W9-AAG-826

Personal Computer Operation and Troubleshooting

second edition

ROGER M. KERSEY

Electronics Chairperson
National Institute of Technology
Livonia, Michigan

Prentice Hall

Englewood Cliffs, New Jersey   Columbus, Ohio

Library of Congress Cataloging-in-Publication Data
Kersey, Roger.
Personal computer operation and troubleshooting / Roger M. Kersey.
—2nd ed.
p. cm.
Includes bibliographical references and index.
ISBN 0–13–656380–5
1. Microcomputers. 2. Microcomputers—Maintenance and repair.
I. Title.
Qa76.5.K415 1996
004.165—dc20 95–38789
CIP

Cover photo: Karageorge/H. Armstrong Roberts
Editor: Charles E. Stewart, Jr.
Production and Design Coordination: Betsy Keefer
Marketing Manager: Debbie Yarnell
Cover Designer: Proof Positive/Farrowlyne Associates, Inc.
Production Manager: Pamela D. Bennett
Illustrations: Carlisle Communications, Ltd.

This book was set in Times Roman by Carlisle Communications, Ltd., and was printed and bound by Quebecor Printing/Semline. The cover was printed by Phoenix Color Corp.

Printed in the United States of America

10 9 8 7 6 5 4 3 2 1

ISBN: 0-13-656380-5

Prentice-Hall International (UK), Limited, *London*
Prentice-Hall of Australia Pty. Limited, *Sydney*
Prentice-Hall of Canada, Inc., *Toronto*
Prentice-Hall Hispanoamericana, S.A., *Mexico*
Prentice-Hall of India Private Limited, *New Delhi*
Prentice-Hall of Japan, Inc., *Tokyo*
Simon & Schuster Asia Pte. Ltd., *Singapore*
Editora Prentice-Hall do Brasil, Ltda., *Rio de Janeiro*

To Regina, Chris, and April
For believing I still exist

Preface

In this ever-changing world of the microcomputer and the information superhighway, it is easy to become weary of changes in technologies and fall behind even though most of what we do is, in some way, controlled by the technologies. This book is intended to help the reader understand the concepts of the hardware technologies in the microcomputer field.

This book deals with the hardware aspects of the IBM PC, XT, and AT compatibles, which account for the majority of the world's microcomputing power. The book has been rewritten so that it may be used in a college or technical school setting or by any technical person with a working knowledge of electronics. The prerequisite for using this book is a working knowledge of basic electronics and solid-state and digital electronics. Although a basic knowledge of microprocessors is helpful, it is not necessary.

Although this book is a hardware book on microcomputer systems, it does cover some of the important concepts of MS-DOS. A understanding of the disk operating system used in the microcomputer is a requirement for today's technicians, and that requirement is demonstrated in this text.

It has been said that if you understand the concepts of a computer system, then you can use those concepts as basic building blocks to configure, troubleshoot, and repair any computer system. This book uses concepts to help the reader with an understanding of the hardware necessary to work effectively on microcomputer systems. Internal block diagrams, pin configurations, and timing diagrams for important integrated circuits are included. The operation of these integrated circuits is the foundation for discussing the operation of each section of the microcomputer system and peripherals. The basic operation of each part of the system is then related to the other sections to provide the reader with an overall concept of the operation of the entire microcomputer system.

Each chapter starts with objectives that provide the user with a road map of the most important topics in the chapter. Either throughout the chapter or at the end of the chapter are troubleshooting tips as well as summaries and review questions.

Chapter 1: This chapter introduces the reader to the basic features and options of microcomputers, ranging from the original XT through the more advanced AT system. Common options and peripherals are discussed to provide the reader with basic knowledge of microcomputer hardware; also discussed in an introductory fashion are some software concepts.

Chapter 2: This chapter provides the reader with the basic knowledge of DOS used in today's microcomputer systems. The depth of this information fulfills the entry requirement for a microcomputer technician.

Chapter 3: In the beginning was the PC, the granddaddy of today's microcomputer systems. Although the 8088 found in the PC is not popular today, understanding the basic concepts gives the microcomputer technician the building blocks to understand the AT systems of today. This chapter provides the reader with an in-depth knowledge of the operation and troubleshooting of this microprocessor and support circuitry. A good understanding of the older technology is essential to understanding the advanced Intel microprocessor.

Chapter 4: This chapter uses the in-depth knowledge presented in Chapter 3 to help the reader understand the more complex operation and troubleshooting of the second major step in microcomputer systems—the 80286. Also covered in this chapter is Intel's first AT chip set, which forms the basis of the chip sets used in today's more advanced microcomputer systems. Troubleshooting and operational concepts for the 80286 are discussed from a hardware perspective.

Chapter 5: This chapter applies the knowledge developed in Chapters 3 and 4 to the next levels of advanced technology—microcomputer systems using the 80386DX and 80386SX microcomputers. This chapter also covers other AT chip sets available and used in the field. Troubleshooting and operational concepts for the 80386DX/SX microprocessors are discussed from a hardware perspective.

Chapter 6: This chapter uses the knowledge developed in Chapters 3 through 5 as the basic building blocks for presenting the operational concepts and troubleshooting of the most advanced Intel microprocessor used today. The 80486 series and Pentium microprocessors are covered, along with advanced I/O architectures, and chip sets are covered from a hardware perspective.

Chapter 7: This chapter is divided into two sections. The first section discusses the circuitry, operation, and troubleshooting of both the XT and AT keyboards and their support circuitry. The second section provides the reader with the basic concepts of the operation and troubleshooting of power supplies used in today's microcomputer systems.

Chapter 8: This chapter dealing with video systems is broken into two sections. First, the operational concepts and troubleshooting of the different types of video display monitors are discussed. Second, the hardware needed to take the data from the computer's data bus and provide the correct signals to display the video images on the video monitor of the computer is examined. Both basic and detailed information is provided about the different types of video adapters and systems available.

Chapter 9: This chapter covers a variety of mass storage devices. Beginning with a basic discussion on floppy disk, hard disk, optical, and tape backup systems, this chapter goes into detail on the floppy disk system. The knowledge gained in understanding the floppy disk system is then applied to the operation of the different types of hard disk systems and interface standards.

Chapter 10: This chapter on I/O systems and devices deals with the basic and detail operation of parallel and serial data communication interfaces. Following I/O adapters, this chapter covers the basic operation of dot matrix, laser, and ink jet printers. The chapter ends with a discussion of the operation of the three different types of mice used with microcomputer systems.

Chapter 11: This chapter is divided into two sections, starting with standard data communications using modems. The basic operational concepts for data communications using the modem and its hardware are discussed along with an introduction to the hardware used to operate software used in LANs.

Chapter 12: Although each chapter in the book contains troubleshooting concepts and tips, the test equipment and troubleshooting chapter discusses the basic operation and use of test equipment in microcomputer systems. Manufacturers' specifications and a logical approach to troubleshooting are also discussed.

ACKNOWLEDGMENTS

Special thanks goes to my wife Gina, son Chris, and daughter April, and the rest of my family and friends for their patience and understanding during the preparation of this book.

I would also like to thank Charles Stewart for sticking up for me and for his encouragement during the writing of this second edition. Thanks also to Betsy Keefer for her help in getting this book in a more readable format and ready for print. Last but not least, thanks to my friend Dale at Lasercomp Inc. for his help and understanding during this time.

I am grateful to the following companies, especially IBM for the Personal Computer and Intel for their great microprocessors and support hardware, for their illustrations, references, and the use of their trademarks in references: International Business Machines Corporation, Intel Corporation, Microsoft Corporation, Micron Technology, Inc., Texas Instruments, Inc., Motorola Corporation, and NEC Electronics, Inc.

R. M. K.

Brief Contents

Chapter 1 *Personal Computer Introduction* 1
Chapter 2 *Disk Operating System* 34
Chapter 3 *PC XT System Board* 119
Chapter 4 *80286 Microprocessor System Board* 242
Chapter 5 *80386 Microcomputer System Board* 300
Chapter 6 *80486/Pentium and Advanced AT Systems* 341
Chapter 7 *Keyboards and Power Supplies* 390
Chapter 8 *Video Systems* 425
Chapter 9 *Mass Storage Systems* 472
Chapter 10 *I/O Systems and Devices* 525
Chapter 11 *Data Communications/Networks* 568
Chapter 12 *Test Equipment and Troubleshooting* 589
Glossary 611
References 617
Answers to Review Questions 618
Index 625

Contents

Preface v

Chapter 1 *Personal Computer Introduction 1*

1.1 Types of Personal Computers 1
1.2 Basic Hardware of a Personal Computer 3
1.3 Description of Computer System Equipment 4
1.4 Video Monitor 18
1.5 Keyboard 20
1.6 Options to Complete the System 23
1.7 Handling, Inserting, and Removing Diskettes 26
1.8 Turning the Computer System On and Off 27
Chapter Summary 29
Review Questions 31

Chapter 2 *Disk Operating System 34*

2.1 What Is DOS and What Does It Do? 35
2.2 How DOS Treats Information on a Disk System 35
2.3 Naming Files 37
2.4 Types of Files 37
2.5 The Root Directory 38
2.6 DOS Wildcards 40
2.7 DOS Syntax Specifications 40
2.8 Special Keys 41
2.9 DOS 5.0, 6.0, and 6.2 41
2.10 Types of DOS Commands 42
2.11 Loading DOS 43
2.12 DOS Command Prompt 44
2.13 Most Common DOS Commands 44
2.14 Batch Files 91
2.15 The Basic AUTOEXEC.BAT File 93
2.16 CONFIG.SYS File 95

2.17 Microsoft Anti-Virus 104
2.18 Defragment (External) 104
2.19 Doublespace (External) 105
2.20 Memory Maker (External) 105
2.21 Fixed Disk (External) 106
2.22 Setting Up Hard Drives 106
Chapter Summary 113
Review Questions 117

Chapter 3 *PC XT System Board 119*
3.1 Microprocessor Subsystem 119
Summary 171
Review Questions 173
3.2 Extended Control Logic 173
3.3 Address Decoding 176
Summary 184
Review Questions 184
3.4 8237 Direct Memory Access Controller (DMA) 185
Summary 199
Review Questions 200
3.5 Read-Only Memory (ROM) 201
Summary 208
Review Questions 208
3.6 Random Access Memory (RAM) 209
Summary 224
Review Questions 224
3.7 Programmable Interval Timer (PIT) 225
Summary 231
Review Questions 231
3.8 Programmable Peripheral Interface (PPI) 231
Summary 237
Review Questions 238
3.9 I/O Interface Bus Channel 238
Review Questions 240

Chapter 4 *80286 Microprocessor System Board 242*
4.1 AT Systems 242
4.2 80286 Microprocessor 246
4.3 80287 80-Bit Numeric Processor Extension 255
4.4 Intel 82230/82231 AT Chip Set 258
4.5 80286 AT/System Board Using the 82230/82231 AT Chip Set 270
Chapter Summary 292
Review Questions 296

Chapter 5 *80386 Microcomputer System Board 300*
5.1 80386 DX/SX Internal Block Diagrams 301
5.2 80386 Operating Modes 302
5.3 80386 Instructions 306
5.4 80386 Microprocessor 307
5.5 80387 DX/SX Numeric Coprocessor 315
5.6 386 System Boards 318
Chapter Summary 335
Review Questions 338

Chapter 6 *80486/Pentium and Advanced AT Systems 341*
6.1 80486DX/DX2/DX4/SX/SX2 MPU 341

6.2 80486 Internal Architecture 341
6.3 80486 Registers 345
6.4 80486 Internal Cache 348
6.5 80486 Instructions 349
6.6 80486 Microprocessor 350
6.7 Pentium Processor 363
6.8 Advanced AT I/O Buses 375
6.9 Extended Industry Standard Architecture (EISA) 378
6.10 Video Electronic Standard Association (VESA) Video Local Bus (VLB) 380
6.11 Peripheral Component Interconnect (PCI) Bus 381
Chapter Summary 383
Review Questions 387

Chapter 7 *Keyboards and Power Supplies 390*
7.1 Keyboards 390
7.2 Power Supplies 407
Chapter Summary 421
Review Questions 422

Chapter 8 *Video Systems 425*
8.1 The Video Display Screen 426
8.2 Producing the Video Image 428
8.3 Video Monitors 429
8.4 Video Adapters 437
8.5 6845 Cathode Ray Tube Controller 440
8.6 IBM Monochrome Video Adapter 450
8.7 IBM Color-Graphics Video Adapter 456
8.8 Video Adapter Outputs 462
Chapter Summary 466
Review Questions 469

Chapter 9 *Mass Storage Systems 472*
9.1 Overview 473
9.2 Floppy Disk Systems 476
9.3 Floppy Disk Adapter 479
9.4 Hard Drive Systems 511
Chapter Summary 518
Review Questions 522

Chapter 10 *I/O Systems and Devices 525*
10.1 Parallel Data Communications 526
10.2 Serial Data Communications 532
10.3 Printers 550
10.4 The Mouse 561
Chapter Summary 563
Review Questions 566

Chapter 11 *Data Communications/Networks* 568
11.1 Data Communications 568
11.2 Networks 574
Chapter Summary 584
Review Questions 585

Chapter 12 *Test Equipment and Troubleshooting 589*
12.1 Test Equipment 589
12.2 Microcomputer Signals and Specifications 594
12.3 Manufacturer's Specifications 596

12.4 Microcomputer Problems 597
12.5 Logical Approach to System Troubleshooting 601
12.6 Repairing Hardware Problems 604
12.7 Troubleshooting Tips 606
Chapter Summary 607
Review Questions 608

Glossary 611

References 617

Answers to Review Questions 618

Index 625

chapter 1

Personal Computer Introduction

OBJECTIVES

Chapter 1 gives a brief introduction to the parts and use of the IBM Personal Computer (PC) or compatible computer system (a microcomputer system that executes most, if not all, software written for the IBM PC. Note that no set standard exists which a computer must follow to be labeled compatible). You should be able to meet the following objectives upon completion of this chapter.

1. List the basic differences between the types of PCs.
2. List and explain the purpose of the three main components in a PC.
3. List and explain the purpose of the most common options in a PC.
4. Explain how to handle and use diskettes.
5. Explain the procedure to start or stop a computer system.
6. Explain the purpose and function of POST.

1.1 TYPES OF PERSONAL COMPUTERS

Since the introduction of the IBM PC in 1981, microcomputer systems have changed to the point that many systems now outperform most minicomputers and reach the performance levels of small mainframe computers of 1981. With the changes in microcomputers, different types of computers have been established: the PC, XT, and AT. The main differences in these PCs are the speed at which they operate, the size of the data they can manipulate, the types of instructions, the addressable memory bus size, and the number of devices they can control.

The original PC from IBM contained 64 kilobytes (KB or, most commonly, K) of DRAM (dynamic random access memory) on the system board, an 8088 16-bit MPU (microprocessor unit) running at 4.77 MHz (megahertz), with an 8-bit data bus, 20-bit address bus (capable of addressing 1 million memory locations), video display system (MDA or CGA), cassette interface, and cassette unit to store and retrieve programs. Standard with the system unit was the language of BASIC programed into ROM (read only memory). Later most PCs were upgraded to hold 256K of DRAM on the motherboard, and an optional 5 1/4-inch SSDD (single-sided double-density) floppy disk system was incorporated.

The next step in the development of the PC was the XT (eXtra Technology) in 1982. The

XT contained an 8088 MPU running at 4.77 MHz, a 5 1/4-inch DSDD (double-sided double-density) floppy disk system, 256K to 640K of DRAM on the motherboard, and a video display system (MDA or CGA). Many XTs also came with an optional hard drive system; with a capacity of 10 MB (megabytes), to increase storage capability. Because IBM decided to open the architecture to their PCs, many other companies started cloning the IBM and interface adapters in the early 1980s. During that time the speed of most compatible XTs was increased to between 8 and 10 MHz, thus extending their usefulness in the computer field. Although today the technology in the XT computer system is old, it still accounts for over one third of the PCs in the field.

Following the XT in 1983 came the first AT (Advanced Technology). The first AT used the Intel 80286 (also known as the 286) 16-bit MPU operating at 6 MHz, a 16-bit data bus, 24-bit address bus (capable of addressing up to 16 million physical memory locations), 1 MB to 4 MB of DRAM on the motherboard, memory management capabilities, the ability to multitask programs, real-time clock, battery–backed-up configuration data buffer, 5 1/4-inch DSHD (double-sided high-density) floppy disk system, and a video display system (MDA, CGA, or EGA). Most 286s were sold with hard disk storage (typically 10 MB to 20 MB). Later the speed of the 286s was increased to 8, 10, 12, 16, and 20 MHz, thus increasing the life of the 286 ATs.

In 1986 the next level of ATs used the Intel 80386 (also known as the 386) 32-bit MPU running at 16 MHz, 32-bit data path, 32-bit address bus (capable of addressing 4 GB (gigabyte) physical memory locations), 1 MB to 8 MB DRAM on the motherboard, increased memory management capabilities, improved multitasking, real-time clock, battery–backed-up configuration data buffer, 5 1/4-inch DSHD floppy disk system, and a video system (MDA, CGA, EGA, PGA, later in 1988 VGA, SVGA, or 8514a). Most 386s were sold with hard disk storage greater than 20 MB. In the late 1980s the speed of the 386 ATs was increased to 20, 25, and 33 MHz. The 386s running at or above 25 MHz normally use an external cache controller (special circuitry used to control the flow of data from slow dynamic RAM to high-speed static RAM and to anticipate the actions of the computer so the MPU always has data ready for it to process without waiting), thus increasing the throughput (overall speed) of the computer.

In 1989 the ATs with the 80386SX MPU from Intel (also called the SX) were introduced. This MPU operates like the standard 80386 except it has a 16-bit data bus and clock speeds of 16 to 33 MHz. By reducing the width of the data bus, SX computers can use the same multitasking and memory management capabilities of the standard 80386 but cost less to build and have a little less throughput because of the smaller data bus.

In 1990 came the ATs with the 80486DX MPU from Intel (also known as the 486). The 80486 MPU is a combination of the 80386 MPU, 80387 math coprocessor (special MPU that performs only math operations at least 10 times faster than the standard MPU), and internal cache controller. Incorporating combinations of the above elements into a single integrated circuit (IC) allows the 486 operating at 25 MHz to have higher throughput than a 386 with a 387 and external cache operating at 33 MHz.

The first 486DX operated at 25 MHz, but that was soon increased to 33 MHz. Intel started producing a 50-MHz version of the 486DX, but speed problems encountered with support circuitry operating at 50 MHz limited the usefulness of that version. Later Intel started producing the overdrive version of the 486DX-25 and 33. In the overdrive 486DXs, the internal clock speed is doubled, thus providing internal clock speeds of 50 MHz and 66 MHz. Intel thus named the new 486DXs 486DX2 (also known as the DX2). In the early 1990s Intel developed the 486DX4, which has an internal operating speed of 99 MHz and an external bus speed of 33 MHz, allowing computer makers to increase the speeds of their systems without changing the support circuitry.

Not long after the 486DX became the dominant MPU, Intel starting producing the 486SX, which is the same as the 486DX except that a math coprocessor is not present. The 486SX performs nonmath operations at about the same speed as the 486DX running at the same clock frequency. The speeds of the 486SX range from 25 MHz to 33 MHz, and 50 MHz or 66 MHz with the 486SX2. The 486SX was developed for use in portable computers; be-

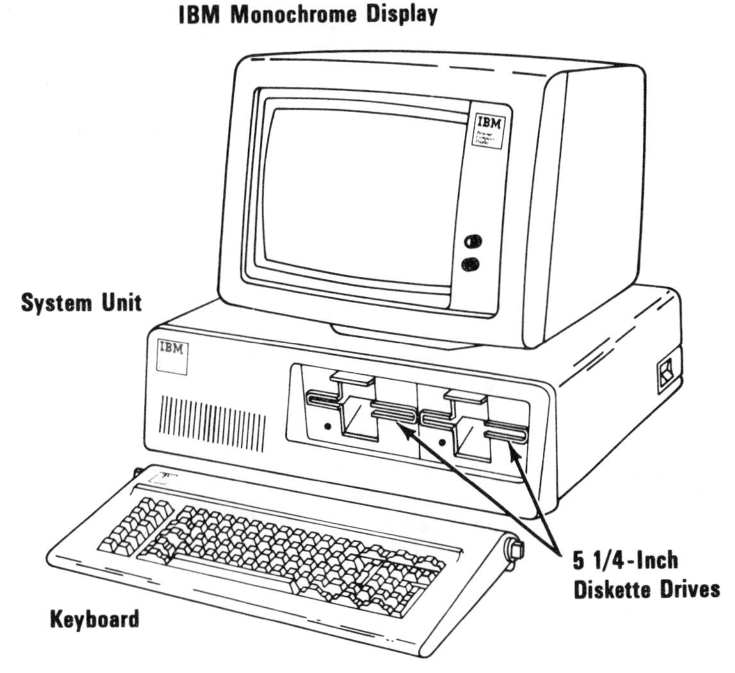

Figure 1-1 The IBM PC system. (Courtesy of IBM)

cause it does not have a math coprocessor, power dissipation is much lower, and some of the versions operate at less than 5 volts.

In 1992 Intel came out with the Pentium MPU to support the high power needs of work stations and power users. Pentiums are now available in 60-MHz, 66-MHz, and 90-MHz versions. The first Pentiums operated at 60 MHz and had a heat problem; cooling fans and heat sinks were added to reduce that problem. The Pentium contains a 64-bit data bus (32-bit data and 32-bit code), a larger internal cache (8K code cache and 8K data cache), an integrated smart superscaler pipeline that allows it to execute two instructions at once, and a floating point unit (FPU, also known as a math co-processor). The Pentium currently represents the top of the line for non-RISC (Reduced Instruction-Set Computing) processors.

1.2 BASIC HARDWARE OF A PERSONAL COMPUTER

Although many different levels of PCs are based on the IBM PC and Intel 80xxx chip family, all systems (Fig. 1-1) basically have a minimum of three hardware sections: the system unit (also known as the base unit), some type of video display, and a keyboard.

The system unit contains the system board (also known as the motherboard, which contains the MPU, ROM, DRAM, I/O (input/output) interface channel, and support hardware circuitry), power supply, some type of mass storage system, video controller, and data communication ports inside a steel or plastic case.

The video display, the main output device, gives the operator a visual indication of the operation the computer is performing. The video display requires a compatible video adapter in the system unit to operate correctly.

The keyboard is an input device that contains 83 or more keys. Each key is a switch that allows the operator to enter data into the computer. The code produced when a key is pressed either represents some type of printable character (letters, numbers, or symbols) or causes the computer to perform a specified operation.

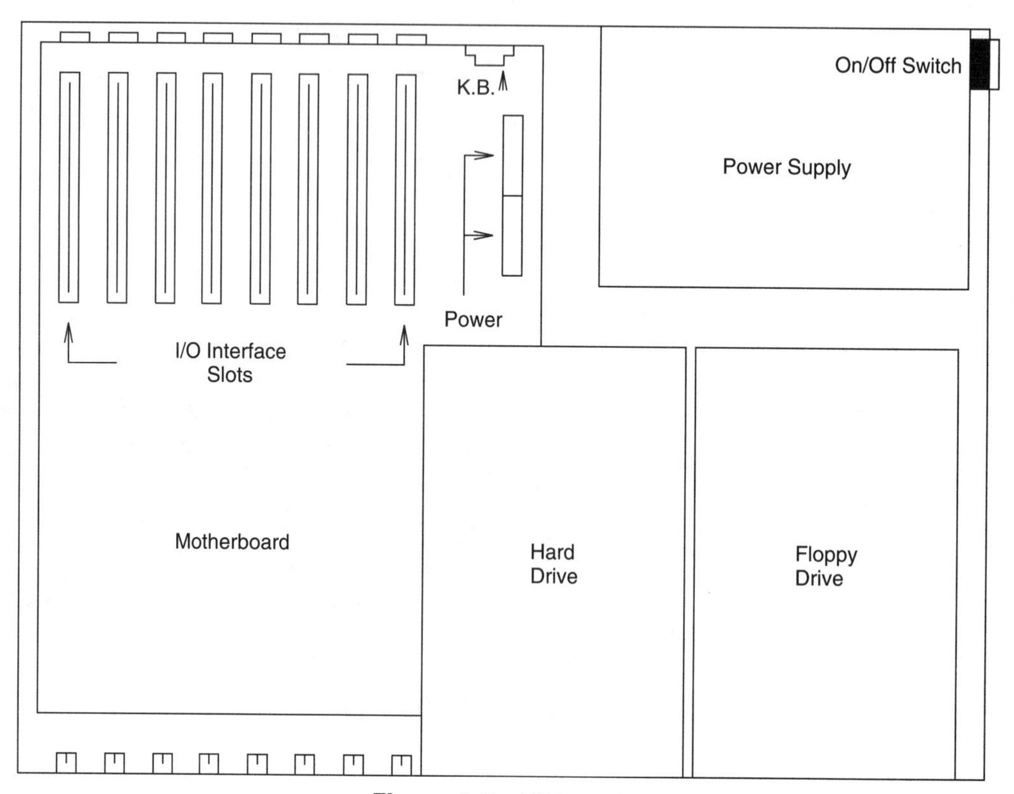

Figure 1-2 XT layout.

1.3 DESCRIPTION OF COMPUTER SYSTEM EQUIPMENT

1.3.1 System Unit

Figure 1-2 shows a typical layout of the equipment found in most XT system units. The motherboard is normally located on the left side of the case, with the I/O interface slots and keyboard (K.B.) input near the rear of the case. The power supply is located in the right rear corner of the case. The floppy drive(s) is located in the front right-hand side and the optional hard drive just to the left of the floppy drive(s). A complete system contains two or more interface cards in the interface slots (cards can be placed in any slot), normally one card for the video adapter and one for the floppy disk adapter.

1.3.2 PC/XT Motherboard

Figure 1-3 shows a typical PC/XT motherboard system board. There are only 62 pin I/O interface slots in this type of MPU (8088 usually, sometimes an 8086). The 40-pin socket next to the MPU is for the 8087 math co-processor IC. Older motherboards had only five I/O interface slots, but most motherboards now contain eight I/O interface slots. The DMA (direct memory access controller), PPI (programmable peripheral interface), CLK (clock generator/driver), PIC (programmable interrupt controller), PIT (programmable interval timer), and remaining ICs in the diagram are used to support the 8088 MPU.

The ROM/EPROM (read only memory/eraseable programmable read only memory) section has one to four ICs. One of the ICs in this section must contain BIOS (basic input output system); this program, which is stored in ROM or EPROM, is a list of instructions telling the computer how to act like a computer. BIOS checks, configures, and controls the equipment in the computer. All software works with BIOS to perform the tasks of the program. Some motherboards contain more than BIOS; the other ICs may contain a programing language, DOS (disk operating system), setup/utility software, or application software, but in most cases empty sockets are available for future expansion.

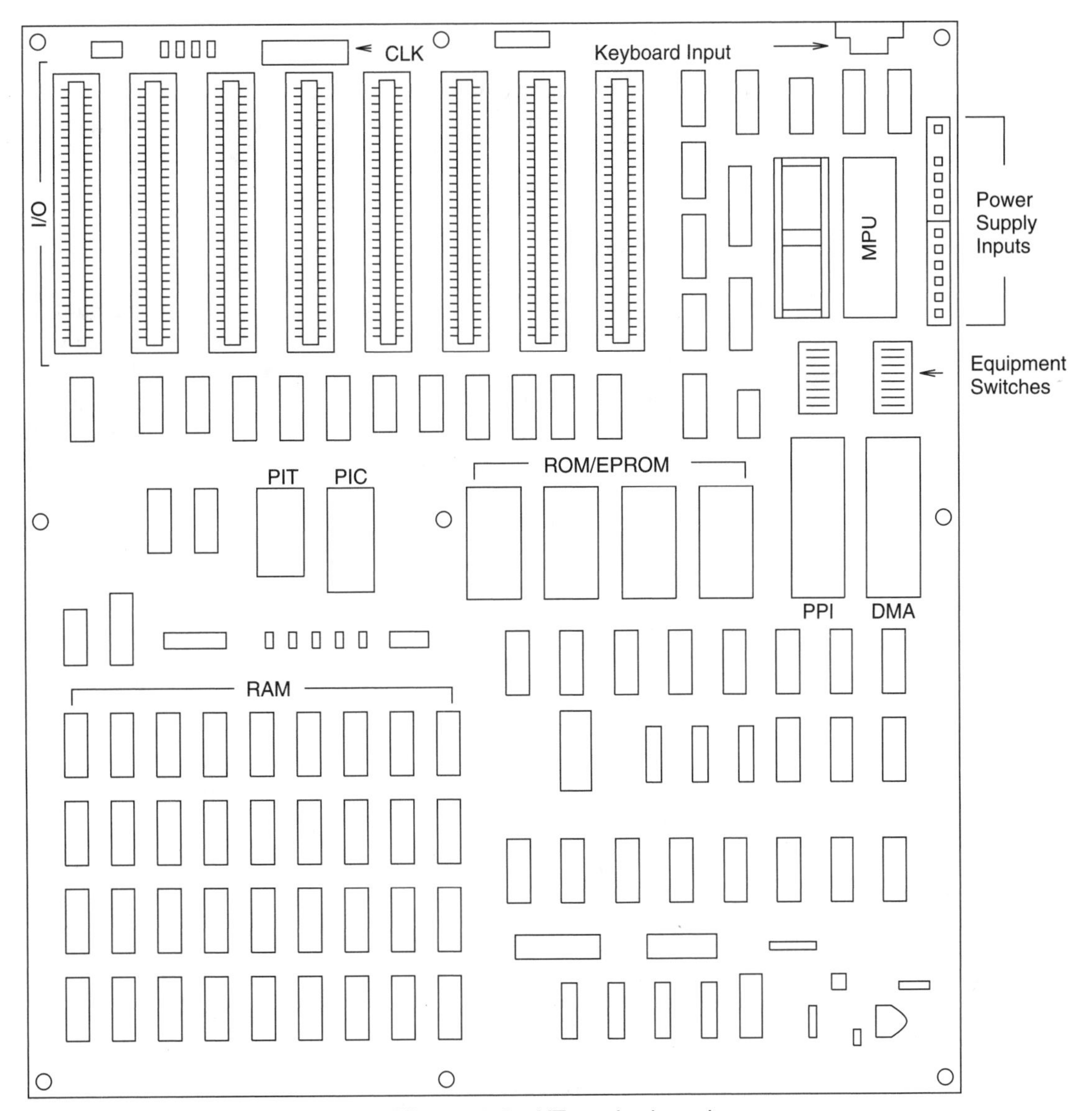

Figure 1-3 XT motherboard.

The 8088 MPU has a 20-bit address bus and can therefore address 1,000,000 (1,048,576 actual) memory locations, but because of how the system memory is mapped and how BIOS and DOS work with the memory, only the first 640K (655,360 actual) memory locations are used for conventional memory (standard user memory). The remaining memory locations are reserved for I/O adapters and ROM.

The RAM section is normally made up of four rows of nine DRAMs each. DRAMs normally are 1 bit wide; therefore eight DRAMs are connected in parallel to supply the data for an 8-bit data bus, one IC for each bit of the data bus. In the PC eight of the ICs are used for the data and the additional IC is used for storing a parity bit (a parity bit is used to check for proper data).

In the original PC the ICs used in the DRAM section were 4116 DRAMs (16K $\times$ 1); thus each row supplied 16K of memory, producing a total of 64K of storage on the motherboard. Any additional memory required an external memory card placed in one of the I/O interface slots. Later motherboards used the 4164 DRAM (64K $\times$ 1), so that each row supplied 64K of memory, producing a total of 256K of storage on the motherboard; additional memory required an external memory card. The latest type of XT motherboard uses two types of ICs, two rows of 41256 DRAMs (256K $\times$ 1) to produce 512K of memory and two rows of 4164

DRAMs (64K × 1) to produce an additional 128K of memory, giving a total of 640K of memory on the motherboard.

The equipment dip switches (or jumpers) tell the computer system (via BIOS) what type of equipment options are in the computer (type of video, amount of memory, math coprocessor present, number of floppy drives). This information is needed by BIOS to configure the devices and check them upon start-up.

1.3.3 AT Motherboard

Figure 1-4 shows a typical 386/486 AT motherboard. There are 8-bit (62-pin I/O interface slot), 16-bit (additional 36-pin I/O interface slot, which is used in conjunction with the 8-bit interface slot), and 32-bit (additional 86-pin preparatory I/O interface slot [for 32-bit memory cards], which is used in conjunction with the 8-bit interface slot) interface slots, and the MPU is an 80386. A 486 AT system board would look very similar, except an 80486 would be used and no socket for an 80387 would be seen. If the motherboard contained only 8-bit

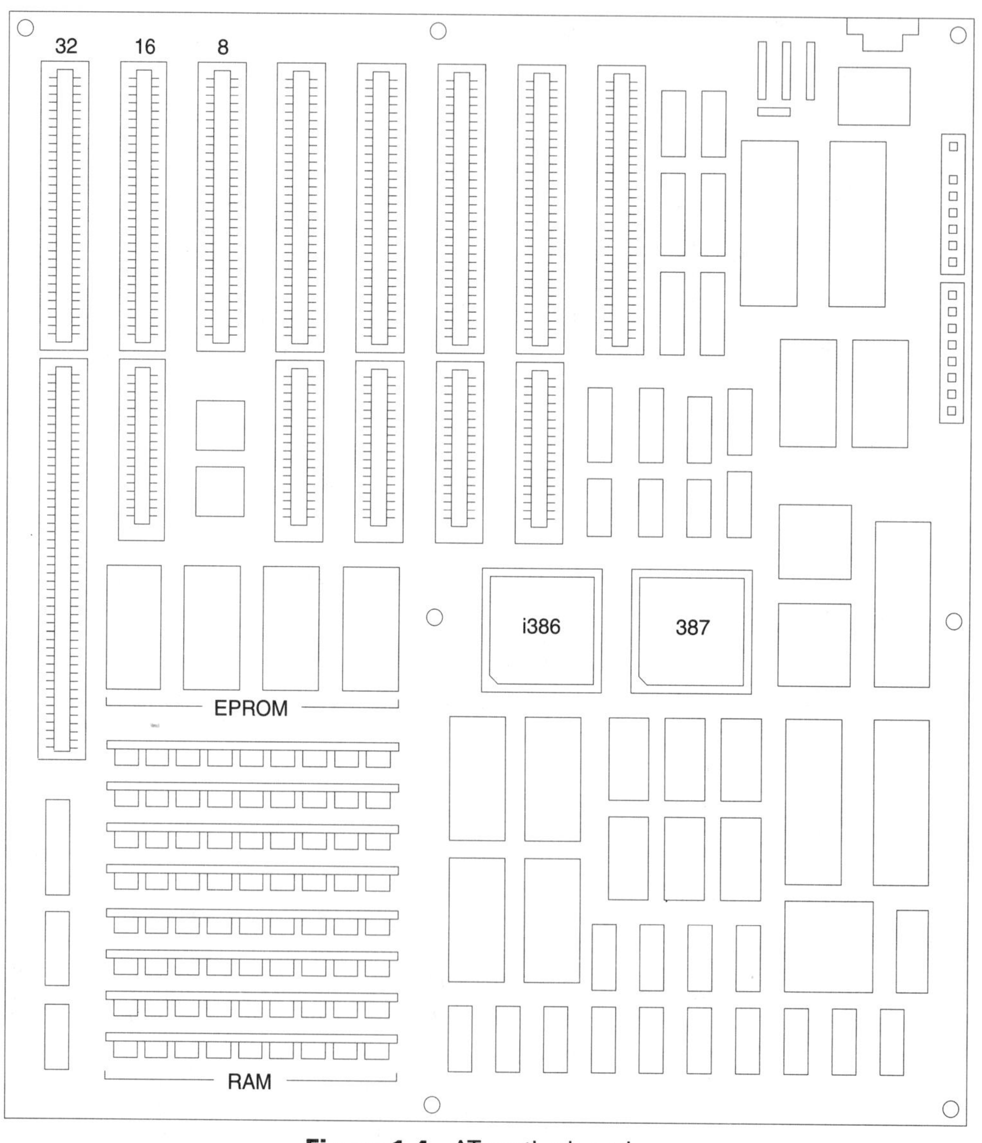

Figure 1-4 AT motherboard.

and 16-bit interface slots, it would probably be either a 286 or 386SX AT motherboard. AT motherboards usually contain eight interface slots of mixed sizes (a 286 or 386SX AT usually contains five 16-bit and three 8-bit slots; a 386 or 486 AT contains one or two 32-bit, four 16-bit, and two 8-bit slots). ATs contain two DMAs, a 8042, a CLK, two PICs, a PIT, and other smaller support ICs.

The ROM/EPROM section may have two or four ICs. One set of the ICs (one for low byte, one for high byte) contains BIOS. Because ATs contain more support ICs, more memory space is needed to perform BIOS functions. ABIOS (AT BIOS) performs the same function as in the XT with the addition of a set-up program; any empty sockets can hold additional software for the system, as in the XT systems.

The 286/386SX MPU has a 24-bit address bus (16,777,216 actual addresses); 386/486 MPUs have a 32-bit address bus (4,294,967,296 actual addresses). If an AT system contains 1MB of DRAM, the first 640K are used as conventional memory, the remaining memory being hardware decoded for the first 384K memory locations above 1 MB (called extended memory). All physical memory at address locations above 1MB is used as extended memory unless otherwise configured. Therefore, an AT system with 2 MB has 640K of conventional memory and 1384K of extended memory. Physical memory locations between 640K and 1 MB (known as upper memory) are reserved for I/O devices and system ROM(s), which help maintain 8088 compatibility. The extended memory that the AT MPUs can address directly can be used only with the right type of software and operating system or can be converted into another type of memory.

The RAM section in older ATs is normally made up of four rows of nine DRAM DIPs (dual in-line packages) each. If the four rows are loaded with 41256 (256K ×1) DRAMs, the board contains 1MB of memory; any additional memory requires an external memory card. If the four rows are loaded with 1M ×1 chips, the total memory is 4 MB; additional memory requires an external memory card.

The latest trend in AT memory is to use SIPs (single in-line packages) or SIMMs (single in-line memory modules). These units of memory contain micro-size or surface mount–technology memory ICs soldered onto a small memory board. This type of memory requires less motherboard space and reduces the number of sockets required for memory, lowering the cost of the motherboard. Each unit of memory contains 256K to 4MB of memory plus a parity bit for each memory location, requiring only about one quarter the space required for DIPs and using about 30% less power. Since the introduction of SIPs and SIMMs, cost per byte has decreased to the point that it is 25% less than the cost of memory using DIPs. The system in Figure 1-4 contains 8MB of memory using SIPs or SIMMs; extra memory requires an external memory board.

In ATs, equipment dip switches are not normally necessary because BIOS accesses a battery–backed-up memory buffer, which contains information about the equipment connected to the computer. Some ATs still use some dip switches for default settings in case the battery–backed-up memory fails.

1.3.4 Integrated Motherboards

The last type of motherboard (Fig. 1-5) is a highly integrated AT motherboard. It contains everything a standard motherboard has, plus a built-in video adapter, I/O ports (printer [parallel], serial, and mouse interface ports), and hard/floppy drive adapter. The integration of these adapters results in a complete computer system for less cost than with external adapters. However, because many of the ICs are custom made, the board is difficult to troubleshoot, and upgrading the board may be impossible without disabling part of the motherboard. These boards can be found in XT, 286, and SX versions, usually not in 386DX and 486DX versions (because of upgradability). Normally only three I/O interface slots are needed because almost everything is built into the motherboard.

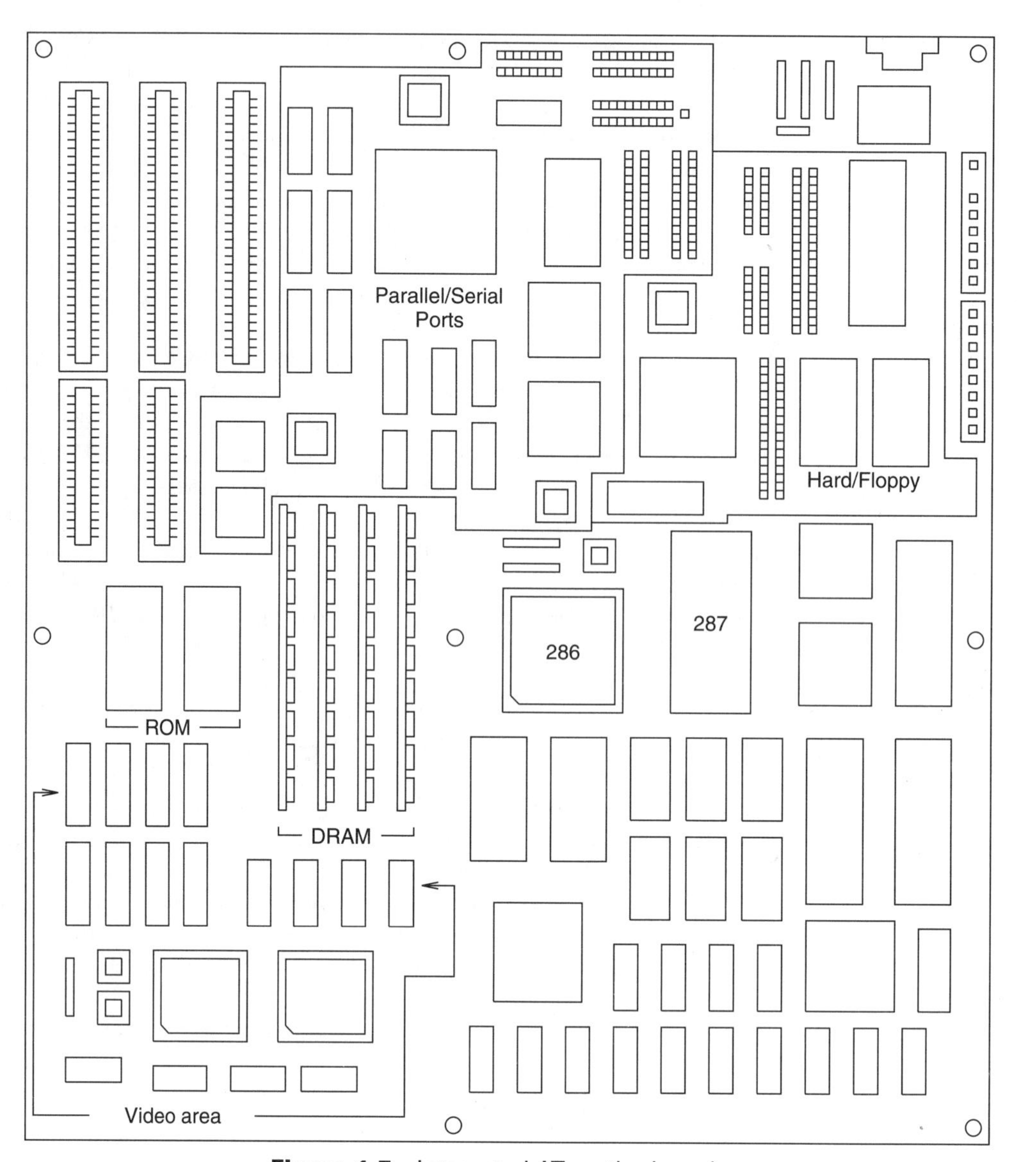

Figure 1-5 Integrated AT motherboard.

1.3.5 Power Supply

The power supplies in XT and AT systems are basically the same except for the power ratings. The power supply is normally located in the right rear corner of a desktop case or in the upper rear corner of a tower case. A typical power supply uses a variable switching voltage regulator, which provides very high efficiency. The voltage output connections are standardized, as are the four different output voltages (+5, −5, +12, and −12 volts), and most power supplies share the same dimensions, so changing a power supply is very easy. One pair of connectors provides power (+5, −5, +12, −12 volts, ground, and power good signal) to the motherboard; the remaining two to four (4-pin molex) connectors supply power (+5, +12 volts, and ground) to any hard/floppy/tape back-up drives. Currently most XT power supplies are rated at 150 watts, whereas most AT power supplies are rated between 200 and 300 watts, depending on the usage of the system.

1.3.6 Mass Storage System

Two types of mass storage systems are available—the floppy disk system and hard disk system. Both allow the user to store and retrieve large amounts of data from the computer sys-

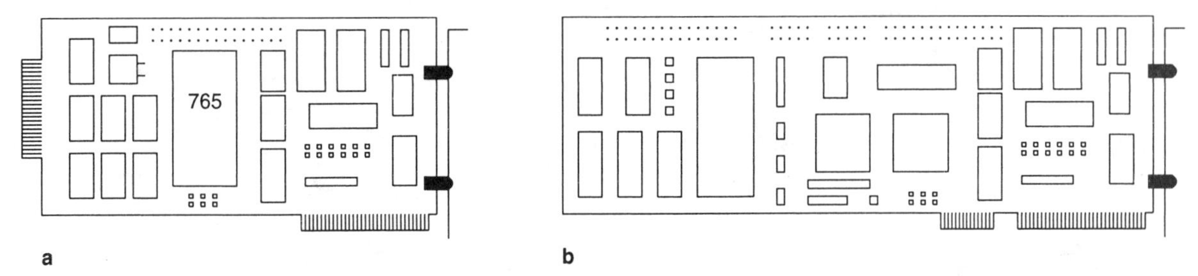

Figure 1-6 **(a)** 8-bit XT floppy disk adapter; **(b)** 16-bit AT floppy/hard disk adapter.

tem and place it on a nonvolatile magnetic medium for later use. The main differences in the two systems are the capacity of the media and the speed at which the data can be accessed.

1.3.7 Floppy Disk System

The floppy disk system consists of four parts—floppy disk adapter, floppy disk drive unit, diskette, and operating system. The floppy disk system is basically the same for both XT and AT systems, except that most XT systems normally cannot control high-density drives.

Floppy Disk Adapter. Figure 1-6(a) shows a typical XT-style floppy disk adapter. This type of adapter can control two double-density floppy disk drives in a 5 1/4-inch (360K) and/or 3 1/2-inch (720K, requires software driver in some cases) form factor through a 34-pin edge connector located at the front of the card. The main controlling element is a 765AC FDFC (floppy disk formatter controller) IC, which takes parallel data from the computer and converts it into serial data for the disk drive unit(s). The device also reads serial data from the disk drive unit(s) and converts it into parallel data for the computer. The 765 controls the motors and sensors of the disk drive unit(s) and is capable of controlling up to four disk drive units, although most adapters are set up to control only two. Figure 1-6(b) shows a typical AT-style floppy/hard disk adapter, which is standard equipment on ATs because they usually contain a hard disk system. The floppy part of the adapter performs the same functions as the floppy disk adapter of the XT, except that data is transferred to and from the adapter using the 16-bit data bus of the computer. Usually the floppy disk controller IC is not a 765 but a more advanced device that can control both double- and high-density drives. High-density 5 1/4-inch drives hold 1.2MB of data and the 3 1/2-inch drives hold 1.44MB of data. The control and data signals are supplied to the disk drive units through a 34-pin header that is normally set up to control two drives.

Floppy Disk Drive Unit. The floppy disk drive units (Fig. 1-7) contain TTL (transitor transistor logic) and analog circuitry, two motors, and assorted sensors. Its electronic circuitry is used to control the speed of the spindle motor that rotates the diskette. A tachometer or servo system is used to maintain the 300-rpm speed of the spindle motor. The physical appearance of double- and high-density drive units is basically the same despite some differences in the circuitry and one of the control signals.

The floppy disk drive unit also controls the location of the read/write (R/W) heads on the diskette. The unit's stepper motor determines the position of the R/W heads on the diskette, where each location of the heads is called a track.

Information provided by different sensors determines the proper operation of the floppy disk drive unit. There is a track zero sensor, which is a mechanical or optical switch. A closed switch indicates that the R/W heads are on track zero, the outermost track of the diskette. Track zero is used as a reference for many disk operations. The write-protect sensor is a light sensor that indicates if the diskette in the drive is write-protected. If light passes through the sensor, the diskette can be written to; if no light passes, the disk drive unit does not attempt to write to the diskette. The last sensor is the index hole; this sensor informs the floppy disk adapter when the R/W head is positioned over the first sector of the diskette.

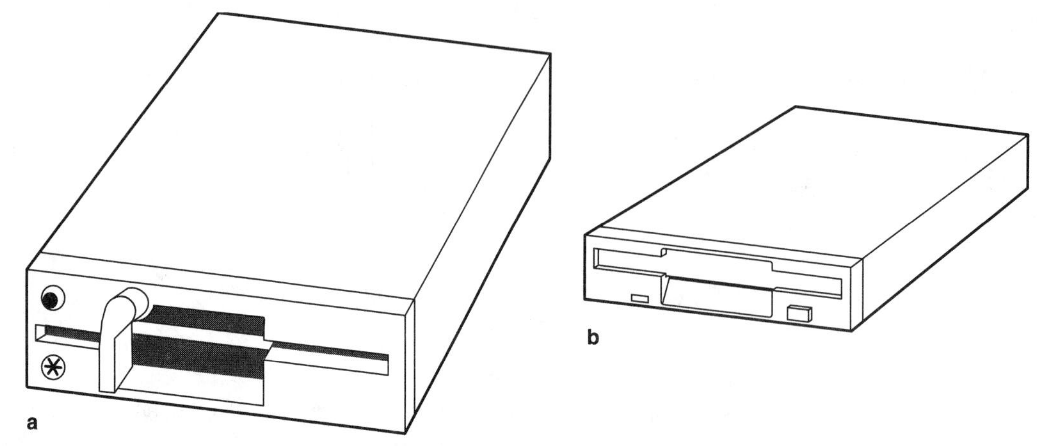

Figure 1-7 **(a)** 5.25-inch floppy disk drive unit; **(b)** 3.5-inch floppy disk drive unit.

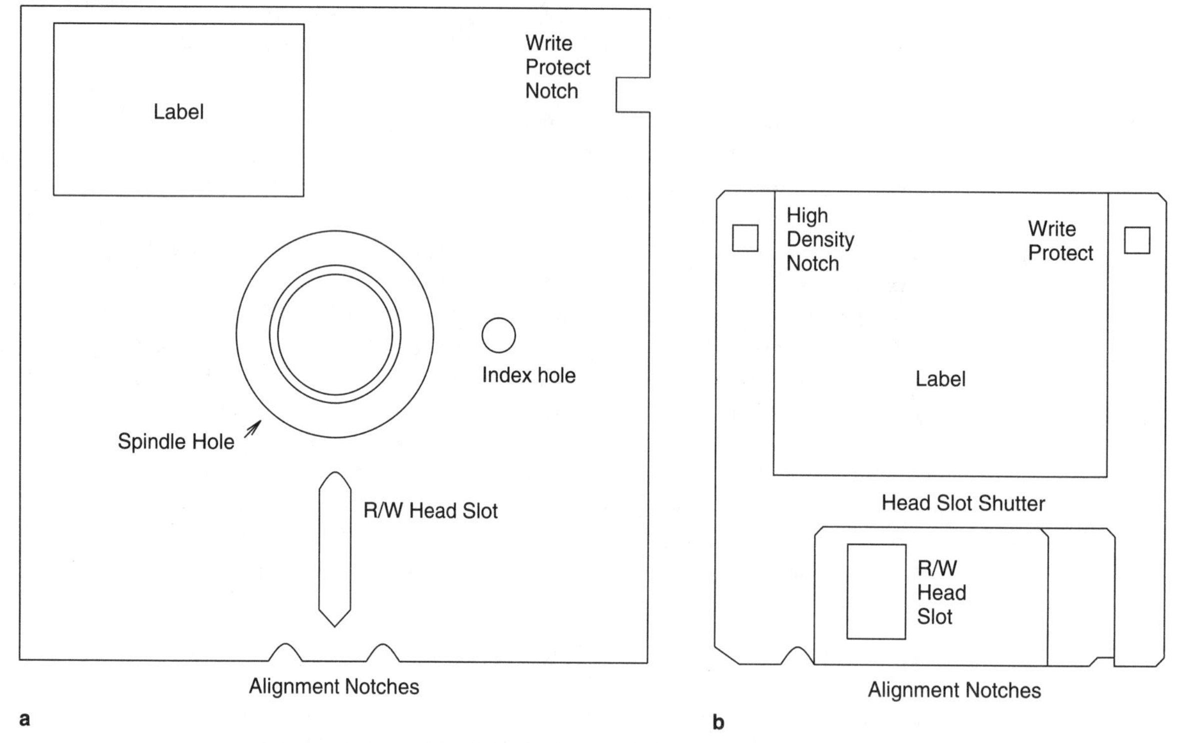

Figure 1-8 **(a)** 5.25-inch floppy diskette; **(b)** 3.5-inch floppy diskette.

Finally, the floppy disk drive unit converts TTL levels from the floppy disk drive adapter into analog levels that are stored on the diskette during a write operation. During a read operation, analog levels from the R/W heads are converted into TTL levels for the floppy disk drive adapter.

Floppy Diskette. Figure 1-8(a) shows a 5 1/4-inch floppy (mini) diskette and identifies its parts. The only physical way in most cases to tell the difference between DSDD, (360K storage) and DSHD (1.2MB storage) diskettes is that the DSHD diskette normally does not have the reinforcement ring (second circle around the spindle hole). The main difference is in the type of magnetic coating on the surface of the diskette (DSHD diskettes have a stronger magnetic coating, allowing high density drive units to store more information in the same physical area).

This type of storage medium gets its name from the fact that the material inside the jacket is very thin (therefore floppy), and the physical shape is round (therefore disk). The 5 1/4-inch dimension is the external size of the protective jacket used to support the disc, protect its surface (from contamination), and allow easy insertion and removal of the diskette from the drive units. The term *mini diskette* is derived from the fact that the first diskettes were 8 inches, and the 5 1/4-inch diskette is smaller.

The label should always be located on side A of the floppy diskette. Side B is identified by the fold around the edges of the diskette.

The R/W head slot allows the R/W heads to access the surface of the diskette. Each position of the R/W heads from the outside to the inside of the diskette is called a track. *Never touch the surface of the diskette, it may cause damage to the information stored on the diskette.*

The hole in the center of the diskette is called the spindle hole. The disk drive unit uses this hole to rotate the disk when writing or reading information to or from it.

The small hole near the spindle hole is called the index hole. When the index hole on the diskette jacket matches the index hole on the disk inside, a sensor in the disk drive unit tells the computer that the R/W heads are at the first sector of the diskette. This signal is used to help determine where the R/W heads are during a rotation.

The two small notches on the edges of the diskette near the head slot are called the alignment notches. These ensure that the diskette is aligned in the drive unit properly.

The notch in the upper right-hand corner of side A is called the write-protect notch. If the notch is covered with some kind of nontransparent tape, the disk drive unit does not write to the diskette. If the notch is not covered, the disk drive unit can perform a write operation on the diskette.

Figure 1-8(b) shows a 3 1/2-inch (micro) floppy diskette and indentifies its parts. The way to physically tell the difference between DSDD (720K storage) and DSHD (1.44MB storage) diskettes is that DSDD diskettes do not have a notch in the plastic opposite the write-protect notch. The important difference, however, is the type of magnetic coating on the surface of the diskette.

Micro-diskettes are now becoming the norm because of their small size, hard plastic case, and high storage capabilities. Because of the way the plastic jacket is made, the label can be placed only on the correct side of the diskette. Side B of the diskette contains the spindle platter, which must not be covered.

The head slot shutter is a spring-activated metal door that opens (slides to the right) when the diskette is placed properly into a drive unit. When the shutter opens, the R/W head slot is exposed, allowing the R/W heads access to the surface of the diskette. Each position of the R/W heads (80 positions on both DSDD and DSHD) is called a track. Because the shutter closes automatically when the diskette is removed from the drive unit, touching the surface of the diskette is normally not possible. Touching it may cause damage to the information stored on the diskette.

On side B of the micro floppy diskette, a 7/8-inch round metal spindle platter is located at the center of the diskette. A small square hole in the center of this platter is used to rotate the disk inside the case via the spindle motor at 300 rpm.

At the edge of the spindle hole is a rectangular hole called the index hole. As the disk rotates, a light is shined on the metal platter; when the index hole passes the sensor, no light is reflected back to the sensor, indicating that the R/W heads are over the first sector. This signal is used as a reference point, and a timing circuit then divides the rotating disk into its other sectors. This signal therefore is used to help determine where the R/W heads are during a rotation.

The notches on the edge of the head slot shutter, called alignment notches, ensure proper alignment of the diskette in the drive unit.

The notch (square hole) in the upper-right corner of side A is called the write-protect notch. A small door slides back and forth over this notch (accessed from side B). If the door covers the notch, the diskette can be read from or written to. If the door does not cover the notch, the drive unit does not perform a write operation on the diskette, but the diskette can still be read.

1.3.8 Operating System

The most popular PC operating system is DOS (disk operating system), developed by MicroSoft. OS2, Unix, and Windows (actually an environment rather than an operating system) are other popular operating systems that are used primarily when multiuser (more than one user of the same software at the same time) or multitasking (more than one software package operating at the same time) is needed.

The operating system is software that completes the disk systems and allows the user easy access to the power of the computer. The operating system is loaded into the computer and works with the monitor program (BIOS in the PC, stored in ROM/EPROM) to control the physical hardware in the computer. The operating system standardizes the operation of the hardware in relation to the software, so that software written to run under DOS has a standard reference point to control the computer. As the name DOS implies, this system deals primarily with floppy and hard disk systems, but DOS also controls other hardware functions in the computer (video, RAM, keyboard, I/O interfaces, and most other peripherals).

Once the computer is turned on and BIOS runs POST (power-on self-test), the computer looks at the disk system to load DOS. As DOS is loaded, it takes more control of the computer, at which point other software may be loaded into the computer to make it perform specific tasks. DOS performs the same types of tasks whether it is controlling a floppy disk or hard disk system. The main difference in the operation of DOS is that a floppy disk system has a smaller data capacity than a hard disk system.

1.3.9 Hard Disk System

The hardware in a hard disk system is not as standardized as in the floppy disk system. The many different types of interfaces in hard drive systems all react the same as far as the computer itself is concerned. The difference lies in how the interfaces control and process the data that go to the disk drive unit. Once the interface is selected, a compatible drive unit must be used for the system to operate. A hard disk system has three parts—hard disk controller, disk drive unit, and operating system. The operating system is the same as for a floppy disk system.

Hard Disk Adapter. Today's complex and large software packages make it difficult and impractical to operate software with only a floppy disk system, even if the system contains two drives. Understanding the operation of a hard drive system and how to diagnose problems is a necessity for today's technician.

Because hard drive systems are much larger, faster, and more complex, the hard disk adapter of a computer requires more circuitry and smarter ICs to control the system. Basically, however, the hard drive adapter performs the same functions as a floppy disk adapter. Western Digital, one of the largest manufacturers of hard disk adapters, has set the standard for how a hard disk adapter should function. As seen in Figure 1-6, an AT floppy/hard disk adapter requires much more circuitry than a floppy disk adapter alone. If an XT uses a hard disk system, it usually requires a separate hard disk adapter, a specialized microprocessor that controls the interfacing between the PC and the hard drive unit. Because a hard drive system requires more control than floppy systems, normally two ribbon cables (one for control signals and one for data) connect the adapter to the hard drive unit. The following is a brief summary of the major types of interfaces used in hard disk systems.

MFM (modified frequency modulation) was the first standardized hard disk interface. Low cost and reliability have been its strong points. Its major weakness is less speed and efficiency than some of the newer interfaces. The most common capacities for MFM systems are 20MB, 40MB, and 80MB.

SCSI (small computer system interface) is used mainly in computers that are not normally PC compatible. Some PCs use SCSI because this interface can control much more than just hard drive systems (tape backup, CDs, and scanners), data transfer rates are very high, and up to seven devices can be controlled by one adapter. Most of the intelligence for an SCSI system is in the device that the adapter controls. The most common capacities for SCSI systems are 20MB, 40MB, 80MB, 100MB, 200MB, 500MB, 1.2GB, and 2GB.

The next interface standard to be developed was RLL (run length limited). Because of the way data is stored on the surface of the disk, this interface allows storage of about 50% more data in the same area than MFM and can transfer data 50% faster. Because of the tight packing of data on the disk, errors are more likely to create problems as the integrity of the disk surface changes. The most common capacities for RLL systems are 30MB, 65MB, and 120MB.

In the quest for faster and more intelligent hard drive adapters, the ESDI (enhanced small device interface) standard was developed. This interface is one of the fastest and most reliable interfaces today but also costs more than twice as much as any other interface. ESDI is used in very large hard drive systems in which the main emphasis is speed. The most common capacities for ESDI systems are 100MB, 155MB, 220MB, 320MB, 650MB, and 1GB.

The latest hard disk interface is IDE (intelligent drive electronics). In this system the adapter is usually nothing more than a buffer to the computer bus; the actual intelligent part of the adapter is built into the disk drive unit itself. Using this type of interface, computer makers can integrate the hard disk adapter into the motherboard or any I/O adapter with very little circuitry. Also, because the adapter is part of the drive unit, custom circuitry can be added to correct for errors and to enhance the speed of the drive without affecting the compatibility of the interface. Most IDE drives use a form of RLL for storing data on the diskette. The most common capacities for IDE systems are 20MB, 40MB, 80MB, 100MB, 200MB, 340MB, 420MB, 540MB, 740MB, and 1GB.

Hard Drive Unit. The hard drive system gets its name from the fact that the disk inside the drive unit is made of a special type of metal about 1/16-inch thick. Today's hard drive units are 5 1/4 inches and 3 1/2 inches (Fig. 1-9). Another name given to hard drive systems is nonremovable disk because in most hard drive systems the disk remains inside the disk drive unit and cannot be removed.

The magnetic coating on the surface of the disk is much denser and allows more data to be stored in a smaller area. To further increase the capacity of hard drive units, more than one disk (platter) is used. Typically a hard drive has at least two platters (disks) and up to eight platters. These disks are stacked like records, and all turn at the same time. An R/W head is required for each side of a disk where data is stored. Therefore, a hard disk drive unit can have as many as 16 R/W heads. They perform the same function as in the floppy disk system but are much more sensitive.

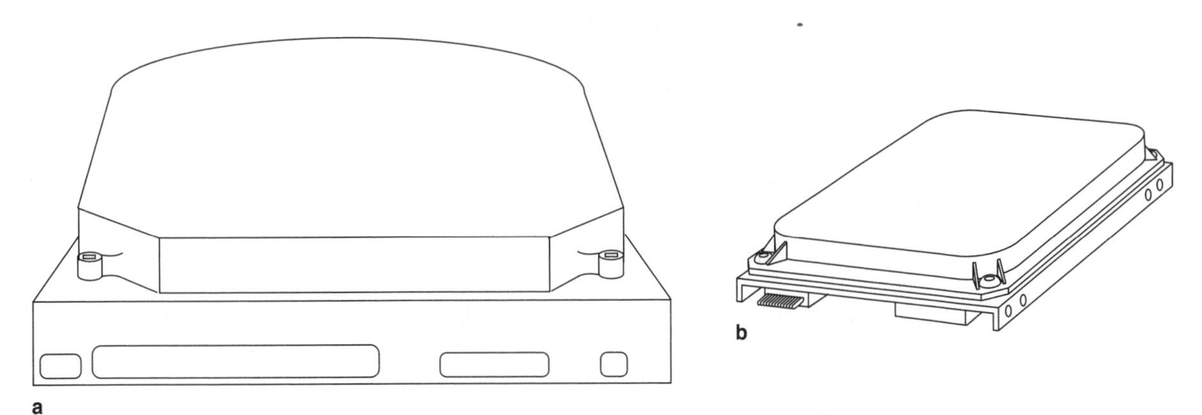

Figure 1-9 **(a)** 5 1/4-inch hard drive unit; **(b)** 3 1/2-inch hard drive unit.

The spindle motor of a hard drive unit rotates all the disks in the drive unit at about 3600 rpm. This higher rate speeds up disk operations in the unit and allows more data to be stored on the diskette.

The R/W heads of the hard drive unit are ganged together as one unit called the head assembly. The head assembly is moved by either a stepper motor or a voice coil. Stepper motors are used in slow, low-cost drives, whereas voice coils are used in fast drives for which cost is not a factor. Each position that an R/W head lands on is called a track, and when talking about the head assembly (with all R/W heads used as one unit), the position is called a cylinder, which represents the total number of tracks. Today's hard drive units contain anywhere from 400 to 1024 cylinders.

Because the disk is not removable in most hard drive systems, there is no need for a write-protect sensor. The track zero sensor performs the same task as in the floppy drive unit. To determine the relative position of the R/W heads as far as rotation is concerned, an index mark sensor is used in some hard drive units, whereas others use the surface of one of the disks to supply servo information to the electronics.

1.3.10 Video Adapter

Video systems have two sections—the video adapter and the video monitor. The many different types of video adapters (MDA, CGA, EGA, VGA, SVAG, and 8514A) all perform basically the same task. Video monitors, like video adapters, differ in ability but all perform basically the same task. The video adapter part of the system is found in the system unit.

The purpose of a video adapter is to convert the data in screen memory into signals the video monitor can use. The video adapter performs the following:

- Controls the conversion of ASCII codes in screen memory into dot codes.
- Determines the location of the electron beam on the screen of the CRT (cathode ray tube) by controlling the vertical and horizontal sync pulses.
- Controls the blanking of the screen during retrace.
- Controls resolution, the number of points that can be controlled on the screen.
- Controls the types of characters and graphics on the screen.
- Controls the colors on the screen.
- Determines cursor position, updates cursor.
- Determines and controls the top of page and scrolls the information on the screen.
- Controls the operation of a light pen in some cases.

MDA (monochrome display adapter) (Fig. 1-10(a)) provides very high resolution characters (characters are made up of a maximum of 63 dots, 7 rows by 9 dots each) in black and white, but under normal conditions does not allow the user to perform graphic commands. The user may use graphic characters to create pictures on the screen. Characters can be displayed only in an 80 $\times$ 25 format (80 characters across the screen with 25 rows of characters per screen). This adapter has 4K of VRAM (video random access memory) with 2K used for character codes and 2K used for attribute information (information regarding how the characters are displayed, such as bold face and underlining). All outputs to the monitor are TTL compatible. A form of the MDA called the Hercules adapter does allow graphic commands to be performed with certain types of software.

CGA (color graphics adapter) provides low-resolution and low-cost color graphics. Characters may be displayed in a 40 $\times$ 25 or 80 $\times$ 25 format, with 1 of 16 different background and border colors and 1 of 8 different character colors. The character resolution is the lowest of all video adapters (7 rows by 7 dots), which makes characters more difficult to see. Character enhancements are shown by different colors. In the graphics mode low resolution is no longer used. Medium resolution is 320 $\times$ 200 pixels (a pixel is the smallest area on the

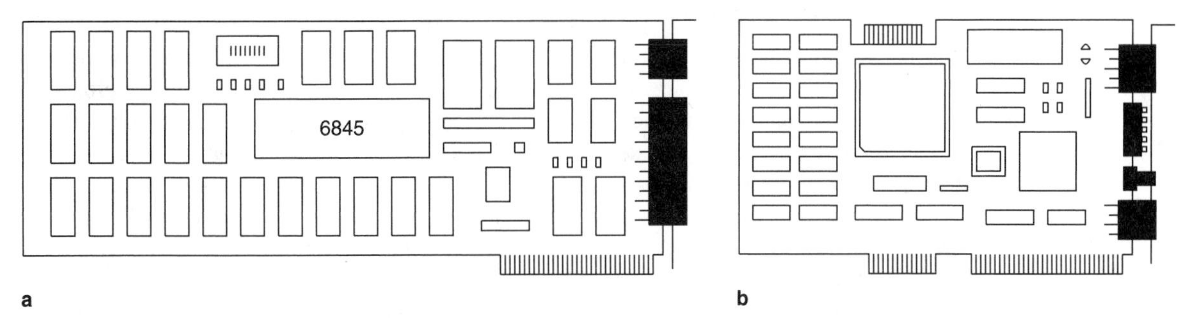

Figure 1-10 **(a)** Monochrome display adapter; **(b)** VGA display adapter.

screen that the adapter can control in the selected resolution mode) in one of four colors, with high resolution being 640 × 200 pixels in one of two colors. There is 16K of video memory to control the characters, graphics, and colors for the adapter. There are a variety of outputs to the monitor. RBG (red, blue, green, which are TTL compatible) provides the best-looking video, and a composite output (which is a combination of all the signals mixed into one complex waveform, applied to a TV-like monitor) produces poor-looking video but is cheap.

EGA (enhanced graphics adapter) provides high resolution in 16 colors and displays characters that look almost as good as those in MDA. Character size is 7 × 9, the same as MDA, but there is one less dot spacing between characters in the row. EGA also provides more resolution in the graphics mode with more colors; maximum resolution is 640 × 350 pixels in 16 colors. EGA adapters are downward compatible (support all MDA and CGA video modes). EGA adapters require between 64K and 256K of VRAM. Outputs of monitors are TTL RGB and composite compatible.

VGA (video graphics array) (Fig. 1-10(b)) provides somewhat higher resolution than the EGA system and offers many more colors. Characters look about the same, but the graphics mode has more control points—640 × 480 pixels—and each pixel can be in one of 256 colors. Some adapters allow the user to produce more than 80 characters per row and 25 rows per page at a lower character resolution. The VGA adapter is downward compatible with MDA, CGA, and EGA. Many older VGA adapters provide both TTL RGB (CGA and EGA type monitors) and analog (VGA monitor) outputs. Most standard VGA cards contain 256K of VRAM.

SVGA (super video graphics array) provides higher resolution than VGA but at a higher cost. Characters look the same, but the graphics mode has more control points—1280 × 1280 pixels—and each pixel can be in one of 16 million colors. An SVGA adapter is downward compatible with MDA, CGA, EGA, and VGA. It normally contains 256K to 4MB of VRAM and provides analog outputs.

The designation 8514A is not the name of a video adapter but is an IBM high-resolution graphics standard that is becoming more popular in CAD applications. An 8514A adapter is usually an enhanced SVAG card that provides outstanding resolution, 1024 × 768 in anywhere from 16 to 65,536 colors, depending on the amount of memory installed (from 512K to 4M). Only an analog output is provided.

1.3.11 External Memory Cards

By the virtue of BIOS and DOS, only the first 640K of memory can be used by standard DOS-based programs as user (conventional) memory, even though the 8088 can address 1MB, the 286/386SX 16MB, and the 386/486 4GB. Memory locations between 640K and 1MB are reserved for I/O (video and hard drive adapters) and system ROM(s); this area of memory is referred to as UMEM (Upper MEMory). Memory above 1MB is called extended memory and is used differently in XT and AT systems; the external memory cards must be compatible with how the memory is used.

In an XT system that uses the 8088, only 1MB can be addressed directly; any memory above 1MB must be addressed indirectly by means of a software driver (memory manager) and a compatible hardware memory card. The current standard was developed by Lotus/Intel/Microsoft (LIM) and set forth in the expanded memory specification (EMS). Because this specification was developed before the 286s were in the market, it does not use the advanced hardware memory management capabilities of the 286 or more advanced MPUs. It was intended for use with XT systems, although it does work on AT systems. The software driver developed after EMS is called an EMM (expanded memory manager) driver; it is loaded in conventional memory and is supplied with a memory card or software. This driver makes DOS look for an unused block of 64K in UMEM and sets up a software page frame. The page frame is divided into four pages of 16K each, and this area of memory is used as a gateway between conventional memory and extended memory. When software compatible with EMM needs information located in extended memory, the EMM driver transfers the page or pages into the expanded memory area, where it can be addressed by the DOS-based software. The EMM driver can map out up to 32MB as expanded memory. The memory card in Figure 1-11(a) is a typical 8-bit memory card that can contain 256K (using four rows of 64K × 1 DRAMs), 1MB (using four rows of 256K × 1 DRAMs), or 4MB (using four rows of 1MB × 1 DRAMs). Note that the bottom right side of the card contains 31 traces on the edge connector (62 actual traces, 31 on each side), which allows this card to support only an 8-bit data bus.

All AT systems use MPUs that can directly address more than 1M memory locations, 16MB in the 286/386SX, and 4GB in the 386/486s. In order to maintain complete compatibility with the 8088/8086 MPUs, these MPUs have different types of operating modes (an operating mode controls how the MPU accesses memory). The 286 has two operating modes (real and protected), and the SX386/486DX/486SX/ Pentium systems have two operating modes: real and protected (virtual 86). These modes are usually switched through software; unless the operating system or software switches the MPU operating mode, it stays in the real (default) mode. The real mode causes these MPUs to emulate the operation of the 8088/8086 family with its limitations (only the lower 20 bits of the address bus function). In the real operating mode they cannot address more than the 1MB memory locations directly. Most standard DOS software (software that was not written to take advantage of AT's operations) that can operate on an XT system requires operation in the real mode. Enhanced DOS software programs (the type that takes advantage of the protected and virtual modes), which are very limited in number, allow use of some if not all the memory that the AT MPUs can address directly. While in these modes the MPU can use conventional memory (first 640K), extended memory (above 1MB), high memory (1MB to 1.088MB), and expanded memory (converted extended memory), or any combination of these types of memory, provided that the required drivers and software are used. The card in Figure 1-11(b) would be used in a 286 or SX system because it fits a 16-bit I/O interface slot. (Note the edge connector on the bottom right-hand corner of the card. This could hold 512K to 8MB, depending on the type of DRAMs used.) A 32-bit memory card for a 386/486 system would have more traces on the edge connector near the center of the board.

In the later version of 386, 486, and Pentium motherboards, the memory is in the form of SIMMs. There are usually eight SIMM slots, each capable of supporting 256K (256K × 9),

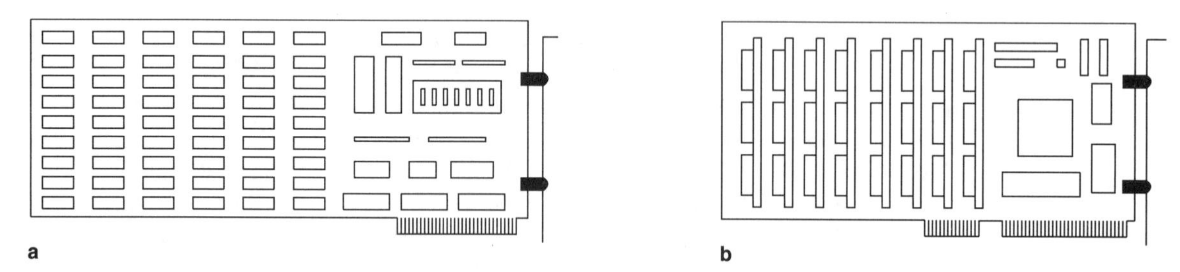

Figure 1-11 **(a)** 8 -bit memory card; **(b)** 16-bit AT memory card

1MB (1MB × 9), or 4MB (4MB × 9) SIMMs, thus allowing up to 32MB of data storage on the motherboard. In the early 1990s some computer companies started using 36-bit SIMM packages which quadruple the amount of memory on each 72-pin SIMM package. With 36-bit SIMMs 128MB of memory can be placed on the motherboard.

1.3.12 Data Communication Ports

The two most popular types of data communication ports found in PCs are serial and parallel. The serial (asynchronous) port is used to communicate with devices that send and receive data one bit at a time. In PCs asynchronous serial communication ports are labeled COM1 (communication port 1), COM2, COM3, and COM4. Typical serial devices are modems, some printers, plotters, laser printers, and mice. The advantages of serial data devices are the need for fewer wires (which lowers cost and reduces the possibly of noise problems) and longer transmission distances. The disadvantages of serial data communication are that it is slow and it requires more circuitry to implement than does parallel communication.

The parallel ports in PCs are used to communicate with printers (Centronic compatible). Because parallel ports are used primarily with printers, the term *printer port* is used. Printer ports are labeled PRN (primary printer port), LPT1 (line printer 1), and LPT2. The advantage of parallel communication is that one byte of data is transferred at a time, making it faster than serial communication. The disadvantages of parallel communication are the need for more wires, sensitivity to noise, and limited transmission distances (because of signal levels).

In the early days of PCs, each type of port required its own interface card and took up one interface slot. This limited the number of devices connected to the computer (because of a limited number of interface slots on the motherboard). Although these types of I/O cards are still available, they are not popular.

Manufacturers of I/O interface cards have now placed parallel and serial communication ports and other interfaces onto one I/O interface card, called a multifunction I/O card (adapter). Figure 1-12(a) shows a typical 8-bit multifunction I/O card that contains one parallel printer port, two serial ports, battery–backed up-real-time clock, and game port. Figure 1-12(b) shows a typical 8-bit multifunction memory I/O card that contains the same ports as the card in (a), plus up to 4MB of extra memory. Other multifunction I/O cards contain parallel printer ports, serial ports, a game port, a clock, a floppy disk adapter, an IDE adapter, and sometimes even a video adapter, all on one card.

1.3.13 System Unit Cases

Many different types of cases are available for PCs (Fig. 1-13). The primary differences are their size (which limits the number of drive units that can be stored internally in the system unit) and how they take up space in an office. Most cases are made of steel to reduce radio frequency interference emitted from the computer. Computer cases made of plastic usually contain a steel shield around the motherboard.

Most cases also provide a control/status panel. The panel usually contains a turbo switch (switches the speed of the computer), reset switch (resets computer without removing power), and sometimes an on/off switch. The status part of the panel usually contains a turbo light, power-on light, and hard disk access light (lights up when the hard disk is being used).

Tower cases are made to stand upright to reduce the amount of desk space needed. Full tower cases provide the most space for drive units (a drive unit normally specifies the space required for a half-height 5 1/4-inch floppy drive unit)(six exposed, two or three hidden) and can be placed on the floor next to the desk. Mini tower cases are meant to stand on the desk and usually provide less space for drive units (three half height and two micro height exposed). Normally full tower cases have 386/486 systems in them, and mini tower cases have 286/SX/386 systems.

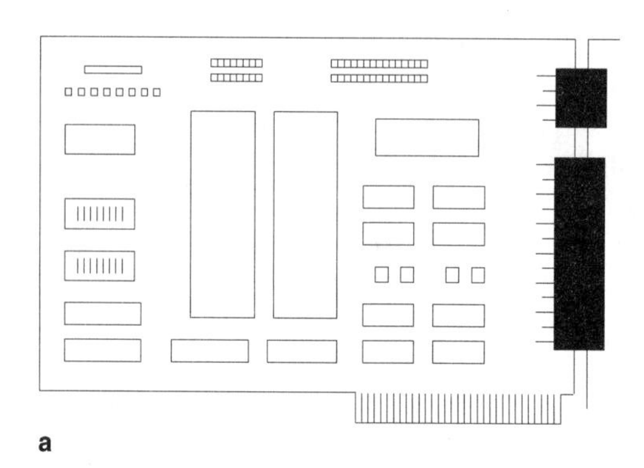

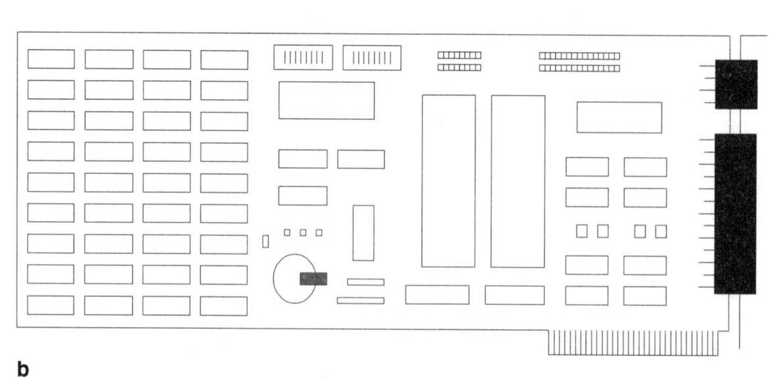

Figure 1-12 **(a)** XT multi-I/O card; **(b)** XT memory multi-I/O card.

There are many different types of desk style cases (full AT, full XT, and mini XT/AT). A full AT case can have any type of XT or AT motherboard in it and normally provides space for three exposed and two hidden drive units. The full XT case usually has an XT motherboard but may also contain a mini size AT motherboard and provides space for two exposed and two hidden drive units. The mini XT/AT case requires an XT or XT-size AT motherboard and provides space for only three exposed drive units.

1.4 VIDEO MONITOR

The video monitor (Fig. 1-14) is an output device (similar to a TV screen) that provides a visual indication of the operation of the computer. The video monitor (also known as the CRT) is controlled by the video adapter, which is located in the system unit. The type of monitor used depends on the video adapter.

Video monitors can display information in either monochrome (colors are black/white, black/green, or black/amber) or color (the number of colors is determined by the video adapter). Video monitors are usually dumb (cannot switch frequencies for different types of video adapters), but some multifrequency video monitors automatically switch frequencies to match the video adapter. Another important specification for monitors is dot pitch (the space between the dots where light is produced); the smaller the value, the sharper the display. Typical dot pitches are 0.39 mm, 0.31 mm (considered a good size), 0.28 mm, and 0.26 mm. The size of the display is specified as a diagonal measurement in inches (12 and 14 inches standard, 16 and 20 inches large) of the displayable area; the larger the size, the more the display costs.

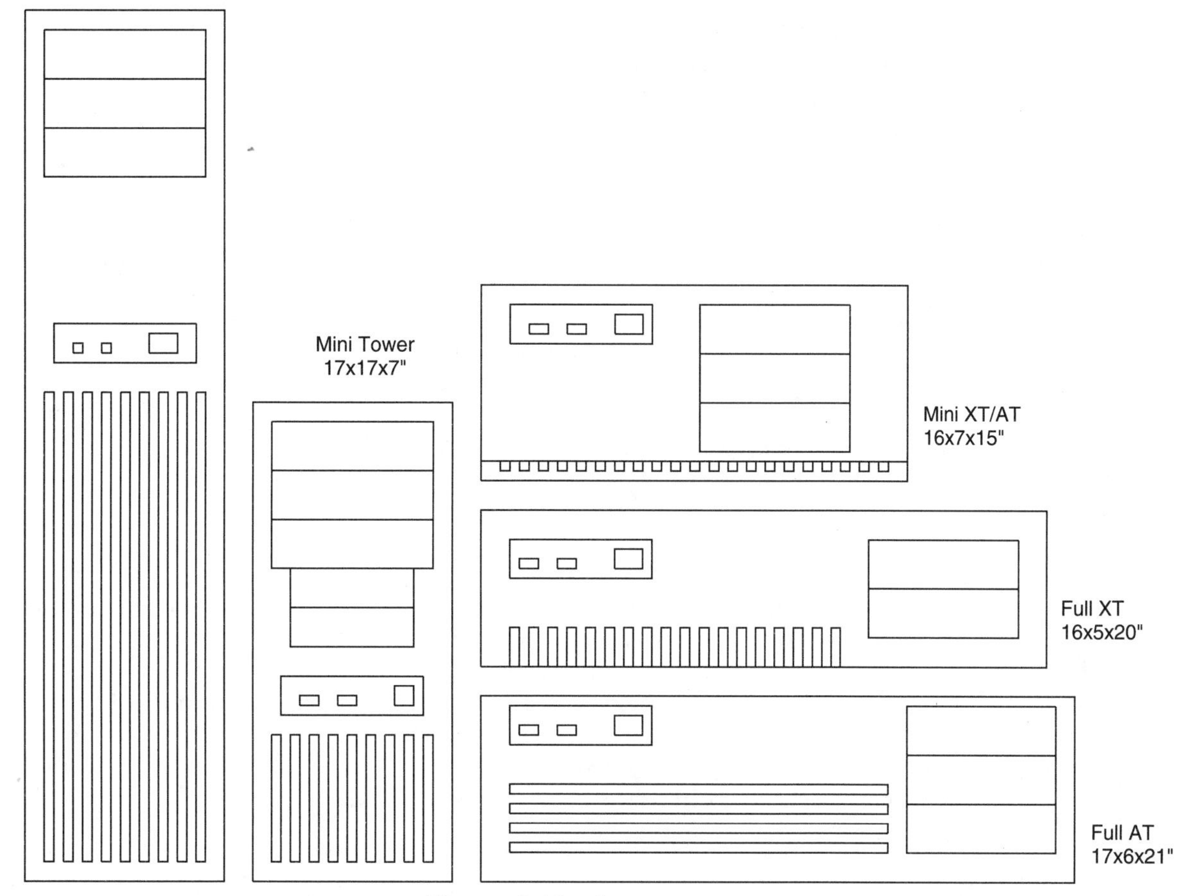

Figure 1-13 Computer cases.

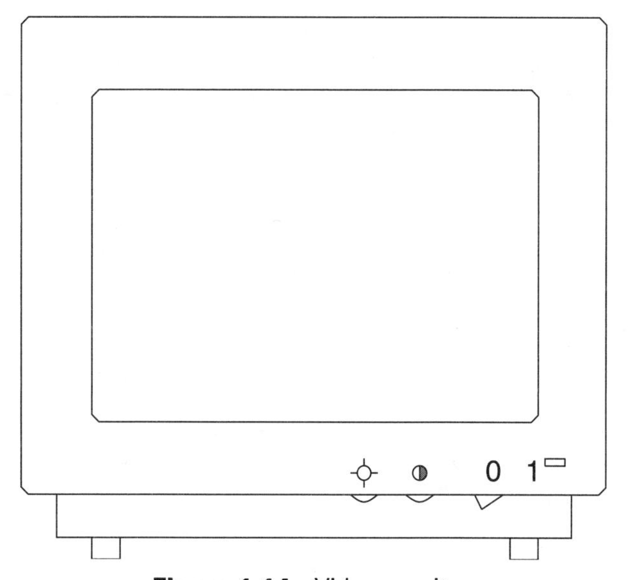

Figure 1-14 Video monitor.

Most monochrome and color displays offer an on/off switch and two front controls that allow the user to adjust the display for maximum visibility. The brightness control (looks like a sun or star) is used to adjust the overall brightness (foreground [character] and background) of the display. The contrast (circle half filled in) is used to adjust the foreground brightness (has little affect on background brightness) in a color monitor. The contrast control is used normally to adjust the background brightness in a monochrome display. Some monitors offer other controls at the front of the monitor.

1.5 KEYBOARD

The most-used part of the computer system is usually the keyboard. There are basically two types of keyboards, the 83-key keyboard (Fig. 1-15) and the 101-key keyboard (Fig. 1-16). The main differences between these keyboards are their layout and the number of keys they contain. When a key is pressed, a key code is sent serially from the keyboard to the keyboard-encoding circuitry on the motherboard, where the key code is processed. Most keyboards are switchable between the key codes for XT and AT computer systems. It is very important that the correct key code format be selected because of how the key codes are sent and processed by the computer's motherboard.

1.5.1 Function Key Area

The function key section of the keyboard has no predefined function; these keys are programmable and perform different functions depending on the software being used.

Ten function keys (labeled F1 through F10) are found on an 83-key keyboard. These keys are located on the far left side of the keyboard. The 101-key keyboard has 12 function keys (labeled F1 through F12). These keys are located along the top of the keyboard in three groups of four keys each. Because most software was written before the 101-key keyboard was developed, function keys F11 and F12 are not normally used.

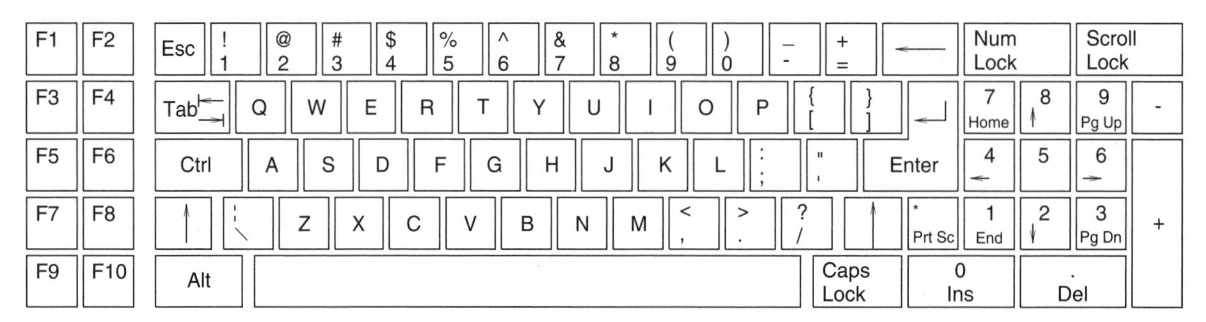

Figure 1-15 83-key keyboard.

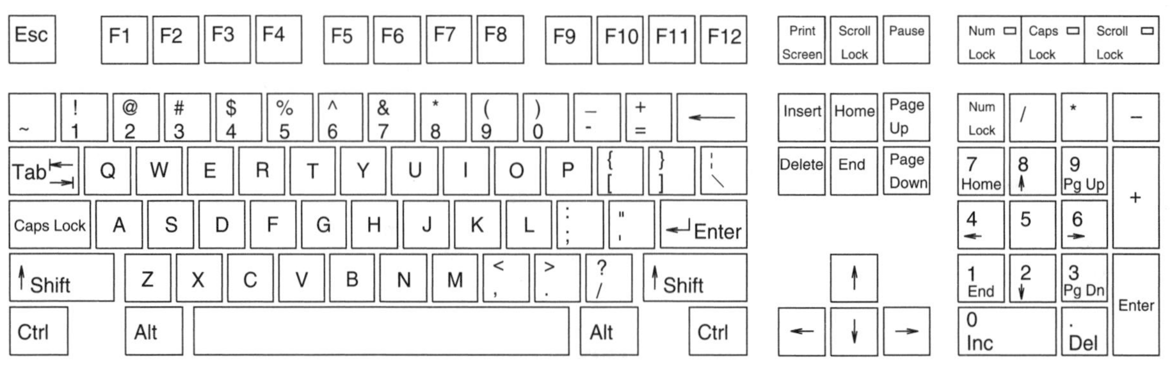

Figure 1-16 101-key keyboard.

1.5.2 Typewriter Area

The typewriter area of both types of keyboards is located in the five lower rows in the center of the keyboard. This area is used the most because it contains the keys that produce the English alphabet, numbers, punctuation marks, and special characters that can be entered into the computer. The following is a list of special keys in the typewriter area and what they do. All other keys in this area perform the same functions as on a typewriter.

The most used key is the *enter* key. This key is located on the right-hand side of the typewriter area (some keyboards use the word *Return* instead of *Enter* on this key). Enter tells the computer that the end of the program line has been reached or that a command should be performed.

The next most used key is probably the *backspace* key, located in the upper right-hand corner of the typewriter area. This key is usually indicated by an arrow pointing left. If the computer is in the overstrike mode, this key erases the character to the left of the cursor and moves the cursor one space to the left without affecting any other character on the screen. If the computer is in the insert mode, this key erases the character to the left of the cursor, moves the cursor one space to the left, and causes all characters to the right of the cursor to move to the left one character position. The function is called a destructive backspace because the character is lost unless retyped.

A very useful key is the *print screen* key. On the 83-key keyboard, (*/PrtSc) is located under the enter key. On the 101-key keyboard (Print/ Screen) is located near F12. When the (*/PrtSc) key is pressed by itself on an 83-key keyboard, the symbol for the asterisk is entered into the computer. When the (Print/Screen) key on the 101-key keyboard or the shift and */PrtSc combination is pressed on an 83-key keyboard, everything on the screen is printed by the printer. Only characters or symbols are printed by the printer unless the graphic command in DOS has been loaded into the computer, in which case everything on the screen is printed on the printer. If the print screen key is pressed along with the control key, everything typed in or printed by the computer is sent to the printer. This function continues until the key sequence is pressed again or the computer is reset.

Just as on a typewriter, there are two *shift* keys. The shift keys are located near the bottom of the typewriter area of the keyboard. The shift keys may be indicated by an arrow pointing up or by the word Shift. When the shift key is held down while another key is pressed, the second function of the key is entered into the computer. For example, if you press the letter S without the shift key, a lower case letter is produced. When the shift key is pressed along with the S, a capital letter is produced.

The IBM PC keyboard has a *caps-lock* key. When this key is pressed once, only upper case letters are produced. This above condition continues until the caps-lock key is pressed again, thereby unlocking the function. Once unlocked all keys produce their lower case characters unless the shift key is used. If the shift key is pressed when the caps-lock key is locked, lower case letters are entered. Some keyboards provide a visual indicator that lights up when the caps-lock key is locked.

As on a typewriter, the bar at the bottom row of the typewriter area of the keyboard is the *space bar*. The space bar has no markings but is the largest key on the keyboard. This key inserts a space into the data being sent to the computer.

The *alternate* key is used with other keys to input a command or function without typing in the full command or function word. This key is located just left of the space bar at the bottom of the typewriter area of the keyboard. Because there is no default value for the Alt key, it may not function all the time.

The *control* key also performs special tasks when used with another key. Its function differs with different types of software. Ctrl is located under the tab key or next to the alternate key.

The *tab* key is located near the upper left-hand side of the typewriter area of the keyboard. Some keyboards have the word Tab on the key and others have arrows pointing left and right. When this key is pressed, the cursor moves to the next tab position to the right. If the shift and tab keys are pressed together, the cursor moves to the next tab location to the left. Tab settings differ from language to language and program to program.

The *escape* key performs different tasks with different languages and application programs. Normally the escape key executes a soft break, causing the main menu to be displayed on the screen. This key normally allows the user to exit or change the work environment without losing data.

On some typewriters the letter O is used instead of the number zero. When using computers this character exchange is not proper and may cause errors in the program. Similarly, on the computer you should not use the letter l instead of the number one.

1.5.3 Numeric Keypad Area

The numeric keypad contains 15 keys on the 83-key keyboard and 17 keys on the 101-key keyboard. On both keyboards the numeric keypad is located on the far right side of the keyboard. The numeric keypad serves two functions on both keyboards. When the keyboard is in the numeric mode, these keys are set up like a calculator and allow the user to enter numbers and symbols into the computer quickly. When the keyboard is in the cursor mode, these keys allow the user to change the position of the cursor on the screen. The 101-key keyboard has another set of 11 keys that always performs the cursor positioning (even if the keyboard is in the numeric lock mode). These keys are located between the typewriter area and the numeric keypad.

The *numeric lock* (NumLock) key is a toggle key that controls the mode of operation of the numeric keypad. When the computer is first powered up, the NumLock key is in the unlocked position (cursor mode)(NumLock light is off), allowing the keys on the keypad to position the cursor on the screen. When it is pressed once, it is in the locked position (numeric mode)(NumLock light is on), and the keys on the keypad input numerical data. The NumLock key is unlocked by pressing it again. With some software, pressing Ctrl and NumLock together freezes the contents of the display. The freeze ceases as soon as any other key is pressed.

The *scroll lock/break* key is a toggle key that affects the contents of the screen scrolling. When listing a program that requires more than one screen, the scrolling process can be stopped by pressing the shift and scroll lock keys (Scroll Lock light goes on). To start the display listing again, simply press the scroll lock/break key again (Scroll Lock light goes off). To stop most running programs that have gotten out of control, press the control key along with Scroll Lock/Break (this sequence issues a control C).

On the 101-key keyboard, the *pause* key (Pause/Break) freezes the contents of the screen until any other key is pressed. If the Ctrl and Pause/Break keys are pressed, a program that has gotten out of control normally stops (this sequence issues a control C).

The plus and minus keys on the keypad are not affected by the NumLock key. Whenever these keys are pressed, a plus or minus sign is entered into the computer. All other keys relate to NumLock; when locked, none of the following keys produces numeric input. Instead, each performs a unique function.

The *insert* key is located in the bottom left corner of the numeric keypad. When NumLock is unlocked, this key controls whether the computer is in the insert or the overstrike mode. Pressing this key once puts the computer into the insert mode; pressing it again changes to the overstrike mode. In the insert mode, each time a key is pressed a new character is entered into the computer at the position of the cursor. All characters to the right of the cursor move to the right one space. In the overstrike mode, any key pressed causes a character to replace the character above the cursor, without affecting any other character on the screen.

The *delete* key is located on the bottom row, just to the right of the *insert* key on the keypad. This key is not affected by the insert or overstrike mode. When NumLock is unlocked, the decimal-delete key deletes the character at the position of the cursor. After the character is deleted all characters to the right of the cursor move one postion to the left. When there are no characters on the line, the blank line is deleted.

When the *home* key is pressed and NumLock is unlocked, the cursor either moves to the top left-hand corner of the present screen or, in some programs, goes to the first screen of text in the document.

When the *up arrow* key is pressed and NumLock is unlocked, the cursor moves upward one character row.

When the *page up* key is pressed and NumLock is unlocked, either the cursor moves to the top of the current screen or the previous page of text is displayed on the screen.

When the *left arrow* key is pressed and NumLock is unlocked, the cursor moves to the left one character position.

The number 5 on the keypad has no function when NumLock is unlocked.

When the *right arrow* key is pressed and NumLock is unlocked, the cursor moves to the right one character position.

When the *end key* is pressed and NumLock is unlocked, the cursor moves either to the last character of the present screen or to the last character of a document.

When the *down arrow* key is pressed and NumLock is unlocked, the cursor moves down one character row.

When the *page down* key is pressed and NumLock is unlocked, the cursor is positioned at the bottom of the current screen page or the next screen of text.

1.6 OPTIONS TO COMPLETE THE SYSTEM

To complete the computer system, at least two additional items are needed. The printer allows the computer to produce a permanent record of the work it performs. Software is a list of instructions that allows the user to perform specific tasks on the computer.

1.6.1 Printer

No computer system is complete without a printer, a device that takes information (character or graphic) from the computer and makes a permanent record on paper (hard copy). There are many different types of printers; some use mechanical force or heat, and others use light to produce the hard copy.

The most popular type of printer is a dot-matrix printer. Impact dot-matrix printers (Fig. 1-17) use mechanical force to produce characters and graphics using a series of dots in rows and columns. The most common printers contain 9 or 24 pins in their print head. The advantages of this type of printer are low cost, near letter quality (24-pin printers), programmable characters, and speed (60 to 400 characters per second). The disadvantages are noise and lack of letter quality (9-pin printers).

Over the past few years, the price of laser printers has decreased to the point that most businesses and professionals can afford them. A laser printer (Fig. 1-18) is a high-resolution dot-matrix printer (300 to 600 per inch). The laser printer uses a small laser and crystal deflection system that places the image on a photosensitive drum (just like a copying machine).

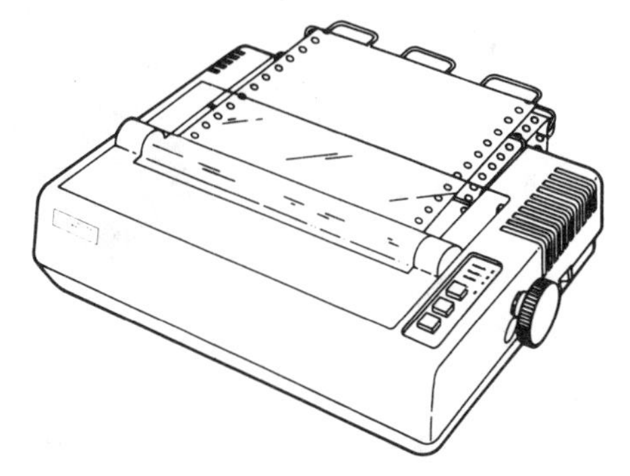

Figure 1-17 IBM graphic printer. (Courtesy of IBM)

The toner (black plastic material) in the printer is transferred to the areas of the drum where the light has hit. Paper is brought in proximity to the drum, and the image is transferred to the paper. The paper is then sent through a heating process, which bonds the toner to the paper, while the drum is electrostaticly erased. The advantages of laser printers are very high resolution, letter quality, and speed (4 to 20 pages per minute). The disadvantages are a somewhat shorter life and cost (less than $700) compared with a dot-matrix printer.

Starting in the 1990s, printer manufacturers redesigned their ink jet printers to be less costly and more reliable. An ink jet printer (Fig. 1-19) is a dot-matrix printer with a print head in a disposable cartridge that contains the ink for the printer. Many small cavities at the bottom of the ink cartridge fill up with ink. When the printer is commanded to print a dot on the paper, a heater in one or more of the small cavities is turned on. As the ink heats, it expands the very small opening in the cavity and causes the ink to leave the cartridge. The voltage applied to the heater and the size of the opening determine the amount of ink that hits the paper and therefore the size of the dot. Most ink jet printers can print 300 to 600 dots per inch, producing laser printer quality text and graphics. Ink jet printers are also very quiet and as fast as most dot-matrix printers (two to four pages per minute in black and white). Because the print head is changed each time the ink cartridge is replaced, the print quality usually does not deteriorate over time. Most ink jet printers also have the ability to print in color. Color ink jet printers may use one cartridge containing up to three colors (red, blue, and yellow), plus one

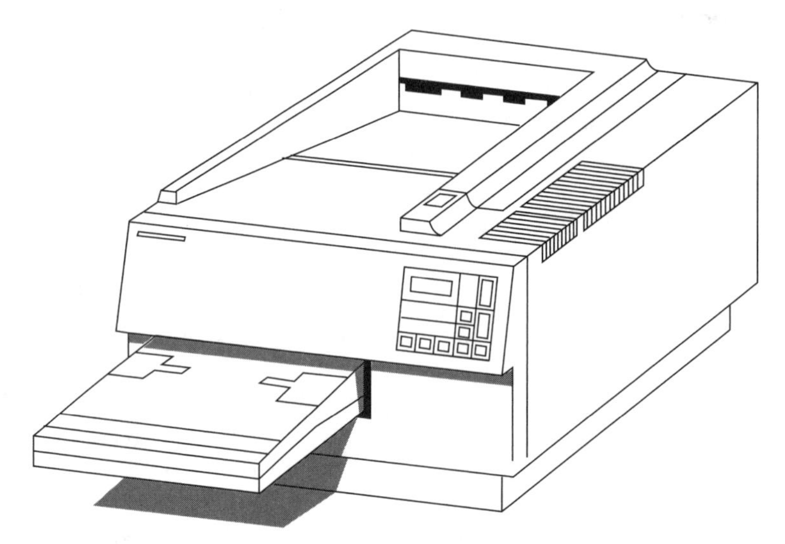

Figure 1-18 Laser printer.

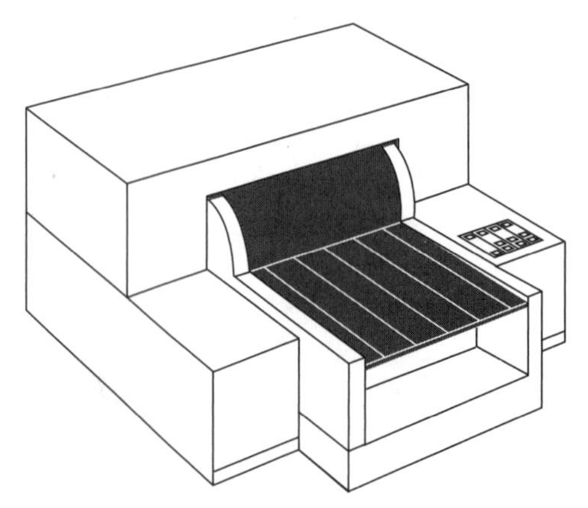

Figure 1-19 Ink jet printer.

cartridge containing only black, whereas other ink jet printers have a separate ink cartridge for each color. By mixing the colors and the amounts of each color, just about any color can be recreated. The advantages of ink jet printers are low cost ($250 to $1000), low noise, high resolution, color printing, and speed. The disadvantages of ink jet printers are the cost of the ink cartridges and the fact that the quality of the print is highly dependent upon the quality of the paper.

1.6.2 Software

A good computer repair technician must not only be proficient in hardware but should have some entry-level knowledge of some software. Such knowledge allows the technician to determine more easily if a problem lies in the software, hardware, or user error. Which software should be learned? That depends on which part of the computer field the technician plans to work in. Therefore, because the scope of this book is to provide hardware knowledge of PCs, not software knowledge, only software common to all PCs is covered.

Software can be divided into three general categories—operating systems, programming languages, and application software. Because the operating system is necessary for any PC to function and perform other tasks through software, this book covers DOS. Programming languages are instructions that a person puts into a list (writing a program) to perform a specific task. Application software is any prewritten program that performs specific tasks.

Operating System. Although there are many different types of operating systems, DOS is by far the most popular for PCs. Basically all other software requires DOS before any type of operation can be performed. Most software is written to work with PCs running under DOS 2.1 or higher. For further information, see discussions of operating systems in Chapter 2.

Windows (MicroSoft) is a graphics environment that runs under DOS to create a GUI (graphics user interface). GUIs are very user friendly because once the user learns how to use Windows, all programs written for Windows use the same type of instructions. Most functions are represented by icons (pictures that indicate the tasks the function performs). Windows also allows the user to share data without converting the data with any software written for Windows. With the right type of hardware, Windows also allows multitasking of software.

Programming Languages. BASIC (beginner's all-purpose symbolic instruction code) is an interpretive high-level computer language that allows the user to develop programs using English-like terms. It was developed at Dartmouth College for use by non–math-oriented computer users. It is easy to learn and comes with most small computers. The version of BASIC that comes with DOS is not very advanced. More advanced versions of BASIC are Professional BASIC, Quick BASIC (by MicroSoft), and Turbo BASIC (by Borland International).

"C" (language compiler) is a high-level compiled system development language that is very efficient. Many application programs and DOS use C. Lattice "C" (Lattice) is one of the standards; others are Quick C (MicroSoft) and Turbo C Prof (Borland International).

Toolbook (Asyetrix) is a Windows programming language that is used to develop software applications and front ends (conversion utilities) for the Windows environment. Toolbook allows the programmer to develop programs by linking graphic routines together to perform specific software applications.

Application Software.

Utilities: Utility software allows the user to perform tasks that enhance other software running in the computer. DOS utility software such as PC-Tools (Central Point Software) and Norton Utilities (Symantec) allows the user to perform DOS functions and functions that should have been available through DOS with a minimum amount of effort. Direct Access (Delta Technology) enables users to develop menus for hard disk systems, allowing the user to press one key to start the software. Sideways (Funk Software) is a program that allows printers to print horizontally (sideways) across the paper.

Spread Sheet: A spread sheet is a program used in business to keep records of transactions involving numbers. This type of software can usually produce graphical charts using the values in the spread sheet for visual representation. Lotus 123 (Lotus), Quattro Pro (Borland International), and Excel (MicroSoft) are the major PC spread sheet programs.

Data Base: A data base management program allows the user to input into and output data from an electronic file cabinet using many different formats. The user can then sort, display, and print the stored data in a variety of ways. Foxbase (Fox Software), R:Base (Microrim), and dBASE (Ashton Tate) are some of the most popular data base managers.

Word Processor: A word processor allows the computer to act as an intelligent memory typewriter. A word processor allows the user to make corrections, reformat, and update the document easily. Most word processors contain a spell checker and some type of thesaurus. The more advanced word processors now allow the user to import graphics (pictures or art) into the document and print it out. Aim Professional (Lotus), WordPerfect (WordPerfect Corp.), and Word for Windows (MicroSoft) are among the most powerful word processors today.

CAD/Graphics: A CAD (computer-aided drafting) or graphics program allows the user to create pictures, schematics, and line art on the computer. The graphics created can then be stored, calibrated, and changed at a later time. Design CAD (American Small Business Co.), Auto CAD (AutoDesk), and PC Paintbrush (Z-Soft) are very popular.

Desktop Publishing: Desktop publishing (DTP) allows the user to integrate graphical and character-based information in one document. The software then allows the user to modify how the information is displayed and printed out. PageMaker (Adobe) and Ventura Publisher (Xerox) are the major players in the market today.

1.7 HANDLING, INSERTING, AND REMOVING DISKETTES

If the following procedures are followed, the integrity of the data and life of the diskette are prolonged, resulting in fewer mass storage problems. The following are instructions for handling, storing, and caring for 5 1/4-inch diskettes. Although 3 1/2-inch diskettes have a more durable jacket, these procedures should still be applied. High-density diskettes are even more susceptible to damage; therefore, these procedures should be followed to the letter.

In addition to the handling procedures described in Figure 1-20, the following procedures should also be followed.

1. Keep diskettes in a dry environment.
2. Do not place diskettes on or near the video monitor.
3. If a diskette has been subjected to extremely cold temperatures, allow the diskette to warm up before using it.
4. Store diskettes in a vertical position, not a horizontal one.

1.7.1 Loading Diskettes

To insert or load a diskette into a disk drive unit, perform the following steps (Fig. 1-21).

1. Lift the load lever (door latch) on the 5 1/4-inch drive (3 1/2-inch drives do not have door latches) only if the LED (light emitting diode, red light) on the drive unit you wish to load is off. If the LED is lighted, the R/W heads are engaged and damage may occur.

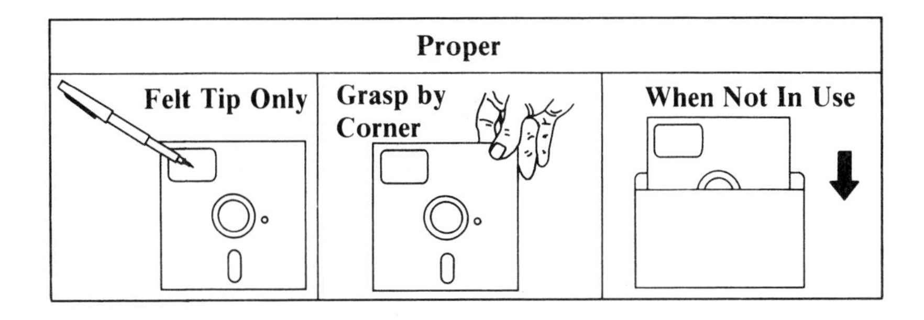

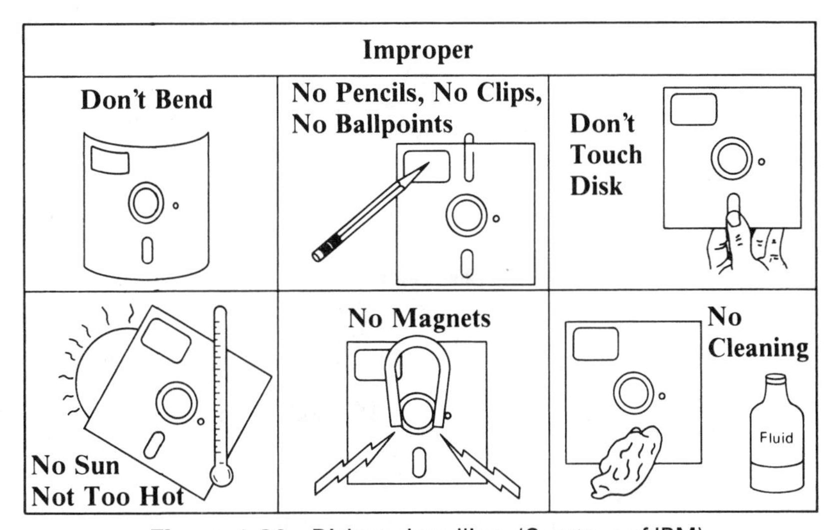

Figure 1-20 Diskette handling. (Courtesy of IBM)

2. Insert the diskette into the slot of the drive unit with the diskette label facing upward, the write-protect notch to the left, and the R/W slot away from you. This slot is the first part of the diskette inserted.
3. Once the 5 1/4-inch diskette has been inserted, pull the load lever down until the latch is closed. Once the 3 1/2-inch diskette has been inserted completely, the diskette should lock down and the eject button should pop out automatically.

1.7.2 Unloading Diskettes

To remove or unload a diskette from a disk drive unit, perform the following steps (Fig. 1-21).

1. Lift the load lever (5 1/4-inch drive) or press the eject button (3 1/2-inch drive; note that the diskette pops out about halfway), provided that the LED on the drive unit is not lighted.
2. Carefully remove the diskette from the drive. Place the 5 1/4-inch diskette in its protective sleeve (jacket).
3. Leave the load lever open.

1.8 TURNING THE COMPUTER SYSTEM ON AND OFF

These procedures refer to starting a computer system with floppy drives only or with a hard drive system in which the user wishes to boot the computer from the floppy drive. If the user wishes to boot the system from the hard drive, no diskette should be placed in floppy

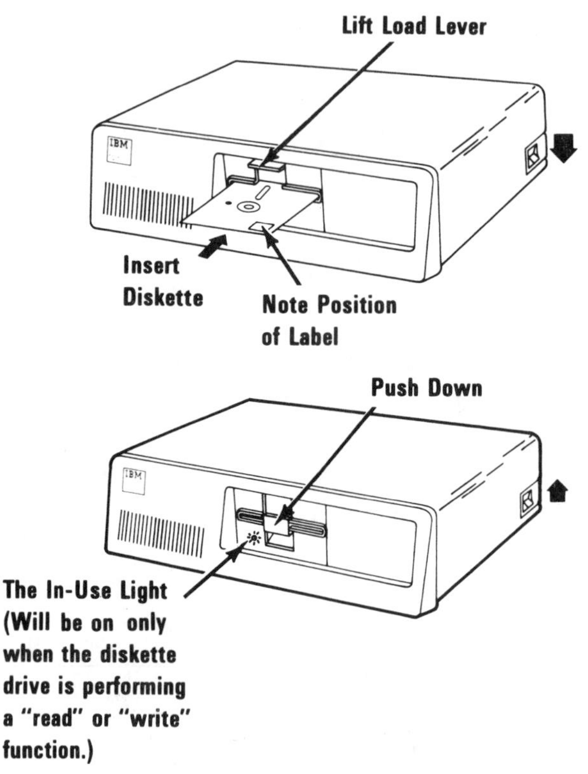

Figure 1-21 IBM PC diskette loading and unloading. (Courtesy of IBM)

drive. If the computer does not see a diskette in the drive, the computer boots from the hard drive.

1.8.1 Start Sequence

1. Turn on all peripheral devices (monitor, printer, and the like).
2. Insert the DOS diskette.
3. Apply power to the computer by using the power supply switch located on the right side of the base unit, toward the back of the unit.
4. The computer performs a POST, which checks the equipment in the system. For more detail on POST, see the Power-On Self-Test section, later in this chapter.
5. Upon successful completion of POST, the computer reads the diskette in drive A. If the read is successful, questions display which you should answer. If the DOS screen does not appear, there may be a problem with the computer or with the diskette. Once DOS is loaded, other languages or programs may be loaded into the computer.

1.8.2 Stop Sequence

1. Exit the application program or language and return to DOS.
2. Remove all diskettes from the base unit, once the LED on the drive units are off. Leave the door latches open. If the door latches are closed with no diskette in the drive, the R/W heads may be damaged.
3. Turn off the computer.
4. Turn off the power to the monitor and all other peripherals.

1.8.3 POST: Power-On Self-Test

The purpose of POST is to determine if the devices in the computer are connected and working properly. POST tests and configures everything on the motherboard, keyboard, disk drive adapters, and video adapter.

When the power is first turned on, BIOS begins POST. The cursor appears on the screen in 4 to 10 seconds. In about 5 to 60 seconds, one short beep indicates that the user memory has been checked. The amount of time to the beep depends on the amount of memory in the system. In most systems the size of the memory being tested is displayed at the upper left of the screen. After POST is complete and no errors are detected, the computer tries to boot DOS.

If a DOS system diskette is not in drive A, the system checks for a hard drive to boot the system. If no hard drive is found in an IBM PC, the computer enters the IBM cassette version of BASIC. In an IBM-compatible system, an error message appears (example: non-system disk). Some computer systems may lock up, and the user must perform a cold start with a system disk. Still other computer systems may allow the user to insert a system diskette and press any key when ready. If the hard drive does contain DOS, the computer boots DOS from the hard drive. If a DOS system diskette is in drive A, the computer boots DOS from drive A and disregards the hard drive.

CHAPTER SUMMARY

1. PC and XT computer systems use the Intel 8088/86 (16-bit MPU, 8-bit data bus, 20-bit address bus). Motherboards contain up to 640K of DRAM. Systems require a video system and a mass storage system.
2. AT computer systems use one of the following:
   - Intel 80286 (16-bit MPU, 16-bit data bus, 24-bit address bus)
   - Intel 80386 (32-bit MPU, 32-bit data bus, 32-bit address bus)
   - Intel 80386SX (32-bit MPU, 16-bit data bus, 32-bit address bus)
   - Intel 80486(32-bit MPU, 32-bit data bus, 32-bit address bus)

   All systems contain 1MB or more of DRAM. All systems require a video system and a mass storage system.
3. A computer system has three basic components:
   - System board (motherboard), which contains the MPU, ROM, DRAM, I/O interface channel, and support hardware; power supply; mass storage system; video adapter; communication ports; and steel case.
   - Video display, a device that gives the operator a visual indication of the operation of the computer.
   - Keyboard, an input device that allows the user to enter data into the computer.
4. PC/XT motherboards contain 8-bit (62-pin) interface slots, 8087 socket, DMA, PPI, CLK, PIC, PIT, ROM/EPROM, DRAM (640K maximum), and equipment dip switches.
5. AT 286/SX motherboards contain 8-bit (62-pin) and 16-bit (additional 36-pins used with the 8-bit slot) interface slots, 80287/80387SX socket, two DMAs, 8042, CLK, two PICs, PIT, ROM/EPROM, DRAM (over 1MB, 640K usable through standard DOS), real-time clock, and battery–backed-up configuration buffer.
6. AT 386/486 motherboards contain 8-bit, 16-bit, and preparatory 32-bit (86-bit) interface slots, 80387 socket two DMAs, 8042, CLK, two PICs, PIT, ROM/EPROM, DRAM (over 1MB, 640K usable through standard DOS), real-time clock, and battery–backed-up configuration buffer.

7. Integrated motherboards contain the standard equipment and built-in video adapter, floppy/hard drive adapter, and data communication ports.
8. Power supply outputs (+5v, −5v, +12v, and −12v), XT power rating (up to 150w), and AT power rating (200w to 300w).
9. Mass storage systems are of two types:
    - Floppy disk system requires a floppy disk adapter (controls floppy disk drive unit and data); floppy disk drive unit (contains motors, sensors, and electronics to control the transfer of data to and from the diskette); 5 1/4-inch mini floppy diskette (DSDD 360K and DSHD 1.2MB) and/or 3 1/2-inch micro floppy diskette (DSDD 720K and DSHD 1.44MB); and DOS (disk operating system), a program that controls the hardware of the computer system and allows access to mass storage.
    - Hard disk system requires a hard disk adapter (performs same tasks as floppy disk adapter), hard drive unit (contains motors, sensors, electronic circuitry, and hard disk magnetic media), and the operating system (DOS).
10. The video adapter converts the data from the computer into the signals necessary to make the monitor display the information on the video screen. Types of video are MDA (monochrome display adapter), CGA (color graphics adapter), EGA (enhanced graphics adapter), VGA (video graphics array), SVGA (super video graphics array), and 8514A video standard.
11. External memory cards allow the user to add more memory to the computer system than can be held on the motherboard. The first 640K is conventional memory, 640K to 1MB is upper memory (reserved for I/O and ROM), and over 1MB is extended memory. XTs cannot use extended memory; it must be converted into expanded memory. ATs can use extended memory, expanded memory, and high memory.
12. Data communication ports allow the computer to communicate with peripherals (in a parallel or serial format).
13. System unit cases are used to protect the electronics in the system unit and reduces radio frequency interference emitted by the computer.
14. The video monitor of the computer system is the device that produces the video image on the screen. The type of video monitor must match the type of video adapter.
15. The keyboard allows the user to input data to the computer. There are three areas of the keyboard.
    - The function key area is composed of the programmable keys that perform special functions depending on how they are programmed.
    - The typewriter area of the keyboard contains the English alphabet, numbers, punctuation marks, and special characters that can be entered into the computer.
    - The numeric keypad allows the user either to enter numerical data into the computer or to position the cursor easily.
16. Most computer systems contain a printer, a device that produces a hard copy of the work the computer does.
17. Software is a list of instructions telling the computer how to perform a specific task. Types of software are the operating system, programming languages, and application software.
18. To properly handle diskettes, keep them in a dry environment, store in a vertical position, use only a felt tip pen to write on the label, grasp diskettes by corners, and replace diskette in the sleeve when not in use.
19. The following are cautions about handling diskettes: Do not place diskettes near a magnetic field, keep out of extreme conditions (cold, heat, and sunlight), do not bend, do not touch a magnetic surface, do not try to clean the diskette surface, and do not use pencils, ballpoint pens, paper clips, or anything else that would alter the protective jacket of the diskette.

20. Never insert or remove a diskette from a drive unit when the LED is on.
21. To start the computer, turn on all peripheral devices, insert the DOS system disk, and apply power to the system unit.
22. To stop the computer, exit the application program (return to DOS), remove all diskettes from the system unit, turn off the computer, and remove the power from system unit.
23. POST (Power-On Self-Test) is a program in BIOS that is used to configure and check all the equipment in the computer system.

REVIEW QUESTIONS

1. How many actual memory addresses can the 8088 MPU access?
   a. 16,384
   b. 65,536
   c. 262,144
   d. 524,288
   e. 1,048,576

2. The Intel ____________ is a 32-bit MPU with a 16-bit data bus.
   a. 80286
   b. 80386DX
   c. 80386SX
   d. 80486DX2
   e. 80486SX

3. Which one of the three main components of the computer system contains the power supply for the computer?
   a. system unit
   b. keyboard
   c. video monitor

4. What part of the PC/XT motherboard informs the computer of the type of equipment in the computer?
   a. DRAM
   b. 8088 MPU
   c. 8259 PIC
   d. CMOS
   e. None of the given answers

5. A 32-bit interface slot would not be found in which type(s) of computer systems?
   a. XT
   b. 386DX AT
   c. 486SX AT
   d. 486DX4 AT
   e. 486DX2 AT

6. What type of a motherboard contains a built-in video adapter, floppy/hard disk adapter, and data communication ports?
   a. motherboard
   b. integrated motherboard
   c. dedicated motherboard
   d. closed-access motherboard
   e. None of the given answers

7. Which of the following voltages are not found in a PC's power supply?
   a. –5v
   b. +5v
   c. +9v
   d. –9v
   e. None of the given answers

8. Which of the following floppy diskettes holds the most data?
   a. DSHD
   b. DSDD
   c. DSSD
   d. SSDD
   e. HSHD
9. Which of the following video adapter(s) cannot perform graphic commands?
   a. MDA
   b. CGA
   c. EGA
   d. VGA
   e. SVGA
10. Which of the following computer systems cannot use extended memory?
   a. XT
   b. AT
11. Which of the following communication formats is used to communicate with printers most of the time?
   a. serial
   b. parallel
12. A monochrome monitor displays information using how many colors?
   a. One plus the background color
   b. 8
   c. 16
   d. 256
   e. 65,536
13. Which key, when pressed, erases the character above the cursor?
   a. backspace
   b. tab
   c. break
   d. escape
   e. delete
14. Which key, when pressed along with another key, produces the second function of the other key pressed?
   a. space bar
   b. insert
   c. shift
   d. page up
   e. scroll lock
15. A word processor is what type of software?
   a. operating system
   b. environment
   c. programming language
   d. application
   e. utility
16. A diskette that has become wet by washing should be used only after it is dry.
   a. True
   b. False
17. You should not place a diskette next to a monitor, but it may be placed next to an electric fan.
   a. True
   b. False
18. Insert the diskette into the drive unit with the label side up.
   a. True
   b. False

19. You cannot remove a diskette from a drive unit with the LED on, but you can insert the diskette with the LED on.

a. True

b. False

20. What is POST stored in?

a. ROM/EPROM

b. A disk file

c. CMOS

d. DRAM

e. None of the given answers

chapter 2

Disk Operating System

OBJECTIVES

Chapter 2 provides the reader with an introduction to the disk operating system used in the IBM PC and compatibles. Upon completion of this chapter, you should be able to meet the following objectives.

1. Identify what the acronym DOS stands for.
2. Discuss the function of DOS.
3. Explain how DOS treats information.
4. Describe how file names and extensions are used.
5. Give the parameters for files names and extensions.
6. Describe the purpose of special extensions used in DOS.
7. Discuss the parameters of the root directory and directory structure.
8. List and describe the use of DOS wild cards.
9. Explain DOS, redirecting, and filtering syntax.
10. Discuss the function of the special key sequences used with DOS.
11. Explain the difference between DOS 5.0 and 6.2.
12. Identify the difference between internal and external DOS commands.
13. Describe how to load DOS.
14. Describe the function of the DOS prompt.
15. Explain the function of the DOS commands in this chapter.
16. Describe the tree structure of DOS and its related commands.
17. Explain and demonstrate the function of batch files.
18. Describe the function of AUTOEXEC.BAT AND CONFIG.SYS files.
19. Explain the function of DOS commands related to hard drive systems.

2.1 WHAT IS DOS AND WHAT DOES IT DO?

The acronym DOS stands for disk operating system. DOS is the most popular operating system for PCs. Two companies write versions of DOS for the IBM PC and compatible computer systems. MicroSoft, the maker of MS-DOS, holds the majority of the world market and sets the standard by which all software writers develop their software. Digital Research, maker of DR-DOS, is not as popular and competes with MicroSoft by coming out with newer versions more quickly, with additional new commands and features sometimes not found in MS-DOS. Although the battle continues, DR-DOS for the most part is compatible with MS-DOS; therefore, this book deals with version 5.0 and 6.2 of MS-DOS. Most of the software written today requires DOS version 2.1 or higher, and some may require DOS 5.0 or higher.

DOS is a collection of programs that tells the computer how to perform specific tasks. When a DOS program (command) is executed, the computer performs a specific operation; therefore, these programs are referred to as commands. DOS works with BIOS (monitor program) to make the hardware perform the selected task. BIOS is a collection of programs stored in ROM/EPROM that controls the hardware in computer and makes different types of hardware function the same. Application software running under DOS uses DOS commands to control the hardware with the help of BIOS. Therefore, DOS is the reference point (standard) that the software uses to control the hardware.

Although DOS's main function is to control the disk drive system (store [input], retrieve [output], and modify [process] information), it also controls through the BIOS, all the equipment of the computer system (motherboard, video adapter, keyboard, I/O ports, and most peripherals). The execution of one DOS command can represent more than 50 hardware instructions to different devices. Without DOS or some other operating system standard, the user would have to issue all the necessary instructions to the hardware to perform the task. This would require that the user understood and was familiar with all aspects of the particular hardware in the computer system, which would certainly limit the number of people using the system.

2.2 HOW DOS TREATS INFORMATION ON A DISK SYSTEM

DOS treats the information stored on a diskette or hard drive as files. A file is a group of related information that is used together as a unit. Each file on a diskette or hard drive has a name (file name) and an optional file extension (usually specifies the type of file).

Before a diskette or hard drive can be used to hold data, it must be formatted (initialized). Some drives require two additional steps; lower-level format (for some hard drives) and partitioning. Formatting puts timing marks, track, sector, head, size, and error-detection information on the disk. This process also maps out the disk, determines bad areas (so that DOS does not use such areas), sets up a boot record, a file allocation table, and a directory, and determines user space.

In the formatting process, which usually occurs on both sides of the disk (both sides of all disks in a hard drive), each R/W head location (from outermost to the innermost edges of the disk) is called a track (number varies with size of media). As the disk rotates, it is divided into areas called sectors (number varies with size of media). In each sector on every track a package of 512 bytes of user data is stored; this data package is called a record (or a sector, because there is one record per sector). The formatted size of the diskette can therefore be determined by multiplying the number of sides, bytes per sector, tracks, and sectors per track together. The amount of user space (storage size), as seen from Table 2-1, is smaller than the formatted space, because each disk formatted under DOS contains three packages of information (described later) that occupy some of the formatted disk space.

TABLE 2-1
Floppy Disk Types

	360K	*1.2M*	*720K*	*1.44M*
Diskette size	5.25 inch	5.25 inch	3.5 inch	3.5 inch
Density	Double	High	Double	High
Recording method	MFM	MFM	MFM	MFM
Number of sides	2	2	2	2
Bytes per sector	512	512	512	512
Number of tracks	40	80	80	80
Number of sectors per track	9	15	9	18
Number of records per disk	720	2,400	1,440	2,880
Formatted size	368,640	1,228,800	737,136	1,474,560
Storage size	362,496	1,222,656	731,136	1,468,416
Type of PC	PC/XT/AT	AT	PC/XT/AT	AT

The boot record is a short program used to produce the nonsystem error message on the screen of the computer. This error is generated whenever a nonsystem DOS disk tries to boot the computer. This program is located in the first two sectors (1024 bytes) of all floppy diskettes and in the first sector of the DOS partition of all hard disks formatted under DOS. The last word of the boot record is called the signature; it is used to check the validity of the media.

The file allocation table (FAT), is the second package of data on each disk. Its size varies with media. The FAT is used by DOS to convert the clusters of a file into logical sector numbers, (physical location of each sector on the disk) and maps all the allocated, bad, and available disk space. Logical sectors are numbered beginning on side A, track 0 of the disk. Side B, track 0 sectors are then numbered, before switching to the next track on side A to continue the numbering. A cluster may contain two or more logical sectors (the size varies with media). Clusters are used by DOS to reduce head movement and speed up the overall operation of the disk system. In a floppy diskette, clusters are two sectors in size. Normally, in a hard drive the cluster size is four sectors. The cluster size determines the minimum amount of space a file or part of a file uses on the disk every time data in a cluster is transferred. A disk with a cluster size of 1024 bytes means that files are stored in increments of 1024 bytes each. If a file contains 530 bytes, it takes the space of one cluster. If a file on the same disk is 2049 bytes, it takes the space of three clusters.

DOS places two copies of the FAT on each disk in case one FAT becomes damaged. The first entry in the FAT represents the size of the disk (number of logical sectors on the entire disk). The second entry represents the type of format (number of logical sectors per cluster). All remaining entries each represent a cluster in the data area of the disk (beginning with cluster 2). Each entry in a floppy system is 12 bits in length, and a 16-bit entry is used in hard drive systems. As DOS transfers the data in a cluster, it goes to that cluster entry in the FAT to determine what to do next. The FAT entry may represent the next cluster number for the file, the last cluster for the file (end of the chain), a reserved cluster, or a bad cluster. As each transfer takes place, the FAT is updated to represent the transfer.

The directory is the third package of data found on a DOS-formatted disk. The directory (root directory) contains information regarding all files on the diskette. The number of file entries that the root directory can hold is fixed (determined by the type of disk). A floppy diskette can have up to 112 (DSDD) or 224 (DDHD) entries, whereas hard drives usually have 512 (varies with the size of drive) entries. Each file entry in the directory requires 32 bytes. Bytes 0 to 7 of each entry represent the file name; bytes 8 to A hex represent the characters in the file extension (if used); byte B hex is the attribute byte, which specifies the characteristics of the file (read-only, hidden, system, volume label, subdirectory, and archive); bytes C to 15 hex are reserved for DOS use; bytes 16 and 17 hex determine time (created or updated); bytes 18 and 19 hex determine date (created or updated); bytes 1A and 1B hex represent the first cluster number of the file, and bytes 1C to 1F hex are the size of the file in bytes.

Whenever DOS needs to perform an operation on a file, it first reads the directory to find the file name and extension. If one of the entries matches, DOS checks the attribute code to see if the operation can take place. If the operation continues, DOS uses the first cluster number of the directory entry to transfer the first cluster of data. The FAT then supplies the remaining cluster location information. Once the operation is complete, the directory is updated (size of file, time, and date). If necessary, DOS also updates the FAT entries.

2.3 NAMING FILES

When naming a file, a unique name should be chosen that represents the purpose, contents, or type of information in the file. If the name of the file is not unique, DOS replaces the data in the old file with the new information or issues an error statement. DOS divides the name of a file into two parts, the file name and the file extension.

The file name usually indicates the contents of the file. A file name (not optional) may contain any combination of up to eight alphanumeric characters (A to Z and 0 to 9) and these special characters: ampersand (&), apostrophe ('), at sign (@), braces ({ }), caret (^), dollar sign ($), exclamation point (!), grave accent (`), hyphen (-), number sign (#), parentheses [()], percent sign (%), tilde (~), and underscore (_). DOS sees all letters in the alphabet as uppercase letters. A space and any other characters are not valid in a file name. The following names are special reserved names that should not be used: AUX, CLOCK$, COMx (x = 1 through 4), CON, LPTx (x = 1 through 3), NUL, and PRN.

The file extension (optional) is separated from the file name by a period. The file extension usually indicates the type of file, which allows the grouping of files that share the same type of data. The file extension may contain up to three valid characters. The file extension uses the same characters that are valid for the file name.

2.4 TYPES OF FILES

The file extension usually indicates the type of data that the file contains. The following are some of the common DOS file extensions. These extensions may not apply when using a programming language or application software.

2.4.1 Command Files

When the extension COM is used on a file, DOS treats the file as a command file. This usually means that the data in the file is used to perform a DOS command task. However, an extension of COM can indicate a command of a programming language or other type of software.

2.4.2 Execution Files

If the extension of EXE follows a file name, the file is called an execution file. Execution files normally contain a compiled program in machine (object) code that can perform a task without help from any language or other software.

2.4.3 BASIC Files

If the BAS extension is used with a file name, the file is written in the BASIC language and can be accessed only if the computer is operating in BASIC. BASIC files are stored in a special binary-coded format and cannot be read by normal means unless BASIC is loaded into the computer.

2.4.4 System Files

If the SYS extension is used with a file name, the file is used to change the configuration or how some of the functions of DOS are performed. Files with a SYS extension are normally controlled by the CONFIG.SYS file. The CONFIG.SYS file is created by DOS during the install or by the user and changes how DOS uses and configures the hardware. When DOS is first booted up, it searches the root directory for the CONFIG.SYS file and executes it. This file may change how memory is used or the amount of memory used for specific functions and allow additional standard and nonstandard devices to be controlled by DOS.

2.4.5 Back-up Files

If the BAK extension is used with a file name, that file is called a back-up file. In some software, especially in text editors, a back-up file is created, which contains the file as it was before the last changes were made. Back-up files are useful if something happens to the file you are working on and the data is destroyed.

2.4.6 Batch Files

A file with the extension BAT is a batch file. A batch file allows a user to create a file that can perform multiple DOS commands when the file is called for execution. To execute a batch file, the user need only type in the file name (extension not needed) and press the enter key. Batch files may also contain messages, help, or text.

A special batch file used in most computers is called AUTOEXEC.BAT (auto execute batch). When the computer is booted and the CONFIG.SYS file is executed (if present; otherwise DOS uses the default configuration), DOS searches the root directory for the optional AUTOEXEC.BAT file and executes the command sequence specified. If there is no AUTOEXEC.BAT file, DOS uses its default start sequence (DATE and TIME) and values to complete the DOS boot-up sequence.

2.4.7 Text Files

Text files have no special extension in DOS. Text files contain character information that can be read by the user and is not used as a program or command. Text files usually contain messages, documentation, help, or information about the program or software. Sometimes a text file has an extension of TXT (text), ASC (ASCII code), DOC (documentation), HLP (help), and INF, READ.ME, or README.1ST (information).

2.4.8 Application Files

Application files have no special extensions except those created by the application program. The application program maintains these files, using DOS commands at the request of the application program command.

2.5 THE ROOT DIRECTORY

When a diskette or hard disk is formatted, DOS creates a directory as specified earlier in this chapter. The directory that DOS creates during formatting is called the root directory and is of fixed size. The fixed size of the root directory limits the number of directory entries (file names and other related information) that can be entered. Because of this limitation, a diskette or hard disk could run out of directory space before the remaining free space on the disk is filled up, if the files are small. Once the maximum number of directory entries is reached, no more files can be stored on the disk, even if space is available.

EXAMPLE:

A DSDD 5 1/4-inch diskette has 112 directory entries available in the root directory; if all the files were one cluster in size, a 360K diskette could theoretically hold 360 files. But because the 360K diskette has space for only 112 files names, only about one-third of the disk space would be used before the directory was full. This problem is even more complex with larger diskettes and hard drives.

The problem with the root directory limitation was solved with DOS 2.1. The approach was taken to allow the user to create special files in the root directory that function as another directory (called a subdirectory). A subdirectory is a directory within another directory or subdirectory. A subdirectory is treated like a file by the parent directory. Therefore, the number of file names in a subdirectory is not fixed. The number of entries is determined by the amount of remaining disk space. Because each subdirectory is treated independently, file operations that occur affect only the files in that subdirectory or working directory unless otherwise specified.

Figure 2-1 shows how a diskette can be divided to allow easy assess to its files in a logical order. It shows how one might organize information obtained from school in a notebook or a drawer of a file cabinet. The symbol for a folder represents a file, and the rectangles represent subdirectories. The lines that connect the root directory, subdirectories, and files together show the path that the user (DOS in the case of a computer) must take to get the information. The reference point for the path is always the root directory.

Through DOS directory commands, the user can make (create), change (name of the working directory), and remove directories and subdirectories. Because each directory or subdirectory is treated separately, the user must be able to determine which directory or subdirectory to use. To perform this task, the user must change the directory to make the specified directory the working (current) directory. To change the directory, the user must specify the path with the change directory command. From that point on, all commands are performed only on that directory unless otherwise specified. To make a new directory, the user needs to specify the path for the new directory. To remove a directory (any directory but the root directory), the directory must be empty (all files and subdirectories in that directory must be deleted or removed). The user specifies a path after the remove directory command. Whenever the

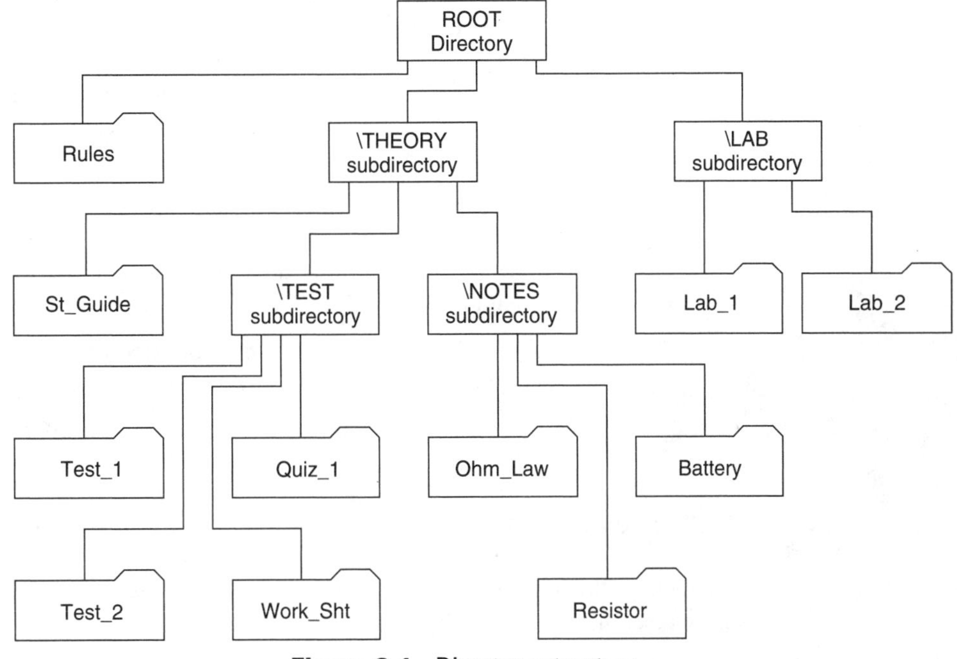

Figure 2-1 Directory structure.

computer boots up, the working directory is the root directory until it is changed. Once the working directory has been changed, the change remains until another directory is selected.

2.6 DOS WILDCARDS

Two wildcards can be used with DOS commands. Both wildcards are used when specifying file names and file extensions.

The question mark (?) wildcard is used to specify that the one character position it is located at can be any valid character. Therefore, a question mark can replace any one character position in a file name or file extension. If more than one character position needs to be wild (any character), more than one question mark needs to be used in that position.

The asterisk (*) wildcard is used to specify that either the entire or all remaining characters from that location in the file name or extension can be any valid character or characters.

2.7 DOS SYNTAX SPECIFICATIONS

Syntax defines the sequence and order that a DOS command requires for the specified command. The syntax also contains required and optional parameters and switches. The following is a list of syntax specifications for MS-DOS.

[brackets] Brackets specify that the parameter is optional (not required). Do not include the brackets in the command.

drive: The drive parameter specifies which disk drive unit is to be used. If this is optional or if a drive is not specified, the computer uses the default drive. Drive units are labeled A to Z and always require a colon (:) after the letter.

path The word *path* specifies that the user has specified a path (directory information) with the command. The path tells the computer how to get to the specified part of the disk (which directory to go to). If a path is not specified, the computer defaults to the current working directory. Each directory (subdirectory) begins with a back slash (\) followed by the name of the directory; a path may contain more than one directory (subdirectory). If a file name follows a path parameter, the user must separate the ending of the path and the beginning of the file name with a back slash (\).

delimiter A delimiter is a character used to separate two pieces of data. In a command, a space is used to separate the command from its parameters, unless otherwise noted. Certain parameters require special delimiters.

... The ellipsis specifies that the parameter may be repeated as many times as needed to achieve the task.

/ The slash is used in a DOS parameter to specify a switch, which causes the DOS command to modify the output of the operation.

2.7.1 Redirecting Syntax

Redirecting allows the user to send output information or get input information from a device or file other than the default input or output device.

> The greater-than sign can be used with most DOS commands to redirect the output to a device other than the default device. To redirect the command, the greater-than sign precedes the device name (prn, con, lpt1, lpt2, lpt3, com1, com2, com3 or com4) or the name of a file (if the file does not exist, DOS creates one).

>> The set (double) of greater-than signs makes DOS append (add) the command output information to another file. The set of greater-than signs must precede the file name.

< The less-than sign causes DOS to get its input for the command from an input other than the keyboard (console). The less-than sign must precede the input name.

2.7.2 Filtering Syntax

Filter commands allow the user to divide, extract, or rearrange the parts of a file or information derived from a DOS command.

< The less-than sign causes the DOS filter command to get its input for a file.

¦ The pipe symbol (¦) is used to cause the DOS filter command to get its input from the preceding DOS command.

2.8 SPECIAL KEYS

The following keys and key sequences are used with DOS to perform specific tasks. The function [F1 to F6] keys on the keyboard perform special tasks in DOS; the function keys that are listed in the section are the keys that are most frequently used with DOS.

The [Ctrl][Alt][Del] key sequence is used to perform a software or warm boot of DOS and the computer system. To initialize this function, the user must press the control [Ctrl] key and hold it down, press the alternate [Alt] key and hold it down, and press the delete [Del] key, then release all keys. The screen blanks out, the computer bypasses the memory test in POST, and the computer boots from drive A or the hard drive.

The [Ctrl][S] key sequence is used to stop the video screen from scrolling. To initialize this function, press the control [Ctrl] key and hold it down, and press the letter [S] key. To start the scrolling again, perform the same sequence.

The [Ctrl][C] or [Ctrl][Break] key sequence is used to cancel a command issued. To initialize this function, press the control [Ctrl] key and hold it down, and press either the letter [C] or [Break] key.

The [F1] or [right arrow] key can be used to copy a single character from the previous DOS command issued to the new command line each time the key is pressed.

The [F3] key is used to repeat the previous DOS command issued to the new command line.

The [F6] or [Ctrl] [Z] keys are used to end a batch file. To initialize the function, either press the [F6] key once or press and hold down the [Ctrl] key and press the letter [Z] key.

2.9 DOS 5.0, 6.0, AND 6.2

Although DOS 6.2 is currently the highest version of DOS, 5.0 is a very popular version of DOS. The main advantages of DOS 6.0 and DOS 6.2 are some of its new utilities (programs that perform high-level operations used to maintain and enhance disk management), better memory management, and some new commands.

Before DOS 5.0, 6.0, or 6.2 can be used in a computer system, the user must install DOS on the disk (floppy or hard) system. During the installation of DOS, the user is prompted to answer questions about the hardware in the computer. This information is used to allow DOS to configure itself automatically to the hardware.

2.10 TYPES OF DOS COMMANDS

The DOS commands that are not used frequently are called external commands; commands that are used very frequently are called internal commands. External commands are not stored in memory when DOS is loaded. To perform these commands, the computer must access the DOS disk to get the information needed to perform the task, load the program into RAM, and execute the program. Once the task is performed, DOS prompts the user to ask if he wishes to execute the command again. If the user answers yes, DOS executes the command again while it is still in memory. If the user answers no, DOS clears the memory of the command and gets ready for the next command.

Internal commands are stored in the memory of the computer when DOS is loaded and are accessed without going to the disk when needed to perform a task. The COMMAND.COM in DOS contains all internal commands. External commands are found within their own separate files. The following is a list of the most common external and internal DOS commands used in this unit of study. (Later in this chapter, there is a list of all DOS commands and a description of how they are used.)

Internal

BUFFERS	CALL	CD
CHOICE	CLS	COPY
DATE	DEL [ERASE]	DEVICE
DEVICEHIGH	DIR	DOS
ECHO	FCBS	FILES
LABEL	LASTDRIV	LH [LOADHIGH]
MD	MENUCOLOR	MENUDEFAULT
MENUITEM	PATH	PAUSE
RD	REM	RENAME
SET	SHELL	STACKS
TIME	TYPE	VERIFY
VOL		

External

CHKDSK	COMP	DBLSPACE
DEBUG	DEFRAG	DELTREE
DISKCOMP	DISKCOPY	DOSSHELL
EMM386	FC	FDISK
FORMAT	GRAPHICS	HELP
HIMEM	MEM	MEMMAKER
MORE	MOVE	MSAV
MSBACKUP	MSCDEX	MSD
PRINT	SCANDISK	SETVER
SMARTDRV	SYS	TREE
UNDELETE	UNFORMAT	VSAFE
XCOPY		

2.11 LOADING DOS

After the computer has executed POST and no problems have been detected, the computer reads drive A. If no diskette is in drive A, the computer goes to the hard drive (usually drive C). If no hard disk is found, BIOS prints a message on the screen and prompts the user to insert a disk with system files into drive A. In some of the more advanced computer systems, the user can set BIOS to access drive C for boot-up before it tries to boot up from drive A.

When the diskette or hard drive is read, the first package of information transferred into the computer is the boot record. Once the boot record is loaded, it takes control and checks the disk to make sure that the IO.SYS and MSDOS.SYS files are the next two files on the disk. The boot record than loads the IO.SYS file into the computer.

IO.SYS (hidden read-only system file) is a short program that enables the low-level interface routines in BIOS. This file is not listed with the directory command because it is hidden. This file loads the MSDOS.SYS file and hands over control of the computer to MSDOS.SYS.

MSDOS.SYS (hidden read-only system file) is a program that enables the upper-level interface routines. These routines control all file management and devices and initialize some user functions. The computer at this time checks for a CONFIG.SYS file; if one exists, the information in the CONFIG.SYS file reconfigures the installable devices. If no CONFIG.SYS file exists, the default device drivers are used. This file then loads the first part of the COMMAND.COM file into the computer and hands over control to the command processor.

The third file is called the COMMAND.COM (command processor) file (a read-only file). This file contains three sections: resident, initialization, and transient. The resident section contains routines to handle some interrupts (control break, terminate address, and critical errors), the reloading routine of the transient portion of the command processor, error messages, and execution routines for command and execution files. Once the resident part of the command processor is complete, the initialization part is loaded and executed.

The initialization part of the command processor determines segment addresses for programs and loads the AUTOEXEC.BAT file routine for the command processor. The transient portion is then loaded and given control.

The transient part of the command processor is loaded into the memory area. This part contains the actual internal commands for DOS, generates the DOS prompt, builds the command line, and loads the routines necessary to execute batch files. If an AUTOEXEC.BAT file exists, the computer loads and executes this file; otherwise, the computer defaults to the internal start-up sequence. If no AUTOEXEC.BAT file exists, the computer displays the following:

2.11.1 MS-DOS Version 5.00 Start-up Screen

```
Current date is Tue  1-01-1980
Enter new date (mm-dd-yy):  [Enter]
Current time is 11:36:52.70p
Enter new time:  [Enter]

Microsft(R) MS-DOS(R) Version 5.00
           (C)Copyright Microsoft Corp 1981-1991.

A>_
```

2.11.2 MS-DOS Version 6.2 Start-up Screen

```
Starting MS-DOS...

Current date is Tue  1-01-1980
Enter new date: mm-dd-yy  [Enter]
```

```
Current time is    0:00.47.89a (the time may vary)
Enter new time:   [Enter]

Microsoft(R) MS-DOS(R) Version 6.2
             (C)Copyright Microsoft Corp 1981-1993.

A:\>_
```

Starting the computer from a powered-down condition is called a cold start. If the computer has power applied to it and the user wishes to restart the computer, the user should press the [Ctrl] [Alt] [Del] (see p. 41). This sequence, called a warm start, restarts the computer without running POST.

Because MS-DOS 6.2 is the latest version of DOS, the examples listed in this chapter use the format found in DOS 6.2. There may be small differences in the way the screen looks between versions 5.0, 6.0, and 6.2 of DOS, but the commands common to all versions function the same.

2.12 DOS COMMAND PROMPT

Once DOS has been loaded into the computer and is ready for the user to issue a command, it displays the DOS prompt. The default DOS prompt in version 6.2 is the current drive designation followed by a colon, followed by backslash (\) and the name of the working directory and the greater-than sign. When no characters follow the backslash, the root directory is selected. In older versions of DOS, the DOS prompt is only the current drive designation followed by a colon and the greater-than sign. The type of prompt can be changed with the prompt command in all versions of DOS.

2.13 MOST COMMON DOS COMMANDS

The following are the most frequently used commands in DOS. They represent about 94% of the commands that a technician ever uses.

Most DOS commands are followed by command parameters. These may or may not be optional, as specified in the description of the command. The commands and parameters may be entered in either uppercase or lowercase letters, because DOS sees all letters as uppercase.

2.13.1 Help (external)

```
HELP
```

The HELP comand provides information regarding any DOS command in MS-DOS 6.2 through a menu system. The user can also print out the help information or view it on the screen.

2.13.2 Date (internal)

```
DATE [mm-dd-yy]
```

The DATE command displays the current date stored inside the computer and allows the user to update the date. If the computer does not have a real-time clock, the date is lost when power is removed. The main purpose of the date is to provide information about the date when the file was created or last modified.

```
A:\>DATE
Current date is Sun 01-10-1993
Enter new date (mm-dd-yy): _
```

If the current date is correct, press the enter key. If you wish to change the date, type in the new date in the following format:

mm	month number, 1-12
-	hyphen
dd	day number, 1-31
-	hyphen
yy	year number, 80-99 or
yyyy	full year number, 1980-2099

Pressing the *enter* key completes the command and updates the date inside the computer.

If the date in the (mm-dd-yy) format is typed in after the date command and the enter key is pressed, the computer updates the date and returns the command prompt and the cursor.

```
A:\>DATE 1-10-93
A:\>_
```

2.13.3 Time (internal)

```
TIME [hours:[minutes:[seconds.[hundredths]]][a or p]]
```

The TIME command displays the current time in military time (24-hour clock). The user has the option of using a nonmilitary time format by using the letter [a] for AM time or [p] for PM time. If the computer does not have a real-time clock, the time is lost when power is removed. The main purpose of time is to provide information about when the file was created or last modified.

```
A:\>TIME
Current time is 7:05:30.45p
Enter new time: _
```

If the current time is correct, press the enter key. If you wish to change the current time in the computer, enter the following information in the following military format:

hh	hour number, 0-23
:	colon [remaining values optional]
mm	minute number, 0-59
:	colon [remaining values optional]
ss	second number, 0-59
.	period [remaining values optional]
xx	hundreds of a second, 0-99

Press the enter key to complete the command.

The user may elect to use the nonmilitary time format by using the following parameters:

hh	hour number, 1-12
:	colon [remaining values optional]
mm	minute number, 0-59
:	colon [remaining values optional]
ss	second number, 0-59
.	period [remaining values optional]
alp	[a] for AM time period. [p] for PM time period.

Press the enter key to complete the command.

If the time in the (hh:mm:ss.xx)(alp) format is typed in after the time command and the enter key is pressed, the computer updates the time and returns the command prompt and the cursor.

```
A:\>TIME 5:23p
A:\>_
```

2.13.4 Clear Screen (internal)

```
CLS
```

The CLS command clears the display screen of all data and places the DOS command prompt and cursor in the upper-left corner of the display screen.

```
A:\>CLS [Enter]
A:\>_      THE VIDEO SCREEN BLANKS; THE DOS COMMAND PROMPT AND
           CURSOR ARE IN UPPER LEFT CORNER OF THE SCREEN.
```

2.13.5 Format (external)

```
FORMAT drive: [/v[:label]] [/q] [/u] [[/f:size] or [/t:tracks
              /n:sectors]] [/b or /s]

FORMAT drive: [/v[:label]] [/q] [/u] [/1] [/4] [/b or /s]

FORMAT drive: [/q] [/u] [/1] [/4] [/8] [/b or /s]
```

The FORMAT command is used on all new diskettes (or hard drive systems) to initialize the surface of the disk system for use. The FORMAT command sets up tracks and sectors and creates a boot record, FAT, and directory. Because FORMAT is an external DOS command, the default disk or working directory must contain the FORMAT.COM file; if a file is not found, an error occurs. The only exception is when FORMAT.COM is in the path statement.

The format drive parameter is required, or an error message is produced: "Required parameter missing -." If the optional switches are not specified, the FORMAT command formats the disk to its old size (if previously formatted) or formats it to the size of the drive specified in the CMOS data table of AT class computers.

Before DOS formats a diskette or hard disk, DOS creates a copy of the FAT and directory and saves this information on the diskette, unless the user informs DOS to format unconditionally. This process allows DOS to reconstruct a previously formatted disk as long as the disk area has not been written over. In DOS 5.0, this was performed by a command called MIRROR, but in DOS 6.0 and 6.2 this is done automatically.

Note: When not to use the FORMAT command:

1. While in a network.
2. On a drive prepared by the SUBST command.
3. On an Interlnk drive.

Format Parameters

drive: The drive parameter is not optional; you must specify the drive letter followed by a colon.

Format Switches

/v:label The /v switch specifies that a volume label (name) should be placed on the disk. The ":" is followed by a volume label (up to 11 characters). This switch is optional; if this switch is not specified, the FORMAT command asks for a volume label after the format is complete. If the volume switch is used, the computer does not request a volume label after the format is complete.

/q The "q" (quick) format switch formats the disk by deleting and writing a new FAT and root directory. This type of format does not rewrite the timing marks or scan the disk for any bad areas. This switch works only with previously formatted disks. The advantage of this switch is that is only takes seconds to format the disk, but the disadvantage is that any bad areas on the disk are not detected.

/u The "u" (unconditional) format switch formats the disk and does not save the old FAT and root directory, so the disk can be unformatted by the UNFORMAT command. All old data on the disk is lost. This format process is quicker because the old format information is not saved or even checked.

/f:size The "f" (fixed) format switch along with the "size" parameter is used to select the fixed format size of diskettes (you should not use this command when formatting a hard drive). The following are the size values supported by DOS 6.0 and 6.2.

5 1/4-inch diskette sizes
160 or 160K or 160KB (SSDD)
180 or 180K or 180KB (SSDD)
320 or 320K or 320KB (DSDD)
360 or 360K or 360KB (DSDD)
1.2 or 1.2M or 1.2MB (DSHD)

3 1/2-inch diskette sizes
720 or 720K or 720KB (DSDD)
1.44 or 1.44M or 1.44MB (DSHD)
2.88 or 2.88M or 2.88MB (DSED)

SSDD: single-sided double-density
DSDD: double-sided double-density
DSHD: double-sided high (quadruple)-density
DSED: double-sided extra–high-density

/b — The "b" (boot) switch reserves space for the IO.SYS and MSDOS.SYS files on the diskette. This switch was used in older versions of DOS to make sure the SYS command would have room on the diskette for these hidden system files; this switch is no longer necessary with DOS 6.0 and 6.2.

/s — The "s" (system) switch is used to copy the DOS system boot-up files to the disk once the format is complete (when using this switch the SYS command is not necessary). Placing the IO.SYS, MSDOS.SYS, and COMMAND.COM files on the disk means that the disk is able to boot up the computer in DOS.

/t:tracks
/n:sectors — The "t" (tracks) and "n" (number of sectors) switches must be used together to specify the number of tracks and sectors per track the FORMAT command should format on the disk. These switches are used for disks that do not have standard format sizes using the /f:size switch.

/1 — The "1" (single-sided) format switch forces the FORMAT command to format only one side of the disk. This switch is used only if the disk is to be used on a very old single-side disk drive unit.

/4 — The "4" (double-density) format switch forces the FORMAT command to format the diskette in a high-density 5 1/4-inch drive as a double-density format standard. This switch is needed only if the user does not use the /f:size switch or to format a DSDD 360K diskette in a high-density drive unit.

/8 — The "8" (eight-sector) format switch forces the FORMAT command to format the 5 1/4-inch diskette to eight sectors per track (160K or 320K) rather than the 180K or 360K format. This command is necessary only if the /f:size switch is not used or when working with MS-DOS versions earlier than 2.0.

The following example shows how to format the 3.5-inch DSDD diskette in drive B to 720K.

```
C:\>FORMAT B: /F:720    [Enter]
Insert new diskette for drive B:
and press ENTER when ready...

Checking existing disk format.
Saving UNFORMAT information.
Verifying 720K
 xx percent completed. (THIS LINE IS REPLACED WITH THE FOLLOW-
                        ING LINE ONCE THE FORMAT IS COMPLETE.)

Format Complete.

Volume label (11 characters, ENTER for none)?_
```

```
     730112 bytes total disk space
     730112 bytes available on disk

       1024 bytes in each allocation unit.
        713 allocation units available on disk.

Volume Serial Number is 2B5C-0DE6

Format another (Y/N)?_
```

The format size (/f:size) was used because the 3 1/2-inch DSDD diskette was in a high-density drive. Because the unconditional switch is not used, DOS defaults to a safe format, which allows the user to unformat the diskette as long as the diskette is not written to.

The next example formats the diskette in drive B with the system files and unconditionally, thus destroying the previous format and information on the diskette. Because the unconditional switch is used, the FORMAT command does not check for or save the old format information, and the formatting occurs more quickly.

```
C:\>FORMAT B: /u /s /F:1.44    [Enter]
Insert new diskette for drive B:
and press ENTER when ready...

Formatting 1.44M
 xx percent completed. (THIS LINE IS REPLACED WITH THE FOLLOW-
                        ING LINE ONCE THE FORMAT IS COMPLETE.)

Format complete.
System transferred

Volume label (11 characters, ENTER for none)? BOOT [ENTER]

   1457644 bytes total disk space
    184320 bytes used by system
   1273344 bytes available on disk

       512 bytes in each allocation unit.
      2847 allocation units available on disk.

Volume Serial Number is 2C3D-0DF4

Format another (Y/N)?_
```

Note that the IO.SYS, MSDOS.SYS, and COMMAND.COM files have used up 184,320 bytes of space; thus the number of available bytes on the diskette is less than the formatted space.

The next example of the FORMAT command formats the diskette in drive A, with 80 tracks and 15 sectors per track. In the results of the format, note that a number of bad sectors were found; these sectors are blocked out and not used to store information.

Note: If a diskette shows bad sectors when formatted for the first time, try to format the diskette again; many times dirt or some other type of material on the diskette is dislodged by the time the second format is complete.

```
C:\>FORMAT A: /t:80 /n:15  [Enter]
Insert new diskette for drive A:
and press ENTER when ready...

Checking existing disk format.
Saving UNFORMAT information.
Verifying 1.2M
 xx percent completed. (THIS LINE IS REPLACED WITH THE FOLLOW-
                        ING LINE ONCE THE FORMAT IS COMPLETE.)

Format complete.

Volume label (11 characters, ENTER for none)?

   1213952 bytes total disk space
      7680 bytes in bad sectors
   1206272 bytes available on disk

       512 bytes in each allocation unit.
      2356 allocation units available on disk.

Volume Serial Number is 1A36-12EE

Format another (Y/N)?
```

In the last example of the FORMAT command, the 5 1/4-inch double-density diskette in a high-density drive unit is formatted to the double-density size, using the quick format switch.

```
C:\>FORMAT A: /q /4   [Enter]
Insert new diskette for drive A:
and press ENTER when ready...

Checking existing disk format.
Invalid existing format.
This disk cannot be QuickFormatted.
Proceed with Unconditional Format (Y/N)?
Formatting 360K
 xx percent completed. (THIS LINE IS REPLACED WITH THE FOLLOW-
                        ING LINE ONCE THE FORMAT IS COMPLETE.)

Format complete.

Volume label (11 characters, ENTER for none)?

    362496 bytes total disk space
    362496 bytes available on disk

      1024 bytes in each allocation unit.
       354 allocation units available on disk.

Volume Serial Number is 3D49-12F4

QuickFormat another (Y/N)?
```

Because the diskette in the drive did not have a previous format on it, the messages "Invalid existing format" and "This disk cannot be QuickFormatted" are displayed. Next DOS asks a question of the user: "Proceed with Unconditional Format (Y/N)?" This allows the user to proceed if "Y" (yes) is selected or to end the format sequence if the answer is no ("N").

2.13.6 Unformat (external)

```
UNFORMAT drive: [/l] [/test] [/p]
```

The UNFORMAT command restores the disk information and old format that were erased by the FORMAT command in MS-DOS 6.0 and 6.2. Restoring of the disk does not occur if the diskette was formatted using the "u" (unconditional) format switch. Any files modified since the last format may be lost.

Parameter

drive: The drive parameter is not optional; specify the drive letter to be unformatted followed by a colon (:).

Switches

/l The "l" (list) switch is used to display all files and subdirectories found by the UNFORMAT command. If this switch is not used, the unformat defaults to displaying only subdirectories and files that are fragmented. A fragmented file is a file that is located in nonadjacent sectors on the diskette.

/test The "t" (test) switch is used to display how the UNFORMAT command would re-create the diskette information. The unformatting of the diskette does not actually occur.

/p The "p" (print) switch pipes the output messages to the printer connected to lpt1.

```
C:\>UNFORMAT A:

Insert disk to rebuild in drive A:
and press ENTER when ready.

Restores the system area of your disk by using the image
file created by the MIRROR command.

    WARNING !!        WARNING !!

This command should be used only to recover from the inad-
vertent use of the FORMAT command or the RECOVER command.
Any other use of the UNFORMAT command may cause you to
lose data! Files modified since the MIRROR image file was
created may be lost.

Searching disk for MIRROR image.

The last time the MIRROR or FORMAT command was used was at
07:51 on 01-10-93.
```

```
The MIRROR image file has been validated.

Are you sure you want to update the system area of your
drive A (Y/N)?

The system area of drive A has been rebuilt.

You may need to restart the system.

C:\>_
```

2.13.7 Label (internal)

```
LABEL [drive:][label]
```

The LABEL command allows the user to change, create, or delete a volume label (name) on a disk. The volume label is part of the root directory and is visible with the DIR (directory) command or the VOL (volume) command.

Label Parameters

drive: This optional parameter is used to specify which drive contains the label that the LABEL command is to process. If no drive is specified, the current drive label is processed.

label This optional parameter allows the user to input the new label immediately without viewing the old label and serial number. If the disk has not been labeled by either the FORMAT command or LABEL command, the volume label is "no label."

In the following example, the diskette in drive A, the default drive, has no label. By typing in "first disk" and pressing the enter key, DOS creates a label for the disk called FIRST DISK. *Note:* DOS does not differentiate between upper- and lowercase letters for the volume label.

```
A:\>LABEL    [ENTER]
Volume in drive A has no label
Volume Serial Number is 105E-0FD3
Volume label (11 characters, ENTER for none)? first disk

A:\>_
```

In the next example, the diskette in drive A already has a label of FIRST DISK, but the enter key is pressed when asking for a new volume label, at which point DOS prompts the user to verify that the label should be deleted. This message is not displayed if another volume label is replacing the old one.

```
A:\>LABEL    [ENTER]
Volume in drive A is FIRST DISK
Volume Serial Number is 105E-0FD3
Volume label (11 characters, ENTER for none)?

Delete current volume label (Y/N)?

A:\>_
```

2.13.8 Volume (internal)

```
VOL [drive:]
```

The VOL label command displays the volume label and serial number of a diskette or hard disk.

VOL Parameters

drive: This drive parameter is optional; if not specified, the volume command displays the volume label of the current drive.

```
A:\>VOL

 Volume in drive A is HELP
 Volume Serial Number is 105E-0FD3

A:\>_
```

2.13.9 System (external)

```
SYS [drive1:][path] drive2:
```

The SYS command creates the MS-DOS system files IO.SYS and MSDOS.SYS (both hidden files) and the COMMAND.COM file for a disk or diskette. This command allows a disk or diskette to boot up the computer in MS-DOS. Once booted with the system files, only internal MS-DOS commands can be performed. External MS-DOS commands require the files for each command used.

SYS Parameters

drive1: This optional parameter specifies which drive contains the system file used to create the new system files. If this parameter is not specified, the current drive is used.

path This optional parameter specifies the path that contains the system file used to create the new system files. If this parameter is not specified, the current working directory is used.

drive2: This required parameter specifies the drive where the system command is to create the system files.

```
C:\>SYS A:
System transferred

C:\>_
```

2.13.10 Directory (internal)

```
DIR [drive:][path][filename] [/p] [/w]
    [/a[[:]attributes]][/o[[:]sortorder]] [/s] [/b] [/l] [/c]
```

The DIR command displays a list of the file names and subdirectories in the current drive and working directory (unless otherwise specified). The listing displays the volume label (name), volume serial number, directory name, file or subdirectory name, file extension, size of file, and date and time the file was created or last modified.

DIR Parameters

drive: This optional parameter specifies the drive for the directory command, which may be other than the current drive.

path This optional parameter specifies the path for the directory command, which may be other than the working directory.

DIR Switches

/p The “p” (page) switch lists only one video page of the directory listing at a time; at the end of each page, DOS prompts the user to press any key to see the next screen of listings.

/w The “w” (wide) page switch lists only the file names and extension in a five-column format.

/a[[:]attributes] The “a” (attribute) switch is used to display or not to display the listing of file names with specified attributes. If the “a” switch is used without any attributes, all files, regardless of their attributes, are displayed. The colon is not necessary. Do not use spaces with the attributes. The attributes are as follows:

Display	Do not display
H (hidden files)	−H (hidden files)
S (system files)	−S (system files)
D (directories)	−D (directories)
A (archived files)	−A (archived files)
R (read-only files)	−R (read-only files)

/o[[:]sortorder] The “o” (order) switch is used to cause the listing of the directory command to be displayed in a specified order. This does not change the actual order of the files in the directory; it just changes the order on the display screen. The colon is not necessary. Do not use spaces with the sort-order parameters. The sort-order parameters are as follows.

N File names in alphabetic order
−N File names in reverse alphabetic order
E Extensions in alphabetic order
−E Extensions in reverse alphabetic order
D By date and time, ascending (earliest first)
−D By date and time, descending (latest first)
S By size, ascending (smallest first)
−S By size, descending (largest first)
G Group directories before files
−G Group directories with files
C By compression ratio (smallest first)
−C By compression ratio (largest first)

/s The “s” (search) switch lists all file information for the specified file name for the directory and all subdirectories on the disk. Each directory or subdirectory name is displayed, along with the file name, extension, size, data, and time the file was created.

The remaining switches are not used very often and are not discussed here.

In addition to these parameters and switches, the DIR command can use DOS wild cards and be redirected and filtered.

The following example displays the files in the default directory (“\” root directory) in the current drive. The C: part of the DOS prompt C:\> specifies that the current drive is drive

C. The \ (backslash with no characters following it) specifies the root directory, which is the current working directory. A name following the backslash would indicate that the current working directory is a subdirectory path.

If the disk has a label, the name is specified on the first line of the directory listing. The volume serial number appears on the second line. The current drive and working directory are listed on the third line.

The leftmost column specifies the file or subdirectory name; the next column to the right specifies the file extension. The next column lists the size of the file, followed by the date and time the file or subdirectory was created or last updated.

The number of files is listed, followed by the total size of the files. The last line of the listing indicates the remaining number of bytes on the disk.

```
C:\>DIR

 Volume in drive C has no label
 Volume Serial Number is 1A97-4AE4
 Directory of C:\

DOS          <DIR>        03-08-93   2:34a
CONFIG   SYS          178 04-23-93   9:23a
AUTOEXEC BAT           62 04-23-93   9:23a
COMMAND  COM        52925 03-10-93   6:00a
MMOUSE   COM        16896 11-29-90  12:00a
WINA20   386         9349 03-10-93   6:00a
        6 file(s)         79410 bytes
                       37269504 bytes free
```

In the following example, the DIR command was performed by drive A, because the drive parameter was used.

```
C:\>DIR A:

 Volume in drive A is HELP
 Volume Serial Number is 105E-0FD3
 Directory of A:\

FIRST    BAT          114 01-10-93   8:04a
COMMAND  COM        52925 03-10-93   6:00a
MMOUSE   COM        16896 11-29-90  12:00a
WINA20   386         9349 03-10-93   6:00a
FORMAT   COM        22717 03-10-93   6:00a
CONFIG   SYS          178 04-23-93   9:23a
AUTOEXEC BAT           62 04-23-93   9:23a
        7 file(s)       102241 bytes
                       1222144 bytes free
```

In the following example, the DIR command was executed using the wide-output listing [/W]. Only the file names and extension are listed. Subdirectories are specified with brackets [].

```
C:\>DIR /W

 Volume in drive C has no label
 Volume Serial Number is 1A97-4AE4
 Directory of C:\
```

```
[DOS]   CONFIG.SYS   AUTOEXEC.BAT  COMMAND.COM  MMOUSE.COM
WINA20.386
         6 file(s)        79410 bytes
                       37269504 bytes free
```

In the following example, the DIR command was issued using the attribute switch. Because no attribute was selected, all files, regardless of their attribute, are listed.

```
C:\>DIR /A

 Volume in drive C has no label
 Volume Serial Number is 1A97-4AE4
 Directory of C:\

IO       SYS        40470 03-10-93   6:00a
MSDOS    SYS        38138 03-10-93   6:00a
DOS          <DIR>        03-08-93   2:34a
CONFIG   SYS          178 04-23-93   9:23a
AUTOEXEC BAT           62 04-23-93   9:23a
COMMAND  COM        52925 03-10-93   6:00a
MMOUSE   COM        16896 11-29-90  12:00a
WINA20   386         9349 03-10-93   6:00a
        8 file(s)       158018 bytes
                      37269504 bytes free
```

In the following example, the DIR command is using a sort option [/O]; the [N] specifies that the output should be displayed sorted in alphabetical order.

```
C:\>DIR /ON

 Volume in drive C has no label
 Volume Serial Number is 1A97-4AE4
 Directory of C:\

AUTOEXEC BAT           62 04-23-93   9:23a
COMMAND  COM        52925 03-10-93   6:00a
CONFIG   SYS          178 04-23-93   9:23a
DOS          <DIR>        03-08-93   2:34a
MMOUSE   COM        16896 11-29-90  12:00a
WINA20   386         9349 03-10-93   6:00a
             6 file(s)        79410 bytes
                           37269504 bytes free
```

2.13.11 More (external)

```
MORE < [drive:] [path] filename
[source] ¦ MORE
```

The MORE filter is used with any type of command that displays information to the video screen. This filter produces the same results as the page switch in the directory command but can be used with other commands as well. The most popular way to use the MORE command is to have it follow the DOS command. To use the MORE command as the output, the pipe sign (¦) must precede the key word MORE. The source is the command, file, or device from which the information is coming.

```
C:\>DIR \DOS\*.EXE ¦MORE

 Volume in drive C has no label
 Volume Serial Number is 1A97-4AE4
 Directory of C:\DOS

NLSFUNC  EXE      7036 03-10-93   6:00a
ATTRIB   EXE     11165 03-10-93   6:00a

***** 14 FILES LISTED IN THIS AREA *****

MOVE     EXE     17823 03-10-93   6:00a
INSTBIN  EXE    135680 03-20-93   8:22a
— More —

** AT THE BOTTOM OF VIDEO PAGE **


** BEGINNING OF NEXT VIDEO PAGE **

SMARTDRV EXE     42073 03-10-93   6:00a
BACKUP   EXE     36092 04-09-91   5:00a

***** 19 FILES LISTED IN THIS AREA *****

SUBST    EXE     18478 03-20-93   6:00a
MSBACKUP EXE      5506 03-10-93   6:00a
— More —

** AT THE BOTTOM OF VIDEO PAGE **


** BEGINNING OF NEXT VIDEO PAGE **

UNDELETE EXE     26420 03-10-93   6:00a
INTERLNK EXE     17197 03-10-93   6:00a

***** 5 FILES LISTED IN THIS AREA *****

SIZER    EXE      7169 03-10-93   6:00a
MSAV     EXE    172198 03-10-93   6:00a
        50 file(s)     2344981 bytes
                      37265408 bytes free
```

2.13.12 Disk Copy (external)

```
DISKCOPY [drive1: [drive2:]] [/1] [/v]
```

The DISKCOPY command makes an exact copy of the disk in the source drive onto the target drive. Exact copy means that all files on both drives after the command is complete are the same in both size and physical locations on each drive. If the target drive is not formatted, DOS formats the disk. If the target drive has bad sectors, DOS produces an error message (Target disk may not be usable). The target disk must be the same size as the source disk.

DISKCOPY Parameters

drive1: This optional parameter specifies the source drive.

drive2: This optional parameter specifies the target drive.

If the drive parameters are not used, DOS uses the current drive as both the source and target. Diskcopy cannot be performed on a nonremovable (hard disk) drive.

DISKCOPY Switches

/1 The "1" (one) switch is used to inform DOS to copy the first side or both drives.

/v The "v" (verify) switch verifies the copy process but slows down the operation.

In the following example, the DISKCOPY command copies the contents of drive A (source) diskette onto drive B (target) diskette. Because two drives are used, there is no need to exchange diskettes.

```
C:\>DISKCOPY A: B:

Insert SOURCE diskette in drive A:

Insert TARGET diskette in drive B:

Press any key to continue . . .

Copying 40 tracks
9 Sectors/Track, 2 Side(s)

Volume Serial Number is 0ACA-1034

Copy another diskette (Y/N)?
```

In the following example, the DISKCOPY command copies the contents of drive A (source) diskette onto a different diskette that will be placed in drive A (called the target diskette). Because only one drive is used, the user must swap the source and target diskette more than once. The number of times the user must swap diskettes depends on the size of the source diskette and the amount of conventional RAM.

```
C:\>DISKCOPY A: A: /V

Insert SOURCE diskette in drive A:

Press any key to continue . . .

Copying 80 tracks
18 sectors per track, 2 side(s)

Insert TARGET diskette in drive A:
```

```
Press any key to continue . . .

Insert SOURCE diskette in drive A:

Press any key to continue . . .

Insert TARGET diskette in drive A:

Press any key to continue . . .

Insert SOURCE diskette in drive A:

Press any key to continue . . .

Insert TARGET diskette in drive A:

Press any key to continue . . .

Volume Serial Number is 09FF-3451

Copy another diskette (Y/N)?
```

2.13.13 Disk Compare (external)

```
DISKCOMP [drive1: [drive2:]] [/1] [/8]
```

The DISKCOMP compares the contents of two disks location by location (track by track) and informs the user of any differences. This command should be performed only on disks that have been duplicated under disk copy; otherwise, errors usually occur, even though the same number and size of files are found on both disks.

DISKCOMP Parameters

drive1: This optional parameter specifies the first disk used in the disk compare.

drive2: This optional parameter specifies the second disk used in the disk compare.

DISKCOMP Switches

/1 The "1" (side one) switch compares only the first sides of the diskettes.

/8 The "8" (eight sector) compares only 8 sectors per track, even if the diskettes contain more than 8 sectors per track.

In the following example, the contents of drive A (FIRST) diskette are compared to the contents of drive B (SECOND) diskette. No error was encountered during the process.

```
A:\>DISKCOMP A: B:

Insert FIRST diskette in drive A:

Insert SECOND diskette in drive B:

Press any key to continue . . .
```

```
Comparing 40 tracks
9 sectors per track, 2 side(s)

Compare OK

Compare another diskettes (Y/N)?
```

In the following example, only one disk drive is available; the computer prompts the user to swap the diskettes during the comparison. One error was detected.

```
C:\>DISKCOMP A: A:

Insert FIRST diskette in drive A:

Press any key to continue . . .

Comparing 80 tracks
18 sectors per track, 2 side(s)

Insert SECOND diskette in drive A:

Press any key to continue . . .

Compare error on
side1, track 26

Insert FIRST diskette in drive A:

Press any key to continue . . .

Insert SECOND diskette in drive A:

Press any key to continue . . .

Compare another diskette (Y/N)?
```

2.13.14 Copy (internal)

```
COPY [drive:][path1]name1 [drive:][path2][name2] [/v]

COPY [/a or /b] source [/a or /b] [+ source [/a or /b] [+
     ...]] [destination [/a or /b]] [/v]
```

The COPY command makes a copy of a file or files as specified. The COPY command reads one file at a time from the source diskette and then writes the file to the destination diskette; after each file is copied, the computer displays the file name on the screen. The copy command must follow these rules:

1. The file(s) cannot have the same name if the destination is on the same drive or directory.
2. The command works on only one directory at a time; no parameter allows multidirectory copies.

3. The command uses the listing of the directory to determine the order in which to copy multiple files. Therefore, even though the number of files may be the same on two different drives, the physical locations of the files may be different.
4. Hidden files cannot be copied with the copy command.
5. The command does not format the diskette.
6. When copying files from one disk to another disk, two drives are required.

The first syntax listing is the most common; therefore, only those parameters and switches are discussed here.

COPY Parameter

drive1: drive2:	Drive1: specifies the source drive and drive2: specifies the destination drive. If a drive parameter is not specified, the current drive is used.
path1 path2	Path1 specifies the source path and path2 is the destination path. If the path is not specified, the working directory is used.
name1 name2	Name1 is the file name(s) of the source and name2 is the file name(s) of the destination. DOS wild cards are valid within these parameters. The destination name is optional and is used if the destination file name needs to be different from the original file name.

COPY Switch

/v	The "v" (verify) switch causes COPY command to verify the copying of each file. When this switch is used, the COPY command takes longer.

```
C:\>COPY *.* A:
CONFIG.SYS
AUTOEXEC.BAT
COMMAND.COM
MMOUSE.COM
WINA20.386
        5 file(s) copied
```

In the previous example, all the files located on the current drive (C) in the current working directory ("\", root directory) were copied to the last working directory of drive A, using the same names. The names of the files copied to drive A are the same as those of the source drive because the pathname2 was not specified.

```
C:\>COPY AUTOEXEC.BAT \BATCH\AUTO.OLD /V
        1 file(s) copied
```

In the previous example, the file (AUTOEXEC.BAT) from the root directory of drive C (C:\, current drive and working directory) is copied into the \BATCH subdirectory of drive C (because a destination drive was not specified). The name of the destination file is stored with the new file name of AUTO.OLD. Because the verify switch was used, the computer verified that the copy occurred correctly.

2.13.15 Verify (internal)

```
VERIFY [on or off]
```

The VERIFY command switches on or off the verify function any time data is written to the diskette. The verify function slows down disk operations. If the VERIFY command does not specify the on or off switch, the status of verify is displayed. Once the VERIFY command is switched on or off, it stays in that condition until the power of the computer is turned off or until the VERIFY command is switched on or off.

```
C:\>VERIFY
VERIFY is off

C:\>VERIFY ON
```

2.13.16 Xcopy (external)

```
XCOPY [drive1:][path1]name1 [drive2:][path2][name2] [/v] [/s]

XCOPY source [destination] [/a or /m]
      [/d:date] [/p] [/s [/e]] [/v] [/w]
```

The XCOPY command copies a file(s) and directory (directories) from one location to another as specified by the DIR command. XCOPY follows the same rules as COPY, except multidirectories can be copied if specified properly. Subdirectories are created on the target disk if the subdirectory does not exist. The main advantage of the XCOPY command is that all or as many of the files and directories of the source as needed are read into memory at one time; then everything is dumped into the target. Because of this advantage, XCOPY copies faster and with less wear on the drives than does the COPY command. The COPY command copies one file at a time between the drives until all files are copied.

The first syntax listing is the most common; therefore, only those parameters and switches are discussed in this text.

The parameters drive1:, path1, name1, drive2:, path2, and name2 follow the same format as the COPY command; the verify switch also works the same as for the COPY command. The "/s" subdirectory switch causes the computer to copy the files in the directory and subdirectories unless they are empty. If the subdirectory switch is not specified, only the contents of the specified directory or subdirectory are copied.

In the following example, all the files in the THEORY subdirectory of drive C (because no drive parameter is specified) are copied to the last working directory of drive A. The computer first reads drive C and copies the files into memory or until memory is filled; then the files are transferred into the working directory of drive A.

```
C:\>XCOPY \THEORY\*.* A:
Reading source file(s) ...
\THEORY\RULES
        1 File(s) copied
```

In the following example, all the files and subdirectories (because the subdirectory switch was specified) within the \THEORY subdirectory of drive C are copied into the \THEORY subdirectory of drive A. Because there is no \THEORY subdirectory on drive A, the computer asks the user whether \THEORY should be a file or a directory. All remaining subdirectories are created automatically, and the files are copied into those subdirectories.

```
C:\>XCOPY \THEORY\*.* A:\THEORY /s
Does THEORY specify a file name
or directory name on the target
(F = file, D = directory)? (PRESS THE LETTER D)
\THEORY\RULES
```

```
\THEORY\TEST\TEST_1
\THEORY\TEST\TEST_2
\THEORY\TEST\QUIZ_1
\THEORY\NOTES\OHM_LAW
        5 File(s) copied
```

The RULES file is copied into the newly created \THEORY subdirectory on drive A. The TEST_1, TEST_2, and QUIZ_1 files are copied into the newly created \THEORY\TEST subdirectory. The OHM_LAW file is copied into the newly created \THEORY\NOTES subdirectory.

2.13.17 Compare (external)

```
COMP [drive1:][path1]name1 [drive1:][path2]name2 [/d] [/a]
     [/l] [/n=number] [/c]
```

The COMP command compares the contents of two files or sets of files and displays different types of messages as a result of the comparison.

The drive1:, path1, drive2:, and path2 parameters follow the same format as the COPY command and are needed only if the file or files are on different drives and/or working directories. Both file names must be specified for this command to function if the files are in the same drive and directory. If name2 is not specified, the computer uses name1 as the default for name2.

The switches specified in the syntax are used to denote special types of comparisons that are not discussed in this book.

In the following example, the first COMP command compares the contents of AUTOEXEC.BAT to AUTO; because the files are the same, the computer responds with the message "Files compare OK." Because the answer to the question "Compare more files (Y/N)?" is "Y" (yes), the computer prompts the user to input the name of the first file to be compared; once typed in and with the enter key pressed, the computer responds with the next question. The name of the second file in the example is located in the \THEORY\TEST subdirectory of drive A, and the name of the second file is AUTO.OLD. Because the file sizes are different, the computer displayed a message to that effect.

```
A:\>COMP AUTOEXEC.BAT AUTO
Comparing AUTOEXEC.BAT and AUTO...
Files compare OK

Compare more files (Y/N)? (ANSWER YES)
Name of first file to compare: AUTO
Name of second file to compare: A:\THEORY\TEST\AUTO.OLD
Option:
Comparing AUTO and A:\THEORY\TEST\AUTO.OLD
Files are different sizes

Compare more files (Y/N)?
```

The COMP command has for the most part been replaced with the FC (file compare) command, which allows the user to specify the comparing of many different types of data in two files.

2.13.18 File Compare (external)

```
FC [/a] [/c] [/l] [/lbn] [/n] [/t] [/w] [/nnnn] [drive1:]
   [path1]filename1 [drive2:][path2]filename2
```

The FC command compares the contents of two files or sets of files. The FC command works similarly to the COMP command except that it allows the user to specify how to compare the contents of the files.

The drive1:, path1, filename1, drive2:, path2, and filename2 parameters follow the same format as for the COMP command.

The different switches are not discussed in this text because they are not normally used. If no switches are used, the computer defaults to the ASCII mode unless the file extension is [bin], [com], [exe], [lib], [obj], or [sys]. If there is a difference in the files, the contents of both files are displayed in ASCII format. If there are no differences in the files, the message "FC: no differences encountered" is displayed.

In the following example, the file AUTOEXEC.BAT is compared with the AUTO file, both of which are located in the current drive and working directory. No differences are found.

```
A:\>FC AUTOEXEC.BAT AUTO
Comparing files AUTOEXEC.BAT and AUTO
FC: no differences encountered
```

In the following example, the file CONFIG.OLD located in the current drive and working directory is compared with the file CONFIG located in the \THEORY subdirectory of the current drive. Because no switches are used and there are differences in the files, DOS displays all the lines in each file.

```
A:\>FC CONFIG.OLD \THEORY\CONFIG
Comparing files CONFIG.OLD and \THEORY\CONFIG
***** CONFIG.OLD
DOS=HIGH
FILES=40
BUFFERS=25
***** \THEORY\CONFIG
DOS=HIGH
FILES=25
BUFFERS=20
*****
```

2.13.19 Delete [Erase] (internal)

```
DEL [drive:][path]filename [/p]

ERASE [drive:][path]filename [/p]
```

The DEL or ERASE command deletes a file(s) on the specified drive and/or directory. The only time the DEL command prompts the user to verify the operation is when the user deletes everything in the drive or directory using the (*.*) wild card or when the [/p] prompt switch is used. The prompt switch causes the computer to confirm each deletion of a file.

The [drive:] and [path] parameters specify the location of the file and are optional. If not used, DOS uses either the current drive or working directory. The file name is not optional and should include the file extension if one exists. DOS wildcards are valid with this command.

```
C:\THEORY>DEL *.*

All files in directory will be deleted!
Are you sure (Y/N)?
```

In the previous example, the DEL command deletes all files in the default drive and THEORY subdirectory only if the user answers yes (Y). If the user answers no (N), no files are deleted. If there are subdirectories within the working directory, they and their files are not deleted. The only time the DEL command prompts the user is when all files are to be deleted; otherwise, the deletion takes place as soon as the enter key is pressed unless the [/p] prompt switch is used.

```
C:\>DEL \THEORY\TEST\T*.* /P

C:\THEORY\TEST\TEST_1,     Delete (Y/N)?y
C:\THEORY\TEST\TEST_2,     Delete (Y/N)?n
```

In the above example, the DEL command deletes all files in drive C in the subdirectory of \THEORY\TEST that begin with the letter "T," disregarding any file extension, only after the user answers the question after each file is found. In this example, the user answered "Y" (yes) to the first file found; thus, it is deleted. The answer to the second file found was "N" (no); therefore, this file is not deleted.

How DOS Deletes Files. When DOS deletes (erases) a file, DOS does not delete the data from the cluster locations for the file. DOS changes the first character in the file name to a special character, which informs DOS that the directory entry for this file has been deleted and is available. Also, DOS goes to all FAT entries related to the deleted file and replaces the value for each cluster entry with a hex 00. The hex 00 means that the disk space the cluster represents is now available (free) for use. The new available space could be used to store a new file on the disk. This means that the old data from the previous file will be overwritten and lost.

Before DOS places a new file on the disk, it looks through the directory to find if the file name has already been used; if not, the first directory entry space that is available (either an entry location that was never used or one that has been deleted) is used for the new directory entry. If this happens, the old file information in the directory is lost, along with the first cluster number in the cluster chain for the old file, and the file cannot be recovered. Even if the deleted directory listing is not used by the new file copied onto the disk, the file may still not be recoverable because one or more of the old clusters related to the deleted file may have been overwritten. This may occur because, when DOS stores files on the disk, it always looks first for available space in the outermost tracks of the disk (track 0) and uses this space first when storing data on the disk. If the new file is large (requiring more clusters than were available before the deleted file), the data in the cluster location is overwritten and the data is lost.

Following the DEL (erase) command is the UNDELETE command, which may be used to recover some if not most deleted files.

2.13.20 Undelete (external)

```
UNDELETE [[drive:][path]filename] [/dt] [/ds] [/dos]

UNDELETE [/list or /all or /purge [drive] or /status or /load
or /unload or /s[drive] /tdrive[-entries]]
```

The UNDELETE command restores files that were previously deleted by using the DEL (ERASE) command. The UNDELETE command has three levels for protecting files against deletion: standard, Delete Tracker, and Delete Sentry.

A word of caution: The UNDELETE command may not always restore (retrieve) a file, especially if there have been any store operations on the disk since the file was deleted. If a file is accidentally deleted, the user should immediately try to undelete the file, in which case the file can be restored. The UNDELETE command cannot restore directories and the deleted files within those removed directories.

UNDELETE Standard Level. The standard level of protection is the least effective, because, if the directory entry has been written over or any of the clusters in the chain have been used, it is not able to recover the file completely. This protection level requires no memory or disk space. When using UNDELETE in this mode (the Delete Tracker or Delete Sentry was not enabled), DOS scans the directory and path specified, looks for entries that have been marked as deleted (first character in file name changed to the delete character), and checks to see if they can be restored. This is determined by checking the data in the entry and verifying that the first cluster listed in the directory entry for that file is still available. If it is, the file or at least the first cluster of information might be restorable. If the file is to be undeleted, the computer prompts the user for the first character in the file name and then reads the number of bytes of the old file from the directory entry; this determines the number of clusters that were associated with the deleted file. DOS then goes to the first cluster number listed in the directory; if only one cluster was related to the deleted file, DOS marks the cluster as being the last cluster for the file and then updates the directory. If the deleted file requires more than one cluster, DOS writes the number value of the next available cluster location into the current cluster entry. This process continues until the total number of clusters marked is equal to the total number of clusters needed by the deleted file. The last cluster in the chain is marked to indicate the last cluster in the file chain. DOS then updates the FAT and the directory to recover the file.

UNDELETE Delete Tracker Level. The intermediate protection level of the UNDELETE command uses the PCTRACKER.DEL file (a hidden file) to store a record of the locations of the deleted file. When the UNDELETE command is used and the Delete Tracker is enabled, DOS uses the data in the PCTRACKER.DEL file to determine the location(s) of the deleted files during the recovery process. This level of protection may allow partial recovery of the deleted file. The Delete Tracker uses 13.5K of memory and 5K or more of disk space, depending on the number of entries (number of deleted files) in the PCTRACKER.DEL file.

UNDELETE Delete Sentry Level. The highest level of protection for the UNDELETE command is the Delete Sentry level. When this level is enabled, it uses 13.5K of memory and a percentage of the hard drive as specified in the UNDELETE.INI file. The Delete Sentry creates a hidden directory named SENTRY; when a file is deleted, DOS moves the directory entry information into the SENTRY directory. When the UNDELETE command is issued and the file to be undeleted is found in the SENTRY directory, DOS moves the file back to its original location. Therefore, when using Delete Sentry, the file is really not deleted, just moved to another location. Because the size of the SENTRY directory is limited, when it is filled the Delete Sentry program deletes enough of the oldest file to make room for the latest deleted file. After a file is in the SENTRY directory for seven (default value that can be changed) days, it is purged.

UNDELETE Parameters

drive:	The drive that contains the file to be undeleted.
path	The path that contains the file to be undeleted.

Switches. Note that, if the Delete Tracker or Delete Sentry switches are not used and loaded into memory, the UNDELETE command operates in the standard protection mode.

/dt	This switch "/dt" (Delete Tracker) causes the UNDELETE command to undelete after prompting only those files listed in the deletion-tracking file.

/ds	This switch "/ds" (Delete Sentry) causes the UNDELETE command to undelete after prompting only those files listed in the SENTRY directory.
/s[drive]	The "/s[drive]" (Sentry) switch is used to enable the Delete Sentry (highest) protection level. The switch also loads the memory-resident portion of the Undelete program into memory and then uses the data found in the UNDELETE.INI file to configure this option. If the optional drive is not specified, the current drive is used as the default.
/tdrive [-entries]	The "/tdrive[-entries]" (Tracker) switch is used to enable the Delete Tracker protection level. This switch also loads the memory-resident portion of the Undelete program into memory using the data found in the PCTRACKR.DEL file. The optional entries switch is used to select the number of file entries to be saved in the PCTRACKR.DEL file. If the entries option is not used, the default value is used.

In the following example, neither the Delete Tracker nor Delete Sentry was enabled (because no Delete Sentry control or Deletion-tracking files are found), so DOS uses the standard MS-DOS method of undeleting files. MS-DOS detects two deleted files; both are able to be undeleted. The computer lists the files one at a time, except that the first character in the file name is a question mark. The computer first asks the user if he or she wishes to undelete the file. If the answer is "Y" (yes), the computer prompts the user to input a character to be used as the first character in the file name. If the undeletion is successful, the computer informs the user. If the answer to the undelete question is "N" (no), the computer displays the next file that can be undeleted.

Note: If there are files that cannot be undeleted, the files are still displayed, but the computer informs the user that the file is unrecoverable.

```
C:\>UNDELETE

UNDELETE - A delete protection facility
Copyright (C) 1987-1993 Central Point Software, Inc.
All rights reserved.

Directory: C:\
File Specifications: *.*

    Delete Sentry control file not found.

    Deletion-tracking file not found.

    MS-DOS directory contains      2 deleted files.
    Of those,       2 files may be recovered.

Using the MS-DOS directory method.

      ?C        171077  8-04-88 11:54p ...A Undelete(Y/N)?y
      Please type the first character for ?C     .    :p
File successfully undeleted.

    ?ELPP           16  1-12-93 8:45p ...A Undelete (Y/N)?n

C:\>_
```

The following example shows how the Delete Tracking memory resident program can be loaded into memory. This program and the tracking of deleted files continue until power is removed from the computer, unless otherwise specified. Only those files deleted while the Delete Tracking program is running are recorded in the PCTRACKR.DEL file. This command can be used in the AUTOEXEC.BAT file for automatic Delete Tracking. In the example, only drive C is tracked for deleted files using the Delete Tracking method.

```
C:\>UNDELETE /TC:

UNDELETE  - A delete protection facility
Copyright (C) 1987-1993 Central Point Software, Inc.
All rights reserved.

UNDELETE loaded.

Delete Protection Method is Delete Tracking.
Enabled for drives: C

C:\>_
```

In the next example of the UNDELETE command, the delete tracking was enabled because of the display of the deletion-tracking file contents and the fact that no Delete Sentry control file was found. Note that the user does not have to enter the first character in the file name, because the information on the file and file name are stored in the PCTRACKR.DEL file.

```
C:\>UNDELETE

UNDELETE - A delete protection facility
Copyright (C) 1987-1993 Central Point Software, Inc.
All rights reserved.

Directory: C:\
File Specifications: *.*

    Delete Sentry control file not found.

    Deletion-tracking file contains    3 deleted files.
    Of those,     3 files have all clusters available,
                  0 files have some clusters available,
                  0 files have no clusters available.

     MS-DOS directory contains 3 deleted files.
     Of those,     3 files may be recovered.

Using the Deletion-tracking method.

     COME              52925  3-10-93  6:00a  ...A Deleted:
     1-13-93  2:07a

All of the clusters for this file are available. Undelete
(Y/N)?n
```

```
    AUTO      OLD       62  4-23-93  9:23A  ...A Deleted:
    1-13-93   2:07a

All of the clusters for this file are available. Undelete
(Y/N)?y

File successfully undeleted.

    TEST                15  1-12-93  6:47p  ...A Deleted:
    1-13-93   2:07a

All of the clusters for this file are available. Undelete
(Y/N)?n

C:>_
```

The following example shows how the Delete Sentry memory-resident program can be loaded into memory. This program and the tracking of deleted files continue until power is removed from the computer, unless otherwise specified. Only those files deleted while the Delete Sentry program is running are moved to the SENTRY directory. This command can be used in the AUTOEXEC.BAT file for automatic Delete Sentry protection. In the example, only drive C is tracked for deleted files using the Delete Sentry method.

```
C:\>UNDELETE /S

UNDELETE - A delete protection facility
Copyright (C) 1987-1993 Central Point Software, Inc.
All rights reserved.

UNDELETE loaded.

Delete Protection Method is Delete Sentry.
Enabled for drives : C

C:\>_
```

In the following example of the undelete command, the Delete Sentry is enabled because of the display of the Delete Sentry control file contents. The computer also displays the Delete-tracking file and MS-DOS directory contents. Note that the user does not have to enter the first character in the file name, because the file has only been moved into the SENTRY directory along with all of its information. The /DS (Delete Sentry only) switch is used to specify files to delete under Delete Sentry control only.

```
C:\>UNDELETE /DS

UNDELETE - A delete protection facility
Copyright (C) 1987-1993 Central Point Software, Inc.
All rights reserved.

Directory: C:\
File Specifications: *.*

     Delete Sentry control file contains 2 deleted files.
```

```
    Deletion-tracking file contains 2 deleted files.
    Of those,      2 files have all clusters available,
                   0 files have some clusters available,
                   0 files have no clusters available.

     MS-DOS directory contains    3 deleted files.
     Of those,     1 file may be recovered.

Using the Delete Sentry method.

     NEW                  178 4-23-93  9:23a  ... A Deleted:
     1-13-93  2:10a
This file can be 100% undeleted. Undelete (Y/N)?y

File successfully undeleted.

     MENU     BAT          40 3-09-93  12:05a  .... Deleted:
     1-13-93  2:10a
This file can be 100% undeleted. Undelete (Y/N)?

C:\>_
```

2.13.21 Move (external)

```
MOVE [drive:][path]filename[,[drive:][path]filename[...]] destination
```

The MOVE command moves one or more specified files to a new destination (location on a disk). Multiple files can be moved by using DOS wildcards or by specifying multiple [[drive:][[path]filenames] separated by commas. The destination follows the same format [[drive:][path]]. If only one file is moved, a new file name for the file can be specified in the destination parameter. Moving a file or group of files overwrites any files in the destination with the same name.

In the following example, the AUTOEXEC.BAT and CONFIG.SYS files are moved to the \THEORY subdirectory.

```
C:\>MOVE AUTOEXEC.BAT,CONFIG.SYS \THEORY
c:\autoexec.bat => c:\theory\autoexec.bat [ok]
c:\config.sys => c:\theory\config.sys [ok]
C:\>_
```

2.13.22 Rename (internal)

```
REN [drive:][path]filename1 filename2
```

The REN command is used to change the name of filename1 and give it the new name filename2. The drive and path are optional unless the file in question is in a directory other than the current drive and working directory. This command does not change the size or location of the file or its location in the directory listing.

In the following example, the MMOUSE.COM file was renamed MOUSE.COM.

```
C:\>REN MMOUSE.COM MOUSE.COM

C:\>_
```

2.13.23 Type (internal)

```
TYPE [drive:][path]filename
```

The TYPE command displays the content of a text file in an ASCII format on the video screen. If a file contains more than one video screen, the MORE pipe allows viewing of one video page at a time. The output of the TYPE command is case sensitive. Nonprintable ASCII characters may not be visible or may cause the computer to perform control functions.

```
C:\>TYPE AUTO.BAT
PATH=C:\DOS
MMOUSE

C:\>_
```

2.13.24 Graphics (external)

```
GRAPHICS [type] [[drive:][path]filename] [/r] [/b] [/lcd]
        [/printerbox:std or /printbox:lcd]
```

The GRAPHICS command is a memory-resident program that allows MS-DOS to print graphical information from the video screen on a printer by means of the graphics adapter (CGA, EGA, and VGA) in a bit-mapped format rather than a text format.

This command is normally found in the AUTOEXEC.BAT file if this type of printing is used often. The most frequently used parameter is the type parameter, which specifies which type of printer is to be used. The type parameter is very important because it configures the data for the selected printer; if the incorrect printer is selected, the output from the printer may not look like the video display screen.

Once this command is loaded into memory, it remains there until the computer is turned off. Once it is in memory, the user can press the print screen key to get the printout. Otherwise, the user needs to specify the file name. When the print screen key is pressed, the cursor starts scanning the video screen, from the upper-left to the lower-right corner of the screen. The printing of graphics requires more time than printing standard text because each location must be converted into bit-mapped graphics. The printout is printed sideways on the paper.

GRAPHICS Parameter

type — The type parameter specifies the printer for the graphics command. If the correct printer is not selected, the output from the printer may not be correct.

The following are the most common types of printers supported by MS-DOS 6.2; for a complete listing, see HELP in MS-DOS.

COLOR1: IBM Personal Color Printer (black ribbon)
COLOR4: IBM Personal Color Printer (multicolor, red/blue/green)
COLOR8: IBM Personal Color Printer (multicolor ribbon, Cyan/Magenta/Yellow/Black)
GRAPHICS: IBM Personal Graphics, Proprinter, or Quietwriter printer
HPDEFAULT: Hewlett-Packard (HP) PCL printer
DESKJET: Hewlett-Packard Desk Jet printer
LASERJET: Hewlett-Packard LaserJet printer
LASERJETII: Hewlett-Packard LaserJet II printer
PAINTJET: Hewlett-Packard PaintJet printer
QUIETJET: Hewlett-Packard QuietJet printer

The following example loads the GRAPHICS program into memory, using the HP LaserJet printer format.

```
C:\>GRAPHICS LASERJET

C:\>_
```

2.13.25 Print (external)

```
PRINT [[drive:][path]filename[ ...]] [/t]

PRINT [/d:device] [/b:size] [/u:ticks1] [/m:ticks2]
      [/s:ticks3] [/q:qsize] [/t] [[drive:][path]filename
      [ ...]] [/c] [/p]
```

The PRINT command causes the contents of an ASCII text file to be printed on a printer. The output port for the printer may be selected by the MODE command. The first example of the syntax above is the most common form for using the PRINT command. The user needs to specify the drive and path if the file name is not in the current drive or working directory. The file name requires a file extension if present. Multiple files can be printed using DOS wildcards or multiple file names. The [/t] terminate switch is used to remove all files currently in the print queue (buffer).

The following example shows the PRINT command being used to print the contents of the CONFIG.SYS file on the printer. The first time the PRINT command is used, once the computer is turned on, the PRINT command prompts the user to select a list device (printer port); once the port is selected, the PRINT command does not prompt the user again as long as the computer is not reset. The options for the list device are PRN (the default printer port), LPT1: (line printer), LPT2:, COM1: (asychronous communication port), COM2:, COM3:, and COM4. Once the resident part of the PRINT command is installed, the printer should start printing the file. If the print queue is not empty, DOS displays a message indicating that certain files are in the queue.

```
C:\>PRINT CONFIG.SYS
Name of list device [PRN]:
Resident part of PRINT installed

  C:\>CONFIG.SYS is currently being printed

C:\>_
```

2.13.26 Check Disk (external)

```
CHKDSK [[drive:][path]filename] [/f][/v]
```

The CHKDSK command reads the directory of the specified drive and path, and analyzes all the files on the disk or directory by reading the FAT. Any problems with files or FAT entries are listed on the screen.

CHKDSK Switches

/f The [/f] (fix) switch causes the check disk command to convert the data found in clusters of the FAT that do not correspond to the FAT sequence. FAT errors do not necessarily mean serious problems with the hardware or software. Lost clusters errors occur when the directory entry specifies a certain number of clusters but the number of clusters in the allocation chain does not match. The data from each converted cluster location is stored in a file called FILEnnnn (where nnnn starts with the number 0000 and increments with each cluster location converted) in the root directory. The user can then use a text editor to view the contents of the files to determine if the data can be used. If the file contains machine code, the data is probably not usable. Once the CHKDSK command fixes the FAT, it is updated so that no more errors are listed until a new error occurs.

/v — The [/v] (display) switch causes CHKDSK to list the names of files in each directory as it analyzes the disk structure.

In the following example, the CHKDSK command analyzes the file structure of drive C and lists the serial number, format size (bytes total disk space), number and size of hidden files, directories, user files, and the size of available space on the disk. This command also displays the cluster (allocation unit) size, the total and available number of allocation units. Finally, the amount of conventional memory is displayed, along with the number of free bytes of conventional memory.

```
C:\>CHKDSK C:
Volume Serial Number is 1A97-4AE4

  42661888 bytes total disk space
     79872 bytes in 2 hidden files
     18432 bytes in 7 directories
   5912576 bytes in 160 user files
  36651008 bytes available on disk

      2048 bytes in each allocation unit
     20831 total allocation units on disk
     17896 available allocation units on disk

    655360 total bytes memory
    615088 bytes free

C:\>_
```

2.13.27 Scan Disk (external)

```
SCANDISK [[drive: [drive:...]/ALL]
```

The SCANDISK command performs a detailed analysis of a disk and allows the user to repair many types of errors to files or the surface of the disk. The SCANDISK command is a new addition to MS-DOS 6.2. SCANDISK should be used instead of CHKDSK because it does a more detailed job of analyzing and repairing disk-related problems. The above syntax shows only the most common parameters; use the DOS HELP function to obtain additional information.

SCANDISK Parameters

drive: — The drive parameter is optional. If a drive parameter is not specified, SCANDISK is performed on the current default drive. If a drive parameter is specified, SCANDISK performs its task on the drive specified by the parameter. The user may optionally specify more than one drive separated by a space.

SCANDISK Switches

/ALL The /ALL switch causes SCANDISK to be performed on all local drives.

When SCANDISK is executed, the user first views a screen that shows five areas of analysis: media descriptor, FAT, directory structure, file system, and scan surface. As each area is processed, a check mark appears to indicate completion of that part of the analysis. If SCANDISK finds a problem, a window appears and prompts the user with a message and asks if it should fix the problem. After the repair is made, SCANDISK continues with the analysis. The last item on the list, "scan surface," may take 15 minutes or longer. SCANDISK prompts the user to either continue and perform the surface scan or exit and view an error log. If the user continues with the surface scan analysis, the screen clears and a graphical screen is displayed. On the left a grid appears with shapes within the grid. The shapes change appearance as SCANDISK scans the surface. On the right a listing indicates the following: the drive being scanned, the total number of clusters on the drive, a changing numerical value that indicates the number of clusters being examined, the number of bad clusters found, the cluster size of each graphical grid, a description of what each type of grid shape represents (unused clusters, some used clusters, used clusters, some bad clusters).

If a repair is performed, the user may be prompted to insert a diskette so that the computer can create an undo file in case the repair does not work. The undo file allows the user to undo the repair so that the data can be recovered by some other third-party utility program.

SCANDISK Example

```
C:\>SCANDISK

C:\>SCANDISK D:

C:\>SCANDISK C: D:

C:\>SCANDISK /ALL
```

In the first example, SCANDISK is performed on the default drive, in this case drive C. In the second example, SCANDISK is performed on drive D. In the third example, SCANDISK is performed on both drives C and D. In the last example, SCANDISK is performed on all local drives.

2.13.28 Memory (external)

```
MEM [/c] [/f] [/p]
```

The MEM command analyzes the memory in the computer system and displays information about how the memory is allocated.

MEM Switches

/c The "/c" (classify) switch displays a listing of programs currently in memory.

/f The "/f" (free) switch displays a listing of available free memory.

/p The "p" (page) switch allows the MEM command to display one page at a time.

```
C:\>MEM

Memory Type           Total  =  Used  +  Free
____________________  _______  _______  _______
Conventional            640K     19K     621K
Upper                     0K      0K       0K
Adapter RAM/ROM         384K    384K       0K
Extended (XMS)         3072K     64K    3008K
____________________  _______  _______  _______
Total memory           4096K    467K    3629K

Total under 1MB         640K     19K     621K

Largest executable program size         621K   (636224 bytes)
Largest free upper memory block           0K        (0 bytes)
MS-DOS is resident in the high memory area.
```

2.13.29 MicroSoft Diagnostics (external)

```
MSD [i] [/f[drive:][path]filename] [/p[drive:][path]filename]
    [/s[drive:][path][filename]]
```

The MSD command provides technical information about the computer in a graphical type of format. Although this program can be controlled by the keyboard, it is best controlled by a mouse.

The most common syntax is used by entering the letters MSD and pressing the enter key; this produces the following graphical menuing system (Figure 2-2).

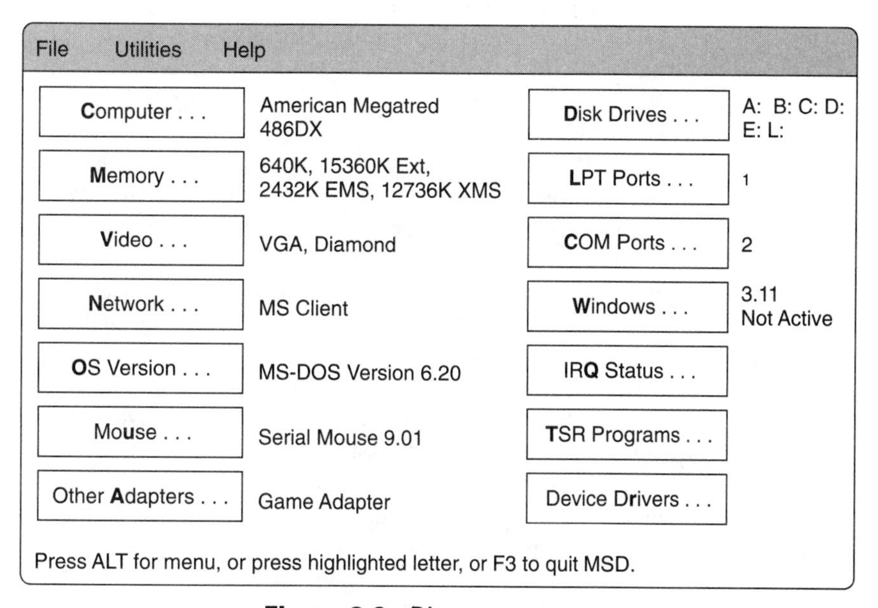

Figure 2-2 Directory structure.

- The three alternate menu functions at the top of the screen use pull-down menus to provide different functions for MSD.
- The File pull-down menu can be accessed either by pressing the ALT key and or using the mouse pointer (a filled–in rectangle) and clicking on the word File. The File pull–down menu allows different file functions.

The following describes the File functions available under the pull-down File menu.

The *Find File. . .* function allows the user to search for a file. The ellipsis (three periods) following the words *Find File* means that when this option is selected the user is prompted to enter data to perform the operation. A window (box) is displayed over a portion of the screen in which the user inputs the data.

The *Print Report. . .* function allows the user to determine how much information from this program should be sent to the printer. The window that opens also prompts the user to select which printer port to use.

The next section of the File pull-down menu lists a number of files that, when selected, are shown in a window on the screen. These are information files for different DOS-related configurations. The number of files listed varies depending on the options selected in DOS.

The last option on the File pull-down menu allows the user to exit the MSD command.

```
C:\>MSD
```

The *Utilities* pull-down menu provides five options.

The *Memory Block Display. . .* option window shows a list of programs (file names) that are loaded in memory and the memory address location and size of each program currently stored in RAM. On the right side of the window, a graphical representation of the computer RAM shows the location in reference to the memory map.

The *Memory Browser. . .* option window shows a list of BIOS memory names, their memory locations, and the size of each BIOS. The right side of the windows shows a graphical representation of the memory map of each BIOS in the computer. Pressing the enter key displays the contents of the BIOS in another window.

The *Insert Command. . .* option window shows a list of DOS command parameters normally listed in the CONFIG.SYS or AUTOEXEC.BAT files and allows the user to modify or insert data into the selected file.

The *Test Printer. . .* option window provides a list of options for printer types, the type of data for the printer, and the port used for the printer.

The *Black and White* option causes the screen to switch between a color screen and a black and white screen.

The HELP pull-down menu lists only one option, called *About. . .*, which, when selected, provides the copyright screen and version number for MicroSoft Diagnostics.

Under the pull-down menu are small boxes (windows) that indicate some part of the computer system. The wording to the right of most of these boxes is a summary of the important information regarding the adjacent window. If the user presses the character that is of a different color for each window title, detailed information regarding the window title is displayed on the screen in another window.

The *Computer. . .* title window, when selected, provides the computer name, BIOS manufacturer, BIOS version, BIOS category, BIOS ID byte, BIOS date, processor, math coprocessor, keyboard, bus type, DMA control, cascaded IRQ2, and BIOS data segment.

The *Memory. . .* title window, when selected, provides a memory map of the RAM in the computer in both a graphical and text format. The size and other types of information are listed for each type of memory (conventional, extended, upper, and expand (XMS) memory).

The *Video. . .* title window, when selected, provides technical data on the video adapter: video adapter type, manufacturer, model, display type, video mode, number of columns, number of rows, video BIOS version, video BIOS date, VESA support installed, VESA version, VESA OEM name, and secondary adapter.

The *Network. . .* title window, when selected, provides technical data on any network cards or software that are in the computer.

The *OS Version. . .* title window, when selected, provides technical data on the current operating system version: operating system, internal revision, OEM serial number, user serial number, OEM version string, DOS location in boot drive, path to program, and environment strings.

The *Mouse. . .* title window, when selected, provides technical data on the mouse if one is connected to the system: mouse hardware, driver manufacturer, DOS driver type, DOS driver version, mouse IRQ, number of mouse buttons, horizontal sensitivity, mouse-to-cursor ratio, vertical sensitivity, threshold speed, and mouse language.

The *Other Adapters. . .* title window, when selected, provides technical data on other types of devices not listed in the title windows; one such device is a game adapter: joystick A–X, Y, button 1, button 2, joystick B–X, Y, button 1, and button 2.

The *Disk Drives. . .* title window, when selected, provides technical data on the disk drives in the computer system: type of drive (floppy or fixed), form factor, size, number of cylinders, number of heads, bytes per sector, number of sectors per track, total size, and amount of free space.

The *LPT Ports. . .* title window, when selected, provides technical data about any line printer ports connected to the computer system: port, port address, on line, paper out, I/O error, time out, busy, and ACK.

The *COM Ports. . .* title window, when selected, provides technical data about any communication ports connected to the computer system: port name, port address, baud rate, parity, data bits, stop bits, carrier detect, ring indicator, data set ready, clear to send, and UART chip used.

The *IRQ Status. . .* title window, when selected, provides technical data about the IRQ (interrupt request) lines in the computer system. Each IRQ number contains the following information: address, description, detected, and handled by.

The *TRS Programs. . .* title window, when selected, provides technical data about any TRS (terminal resident) programs loaded in memory: program name, address, size, and command line parameters.

The *Device Drivers. . .* title window, when selected, provides technical data about any device drivers currently loaded in memory: device, file name, units, header, and attributes.

DOS Directory Commands. As discussed earlier in this chapter under Root Directory, when a disk is formatted under MS-DOS, DOS creates a directory for the disk. Within this directory created by the FORMAT command is a listing of all file names and file-related information. Because such a directory has a fixed size, the number of file names that can be stored on the disk is limited.

The directory created by the FORMAT command is called the Root Directory ("\", the backslash with no character following the symbol).

Because of the limited number of root directory entries, beginning with MS-DOS 2.0, MicroSoft created a group of commands that allows DOS users to create and manage multiple directories located on the same physical disk. Each of these directories is called a subdirectory (many times directory, for short). A subdirectory acts as a file to the directory in which it is located but as the directory to the files located within the subdirectory. Because the subdirectory looks like a file to the directory, the number of file name entries is limited only by the amount of disk space.

Each subdirectory acts independently of any other subdirectory; therefore, deleting all the files in one subdirectory does not alter the contents of the other subdirectories. For MS-DOS to work with multidirectory disks, DOS must be told in which subdirectory the command should be performed or the location of the file in question. Therefore, if the file or command is located in a different directory or subdirectory than the one in which the computer is currently working (called the working directory), the user must specify a path parameter that tells DOS how to get the information to and from the location specified. The base of the path is always the root directory "\"; a subdirectory would then contain a name following the backslash. Each subdirectory name is separated by the backslash (\).

The following six DOS commands are related to directory functions. Each is explained and examples are given using the directory structure in Figure 2-1.

2.13.30 Make Directory (internal)

```
MD [MKDIR][drive:]path
```

The first step in setting up a subdirectory(ies) is to make a directory (subdirectory). The drive parameter is needed only if the directory being made is on a drive other than the current drive. The path is the name of the directory preceded by the backslash. Because a subdirectory acts as a file to the directory it is in, the names given to subdirectories follow the parameters used in naming a file (eight valid characters maximum for the subdirectory name and three optional valid characters maximum for the subdirectory extension). Note that most subdirectories do not contain extensions.

2.13.31 Change Directory (internal)

```
CD [CHDIR][drive:]path

CD[..]
```

The CD command allows the user to change the current working directory. The working directory is the directory DOS uses to find a file, command, or executable file unless otherwise specified within the command syntax. When the computer is first booted up, the working directory is the root directory of the drive the computer was booted up on. If the file or command is not located in the working directory, DOS responds with an error message: "File not found -XXXX (where XXXX is the file name)".

2.13.32 Remove Directory (internal)

```
RD [RMDIR][drive:]path
```

The RD command allows the user to remove a directory from the disk; directories cannot be deleted using the DEL command. Certain conditions must be met before the specified directory can be removed. (1) There must be no files or subdirectories within the directory the user wishes to remove. (2) The user cannot be in the directory that is to be removed or one of its subdirectories. (3) The user cannot remove the root directory.

2.13.33 Path (internal)

```
PATH [[drive:]path[;...]]
```

PATH commands DOS to search a directory or multiple directories for the executable file. DOS always searches the working directory first; if the file is not found, it begins searching the paths (directory names) listed in the PATH command syntax. The PATH command is often found in the AUTOEXEC.BAT file or may be used on the command line of DOS.

2.13.34 Tree (external)

```
TREE [drive:][path] [/f] [/a]
```

The TREE command displays the tree (directory) structure of the directories in a drive. If the drive parameter and path are not used, DOS displays the directories on the current drive beginning with the root directory. The "/f" (file) switch, when specified, causes the file names within each directory to be listed, along with the directories. The "/a" (ASCII) switch, when specified, causes the TREE listing to be seen in an ASCII format.

2.13.35 Delete Tree (external)

```
DELTREE [/y] [drive:]path
```

The DELTREE (new with DOS 6.0) command is used to delete a directory and all files and subdirectories within the specified directory. In reality, DOS first deletes all the files within directories specified and then removes the directories. DOS prompts the user to verify the DELTREE function unless the "/Y" (yes) switch is used.

The following example shows how to create a directory structure as displayed in Figure 2-3. The DIR command is used throughout the examples to verify that the operations are taking place. The end of the example demonstrates how to get the disk back to its original condition. The example assumes that the computer has a hard drive with DOS loaded, with the PATH set for the DOS directory with all the files for the example in the root directory of drive C.

```
A:\>DIR                                              [Step 1]

 Volume in drive A is EXAMPLE
 Volume Serial Number is 2B5C-0DE6
 Directory of A:\

File not found

A:\>MD\THEORY                                        [Step 2]
```

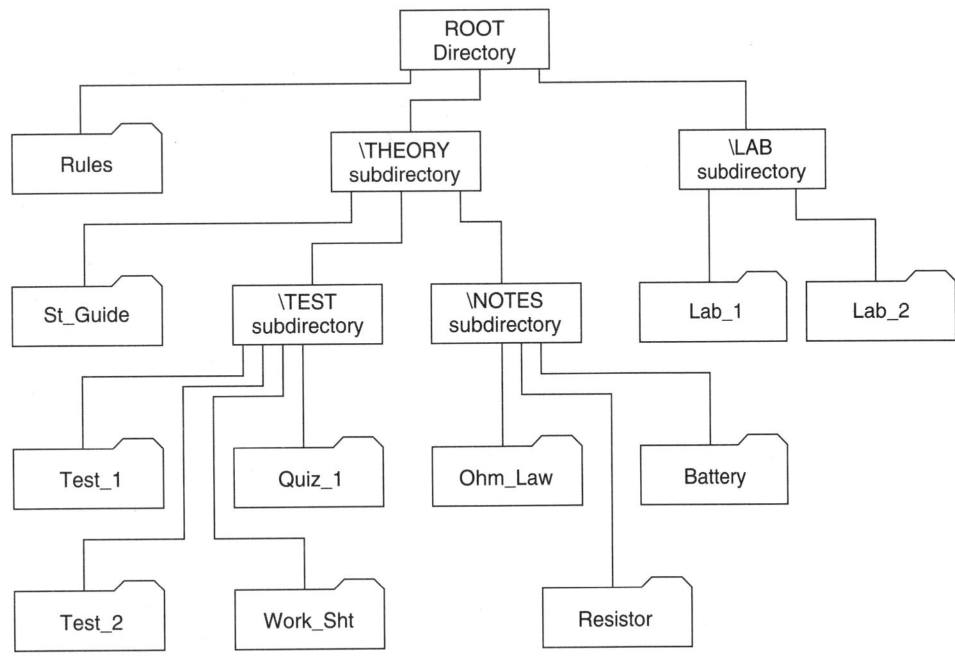

Figure 2-3 Directory structure.

```
A:\>MD\LAB                                              [Step 3]

A:\>DIR                                                 [Step 4]

 Volume in drive A is EXAMPLE
 Volume Serial Number is 2B5C-0DE6
 Directory of A:\

THEORY       <DIR>      01-16-93   2:43p
LAB          <DIR>      01-16-93   2:43p
        2 file(s)          0 bytes
                      728064 bytes free

A:\>COPY C:\RULES A:                                    [Step 5]
        1 file(s) copied

A:\>DIR                                                 [Step 6]

 Volume in drive A is EXAMPLE
 Volume Serial Number is 2B5C-0DE6
 Directory of A:\

THEORY       <DIR>      01-16-93   2:43p
LAB          <DIR>      01-16-93   2:43p
RULES          154      01-15-93   2:15a
        3 file(s)        154 bytes
                      727040 bytes free

A:\>CD\THEORY                                           [Step 7]

A:\THEORY>DIR                                           [Step 8]

 Volume in drive A is EXAMPLE
 Volume Serial Number is 2B5C-0DE6
 Directory of A:\THEORY

.            <DIR>      01-16-93   2:43p
..           <DIR>      01-16-93   2:43p
        2 file(s)          0 bytes
                      727040 bytes free

A:\THEORY>DIR \LAB                                      [Step 9]

 Volume in drive A is EXAMPLE
 Volume Serial Number is 2B5C-0DE6
 Directory of A:\LAB

.            <DIR>      01-16-93   2:43p
..           <DIR>      01-16-93   2:43p
        2 file(s)          0 bytes
                      728064 bytes free
```

```
A:\THEORY>MD\THEORY\TEST                                    [Step 10]

A:\THEORY>MD\THEORY\NOTES                                   [Step 11]

A:\THEORY>DIR                                               [Step 12]

 Volume in drive A is EXAMPLE
 Volume Serial Number is 2B5C-0DE6
 Directory of A:\THEORY

.            <DIR>      01-16-93   2:43p
..           <DIR>      01-16-93   2:43p
TEST         <DIR>      01-16-93   2:52p
NOTES        <DIR>      01-16-93   2:52p
        4 file(s)            0 bytes
                        724992 bytes free

A:\THEORY>COPY C:ST_GUIDE A:\                               [Step 13]
        1 file(s) copied

A:\THEORY>CD                                                [Step 14]
A:\THEORY

A:\THEORY>CD\                                               [Step 15]

A:\>COPY C:TEST_1 A:\THEORY\TEST                            [Step 16]
        1 file(s) copied

A:\>COPY C:TEST_2 A:\THEORY\TEST                            [Step 17]
        1 file(s) copied

A:\>COPY C:QUIZ_1 A:\THEORY\TEST                            [Step 18]
        1 file(s) copied

A:\>COPY C:WORK_SHT A:\THEORY\TEST                          [Step 19]
        1 file(s) copied

A:\>COPY C:OHM_LAW A:\THEORY\NOTES                          [Step 20]
        1 file(s) copied

A:\>COPY C:RESISTOR A:\THEORY\NOTES                         [Step 21]
        1 file(s) copied

A:\>COPY C:BATTERY A:\THEORY\NOTES                          [Step 22]
        1 file(s) copied

A:\>COPY C:LAB_1 A:\LAB                                     [Step 23]
        1 file(s) copied

A:\>COPY C:LAB_2 A:\LAB                                     [Step 24]
        1 file(s) copied

A:\>DIR                                                     [Step 25]
```

```
 Volume in drive A is EXAMPLE
 Volume Serial Number is 2B5C-0DE6
 Directory of A:\

THEORY       <DIR>       01-16-93   2:43p
LAB          <DIR>       01-16-93   2:43p
RULES                154 01-15-93   2:15a
        3 file(s)          154 bytes
                        712704 bytes free

A:\>DIR \THEORY                                        [Step 26]

 Volume in drive A is EXAMPLE
 Volume Serial Number is 2B5C-0DE6
 Directory of A:\THEORY

.            <DIR>       01-16-93   2:43p
..           <DIR>       01-16-93   2:43p
TEST         <DIR>       01-16-93   2:52p
NOTES        <DIR>       01-16-93   2:52p
ST_GUIDE             252 01-15-93   2:17a
        5 file(s)          252 bytes
                        712704 bytes free

A:\>DIR \THEORY\TEST                                   [Step 27]

 Volume in drive A is EXAMPLE
 Volume Serial Number is 2B5C-0DE6
 Directory of A:\THEORY\TEST

.            <DIR>       01-16-93   2:43p
..           <DIR>       01-16-93   2:43p
TEST_1               179 01-15-93   2:19a
TEST_2               129 01-15-93   2:22a
QUIZ_1                39 01-15-93   2:46p
WORK_SHT             178 04-23-93   9:23a
        6 file(s)          525 bytes
                        712704 bytes free

A:\>DIR \THEORY\NOTES                                  [Step 28]

 Volume in drive A is EXAMPLE
 Volume Serial Number is 2B5C-0DE6
 Directory of A:\THEORY\NOTES

.            <DIR>       01-16-93   2:43p
..           <DIR>       01-16-93   2:43p
OHM_LAW              119 01-15-93   2:48p
RESISTOR             309 01-15-93   2:50p
BATTERY              609 01-15-93   2:50p
        5 file(s)         1118 bytes
                        712704 bytes free

A:\>DIR \LAB                                           [Step 29]
```

```
 Volume in drive A is EXAMPLE
 Volume Serial Number is 2B5C-0DE6
 Directory of A:\LAB

.            <DIR>       01-16-93   2:43p
..           <DIR>       01-16-93   2:43p
LAB_1             1250 01-15-93   2:51p
LAB_2             2041 01-15-93   2:52p
        4 file(s)        3291 bytes
                       712704 bytes free
```

The following is a description of the preceding example and all the steps it took to achieve the directory structure of Figure 2-3.

Step 1 — The DIR command contains no specified drive or path parameter; therefore, DOS uses the current drive (A:) and the working directory (\) root directory. Note that the third line in the listing states "Directory of A:\," which means the root directory of drive A. Because the disk contains no file or subdirectories, the message "File not found" is displayed. After the listing of the directory, the DOS prompt "A:\ >" was displayed.

Step 2 — The MD command is used to create a subdirectory under the root directory called THEORY (\THEORY).

Step 3 — The MD command is used to create a subdirectory under the root directory called LAB (\LAB).

Step 4 — The DIR command is issued again, using the current drive and working directory. Note that now THEORY and LAB appear in the listing. They are identified by the word DIR placed between the less-than and greater-than signs. Also note that the sizes of the THEORY and LAB directories are not listed. Also note that there are zero bytes in the files (directories are listed as files with no file size specified).

Step 5 — The COPY command is used to copy the file RULES from the root directory of drive C into the working directory of drive A (in this example, the root directory), as specified by the DOS prompt.

Step 6 — The DIR command is given again to verify the COPY command. This listing now lists three files, THEORY and LAB, which are subdirectories, and the RULES file.

Step 7 — The CD command changes the working directory from the root directory of drive A to the THEORY subdirectory of drive A. The drive does not change because no drive parameter is specified. Once this command is processed, the DOS prompt changes to "A:\THEORY," which means the working directory is THEORY with the current drive as drive A. The path does not change until the computer is reset or the path is changed with the CD command.

Step 8 — The DIR command is performed on the working directory (\THEORY in this example) in drive A. Note that the third line of the listing displays "Directory of A:\THEORY," which specifies the directory being listed. The two files "." and ".." are created and used by DOS to maintain the path. These cannot be deleted and do not affect the Remove Directory command.

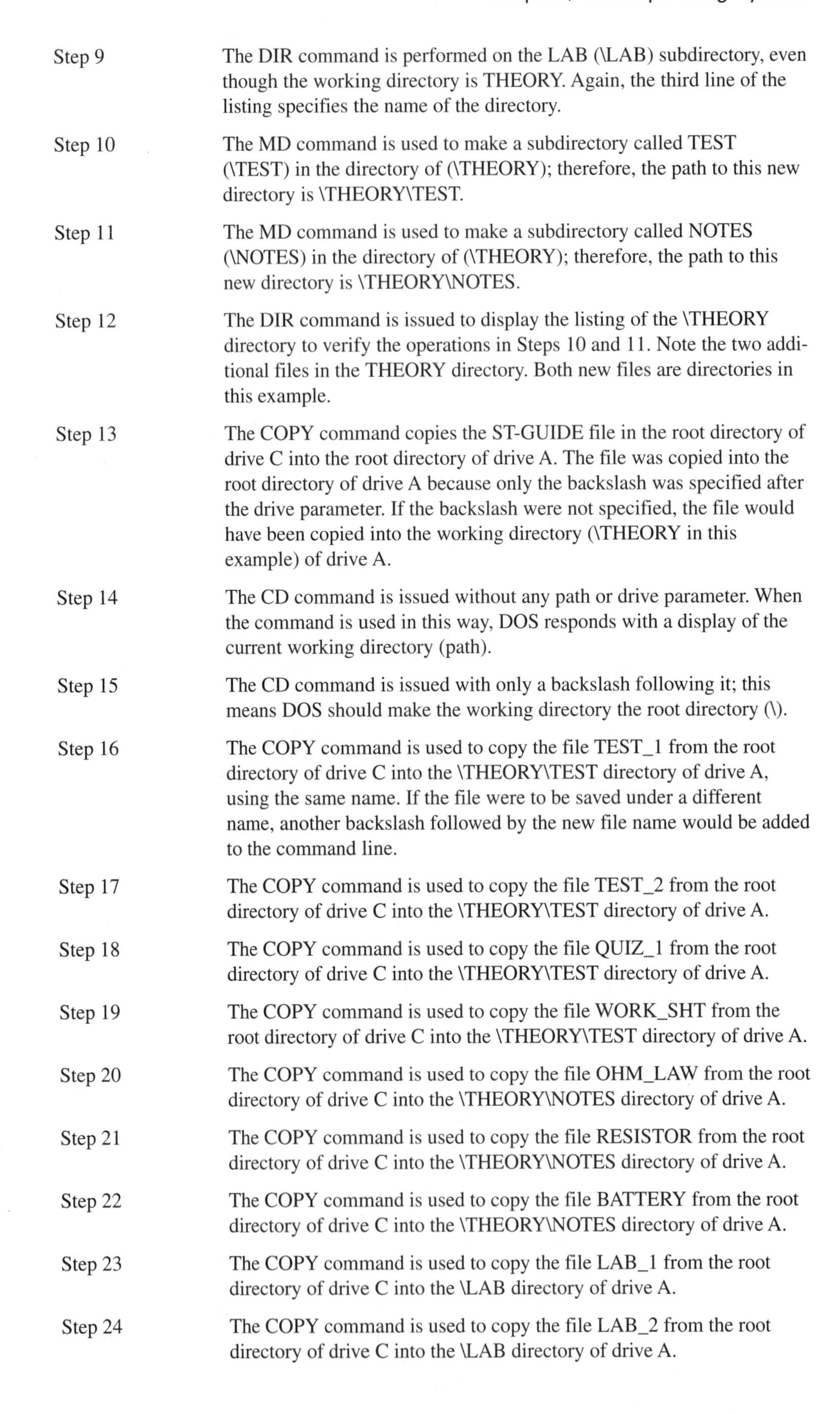

Step 9 The DIR command is performed on the LAB (\LAB) subdirectory, even though the working directory is THEORY. Again, the third line of the listing specifies the name of the directory.

Step 10 The MD command is used to make a subdirectory called TEST (\TEST) in the directory of (\THEORY); therefore, the path to this new directory is \THEORY\TEST.

Step 11 The MD command is used to make a subdirectory called NOTES (\NOTES) in the directory of (\THEORY); therefore, the path to this new directory is \THEORY\NOTES.

Step 12 The DIR command is issued to display the listing of the \THEORY directory to verify the operations in Steps 10 and 11. Note the two additional files in the THEORY directory. Both new files are directories in this example.

Step 13 The COPY command copies the ST-GUIDE file in the root directory of drive C into the root directory of drive A. The file was copied into the root directory of drive A because only the backslash was specified after the drive parameter. If the backslash were not specified, the file would have been copied into the working directory (\THEORY in this example) of drive A.

Step 14 The CD command is issued without any path or drive parameter. When the command is used in this way, DOS responds with a display of the current working directory (path).

Step 15 The CD command is issued with only a backslash following it; this means DOS should make the working directory the root directory (\).

Step 16 The COPY command is used to copy the file TEST_1 from the root directory of drive C into the \THEORY\TEST directory of drive A, using the same name. If the file were to be saved under a different name, another backslash followed by the new file name would be added to the command line.

Step 17 The COPY command is used to copy the file TEST_2 from the root directory of drive C into the \THEORY\TEST directory of drive A.

Step 18 The COPY command is used to copy the file QUIZ_1 from the root directory of drive C into the \THEORY\TEST directory of drive A.

Step 19 The COPY command is used to copy the file WORK_SHT from the root directory of drive C into the \THEORY\TEST directory of drive A.

Step 20 The COPY command is used to copy the file OHM_LAW from the root directory of drive C into the \THEORY\NOTES directory of drive A.

Step 21 The COPY command is used to copy the file RESISTOR from the root directory of drive C into the \THEORY\NOTES directory of drive A.

Step 22 The COPY command is used to copy the file BATTERY from the root directory of drive C into the \THEORY\NOTES directory of drive A.

Step 23 The COPY command is used to copy the file LAB_1 from the root directory of drive C into the \LAB directory of drive A.

Step 24 The COPY command is used to copy the file LAB_2 from the root directory of drive C into the \LAB directory of drive A.

Step 25 The DIR command is issued with no parameters; therefore, the working directory and current drive are used as a default by DOS. Note that this directory listing is for the root directory of drive A.

Step 26 The DIR command is issued to display a listing of the \THEORY directory of drive A.

Step 27 The DIR command is issued to display a listing of the \THEORY\TEST directory of drive A.

Step 28 The DIR command is issued to display a listing of the \THEORY\NOTES directory of drive A.

Step 29 The DIR command is issued to display a listing of the \LAB directory of drive A.

In the following example, the TREE command is performed on drive A and lists only the directory names. The second example of the TREE command also includes the files in each of the listed directories. Note that TREE is an external command; therefore, the TREE command file either must be located in the path for the computer or must be in the working directory itself.

```
A:\>TREE A:
Directory PATH listing for Volume EXAMPLE
Volume Serial Number is 2B5C-0DE6
A:.
├───────THEORY
│       ├─────TEST
│       └─────NOTES
└───────LAB


A:\>_
```

```
A:\>TREE A: /F
Directory PATH listing for Volume EXAMPLE
Volume Serial Name is 2B5C-0DE6
A:.
│       RULES
│
├───────THEORY
│       │     ST_GUIDE
│       ├─────TEST
│       │          TEST_1
│       │          TEST_2
│       │          QUIZ_1
│       │          WORK_SHT
│       └─────NOTES
│                  OHM_LAW
│                  RESISTOR
│                  BATTERY
│
└───────LAB
              LAB_1
              LAB_2
A:\>_
```

In the following example, the RD command is issued to remove the directory called \LAB. But because the directory contains files, the command cannot be performed. In this case the user might perform a DIR command on the directory in question to determine the problem. The user must delete all the files in the directory in question and remove any subdirectories and files within those subdirectories before this command can be performed.

```
A:\>RD\LAB
Invalid path, not directory,
or directory not empty

A:\>_
```

The first step is to delete all the files in the directory in question. If the answer is yes, all the files in the \LAB directory are deleted. If the RD command is issued again and the same error occurs, the user should check to make sure that there are no subdirectories within the \LAB directory. Otherwise, the RD command should work.

```
A:\>DEL \LAB\*.*
All files in directory will be deleted!
Are you sure (Y/N)?y

A:\>RD\LAB

A:\>_
```

Because the computer responds with the DOS prompt in the above example of the RD command, we know that the directory was removed from the disk; otherwise, an error message would have been displayed.

A more efficient way to remove a directory along with all its subdirectories and files is to use a new MS-DOS 6.0 command called DELTREE. The user simply specifies the directory name preceded by a backslash after the DELTREE command word. Answer "y" (yes) to the verification question.

```
A:\>DELTREE \THEORY
Delete directory "\theory" and all its subdirectories? [yn] y
Deleting \theory...

A:\>_
```

2.13.36 DOSshell (external)

```
DOSSHELL [/t] [/b] [/g]
```

The DOSSHELL command causes the computer to bring up a graphical menu interface that allows the user to perform common DOS commands. Three optional switches can be used with the DOSSHELL command. The [/t] (text) switch causes the DOSSHELL to appear in a text mode; this is the default mode if no switch is provided. The [/b] (black–white) switch causes the DOSSHELL to appear in black and white even on a color monitor. The [/g] (graphics) switch causes the DOSSHELL to appear in a more graphical format using icons similar to Windows.

The Areas of DOSSHELL

Menu Bar: The menu bar at the top of the screen contains four options. There are three methods to select one of the options from the menu bar. Using the keyboard, the user can press the

F10 key, then press the highlighted letter associated with the option or use the cursor arrow keys to highlight the option desired, and then press the enter key to enable the pull-down menu for the option. Using a mouse, the user moves the mouse cursor to the wording of the option desired and presses the left mouse button to activate the pull-down menu. In the case of DOSSHELL, the mouse is the easier method of operating the SHELL.

The first option of the menu bar is the *File* option. When this option is selected, a pull-down menu appears on the screen with a list of the different commands that can be performed. Some of the commands are visible and some are barely visible. Those commands that are clearly visible can be processed. The commands that are not clear are commands for which the user must first select a file. The commands that are clearly visible can be performed by simply pressing the proper key (specified following the command) or the highlighted character in the command name, or the user may be required to position the cursor (keyboard or mouse) at the location on the menu for the command and to select the command (press the enter key or press the left mouse button). If the command is followed by three periods (called an ellipsis), a window appears on the screen requesting additional information needed to perform the command.

The commands that are active when the File menu is active are Open, Run. . ., Print, Associate. . ., Search. . ., View File Contents, Move. . ., Copy. . ., Delete. . ., Rename. . ., Change Attributes. . ., Create Directory, Select All, Deselect All, and Exit.

The *Options* area of the menu bar is selected in the same way as the File area. The following options are available in the Options area of the menu bar: Confirmation. . ., File Display Options, Select Across Directories, Show Information, Enable Task Swapper, Display . . ., and Colors.

The *View* option from the menu bar allows the user to set up different formats in which DOSSHELL can display information. The following is a list of the View options: Single File List, Dual File Lists, All Files, Program/File Lists, Program List, Repaint Screen, and Refresh.

The *Tree* option from the menu bar allows the user to select the way the tree structure is displayed in DOSSHELL. The options for Tree are Expand One Level, Expand Branch, Expand All, and Collapse Branch.

The last option on the menu bar is *Help;* this option displays help for the user while using DOSSHELL. The options of Help are Index, Keyboard, Shell Basics, Commands, Procedures, Using Help, and About Shell.

Current Directory Listing: The first line under the menu bar lists the drive and directory that are currently selected. As the user selects different drives and directories, the listing changes.

Switching between Areas: The tab key allows the user to switch between the Drive Icons area, Directory Tree area, Files Listing area, and Program List area. As each area is selected (made active), the title is highlighted.

Drive Icons: Under the line that lists the current drive and working directory are the disk drive icons; the number of drive icons depends on the number of drives in the computer. Each drive is specified by the drive letter, followed by a colon, inside brackets. The active drive icon is highlighted. A different drive can be selected by using the mouse or cursor arrow keys and pressing the enter key or left mouse button.

Directory Tree: The Directory Tree area of the screen displays the tree structure of the selected drive. When this area is active, the user can use the arrow keys or mouse to make a new directory active. Once the directory is made active, the file listing on the opposite side of the screen displays the file within the active directory.

File Listing: The File Listing area of the screen displays the file names, file extensions, size, and date the file was created. Additional information can be listed by changing the View in the menu bar. The highlighted file name is the active file. If the file is large, the user can scroll through the listing of file names with the arrow keys.

Main Area: The Main area is located near the bottom of the screen. In this area is a listing of Program Files that can be executed from DOSSHELL. DOS sets up some of the DOS commands, but the user can customize this area to provide quick access to the most-used programs.

Status Bar: At the very bottom of the screen is the status bar, which tells the user how to get to the menu bar (F10) and to temporarily exit DOSSHELL and lists the current time.

2.13.37 Back Up (external)

```
MSBACKUP [setup_file] [/bw] [/lcd] [/mda]
```

MS-DOS has improved the functions of backing up and restoring the data on a hard drive with the menu-driven MSBACKUP command utility. MS-DOS still contains the older BACKUP and RESTORE commands that were used in MS-DOS 5.0 and earlier. If the files of a backup have been backed up on an older version of DOS, the BACKUP and RESTORE commands should be used. Otherwise, the user should use the MSBACKUP command utility, which is faster and more accurate.

When installing MS-DOS 6.2, the user is asked whether or not MS-DOS should install either the DOS or WINDOWS or both versions of MSBACKUP. Installing both versions means more hard disk space is used.

The first time MSBACKUP is executed, DOS prompts the user to configure the hardware so that MSBACKUP can use the hardware in the most efficient manner. Once this occurs, the user is given four options from a pop-up menu.

The Backup option allows the user to select the drive, directory, and/or file to be backed up, the type of back-up drive, and the type of back-up. The Compare option allows the user to compare the data on the backed-up floppies with the files on the hard drive. The Restore command allows the user to select the back-up drive, type of restore, and location of original files. The last option ends the execution of the MSBACKUP command.

MSBACKUP removes all TSR (terminal stay resident) or memory-resident programs from memory before the back-up occurs. There must also be at least 512K of memory free in the computer.

2.13.38 Debug (external)

```
DEBUG [[drive:][path][filename [testfile]parameters]]]
```

The DEBUG command is a single-pass assembler language program. This program allows the user to test and debug executable files, but it can also be used to assemble small assembler language programs into machine code. Its options are very limited and, because of this, it is not used often. Most assembly language programs are written using more advanced assemblers. Therefore, only a brief description of the DEBUG command is given in this text; refer to the MS-DOS manual under Debug if more information is needed.

To enter Debug the user must specify the command and should include the file name. If no other parameters (drive:, path, testfile) are given, DOS uses the defaults.

```
A:\>DEBUG LETTER
-
```

In the first example, Debug is loaded from drive A (current drive). Once loaded, Debug looks for a file in the current drive and working directory called letter. Because the Debug prompt is displayed, Debug found the file called LETTER and named the memory location (LETTER) used by Debug. This does not load the LETTER file into memory; it simply labels the memory locations used by Debug until Debug is exited or the Debug *Name* command changes the name. For the file named LETTER to be transferred into memory, the Debug command *Load* must be used.

```
A:\>DEBUG LETTER
File not found
-
```

In this example, the file LETTER is not found on the current drive and working directory. Debug then names the memory it is using.

Debug Commands

A	The *Assemble* command is used to assemble the source code entered into the computer into object (machine) code.
C	The *Compare* command compares the contents of two areas of memory.
D	The *Dump* command is used to display the contents of a memory location or group of memory locations. The Dump command provides a hexadecimal and ASCII equivalent for each memory location displayed.
E	The *Enter* command is used to enter 1 byte of data into the computer at a time. The current byte of data is displayed first, and then Debug allows the user to change the value.
F	The *Fill* command allows the programmer to replace the values in a range of memory locations with the values within the list parameters.
G	The *Go* command allows the programmer to execute the Debug program currently in memory.
H	The *Hex* command is used to determine the difference between two hexadecimal values and displays the sum and differences.
I	The *Input* command allows the transfer of data on the specified input port to be read into the Debug program.
L	The *Load* command loads the specified file from disk into memory.
M	The *Move* command moves the contents memory from one location to another location, as specified by the range.
N	The *Name* command specifies the name of the file Debug will use when writing or reading data to or from the disk.
O	The *Output* command transfers the value from memory into an output port.
P	The *Proceed* command executes a loop, string instruction (prefixes), software interrupts, subroutines, and trace instructions.
Q	The *Quit* command causes the computer to exit Debug and gives control of the computer back to DOS.
R	The *Register* command allows the programmer to display and change the contents of any register inside the MPU.
S	The *Search* command searches the specified memory location(s) for one or more bytes in a data pattern.

T	The *Trace* command allows the user to execute one or more instructions in the Debug program.
U	The *Unassemble* command converts the machine code located in the memory of the computer into source (mnemonic) code format. The source code is not saved, but is merely displayed on the video screen.
W	The *Write* command is used to store on a disk drive a Debug program that is currently in memory.
XA	The *Allocate Expanded Memory* command is used to specify the number of pages assigned to expanded memory.
XD	The *Deallocate Expanded Memory* command is used to deactivate expanded memory.
XM	The *Map Expanded Memory Pages* command is used to map logical pages to handle physical pages of expanded memory.
XS	The *Display Expanded Memory Status* command displays data regarding the status of expanded memory.

2.13.39 Edit (external)

```
EDIT [[drive:][path]filename] [/b] [/g] [/h] [/nohi]
```

The EDIT command is a menu-driven text editor that allows the user to create and edit text (ASCII) files. This command replaces the EDLIN command in the older versions of DOS, although EDLIN is still provided with MS-DOS 6.0; because of its lack of functions, EDLIN is not used very often. The most-used command syntax is typing the word EDIT and pressing the enter key.

Press the ESC key to start editing. The ALT key is used to activate the menu bar at the top of the screen. Pressing F1 provides help for the user. The menu bar contains four options in the form of pull-down menus. The EDIT menu can be controlled by either the keyboard or the mouse, just like the DOSSHELL command.

Once the menu bar is made active by pressing the ALT key, the first letter in each menu option is highlighted, at which point the user can use the mouse, the arrow keys, or the letter that is highlighted to open the pull-down menu.

Under the File option, the user can create a New file, Open... an existing file, Save the file under the same name, Save As... save the file under a different name or location, Print... the contents of the file, or Exit the EDIT command.

Under the Edit option, the user can Cut selected areas of the file, Copy selected areas of the file, Paste the data in the memory to an area of the file, or Clear a selected area of the file. The selection of the file area is done in one of two ways: (1) placing the mouse cursor at the beginning of the area to be selected and holding down the left mouse button until the end of the area selected is reached by the mouse cursor, or (2) pressing and holding down the Shift key while using the cursor arrow keys until the proper sections of the file are selected.

Under the Search option, the user can Find. . . a character or group of characters specified, Repeat Last Find sequence, or Change. . . the character(s) found.

Under Options, the user can change the Color of the display or specify the Help file path.

In the lower-right corner of the screen, the status of the cursor is displayed. The position of the cursor vertically (which row) is displayed, followed by a colon. The number to the right of the colon specifies the cursor position in the current line.

2.14 BATCH FILES

A batch file contains a list of commands or instructions (DOS or other software instructions) that are performed as soon as the file name is typed into the computer and the enter key is pressed. A batch file can be edited with the EDIT, EDLIN, or any type of software that allows ASCII (text) editing.

The most common text file, found in most computer systems, is the AUTOEXEC.BAT file. Once the computer boots up DOS and the CONFIG.SYS file changes the parameters of the computer, DOS searches the drive for the AUTOEXEC.BAT (auto execution batch) file to execute. The status of the computer after the execution of the AUTOEXEC.BAT file depends on the command lines listed in the AUTOEXEC.BAT file.

The following commands are normally related to batch-type files and their most commonly used parameters and switches.

2.14.1 Call (internal)

```
CALL [drive:][path]filename
```

The CALL command allows a new batch program to take control without stopping the first batch program.

2.14.2 Choice

```
CHOICE [/C[:]keys][/N][/S][/T[:]c,nn][text]
```

The CHOICE command prompts the user for a choice during the execution of a batch program. The command is only valid during the execution of a batch file.

CHOICE Switches

/c:keys	This switch is used to specify the keys that can be used with the CHOICE command.
/n	This switch is used to disable the displaying of the prompt.
/s	This switch forces the CHOICE command to be case sensitive.
/t:c,nn	This switch sets a time period for the choice to be made, where "c" specifies the default key and "nn" specifies the number of seconds before the default is selected.

2.14.3 Echo (internal)

```
ECHO [on or off]
```

```
@ECHO
```

The ECHO command is used to display or hide the text command lines in a batch program or to display messages on the video screen.

If the word *off* follows the ECHO command, then the contents of the remaining lines in the batch file are not seen unless the word ECHO starts the line. The ECHO *on* command line turns on the listing of each batch command line as the line is being executed in the batch file. If the @ (at sign) precedes the ECHO command, that line is not displayed or executed.

2.14.4 Loadhigh (external)

```
LH [drive:][path]filename [parameters]
```

The LH command is a program that allows the contents of the file name specified to move into upper memory if available, which frees up conventional memory.

Although special switches can be used with LH command which are not specified in the above syntax, those switches are normally specified only when the MEMMAKER command is issued. If the user tries to develop the value for the special switches, the computer's memory may be overwritten and the system will lock up.

2.14.5 MSCDEX (external)

```
MSCDEX /d:drive [/d:drive2...] [/e] [/s] [/v] [/l:letter]
        [/m:number]
```

The MSCDEX command is a program that interfaces the data from the CD driver supplied by the CD manufacturer to MS-DOS and MS-WINDOWS application programs. This program must be used with MS-DOS 6.0 or an error will occur. The user may use the MSCDEX command supplied with their CD drive in versions of MS-DOS 6.0 or older.

MSCDEX Parameters

/d:drive
/d:drive2 — The drive parameter specifies the driver name (signature) from the CD-ROM device driver supplied by the manufacturer. Some CD-ROMs contain more than one device driver. The drive (signature) name must match the one used in the CONFIG.SYS file.

/e — This parameter makes MSCDEX use expanded memory if it is available.

/s — This parameter enables the sharing of data from the server in a network environment (MS-NET or Windows for Workgroups).

/v — This parameter displays memory statistics when MSCDEX is started.

/l:letter — This parameter specifies a drive letter for the CD-ROM drive. If this parameter is not used, MS-DOS assigns its own drive letter based on the last available physical drive in the system.

/m:number — This parameter specifies the number of sector buffers to be used with the CD-ROM. The greater the number of buffers, the faster the CD-ROM operates. The number of buffers has a limit above which increasing the number of buffers no longer helps speed up the system. Usually, this value is between 8 and 32.

2.14.6 Path (internal)

```
PATH [[drive:][path[;...]]
```

The PATH command specifies the directories that MS-DOS should search to find an executable file. The order in which paths are specified is the order in which MS-DOS searches and executes. The PATH command is normally located in the AUTOEXEC.BAT file.

2.14.7 Pause (internal)

```
PAUSE
```

The PAUSE command causes the computer to suspend processing of the batch file until the user presses any key. In most cases a batch line that displays a message precedes the PAUSE command.

2.14.8 Remark (internal)

```
REM [string]
```

The REM command is used to place in a batch file messages that will not be executed or displayed during the execution of the batch file. A space follows the REM command, which starts the string; the string of characters ends when the enter key is pressed, thus ending the line.

2.14.9 Set (internal)

```
SET [variable=[string]]
```

The SET command is used to set the environment variables of the batch files or programs that control the way MS-DOS works. The variable parameter is the variable that is to be modified or set. The string is the data associated with the variable.

2.14.10 Smart Drive (external)

```
[drive:][path]SMARTDRV

[drive:][path]SMARTDRV [[drive[+ or −]]...] [/e:elementsize]
[initcachessize] [wincachesize] [/b:<bufferssize] [/c]
[/r] [/l] [/q]
```

The SMARTDRV command is a program used to create and maintain a disk cache (buffer) system. This system is located in extended memory and buffers the fast data from memory for the disk system until the drives can catch. This software also allows the user to regain control of the computer system without waiting for the drive to catch up to the data transfers. SMARTDRV is available in both MS-DOS and MS-WINDOWS; because MS-DOS 6.0 and 6.2 are newer than MS-WINDOWS 3.1, users should specify the MS-DOS version. The command is normally found in the AUTOEXEC.BAT file in MS-WINDOWS 3.1 and MS-DOS 6.0 and 6.2; in an earlier version of both, this command is used as a driver and is found in the CONFIG.SYS file.

The most common syntax of the Smart Drive is to specify the drive and path for SMARTDRV and allow SMARTDRV to set all the parameters automatically.

2.14.11 Anti-virus (external)

```
VSAFE
```

The VSAFE command loads up the memory-resident part of the MSAV (MicroSoft Anti-Virus) command and monitors the computer system for viruses. When it finds a possible virus, VSAFE warns the user. This program uses 22K of memory and should not be loaded when running Windows.

2.15 THE BASIC AUTOEXEC.BAT FILE

The following is typical basic information found in an AUTOEXEC.BAT file. This file can be created or edited with MS-DOS EDIT, EDLIN, or any other type of text or word processor that can save a text file in ASCII. This is the default AUTOEXEC.BAT file for MS-DOS 6.0.

```
@ECHO OFF            [1]

PROMPT $p$g          [2]

PATH C:\DOS          [3]

SET TEMP=C:\DOS      [4]
```

- Line 1 is used to disable the displaying of any command line in the file as it is being executed.
- Line 2 sets the type of MS-DOS prompt; the "$p" switch specifies including the path in the prompt, and the "$g" switch specifies using the ">" (greater-than) sign to end the prompt.
- Line 3 makes the computer search the DOS directory if the command (executable) file is not in the root directory.
- Line 4 sets a temporary environment in the DOS directory of drive C.

The following is a typical example of an AUTOEXEC.BAT file with more than standard equipment.

```
C:\DOS\SMARTDRV.EXE                                    [1]

PATH C:\MENU;C:\PWPLUS;C:\WINDOWS;C:\DOS;              [2]

SET GMKWSPC=C:\PWPLUS                                  [3]

LH C:\MMOUSE                                           [4]

SET TEMP=C:\DOS                                        [5]

SET SOUND=C:\SBPRO                                     [6]

SET BLASTER=A:220 I10 D1 T4                            [7]

C:\SBPRO\SBP-SET /M:12 /VOC:12 /CD:12 /FM:12           [8]

LH C:\DOS\MSCDEX.EXE /D:MSCD001 /E /V /L:L /M:15       [9]
```

- Line 1 sets up a disk caching system from MS-DOS called Smart Drive.
- Line 2 specifies a multiple path for MS-DOS to search for executable files. If the file is not found in the current drive and working directory, MS-DOS first searches the MENU directory, followed by PWPLUS, WINDOWS, and finally the DOS directory.
- Line 3 sets up an environment called GMKWSPC (grammar checker program) in the PWPLUS directory.
- Line 4 loads a mouse command (a program that allows the computer to communicate with the mouse) into high memory, if available; if not, the mouse command is loaded in conventional memory.
- Line 5 sets up a temporary environment (a data area) in the DOS directory.

- Line 6 sets up an environment (a data area) in the sound blaster professional (SBPRO) directory.
- Line 7 sets up data in an environment called BLASTER that specifies data (address location, interrupt value, DMA channel used, and number of voices) for the sound blaster sound board.
- Line 8 configures the SBP-SET file to specify the default values for the sound blaster mixer program.
- Line 9 loads the MSCDEX CD-ROM program from MS-DOS into memory with the parameters following the program name. This command is loaded high if high memory is available; if no high memory is available, it is loaded in conventional memory.

2.16 CONFIG.SYS FILE

Once the last part of COMMAND.COM is loaded into memory during the booting up of MS-DOS, the computer searches the root directory of the boot drive for a file called CONFIG.SYS (configuration system) file. This file allows the software to configure the hardware in a fashion that allows the computer either to operate or to operate more efficiently. The CONFIG.SYS file is very important in determining how memory is used in the computer system.

The following are the most common standard commands found in the CONFIG.SYS file.

2.16.1 Buffers (internal)

```
BUFFERS=n[,m]
```

The BUFFER command specifies the number of memory buffers that will be used for the disk operations. If the number of buffers is set too small, the software operates very slowly; if set too high, conventional memory may not be sufficient to operate the software. Many programs check the CONFIG.SYS file to make sure that the number of buffers is sufficient to operate the software being installed and makes changes in the file if the value is too small.

The "n" parameter specifies the number of buffers allocated, in a range from 1 to 99, with 15 buffers being the default value for a computer with 512K or more of memory.

The "m" parameter specifies the number of buffers allocated in a secondary buffer. In most computer systems, the secondary is set for 0.

Note that if the computer is going to use Smart Drive, the number of buffers can be reduced.

2.16.2 Device Driver (internal)

```
DEVICE=[drive:][path]filename [dd-parameters]
```

The DEVICE command is used to load a device driver (a program that configures hardware devices, usually relating to I/O operations) into memory. Once loaded, the device remains configured until the computer is reset or rebooted.

The drive and path parameters specify where the file name is located in the drive and should be used if the file name is not in the root directory of the boot drive. The parameters that follow the file name are those associated with the specific device driver.

2.16.3 Device High (internal)

```
DEVICEHIGH=[drive:][path]filename [dd-parameters]
```

The DEVICEHIGH command performs the same function as LOADHIGH except that the computer loads the device driver information into upper memory. If not enough upper memory is available, the device driver is loaded into conventional memory. The MEMMAKER program adds additional switches to this command to make it more efficient.

2.16.4 DOS (internal)

```
DOS=[high or low][,umb or noumb]
```

The DOS command allows MS-DOS to load part of itself into high memory and determine whether it should maintain a link in the upper memory area.

The "high" or "low" parameter specifies where to load part of DOS; "high" means to load part of DOS into high memory. If the low parameter is specified, all of DOS is loaded into conventional memory.

The [umb or noumb] (upper memory block) parameter specifies whether or not DOS should maintain a link into the upper memory block.

2.16.5 File Control Blocks (internal)

```
FCBS=x
```

The FCBS command is used to specify the number of file control blocks that can be opened at the same time under MS-DOS. The value of "x," the number of control blocks, can range from 1 to 255, with 4 being the default.

2.16.6 Files (internal)

```
FILES=x
```

The FILES command specifies the number of files that MS-DOS can have open (access) at any one time. The value of "x" ranges from 8 to 255, with a default of 8. Most programs that use the disk heavily use a value two to three times higher than the default. During the installation of this type of software, most software packages check the CONFIG.SYS file to make sure that the number of files is large enough.

2.16.7 Last Drive (internal)

```
LASTDRIVE=x
```

The LASTDRIVE command specifies the maximum number of drives that MS-DOS accesses. The value of "x" is a letter from A to Z. By keeping the value of "x" low, MS-DOS normally operates somewhat faster. The "x" value should be at least one drive letter higher than the highest physical drive in the system.

2.16.8 Shell (internal)

```
SHELL=[[drive:]path]filename [parameters]
```

The SHELL command specifies the location and name of the command interpreter (COMMAND.COM) for MS-DOS. The parameters that follow the filename can configure the size of the environment (/e:xxxx), and the (/p) parameter is used to cause MS-DOS to make the command interpreter permanent.

2.16.9 Stacks (internal)

```
STACKS=n,s
```

The STACKS command sets up the number of data stacks used to handle hardware interrupts, where "n" specifies the number of stacks (range from 8 to 64) and "s" specifies the size of each stack in bytes (range from 32 to 512).

The following is a short list of the most common device drivers found in the CONFIG.SYS file. If different device drivers are needed, refer to the MS-DOS manual. Device drivers are programs that configure the hardware; once the hardware is configured, it remains configured until the computer is reset.

2.16.10 EMM386 Device Driver (external)

```
DEVICE=[drive:][path]EMM386.EXE [on or off or auto] [memory]
[RAM=mmmm-nnnn] [noems] [highscan]
```

The EMM386 device driver allows MS-DOS to access the upper memory area and extended memory so that it can simulate expanded memory. This device drive requires a 386 MPU or higher.

- The [drive:] and [path] parameters specify the drive and path for the EMM386.EXE file.
- The [on or off or auto] determines whether the EMM386 device driver is on all the time, off all the time, or turned on (auto) only when software requires expanded memory.
- The [memory] parameter specifies in kilobytes the amount of memory to be used for expanded memory when enabled.
- The [RAM=mmmm-nnnn] parameter specifies the segment address to be used for upper memory blocks; if the size is not specified, all available space is used.
- The [noems] parameter allows access to the upper memory area and does not allow access to expanded memory.
- The [highscan] parameter checks the availability of upper memory when using upper memory blocks and the EMS windows.

2.16.11 HIMEM Device Driver (external)

```
DEVICE=[drive:][path]HIMEM.SYS
```

The HIMEM.SYS device driver allows MS-DOS to manage extended memory in order to free up conventional memory to run certain programs. HIMEM.SYS should be loaded near the top of the CONFIG.SYS file so that it can manage extended memory for the other device drivers and configuration commands.

2.16.12 Set Version (external)

```
SETVER [drive:][path]filename
```

The SETVER command is used to set up and display the DOS version table and then report the version number to device drivers and programs that were written to operate on earlier versions of MS-DOS.

The following is a typical listing of the CONFIG.SYS file created by MS-DOS during installation. This file can be edited the same way as a batch file.

```
DEVICE=C:\DOS\SETVER.EXE                        [1]

DEVICE=C:\DOS\HIMEM.SYS                         [2]

DOS=HIGH                                        [3]

FILES=30                                        [4]

SHELL=C:\DOS\COMMAND.COM C:\DOS\ /P             [5]
```

- Line 1 sets the version table for device drivers and programs running in the computer.
- Line 2 loads the extended memory management drivers into memory, which allows certain programs to use extended memory.
- Line 3 loads part of COMMAND.COM into high memory.
- Line 4 sets to 30 the number of files MS-DOS can access at any one time.
- Line 5 specifies the location of the COMMAND.COM interpreter and makes it permanent.

The following example is a typical listing of the CONFIG.SYS file with more than the minimum configuration.

```
DEVICE=C:\DOS\SETVER.EXE                                     [1]

DEVICE=C:\DOS\HIMEM.SYS                                      [2]

BUFFERS=50,0                                                 [3]

FILES=50                                                     [4]

LASTDRIVE=M                                                  [5]

DOS=UMB                                                      [6]

DOS=HIGH                                                     [7]

DEVICE=C:\DOS\EMM386.EXE NOEMS HIGHSCAN                      [8]

FCBS=4,0                                                     [9]

SHELL=C:\DOS\COMMAND.COM /E:2048 C:\DOS\ /P                  [10]

STACKS=9,256                                                 [11]

DEVICEHIGH=C:\SBPRO\DRV\SBPCD.SYS /D:MSCD001 /P:220   [12]
```

- Lines 1, 2, 4, 7, and 10 perform the same functions as they did in the first example of the CONFIG.SYS file.
- Line 3 sets the number of buffers to 50.

- Line 5 sets the maximum number of accessible drives for MS-DOS to 13.
- Line 6 causes MS-DOS to maintain a link to the upper memory block area.
- Line 8 loads the EMM386 driver and uses it to manage the upper memory area without creating any expanded memory.
- Line 9 configures MS-DOS to have four file control blocks.
- Line 11 sets up nine stacks for interrupts that are 256 bytes in size.
- Line 12 loads the device driver SBPCD.SYS for a CD-ROM and names the device MSCD001.

Beginning with MS-DOS 6.0, MicroSoft added some internal commands that allow the user to create a CONFIG.SYS file that displays a menu (Fig. 2-4) to allow the user to choose the configuration that best suits the software to be used. Menu commands can be used only in CONFIG.SYS. The following are the most common menu commands and an example of how the menu command can be used to create a multiple-configuration CONFIG.SYS file. If the menu commands are used in the CONFIG.SYS file, the AUTOEXEC.BAT file reflects the menu choices made.

2.16.13 Menu Color (internal)

```
Menucolor=x[,y]
```

The MENUCOLOR command allows the user to define the color of the text and background color of the menu screen of CONFIG.SYS.

MENUCOLOR Parameters

x	The “x” parameter is not optional; this parameter specifies the color of the text used when the menu appears on the screen.
y	The optional “y” parameter specifies the background color for the menu. If the “y” parameter is not specified, the background is black.

MS-DOS 6.2 Startup Menu

1. WINDOWS V-3.1
2. DOS and GAME PROGRAMS
3. CLEAN DOS BOOT

Enter a choice: 2 xx (THE LETTERS "XX" REPRESENT A NUMBER THAT COUNTS DOWN TO 0, AT WHICH TIME THE DEFAULT NUMBER 2 IS AUTOMATICALLY SELECTED.)

F5=Bypass startup files F8=Confirm each line of CONFIG.SYS and AUTOTEXEC.BAT [N]

Figure 2-4 MS-DOS startup menu.

Color Parameters

0 =black	6 =brown	12 =bright red
1 =blue	7 =white	13 =bright magenta
2 =green	8 =gray	14 =yellow
3 =cyan	9 =bright blue	15 =bright white
4 =red	10 =bright green	
5 =magenta	11 =bright cyan	

2.16.14 Menu Default (internal)

```
MENUDEFAULT=blockname[,timeout]
```

The MENUDEFAULT command defines the default configuration when the user does not make a different selection from the menu. The optional time-out parameter defines how long the user has before the default block name is automatically selected.

MENUDEFAULT Parameters

blockname — The block name specifies the menu item block that is used as the default if a different selection is not made within the given time period.

timeout — The optional time-out parameter is used to specify how long DOS waits before the default menu is processed. The time-out value can range from 0 to 90 seconds; if a time-out value is not specified, the menu remains on the screen until a key is pressed.

2.16.15 Menu Item (internal)

```
MENUITEM=blockname[,menu_text]
```

The MENUITEM command specifies a block of CONFIG.SYS lines to be executed when a menu selection is made. Each specified block can contain lines that are found in other blocks; the difference is the parameters of the lines. The different parameters are used to configure the hardware in ways that allow DOS to better use the resources of the computer system.

MENUITEM Parameters

blockname — The block name specifies a group of CONFIG.SYS command lines that configure the computer in a particular fashion. The block name may contain up to 70 characters, with the exception of backslash, comma, equal sign, forward slash, space, and square brackets.

menu_text — The menu text provides the user with messages that explain the purpose for each menu choice. Menu text can contain up to 70 characters.

2.16.16 CONFIG.SYS File Listing

```
[MENU]                                          [1]
MENUCOLOR=7,1                                   [2]
MENUITEM=WINDOWS, WINDOWS V-3.1                 [3]
MENUITEM=DOS, DOS and GAME PROGRAMS             [4]
MENUITEM=CLEAN, CLEAN DOS BOOT                  [5]
MENUDEFAULT=DOS,30                              [6]
```

```
[COMMON]                                                    [7]
DEVICE=C:\DOS\HIMEM.SYS                                     [8]
DEVICE=C:\DOS\SETVER.EXE                                    [9]
DOS=HIGH                                                   [10]
DOS=UMB                                                    [11]
LASTDRIVE=Z                                                [12]
FCBS=16,8                                                  [13]
STACKS=32,256                                              [14]
FILES=50                                                   [15]
BUFFERS=30                                                 [16]
SHELL=C:\DOS\COMMAND.COM C:\DOS/P/E:1024                   [17]

[WINDOWS]                                                  [18]
DEVICE=C:\DOS\EMM386.EXE AUTO 512                          [19]
DEVICEHIGH=C:\VL201.SYS                                    [20]
DEVICEHIGH=C:\DEV\XCDAE.SYS/D:MSCD001/P:300/M:16/T:7/I:11/X [21]

[DOS]                                                      [22]
DEVICE=C:\DOS\EMM386.EXE 4096 RAM                          [23]
DEVICEHIGH=C:\VL201.SYS                                    [24]
DEVICEHIGH=C:\DEV\XCDAE.SYS/D:MSCD001/P:300/M:16/T:7/I:11/X [25]

[CLEAN]                                                    [26]
DEVICE=C:\DOS\EMM386.EXEAUTO 2048                          [27]

[COMMON]                                                   [28]
```

The screen in Figure 2-3 is displayed after the message starting MS-DOS.

CONFIG.SYS Line Specifications

- Line 1 tells DOS to use the menu commands and create a menu at boot up.
- Line 2 specifies the color of the menu screen—a blue background with white text.
- Line 3 specifies the first menu item selection, WINDOWS V-3.1
- Line 4 specifies the second menu item selection, DOS and GAME PROGRAMS.
- Line 5 specifies the third menu item selection, CLEAN DOS BOOT.
- Line 6 specifies the default menu selection to be used if no different selection is made. A blank line separates each menu block.
- Line 7, the [COMMON] block, defines parameters common to all menu selections.
- Line 8 loads the extended memory management driver into memory, allowing certain programs to use extended memory.
- Line 9 sets the version table for device drivers and programs running in the computer.
- Line 10 loads part of COMMAND.COM into high memory.
- Line 11 sets up upper memory blocks in high memory.
- Line 12 specifies the last drive for DOS to equal drive letter Z.
- Line 13 specifies the number of file control blocks that can be opened at the same time under MS-DOS.
- Line 14 sets up the number of data stacks used to handle hardware interrupts.
- Line 15 specifies the maximum number of files that can be opened at one time.
- Line 16 specifies the number of memory buffers to be used for a disk operation.

- Line 17 specifies the location and name of the command interpreter for MS-DOS. A blank line is used to separate the COMMON block from the WINDOWS block.
- Line 18, [WINDOWS], specifies the beginning of the WINDOWS block; these lines are executed only when WINDOWS V-3.1 is selected.
- Line 19 allows up to 512K of memory to be used for expanded memory, but only if the software requires the memory. The value is set low because WINDOWS does not like to use expanded memory.
- Line 20 loads into high memory the software driver for a local bus hard drive adapter.
- Line 21 loads into high memory a device driver for a CD-ROM. A blank line is used to separate the WINDOWS block from the DOS block.
- Line 22, [DOS], specifies the beginning of the DOS block; these lines are executed only when DOS and GAME PROGRAMS are selected.
- Line 23 sets up 4096K of memory to be used for expanded memory. Most DOS and GAME programs use expanded memory rather than extended memory. This line is the main difference between the WINDOWS and DOS blocks.
- Line 24 loads into high memory the software driver for a local bus hard drive adapter.
- Line 25 loads into high memory a device driver for a CD-ROM. A blank line is used to separate the DOS block from the CLEAN block.
- Line 26 specifies the beginning of the CLEAN block; these lines are executed only when CLEAN DOS BOOT is selected.
- Line 27 allows up to 2048K of memory to be used as expanded memory, but only if the software requires expanded memory. A blank line is used to separate the CLEAN block from the ending COMMON block.
- Line 28 ends the CONFIG.SYS file, using the menu option.

2.16.17 AUTOEXEC.BAT File Listing

```
@ECHO OFF                                               [1]
PROMPT $P$G                                             [2]
MOUSE                                                   [3]
GOTO %CONFIG%                                           [4]

:WINDOWS                                                [5]
C:\DOS\SMARTDRV.EXE                                     [6]
PATHC:\WINDOWS;C:\DOS;                                  [7]
LH C:\DOS\MSCDEX.EXE/D:MSCD001/E/M:20/V/L:L             [8]
SET TEMP=C:\TEMP                                        [9]
WIN                                                    [10]
GOTO END                                               [11]

:DOS                                                   [12]
C:\DOS\SMARTDRV.EXE                                    [13]
PATH C:\DOS;C:\GAMES;C:\UTILITY;C:\TAPE;               [14]
LH C:\DOS\MSCDEX.EXE/D:MSCD001/E/M:20/V/L:L            [15]
SET TEMP=C:\TEMP                                       [16]
MENU                                                   [17]
GOTO END                                               [18]
```

```
:CLEAN                                        [19]
PATH C:\DOS;                                  [20]
SET TEMP=C:\TEMP                              [21]
GOTO END                                      [22]

:END                                          [23]
CLS                                           [24]
```

AUTOEXEC.BAT Specifications

- Lines 1 through 4 are common to all menu selections, many from the CONFIG.SYS file using the menu option.
- Line 1 causes DOS not to display the commands on the screen as they are being executed in the AUTOEXEC.BAT file.
- Line 2, the prompt command, causes the DOS prompt to be the current path followed by the ">" greater-than sign.
- Line 3 loads the mouse driver to enable the mouse.
- Line 4 uses the name of the menu selection made when the CONFIG.SYS file is executed to determine which block of parameters is to be used during the execution of the AUTOEXEC.BAT file. A blank line separates each block selection from the CONFIG.SYS menu.
- Line 5 specifies the WINDOWS command line block. The command lines following this block name are executed when the WINDOWS selection is made in the CONFIG.SYS file.
- Line 6 sets up the disk-caching system from MS-DOS.
- Line 7 specifies a multiple path from MS-DOS to search for executable files.
- Line 8 loads into extended memory the MSCDEX CD driver supplied with MS-DOS. This driver is required to run multimedia programs.
- Line 9 sets up a temporary environment in the TEMP directory.
- Line 10 causes the computer to go into WINDOWS.
- Line 11 causes the computer to go to the :END block once the user exits WINDOWS.
- Line 12 specifies the DOS command line block. The command lines following this block name are executed when the DOS selection is made in the CONFIG.SYS file.
- Line 13 sets up the disk-caching system from MS-DOS.
- Line 14 specifies a multiple path from MS-DOS to search for executable files.
- Line 15 loads into extended memory the MSCDEX CD driver supplied with MS-DOS. This driver is required to run multimedia programs.
- Line 16 sets up a temporary environment in the TEMP directory.
- Line 17 causes DOS to execute a menu program. (*Note:* The menu program is not part of DOS.)
- Line 18 causes the computer to go to the :END block once the user exits the menu program.
- Line 19 specifies the CLEAN command line block. The command lines following this block name are executed when the CLEAN selection is made in the CONFIG.SYS file.
- Line 20 specifies a multiple path from MS-DOS to search for executable files.
- Line 21 sets up a temporary environment in the TEMP directory.
- Line 22 causes the computer to go to the :END block.
- Line 23 is used to specify the ending block of the AUTOEXEC.BAT file.
- Line 24 causes DOS to clear the screen.

2.17 MICROSOFT ANTI-VIRUS

```
MSAV [drive:]
```

The MSAV command is used to load the MicroSoft Anti-Virus program into memory. This program allows the user to scan and clean different system drives for known viruses. The program is menu driven and can be controlled by the keyboard. To have virus protection always operating, use the VSAFE command. *Caution:* VSAFE should not be loaded into memory when Windows is running.

The main menu of MicroSoft Anti-Virus gives the user five options. The choices can be accessed by using the cursor arrow keys, by pressing the highlighted letter in the options, by using the function keys listed at the bottom of the screen, or by using the mouse. As each menu choice is highlighted within the main menu, a description of the function is listed in a window on the right side of the screen.

The Detect option causes the program to scan memory and then the selected disk for any known viruses. At the end of the scan, the program displays the status of the scan. This window displays the drive checked, the types of files checked, the time scanned, the number of infections, and the number of viruses cleaned. If the computer finds a virus, the program prompts the user to allow the cleaning of the virus.

The Detect and Clean option causes the program to perform the same function as the Detect option, except that the computer does not prompt the user before cleaning any viruses found.

The Select new drive option is used to select which drive is to be scanned and or cleaned.

The Options option allows the user to configure how the antivirus program works with files. The following are the options: Verify Integrity, Create New Checksums, Create Checksums on Floppy, Disable Alarm Sound, Create Backup, Create Report, Prompt While Detect, Anti-Stealth, and Check All Files.

The Exit option allows the user to exit the antivirus program and return control to MS-DOS.

2.18 DEFRAGMENT (EXTERNAL)

```
DEFRAG
```

The DEFRAG command loads the graphical menu-driven defragment program into memory. A file becomes fragmented when the available adjacent clusters on a disk drive are smaller than the size of the file to be stored. DOS then stores part of the file in one area of the disk and the remaining part of the file in one or more other areas of the disk, depending on the size of the file and the available adjacent clusters. The defragment program rearranges the file locations into adjacent clusters.

It is important to defragment a disk drive because the more files are fragmented, the more different locations (tracks) the R/W heads must access to get or output the data on the drive. Besides making the disk drive unit work harder, fragmented files also take longer to access. On average, the user should defragment a hard drive once every month. When defragmenting is done often, the process does not take very long because the files are not as badly fragmented.

Once the enter key is pressed, and the defragment program is loaded; the program then asks the user to select the drive to be optimized (defragmented). The user can use the mouse or the keyboard to make the selection. Once the drive is selected, the computer scans the disk in question to determine which type of optimization method should be used. The user then has the option of optimizing the drive using the method selected or reconfiguring the defragment program.

Once the optimization process begins, the user sees a graphical representation of the defragmenting work being done on the drive. At the bottom of the screen in the Status area, the user sees a bar that shows (in percent) how complete the operation is at any time during the operation. At the very bottom of the screen, the FAT and directory information is written, read, and updated.

Under the Optimize option on the menu bar, the user has the following choices: Begin optimization, Drive (select drive), Optimization Method... (full optimization or unfragment files only), Sort (determine the order of sorting: unsorted, by name, by extension, by date and time, by size, in ascending or descending order), Map legend. . . (describes what the symbol means in each area of the graphical map of the drive), About Defrag. . ., and Exit the defragment program.

2.19 DOUBLESPACE (EXTERNAL)

```
DBLSPACE
```

The DBLSPACE command is used to install and configure the DBLSPACE files on a disk drive. If this is the first time DBLSPACE is used, the computer installs the files and prepares the disk drive to work with the Doublespace system. Doublespace increases the amount of free disk space by compressing files on a special area of the disk drive. These compressed files require the Doublespace system to be loaded into the computer before the compressed files can be used. The compression of files yields about 50% more disk space than does storing files in the standard format.

The problem with compressed files or files stored on a compressed drive is that some programs may not function correctly. Also, some programs tend to operate much more slowly when they are stored on a compressed drive. Through its install program, MS-DOS has tried to anticipate problems with programs running on a compressed drive, but it is not always accurate, and the program may lock up the system. Therefore, be careful when storing programs on a compressed drive.

The installation is menu driven and supplies on-screen help with the decisions the user must make. The user has two choices in installing Doublespace: (1) The Express Setup compresses drive C and determines the compression ratio setting automatically. (2) The Custom Setup requires the user to answer questions about how MS-DOS is to set up Doublespace. Even though the Express Setup is recommended, it leaves very little for the user to select from; therefore, the Custom Setup should be the choice for anyone who knows something about MS-DOS and the hardware in the computer.

Upon selecting the Custom Setup, the user is asked to either compress an existing drive or Create a new, empty compressed drive. In most cases, it is better to create a new, empty compressed drive (so that the user can experiment with software to determine which software operates correctly on the compressed drive). The computer displays the disk drives that are available. Once the drive is selected, the computer informs the user how much free space Doublespace will leave on the original drive, the comparison ratio of the new drive, and the letter associated with the new drive. The amount of time it takes to set up the hard drive varies depending on the size of the drive and the number of files to be compressed. After 1 to 10 minutes on average, a status screen appears listing the following information: space used from drive C, free space on new drive, compression ratio, and total time to create the new drive. After the enter key is pressed for the status screen, the computer updates the AUTOEXEC.BAT and CONFIG.SYS files to enable Doublespace to function as soon as the computer is booted.

2.20 MEMORY MAKER (EXTERNAL)

```
MEMMAKER
```

The MEMMAKER program is used to optimize memory configurations in the CONFIG.SYS and AUTOEXEC.BAT files. This is done by moving device drivers and memory-resident programs into upper memory. MEMMAKER requires a computer system with a 386 or higher processor and extended memory. Once MEMMAKER is typed into the computer and the enter key is pressed, the program prompts the user for information regarding her or his system and then, based on the information provided, modifies the memory configurations of device drivers and memory-resident programs. During this process, the computer reboots itself and has the user verify that the computer is operating correctly. If the user confirms the proper operation of the computer system, the temporary changes in the CONFIG.SYS and AUTOEXEC.BAT files are made permanent.

2.21 FIXED DISK (EXTERNAL)

```
FDISK
```

The FDISK program is used to configure a hard disk to be used with MS-DOS. This program allows the user to set up partitions on the drive, basically allowing the computer to define how the physical disk is to be used by MS-DOS. The user is prompted with options on how the hard drive can be used. Note that if FDISK is performed on a hard drive that is already formatted, the original data is lost. Earlier versions of MS-DOS (3.30 and earlier) limited a single hard drive to 32MB, but MS-DOS now allows partitions of as much as 2GB.

The first screen that appears is the Options screen, which specifies which drive is currently selected. Next, the user must choose among four options: (1) Create the DOS partition or Logical DOS Drive. (2) Set which drive is the active partition. (3) Delete partition or Logical DOS drive. (4) Display partition information.

If the drive has not been formatted or used before, the user should select option 1, which brings up a new screen of options, called the Create DOS Partition or Logical DOS Drive screen, which allows the user to (1) Create primary DOS partition, (2) create extended DOS partition, or (3) create logical DOS drive(s) in the extended DOS partition.

If the drive has been formatted or used, the user should select option 3, which brings up a new screen of options, called the Delete DOS Partition or Logical DOS Drive screen, which allows the user to (1) delete primary DOS partition, (2) delete extended DOS partition, (3) delete logical DOS drive(s) in the extended DOS partition, or (4) delete non-DOS partition. The partition must be deleted before the new partition can be created.

If more than one partition is on the hard drive, the user can set which partition is active (bootable) by choosing option 2.

If option 4 is selected, a new screen displays all partition information regarding the selected drive.

2.22 SETTING UP HARD DRIVES

To set up a hard drive, the user must perform the following steps:

1. If the hard drive is new to the system and the system is in an AT class computer, the user must know the hard drive type number (normally a value between 1 and 47); each drive type number specifies the number of tracks (Cylncylinders), number of heads, and sectors per track (sometimes, depending on the drive, a write precompensation value and landing zone value are provided). If the hard drive is not within the listed hard drive type, the last number in most BIOS allows the user to define values. Therefore, the user should know these values before installing the drive.
2. Some hard drives require the user to perform a lower–level format (this no longer occurs very often). If the drive being installed does not require a lower-level format, proceed to step 3. The installation guide should inform the user how to perform a lower-level format; many drives that require such a format provide a program that

performs all the necessary functions. A lower-level format places timing marks on the drive and verifies the surface of the drive, masking out any bad areas.

3. Once all the cables and power are connected and the hard drive type number is stored in BIOS, the user is ready for the next step. The FDISK command must be executed by typing FDISK, pressing the enter key, and following the instructions on the screen.
4. After FDISK is complete and the computer has been booted up from the floppy drive, the next step is to format the drive with the system if only one drive is being used. If the drive is not the bootable hard drive, do not load the system files on the drive.
5. After the hard drive is formatted, MS-DOS is installed using the setup file on the setup disk. Simply make the drive that contains the MS-DOS disks the current drive, type the word SETUP, and press the enter key. The user is instructed what to do to complete the installation of MS-DOS.

The following example shows how to partition a hard drive using FDISK, format the hard drive with the system, and install MS-DOS 6.0 or 6.2. This example assumes only one hard drive, which is the MS-DOS boot drive, with only one DOS partition.

```
A:\>FDISK
```

```
                         MS-DOS Version 6

                      Fixed Disk Setup Program

              (C)Copyright Microsoft Corp. 1983  -  1993

                           FDISK Options

   Current fixed disk drive: 1

   Choose one of the following:

   1.  Create DOS partition or Logical DOS Drive

   2.  Set active partition

   3.  Delete partition or Logical DOS Drive

   4.  Display partition information


   Enter choice: [1]    **PRESS THE ENTER KEY***


   Press Esc to exit FDISK
```

```
            Create DOS Partition or Logical DOS Drive

Current fixed disk drive: 1

1.  Create Primary DOS Partition

2.  Create Extended DOS Partition

3.  Create Logical DOS Drive(s) in the Extended DOS Partition

Enter choice: [1] **PRESS THE ENTER KEY**




Press Esc to return to FDISK Options
```

```
                    Create Primary DOS Partition

Current fixed disk drive: 1

  Do you wish to use the maximum available size for a Primary
  DOS Partition and make the partition active (Y/N)..........
  ..........? [Y]




  Press Esc to continue
```

```
  System will now restart

  Insert DOS system diskette in drive A:

  Press any key when ready . . .  **PRESS ENTER KEY**
```

```
Starting MS-DOS...

Current date is Sun 01-17-1993

Enter new date (mm-dd-yy): **PRESS ENTER KEY**

Current time is   7:33:44.25p

Enter new time: **PRESS ENTER KEY**


Microsoft(R) MS-DOS(R) Version 6

              (C)Copyright Microsoft Corp 1981-1993.

A:\>FORMAT C: /S


WARNING: ALL DATA ON NONREMOVABLE DISK

DRIVE C: WILL BE LOST!

Proceed with Format (Y/N)?y


Formatting 40.78M             < (THESE TWO LINES ARE REPLACED WITH
 xx percent completed.        < THE FOLLOWING THREE LINES ONCE THE
                                FORMAT IS COMPLETE, WHERE "XX" IS A
Formatting 40.78M               NUMBER THAT INCREMENTS TOWARD 100
Format complete.                UPON COMPLETION.)

System transferred

Volume label (11 characters, ENTER for none)? *press enter
key*

  42661888 bytes total disk space
    186368 bytes used by system
  42475520 bytes available on disk

      2048 bytes in each allocation unit.
     20740 allocation units available on disk.

Volume Serial Number is 2555-1AEF

A:\>
```

```
A:\>SETUP **PLACE MS-DOS SETUP DISK IN DRIVE A**
(THE SCREEN WILL CLEAR TO DISPLAY THE NEXT SCREEN.)
```

```
Microsoft MS-DOS 6 Setup


            Please wait

  Setup is checking your system configuration.
```

```
Microsoft MS-DOS 6 Setup

Welcome to Setup.

The Setup program prepares MS-DOS 6 to run on your

computer.

*  To set up MS-DOS now, press ENTER.

*  To learn more about Setup before continuing, press F1.

*  To quit Setup without installing MS-DOS, press F3.




ENTER=Continue   F1=Help   F3=Exit   F5=Remove Color
```

```
Microsoft MS-DOS 6 Setup
========================

          During Setup, you will need to provide and label
          one or two floppy disks.   Each disk can be
          unformatted or newly formatted and must work in
          drive A.    (If you use 360K disks, you may need
          two disks; otherwise, you need only one disk.)

          Label the disk(s) as follows:

               UNINSTALL #1

               UNINSTALL #2 (if needed)

          Setup saves some of your original DOS files on
          the UNINSTALL disk(s), and others on your hard
          disk in a directory named OLD_DOS.x. With these
          files, you can restore your original DOS if nec-
          essary.

               * When you finish labeling your UNINSTALL
                 disk(s), press ENTER to continue Setup.

ENTER=Continue   F1=Help   F3=Exit
```

```
Microsoft MS-DOS 6 Setup
========================

      Setup will use the following system settings:

 +----------------------------------------------------------+
 |                  DOS Type: MS-DOS                        |
 |                  MS-DOS Path: C:\DOS                     |
 |                  Display Type: VGA                       |
 |                                                          |
 |                  The settings are correct.               |
 +----------------------------------------------------------+

If all the settings are correct, press ENTER.

To change a setting, press the UP ARROW or DOWN ARROW key
until the setting is selected. Then press ENTER to see alter-
natives.

ENTER=Continue   F1=Help   F3=Exit
```

```
ENTER=Continue    F1=Help    F3=Exit

Microsoft MS-DOS 6 Setup

        The following programs can be installed on your
        computer.

                        Program for              Bytes used

        Backup:         MS-DOS only               901,120
        Undelete:       MS-DOS only                32,768
        Anti-Virus:     MS-DOS only               360,448

        Install the listed programs.

        Space required for MS-DOS and programs: 5,494,336
        Space available on drive C: 42,475,520

        To install the listed programs, press ENTER. To see a
        list of available options, press the UP or DOWN ARROW
        key to highlight a program, and then press ENTER.

ENTER=Continue F1=Help F3=Exit
```

```
Microsoft MS-DOS 6 Setup

        Setup is ready to upgrade your system to MS-DOS 6.
        Do not interrupt Setup during the upgrade process.

        * To install MS-DOS 6 files now, press Y.

        * To exit Setup without installing MS-DOS, press F3.

F3=Exit    Y=Install MS-DOS
```

```
Microsoft MS-DOS 6 Setup

          XXXXXXXXXXXXXXXXXXXXXXXXXXXXXXXXXXXXXXXXXXXXXXXXXXX
          XXXXXXXXXXXXXXXXXXXXXXXXXXXXXXXXXXXXXXXXXXXXXXXXXXX
          XXXXXXXXXXXXXXXXXXXXXXXXXXXXXXXXXXXXXXXXXXXXXXXXXXX
          XXXXXXXXXXXXXXXXXXXXXXXXXXXXXXXXXXXXXXXXXXXXXXXXXXX
          XXXXXXXXXXXXXXXXXXXXXXXXXXXXXXXXXXXXXXXXXXXXXXXXXXX
          XXXXXXXXXXXXXXXXXXXXXXXXXXXXXXXXXXXXXXXXXXXXXXXXXXX
          XXXXXXXXXXXXXXXXXXXXXXXXXXXXXXXXXXXXXXXXXXXXXXXXXXX
          XXXXXXXXXXXXXXXXXXXXXXXXXXXXXXXXXXXXXXXXXXXXXXXXXXX
          XXXXXXXXXXXXXXXXXXXXXXXXXXXXXXXXXXXXXXXXXXXXXXXXXXX

           nn% complete
```

```
ENTER=Continue  F3=Exit          ooooooo    fffffffffffff
```

The messages in the "xx" section change as the installation of MS-DOS proceeds. Also, as the installation continues, the value of "nn" increases toward 100. In the lower-right corner of the screen the "ooooo" operation function shows which type of operation is being performed. The "f f f f f f" area lists the names of the files being installed.

At different times, MS-DOS opens a window that prompts the user to change the disk.

The second to last screen prompts the user to remove the floppy disk from the drive and press the enter key to continue.

The last screen informs the user that the AUTOEXEC.BAT and CONFIG.SYS files have been modified and to press any key when ready.

SUMMARY

1. DOS is the acronym for Disk Operating System.
2. Microsoft is the main program writer for DOS (MS-DOS).
3. The main purpose of DOS is to act as an interface that allows application software written for DOS to control the hardware (mainly the disk drive system).
4. DOS treats the information stored on the disk system as files.
5. All disks formatted under MS-DOS contain a minimum of a boot record, file allocation table, and directory.
6. File names can have a maximum of eight valid characters; a file extension can have a maximum of three valid characters. A space is not a valid character. File names must be unique to the working directory.
7. File extensions normally specify the type of data in a file.
8. The directory created when a disk is formatted under DOS is called the root directory and is limited in the number of file names it can contain.
9. There are two wild cards in DOS, the question mark (?) and the asterisk (*).

10. Syntax defines the sequence and order that a DOS command requires for the specified command.
11. The [Ctrl] [Alt] [Del] key sequence is used to perform a software or warm boot.
12. Internal DOS commands are located in the COMMAND.COM file and are loaded into memory when DOS is booted. External DOS commands are located in their own files and are loaded into memory only when the user calls up the command for execution.
13. The default DOS command prompt is the current drive followed by a colon, followed by the working directory, and ending with the greater-than sign.
14. The following is a list of the most-used DOS commands:

HELP	Provides help information on all DOS commands.
DATE	Allows the user to change or view the current date.
TIME	Allows the user to change or view the current time.
CLS	Clears the video screen of data and places the cursor in the upper-left corner of the screen.
FORMAT	Initializes a new disk or reinitializes a used disk.
UNFORMAT	Restores the disk information that was erased by the format command.
LABEL	Allows the user to change, create, or delete the volume label name on a disk.
VOL	Displays the volume label.
SYS	Transfers the system files to a disk, making it bootable.
DIR	Displays a list of the files and other related file information.
MORE	The more filter is used to display information one video page at a time.
DISKCOPY	Makes an exact copy of another disk.
DISKCOMP	Compares the information on two disks.
COPY	Transfers a copy of a file(s) to another location.
VERIFY	Controls the condition of the VERIFY switch.
XCOPY	Transfers a copy of a file(s) to another location, including files within subdirectories.
COMP	Compares the contents of two files.
FC	Compares the contents of two files.
DEL	Deletes a file(s) from a disk.

ERASE	Deletes a file(s) from a disk.
UNDELETE	Restores files that were previously deleted by using the DEL (ERASE) command.
MOVE	Moves a file(s) from one location to another location.
RENAME	Changes the name of a file.
TYPE	Displays the ASCII characters in a file on the video screen.
GRAPHICS	Allows MS-DOS to print graphical information on the printer.
PRINT	Prints the contents of a file on a printer.
CHKDSK	Analyzes the files on the disk and produces a report.
SCANDISK	Performs a detailed analysis of a disk and allows the repair of many different types of disk problems.
MEM	Displays a listing of programs in memory.
MSD	Enables the MicroSoft Diagnostics program.
MD	Makes a subdirectory within a directory.
CD	Changes the working directory.
RD	Removes a directory.
PATH	Specifies a path for DOS to search when calling up an executable file.
TREE	Displays the directory structure.
DELTREE	Deletes, files, and removes the specified directory(ies).
DOSSHELL	Allows the user to process most common DOS commands using a menu interface.
MSBACKUP	A menu program that controls the backing up of the disk system.
DEBUG	A single-pass assembler.
EDIT	A menu-driven text processor.
CALL	A command that allows a new batch program to take control without stopping the first batch program.
ECHO	Controls whether text in a batch file is displayed or not.
LH	Loads the program file into upper memory.

MSCDEX	A MicroSoft CD-ROM command interface.
PAUSE	Causes the computer to suspend processing of a batch file until the user responds.
REM	Allows the user to place remarks in a batch file.
SET	Sets the environment variables in a batch file.
SMARTDRV	MicroSoft's disk-caching system.
VSAFE	A memory-resident antivirus monitor.
BUFFERS	Specifies the number of buffers used in a disk operation.
DEVICE	A command that loads a device driver into memory.
DEVICEHIGH	A command that loads a device drive into upper memory.
DOS	Loads part of MS-DOS into high memory.
FCBS	Specifies the number of file control blocks.
FILES	Specifies the number of files DOS can access at one time.
LASTDRIV	Specifies the maximum number of drives in DOS.
SHELL	Specifies the location of the command interpreter.
STACKS	Specifies the number of data stacks.
EMM386	Expanded memory manager for DOS.
HIMEM	Extended memory manager for DOS.
SETVER	Sets up the DOS version table.
MENUCOLOR	Allows the user to select the color of the CONFIG.SYS menu system.
MENUDEFAULT	Allows the user to define a default selection from the CONFIG.SYS menu system.
MENUITEM	Allows the user to define a menu control block.
MSAV	Enables the MicroSoft Anti-Virus program.
DEFRAG	Allows DOS to defragment a disk.
DBLSPACE	Compresses the files on a disk to gain space.
MEMMAKER	Optimizes memory configurations.
FDISK	Configures a hard drive for use with MS-DOS.

15. A batch file contains DOS or other commands that are executed as the batch file is executed.
16. The CONFIG.SYS file is used to configure the hardware and I/O devices in the computer system.

REVIEW QUESTIONS

1. DOS is the acronym for what?
   a. Disk Operation Sequence
   b. Disk Operating System
   c. Drive Option System
   d. Disk Option System
   e. None of the given answers.

2. What part of the disk contains a listing of the files on the disk?
   a. boot record
   b. file allocation table
   c. directory
   d. COMMMAND.COM
   e. None of the given answers.

3. What part of the disk contains a map of the files on the disk?
   a. boot record
   b. file allocation table
   c. directory
   d. COMMAND.COM
   e. None of the given answers.

4. Is "FILE 01.BAT" a valid file name?
   a. True
   b. False

5. Is the semicolon a part of the drive parameter?
   a. True
   b. False

6. Which key, when pressed once, repeats the last DOS command line?
   a. Enter
   b. F1
   c. F2
   d. F3
   e. None of the given answers.

7. DIR is an internal DOS command.
   a. True
   b. False

8. Which type of DOS command is located in its own file?
   a. internal
   b. external

9. The name of the working directory in the DOS prompt "A:\>" is the root directory.
   a. True
   b. False

10. The command processor is found in which DOS file?
   a. directory
   b. root record
   c. IOSYS
   d. COMMAND.COM
   e. None of the given answers.

11. To change the current drive, the user must type the drive letter followed by a colon and then press the enter key.
   a. True
   b. False

12. The DATE command allows the user to change the default time in the computer.
a. True
b. False

13. The XCOPY command is used to initialize a disk.
a. True
b. False

14. Nonremovable disks cannot be formatted.
a. True
b. False

15. Which format switch is used to make a disk bootable?
a. /U
b. /S
c. /F
d. /B
e. None of the given answers.

16. Which DOS command only displays the label of the disk?
a. LABEL
b. DIR
c. FORMAT
d. VOL
e. None of the given answers.

17. Which DOS filter is used to cause the output display to show only one video page at a time?
a. /W
b. MORE
c. /P
d. /S
e. None of the given answers.

18. The FC command compares the entire contents of two disks file for file and location for location.
a. True
b. False

19. Which copy command produces less wear and tear on the drive unit?
a. COPY
b. XCOPY

20. Which compare command allows the user more control when comparing two or more files?
a. COMP
b. FC

21. A deleted file can always be recovered when using the standard DOS undelete protection level.
a. True
b. False

22. The RENAME command allows the user to rename a file and creates a backup file using the old name.
a. True
b. False

23. Which command displays how memory is being used in the computer?
a. MEM
b. CHKDSK
c. EDIT
d. DEBUG
e. None of the given answers.

24. MSAV is used to remove viruses from a disk.
a. True
b. False

25. The DIR command displays the directory structure on the disk.
a. True
b. False

chapter 3

PC XT System Board

OBJECTIVES

Upon completion of this chapter, you should be able to perform the following tasks:

1. Explain the function of each section of the IBM PC system board and any differences in a generic non–IBM PC system board.
2. Discuss the operation of each of the ICs on the IBM PC system board.
3. Read and explain the functions of timing diagrams.
4. Make logical troubleshooting decisions on paper.

The IBM PC system board (motherboard) is divided into nine sections. Each section is found on one or more schematics. The following are the sections of the IBM PC system board.

Microprocessor subsystem (1 of 10)
Extended control logic section (2 of 10)
Address decoding section (3 of 10)
DMA section (4 of 10)
ROM section (5 of 10)
RAM sections (6/7 of 10)
Programmable timer section (8 of 10)
Peripheral interface section (9 of 10)
I/O interface bus channel (10 of 10)

3.1 MICROPROCESSOR SUBSYSTEM

The microprocessor subsystem contains the integrated circuits (ICs) of the main controlling elements of the IBM PC. If any one of the components in the subsystem fails, the computer may not boot. Figure 3-1 shows the schematic of the microprocessor subsystem for an IBM PC. The following ICs are included:

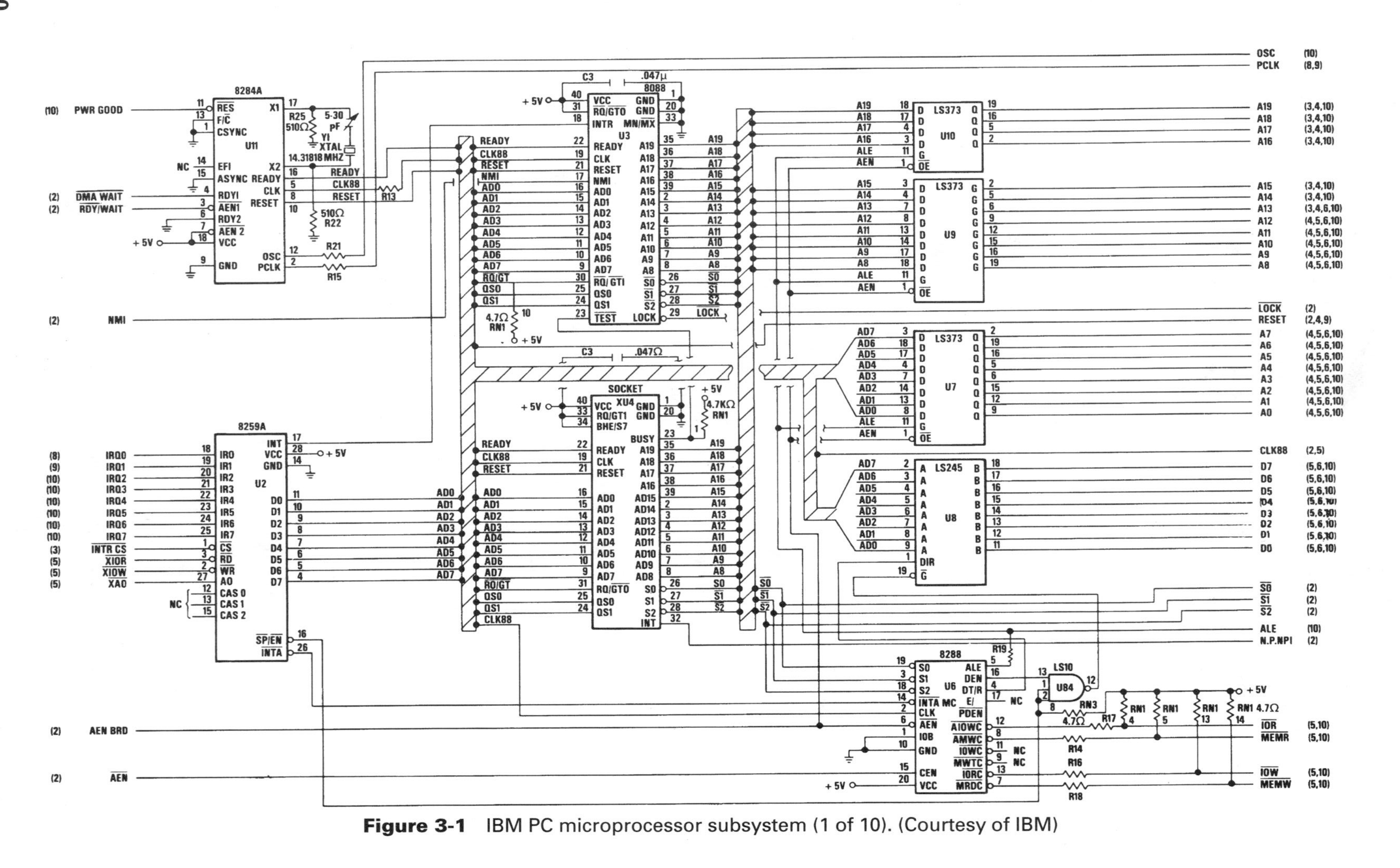

Figure 3-1 IBM PC microprocessor subsystem (1 of 10). (Courtesy of IBM)

Intel 8284 clock generator and driver
Intel 8088 16-bit *N*-channel HMOS MPU
Intel 8087 math coprocessor (optional equipment)
Intel 8288 bus controller
Intel 8259 programmable interrupt controller
74LS373 TTL address latches
74LS245 TTL data bus transceiver

3.1.1 Troubleshooting Note

All troubleshooting checks assume that the technician has verified that the power supply voltage being applied to the IC or ICs in question is correct.

3.1.2 8284A Clock Generator and Driver

The 8284A clock generator and driver chip from Intel is responsible for producing the necessary clock signals for the system board (see Figure 3-2). The 8284 contains three control sections, which operate independently of each other except when they are updated. Of the five outputs, three produce frequencies for timing and the remaining two produce control signals. (*Note:* Throughout this book, if an input or output line for an IC contains a negation bar above it, that line is an active low; otherwise, the line is an active high.)

The oscillator (OSC) outputs the crystal frequency to the I/O bus channel. The second output is the CLK (known as the CLK88 system clock) output; this frequency is supplied to the 8088 MPU and any other IC that must be synchronized with bus operations. The third frequency is the PCLK (peripheral clock) and is used by the timer and interface sections of the system board. The READY output indicates that the bus is ready to start a bus cycle or complete its current bus cycle. The RESET output resets all smart ICs in the system whenever any of the power-supply voltages are not correct.

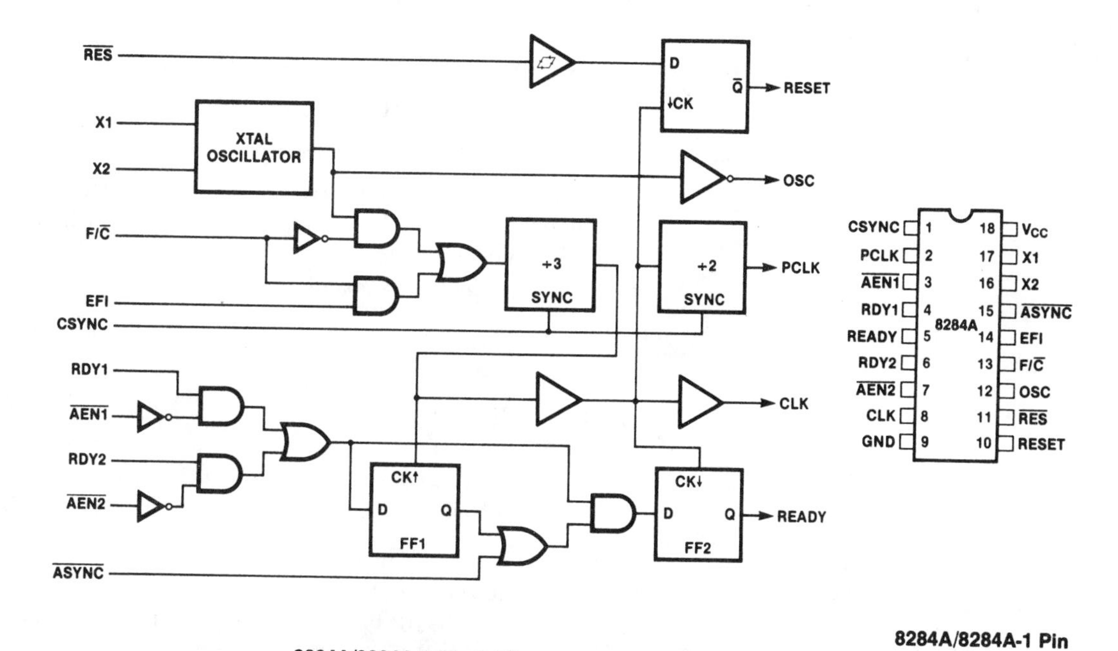

8284A/8284A-1 Block Diagram

8284A/8284A-1 Pin Configuration

Figure 3-2 8284A Block diagram and pin configuration. (Courtesy of Intel)

3.1.3 8284 Pin Configuration

X1–X2

The crystal inputs are used to produce the OSC output for the 8284. Attached to these inputs is a 14.31818-MHz crystal in series with a 5.30-pF and two 510-Ω resistors. The main purpose of this filter is to help filter out unwanted frequencies.

$\overline{\text{RES}}$

The $\overline{\text{RESet}}$ input generates the RESET output signal for the system. This signal is derived from the power-good output from the power supply. Whenever this input goes low, it indicates a problem with the power supply (below 10% to 19% to above 10% to 19% of the rated output voltages). The circuitry in the power supply determines when and under which conditions the power-good line changes. Some power supplies monitor only high or low voltage levels, whereas others monitor a low to high voltage range. The internal Schmitt trigger on the input ensures that only proper levels cause the RESET output condition to occur, not noise. A low on the $\overline{\text{RES}}$ input causes the RESET output signal to go high (reset mode); a high causes the RESET output signal to go low (normal operating mode). Either change occurs within one CLK cycle.

F/$\overline{\text{C}}$

The Frequency/$\overline{\text{Crystal}}$ input selects which input frequency generates the CLK and PCLK output frequencies of the 8284. When high, the frequency applied to the EFI controls the outputs when low, the crystal frequency input is selected and controls them. (*Note:* The OSC output is not affected by the level on this input.)

IBM PC: Because the original IBM PC operated at only one frequency, this input is grounded.

xxx PC: Most generic PCs have a turbo mode, which allows the PC to operate at two different frequencies. In this type of PC, this input may be tied to a turbo switch or an I/O port (for keyboard control or a combination of both). By changing the level on this input, the PC can operate in either the standard mode (4.772727 MHz) or turbo mode (8 or 10 MHz). (*Note:* Some hardware and software may not operate normally in the turbo mode.)

EFI

The frequency applied to the External Frequency Input (if any) is used to develop the CLK and PCLK output frequencies, when the F/$\overline{\text{C}}$ input is high. The input frequency must be TTL compatible. When the F/$\overline{\text{C}}$ input is low, this input has no effect on the 8284A.

IBM PC: Because the original IBM PC operated at only one frequency, the EFI pin was left open.

xxx PC: If the generic PC has a turbo mode, a frequency three times higher than the turbo frequency is applied to this input from a TTL-compatible oscillator circuit. Whether this input affects the CLK and PCLK is determined by the F/$\overline{\text{C}}$ input.

$\overline{\text{AEN1}}$
$\overline{\text{AEN2}}$

The $\overline{\text{Address ENable 1/2}}$ inputs together with the RDY 1/2 inputs determine the readiness of the bus for MPU operations. Both $\overline{\text{AEN}}$ lines are active lows. In the IBM PC and clones, the $\overline{\text{AEN2}}$ is tied high, disabling it. The $\overline{\text{AEN1}}$ line gets its signal from the RDY/$\overline{\text{WAIT}}$ line, controlled by the external control logic section of the system board. This input indicates whether the device that is being addressed has been addressed or the bus is ready to start a bus cycle. When this line goes high during a bus cycle, the selected device is in the process of being accessed (addressed). If this line is high during the absence of a bus cycle, it may indicate a problem with the system board. When this line goes low during a bus cycle, the selected device has been addressed properly and the instruction is then completed (the transfer takes place). If this line is low during the absence of a bus cycle, the system board is operating normally.

RDY1
RDY2

The ReaDY 1/2 inputs together with the $\overline{\text{AEN}}$ 1/2 inputs determine the readiness of the bus for MPU operations. Both RDY lines are active highs. In the IBM PC and clones, RDY2 is tied to ground, disabling it. The RDY1 line gets its signal from the $\overline{\text{DMAWAIT}}$ line, controlled by the external control logic section of the system board. This input indicates whether the other bus master (DMA) is using the bus. Only one bus master may control the bus (either the MPU via the 8288 bus controller or the DMA). Whenever this line is low, the DMA is using the bus and the MPU goes into a wait state while the bus controller disables its command outputs (read and write lines). When this line is high, the DMA is not using the bus, and the MPU via the bus controller has control of the bus.

$\overline{\text{ASYNC}}$

The READY $\overline{\text{ASYNChronization}}$ input allows external synchronization of the READY output in the 8284 with other 8284s. This input is ORed (OR Gate) internally with the output of $\overline{\text{AEN}}$1/2 and RDY1/2 inputs to produce a READY output. In the IBM PC and clones, this input is tied low, disabling it.

CSYNC

The Clock SYNChronization input allows multiple 8284s to operate in the system and have their clock outputs synchronized. In the IBM PC and clones, this pin is tied low, disabling it.

Vcc

This supplies the proper voltage potential from the power supply, allowing the chip to operate properly.

GND

GrouND supplies the current necessary to operate the chip and acts as a common reference point. (*Note:* Because all ICs use Vcc and GND, the function of these inputs is not listed again. But the levels on these pins should always be checked if a problem develops with the specific IC.)

RESET

The RESET output causes the 8088 MPU and all smart ICs to RESET to their starting values. This output is controlled by the $\overline{\text{RES}}$ input (from the power–good signal supplied by the power supply) and is updated within 40 ns of the falling edge of the CLK output.

Condition: RESET output is low.
Result: Normal operation.

Condition: RESET output is high.
Result: All smart ICs reset, all computer operations halt.
Cause: Check power-good line ($\overline{\text{RES}}$ input) from power supply. If low, check output voltages of power supply (see power supply troubleshooting section). If $\overline{\text{RES}}$ input is high, change IC.
Cause: Check CLK output frequency and duty cycle. If bad, check OSC output, crystal inputs, F/$\overline{\text{C}}$ input, and EFI input for correct levels; if bad, correct problem; otherwise, change IC.

OSC

The OSCillator output produces the crystal frequency for any device in the system that needs it. In the IBM PC and most clones, this frequency is 14.31818 MHz. This output is applied only to the I/O channel bus, not to any device on the system board. The output frequency has additional drive current supplied through an internal buffer-inverter.

Condition: TTL-compatible output, at crystal frequency with 50% duty cycle.
Result: Normal operation.
Condition: Non–TTL-compatible output and/or 50% duty cycle not present.
Result: Any device that uses OSC output may not function correctly. In some older CGA video adapters, the OSC frequency is used to create the color burst frequency; if not, proper colors are absent or incorrect on a composite monitor.
Cause: A device that uses the OSC output may be loading down OSC output. Disconnect all I/O adapters that use the OSC output and check OSC levels. If OSC levels return, troubleshoot device, loading down output. If OSC levels do not return, change 8284.
Condition: Incorrect OSC output frequency.
Result: Any device that uses OSC output does not function correctly, and the computer may stop operating.
Cause: See first cause for non–TTL-compatible output and/or 50% duty cycle not present.
Cause: Check crystal and high-pass filter on the crystal inputs of 8284; if bad, change component. If good, change 8284.

CLK

The CLocK output (called CLK88 in the IBM PC and most clones) supplies the necessary clock frequency for the 8088, I/O channel bus, and ICs that require synchronization with the 8088 bus operations. This output is TTL compatible. Internally, the 8284 divides the oscillator (crystal) frequency or external frequency input by 3 to produce this output, which has a duty cycle of 33%. Internally, this signal is also used to update the RESET and READY outputs and determines the frequency of the PCLK output. The CLK output changes its levels internally on the rising edge of the oscillator or external frequency input.

IBM PC: CLK88 output operates at 4.772727 MHz.
xxx PC: CLK88 output operates at 4.772727 MHz (nonturbo mode) or 8 to 10 MHz (turbo mode).

Condition: TTL-compatible output, at one-third crystal frequency (nonturbo mode) or one-third EFI frequency (turbo mode) with 33% duty cycle.
Result: Normal operation.
Condition: Non–TTL-compatible output and/or 33% duty cycle not present.
Result: Any I/O device or IC that uses CLK output does not function correctly, and the computer stops operating.
Cause: I/O device that uses the CLK output may be loading down CLK output. Disconnect all I/O adapters that use the CLK output and check CLK levels. If CLK levels return, troubleshoot device. Go to next cause.
Cause: Check the CLK inputs of the following system board ICs: 8088 MPU, 8087 coprocessor, 8288 bus controller, DCLK output from the external control logic, DMAC 8237, and 74244 line driver in ROM section. If bad, troubleshoot; otherwise, replace 8284.
Condition: Incorrect CLK output frequency.
Result: Any device that uses CLK output does not function correctly, and the computer may stop operating.

Cause: See first and second cause for non–TTL-compatible output and/or 33% duty cycle not present.

Cause: If computer does not have a turbo mode, check OSC output. If not correct, go to troubleshooting OSC output. If good, check the level on the F/$\overline{C}$ line and correct the problem; otherwise, replace 8284.

Cause: If computer has turbo mode, toggle between normal and turbo modes while checking for frequency changes. If the F/$\overline{C}$ lines change (high for normal mode and low for turbo mode), make the following checks; otherwise, check the circuitry controlling F/$\overline{C}$ input. If incorrect frequency occurs during normal (nonturbo) mode, go to second cause and troubleshoot. If incorrect frequency occurs during turbo mode, check the circuitry producing the EFI input frequency and correct the problem; otherwise, replace 8284.

PCLK

The Peripheral CLocK output is a TTL-compatible frequency that is supplied to the timer counter and peripheral interface sections of the system board. Internally, the 8284 divides the CLK signal by 2 to produce this frequency, which has a duty cycle of 50%. This output changes its level shortly after the falling edge of the CLK and before the falling edge of the next oscillator signal.

IBM PC: PCLK output operates at 2.3863633 MHz with a 50% duty cycle.

xxx PC: PLCK output operates at 2.3863633 MHz (nonturbo mode) or 4 to 5 MHz (turbo mode) with a 50% duty cycle.

Condition: TTL-compatible output, at one-half CLK output frequency with 50% duty cycle.

Result: Normal operation.

Condition: Non–TTL-compatible output and/or 50% duty cycle not present.

Result: Any IC that uses PCLK output does not function correctly, and the computer may stop operating.

Cause: Check the PCLK inputs of the following system board ICs: CLK inputs of 8253/8254 PIT and CLK input of 74322 keyboard shift register. If bad, troubleshoot; otherwise, replace 8284.

Condition: Incorrect PCLK output frequency.

Result: Any device that uses PCLK output does not function correctly, and the computer may stop operating.

Cause: See first cause for non–TTL-compatible output and/or 50% duty cycle not present.

Cause: If the computer does not have a turbo mode, check CLK output. If not correct, go to troubleshooting CLK output. If good, check the level on the F/$\overline{C}$ line and correct the problem; otherwise, replace 8284.

Cause: If the computer has a turbo mode, toggle between normal and turbo modes while checking for frequency changes. If the F/$\overline{C}$ lines change (low for normal mode and high for turbo mode), make the following checks; otherwise, check circuitry controlling F/$\overline{C}$ input. If incorrect frequency occurs during normal (nonturbo) or turbo modes, go to second cause and troubleshoot; otherwise, replace 8284.

READY

The READY output indicates the readiness of the address data bus. When low, the bus is not ready. When high, the bus is ready. In both the IBM PC and clones, this output is connected to the READY input of the MPU. This line is controlled by two input lines, and the output is updated as follows: inactive state maximum of 8 ns prior to or active state maximum of 15 ns prior to and up to 50 ns after the falling edge of the CLK output. The $\overline{\text{AEN1}}$ lines must be low (indicating the device being addressed has been given enough time to be addressed), and the RDY1 line must be high (indicating that another bus master is not using the bus) before the bus is ready.

Condition: READY line is high before a bus cycle.
Condition: READY line low while DMA is using the bus.
Condition: READY line is high at the beginning of a bus cycle, then goes low during addressing of the device, followed by the READY line going high.
Result: Normal operation.

Condition: READY line stays high all the time.
Result: Computer halts; memory errors may occur.
Cause: Check the $\overline{\text{AEN1}}$ line; this line should be low unless a bus operation is occurring, in which case the line should toggle low–high–low. If this line stays high, check external control logic circuitry and the level on the I/O Channel Ready line.
Cause: Check the RDY line; this line should be high unless the DMA is using the bus. If this line stays low, check the DMA circuitry and the following lines in the external control logic section: $\overline{\text{LOCK}}$, $\overline{\text{HRQ DMA}}$, $\overline{\text{S0}}$, $\overline{\text{S1}}$, $\overline{\text{S2}}$, and CLK.

3.1.4 8284 Timing Diagrams

The timing diagrams in Figures 3-3 through 3-6 show the operation of the 8284 clock generator-driver. Using the internal block diagram, note the relationships between the three sections of the 8284 as it relates to the IBM PC and clones. Each section works independently of the others, except for the updating of the RESET and READY outputs.

Clock Section (Figs. 3-3 and 3-4). The OSC output always represents the crystal frequency regardless of the condition of F/$\overline{\text{C}}$ line. The CLK output is one-third of either the OSC frequency or the external frequency input frequency, depending on the condition of the F/$\overline{\text{C}}$ line. This output is positive edge triggered from the OSC output and produces a duty cycle of 33%. If the

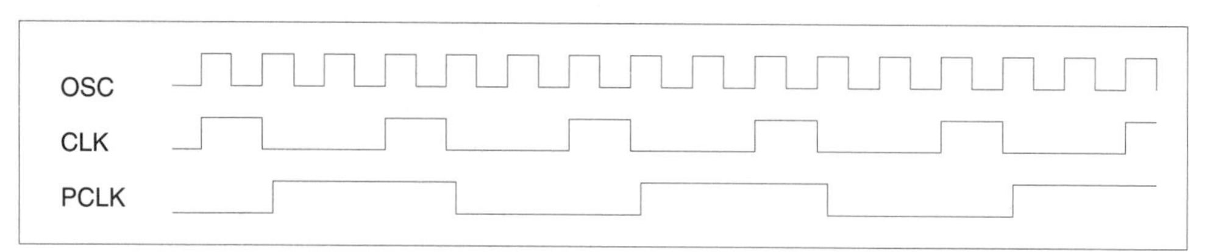

Figure 3-3 8284A nonturbo-mode timing.

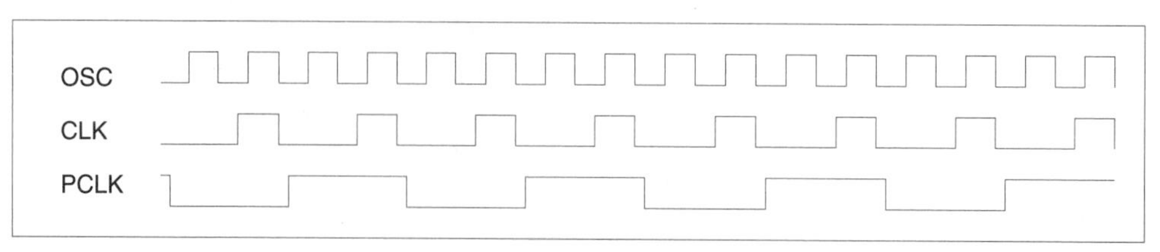

Figure 3-4 8284A turbo timing.

CLK output is not one-third the OSC output, check the EFI input and make sure the CLK output is one-third its frequency. The PCLK output is one-half the frequency of the CLK output, regardless of the condition of the $F/\overline{C}$ line. This output is positive edge triggered from the OSC output and produces a duty cycle of 50%.

Reset Section (Fig. 3-5). The $\overline{RES}$ input (derived from the power-good line from the power supply) is used to determine the level on the RESET output. If low for more than one CLK cycle, the RESET output goes high, placing the system in a reset mode. If high for more than one CLK cycle, the RESET output goes low, allowing for normal operation. The RESET output is updated up to a maximum of 40 ns after the falling edge of the CLK; therefore, if the CLK signal is lost, the output does not change.

Ready Section (Figs. 3-6(a) DMA operation and 3-6(b) MPU operation). Because the IBM PC and clones disable the $\overline{AEN2}$ and RDY2 lines, these lines are not shown in the timing diagram. If a problem develops with the READY output, these lines should be checked for proper levels ($\overline{AEN2}$=low and RDY2=high). The $\overline{AEN1}$ line must be low (indicating that the device being addressed has been given enough time to be addressed), and the RDY1 line must be high (indicating that the DMA is not using the bus) before the bus is ready. The READY output is an active high output and is updated on the falling edge of the CLK. When the $\overline{AEN1}$ line is high, a device is being addressed and the bus is not ready. When the $\overline{AEN1}$ line is low, the device has been addressed and the cycle can continue, provided that the RDY1 line is at its correct level. When the RDY1 line is low, the DMA is using the bus, and the bus is not ready for the MPU to use. When the RDY1 line is high, the DMA is not using the bus; therefore, the MPU can use the bus, provided that the $\overline{AEN1}$ line is at its correct level.

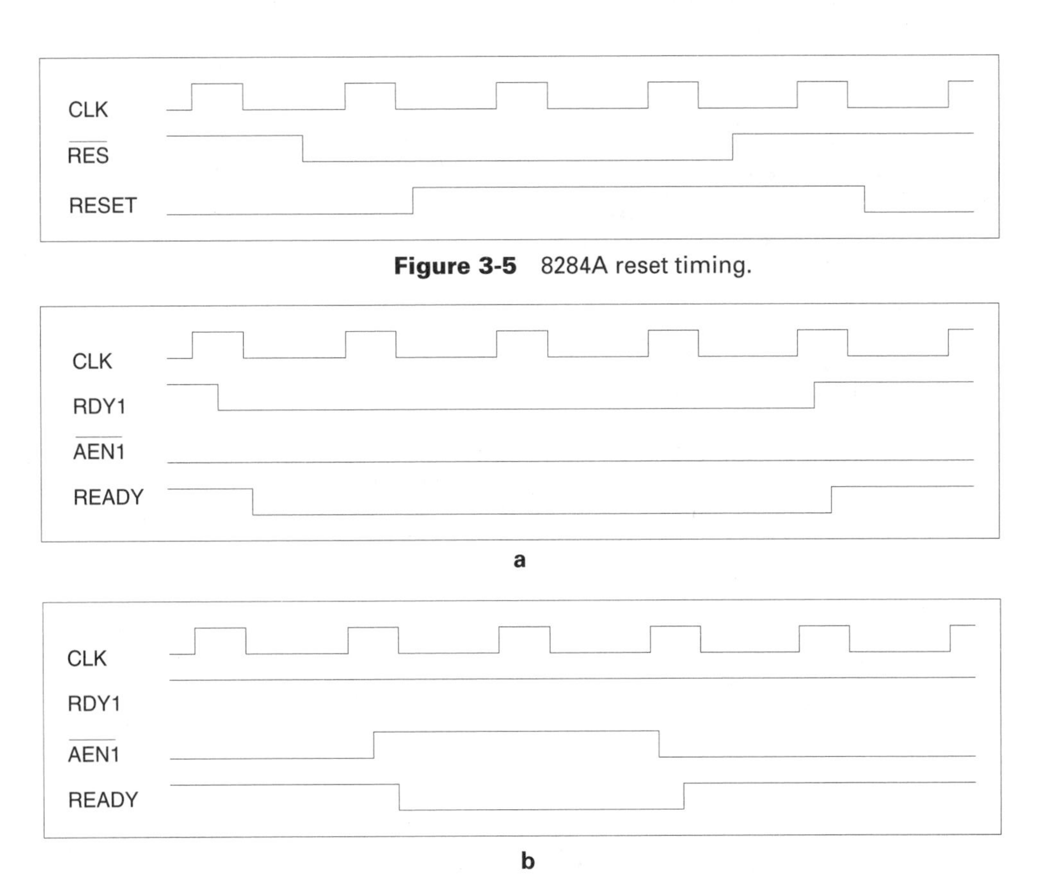

Figure 3-5 8284A reset timing.

Figure 3-6 **(a)** 8284A DMA operation timing; **(b)** 8284A MPU operation timing.

3.1.5 8088 Microprocessor

The 8088 is an HMOS (high-performance *N*-channel MOS) MPU produced by Intel and is software compatible with Intel's 8086. The 8088 MPU is available only in a 40-pin package, with two operating frequencies and temperature ranges. In 1980, Intel developed the 8088, which is an 8-bit data bus version of the 8086. The 8088 has the same instruction set as the 8086 but uses a less expensive 8-bit data bus for memory and I/O devices. After the introduction of the IBM PC in 1981, the popularity of the 8088 microprocessor increased to the point that it was one of the most used microprocessors in the world, only now replaced by the more advanced Intel xxx86 (80286, 80386, and 80486) family of microprocessors. Even though this IC is not as popular as it was in the early to mid 1980s, many of the basic pin functions and operations of the 8088 have been carried over with the rest of Intels xxx86 family. Therefore, it is important for the technician to learn the basics of this IC so that he or she can apply this knowledge to the more advanced Intel MPUs that have now taken the lead.

Figure 3-7 shows the internal block diagram and pin configuration of the 8088 MPU. A brief description of the internal registers and sections of the MPU follows.

3.1.6 8088 MPU Bus

16-Bit Internal Architecture; 16-Bit Internal Data Bus, ALU (Arithmetic Logic Unit), and Registers. Because the 8088 uses 16-bit architecture, the MPU acts as a 16-bit microprocessor once the data have been entered.

20-Bit External Multiplexed Address/Data/Status Bus. Because the 8088 has an 8-bit data bus, 20-bit address bus, and 4-bit status bus, in a 40-pin DIP (Dual Inline Package) the functions of these lines are multiplexed. During certain times of a bus cycle (the sequence necessary to cause a data transfer), these lines are used to produce the address, while at other times some of the same lines act as the data bus or status bus.

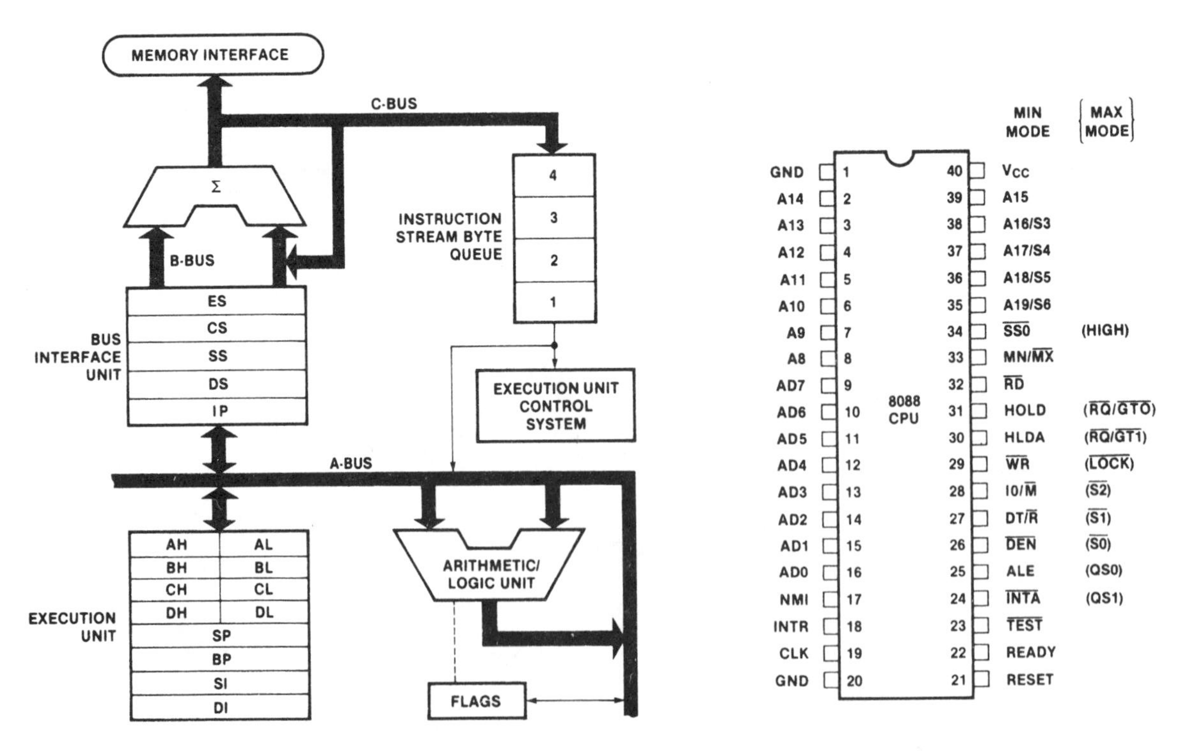

Figure 3-7 8088 block diagram and pin configuration. (Courtesy of Intel)

Operation

1. The data enters and exits the MPU in 8-bit packages; therefore, the data bus is 8 bits wide. A 16-bit operation causes the MPU to transfer two 8-bit data packages sequentially.
2. The 8088 produces a 20-bit external address, which allows the MPU direct access to 1,024,000 memory locations. The address is developed by using two 16-bit values by the summation circuitry of the MPU.
3. The status bus signal is a 4-bit value used to provide certain Intel control ICs with information on which segment register is being used. The status bus information is not used in the IBM PC or compatibles.

3.1.7 8088 MPU Architecture

The 8088 is a third-generation MPU that uses an instruction pipeline that consists of two separate control sections, called the execution unit (EU) and the bus interface unit (BIU). The control units operate independently of each other, which allows the BIU to prefetch data while the EU is performing an operation on the data already inside the MPU, rather than waiting for the EU to perform its task before fetching another byte of data.

Execution Unit. The EU contains the ALU and nine general data registers. All data manipulation within the MPU is handled by the EU; this section has no control over how data is transferred to and from the MPU.

The ALU section of the 8088 handles all arithmetic/logic operations of the MPU once the data is already inside the MPU. The operation performed is determined by the EU control system, which is part of the BIU. This area holds and decodes the current instruction and determines the operational sequence. The ALU can process two 16-bit values and produce a 16-bit result, with additional information about the ALU operation sent to the FLAGs register.

Register A (Accumulator) is a 16-bit register that can be used as two 8-bit registers (AX = 16 bit; AH = high byte of accumulator; AL = low byte of accumulator). This is a general-purpose register that can be used to load or store data and is used in many instructions for holding the results of the operation.

The Base register is a 16-bit register that can be used as two 8-bit registers (BX = 16 bit; BH = high byte; BL = low byte). This is a general-purpose register that can be used to load or store data and is used as an address offset in certain addressing modes.

The Count register is a 16-bit register that can be used as two 8-bit registers (CX = 16 bit; CH = high byte; CL = low byte). This is a general-purpose register that can be used to load or store data and is used by some instructions as a counter.

The Data register is a 16-bit register that can be used as two 8-bit registers (DX = 16 bit; DH = high byte; DL = low byte). This is a general-purpose register that can be used to load or store data and is used by some instructions as a temporary storage register for the results of the operation.

The Stack Pointer (SP) is a 16-bit–only register that is used to point to an address location where the internal environment will be stored during an interrupt or branch.

The Base Pointer (BP) is a 16-bit–only register that is used to point to an address location for certain instructions or that can be used as a general-purpose register.

The Source Index (SI) register is a 16-bit–only register that is used with string instructions for determining the address of the source (where the data is coming from).

The Destination Index (DI) register is a 16-bit–only register that is used with string instructions for determining the address of the destination (where the data is going to).

The Flags register is a 16-bit register, of which 9 bits are defined for use by the ALU.

Carry Flag (CF), bit 0: This flag sets if a carry or borrow is produced in the most significant bit of a byte or word operation.

Parity Flag (PF), bit 2: This flag sets if the number of 1s in the lower byte of the result is even.

Auxiliary Flag (AF), bit 4: This flag sets if a half-carry is produced from the lower nibble to the upper nibble in a byte operation or a carry from the lower byte to the upper byte in a word operation.

Zero Flag (ZF), bit 6: This flag sets if the result of an operation (byte or word) is zero.

Sign Flag (SF), bit 7: This flag sets if the most significant bit in the result is set (byte or word), which indicates a negative signed number.

Trap Flag (TF), bit 8: This flag sets when the computer is in the single-step mode, which allows the MPU to generate an internal interrupt after each instruction is performed.

Interrupt-enable Flag (IF), bit 9: This flag sets when interrupts are enabled, allowing a maskable, external interrupt request to occur. This flag does not affect any internal or nonmaskable external interrupts.

Direction Flag (DF), bit 10: This flag determines whether string instructions autodecrement or autoincrement. If set, the instruction decrements, and, if clear, the instruction increments.

Overflow Flag (OF), bit 11: This flag sets if the signed result is larger than the number of bits that the operand can hold.

Bits 1 and 12 through 15 are always set, and bits 3 and 5 are always reset.

Bus Interface Unit. The main purpose of the BIU is to control the flow of data between the outside world and the internal environment of the MPU. This section contains the memory interface, instruction stream byte queue, EU control system, summation circuitry, and segment registers.

The memory interface circuitry controls the multiplexed address/data/status bus; this circuitry gets its command sequence from the EU control system, based on the current instruction.

The instruction stream byte queue is a 4-byte buffer that allows the MPU to prefetch up to 4 bytes of an instruction while the EU is processing the current instruction or a different instruction. By buffering the bytes of the instruction, both the BIU and EU can function simultaneously the majority of the time. This buffer acts as a FIFO (first in, first out) buffer. The bus interface stops prefetching data when the queue becomes full and continues prefetching data as the queue begins to empty.

The purpose of the EU control system is to hold and decode the current instruction until it can be processed. This area of the microprocessor is also responsible for controlling the sequence of events inside the 8088 as specified by the instruction. The EU control system is the control logic element of the 8088.

The five 16-bit–only registers of the BIU are used to provide the data values necessary to develop the effective address (EA) sent out on the address bus.

The Instruction Pointer (IP) is a 16-bit register that points to the next instruction byte that is not in the instruction queue and normally represents the offset of the specified segment. The IP automatically increments as each byte of the instruction is prefetched into the instruction queue.

The Code Segment (CS), Data Segment (DS), Stack Segment (SS), and Extra Segment (ES) are 16-bit–only registers used normally to define a 64K block of memory. The segment value used is determined by the type of data transfer unless overridden by software control. The code segment value is the default segment value and is used with all instruction prefetching. The stack segment value is used with stack operations and most operations involving the base pointer. The data segment value is used during data operations and string operations. The extra segment register is used like the data segment register and is referenced to external data. Memory segments are defined by software and not by hardware; segments can be sequential or random or can overlap each other, depending on the requirements of the software.

The summation circuitry is used to develop the address that is sent to the memory interface and the outside world. This section uses two 16-bit values to develop the 20-bit address for the MPU.

3.1.8 Effective Address

The effective address is the actual address value that is sent to the outside world to determine which device will be accessed. The following sequence describes how the 8088 develops the effective address.

1. The contents of one of the segment registers or a 16-bit value representing a segment value is transferred to the summation circuitry.
2. Four binary zeros are arithmetically shifted into the least-significant position of the segment value located in the summation circuitry. This process produces a shifted segment value that is 20 bits in size.
3. The contents of the IP or a 16-bit value representing the IP value or offset is transferred to the summation circuitry.
4. The summation circuitry adds the value of the shifted segment value and the IP value located inside the summation circuitry. The 20-bit result is used as the effective address and is sent to the memory interface. Any carry produced by the most significant bit (bit 19) is disregarded and not used.

	Summation Circuitry
Segment value CS (100E hex) →	1 0 0 E hex
Shifted segment value →	1 0 0 E O hex
IP value (A3E1 hex) →	A 3 E 1 hex
Effective address →	1 A 4 C 1 hex

Here the hex value of 1 is the segment (page or bank) value of the address and the hex value of A4C1 is the memory offset (the location in the selected memory segment that will be accessed).

3.1.9 8088 Data Formats

The 8088 can process 8-bit (byte) or 16-bit (word) data as specified by the instruction. This MPU processes not only numerical data but also string data (characters, data that has no numerical value).

Unsigned binary numbers range from 0 to 255 decimal for byte operations (0000 0000 through 1111 1111) and are seen as positive values only. Unsigned binary words range from 0 to 65,535 decimal (0000 0000 0000 0000 through 1111 1111 1111 1111) and are seen as positive values only.

Signed binary numbers range from −128 to +127 decimal for byte operations (0000 0000 through 1111 1111) and are seen as positive or negative values depending on whether the most significant bit is set or reset. If the most significant bit is set, the number is negative; if clear, the number is positive. Signed binary words range from −32,768 to 32,767 decimal (0000 0000 0000 0000 through 1111 1111 1111 1111) and are seen as positive or negative numbers depending on whether the most significant bit is set or clear.

Packed decimal numbers (ASCII numbers) range from 00 to 99 (the largest number in either nibble ranges from 0000 through 1001), with each nibble representing each digit, and are positive only.

Unpacked decimal numbers (ASCII numbers) range from 0 to 9 (the high nibble is zeroed, and the low nibble ranges from 0000 through 1001), with the low nibble representing the only digit, and are positive only.

String data has no numerical value; they can be processed in either a byte or word format. String data normally represent some form of character data.

3.1.10 8088 Addressing Modes

There are 24 combinations of address modes in the 8088. The address mode defines how the data is accessed and processed.

3.1.11 8088 Instructions

The 8088 contains the same instruction set as the 8086; the difference between the two MPUs is that data in the 8086 is transferred in 16-bit packages, whereas the 8088 transfers data in 8-bit packages. The following is a listing of the different formats of instructions. Most instructions can be performed on either 8- or 16-bit data packages. An instruction may be 1 to 6 bytes in size.

14	Data transfer instructions
20	Math instructions
12	Flags and processor control instructions
12	Bit-manipulation instructions
4	Transfer of processor control instructions
24	Jump instructions
4	String instructions

3.1.12 8088 Pin Configurations (Fig. 3-7)

Multiplexed Address/Data Bus. Because the 8088 is an 8-bit data bus version of the 8086, the 8088 uses a multiplexed address/data bus just like the 8086. At the time the 8086 was developed, it was not practical to produce an IC that contained more than 40 pins, and because of the number of address, data, and control lines needed to make the 8086 functional, it was decided to multiplex the address, data, and status buses. This is why it takes a minimum of four MPU clocks to perform one bus operation (bus cycle). In the schematic of the IBM PC, the multiplexed address/data bus is known as the local bus. Once the multiplexed bus is demultiplexed into separate address and data buses, it is known as the external address and data bus.

AD0-AD7

The Address/Data lines serve two different functions. AD0 to AD7 act as the lower 8-bits of the address bus during clock T1 of the bus cycle. Address lines act as outputs from the MPU and inputs to all other devices. Throughout clock cycles T2, T3, Tw (wait clocks, if necessary) and T4 of the bus cycle, AD0 to AD7 act as bidirectional data lines. Each bus cycle requires a minimum of four MPU clocks (CLK88). During an interrupt or local bus hold acknowledge, these lines go into tristate and float.

IBM PC: Because the original IBM PC operated at the 4.772727-MHz clock speed of the 8088, it could perform a maximum of 1,193,100 bus operations per second (data transfers), but less if wait clocks were needed.

xxx PC: A generic PC that operates in the turbo mode of 8 to 10 MHz can perform a maximum of 2 to 2.25 million bus operations per second, but less if wait clocks are needed. This, in theory, makes the turbo PC seem to operate 2.0 to 2.25 times faster than the IBM PC, but this is not normally the case. Because most turbo PCs use the same ICs found in the IBM PC, more wait clocks are normally generated, reducing the overall performance of the turbo to about 1.5 to 1.8 times faster.

Condition: AD0 to AD7 produce the lower 8 bits of address during T1 and represent the data on the data bus during T2, T3, Tw, and T4 of the bus cycle.

Result: Normal operation.

Condition: AD0 to AD7 do not represent the lower 8 bits of address during T1 of the bus cycle.
Result: Computer may halt.
Cause: If AD0 to AD7 stay low constantly, check the operation on all ICs in the microprocessor system that are connected to these lines (8087 math coprocessor, 8259 PIC, 74245 bus transceiver, and 74373 address latch). If all devices check okay, change 8088. If AD0 to AD7 stay high constantly, check for interrupt problems.

Condition: AD0 to AD7 do not produce the correct data for the data bus, throughout T4 of the bus cycle, during a write operation.
Result: Computer may operate for a short time, but then halts or produces errors.
Cause: If the software program is correct, check for the proper operation of the following ICs in the microprocessor subsystem that are connected to these lines: 8087 math coprocessor, 8259 PIC, 74245 bus transceiver, and 74373 address latch. Also check for proper operation of the 8288 bus controller. If all devices check out okay, change the 8088.

Condition: AD0 to AD7 fails to accept the data on the data bus, throughout T4 of the bus cycle, during a read operation.
Result: Computer may operate for a short time, but then halts or produces errors.
Cause: See the cause of the write operation.

A8–A15

In the 8086, these lines perform the same function as lines AD0 to AD7 (supplying D8 to D15 for the data bus), but in the 8088 these lines act only as address lines (A8 to A15, upper byte of offset value). These lines function as address lines during the entire bus cycle, but their levels are used only during T1 of the bus cycle, because that is the only time AD0 to AD7 lines represent the true address. These lines act the same as AD0 to AD7 during a local bus hold and interrupt acknowledge.

Condition: A8 to A15 produce the high byte of the address offset during the entire bus cycle.
Result: Normal operation.

Condition: A8 to A15 produce the high byte of the address offset only during T1 of the bus cycle.
Result: No adverse problem is detected for the PC because the address is latched in only during T1 of the bus cycle.

Condition: A8 to A15 do not produce the high byte of the address offset during T1 of the bus cycle.
Result: Computer may operate for a short time but then halts and produces errors.
Cause: Make sure one of the following ICs is not loading down the bus: 74373 address latch or 8087 math coprocessor.

A16–A19

These output lines act as the four most significant bits (segment value) of the address bus during T1 of the bus cycle. During the rest of the bus cycle, two of the lines are encoded to indicate which segment register is currently being used. During an I/O bus cycle, these lines remain low. In the IBM PC, the segment function of these lines is not used by any device other than the 8087 math coprocessor, which monitors them when the 8088 is in control of the bus. These lines go into tristate and float during a local bus hold acknowledge.

Condition: A16 to A19 produce the high nibble of the address (segment value) during T1 of the bus cycle and then display the segment register status during the remaining time of the bus cycle.
Result: Normal operation.
Condition: A16 to A19 do not produce the high nibble of the address offset during T1 of the bus cycle.
Result: Computer may operate for a short time but then halts and produces errors.
Cause: Make sure one of the following ICs is not loading down the bus: 74373 address latch or 8087 math coprocessor.

Control/Status Bus. The second bus of the 8088 MPU is active during the entire bus cycle. These lines are used to control and determine the status of the external buses in the computer system. This bus determines when the MPU accesses various devices and how the MPU should proceed with the bus operation.

The 8088 MPU uses T1 of each bus cycle to output the address on the multiplexed address/data bus and develop the address on the external address bus. T2 is used to set up the direction and start data flowing. T3, Tw (if necessary), and T4 are used to transfer the data, with the end of the transfer occurring during T4. During wait clocks, the MPU performs internal housekeeping tasks.

READY

The READY input signal is developed by the 8284 clock generator. This signal indicates if the MPU can complete its current bus cycle or if it can access the bus for its next bus cycle. This line must be high at the beginning and end of the bus cycle.

During a typical bus cycle, the READY line goes low during the accessing of the selected device. If by the end of T3 of the bus cycle the READY line is not high, the MPU performs a wait clock. During any wait clocks, the MPU performs internal housekeeping tasks and maintains the selected status signal to the bus controller. The MPU checks the READY line near the end of each wait clock. Once the READY line goes high, the MPU signals the bus controller that the bus is ready to complete the bus cycle by sending a passive status signal to the bus controller through the status lines ($\overline{S0}$ to $\overline{S2}$). This action causes the bus controller to complete the transfer within the next clock cycle (clock T4 of the bus cycle).

Condition: At the end of T1 or at some time in T2 of the bus cycle, READY goes low (indicating that the bus is being accessed). It then returns high after one or more clock cycles, depending on the operation of the MPU, to complete its fourth and final clock of the bus cycle.
Result: Normal operation. (*Note:* Most bus cycles contain 4 to 6 clock cycles. DMA cycles may contain more.)
Condition: READY = 0, prior to beginning of bus cycle.
Result: Bus cycle cannot start, bus controller is in idle (passive) state, no changes occur, and computer halts (waits).
Cause: If the low is produced by the $\overline{\text{AEN1}}$ ($\overline{\text{RDY/WAIT}}$) line being high, the problem is probably located in a device on the external bus (check the I/O CH RDY, $\overline{\text{DACK 0 BRD}}$ [see p. 224] and CLK lines).
Cause: If the low is produced by the RDY1 ($\overline{\text{DMAWAIT}}$) line being low, the DMA is accessing the external bus, which does not indicate a problem unless the line stays high for an extended period of time while the computer is not performing any disk tasks.

Condition: READY = 0, after a bus cycle begins and never returns high.
Result: Bus cycle cannot end, bus controller continues to control the bus, and computer halts.
Cause: If the low is produced by the $\overline{\text{AEN1}}$ ($\overline{\text{RDY}}$/WAIT) line being high, the problem is probably located in a device on the external bus (check the I/O CH RDY, $\overline{\text{DACK 0 BRD}}$, and CLK lines).
Cause: If the low is produced by the RDY1 ($\overline{\text{DMAWAIT}}$) line being low, the DMA is accessing the external bus, which does not indicate a problem unless the line stays high for an extended period of time while the computer is not performing any disk tasks.

Condition: The READY line never goes low after the beginning of a bus cycle.
Result: Computer most likely halts.
Cause: Check the input levels on the $\overline{\text{AEN1}}$ and RDY1 lines of the 8284A. If they indicate that the READY line should go low during a bus cycle, change the 8284A. If these lines never indicate that the READY line should go low, go to the next cause condition.
Cause: Check the status inputs of the 8288 bus controller. If good, check the output controls of the bus controller. If good, the problem is in the extended control logic section. If the status inputs of the 8288 bus controller are bad, check the status outputs of the 8088 MPU.

INTR

The INTerrupt Request (maskable interrupt) input of the 8088 MPU informs the MPU that the programmable interrupt controller has requested the servicing of an interrupt, when this lines goes high. This line is sampled during the last clock of each instruction. If the INTR is high on the last clock of the instruction, the MPU begins the interrupt sequence. The MPU then issues two successive interrupt acknowledge signals to the bus controller through its output status lines. The MPU also locks out any other bus master (using the $\overline{\text{LOCK}}$ output [see p. 162]) from T2 of the first interrupt acknowledge to T2 of the second. No other interrupt can be acknowledged until the current interrupt is serviced.

A typical interrupt sequence occurs as follows:

1. Upon receiving an interrupt request by one of the hardware inputs on the programmable interrupt controller (8259), the 8259 causes its INT (interrupt output) line to go high. The INT output from the 8259 is connected to the INTR input of the MPU.
2. On the last clock of the bus cycle, the MPU samples the INTR line. Because in this example the INTR line is high, the MPU sends the first of two successive interrupt acknowledge signals to the bus controller.
3. The MPU sends the first interrupt acknowledge signal to the bus controller through the status lines ($\overline{\text{S0}}$ to $\overline{\text{S2}}$). This signal causes the bus controller to disable all command outputs (all read and write lines are high). The bus controller then causes the $\overline{\text{INTA}}$ (INTerrupt Acknowledge, see p. 162) output line to pulse low. This signal is applied to the $\overline{\text{INTA}}$ input of the interrupt controller (8259). Upon receiving this acknowledge signal, the interrupt controller readies the interrupt vector code. During T2 of this first interrupt, the MPU locks out any other bus master (DMA) by bringing the $\overline{\text{LOCK}}$ out low. This concludes the first $\overline{\text{INTA}}$ signal (four clocks).

4. After four clocks, the MPU sends the second $\overline{INTA}$ signal to the bus controller. This signal causes the bus controller to keep its command outputs disabled. The bus controller then sends a pulse on the $\overline{INTA}$ line to the interrupt controller. Upon receiving the second $\overline{INTA}$ signal, the interrupt controller releases its interrupt vector code (8 bits; range 08_{16} to $0F_{16}$ on a PC XT) on the data bus and resets the INT line. During T2 of the second $\overline{INTA}$ signal, the MPU unlocks requests from other bus masters (DMA) by making the $\overline{LOCK}$ line go high. The MPU then reads the interrupt vector code (in T4 of the second interrupt), which concludes the second $\overline{INTA}$ signal.
5. The MPU multiplies the vector code by 4 to determine the first of four address locations. The contents of the four address locations are used to determine an effective address for the interrupt service routine. The MPU then executes the service routine.
6. At the end of the interrupt service routine, an interrupt return instruction restores the contents of the flags register, CS and IP, from its location on the stack and sets the interrupt flag, allowing the INTR input to accept a new request.

Condition: The INTR line goes high during interrupt request and goes low after the MPU reads the vector code. The INTR line may go high again after the vector code is read, but the MPU does not process the interrupt request until the current interrupt is completely serviced.
Result: Normal operation.

Condition: The INTR line never goes high.
Result: Computer probably halts because interrupts are generated even though a program is not being executed (see the programmable interrupt controller section in this chapter for more information).
Cause: Check the request inputs of the interrupt controller and make sure they are operational. If bad, backtrack to the devices controlling the request inputs and troubleshoot.

Condition: INTR line is always high.
Result: Computer halts or stops the execution of the program. Although interrupts occur, they should not be continuous.
Cause: Check the request inputs of the interrupt controller. If an input is stuck high, troubleshoot the device(s) controlling that input.
Cause: Check the MPU, bus controller, and interrupt controller, making sure the $\overline{INTA}$ signal is being received in the correct sequence.

NMI

The NonMaskable Interrupt input is leading edge triggered and has a higher priority than the INTR input. If after the leading edge transition the line remains high for two clocks, the 8088 performs a type 2 interrupt. A type 2 interrupt causes the 8088 to read four memory locations beginning at effective address 00008_{16}; the contents of the four memory locations are used by the MPU to create an effective address which indicates the beginning address of the interrupt service routine. The MPU then vectors to the address and executes the service routine. A type 2 interrupt is used to indicate some type of parity error in the IBM PC. This input is controlled by the external control logic on the system board. Parity errors either in memory or I/O are generated only after read operations.

Generally, there are two types of parity errors; both are created when the total number of logical 1s in a memory or I/O location is not correct. Parity error 1, or parity check 1, means the problem is on the system board. Parity error 2, or parity check 2, indicates a parity problem on an external interface board.

Often, parity errors are encountered but do not produce any major problems in the computer other than needing to reboot the system. If parity errors occur continuously, the following are common causes in the order of importance.

1. Power supply noise
2. Fluctuating supply voltages
3. DRAM falling out of tolerance, not retaining data charge levels before refresh occurs
4. Programmable interval timer not programmed or working properly
5. DAM not programmed or working properly
6. Refresh gating in the address decoding section not working properly

IBM PC: The IBM PC always checks for parity errors.

xxx PC: Generic PCs allow the user to determine whether parity will be checked and generated by setting the equipment dip switches on the computer. If parity is disabled, parity errors do not produce any messages for the user.

Condition: The NMI line never goes high; program runs normally.
Result: Normal operation.

Condition: The NMI line never goes high, program does not operate normally, and data is not correct.
Result: Computer halts or operates in an unpredictable fashion.
Cause: Parity errors are occurring; if parity is disabled, enable parity checking to determine problem.
Cause: The NMI signal is being received but MPU does not respond. Change MPU; otherwise, see the causes below.
Cause: Parity circuitry in DRAM section is bad or not receiving updated signal at the proper times. See DRAM troubleshooting section.
Cause: The NMI register in the external control logic section is bad or not receiving updated signals at the proper times. See DRAM troubleshooting section.

Condition: The NMI line never goes low, or computer does not boot and produces a memory error code.
Result: Computer halts or does not fully boot.
Cause: Check DRAM with external memory tester.
Cause: Perform the last three checks on the causes for the previous condition.

RESET

The RESET input, when high for at least four clock cycles, causes the 8088 to immediately terminate all operations. This signal is generated by the 8284 clock generator from the power-good line from the power supply. When this line is high, one or more supply voltages from the power supply are out of tolerance. The operations of the 8088 begin in the order specified in the 8088 reference manual as soon as this line goes low.

Condition: RESET input stays low; computer operates normally.
Result: Normal operation.

Condition: RESET input stays low; computer fails to operate.
Result: Computer halts, or program fails to operate.
Cause: Check power supply for proper +5-, −5-, +12- and −12-V voltage levels. If any are bad, repair or replace power supply.
Cause: If voltage levels are good, check the power-good output of the power supply ($\overline{\text{RES}}$ input of 8284A). If high, everything up to this point is good. If low, check the wiring (look for broken wire or bad connection); otherwise, repair or replace the power supply.
Cause: If $\overline{\text{RES}}$ input to the 8284 is high and the RESET output is high, replace the 8284. If $\overline{\text{RES}}$ input to the 8284 is low and the RESET output is low, check the following ICs for a loading problem: 8088 MPU, 8087 math coprocessor, and external control logic.

Condition: RESET input stays high.
Result: Computer halts or does not attempt to boot.
Cause: Check power supply for proper +5-, −5-, +12- and −12-V voltage levels. If any are bad, repair or replace power supply.
Cause: If voltage levels are good, check the power-good output of the power supply ($\overline{\text{RES}}$ input of 8284A). If high, everything up to this point is good. If low, check the wiring (look for broken wire or bad connection); otherwise, repair or replace the power supply.
Cause: If the $\overline{\text{RES}}$ input to the 8284 is high and the RESET output is high, replace the 8284. If the $\overline{\text{RES}}$ input to the 8284 is low and the RESET output is high, recheck the power supply.

$\overline{\text{RD}}$ When low, the ReaD output indicates that the microprocessor is performing a read operation while in the minimum operational mode. Because the IBM PC and clones operate in the maximum operational mode only, this pin is not used.

CLK The CLocK input receives its signal from the 8284 clock generator. This signal (CLK88) is the system clock for the PC and is used to synchronize all bus operations.

IBM PC: In the original IBM PC, this signal has a frequency of 4.772727 MHz with a duty cycle of 33%.

xxx PC: If the generic computer has a turbo mode of operation, the clock frequency may range from 8 to 10 MHz, with a 33% duty cycle. The nonturbo mode remains at 4.772727 MHz at 33%.

Condition: Clock frequency and duty cycle are proper values (actual values depend on machine).
Result: Normal operation.

Condition: Clock frequency either more or less than its proper value.
Result: Depending on how far off the frequency is, the system probably operates either faster or more slowly than anticipated. Some software may experience problems.
Cause: Check clock output frequency from the 8284. It should be one-third the frequency of either the crystal or external fre-

quency input frequency, depending on the F/$\overline{C}$ input. See troubleshooting section on the 8284A.

MN/$\overline{MX}$ The $\overline{\text{MiNimum/MaXimum}}$ mode input sets the operational mode of the 8088. If high, the 8088 is in the minimum operation mode. This means no coprocessor can be used and the 8088 directly controls all bus operations, decreasing the throughput. (Throughput is the number of operations that a MPU can perform in a given amount of time.) All IBM and generic PCs use the 8088 in the maximum mode of operation; therefore, this pin is grounded. In this mode, the MPU issues status signals to a bus controller, which in turn controls the bus operation selected. The maximum mode of operation also allows the 8088-based system to use a math coprocessor to speed up math operations.

IBM PC: The level on this input produces the same type of result in either the IBM PC or clone.

Condition: The level of the MN/$\overline{MX}$ pin is low.
Result: Computer operates normally.

Condition: The level of the MN/$\overline{MX}$ pin is high.
Result: Computer halts; no valid bus sequence is occurring.
Cause: Bad, broken, or dirty connection to the MN/$\overline{MX}$ pin, causing the input to float high. (Correct problem and make sure that there is a valid logic low on the input.)

$\overline{SS0}$ This output line indicates the status of the 8088 when in the minimum mode of operation. In the maximum mode, this output is always high. Because all PCs operate in the maximum mode, this line is not used or connected to anything.

$\overline{S0}$–$\overline{S2}$ These three status outputs (Table 3-1) are used by the 8088 in the maximum operation mode to inform the bus controller (8288) which type of operation is to be performed and when the bus cycle begins and ends. These signals are active beginning at the end of T4 of the previous bus cycle, and they stay active throughout T1, T2, and the end of T3 during a minimum bus cycle. If the bus is not ready by the end of T3, the status lines remain active until the ready line goes high at the end of one of the wait clocks (Tw). Once the ready lines go high, the status lines go into a passive state to end the bus cycle. A change in levels during the high clock of T4 of the bus cycle indicates the beginning of the next bus cycle.

TABLE 3-1
8088 Status Bus Table

$\overline{S2}$	$\overline{S1}$	$\overline{S0}$	*Operation*
0	0	0	Interrupt acknowledge
0	0	1	Read I/O port
0	1	0	Write I/O port
0	1	1	Halt
1	0	0	Code access
1	0	1	Read memory
1	1	0	Write memory
1	1	1	Passive

Condition: Status lines are active during the end of T4; stay active throughout T1, T2, and the end of T3 (Tw in a nonminimum bus cycle) of the bus cycle; and go passive when the bus is ready.
Result: Normal operation.

Condition: Status lines never go active when an instruction is to be executed.
Result: Computer halts; no operations are performed.
Cause: Check bus controller (8288) and verify that it is not loading down the status lines. If the 8288 is good, replace the 8088.

Condition: Status lines are always active and never go passive to end the bus cycle.
Result: Computer halts; no operation is completed.
Cause: Check READY line; if high, change the 8088 or check the 8288 bus controller. If low, check the troubleshooting section of the 8284A clock generator driver.

$\overline{\text{LOCK}}$

If low, the $\overline{\text{LOCK}}$ output pin locks out control of any other bus master in the system (DMA). Instructions with a prefix of LOCK activate this output, which remains active until the next instruction. This pin prevents the DMA from requesting use of the bus during certain MPU operations (mostly during interrupts).

Condition: The $\overline{\text{LOCK}}$ line goes low shortly after the status is sent and goes high near the end of most operations. The $\overline{\text{LOCK}}$ line goes high halfway through the second $\overline{\text{INTA}}$ signal during an interrupt sequence.
Result: DMA cannot request permission to use the bus; normal operation.

Condition: The $\overline{\text{LOCK}}$ line remains low for extended periods of time.
Result: The computer halts and produces different types of errors. Because the DMA is responsible for memory refresh, parity errors may occur. Also, the DMA is responsible for the transfer of data during floppy- and hard-drive operations; transfer errors may occur (the exact type of error message varies from system to system).
Cause: Software or firmware glitch or bug.
Cause: Short in the extended control logic section, which causes the line to stay at ground level.

Coprocessor Control Status Pins

$\overline{\text{TEST}}$

The $\overline{\text{TEST}}$ input indicates whether the co-processor is connected and allows synchronization if the system is ready or busy. This input affects only the 8088 operating in the maximum operation mode. If low, this level, along with the wait-for-test instruction, produces a wait state in the 8088. If high, the 8088 continues normal execution of its program.

$\overline{\text{RQ}}/\overline{\text{GT0}}$
$\overline{\text{RQ}}/\overline{\text{GT1}}$

The $\overline{\text{ReQuest}}/\overline{\text{GranT 0}}$ (HOLD=min) and $\overline{\text{ReQuest}}/\overline{\text{GranT 1}}$ lines (HLDA=min) allow communication between the 8088 and 8087. This communication is necessary because both ICs share the same local multiplexed address/data bus. The 8088 has $\overline{\text{RQ}}/\overline{\text{GT0}}$ tied high, which indicates that the processor is the master processor. $\overline{\text{RQ}}/\overline{\text{GT1}}$ is the bidirectional handshake line that allows the 8088 to communicate with the 8087. It allows the 8088 to receive a request or grant permission for the 8087 to use the local bus. The handshaking process is described in the 8087 section of this chapter.

TABLE 3-2
8088 Coprocessor Status Table

QS1	*QS0*	*Status*
0	0	No operation in queue
0	1	First byte of operations code from queue
1	0	Empty the queue
1	1	Subsequent byte from queue

QS0-QS1 The Queue Status output (Table 3-2) pins provide the status of the instruction queue to the math coprocessor in the system, if present. QS0 to QS1 are valid only during the clock cycle after a queue operation is performed. The 8087 needs the queue status along with the bus status to process escape instructions.

3.1.13 8088 Timing Diagrams (Fig. 3-8(a) and (b))

In the following, we discuss timing diagrams of the 8088 operating in the maximum operational mode. Remember that in the maximum mode the status lines ($\overline{S0}$, $\overline{S1}$, and $\overline{S2}$) are used to inform the bus controller which type of data transfer to perform and when it begins and ends.

The first timing diagram (Figure 3-8a) shows a typical minimum operation, requiring only four clocks to perform the bus operation. Normally, a minimum operation occurs when a read or write operation occurs with the memory on the system board.

At the end of T4 of the previous bus cycle, the following occurs. The status lines must be passive during this time to allow the previous bus cycle to end.

1. The 8088 places status on the status lines ($\overline{S0}$ to $\overline{S2}$).
2. The 8088 begins placing the address of the next instruction on the multiplexed address/data/status bus.

T1 of the Current Bus Cycle

1. Active status is now valid on the status lines (which informs the bus controller which operation is to be performed).
2. A valid address is now on the address lines and stays valid until near the end of T1.

T2 of the Current Bus Cycle

1. Status stays valid.
2. Address/data lines AD0 to AD7 switch from producing the valid address into bidirectional data lines. The direction and when data appear on these lines are determined by the status lines (type of operation), bus controller, and support circuitry.
3. The levels on address lines A8 to A15 remain valid.
4. The levels on address/status lines A16 to A19 now indicate which of the segment registers are currently being used (S3 to S6). Because no device in the IBM PC or clone uses these values, they have no effect on the computer system.
5. The 8284 clock generator/driver, through the extended control logic at some time during T2, causes the READY line input to the 8088 to go low during memory operations. This indicates that the device in question is being accessed. Under certain operations (I/O operations), the READY line may not go low until some time during T3.

T3 of the Current Bus Cycle

1. Status lines remain valid up to the time when the READY line goes high. Once the READY line goes high, the status lines go into a passive state (all lines high).

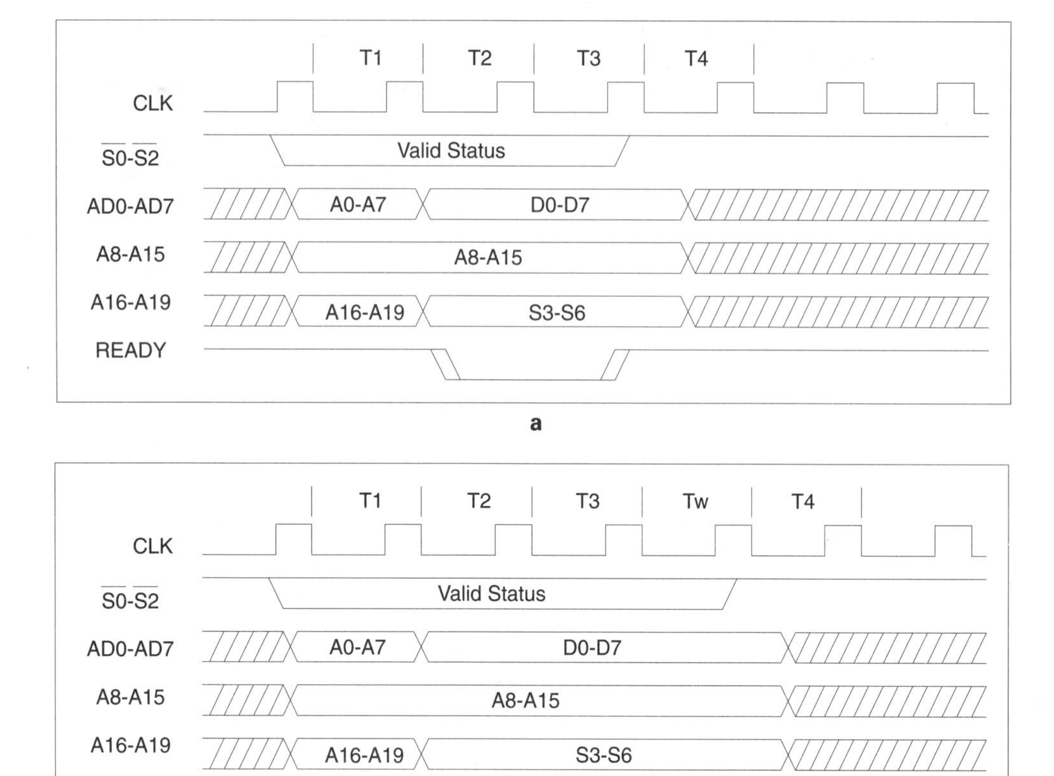

Figure 3-8 **(a)** 8088 timing (minimum); **(b)** 8088 timing (non-minimum).

2. The data on AD0 to AD7 should become valid and stay valid prior to the time the READY line goes high and stay valid throughout the rest of T3.
3. A8 to A15 continue producing that part of the address.
4. A16 to A19 continue producing the segment register status (S3 to S6).
5. When a minimum transfer occurs, the READY line goes high before the end of T3. Because the READY line went high by the end of T3, the 8088 causes the status lines to all go high (passive state), indicating to the bus controller that it must complete the data transfer within one complete clock to end the bus cycle.

T4 of the Current Bus Cycle

1. Status lines stay passive to the end of the transfer (at least halfway through T4). These lines may go active near the end of T4 if another operation is ready to take place.
2. AD0 to AD7 must still represent the valid data through the end of the transfer (at least halfway through T4). These lines may then switch to represent the low byte of the next valid address offset near the end of T4 if another operation is ready to take place.
3. A8 to A15 still represent those bits of the valid address through the end of the transfer (at least halfway through T4). These lines may then switch to represent the high byte of the offset part of the next address near the end of T4 if another operation is ready to take place.
4. A16 to A19 still represent the segment register status through the end of the transfer (at least halfway through T4). These lines may then switch to represent the segment part of the next valid address near the end of T4 if another operation is ready to take place.
5. The READY line should remain high throughout T4.

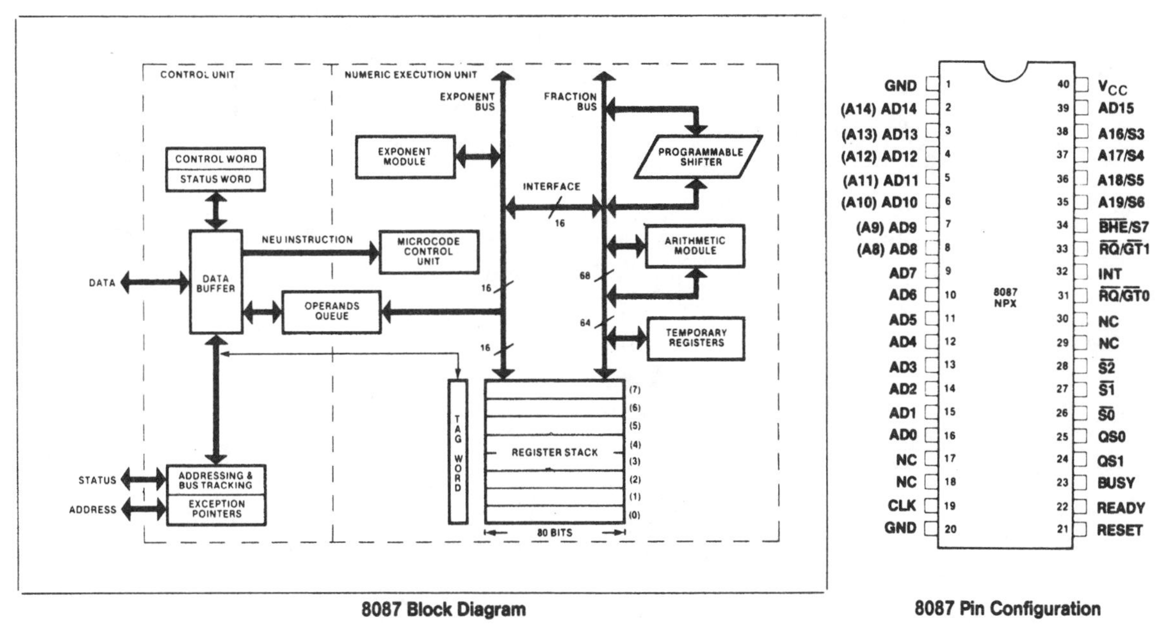

Figure 3-9 8087 Block diagram and pin configuration. (Courtesy of Intel)

The only difference between the first and second (Figure 3-8(b)) timing diagrams is that the READY line in the second is not high by the end of T3, causing the MPU to wait one clock cycle and not make any external changes (internal housekeeping is performed during this time). At the end of the first wait clock, the READY line is checked; if high, the status lines go passive, which ends the bus cycle within one clock (see T4 of the minimum operation). If the READY line is still low, another wait clock is generated, and the process continues until the READY line goes high.

3.1.14 8087 Math Coprocessor

The empty 40-pin socket normally located next to the 8088 MPU is the socket for the 8087 Numeric Data Coprocessor (Fig. 3-9). When used with the right software, this allows the 8087 to process complex math operations up to 50 times faster than the 8088 MPU. The 8087 has the ability to process numeric data only if the software uses the 8087 math coprocessor instructions. This one IC costs more than the entire motherboard, which means that the 8087 is used only when high-speed number crunching is necessary and price is of little consequence. Current dealer prices range from $60 to $100.

Because the 8087 is not as popular as the math coprocessor in AT class computers, the discussion of it is limited in this chapter.

The 8087 math coprocessor is wired in parallel with the 8088; this allows a mirror image of the 8088 instruction queue to be present in the 8087. When a math coprocessor escape code is read into the instruction queue, the 8087 requests permission to use the bus. Once in control, the 8087 keeps control over the local bus until the math operation is complete.

8087 Features

80-Bit number data coprocessor

Type of math operations in addition to 8088 instructions:

- Exponential operations
- Logarithmic operations
- Trigonometric operations

Data types:
- 18-bit BCD
- 32-, 64-, 80-bit floating point
- 16-, 32-, 64-bit integers

Eight 80-bit registers:
- Bit 79 is the sign bit.
- Bits 64 to 78 are the exponent field.
- Bits 0 to 64 are the significant field.

3.1.15 8087 Pin Configurations

Multiplexed Address/Data/Status Bus

AD0–AD15 — When the 8088 controls the bus, these lines are not active and have no effect on the bus, except that AD0 to AD7 monitor the data during the time the data is valid. There are 16 address/data lines that operate like the address/data line in the 8088. If the 8087 is used with the 8086 MPU, these lines act as address lines during T1 and bidirectional data lines during the remaining part of the bus cycle. When the 8087 is used with the 8088, only AD0 to AD7 are used as the low byte of the offset address during T1 of the bus cycle; during the remaining part of the bus cycle, these lines act as the data bus. AD8 to AD15 produce the high byte of the address during the entire bus cycle because the BHE/S7 (Byte high enable) is tied high (which disables the data functions of AD8 to AD15 when used in an 8088 system). Note that AD8 to AD15 are connected to the A8 to A15 lines of the subsystem.

A16–A19
S3–S6 — When the 8087 controls the bus, these lines supply the high nibble of the 20-bit address to the address bus during T1 of the bus cycle and segment status information during the remaining part of the bus cycle. When the 8087 is not in control of the bus, the 8087 monitors these lines.

$\overline{\text{BHE}}$/S7 — In the IBM PC and compatibles that use the 8088, the Byte High Enable line is tied high. The high on $\overline{\text{BHE}}$ causes the 8087 to use only AD0 to AD7 for data transfers while disabling the data functions on AD8 to AD15.

$\overline{\text{S0}}$–$\overline{\text{S2}}$ — The output status lines are active only when the 8087 controls the bus. During this time, only three status functions are valid, and they follow the same coding as the status lines in the 8088. The three valid status codes are read and write memory and passive.

$\overline{\text{RQ}}/\overline{\text{GT0}}$ — The $\overline{\text{Request/Grant Zero}}$ line allows the 8087 to gain control of the local bus from the 8088. These lines are bidirectional; this allows handshaking between the 8087 and the 8088. Handshaking is the process that allows two devices to communicate with each other over the same control lines. In the IBM PC and compatibles, this line is connected to the $\overline{\text{RQ}}/\overline{\text{GT1}}$ line of the 8088. This means that the 8087 is the slave and the 8088 is the original bus master. The 8087 gains control over the bus by the following process:

1. When the math coprocessor escape code is decoded by the 8088 and 8087, the 8087 performs the following sequence.
2. The 8087 causes the $\overline{\text{RQ}}/\overline{\text{GT0}}$ line to pulse low for one clock cycle (request pulse), which is applied to the $\overline{\text{RQ}}/\overline{\text{GT1}}$ line of the 8088. This process informs the 8088 that the 8087 is requesting control over the bus.

3. The 8087 waits until the 8088 pulses its $\overline{\text{RQ}}/\overline{\text{GT1}}$ line low (grant pulse), at which time the 8088 goes into a wait state, and the 8087 begins to control the bus and begins processing the math operation.
4. Upon completion of the math operation, the 8087 pulses the $\overline{\text{RQ}}/\overline{\text{GT0}}$ low (release pulse), which gives control back to the 8088.

$\overline{\text{RQ}}/\overline{\text{GT1}}$	The $\overline{\text{Request}}/\overline{\text{Grant 1}}$ line in the IBM PC and compatibles is tied high, which disables the function of this pin.
QS0–QS1	The Queue Status input lines of the 8087 allow the 8087 to monitor the queue status of the 8088. This information is necessary for the 8087 to gain control over the bus at the proper time.
INT	The INTerrupt output line, when high, indicates that the 8087 has received an unmasked exception during a numeric instruction. This output can produce an NMI in the 8088.
BUSY	The BUSY output line, when high, informs the 8088 that the 8087 is currently using the bus. This line is connected to the $\overline{\text{TEST}}$ input of the 8088 and provides synchronization.
READY	The READY input is used to indicate when the bus is ready to complete its bus cycle to the 8087. The signal applied to this line is produced by the READY output of the 8284 to provide synchronization of the bus.
RESET	The active high input produced by the RESET output of the 8284 is used to reset the 8087.
CLK	The CLocK input is produced by the CLK88 output of the 8284, which synchronizes the 8087 with the bus cycle.

3.1.16 8087 Troubleshooting

When an 8087 is purchased, the manufacturer normally supplies a test program to verify the operation of the 8087 once installed. Also, most diagnostic software tests the 8087. If the diagnostic software fails to recognize the 8087, the problem is normally that the 8087 is not installed correctly. Remove the IC and then verify proper installation; if correct, change the math coprocessor. If changing the 8087 does not correct the problem, use the following checks.

1. Verify Vcc and GND levels.
2. Check the BUSY line; it should be a logic 1 during a 8087 operation.
3. Check the $\overline{\text{RQ}}/\overline{\text{GT1}}$ and $\overline{\text{BHE}}$/S7 lines; they should be a logic 1.
4. Check the $\overline{\text{TEST}}$ line of the 8088; it should be a logic 1.
5. Check the $\overline{\text{RQ}}/\overline{\text{GT0}}$ lines of the 8088; they should be a logic 1.

If the diagnostic software indicates that the 8087 is operational but the software does not function, make sure that the software supports a math coprocessor and that the math coprocessor functions are installed.

3.1.17 8288 Bus Controller

The 8288 bus controller generates the control signals for the bus cycle once the 8088 indicates the type of transfer. This IC controls the operational sequence on both the multiplexed local bus (address/data/status) and the demultiplexed external bus (address, data, control, and status buses). This IC is required by the IBM PC because the 8088 operates in the maximum mode, which means that the 8088 does not directly control the signals. The bus controller causes the 8088's throughput to increase, allowing the IBM PC to operate more efficiently. The 8288

(Fig. 3-10) acts as a buffer between the local and external address/data. Buffering allows other bus masters to use this without interfering with the operation of the 8088.

3.1.18 8288 Pin Configurations

Status Decoder

$\overline{S0}$–$\overline{S2}$ The Status input pins are used by the 8088 to communicate with the bus controller. The level on these inputs determines the type of data transfer that takes place (Tables 3-3 and 3-4). The level on these inputs also determines when the bus cycle begins and ends. The bus cycle begins once the status lines go from a passive state to one of the active states. The bus cycle ends within one more clock cycle when the status lines go from active to passive.

Condition: Status lines go from passive to active, stay active at least until the end of T3 or at the end of one of the wait clocks (Tw) before returning to a passive state.

Result: Normal operation of a bus cycle.

Condition: Status lines go from passive to active but never return to a passive state.

Result: Computer halts or hangs up; data transfer is never completed.

Cause: Check for a CLK88 signal on the CLK input of both the 8288 bus controller and the 8088 MPU. If bad or missing, troubleshoot the 8284A clock generator driver IC or check for broken traces on the printed circuit board.

Cause: Check status outputs of 8088; if the same, go to troubleshooting the READY input line of the MPU. If the status is not the same, check for broken traces on the printed circuit board.

CLK The CLocK input is derived from the 8284 CLK88 line. The frequency and duty cycle are the same as the frequency and duty cycle supplied to the CLK input of the 8088 MPU. This line allows the 8288 to synchronize its command sequence with the MPU.

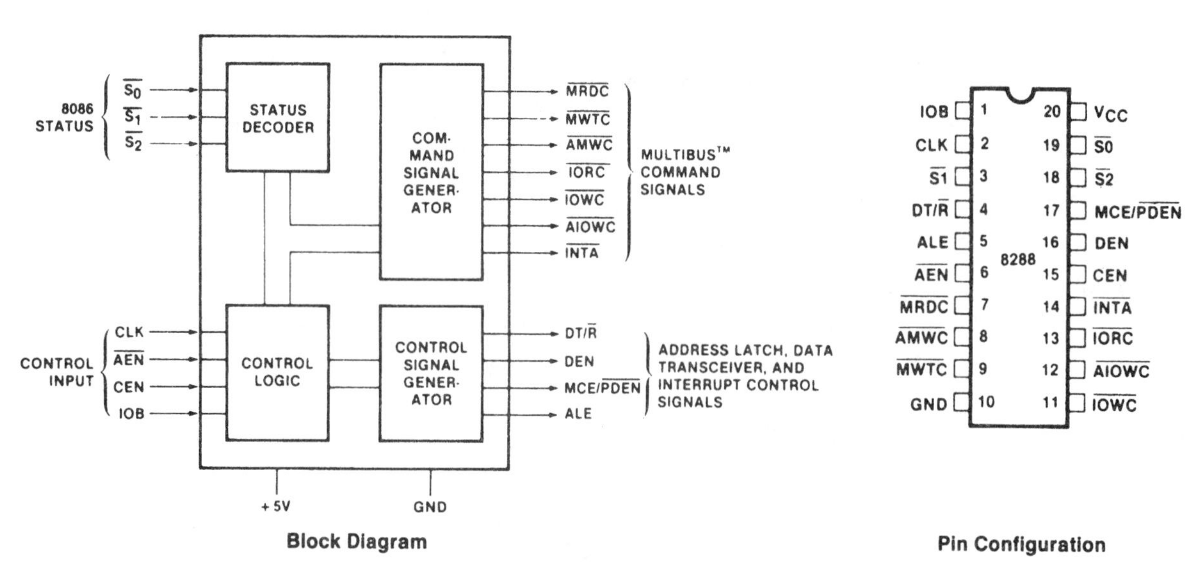

Figure 3-10 8288 block diagram and pin configuration. (Courtesy of Intel)

Condition: Frequency and duty cycle are the same as the CLK output of the 8284A and the CLK input of the 8088.
Result: Normal operation.

Condition: Frequency and/or duty cycle is not correct but is the same as the CLK output of the 8284A and CLK input of the 8088.
Result: Depending on the severity of the problem, the computer may lock up (halt) or may operate too slow or too fast.
Cause: Check the 8284A out of the circuit or replace the 8284A with a known good 8284A; follow the troubleshooting process for the 8284A.
Cause: The 8088, 8087, 8288, or external control logic may be loading down the clock signal. Normally, the loading of a signal causes the voltage level of the signal to decrease, depending on how much loading is present.

$\overline{\text{AEN}}$ The Address ENable input is used to enable (high=not active or low=active) the command outputs ($\overline{\text{MRDC}}$, $\overline{\text{MWTC}}$, $\overline{\text{IORC}}$, $\overline{\text{IOWC}}$, $\overline{\text{AMWC}}$, $\overline{\text{AIOWC}}$, and $\overline{\text{INTA}}$ lines) of the 8288, 115 ns after the line goes low. In the PC this line is derived from the AEN BRD line in the extended control logic section. This line is left low, except during a HOLDA (hold acknowledge) when the DMA is using the bus, when it changes to a high. During a high, the command outputs go into tristate, which eliminates any conflicts between the bus controller and DMA.

Condition: $\overline{\text{AEN}}$ is low during an MPU bus cycle.
Condition: $\overline{\text{AEN}}$ stays low except during a DMA cycle (HOLDA, high).
Result: Normal operation; because the extended control logic controls this line, it normally is always low except during a DMA bus cycle.

TABLE 3-3
8288 Status Bus Table

$\overline{S2}$	$\overline{S1}$	$\overline{S0}$	*Input Function*	*Output Function*
0	0	0	Interrupt acknowledge	$\overline{\text{INTA}}$
0	0	1	Read I/O port	$\overline{\text{IORC}}$
0	1	0	Write I/O port	($\overline{\text{IOWC}}$),$\overline{\text{AIOWC}}$
0	1	1	Halt	None
1	0	0	Code access	$\overline{\text{MRDC}}$ (normally ROM)
1	0	1	Read memory	$\overline{\text{MRDC}}$ (RAM or ROM)
1	1	0	Write memory	($\overline{\text{MWTC}}$),$\overline{\text{AMWC}}$
1	1	1	Passive	None

TABLE 3-4
8288 Output Commands Functions Used in the IBM

$\overline{\text{INTA}}$	Interrupt acknowledge (no read or write command lines are selected)
$\overline{\text{IORC}}$	I/O read command
$\overline{\text{AIOWC}}$	Advanced I/O write command
$\overline{\text{MRDC}}$	Memory read command code access (ROM)
$\overline{\text{AMWC}}$	Advanced memory write command
$\overline{\text{None}}$	During a passive state, no outputs are selected

Condition: The $\overline{\text{AEN}}$ line stays low all the time.
Result: Computer halts; memory is not refreshed.
Cause: Check the HOLDA line on the 8237 DMA; it should go high once every 72 clocks. If a high does not occur, troubleshoot the DMA section.
Cause: Make sure one-half of the PCLK signal is applied to all the 8254 PIT CLK inputs.
Cause: Make sure the CLK signal is supplied to the CLK inputs of the 8288, 8088, and 8237.

CEN

The Command ENable input, when high, activates the selected command output while enabling data flow. When low, all command outputs, DEN (data enable, see p. 149) and PDEN (peripheral DEN) lines are deactivated (logic high); no read or write operation occurs and no data flows. This line acts as a qualifier for $\overline{\text{AEN}}$ to determine when data flows. In the PC, the CEN signal is derived from the $\overline{\text{AEN}}$ line in the extended control logic section and is normally high except during a DMA cycle (HOLDA).

Condition: CEN is high during an MPU bus cycle.

Condition: CEN stays high except during a DMA cycle (HOLDA, high).
Result: Normal operation; because the extended control logic controls this line, it normally is always high except during a DMA bus cycle.

Condition: CEN line stays high during all operations.
Result: Computer halts; memory is not refreshed.
Cause: Check the HOLDA line on the 8237 DMA; it should go high once every 72 clocks. If a high pulse does not occur, check the DMA circuit.
Cause: Make sure one-half of the PCLK signal is applied to all the 8254 PIT CLK inputs.
Cause: Make sure the CLK signal is supplied to the CLK inputs of the 8288, 8088, and 8237.

IOB

The Input/Output Bus-mode input determines whether the 8288 is the I/O bus master or the system bus master. In the IBM PC, this line is grounded, which makes the 8288 the system bus master. This grounding also causes a 155-ns delay in the active command output after AEN becomes enabled during an MPU bus cycle.

Condition: IOB is at ground level.
Result: Normal operation.

Condition: IOB is not at ground level.
Result: Computer halts intermittently.
Cause: Broken connection on printed circuit board or bad solder connection.

Local Bus Controls. The local bus control lines are used to control how the multiplexed address/data/status bus interfaces with the demultiplexed external address and data buses.

DT/$\overline{\text{R}}$

The Data Transmit/$\overline{\text{Receive}}$ output controls the direction of data flow through the 74245 octal data bus transceiver. DT/$\overline{\text{R}}$ goes high during a write (I/O or memory) operation, which allows data to flow from the local bus to the external bus when the 74245 is enabled. This line goes low during a read (I/O or memory) operation, which allows data to flow from the external bus to the local bus when the 74245 is enabled. During an interrupt acknowledge, this line functions like a memory read operation.

Condition: DT/$\overline{R}$ sets during T1 and stays set throughout the write operation (data transfer from the local to the external data bus).
Condition: DT/$\overline{R}$ resets during T1 and stays reset throughout the read operation (data transfer from the external to the local data bus).
Result: Normal operation.
Condition: DT/$\overline{R}$ is at the proper level but changes before the end of the operation.
Result: Computer halts intermittently.
Cause: Replace the 74245; it may be loading down the DT/$\overline{R}$ output.
Cause: DT/$\overline{R}$ output may not be able to supply enough current for the 74245; change the 8288.
Condition: Proper DT/$\overline{R}$ level is not selected during T1 and throughout the operation.
Result: Computer halts intermittently.
Cause: Check the status inputs; if correct and the DT/$\overline{R}$ line is not correct, check the CLK input to the 8288.
Cause: Check the status inputs; if correct and the DT/$\overline{R}$ line is not correct, change the 74245 and/or 8288.

DEN

The Data ENable output allows data to pass through the 74245 when high. This line goes active (high) during the beginning of T2; after the ALE (address latch enable, see below) is pulsed, the level of the DT/$\overline{R}$ line is selected, and the selected read or write goes active. During an advanced write operation, this line goes active at the beginning of T2. During a read operation, this line goes active about half way through T2. This line functions as a memory read during an interrupt acknowledge.

The direction depends on the level on the DIR pin of the 74245. In the PC, this signal is applied to a NAND inverter; in most clones, an inverter is used. The NAND inverter used in the IBM PC is also used to disable any data transfers between the local and external data bus whenever the 5-V supply goes below its lower voltage range. The output of the NAND inverter is applied to the $\overline{G}$ (enabling) input of the 74245 bus transceiver.

Condition: DEN becomes active (high) sometime in T2 and stays active throughout the first third of T4.
Result: Normal operation.
Condition: DEN becomes active (high) sometime in T2 and stays active throughout the first third of T4, but data is not transferred between the local and external data buses.
Result: Computer halts or fails to execute program properly.
Cause: Because DEN is high, check for the proper operation of the NAND gate that is connected between the DEN output and the $\overline{G}$ (enabling) input of the 74245. This gate should act as an inverter; correct if the problem is bad. In a clone, the NAND gate is sometimes replaced by an inverter.
Cause: If DEN is high and the $\overline{G}$ (enabling) input of the 74245 is low, verify that the data on the inputs and outputs of the 74245 are the same (use a logic analyzer) during T3 through the first third of T4. If not the same, change the 74245. If the same, the problem is not hardware.
Condition: DEN does not go high at the proper time during the bus cycle.
Result: Computer halts or fails to execute program properly.

Cause: Check the CLK input to the 8288. If good, go to next cause; otherwise, troubleshoot the CLK88 signal.
Cause: Check the status inputs of the 8288. If good, replace the 8288; otherwise, troubleshoot the status lines of the 8088.

MCE/$\overline{\text{PDEN}}$ The Master Cascade Enable/$\overline{\text{Peripheral Data ENable}}$ output enables any cascaded bus controllers in the system during an interrupt. This output also enables any special I/O bus controllers during a DEN. The IBM and clones do not use the MCE/$\overline{\text{PDEN}}$ output; therefore, no connection is made.

ALE The Address Latch Enable output latches the levels on the inputs of the local bus address latches (AD0 to AD7, A8 to A15, and A16 to A19) on the falling edge of the pulse. The ALE output is applied to the clock inputs of all address latches on the local bus. The ALE line is also applied to the expansion I/O card slots of the system board for controlling external memory. This line functions normally during an interrupt acknowledge.

Condition: The ALE line is pulsed from low to high to low within T1 of the bus cycle and while the multiplexed address/data/status bus is producing the address.
Result: Normal operation.
Condition: The ALE line is pulsed in a clock cycle other than T1 of the bus cycle.
Result: Computer halts; incorrect address is accessed.
Cause: Check CLK input of 8288. If good, go to next cause; otherwise, troubleshoot the CLK88 signal.
Cause: Check the status lines of the 8288. If good, replace the 8288; otherwise, troubleshoot the status lines of the 8088.

External/Local Bus Command Signals. The signals developed in this section of the 8288 bus controller are used to enable operations on the external bus. Only one of these lines ever goes active (low) at any one time.

$\overline{\text{MRDC}}$ The $\overline{\text{Memory ReaD Command}}$ output enables read operations in RAM and ROM. When MRDC goes low, the device being addressed applies a copy of its data to the data bus. This line goes active (low) after ALE is pulsed and the DT/$\overline{\text{R}}$ line is selected.

$\overline{\text{AMWC}}$ The $\overline{\text{Advance Memory Write Command}}$ output enables write operations in RAM. When $\overline{\text{AMWC}}$ goes low, the device being addressed accepts the data and places a copy into the location selected. This line goes active (low) after ALE is pulsed and the DT/$\overline{\text{R}}$ line is selected.

$\overline{\text{IORC}}$ The $\overline{\text{Input/Output Read Command}}$ output enables read operations to occur in I/O devices. When $\overline{\text{IORC}}$ goes low, the I/O device being addressed applies a copy of its data to the data bus. This line goes active (low) after ALE is pulsed and the DT/$\overline{\text{R}}$ line is selected.

$\overline{\text{AIOWC}}$ The $\overline{\text{Advanced Input/Output Write Command}}$ output enables write operations to occur in I/O devices. When $\overline{\text{AIOWC}}$ goes low, the I/O device being addressed accepts the data on the data bus and places a copy into the location selected. This line goes active (low) after ALE is pulsed and the DT/$\overline{\text{R}}$ line is selected.

Condition: Depending on the selected operation (determined by the status inputs of the 8088), the selected output ($\overline{\text{MRDC}}$, $\overline{\text{AMWC}}$, $\overline{\text{IORC}}$, or $\overline{\text{AIOWC}}$) is made active at the end of T1 or beginning of T2. The line remains active through the first part of T4 and must go inactive before DEN goes low.

Result: Normal operation.

Condition: The correct command output line is not selected at any time during the bus cycle.

Result: Computer halts with possible damage to support circuitry.

Cause: Change the 8288 bus controller.

Condition: The selected command output does not stay active for the proper time period.

Result: Computer halts; valid data may not be transferred.

Cause: Check the CLK input of the 8288. If good, go to the next cause; otherwise, troubleshoot the CLK88 signal.

Cause: Change the 8288 bus controller.

$\overline{\text{IOWC}}$ $\overline{\text{MWTC}}$ The $\overline{\text{Input/Output Write Command}}$ and $\overline{\text{Memory WriTe Command}}$ output lines are not used and therefore are not connected.

$\overline{\text{INTA}}$ The $\overline{\text{INTerrupt Acknowledge}}$ output acknowledges a system interrupt. When the status inputs select the $\overline{\text{INTA}}$ output, all other command outputs go inactive (high). The high on all the read and write lines prevents data from being placed on the external data bus. The $\overline{\text{INTA}}$ signal is applied to the $\overline{\text{INTA}}$ input of the 8259 to inform the 8259 that the external bus is disabled.

3.1.19 8288 Timing Diagrams

The following timing diagrams show the operation of the five most used commands regarding bus operations.

Interrupt Acknowledge Timing (Fig. 3-11). When the interrupt request is acknowledged by the programmable interrupt controller (8259 PIC), the PIC informs the 8088 that an interrupt needs to be serviced. Once the 8088 acknowledges the interrupt, it sends out two consecutive interrupt acknowledge status signals from its status output to the bus controller (8288). Figure 3-11 shows how the 8288 bus controller processes the signal.

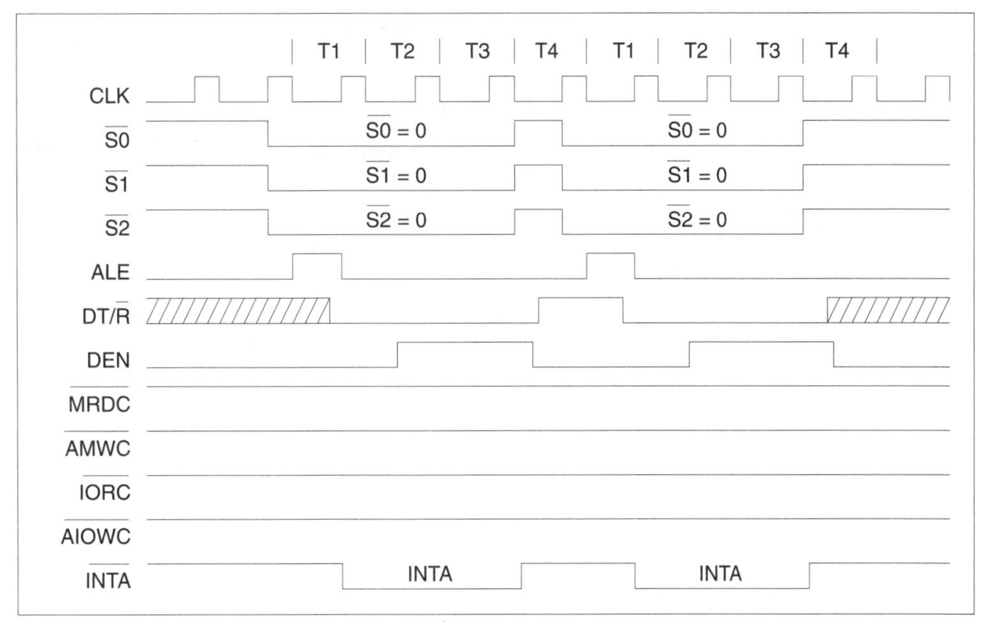

Figure 3-11 8288 interrupt acknowledge timing.

T4 of the Previous Bus Cycle

1. The 8088 sends out the interrupt acknowledge status ($\overline{S0}$=0, $\overline{S1}$=0, and $\overline{S2}$=0).
2. The 8288 bus controller receives the status.

T1 Interrupt Acknowledge (First of Two from 8088)

1. ALE is pulsed; this causes the address on the multiplexed address/data/status (local) bus to be loaded into the address latches.
2. Direction is selected, during an interrupt acknowledge, the DT/$\overline{R}$ line is always low (receive or read operation).
3. The $\overline{INTA}$ output is brought from a high to a low level. This informs the PIC that interrupt acknowledge is being processed by the bus controller. Upon receiving the first of two $\overline{INTA}$ signals from the bus controller, the PIC prepares to release its vector code.

T2 Interrupt Acknowledge (First of Two from 8088)

1. The DEN line goes high, enabling the flow of data from the external to the local bus. *Note:* Because none of the read or write lines go low, no device (IC) produces or accepts the data; therefore, no transfer takes place. The 8088 MPU does not accept or produce data during the first interrupt acknowledge.

T3 Interrupt Acknowledge (First of Two from 8088)

1. Status lines go passive; this informs the bus controller that it must end the sequence within one more clock cycle.

T4 Interrupt Acknowledge (First of Two from 8088)

1. $\overline{INTA}$ sets to inform PIC of the end of the first interrupt acknowledge.
2. DEN resets to disable the data path between the local and external buses.
3. DT/$\overline{R}$ sets to its normal state.
4. A second interrupt acknowledge is received from the 8088 on the status lines ($\overline{S0}$, $\overline{S1}$, and $\overline{S2}$).

T1 Interrupt Acknowledge (Second of Two from 8088)

1. ALE is pulsed; this causes the address on the multiplexed address/data/status (local) bus to be loaded into the address latches.
2. Direct is selected; during an interrupt acknowledge, the DT/$\overline{R}$ line is always low (receive or read operation).
3. The $\overline{INTA}$ output is brought from a high to a low level. This informs the PIC to release its vector code onto the data bus at the end of this interrupt acknowledge cycle. The vector code remains on the local data bus for the rest of the interrupt cycle so that the 8088 can read the vector code.

T2 Interrupt Acknowledge (Second of Two from 8088)

1. The DEN line goes high, enabling the flow of data from the external to the local bus. *Note:* Because none of the read or write lines go low, no device (IC) produces or accepts the data; therefore, no transfer takes place. The exception is that the PIC, upon receiving its second $\overline{INTA}$ signal, has already released its vector code onto the data bus. The 8088 MPU reads the vector code during the second $\overline{INTA}$ signal.

T3 Interrupt Acknowledge (Second of Two from 8088)

1. Status lines go passive; this informs the bus controller that it must end the sequence within one more clock cycle.

T4 Interrupt Acknowledge (Second of Two from 8088)

1. $\overline{\text{INTA}}$ sets to inform the PIC of the end of the first $\overline{\text{INTA}}$ signal.
2. DEN resets to disable the data path between the local and external buses.
3. The 8088 stops reading the vector code on the data bus.
4. DT/$\overline{\text{R}}$ sets to its normal state.

This concludes the $\overline{\text{INTA}}$ sequence as far as the bus controller is concerned. The servicing of the interrupt is now up to the 8088.

IORC: Input/Output Read Command (Fig. 3-12). The $\overline{\text{IORC}}$ sequence is performed any time the 8088 needs information from an I/O device; this information may be data or status information. The following are I/O devices on the motherboard: 8259 programmable interrupt controller, 8237 direct memory access controller, 8254 programmable interval timer, and 8255 programmable peripheral interface. Other I/O devices are any adapter card connected to the I/O bus channel; typical devices are a video adapter, floppy diskette adapter, printer adapter, and asynchronous adapter. I/O read and write bus cycles normally take more than four clocks, because the signals (data and addressing) must physically pass through more devices to perform the transfer; each device produces more time delays.

T4 of the Previous Bus Cycle

1. The 8088 places status on its output status lines.
2. The 8288 bus controller reads the levels on the status lines to determine operation.

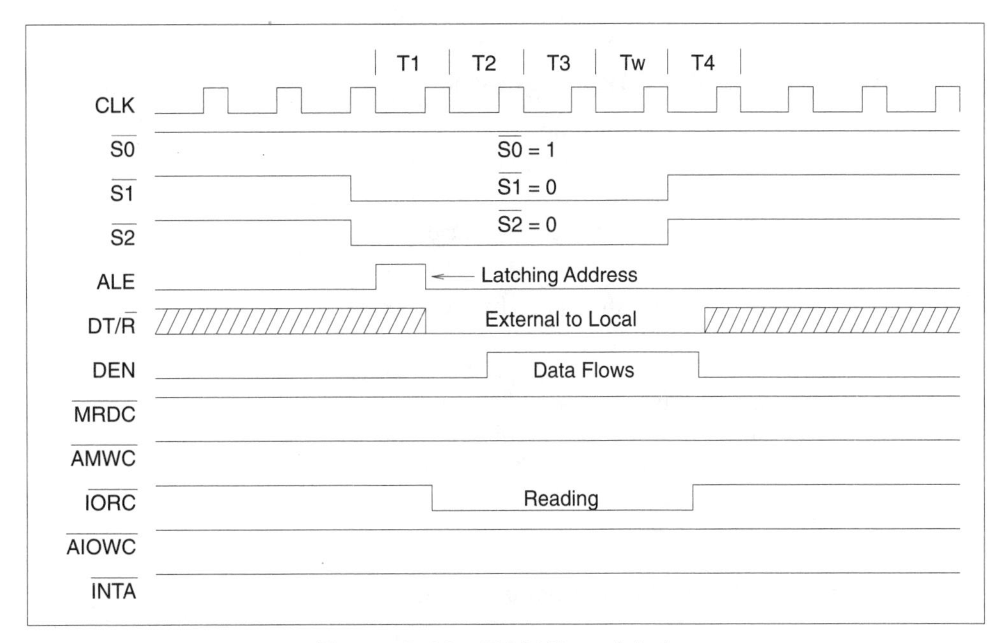

Figure 3-12 8288 I/O read timing.

T1 of I/O Read Bus Cycle

1. The ALE is pulsed; this causes the address on the multiplexed address/data/status (local) bus to be loaded into the address latches. The address is decoded and accesses the device in question.
2. Direction is selected; the DT/$\overline{\text{R}}$ line is always low during a read operation. This allows data to flow from the external data bus to the local data bus once the data bus transceiver is enabled.
3. The $\overline{\text{IORC}}$ line goes low. This informs the device being addressed to place its data on the data bus.

T2 I/O Read Bus Cycle

1. The DEN line goes high, enabling the flow of data from the external to the local bus. Data starts to flow from the external to the local data bus.

T3 I/O Read Bus Cycle

1. The 8088 reads its READY input line near the end of T3. As with most I/O operations, in this example the READY line is low; therefore, the 8088 MPU maintains the levels on the status line. Thus, no changes occur in the 8288 bus controller; the data is not ready to be transferred.
   Note: If the data was ready, the MPU would place that status bus in a passive state, which would cause the bus controller to end the cycle within one more clock.

Tw I/O Read Bus Cycle

1. Because the MPU did not bring the status lines to a passive state in T3 of the I/O ready bus cycle, the MPU reads the READY line at the end of each Tw (wait clock) until the READY line is high. When the MPU reads a high on the READY line, it causes the status lines to go passive, which informs the bus controller to end the current cycle within one more clock.

   In this example the READY line was high near the end of the first Tw, which caused the MPU to place the status line in a passive state. This change in status was sent to the bus controller, informing it to conclude the operation within one more clock.

T4 I/O Read Bus Cycle

1. The $\overline{\text{IORC}}$ line sets, which informs the device producing the data to stop sending data. The MPU has read the data.
2. The DEN resets to disable the data path between the local and external buses. This ends the operation, the data starts becoming invalid.
3. The DT/$\overline{\text{R}}$ sets its normal level.
4. Near the end of the cycle, a new valid status may appear on the status lines.

AIOWC: Advanced Input/Output Write Command (Fig. 3-13). The $\overline{\text{AIOWC}}$ sequence is performed whenever the 8088 sends information to an I/O device; this information may be data or control information. The $\overline{\text{AIOWC}}$ sequence accesses the same devices as the I/O ready sequence. The advanced output is used on the bus controller to allow more time for the data to reach the selected I/O device because it normally takes more time to write information than to read information.

T4 of the Previous Bus Cycle

1. The 8088 places status on its output status lines.
2. The 8288 bus controller reads the levels on the status lines to determine operation.

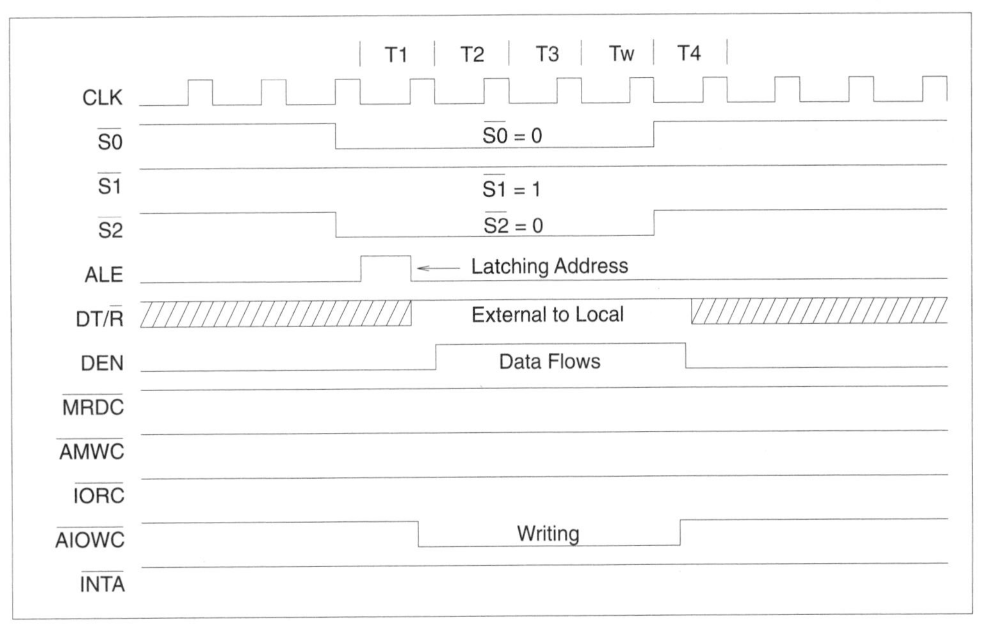

Figure 3-13 8288 I/O write timing.

T1 I/O Write Bus Cycle

1. The ALE is pulsed; this causes the address on the multiplexed address/data/status (local) bus to be loaded into the address latches. The address is decoded and accesses the device in question.
2. Direction is selected; the DT/$\overline{R}$ line is always high during a write operation. This allows data to flow from the local data bus to the external data bus once the data bus transceiver is enabled.
3. The $\overline{AIOWC}$ lines go low. This informs the device being addressed to begin accepting the data on the data bus.

T2 I/O Write Bus Cycle

1. The DEN line goes high at the beginning of T2, enabling the flow of data from the local to the external bus, and data starts to flow.

T3 I/O Write Bus Cycle

1. The 8088 reads its READY input line near the end of T3. As with most I/O operations, in this example the READY line was low; therefore, the 8088 MPU maintains the levels on the status line. Thus, no changes occur in the 8288 bus controller; the data is not ready to be transferred.
   Note: If the data was ready, the MPU would place that status bus in a passive state, which would cause the bus controller to end the cycle within one more clock.

Tw I/O Write Bus Cycle

1. Because the MPU did not bring the status lines to a passive state in T3 of the I/O ready bus cycle, the MPU reads the READY line at the end of each Tw until the READY line is high. When the MPU reads a high on the READY line, it causes the status lines to go passive, which informs the bus controller to end the current cycle within one more clock.

   In this example the READY line was high near the end of the first Tw, which caused the MPU to place the status line into a passive state. This change in status was sent to the bus controller, informing it to conclude the operation within one more clock.

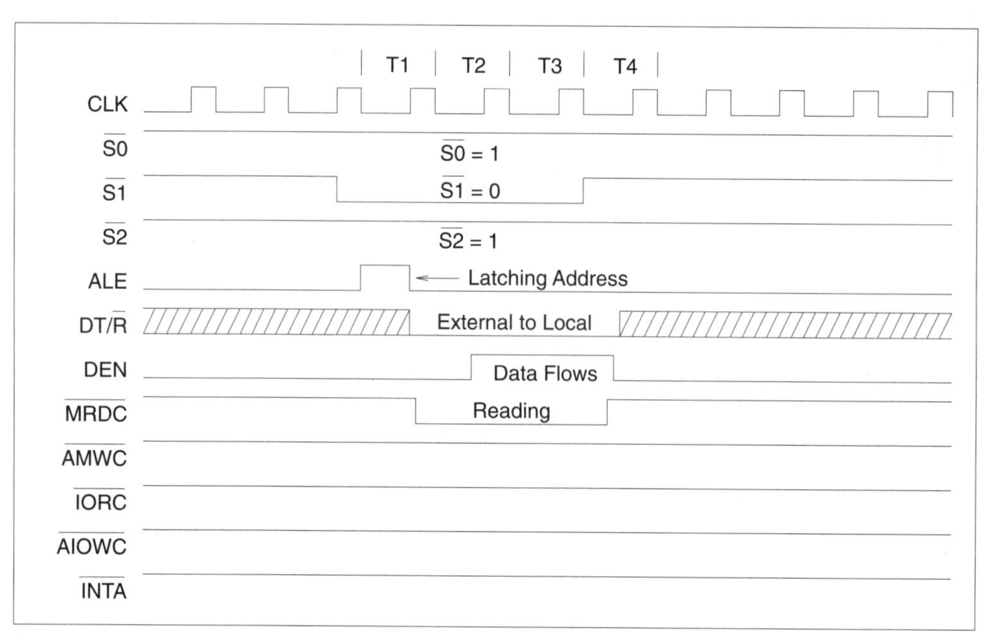

Figure 3-14 8288 memory read timing.

T4 I/O Write Bus Cycle

1. The AIOWC line sets, which informs the device accepting the data to stop accepting data.
2. The DEN resets to disable the data path between the local and external buses. This ends the operation. The data starts becoming invalid.
3. Because the DT/R was set, no change occurs.
4. Near the end of the cycle, a new valid status may appear on the status lines.

MRDC: Memory Read Command (Fig. 3-14). The MRDC sequence is performed whenever the 8088 needs information from memory (RAM or ROM). Memory reads and writes normally do not require any wait clocks for the bus cycle. In some add-on memory cards, wait clocks are produced because of the extra time delays associated with the additional circuitry required by these cards. Also, in many ATs that have MPUs operating at very high speeds, wait states may be inserted into memory operations if the computer is using slow, less expensive memory. The sequence is the same as in the I/O read bus cycle, except that normally no wait clocks are generated and the MRDC line changes instead of the IORC line.

AMWC: Advanced Memory Write Command (Fig. 3-15). The AMWC sequence is performed whenever the 8088 sends (stores) information to memory (RAM only). The sequence is the same as in the I/O write bus cycle, except that normally no wait clocks are generated and the AMWC line changes instead of the AIOWC line.

3.1.20 8259 Programmable Interrupt Controller

The purpose of the 8259 (Figure 3-16) is to accept and encode the priority of eight different hardware interrupts. The PC needs a programmable interrupt controller (PIC) because the PC can be configured with many different types of devices, each requiring an interrupt. Eight 8259s can be cascaded together to supply up to 64 different levels of interrupts, but only one is used in the PC, whereas ATs normally contain two.

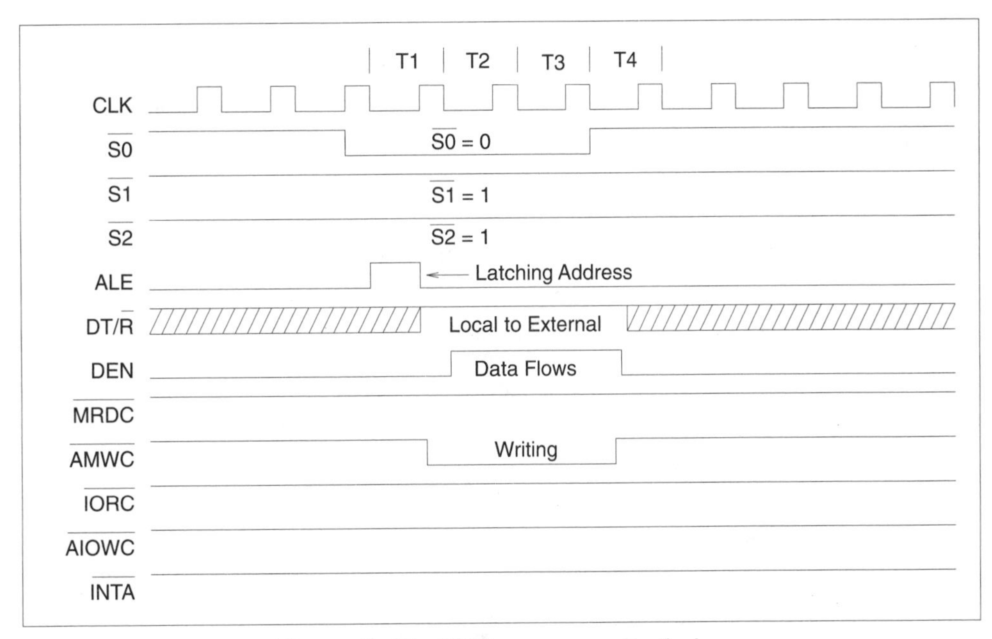

Figure 3-15 8288 memory write timing.

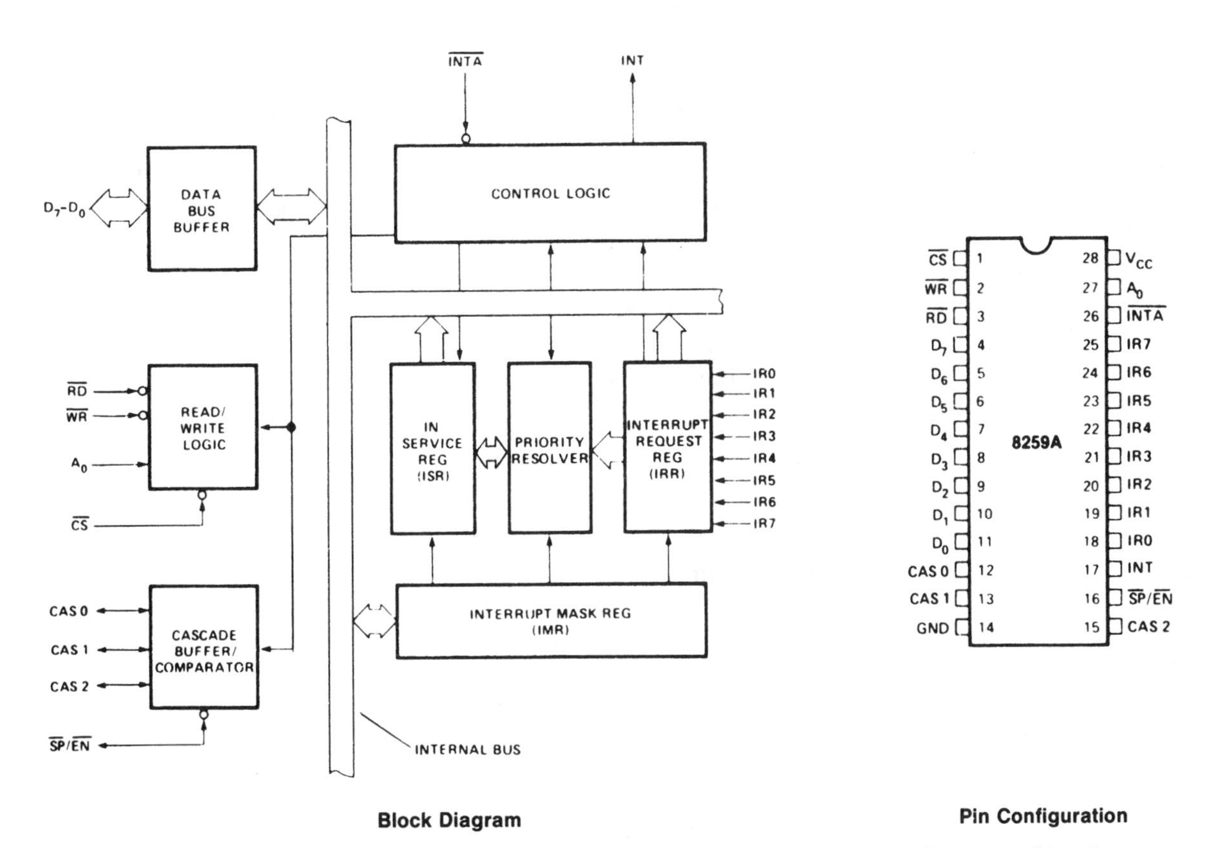

Figure 3-16 8259 block diagram and pin configuration. (Courtesy of Intel)

8259 Block Functions

CBC — The Cascade Buffer/Comparator block allows multiple programmable interrupt controllers to act as one large interrupt controller. This block determines the priority of each controller; the individual controller determines the priority of each input. A maximum of eight controllers can be cascaded together. This block is disabled in the PC because only one 8259 is present in the system.

RWL
The Read/Write Logic block controls the interfacing between the PIC and the 8088. This section of the PIC is used whenever the 8088 programs or reads the status of the PIC. This block is not enabled during the interrupt sequence.

DBB
The Data Bus Buffer block, under the control of the RWL and CL, controls the flow of data between the external and internal data buses of the PIC.

CL
The Control Logic block determines the sequence of operations inside the 8259 during an interrupt sequence. The CL block uses information from the interrupt request register, priority resolver, RWL, and CBC to determine the operation. This section of the PIC is also responsible for the handshaking signals between itself and the 8088 MPU and 8288 bus controller to perform the interrupt sequence.

IRR
The Interrupt Request Register block is an 8-bit buffered register. As one or more of the interrupt request inputs go high, one or more bits are set, which corresponds to the inputs that went high. The bits reset only after their interrupt has been serviced.

PR
The Priority Resolver block reads the value of the IRR and determines which interrupt has the highest priority. The PR sends a code to the in-service register, which represents the interrupt with the highest priority. The PR code is sent to the in-service register only if it is empty or after the interrupt being serviced is complete. Therefore, once an interrupt sequence begins, that interrupt must be serviced before another interrupt request can be serviced, even if the latter interrupt request is higher.

ISR
The In-Service Register holds the interrupt code currently being serviced. When the PIC receives the first $\overline{\text{INTA}}$ signal, the ISR bit corresponding to the IRR bit with the highest priority sets, which then resets the IRR bit, indicating that the interrupt sequence has begun. After the second $\overline{\text{INTA}}$ signal is received and the vector code is released, the ISR bit resets.

IMR
The Interrupt Mask Register block holds the preprogramed interrupt code numbers for each interrupt request input. These code bits are used to mask out selected interrupt inputs in the IRR.

3.1.21 8259 Pin Configurations

D0–D7
The eight bidirectional tristate data lines are used to program the chip and check the status of the chip; a chip select signal must be used during either of these processes. These lines are also used to place the interrupt vector on the data bus at the end of the interrupt acknowledge sequence (no chip select signal is needed).

Condition: Active during programming the PIC, the read status of the PIC, and the end of the second $\overline{\text{INTA}}$ signal of an interrupt sequence.
Result: Normal operation.
Condition: Not active when the computer is first booted up, when $\overline{\text{CS}}$ is low, and when $\overline{\text{WR}}$ is low.
Result: The PIC is not being programed; computer may halt when an interrupt occurs.

Cause: Verify the proper operation of the bus controller 8288.
Cause: Verify the proper operation of the bus transceiver 74245.
Cause: Verify that the data bus is not being loaded down.

Condition: Not active when the computer is reading the status of the PIC, when $\overline{\text{CS}}$ is low, and when $\overline{\text{RD}}$ is low. Computer may halt.
Result: The PIC is not producing status information for the computer.
Cause: See causes above.

Condition: Not active during the end of the second $\overline{\text{INTA}}$ signal.
Result: Computer may halt when an interrupt occurs because of a wrong vector (INT) code.
Cause: Verify programming of PIC.
Cause: Verify bus controller operations.
Cause: Change PIC.

$\overline{\text{CS}}$

The $\overline{\text{Chip Select}}$ input, when low, enables the IC so that the 8088 can program information into the PIC or read its status. This line is not used during an interrupt sequence. The input is controlled by the $\overline{\text{INTR CS}}$ line from the address-decoding section of the system board.

Condition: The $\overline{\text{CS}}$ line is low during programming and reading the status of the PIC.
Result: Normal operation.

Condition: The $\overline{\text{CS}}$ line does not go low during programming and reading the status of the PIC.
Result: Computer halts; PIC is not programmed correctly.
Cause: Verify that the address being used is within the memory map range of the PIC.
Cause: Verify the operation of the bus controller.
Cause: Troubleshoot address-decoding section.

$\overline{\text{WR}}$

The $\overline{\text{WRite}}$ input pin is used to cause the PIC to accept data (write operation) from the data bus for the purpose of programming when $\overline{\text{CS}}$ and A0 are at the proper levels. This input has no effect on the interrupt sequence. The 8088 via the program located in BIOS ROM must program this IC before the PC can use this chip.

Condition: The $\overline{\text{WR}}$ line is low during programming of the PIC.
Result: Normal operation.

Condition: The $\overline{\text{WR}}$ line does not go low during programming of the PIC.
Result: Computer halts when an interrupt occurs. The PIC is not programmed correctly.
Cause: Verify the operation of the bus controller.

$\overline{\text{RD}}$

The $\overline{\text{ReaD}}$ input pin is used to cause the PIC to produce data (read operation) for the data bus, indicating the status of the registers selected by the level on A0 and when $\overline{\text{CS}}$ is low. This input has no effect on the interrupt sequence.

Condition: The $\overline{\text{RD}}$ line is low while reading the status of the PIC.
Result: Normal operation.

Condition: The $\overline{\text{RD}}$ line does not go low while reading the status of the PIC.
Result: Computer may halt when an interrupt occurs. The PIC may not be programmed correctly.
Cause: Verify the operation of the bus controller.

A0 — The Address Zero input, along with the $\overline{CS}$ signal, and the read and write lines determine which of the internal registers of the 8259 is to be accessed.

IR0-7 — The Interrupt Request inputs (IR0 to 7) allow up to eight different interrupts to be interpreted and processed. Each interrupt input has a different priority and a different 1-byte INT value (vector code) for the 8088. The priority of each input is programmed by BIOS when the PC is powered up. In the PC, IR0 has the highest priority and IR7 the lowest priority.

Table 3–5 lists the eight different IR inputs and their priority levels, interrupt numbers, and types for a PC.

Condition: One or more of the IRx lines goes high when a device requires an interrupt service routine.
Result: Normal operation.

Condition: Corresponding IRx line does not go low when the device is requesting an interrupt service routine.
Result: The device requesting the interrupt cannot be serviced and the computer may halt.
Cause: Verify that the device is actually producing the interrupt request level (high).
Cause: Check the configuration of the software and verify that the correct device is being used.

INT — The INTerrupt output informs the 8088 to start the interrupt service routine process. This output stays high until the second $\overline{INTA}$ pulse.

Condition: The INT output goes high once one or more of the IR lines go high and resets during the second $\overline{INTA}$ signal.
Result: Normal operation.

Condition: The INT output does not go high once one or more of the IR lines go high.
Result: Computer may halt.

TABLE 3–5
8259 Interrupt Table

IRQ	*Priority*	*Vector No.*	*Type*
IR0	0	08H	Timer (time of day)
IR1	1	09H	Keyboard
IR2	2	0AH	Reserved (I/O ready)
IR3	3	0BH	Communications SDLC communications BSC communications Primary cluster
IR4	4	0CH	Asynchronous communications printer (COM1) SDLC Communications
IR5	5	0DH	Disk (hard disk)
IR6	6	0EH	Diskette (floppy disk)
IR7	7	0FH	Printer Alternate cluster

Cause: Change PIC; INT output may not be producing enough current for the 8088.
Cause: The INT input of the 8088 MPU may be loading down the INT output of the PIC.

$\overline{\text{INTA}}$ The $\overline{\text{INTerrupt Acknowledge}}$ input indicates that the 8088 has acknowledged the interrupt via the 8288 bus controller. The 8088 informs the bus controller, which informs the interrupt controller, that the interrupt is acknowledged by the 8088.

Condition: The $\overline{\text{INTA}}$ line goes low during each interrupt acknowledge from the 8088 via the bus controller.
Result: Normal operation.
Condition: The $\overline{\text{INTA}}$ line does not go low during an interrupt acknowledge from the 8088 via the bus controller.
Result: Computer may halt; interrupt cannot be serviced because the interrupt sequence is not correct.
Cause: Verify the $\overline{\text{INTA}}$ from the 8088; check status lines $\overline{\text{S0}}$ through $\overline{\text{S2}}$.
Cause: Verify that the bus controller is operating correctly during an interrupt acknowledge.

Cascaded Control Lines

CAS0-2 The CAScade inputs determine the order of priority of each 8259 interrupt controller in the system, if more than one exist. Because the IBM PC has one 8259, these lines are not connected.

$\overline{\text{SP}}/\overline{\text{EN}}$ The $\overline{\text{Slave Program}}/\overline{\text{ENable}}$ buffer input serves two functions. In the buffered mode, this input determines the operation of the data buffers. When the 8259 operates in the nonbuffered mode, the following conditions exist: If low, it indicates a 8259 slave. In the IBM PC this input is high, indicating that the 8259 is the master interrupter controller.

3.1.22 8259 Timing Diagram

Figure 3–17 is a timing diagram of a typical interrupt sequence, and a listing of what actually occurs during an interrupt follows. The sequence begins with one or more of the interrupt request lines going high.

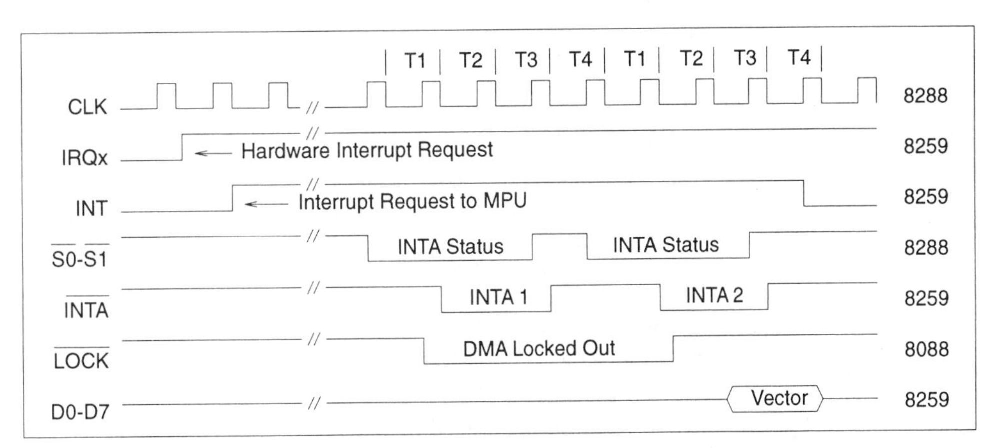

Figure 3-17 8259 interrupt timing.

1. One or more of the interrupt request lines go high. This indicates an interrupt request. The IRR sets the appropriate bit or bits.
2. The PR determines which request input has the highest priority. The INT output goes high.
3. The 8088 sends two successive $\overline{INTA}$ codes to the 8288. The amount of time it takes for the 8088 to send the first $\overline{INTA}$ code varies with the operation currently being executed by the 8088. If the 8088 is idle, the first $\overline{INTA}$ code may be sent out within one clock. If the 8088 is processing an instruction, the first $\overline{INTA}$ code is not issued to the bus controller until the instruction is complete. The 8088 at this time also locks out any other bus master (DMA) by bringing the $\overline{LOCK}$ line low.
4. The first $\overline{INTA}$ code disables the external bus by not allowing any read or write command lines to go low. Then the 8288 sends an $\overline{INTA}$ signal (low) to the 8259.
5. Upon receiving the first $\overline{INTA}$ signal from the 8288, the 8259 sets the ISR bit determined by the PR and resets the bits in the IRR. The first interrupt acknowledge takes a minimum of four clocks.
6. After receiving the second $\overline{INTA}$ signal from the 8088, the 8288 keeps the external bus disabled. The 8288 then sends out another $\overline{INTA}$ signal (low) to the 8259. During T2 of the second interrupt acknowledge, the 8088 unlocks the $\overline{LOCK}$ line (high).
7. Upon receiving the second $\overline{INTA}$ signal, the 8259 releases the 8-bit vector code to the data bus. At the end of the second $\overline{INTA}$ signal, the ISR bit resets, the INT output resets, and the 8088 reads the vector code. This completes the interrupt sequence as far as the 8259 is concerned.
8. Once the 8088 reads the 8-bit vector code, the code is used internally to generate an effective address.

   Note: This 8088 is set up internally to process up to a maximum of 256 interrupts; the vector code that is read acts as an INT for the 8088. Each INT or vector code represents four memory locations (the first 1K of DRAM) in the first page of memory (page 0).

   The first eight interrupts in the 8088 are internally predefined; all remaining interrupts are generated by either the 8259 or software.

   Once the INT signal is developed, the computer pushes the contents of the IP and CS register onto the stack. The 8088 then multiplies the interrupt number by 4 and uses that value to read the 4 bytes of data from memory (page 0). The first byte read is transferred into the lower byte of the IP, and the next byte is transferred into the upper byte of the IP. The third byte is transferred into the lower byte of the CS register, while the fourth byte is transferred into the high byte of the CS register. Once the data is in these registers, the 8088 uses this effective address to begin processing the interrupt service routine.

EXAMPLE 3-1

IR1 = 1, vector = 09 hex, device = keyboard:

$09 \times 04 = 36$ memory locations

Memory location 36 = 23 in hex

MLoc 23 data into low byte of IP
MLoc 24 data into high byte of IP
MLoc 25 data into low byte of CS
MLoc 26 data into high byte of CS

The amount of time it takes the 8088 to complete the interrupt service routine depends on the specific routine. The 8088 does not accept any other INT for processing until the 8088 completes the current interrupt service routine. At the end of the routine, the hardware requesting the interrupt is reset, and the 8088 sets the interrupt enable flag and returns the status of CS and IP to their values before the interrupt.

Table 3-6 lists the 8259 interrupts, their vector numbers, and the memory locations that contain the starting address values for the interrupts.

3.1.23 74373 Address Latch

The 74373 (Figs. 3-18 and 3-19) is a TTL octal D-type transparent latch. It holds the address produced by the multiplexed local bus for the demultiplexed external bus. The PC requires this function because the local bus produces the complete address only during T1 of each bus cycle, even though that valid address is needed during the complete bus cycle. The inputs of the latches are connected to the multiplexed local bus; the outputs of the latches produce the address on the demultiplexed external bus.

Whatever level is on each of the data inputs is allowed into the latch as long as the enable C line is high. As soon as the enable C line goes low, the level that was last on the data inputs is latched (stored) into the data latch and does not change until the enable C line goes high again or power is removed. The bus controller controls the latching by the use of the ALE line, which is pulsed low–high–low during T1 of the bus cycle.

The $\overline{OC}$ (Output Control) line, when low, enables the tristate output of the data latch, allowing the data stored inside the latch to be applied to the output of the latch. In the IBM PC and clones, this line is developed by the external control logic section and is called the AEN BRD (Address ENable BoaRD). This line stays low unless another bus master is in control of the bus. Therefore, as soon as the enable C line goes high, the outputs change. If another bus

TABLE 3-6
8259 Interrupt Vector Table

Interrupt	*Vector No.*	*Address Range*
IRQ0	08 hex	00020–00023 hex
IRQ1	09 hex	00024–00027 hex
IRQ2	0A hex	00028–0002B hex
IRQ3	0B hex	0002C–0002F hex
IRQ4	0C hex	00030–00033 hex
IRQ5	0D hex	00034–00037 hex
IRQ6	0E hex	00038–0003B hex
IRQ7	0F hex	0003C–0003F hex

'LS373, 'S373
FUNCTION TABLE

OUTPUT ENABLE	ENABLE LATCH	D	OUTPUT
L	H	H	H
L	H	L	L
L	L	X	Q_0
H	X	X	Z

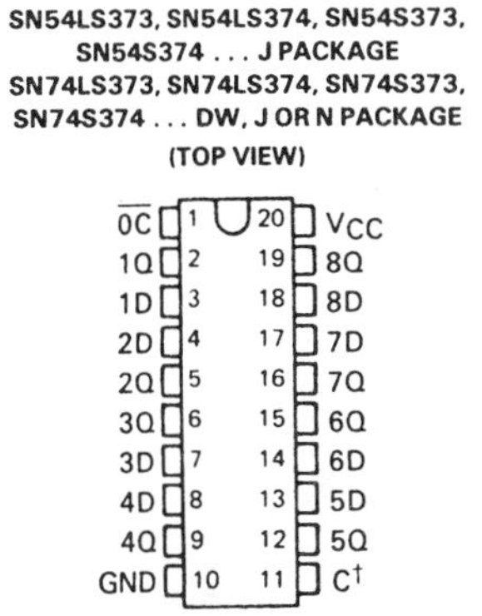

SN54LS373, SN54LS374, SN54S373,
SN54S374 . . . J PACKAGE
SN74LS373, SN74LS374, SN74S373,
SN74S374 . . . DW, J OR N PACKAGE
(TOP VIEW)

Pin	Name	Pin	Name
1	$\overline{OC}$	20	VCC
2	1Q	19	8Q
3	1D	18	8D
4	2D	17	7D
5	2Q	16	7Q
6	3Q	15	6Q
7	3D	14	6D
8	4D	13	5D
9	4Q	12	5Q
10	GND	11	C†

Figure 3-18 74373 truth table and pin configuration. (Reprinted by permission of copyright holder, Texas Instruments, Incorporated, 1985)

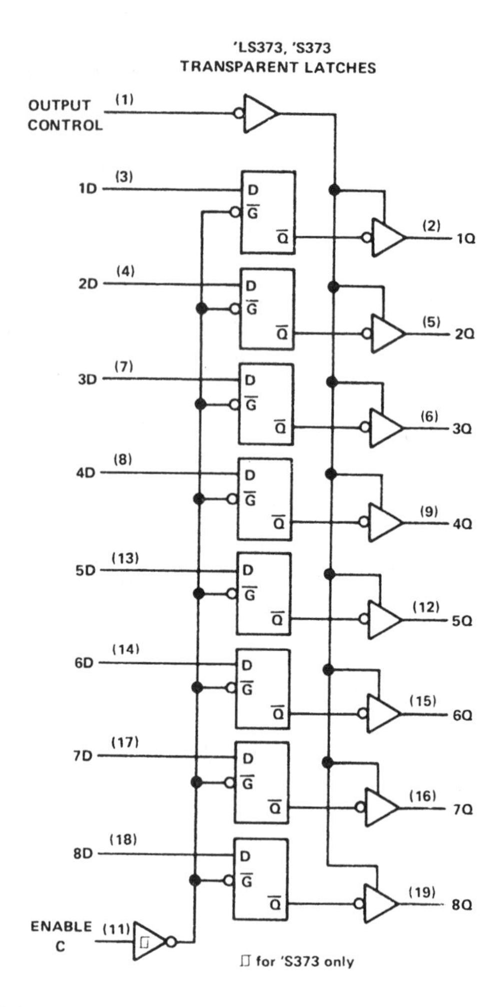

Figure 3-19 74373 block diagram. (Reprinted by permission of copyright holder, Texas Instruments, Incorporated, 1985)

master is controlling the bus, the $\overline{OC}$ line is high, which causes the outputs of the data latch to go into tristate, making the outputs transparent to the external address bus.

See the timing diagrams for the read and write bus cycles for the operation of this IC.

3.1.24 74245 Data Bus Transceiver

The 74245 (Figs. 3-20 and 3-21) octal bus transceiver determines the direction and timing of data flow between the external and local buses. In the PC the multiplexed AD0 to AD7 bus is connected to the eight A inputs/outputs of the bus transceiver. The eight B inputs/outputs of the bus transceiver are connected to the demultiplexed external data D0 to D7 bus.

The DIR (DIRection) input of the 74245 is used to select the direction of data (current) flow. When high, the direction of flow is from side A to side B once the transceiver is enabled. When low, the direction of flow is from side B to side A once the transceiver is enabled. In the PC, the DIR input is controlled by the DT/$\overline{R}$ (data transmit/receive) output from the bus controller. During a read operation, data may flow from the external to the local bus when the transceiver is enabled (MPU is receiving data). During a write operation, data may flow from the local to the external bus when the transceiver is enabled (MPU is transmitting data). Remember that the DIR input only selects the direction of data once the transceiver is enabled. The $\overline{G}$ (Gating or enabling) input is used to enable the data flow once a direction has been selected. When low, data is allowed to flow in the selected direction. When high, no data flows and the transceiver becomes transparent to the multiplexed local bus (AD0 to AD7) and the external data bus (D0 to D7). In the PC, the DEN output is inverted by a NAND inverter and is applied to the $\overline{G}$ input of the bus transceiver. The NAND inverter is used to keep the bus

SN54LS245 . . . J PACKAGE
SN74LS245 . . . DW, J OR N PACKAGE
(TOP VIEW)

Pin	Name	Pin	Name
1	DIR	20	VCC
2	A1	19	$\overline{G}$
3	A2	18	B1
4	A3	17	B2
5	A4	16	B3
6	A5	15	B4
7	A6	14	B5
8	A7	13	B6
9	A8	12	B7
10	GND	11	B8

FUNCTION TABLE

ENABLE $\overline{G}$	DIRECTION CONTROL DIR	OPERATION
L	L	B data to A bus
L	H	A data to B bus
H	X	Isolation

H = high level, L = low level, X = irrelevant

Figure 3-20 74245 truth table and pin configuration. (Reprinted by permission of copyright holder, Texas Instruments, Incorporated, 1985)

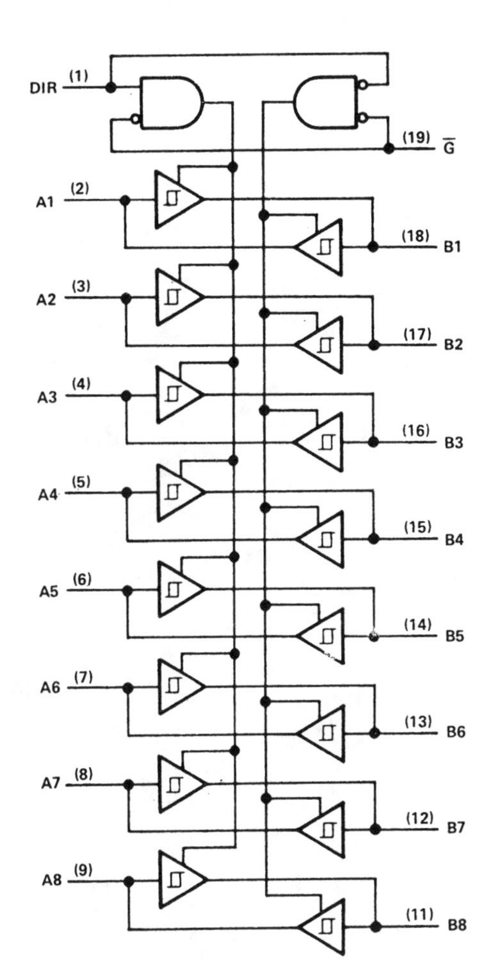

Figure 3-21 74245 block diagram. (Reprinted by permission of copyright holder, Texas Instruments, Incorporated, 1985)

transceiver disabled in the event that the +5-V supply is not within tolerance, thus allowing no data to pass when the IC may be on the borderline of operating voltages. When the DEN line of the bus controller goes high, a low is produced on the $\overline{G}$ input, which enables the bus transceiver, allowing data to flow in the selected direction. When the DEN line is low, a high is produced on the $\overline{G}$ input, which places the bus transceiver in tristate, which effectively blocks any transfer of data between the local and external buses.

3.1.25 Microprocessor Subsystem Timing Diagrams

Each bus cycle requires at least four clock cycles from the 8284 clock generator-driver. If the device being accessed is slow, additional clock cycles are added to the bus cycle. A bus cycle relates to one data transfer in either the read or write mode of operation. Memory operations normally take only four clocks to perform a bus cycle, whereas bus cycles involving I/O devices usually take more than four clock cycles. The clocks in each bus cycle are labeled T1, T2, T3, Tw, and T4, where T1 is the first clock of the bus cycle, T4 is always the last clock cycle, and Tw stands for any wait clocks generated by the 8088 when the device being accessed is slow. Minimum bus cycles contain only four clocks, and nonminimum bus cycles contain the four standard clocks plus one or more wait clocks.

Minimum Memory Write Bus Cycle (Fig. 3-22). Normally, memory write cycles require only four clocks to produce the transfer. The data from the local bus is transferred to the external data bus during this operation.

T1 The 8088 sends out a status signal (memory write) to the bus controller (8288) prior to T1 of the current bus cycle. The levels on the status must be stable by the beginning of T1 and stay valid to at least T3.

While the 8088 is placing the status on the status lines, it places the address on the multiplexed local bus (AD0 to AD7, A8 to A15, and A16 to A19). The levels representing the address must be stable during T1 of the bus cycle.

At the beginning of T1, the 8288 brings the ALE line high. Because the external control logic keeps the outputs of the address latches enabled except when another bus master is using the bus, the levels on the

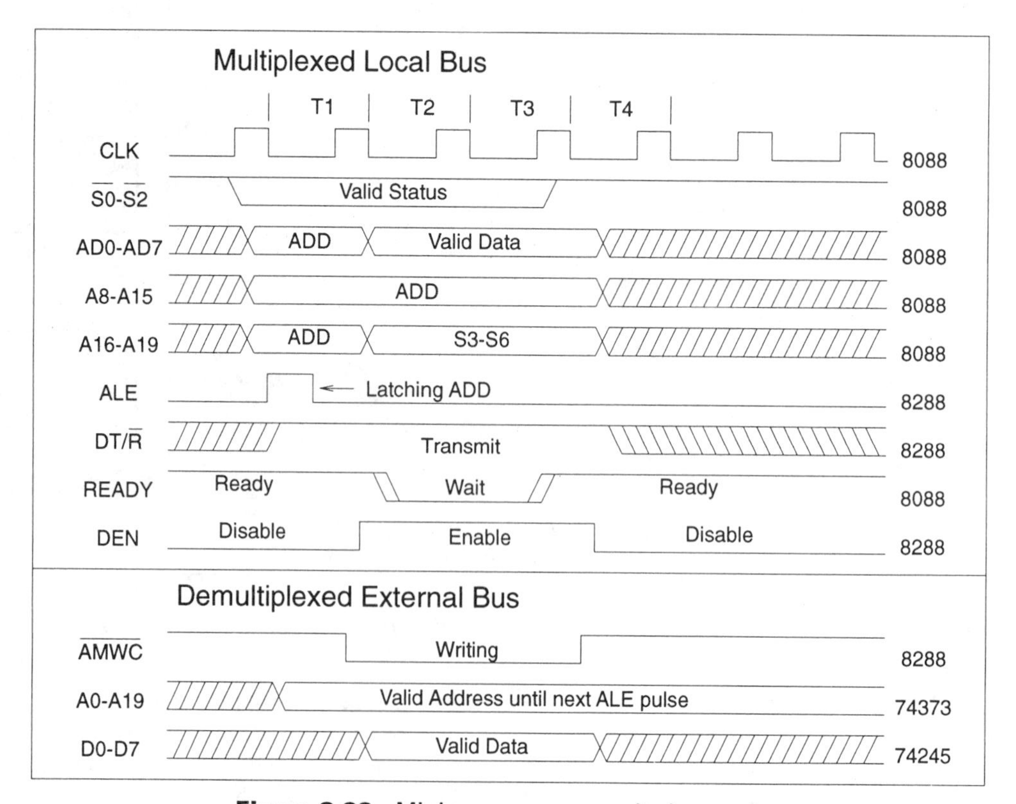

Figure 3-22 Minimum memory write bus cycle.

multiplexed local bus (currently representing the address) are loaded into the address latches and displayed on the demultiplexed external address bus. The levels on the multiplexed local bus remain unchanged until the end of T1. Prior to the end of T1 (about 115 ns), the 8288 brings the ALE line low; during this transition, whatever levels were on the multiplexed local bus (current address) are latched into the address latches.

Shortly after the 8288 brings the ALE line high, it brings the DT/$\overline{R}$ line to its proper level. In this example, the DT/$\overline{R}$ line must go high and stay high throughout the bus cycle. The direction of data flow will be from the local to the external bus; no data flows, however, until the bus transceiver is enabled by the DEN line.

At the end of T1 or beginning of T2, the 8288 brings the $\overline{AMWC}$ line low; this line stays low throughout the bus cycle. This line informs the device being accessed that it should accept the data on the data bus.

Also at the end of T1, AD0 to AD7 of the local bus begins to change from producing the lowest 8 bits of the address to representing data from the 8088. Also, A16 to A19 changes from representing the segment value of the address to segment register status. Because no device in the PC uses this information, it is of no importance.

T2

The 8288 brings the DEN line high, which places a low on the $\overline{G}$ input of the bus transceiver. This low causes the bus transceiver to enable data flow in the direction that was selected by the DT/$\overline{R}$ line. The data now flows from the local bus (AD0 to AD7) to the external data bus (D0 to D7) into the device being accessed.

The READY signal produced by the 8284A is applied to the 8088 and goes low somewhere during T2. When low, this line indicates to the 8088 that the process of accessing the selected device is in progress but is not complete. The levels on all other lines in this example remain constant.

T3

Because in this sample timing diagram the READY line goes high before the end of T3, only four clocks are required to complete this bus cycle.

The 8088 reads the READY line at the end of T3; this line is high, indicating to the 8088 that the device being accessed is ready to complete the bus cycle.

Once the 8088 reads the READY line and it is high, the 8088 brings the status lines ($\overline{S0}$ to $\overline{S2}$) to a passive state (all logic highs). This passive state informs the 8288 bus controller that it must complete the bus cycle within one more clock.

T4

Near the beginning of T4, the bus controller causes the $\overline{AMWC}$ line to set. This causes the device being accessed to stop accepting data from the data bus. At this time, whatever is on the data bus is stored into the selected device. This basically concludes the write operation.

Although the write operation has ended, the 8288 bus controller brings the DEN line low, which disables the data path between the local bus and external data bus. This is done to protect the devices on both the local and external buses.

The following lines may or may not change in T4 depending on what, if any, the next operation is: AD0 to AD7, A8 to A15, A16 to A19, and

DT/$\overline{R}$. If another operation is pending at the end of T4, AD0 to AD7, A8 to A15, A16 to A19, and the status lines ($\overline{S0}$ to $\overline{S2}$) may change.

Nonminimum I/O Write Bus Cycle (Fig. 3-23). I/O write cycles may or may not contain one or more wait clocks (Tw). The number of wait clocks required, if any, depends on the device being accessed and the speed of that device. As in the memory write cycle, the data is transferred from the local bus to the external data bus. All I/O devices are found in page 0 of the memory map. When performing an I/O operation, the 8088 always sets the segment value to zero.

T1 — The same sequence occurs except that the status indicates an I/O write instead of a memory write, the $\overline{AIOWC}$ line goes low instead of the $\overline{AMWC}$ line going low, and the segment value of the address is always zero.

T2 — The same sequence occurs.

T3 — In this example, the READY line does not go high before the end of T3. When the 8088 reads the READY line at the end of T3 and it is still low, it indicates to the 8088 that the device being accessed is not ready to complete the transfer. Therefore, the 8088 does not change the status lines to a passive status. Instead, the 8088 keeps the status valid and does nothing for one complete clock. This time period during which the 8088 is doing nothing is called a wait clock (Tw).

Tw — At the end of the first wait clock, the 8088 reads the READY line again. If the READY line is still low, the 8088 waits for one more clock period without changing the status lines and continues this sequence until the READY line goes high.

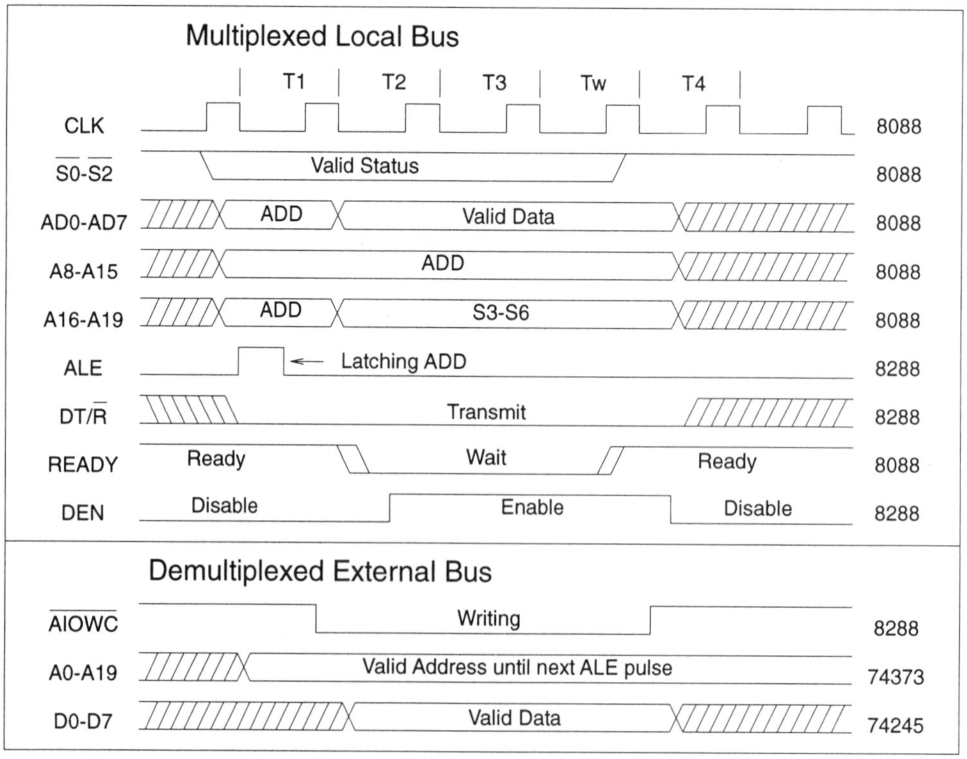

Figure 3-23 Nonminimum I/O write bus cycle.

Since, in this example, the READY line is high by the end of the Tw, the 8088 knows that the device being accessed is ready to complete its transfer.

The 8088 brings the status lines into a passive state, which informs the bus controller to end the bus cycle within one more clock. The final clock of the bus cycle is always labeled T4, no matter how many clocks make up the bus cycle.

T4 — The same sequence occurs except that the $\overline{\text{AIOWC}}$ line goes high, rather than the $\overline{\text{AMWC}}$ line; this ends the write operation.

Minimum Memory Read Bus Cycle (Fig. 3-24). The read bus cycle operates in much the same way as the write bus cycle except that the data flow from the external bus to the local bus. Again, memory operations normally require only four clocks to complete the transfer. The following explains the differences between the memory write bus cycle and the memory read bus cycle.

T1 — Same sequence as for the memory write bus cycle, with the following exceptions:

The status line shows a memory read (or code access) instead of a memory write status.

The DT/$\overline{\text{R}}$ line goes low instead of high. The direction allows data to flow from the external data bus to the local bus when the bus transceiver is enabled.

The $\overline{\text{MRDC}}$ line goes low instead of the $\overline{\text{AMWC}}$.

T2 — Same sequence as for the memory write bus cycle, except that the DEN line does not go high until about halfway through T2.

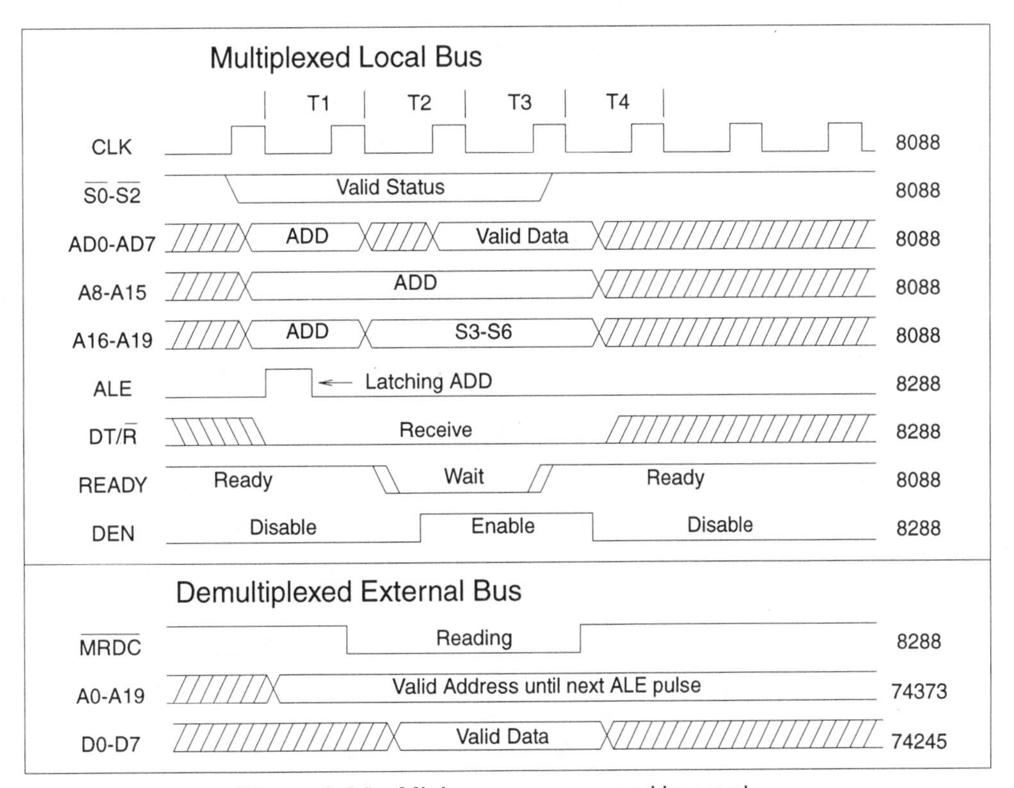

Figure 3-24 Minimum memory read bus cycle.

T3 Same sequence as for the memory write bus cycle.

T4 Same sequence as for the memory write bus cycle, with the following exception: The $\overline{MRDC}$ line goes high to end the read operation.

Nonminimum I/O Read Bus Cycle (Fig. 3-25). As with I/O write cycles, an I/O read cycle may or may not contain one or more wait clocks. The number of waits depends on the device being accessed. The following lists the differences between I/O write and I/O read cycles.

T1 The same sequence occurs as in the I/O write cycle except that status indicates an I/O read instead of an I/O write. The $\overline{IORC}$ line goes low instead of the $\overline{AIOWC}$ line, and the DT/$\overline{R}$ line goes low instead of going high.

T2 Same sequence as for the I/O write bus cycle, except DEN goes high halfway through T2.

T3 Same sequence as for the I/O write bus cycle.

Tw Same sequence as for the I/O write bus cycle.

T4 The same sequence occurs as in the I/O write bus cycle except that the $\overline{IORC}$ line goes high, rather than the $\overline{AIOWC}$ line, which ends the read operation.

3.1.26 Troubleshooting the Microprocessor Subsystem

The following is a list of key signals in the microprocessor subsystem and the effect that each has on the computer system. Remember to check the output of the IC for the signal and, if it is missing, bad, or improper, to check the input signals necessary for the signal.

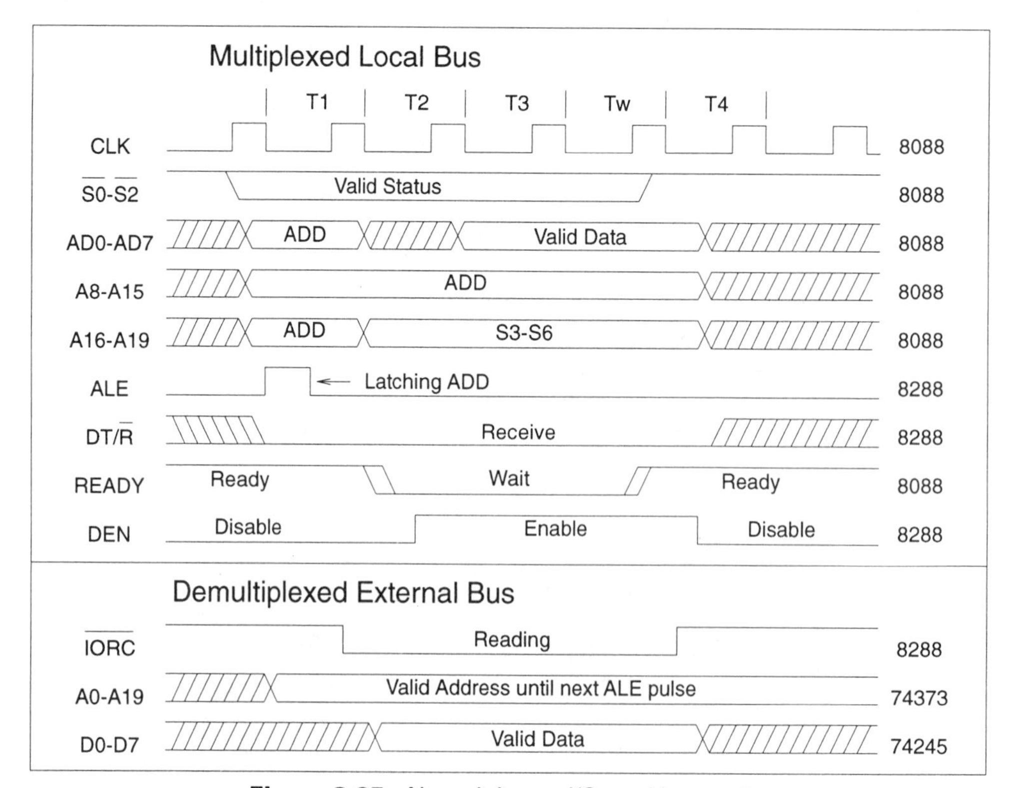

Figure 3-25 Nonminimum I/O read bus cycle.

If the computer fails to boot, the signal is probably completely bad or is stuck at the wrong level. A logic probe and/or DMM (digital mutimeter) can be used to check the levels.

If the computer boots but fails to operate properly, the signal is probably incorrect and not completely bad. An oscilloscope or logic analyzer would be the best piece of test equipment to use for this type of check.

Computer Fails to Boot

1. Bad Vcc.
2. Bad CLK88; check 8284.
3. Bad PCLK; check 8284.
4. Bad READY; check 8284.
5. Bad RESET; check 8284.
6. Bad ALE; check 8288.
7. Bad DT/$\overline{\text{R}}$; check 8288.
8. Bad CEN; check 8288.
9. Bad DEN; check 8288.
10. Bad $\overline{\text{AEN}}$; check AEN 0 BRD.
11. Bad $\overline{\text{INTA}}$; check 8288.
12. Bad $\overline{\text{IORC}}$, $\overline{\text{AIOWC}}$, $\overline{\text{MRDC}}$, or $\overline{\text{AMWC}}$; check 8288.
13. Bad INT; check 8259.

Computer Boots but Fails to Operate

Check numbers 2 to 13 above.

MICROPROCESSOR SUBSYSTEM SUMMARY

1. The 8284A develops the following important signals:
   a. The RESET signal for the computer; when high, all smart ICs in the computer reset. When low, the computer system operates normally.
   b. The OSC output is the frequency of the crystal applied to the crystal inputs to the 8284A, with a 50% duty cycle. Levels of the OSC are TTL compatible. This output is the only supply to the I/O channel bus and is not used by any IC on the motherboard.
   c. The CLK output is one-third of the frequency of either the crystal frequency (non-turbo mode) or the EFI input (turbo mode), with a 33% duty cycle. This output is supplied to all ICs that must be synchronized with bus operations (8088, 8087, 8288, and 8237).
   d. The PCLK output has a frequency one-half of the CLK frequency, with a 50% duty cycle. This output is supplied to the 8254 programmable interval timers section of the computer.
   e. The READY output indicates the readiness of the bus to start a new instruction or complete the current instruction. The READY line, when low, places the MPU in a wait state.
2. The 8088 MPU is the main microprocessor and the main bus master in the computer. If this IC goes bad, the computer cannot boot. The 8088 is responsible for processing data for the computer system. The following is a brief description of some of the more important pin functions:

a. AD0 to AD7: The address/data lines supply the low byte of the address offset beginning at the end of T4 and throughout T1. During the remaining part of the bus cycle, these lines act as the bidirectional data lines.
b. A8 to A15: These address lines produce the high byte of the address offset throughout the entire bus cycle.
c. A16 to A19: These address lines produce the high nibble of the address (segment value) beginning at the end of T4 and throughout T1. During the remaining part of the bus cycle, these lines produce segment status (not used by the IBM).
d. The READY input indicates the readiness of the bus to start a new cycle or end the current bus cycle. When high, the bus is READY; when low, the bus is in the process of accessing a device.
e. The INTR (interrupt request) informs the 8088 that a maskable interrupt needs to be serviced. This signal is developed by the 8259 PIC.
f. The NMI (nonmaskable interrupt) informs the 8088 that a nonmaskable interrupt has occurred (parity error).
g. The RESET line resets the 8088 when high for four clock cycles.
h. The CLK input supplies the synchronizing frequency to the 8088.
i. The $\overline{S0}$, $\overline{S1}$, and $\overline{S2}$ lines allow the 8088 to inform the 8288 bus controller which type of bus operation to perform and when to start and stop the bus cycle.
j. The $\overline{LOCK}$ output locks out the DMA from asking permission to use the bus.

3. The 8087 is the math coprocessor that can be used in the PC. The numeric data coprocessor performs operations only on special math instructions. If this IC is used in the computer system, math operations can be processed up to 50 times faster than with the 8088, but the cost of the speed is high (the cost of the IC is as much as for the entire motherboard).
4. The 8288 bus controller is used by the 8088 to control all bus operations. The bus controller allows the 8088 to operate more effectively. The following is a brief description of pin functions.
   a. The CLK input supplies the synchronizing clock to the 8288; this is the same signal supplied to the 8088.
   b. The $\overline{AEN}$ line is used to determine whether the local or external bus is enabled or disabled.
   c. The $\overline{S0}$, $\overline{S1}$, and $\overline{S2}$ lines are inputs that tell the bus controller which type of bus operation needs to be performed and when to start and stop the bus cycle.
   d. The CEN line is used to disable the command outputs (I/O and memory, read and write lines) during certain bus operations.
   e. The DT/$\overline{R}$ line determines the direction of data flow once the bus is enabled.
   f. The DEN line determines when the data will flow.
   g. The ALE line is used to latch the address on the local bus into the address latches in the microprocessor subsystem, which supplies the address to the external address bus.
   h. The $\overline{MRDC}$ and $\overline{IORC}$ lines, when low, cause the addressed device on the external bus to output its data to the 8088.
   i. The $\overline{AMWC}$ and $\overline{AIOWC}$ lines, when low, cause the address device on the external bus to accept the data from the 8088.
   j. The $\overline{INTA}$ line is used to acknowledge an interrupt sequence.
5. The 8259 programmable interrupt controller is used to process up to eight different hardware interrupts for the PC. The main purpose of this IC is to monitor the hardware interrupt request lines and, once an interrupt signal is seen, to determine which interrupt has the highest priority and start the interrupt sequence. During the interrupt sequence with the 8088, the bus controller disables the external bus. Once the external bus has been disabled, the 8259 releases the interrupt vector code on the data bus, which the 8088 reads, and begins processing the selected interrupt.
6. The 74373s are the address latches for the microprocessor subsystem. These ICs are used to hold the address from the local multiplexed address/data bus for the external address bus until the operation is complete.

7. The 74245 is the data bus transceiver used to control the direction and timing of data flow between the local and external data buses.

MICROPROCESSOR SUBSYSTEM REVIEW QUESTIONS

1. Which one of the following 8284A outputs is not used by any IC on the motherboard?
   a. CLK
   b. OSC
   c. PCLK
   d. READY
   e. RESET

2. If the power-good line is high and the RESET output of the 8284A is low, what is wrong with the computer?
   a. One or more of the power-supply voltages are missing or bad.
   b. The 8284A is bad, because the RESET output should be high.
   c. The power-good line is bad, because this line should be low if power is good.
   d. Nothing is wrong with the computer; this is a normal condition.

3. If the frequency of the clock input to the 8088 changes from 4.772727 to 5.0 MHz, what happens to the operation of the computer?
   a. The computer operates normally but a little faster.
   b. The computer locks up (halts) because of the higher speed.
   c. The 8284A is bad.
   d. The 8087 is loading down the CLK88 line.

4. A parity error causes which one of the following 8088 lines to go high?
   a. READY
   b. INTR
   c. NMI
   d. IRQ

5. If the MN/$\overline{\text{MX}}$ pin of the 8088 is always high, the computer operates normally.
   a. True
   b. False

6. During T1 of any bus cycle, AD0 to AD7 represent the data on the data bus.
   a. True
   b. False

7. How does the 8288 bus controller know when to end the current bus cycle?
   a. The READY line to the 8288 goes high.
   b. The status lines from the 8088 go passive.
   c. The status lines from the 8088 go active.

8. The 8288 disables the external bus during an interrupt acknowledge by not allowing any of the read or write command lines to go low.
   a. True
   b. False

9. The only time the $\overline{\text{CS}}$ line of the 8259 goes low is when the vector code is being produced.
   a. True
   b. False

10. The DIR pin of the 74245 goes low during an I/O write operation.
   a. True
   b. False

3.2 EXTENDED CONTROL LOGIC

The extended control logic section of the IBM PC in Figure 3-26 produces 12 high-level control signals. This section is responsible for producing most of the signals necessary to inform

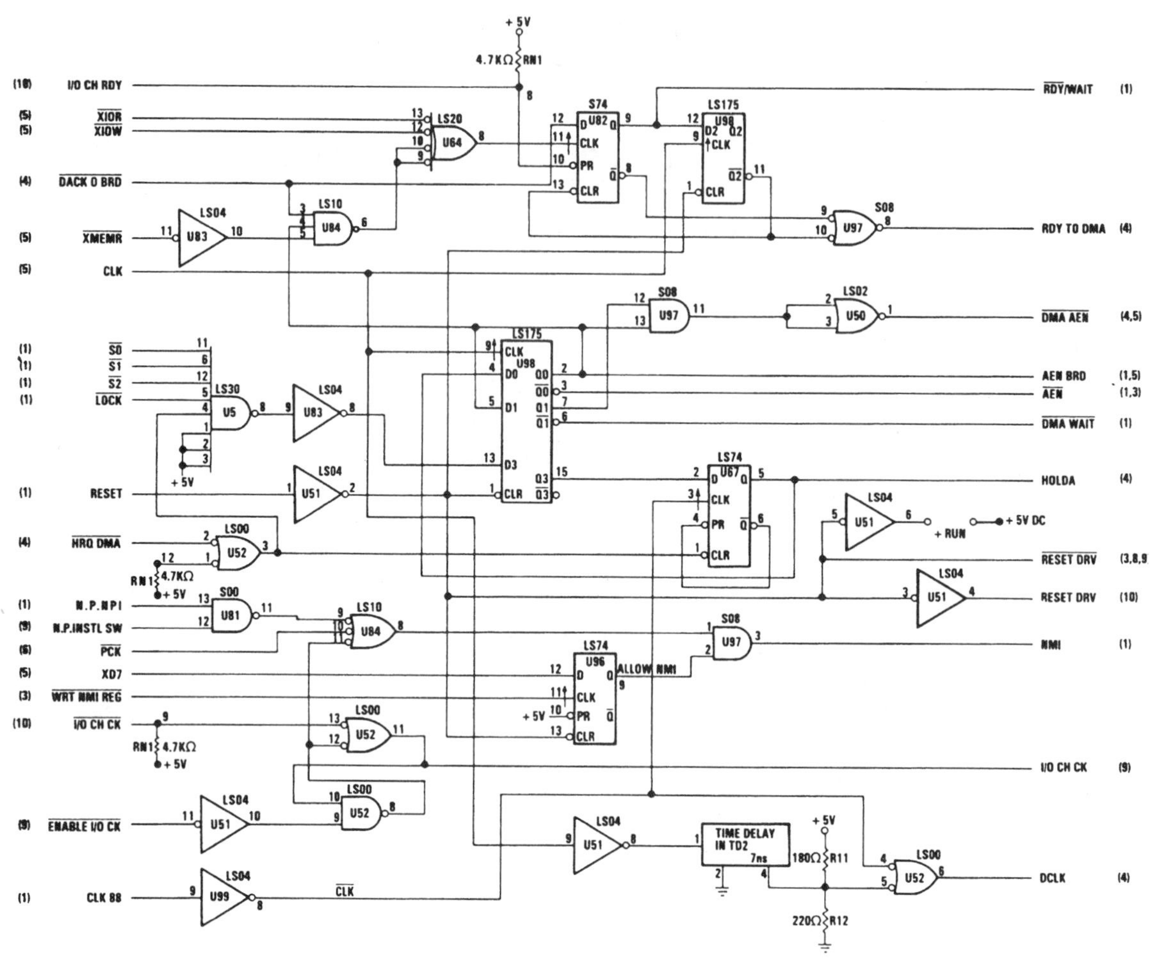

Figure 3-26 IBM PC extended control logic (2 of 10). (Courtesy of IBM)

the MPU of the condition of the bus, wait-state logic, and NMI logic (indicating parity errors). Because each motherboard manufacturer takes a different approach to producing the signals in this section, only the purpose of each output control line is explained.

RDY/WAIT. The ReaDY/WAIT control line is connected to the AEN1 input of the 8284 clock generator. When high, during a valid bus cycle, this line indicates that the memory or I/O device selected for the operation is in the process of being accessed. The high on this input causes the READY output of the 8284 to go low, which causes the 8088 to produce internal wait clocks. During the wait clocks, the status output lines of the 8088 do not change until the READY input line goes high. If wait clocks are produced, the 8088 checks the READY line at the end of each wait clock until the READY line goes high, after which it informs the bus controller to conclude the bus cycle by changing the levels on the output status lines.

The circuitry used to produce this output automatically causes the RDY/WAIT line to go high just after the selected read or write command output goes low. About 300 ns in an IBM PC and 100 to 200 ns in PCs into the memory cycle, this line goes low, indicating that the memory device is ready to complete the data transfer. The time delay is fixed and is determined by the gating and timing in this section of the system board. The reason for the delay is that no memory device on the system board has an access time greater than 300 ns; in fact, the slowest device has a 250 ns (EPROM is used) maximum access time.

If the device being accessed uses the I/O channel ready line, the RDY/WAIT line stays high (wait state) as long as the I/O channel ready line is low. When the I/O channel ready line goes high, the $\overline{\text{RDY}}$/WAIT line goes low, indicating that the I/O device in question is ready to complete the transfer. In most cases, I/O devices are responsible for wait clocks being generated.

The $\overline{\text{LOCK}}$ output pin from the 8088 also keeps the wait generator logic from being triggered again until the 8088 finishes its bus cycle, after which the wait generator logic may be triggered again, allowing the DMA to gain control over the bus.

RDY TO DMA. The ReaDY TO DMA control line indicates the readiness of the bus to the DMA while under a DMA read/write cycle. This signal is developed in the same way as the $\overline{\text{RDY}}$/WAIT signal except that it is inverted. If high, the bus is ready to complete the DMA transfer. If low, it indicates to the DMA that the device it is accessing is not ready to complete the DMA transfer.

$\overline{\text{DMA AEN}}$. The $\overline{\text{DMA Address ENable}}$ control line enables the outputs of the address latches in the DMA section of the IBM PC. If low, the address latches in the DMA section become enabled and control the levels on the external address bus. During this time, the outputs of the address latches in the microprocessor subsystem are transparent (in tristate). If high, the DMA address latches are disabled, and the DMA has no control over the external address bus.

AEN BRD. The Address ENable BoaRD control line enables or disables the address latches in the microprocessor subsystem. If low, the outputs of the address latches in the microprocessor subsystem control the levels on the external address bus (address latches are enabled). If high, the outputs of the address latches in the microprocessor subsystem do not control the levels on the external address bus (latch becomes transparent [tristate]); the other bus master controls the address bus. In the PC, this line is always low except when the DMA is using the bus. Basically, this line controls whether the external address bus is controlled by the microprocessor subsystem or the DMA.

$\overline{\text{AEN}}$. The $\overline{\text{Address ENable}}$ control line enables the bus controller's command (read/write) outputs. If high, the command outputs of the bus controller are enabled, and the bus controller controls the bus operation. If low, the command outputs for the bus controller are disabled, and the other bus master controls the bus operation. In the PC, this line is always high except during a DMA cycle.

$\overline{\text{DMA WAIT}}$. The $\overline{\text{DMA WAIT}}$ control line is connected to the RDY1 input of the 8284A. If low, the DMA is using the bus, and the bus 8088 goes into a wait state until the DMA cycle is complete. When high, the bus is not being used by the DMA, and the 8088 can use the bus.

HOLDA. The HOLD Acknowledge control line indicates to the DMA that the 8088 has given up control of the bus. If high, the DMA takes control of the bus. If low, the DMA cannot take control of the bus.

$\overline{\text{RESET DRV}}$. The $\overline{\text{RESET DRiVe}}$ control line initializes all smart devices in the IBM PC. If low, all smart devices requiring a low reset go into a reset mode (halt); the ICs begin to restart after this line goes high. This line stays high for normal computer operations.

RESET DRV. The RESET DRiVe control line initializes all smart devices in the IBM PC. If high, all smart devices requiring a high reset go into a reset mode (halt); the ICs begin to restart after this line goes low. This line stays low for normal computer operations.

NMI. The NonMaskable Interrupt control line indicates to the 8088 when a memory parity error has occurred. If high, a memory parity error exists and the computer services the interrupt. If low, the system operates normally, and no memory parity error is indicated.

I/O CH CK. The Input/Output CHannel ChecK control line controls whether I/O parity error checking is enabled or disabled on the I/O channel bus. If high, I/O parity error checking is enabled. If low, no I/O parity error checking is performed.

DCLK. The Delay CLocK control line is the clock for the DMA. This signal operates at the same frequency as the CLK88, except that it is delayed by approximately 7 ns plus the delay in the one 7404 and one 74244. The delay is necessary to allow the DMA time to gain access to the bus and synchronize properly.

3.2.1 Troubleshooting the Extended Control Logic

The following are the key signals for the extended control logic section of the IBM PC. A logic analyzer or oscilloscope must be used because these signals change quickly. A latching digital probe may also be used, but this does not give as much detail.

Condition: One or more of the extended control logic lines are not at the correct levels at the proper times.
Result: Computer fails to boot or locks up shortly after booting.
Cause: Check $\overline{\text{RDY}}$/WAIT line; this line should be changing during an 8088 memory or I/O operation.
Cause: Check RDY TO DMA; this line should change levels only during a DMA operation. Try to perform a floppy or hard disk operation.
Cause: Check $\overline{\text{DMA AEN}}$; it should go low once every 15 μs for approximately 1 μs; this proves that memory refresh is working.
Cause: Check AEN BRD; if no software is being executed, this line should go high for about 1 μs every 15 μs.
Cause: Check $\overline{\text{AEN}}$; if no software is being executed, this line should go low for about 1 μs every 15 μs.
Cause: Check $\overline{\text{DMA WAIT}}$; if no software is being executed, this line should go low for about 1 μs once every 15 μs.
Cause: Check HOLDA; if no software is being executed, this line should go high for about 1 μs every 15 μs.
Cause: Check if NMI is stuck high; it should normally be low unless a parity error has occurred.
Cause: Check the DCLK; it should lag the CLK88 clock and be at the same frequency and duty cycle.

3.3 ADDRESS DECODING

The third section of the system board is called the address decoding section (Fig. 3-27). An address (memory location) is a numerical value (binary) that indicates the location of a device (smart ICs, memory, and any programmable IC) connected to the address bus (an IC connected to the address bus is normally connected also to the data bus). All devices connected to the address bus have an address associated with them, except the default bus master in the 8088. The 8088 determines the address for the operation to be performed either directly or indirectly by programming another smart IC that can be used as a bus master. To make hardware and software compatible in different PCs, all addressable devices are located in a standardized memory map, which specifies the memory location for each device.

Addressing a device begins with the address becoming valid on the external address bus; the logic levels on some of the address lines are applied to the address inputs of devices connected to the address bus and to the address decoding section of the system board. (The number of address lines connected to a device varies with the number of addressable locations inside the device.) The address decoding section enables or selects (takes the device out of a tristate and places it in an active mode) the device with a control line called a chip select ($\overline{\text{CS}}$). This is the first step in accessing (addressing) the device for the bus cycle. $\overline{\text{CS}}$ in computer systems are active low; an active-low device draws the most current when the device is on (enabled or in active mode) and the least amount of current when the device is off (disabled or in tristate). By lim-ber devices accessed (enabled) by the current bus cycle, the loading on the address latches in the microprocessor system is reduced. The reduction in loading reduces the amount of power needed and heat produced by the system board, which allows the manufacturer to reduce the size and cost of the computer.

Once the $\overline{\text{CS}}$ has enabled the selected device, the address lines directly connected to the device select the internal location to be used for the bus operation. The levels on the read and write lines then determine whether the data enters or exits the selected device and selected location within the device. The transfer is complete when the $\overline{\text{CS}}$ deselects the device to end the bus cycle.

The address decoding section is made up primarily of a 74138 (3-to-8 demultiplexer) along with some support gating. Its purpose is to set up the memory map and produce the $\overline{\text{CS}}$s of all addressable smart ICs [8237 (DMA), 8259 (PIC), 8253 (PIT), and 8255 (PPI)], the NMI register, and the DMA page register of the system board. This section also produces the $\overline{\text{ROM ADDR SEL}}$ (ROM ADDRess SELect), $\overline{\text{RAM ADDR SEL}}$ (RAM ADDRess SELect), ADDR SEL (ADDRess SELect), the chip selects for RAM ($\overline{\text{RAS0/CAS0}}$, $\overline{\text{RAS1/CAS1}}$, $\overline{\text{RAS2/CAS2}}$, $\overline{\text{RAS3/CAS3}}$; $\overline{\text{RAS}}$ = Row Address Strobe, $\overline{\text{CAS}}$ = Column Address Strobe), and the chip selects for ROM ($\overline{\text{CS0}}$ to $\overline{\text{CS7}}$). The last major responsibility of the address decoding section is to set up the timing sequence for addressing RAM.

All I/O interface adapters and memory adapters that are not part of the system board contain their own address decoding to produce the chip selects necessary for access to the ICs contained on the adapter.

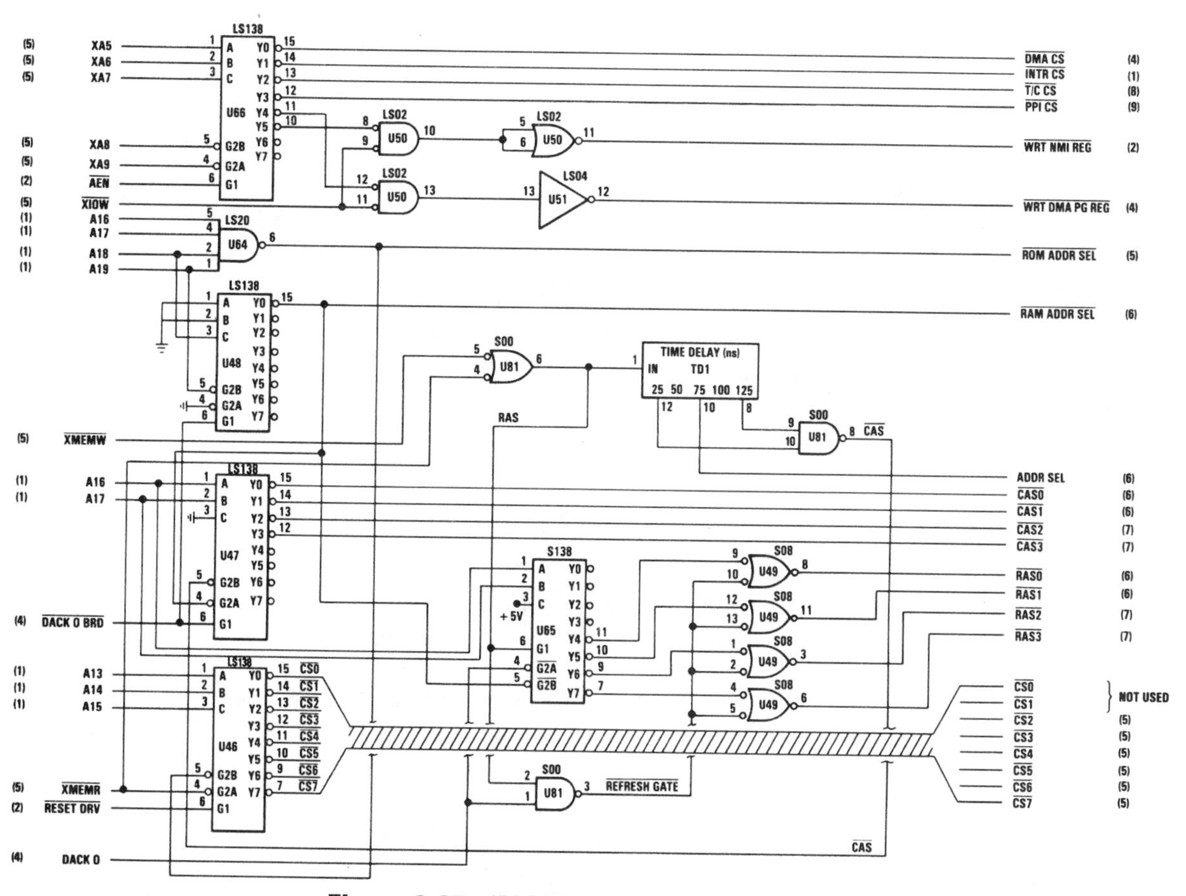

Figure 3-27 IBM PC address decoding section (3 of 10). (Courtesy of IBM)

3.3.1 74138 Pin Configuration (Figs. 3-28 and 3-29)

A/B/C

The A/B/C input lines are used to determine which output is selected. Only one output is selected at any one time. An output can be selected only once the enabling lines are at their proper levels. The select output goes low, while all unselected outputs remain high. The A input controls the LSB (least significant bit), while the C input controls the MSB (most significant bit). The binary count on these three inputs determines which output is selected [A = 0, B = 1, C = 0, a binary count of 010 (two in decimal) means output $\overline{Y2}$ is selected].

G1, $\overline{G2A}$, $\overline{G2B}$

To select any one of the outputs of this IC, the enabling (Gating) lines must be at their proper levels. Three enabling lines must be at the following levels to enable the selected output: G1 = 1, $\overline{G2A}$ = 0, and $\overline{G2B}$ = 0.

$\overline{Y0}$–$\overline{Y7}$

There are eight active-low output lines on this IC. Only one output may be selected at any one time; the selected output goes low, while all other outputs remain high. If the gating inputs are not all at their proper levels, all outputs are high and no output is selected.

Pin	Name	Pin	Name
1	A	16	VCC
2	B	15	Y0
3	C	14	Y1
4	$\overline{G2A}$	13	Y2
5	$\overline{G2B}$	12	Y3
6	G1	11	Y4
7	Y7	10	Y5
8	GND	9	Y6

INPUTS					OUTPUTS							
ENABLE		SELECT										
G1	$\overline{G2}$*	C	B	A	Y0	Y1	Y2	Y3	Y4	Y5	Y6	Y7
X	H	X	X	X	H	H	H	H	H	H	H	H
L	X	X	X	X	H	H	H	H	H	H	H	H
H	L	L	L	L	L	H	H	H	H	H	H	H
H	L	L	L	H	H	L	H	H	H	H	H	H
H	L	L	H	L	H	H	L	H	H	H	H	H
H	L	L	H	H	H	H	H	L	H	H	H	H
H	L	H	L	L	H	H	H	H	L	H	H	H
H	L	H	L	H	H	H	H	H	H	L	H	H
H	L	H	H	L	H	H	H	H	H	H	L	H
H	L	H	H	H	H	H	H	H	H	H	H	L

*$\overline{G2}$ = $\overline{G2A}$ + $\overline{G2B}$

H = high level, L = low level, X = irrelevant

Figure 3-28 74138 pin configuration and truth table. (Reprinted by permission of copyright holder, Texas Instruments, Incorporated, 1985)

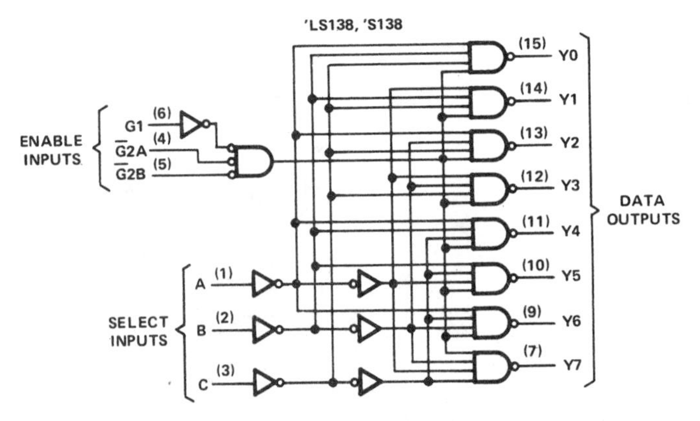

Pin numbers shown on logic notation are for D, J or N packages.

Figure 3-29 74138 logic diagram. (Reprinted by permission of copyright holder, Texas Instruments, Incorporated, 1985)

3.3.2 Determining Address Decoding for the System Board

To determine the address ranges of any $\overline{CS}$, you must look at which control and address lines are controlling the inputs of the selected 74138 and, using the truth table, develop the addresses that enable the selected output. The following is an example (using Fig. 3-27) of how to determine the address ranges for the $\overline{DMACS}$.

For the $\overline{DMA\ CS}$ to go low, the following conditions must be met on IC U66 (74138): $\overline{AEN}$ = 1, XA9 = 0, XA8 = 0, XA7 = 0, XA6 = 0, and XA5 = 0. $\overline{AEN}$ must be high because it is connected to the G1 input of U66. When $\overline{AEN}$ is high, the command outputs of the bus controller are enabled (8088 has control over the bus). The $\overline{DMA\ CS}$ enables only the DMA; the remaining lines connected to the DMA determine which operation is performed. Any address, data, or control line that is preceded by the letter X means that the line can be controlled by either the 8088 MPU or the 8237 DMA.

Only address lines 5 to 9 are used, and to select the $\overline{Y0}$ output of the U66, these lines must all be low (Table 3-7). Because the other address lines are not connected to the U66, they act as "don't cares" in determining the address range of the $\overline{CS}$ used. To determine the address ranges of the $\overline{CS}$, you must assume the condition where all don't cares are high or low.

The first assumption is when address lines 10 to 19 are all low and address lines 0 to 4 are all low. The address is 00000, which indicates the lower end of the first address range for $\overline{Y0}$.

The second assumption is when address lines 10 to 19 are all low and address lines 0 to 4 are all high. The address is 0001F, which indicates the upper end of the first address range for $\overline{Y0}$.

The first address range is from 00000 to 0001F.

The third assumption is when address lines 10 to 19 are all high and address lines 0 to 4 are all low. The address is FFC00, which indicates the lower end of the second address range for $\overline{Y0}$.

The fourth assumption is when address lines 10 to 19 are all high and address lines 0 to 4 are all high. The address is FFC1F, which indicates the upper end of the second address range for $\overline{Y0}$.

The second address range is from FFC00 to FFC1F.

Tables 3-8 through 3-13 show the inputs, outputs, and addresses for the address decoding section, using the truth table of the 74138 above. EA refers to the effective address range for the $\overline{CS}$.

Note: Although the addressable ICs on the system board have two address ranges that cause the $\overline{CS}$s to go low, only the lower address range is used when the IC is accessed. This is because all addressable ICs in the previous table use I/O read and write lines connected to the IC. Whenever the 8088 performs an I/O operation, the high nibble of the address bus is always low during this type of operation.

3.3.3 RAM Address Decoding Timing (Tables 3-11 and 3-12)

The address decoding section is responsible for the timing sequence of RAM in the computer. The address decoding section allows only one of the $\overline{RAS}$ and one of the $\overline{CAS}$ lines to go low at any one time, except during a memory refresh cycle, when all $\overline{RAS}$ lines go low at the same

TABLE 3-7
Address Lines for $\overline{Y0}$, U66

No.	*19 18 17 16*	*15 14 13 12*	*11 10 9 8*	*7 6 5 4*	*3 2 1 0*
	× × × ×	× × × ×	× × 0 0	0 0 0 ×	× × × ×
1	0 0 0 0	0 0 0 0	0 0 0 0	0 0 0 0	0 0 0 0
2	0 0 0 0	0 0 0 0	0 0 0 0	0 0 0 1	1 1 1 1
3	1 1 1 1	1 1 1 1	1 1 0 0	0 0 0 0	0 0 0 0
4	1 1 1 1	1 1 1 1	1 1 0 0	0 0 0 1	1 1 1 1

TABLE 3-8
U66 74138 Addressable ICs of the System Board

Enabled output	*Inputs*						
	$\overline{AEN}$	*XA9*	*XA8*	*XA7*	*XA6*	*XA5*	*$\overline{XIOW}$*
$\overline{\text{DMA CS}}$	1	0	0	0	0	0	NA
EA	(00000–0001F)			and	(FFC00–FFC1F)		
$\overline{\text{INTR CS}}$	1	0	0	0	0	1	NA
EA	(00020–0003F)			and	(FFC20–FFC3F)		
$\overline{\text{T/C CS}}$	1	0	0	0	1	0	NA
EA	(00040–0005F)			and	(FFC40–FFC5F)		
$\overline{\text{PPI CS}}$	1	0	0	0	1	1	NA
EA	(00060–0007F)			and	(FFC60–FFC7F)		
$\overline{\text{WRT NMI REG}}$	1	0	0	1	0	1	0
EA	(000A0–000BF)			and	(FFCA0–FFCBF)		
$\overline{\text{WRT DMA PG REG}}$	1	0	0	1	0	0	0
EA	(00080–0009F)			and	(FFC80–FFC9F)		

TABLE 3-9
U84 74LS20: ROM Address Select

Enable Output	*A19*	*A18*	*A17*	*A16*
$\overline{\text{ROM ADDR SEL}}$	1	1	1	1
EA	(F0000–FFFFF)			

TABLE 3-10
U48 74138: RAM Address Select (on Board)

Enabled output	*Inputs*				
	A19	*A18*	*$\overline{DACK\ 0\ BRD}$*	*$\overline{XMEMW}$*	*$\overline{XMEMR}$*
$\overline{\text{RAM ADDR SEL}}$	0	0	1	0	0
EA	(00000–3FFFF)				

TABLE 3-11
U47 74138: RAM Column Address Select

Enabled output	*Inputs*					
	A17	*A16*	*$\overline{RAM\ ADDR\ SEL}$*	*$\overline{DACK\ 0\ BRD}$*	*$\overline{XMEMW}$*	*$\overline{XMEMR}$*
$\overline{\text{CAS0}}$	0	0	0	1	0/1	1/0
EA	(00000–0FFFF)					
$\overline{\text{CAS1}}$	0	1	0	1	0/1	1/0
EA	(10000–1FFFF)					
$\overline{\text{CAS2}}$	1	0	0	1	0/1	1/0
EA	(20000–2FFFF)					
$\overline{\text{CAS3}}$	1	1	0	1	0/1	1/0
EA	(30000–3FFFF)					

TABLE 3-12
U65 74138: RAM Row Address Select

Enabled output	*Inputs*					
	A17	*A16*	*$\overline{\text{RAM ADDR SEL}}$*	*$\overline{\text{DACK 0}}$*	*$\overline{\text{XMEMW}}$*	*$\overline{\text{XMEMR}}$*
$\overline{\text{RAS0}}$ EA	0 (00000–0FFFF)	0	0	0	0/1	1/0
$\overline{\text{RAS1}}$ EA	0 (10000–1FFFF)	1	0	0	0/1	1/0
$\overline{\text{RAS2}}$ EA	1 (20000–2FFFF)	0	0	0	0/1	1/0
$\overline{\text{RAS3}}$ EA	1 (30000–3FFFF)	1	0	0	0/1	1/0

In U47 and U65, either $\overline{\text{XMEMW}}$ or $\overline{\text{XMEMR}}$ must be high.

TABLE 3-13
TD1 Address Time Delay

ADDR SEL goes high 75 ns after $\overline{\text{XMEMW}}$ or $\overline{\text{XMEMR}}$ goes low.
$\overline{\text{CAS}}$ goes low to enable the U47 125 ns after $\overline{\text{XMEMW}}$ or $\overline{\text{XMEMR}}$ goes low.

TABLE 3-14
U46 74138: ROM Chip Selects

Enabled output	*Inputs*					
	A15	*A14*	*A13*	*$\overline{\text{ROM ADDR SEL}}$*	*$\overline{\text{XMEMR}}$*	*$\overline{\text{RESET DRV}}$*
Not used	0	0	0	0	0	1
Not used	0	0	1	0	0	1
$\overline{\text{CS2}}$ EA	0 (F4000–F5FFF)	1	0	0	0	1
$\overline{\text{CS3}}$ EA	0 (F6000–F7FFF)	1	1	0	0	1
$\overline{\text{CS4}}$ EA	1 (F8000–F9FFF)	0	0	0	0	1
$\overline{\text{CS5}}$ EA	1 (FA000–FBFFF)	0	1	0	0	1
$\overline{\text{CS6}}$ EA	1 (FC000–FDFFF)	1	0	0	0	1
$\overline{\text{CS7}}$ EA	1 (FE000–FFFFF)	1	1	0	0	1

time and none of the $\overline{\text{CAS}}$ lines goes low. This is referred to as RAS-only refresh and is discussed under the DRAM section of the PC.

1. Once a valid RAM address and a memory read or write operation are selected, the address decoding section causes one of the $\overline{\text{RAS}}$ lines to go low. This process causes the row part of the address to be latched into the DRAM of the page of memory selected. The selected $\overline{\text{RAS}}$ line stays low until the end of the memory operation.

2. Seventy-five nano seconds after either the $\overline{XMEMW}$ or $\overline{XMEMR}$ line goes low and if a valid RAM address is on the address bus, the ADDR SEL line goes high (this level remains high throughout the memory operation and then resets). The ADDR SEL line, when high, causes the column part of the address to be placed on the address inputs of the DRAM.
3. Fifty nano seconds after the ADDR SEL line goes high, the $\overline{CAS}$ line for the selected memory page goes low (the same $\overline{RAS}$ and $\overline{CAS}$ lines must go low for the memory cycle to be completed). This line stays low until the end of the memory cycle (the cycle ends when either the $\overline{XMEMR}$ or $\overline{XMEMW}$ line that went low goes high again). This line going low causes the column part of the address currently on the address inputs of the DRAM to latch in the column part of the address into memory. Once this occurs, the memory location is selected. At the end of the memory cycle, the $\overline{CAS}$ and $\overline{RAS}$ lines go high and the ADDR SEL line goes low.

3.3.4 IBM PC Memory Maps (Tables 3-15 to 3-17)

3.3.5 Differences in Address Decoding

Tables 3-18 and 3-19 list the differences between the IBM PC and PC memory maps. Any address range and function not included in the tables are the same for both the IBM PC and xxxPC.

3.3.6 Address Decoding Troubleshooting

To determine if the address decoding section is causing the problem, the technician needs to monitor the $\overline{CS}$ to the IC in question and then perform some type of operation that causes the 8088 to

TABLE 3-15
Addressable ICs on the System Board

Hex Address	*Function*
00000–0000F	8237 DMA controller
00020–00021	8259 Programmable interrupt controller
00040–00043	8253 Programmable timer
00060–00063	8255 PPI
00080–00083	Page Register of the DMA
000A0–000A1	NMI mask bit register

TABLE 3-16
I/O Devices

Hex Address	*Function*
00200–0020F	Game port
00210–00217	Expansion unit adapter
00278–0027F	Secondary parallel printer port
002FB–002FF	Secondary serial communications port
00300–0031F	Prototype adapter
00320–0032F	Hard disk adapter
00378–0037F	Primary parallel printer port
00380–0038C	Secondary binary synchronous interface
003A0–003A9	Primary binary synchronous interface
003B0–003BF	Monochrome adapter/printer port (PRN)
003D0–003DF	Color/graphics adapter
003F0–003F7	Mini floppy diskette adapter
003F8–003FF	Primary serial communications port

TABLE 3-17
System Memory

Hex Address	*Function*
00000–0FFFF	Page 0, user RAM (64K)
10000–1FFFF	Page 1, user RAM (64K)
20000–2FFFF	Page 2, user RAM (64K)
30000–3FFFF	Page 3, user RAM (64K)
40000–4FFFF	Page 4, user RAM (64K)
50000–5FFFF	Page 5, user RAM (64K)
60000–6FFFF	Page 6, user RAM (64K)
70000–7FFFF	Page 7, user RAM (64K)
80000–8FFFF	Page 8, user RAM (64K)
90000–9FFFF	Page 9, user RAM (64K)
A0000–AFFFF	Reserved (64K), page 10
B0000–B3FFF	Monochrome screen memory (16K), page 11
B4000–B7FFF	Reserved (16K), page 11
B8000–BBFFF	Color/graphics memory (16K), page 11
BC000–BFFFF	Reserved (16K), page 11
C0000–C7FFF	Reserved expansion ROM (32K), page 12
C8000–CBFFF	Fixed-disk ROM (16K), page 12
CC000–CFFFF	Reserved expansion ROM (16K), page 12
D0000–DFFFF	Reserved expansion ROM (64K), page 13
E0000–EFFFF	Reserved expansion ROM (64K), page 14
F0000–F3FFF	Reserved (16K), page 15
F4000–F5FFF	BASIC ROM1 (8K), page 15
F6000–F7FFF	BASIC ROM2 (8K), page 15
F8000–F9FFF	BASIC ROM3 (8K), page 15
FA000–FBFFF	BASIC ROM4 (8K), page 15
FC000–FDFFF	BASIC ROM5 (8K), page 15
FE000–FFFFF	BIOS ROM (8K), page 15

TABLE 3-18
System Memory for IBM PC

Hex Address	*Function*	*IBM PC Chip Select*
00000–0FFFF	User RAM, page 0 (64K)	$\overline{RAS0}$/$\overline{CAS0}$
10000–1FFFF	User RAM, page 1 (64K)	$\overline{RAS1}$/$\overline{CAS1}$
20000–2FFFF	User RAM, page 2 (64K)	$\overline{RAS2}$/$\overline{CAS2}$
30000–3FFFF	User RAM, page 3 (64K)	$\overline{RAS3}$/$\overline{CAS3}$
F4000–F5FFF	BASIC ROM1 (8K)	$\overline{CS2}$
F6000–F7FFF	BASIC ROM2 (8K)	$\overline{CS3}$
F8000–F9FFF	BASIC ROM3 (8K)	$\overline{CS4}$
FA000–FBFFF	BASIC ROM4 (8K)	$\overline{CS5}$
FC000–FDFFF	BASIC ROM5 (8K)	$\overline{CS6}$
FE000–FFFFF	BIOS ROM (8K)	$\overline{CS7}$

access the device. If the $\overline{CS}$ line does not go low, the inputs to the address decoding IC must be monitored to make sure the proper inputs are applied. If the output is not correct and the inputs are correct, change the IC. If the output is not correct and the input is not correct, backtrack to the device supplying the input and troubleshoot that device. The following indicates the kind of operation to perform to check if the address decoding section is functioning correctly.

- To check the $\overline{DMA\ CS}$, monitor the $\overline{CS}$ line on the 8237 DMA and watch for a low pulse (about four clocks) when performing a DIR command on the floppy diskette or hard drive.

TABLE 3-19
System Memory for xxxPC

Hex Address	*Function*	×××*PC*
00000–3FFFF	User RAM, pages 0–3 (256K)	$\overline{RAS0}/\overline{CAS0}$
40000–7FFFF	User RAM, pages 4–7 (256K)	$\overline{RAS1}/\overline{CAS1}$
80000–8FFFF	User RAM, page 8 (64K)	$\overline{RAS2}/\overline{CAS2}$
90000–9FFFF	User RAM, page 9 (64K)	$\overline{RAS3}/\overline{CAS3}$
F4000–F5FFF	Available ROM space (8K)	$\overline{CS2}$
F6000–F7FFF	Available ROM space (8K)	$\overline{CS3}$
F8000–F9FFF	Available ROM space (8K)	$\overline{CS4}$
FA000–FBFFF	Available ROM space (8K)	$\overline{CS5}$
FC000–FDFFF	Available ROM space (8K)	$\overline{CS6}$
FE000–FFFFF	BIOS ROM (8K)	$\overline{CS7}$

- To check the $\overline{\text{INTA CS}}$, monitor the $\overline{\text{CS}}$ line on the 8259 PIC and watch for a low pulse (about four clocks) when performing a warm boot on the computer.
- To check the $\overline{\text{T/C CS}}$, monitor the $\overline{\text{CS}}$ line on the 8253 PIT and watch for a low pulse (about four clocks) when causing the computer to beep. In the programming language of BASIC, this can be done by typing the command BEEP and pressing the enter key.
- To check the $\overline{\text{PPI CS}}$, monitor the $\overline{\text{CS}}$ line on the 8255 PPI and watch for a low pulse (about four clocks) when pressing a key on the keyboard.
- To check any one of the $\overline{\text{CS}}$s of RAM, monitor the $\overline{\text{RAS}}$ and $\overline{\text{CAS}}$ lines on any one of the DRAM ICs in the bank of RAM to be checked. The $\overline{\text{RAS}}$ line should go low and stay low, followed by the $\overline{\text{CAS}}$ line going low for at least 150 ns. A program must be written to access the selected page of memory for testing the $\overline{\text{RAS}}$ and $\overline{\text{CAS}}$ lines in question.
- To check the chip select of ROM BIOS, monitor the $\overline{\text{CS}}$ of BIOS ROM and perform a warm boot; this line should be pulsed low (about four clocks).

ADDRESS DECODING SUMMARY

1. The address decoding section is responsible for producing the chip selects for the addressable ICs on the system board. All I/O interface adapters contain their own address decoding section, which sets up the $\overline{\text{CS}}$s for their own ICs.
2. All $\overline{\text{CS}}$s are active low.
3. A $\overline{\text{CS}}$ goes low whenever the address on the address bus is within the address range of the $\overline{\text{CS}}$ and any other control line levels are correct for the $\overline{\text{CS}}$.
4. The address decoding section is also responsible for controlling the timing sequence for DRAM (read, write, or refresh).
5. The memory locations for the addressable ICs in the computer are found in the memory map of the computer. All address decoding needs to follow the memory map parameters of the computer.

ADDRESS DECODING REVIEW QUESTIONS

1. To select one of the outputs of the 74138, what levels must be on the enable inputs?

a. G1 = 0, $\overline{\text{G2A}}$ = 0, $\overline{\text{G2B}}$ = 0

b. G1 = 0, $\overline{\text{G2A}}$ = 0, $\overline{\text{G2B}}$ = 1

c. G1 = 1, $\overline{\text{G2A}}$ = 1, $\overline{\text{G2B}}$ = 0

d. G1 = 1, $\overline{\text{G2A}}$ = 1, $\overline{\text{G2B}}$ = 1

e. None of the above answers is correct.

2. When one of the outputs of the 74138 is selected, the output will be which logic?

a. high

b. low

3. To select the $\overline{\text{Y3}}$ output of the 74138, the enable lines must be at their proper levels. Which of the following levels must be on inputs A, B, and C?

a. A = 1, B = 1, C = 0

b. A = 0, B = 1, C = 1

c. A = 1, B = 0, C= 1

d. A = 0, B = 0, C = 1

e. None of the above answers is correct.

4. Which one of the following lines goes low with an address on the address bus of 00082 hex?

a. $\overline{\text{DMA CS}}$

b. $\overline{\text{INTR CS}}$

c. $\overline{\text{T/C CS}}$

d. $\overline{\text{PPI CS}}$

e. None of the above answers is correct.

5. The $\overline{\text{CAS0}}$ line goes low before the $\overline{\text{RAS0}}$ lines goes low.

a. True

b. False

6. At the end of the memory cycle, the $\overline{\text{RAS}}$ line goes high.

a. True

b. False

7. When the address 1022E hex is on the address bus, which of the following lines goes low?

a. $\overline{\text{RAS0}}$ and $\overline{\text{CAS0}}$

b. $\overline{\text{RAS1}}$ and $\overline{\text{CAS1}}$

c. $\overline{\text{RAS2}}$ and $\overline{\text{CAS2}}$

d. $\overline{\text{RAS3}}$ and $\overline{\text{CAS3}}$

8. Which $\overline{\text{CS}}$ line with an address of FE012 hex goes low when a memory read operation is performed?

a. $\overline{\text{CS3}}$

b. $\overline{\text{CS4}}$

c. $\overline{\text{CS5}}$

d. $\overline{\text{CS6}}$

e. $\overline{\text{CS7}}$

9. Which I/O device is accessed when an address of 00322 hex is on the address bus?

a. Game port

b. Floppy disk adapter

c. Hard disk adapter

d. Primary serial communications port

e. None of the above answers is correct.

10. The CGA video card is located at memory locations 003B0 through 003BF hex.

a. True

b. False

3.4 8237 DIRECT MEMORY ACCESS CONTROLLER (DMA) (FIG. 3-30)

Thus far the 8259 (interrupt controller) has been discussed; this IC is called a smart IC because it can be programmed, and once programmed it operates and performs its functions without

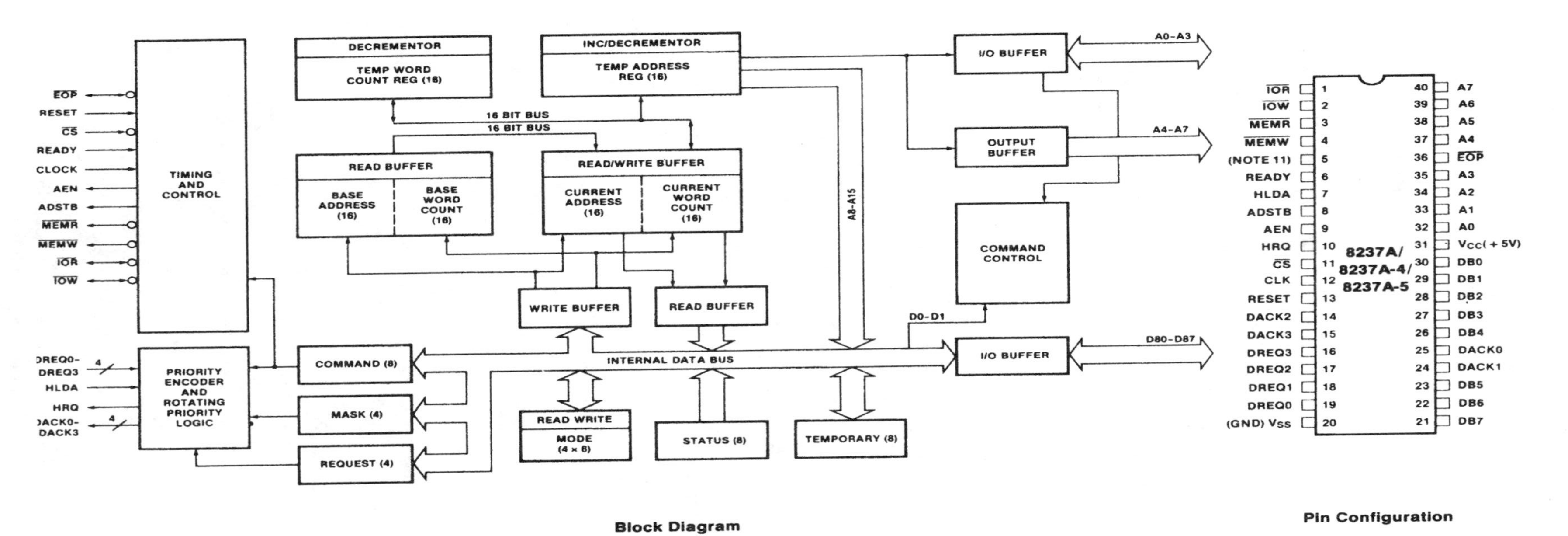

Figure 3-30 8237 Block diagram and pin configuration. (Courtesy of Intel)

any other help from the MPU. Smart ICs are used to cut down on the amount of unnecessary work the MPU has to perform, thereby speeding up the total operation of the computer. In this section we examine another smart IC that helps the MPU to run more efficiently. The main function of the 8237 DMA is to control the transfer of data between the floppy or hard disk adapters and memory, and it is used to refresh dynamic RAM in the IBM PC. The DMA can be used for the following tasks.

1. Dynamic RAM refresh
2. High-speed I/O to memory data transfers
3. High-speed memory to I/O data transfers
4. Memory to memory data transfers (used seldom)

The DMA is simply an I/O controller (specialized microprocessor) that transfers data from I/O to memory or from memory to I/O devices. Normally, data enters the DMA only when the DMA is being programmed. During a DMA data transfer, the DMA simply enables the devices in the memory map and allows the transfer to occur. During most DMA cycles, the data never enters or leaves the DMA.

The DMA can transfer data more quickly than the MPU because the DMA causes a read operation to occur (device places data on the data bus) and, while the read operation is still in progress, the DMA causes a write operation to occur (device accepts data from the data bus). This can take place only if the operation involves a transfer from memory to I/O or I/O to memory. Therefore, if a memory read is performed, only an I/O write can take place, and if an I/O read is performed, only a memory write can take place. A simultaneous read and write operation cannot be performed on either memory or I/O. If the MPU were to perform the same data transfer, it would require two bus cycles and at least eight clock cycles. Therefore, the DMA performs the operation at least twice as fast as the MPU.

If the DMA cycle requires programming other than the default programming loaded into the DMA at the time the computer is booted up, the 8088 under the control of software formats the DMA. After the DMA is formatted, the following sequence can occur. The DMA is informed that a DMA cycle is needed either by one of the DMA channel request lines (hardware request) or by the 8088 programming the proper internal register. Once the DMA cycle is requested, the DMA first asks for permission to use the bus from the external control logic section. Once permission is granted to the DMA (8088 is placed in a wait state and the 8288 bus controller and address latches in the microprocessor subsystem are disabled), the DMA takes control of the bus. The DMA does not give control of the bus back to the 8088 until the cycle is complete unless the DMA is reset. Once the DMA cycle is complete, the 8088 regains control of the bus.

3.4.1 DMA Internal Registers and Control Logic

There are 27 registers inside the DMA and three control logic sections. There are five groups of four registers each; each register in these groups controls one of the four channels of the DMA.

Internal Registers

BAR — There are four 16-bit Base Address Registers, one for each channel in the DMA. These registers hold the initial address value of the selected channel for the DMA cycle. These values change only when the MPU programs a different value into them.

BWCR — There are four 16-bit Base Word Count Registers, one for each channel in the DMA. These registers hold the initial number of DMA transfers to occur in the DMA cycle. These values change only when the MPU programs a different value into them.

CAR — There are four 16-bit Current Address Registers, one for each channel in the DMA. These registers hold the address currently being accessed by the DMA channel. The value of these registers automatically either decrements or increments after each DMA transfer.

CWCR — There are four 16-bit Current Word Count Registers, one for each channel in the DMA. These registers hold the current number of transfers remaining in the DMA cycle and decrement after each transfer. The number of transfers is the value of this register plus 1.

MR — There are four 6-bit Mode Registers, one for each channel in the DMA. These registers indicate which of the four operational modes the DMA is in and which channel is being accessed.

TAR — There is one 16-bit Temporary Address Register, which the selected channel uses. This register temporarily holds address bits A8 to A15 during a DMA transfer.

TWCR — The Temporary Word Count Register is a 16-bit register that temporarily holds the current count before its decrement.

SR — There is one 8-bit Status Register used by all four channels in the DMA. This register shows which of the four channels has requested service and if the selected channel has completed its cycle.

CR — There is one 8-bit Command Register used by all four channels in the DMA. This channel holds the command word, which will be performed. The command word selects the type of operation and how it is to be performed.

TR — There is one 8-bit Temporary Register that holds data from the data bus under certain DMA operations.

MSKR — There is one 4-bit MaSK Register used by the DMA to disable (mask out) any of the DMA input requests.

REQR — There is one 4-bit REQuest Register that allows software to produce a DMA cycle in one of its channels. This register allows software request instead of hardware request to start a DMA cycle.

Note: The I/O buffer holds the lowest 4 bits of the address during a DMA cycle. This buffer automatically increments to the next address at the end of each data transfer. The output buffer holds the next lowest 4 bits of the address during a DMA cycle. This buffer automatically increments to the next address at the end of each data transfer.

Control Logic

T/C — The Timing and Control logic in the DMA generates all internal and external control signals in and for the DMA. The signals generated include read/write, address enables, and timing signals.

PERPL — The Priority Encoder and Rotating Priority Logic determines the priority of the DMA request and generates a DMA acknowledge. This logic section uses the information in the command, mask, and request registers to determine priority.

CC — The Command Control logic determines which command word will be performed.

3.4.2 8237 DMA Pin Configuration

RESET	The RESET input resets all internal registers of the DMA to their starting default values when this input goes high. A reset produces an idle state in the DMA.
$\overline{\text{CS}}$	When low, the $\overline{\text{Chip Select}}$ input enables the DMA for programming by the MPU. The level on this pin is derived from the address decoding section of the IBM PC.
A0–A7	Address lines A0 to A3 are bidirectional. If $\overline{\text{CS}}$ is low, A0 to A3 act as inputs, allowing the 8088 to address one of the 24 internal registers of the 8237. When $\overline{\text{CS}}$ is high and the DMA is in its cycle, all address lines (A0 to A7) act as outputs. During a DMA cycle, A0 to A7 are used as the lower 8 bits of the address that the DMA is accessing. During an MPU cycle, address lines A4 to A7 are in a high-impedance state.
D0–D7	The Data lines are bidirectional I/O lines. When the 8088 has control over the bus, the data lines input data to program the internal registers of the DMA or output data when they read the status of one of the internal registers. During a DMA cycle, these lines act as the eight most significant bits of the address bus (A8 to A15). These bits are strobed into an external address latch by the address strobe line of the DMA during a DMA cycle.
$\overline{\text{EOP}}$	The $\overline{\text{End Of Program}}$ pin is bidirectional. In the PC, this line is tied high through a pull-up resistor and acts only as an output. This line is pulsed low when the terminal count is reached during the DMA cycle, that is, when the current word count register has completed all remaining transfers.
READY	The READY input is used to slow down DMA transfers in slow I/O devices that use the I/O CH CK line. When this line is high, the device being accessed is ready to complete its data transfer. When low, the ready line causes the production of wait clocks in the DMA. The READY line gets its signal from the external control logic RDY TO DMA.
CLOCK	The CLOCK input is derived from DCLK in the external control logic section. The input frequency is 4.772727 MHz (209.52 ns), giving a maximum possible transfer rate of 1.6 Mb/sec for the DMA, and is delayed from the CLK88 output.
AEN	The Address ENable output latches the 8 MSB of the address into the address latches. This output is not used in the IBM PC, so it is not connected.
ADSTB	The ADdress STroBe output is used to latch the data on the data bus (which represents address bits A8 to A15) into the address latches at the proper time during a DMA cycle.
$\overline{\text{MEMR}}$	The $\overline{\text{MEMory Read}}$ output enables the read function in a DMA cycle (DMA read or DMA memory-to-memory transfer). During an MPU cycle, this line is in tristate.
$\overline{\text{MEMW}}$	The $\overline{\text{MEMory Write}}$ output enables the write function in a DMA cycle (DMA write or DMA memory-to-memory transfer). During an MPU cycle, this line is in tristate.

$\overline{\text{IOR}}$ The Input/Output Read line is a bidirectional line that has two functions. Under MPU control, this line is an input and is used to read the status of the selected register of the DMA. Under DMA control, this line acts as an output and is used to control the read operation in the peripheral during a DMA cycle.

$\overline{\text{IOW}}$ The Input/Output Write line is a bidirectional line that has two functions. Under MPU control, this line is an input and is used to write information into the selected register of the DMA. Under DMA control, this line acts as an output and is used to control the write operation in the peripheral during a DMA cycle.

HRQ The DMA uses its Hold ReQuest output to ask the external control logic for permission to use the bus. When high, the request is made. When the DMA has control over the bus, this line is in a high-impedance state.

HLDA The external control logic uses the HoLD Acknowledge input line to grant permission to the DMA to use the bus. The external control logic causes the 8088 to go into a wait state after it completes its current bus cycle. The external control logic also places the bus controller command outputs in tristate along with the address latches in the microprocessor subsystem. This line is an active high and remains high until the DMA ends its cycle.

DREQ0–DREQ3 The Dma REQuest lines are inputs used to inform the DMA of the need for a DMA cycle to begin. The request is made when the line goes high. The priority of each channel is programmed by BIOS. The DREQ0 line has the highest priority, and DREQ3 has the lowest priority. DREQ0 (channel 0) is used to inform the DMA that a DMA refresh cycle needs to be performed. This line is connected to the PIT section of the PC. DREQ1 is not dedicated to any device in the PC; this line is available for use on the I/O channel bus. DREQ2 is used to inform the DMA that the floppy disk adapter requires a DMA cycle. This line is connected to the floppy disk adapter through the I/O channel bus. DREQ3 (channel 3) is used to inform the DMA that the hard (fixed) disk adapter requires a DMA cycle. This line is connected to the hard disk adapter through the I/O channel bus. These lines only go high when the device requesting the DMA cycle is ready for the transfer.

$\overline{\text{DACK0–DACK3}}$ The Dma ACKnowledge output lines inform the device connected to the channel requesting a DMA cycle that the cycle is in progress. The actual 8237 has active-high $\overline{\text{DACKx}}$ output lines, but when used in the IBM PC, the DMA is programmed to make these output lines act as active-low outputs. The acknowledgment occurs when the selected line is low. There is one line for each DMA channel. $\overline{\text{DACK0}}$ is connected to the refresh gates in the address-decoding section of the IBM PC so that all pages of RAM are refreshed at the same time. This line is also available on the I/O channel bus so that it can be used to refresh any memory card that is connected to the I/O channel bus. $\overline{\text{DACK1}}$ is not dedicated to any PC device but is available on the I/O channel bus. $\overline{\text{DACK2}}$ is used to select the floppy disk adapter and is connected to the adapter through the I/O channel bus. $\overline{\text{DACK3}}$ is used to select the hard disk adapter and is connected to the adapter through the I/O channel bus.

3.4.3 DMA Operational Modes

Active Cycles. The 8237 DMA has four operating modes when active. The DMA is active when it has control over the bus system (acting as the bus master). The following are the four operating modes and brief descriptions of their functions.

STM — The Single Transfer Mode allows only 1 byte of data to transfer for each DMA requested. In this mode, the DMA always uses the value of the base address value upon receiving a DMA request. The current word count register is decremented after each transfer.

BTM — The Block Transfer Mode allows up to a 64K-byte block of data to transfer every time a DMA request occurs. The DMA uses the value of the current address register for the address of each transfer during the DMA cycle. After each transfer, the current word count register is decremented, and the current address register is either decremented or incremented. This process continues until all data transfers for the DMA cycle have taken place (number of transfers equal to the value in the base word count register). Some dedicated high-speed adapters (tape backup adapters) use this transfer mode. The problem with this transfer mode is that no other function can take place until the entire block of data has been transferred. This may cause errors in memory, because the memory may not be refreshed soon enough to retain its data.

DTM — The Demand Transfer Mode allows the DMA to transfer data until the DREQ line goes low (inactive) or the terminal count is reached. Therefore, this transfer mode is similar to the block transfer mode except that it can be interrupted. The DMA uses the value in the current address register to determine the address used during the DMA cycle. After the transfer takes place, the current word count register is decremented and the current address register is either decremented or incremented for the next DMA cycle. The PC uses this mode to perform refresh of memory and floppy and hard disk transfers.

CM — The Cascade Mode allows more than one DMA to share the bus with other DMAs. The IBM PC does not use this mode.

Idle Cycle. The DMA is in its idle state when it does not have control over the bus. While in the idle cycle, the DMA tristates its output lines, making it transparent to the bus. The DMA checks its DMA request lines during each clock.

3.4.4 Types of DMA Transfers

READ — The READ transfer occurs when data transfer from memory into an I/O device. The $\overline{\text{MEMR}}$ line go low, prior to bringing the $\overline{\text{IOW}}$ line low. The PC uses the READ transfer to RAM to refresh memory and to transfer data to the floppy and hard disk adapters.

WRITE — The WRITE transfer occurs when data transfer from an I/O device into memory. The $\overline{\text{IOR}}$ lines go low prior to bringing the $\overline{\text{MEMW}}$ line low. The PC uses the WRITE transfer to transfer data from the floppy and hard disk adapter into memory.

VERIFY — The VERIFY transfers simulate data transfers, meaning that the DMA generates an address but keeps the read and write (in both I/O and memory) high (inactive). Therefore, only the address is placed on the address bus. During this type of transfer, the ready input to the DMA is not checked. The VERIFY transfer allows the DMA to control the addressing of the address bus, while a special interface adapter controls the read and write lines.

3.4.5 DMA Section of the IBM PC (Fig. 3-31)

The ICs in this section of the IBM PC are used to hold the address for the address bus during a DMA cycle. During an MPU cycle, the outputs of these ICs are in tristate, making them transparent to the address bus. The DMA controls only data transfers on the external bus, usually I/O to memory and memory to I/O data transfers. (The DMA does have the ability to perform memory-to-memory transfers, but it is seldom used.) Data transfers between the local (multiplexed) bus and the external bus are controlled by the 8288 bus controller in the microprocessor subsystem.

8237 — The main purpose of the 8237 DMA is to control data transfers between memory and the floppy or hard disk adapters. The DMA is also used to refresh dynamic RAM.

74LS244 — The 74LS244 octal buffer/driver supplies the low byte of the address offset to address lines XA0 to XA7 only during a DMA cycle. During an MPU cycle, the outputs of this IC are in tristate, making them transparent to the address.

74LS670 — The 74LS670 is a 4- by 4-bit data register; it is referred to as the DMA page register. The IC can store four 4-bit values that represent the page value of the address (A16 to A19) bus. A different 4-bit value is used for each DMA channel. The reason for this IC is that the 8237 DMA can directly control only 16 address lines (A0 to A15) during a DMA cycle. Before the DMA can take control over the bus, the MPU loads the DMA page register with the page value for the memory location to be used for the data transfer, unless the value in the DMA page register is already loaded with the page value for the transfer.

Writing to the DMA Page Register

1. The MPU places the address for the DMA page register on the address bus.
2. The address decoding section causes the $\overline{\text{WRT DAM PG REG}}$ line to go low, which selects an address range from 00080 through 00083 hex. This line stays low until the end of the bus cycle that writes to the DMA page register.
3. The two LSBs of the address XA0 and XA1 select which of the 4-bit memory locations in the 74LS670 accepts the data on the address bus. XA0 is connected to WA (Write A) and XA1 is connected to WB (Write B); the levels on these two inputs select which 4-bit location is used. Table 3-20 shows which 74LS670 memory location is selected by the address on the address bus and which DMA channel uses the data at the memory location.
4. The four LSBs of the data bus place the data in the selected memory location of the 74LS670.
5. The bus cycle ends when the $\overline{\text{WRT DMA PG REG}}$ line goes high again, which writes the data on the data bus into the selected memory location inside the 74LS670.

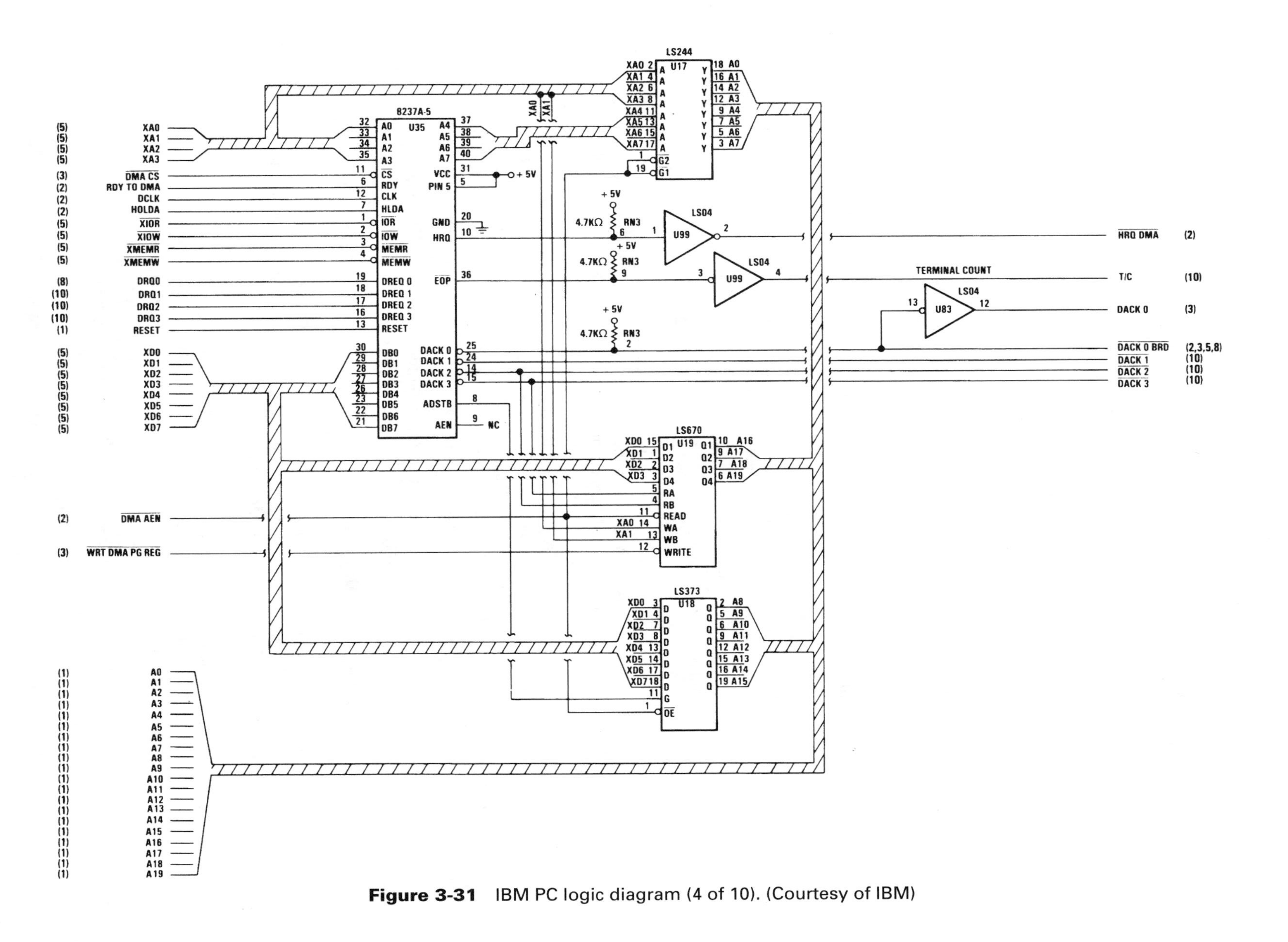

Figure 3-31 IBM PC logic diagram (4 of 10). (Courtesy of IBM)

TABLE 3-20
DMA Page Register Addressing

XA1	*XA0*	*Address*	*DMA Channel*
0	0	00080	No DMA channel
0	1	00081	DMA channel 2
1	0	00082	DMA channel 3
1	1	00083	DMA channel 0 or 1

Reading the DMA Page Register

1. Reading of the contents of the DMA page register occurs only during a DMA cycle and is controlled by the DMA.
2. Once one of the $\overline{\text{DACKx}}$ lines goes low, one of the 4-bit memory locations of the 74LS670 is selected. See the previous table for which location is used. $\overline{\text{DACK2}}$ is connected to RB (Read B) and $\overline{\text{DACK3}}$ is connected to RA (Read A). Because only one of the Dma ACK lines ever goes low at any one time, only address locations 00081, 00082, and 00083 are ever valid. DMA channel 2 uses the data at memory location 00081, DMA channel 3 uses 00082, and DMA channel 0 or 1 could use 00083.
3. The data from the selected memory location do not appear on the outputs of the 74LS670 until the external control logic makes the $\overline{\text{DMA AEN}}$ line go low; this line stays low until the end of the DMA cycle.
4. The outputs of the 74LS670 go into tristate once the $\overline{\text{DMA AEN}}$ line goes high at the end of the DMA cycle.

74LS373 — The 74LS373 octal latch is used to hold the high byte of the offset address (A8 to A15) during the DMA cycle. This is required because the DMA contains only eight address lines. So during the first part of the DMA cycle, the DMA releases the value of A8 to A15 on the data bus. During this time, the DMA pulses high the ADSTB line, which latches the levels on the data bus into the 74LS373 address latch. Once the outputs of the 74LS373 become enabled, these data become the high byte of the offset address on the address bus. The outputs of the 74LS373 become enabled only when the $\overline{\text{DMA AEN}}$ line goes low and remains low throughout the DMA cycle.

74LS04 — These inverters are used to buffer and invert the HRQ DMA, T/C, and $\overline{\text{DACK0}}$ lines for the system board.

3.4.6 DMA Timing Diagrams

The following is an explanation of a DMA cycle in the IBM PC using the timing diagram of the DMA in Figure 3-32 and 3-33. Note that the timing diagram shows only the function of the DMA; in this explanation we describe everything that happens.

1. If the operation that uses the DMA requires that the DMA be formatted with values other than the default values loaded into the DMA when the computer was booted up or from its last cycle, the MPU formats the DMA with its new values. The MPU loads the necessary registers to determine the transfer mode and transfer type to be used during the next DMA cycle. The MPU also loads the page value for the DMA cycle into the DMA page register, if a different page value is needed, from either the default or last values loaded. Once this process is complete, the following sequence occurs. This process is necessary before most DREQx lines request a DMA cycle. Channel 0 (dynamic RAM refresh) does not require this step to generate the DREQ0.

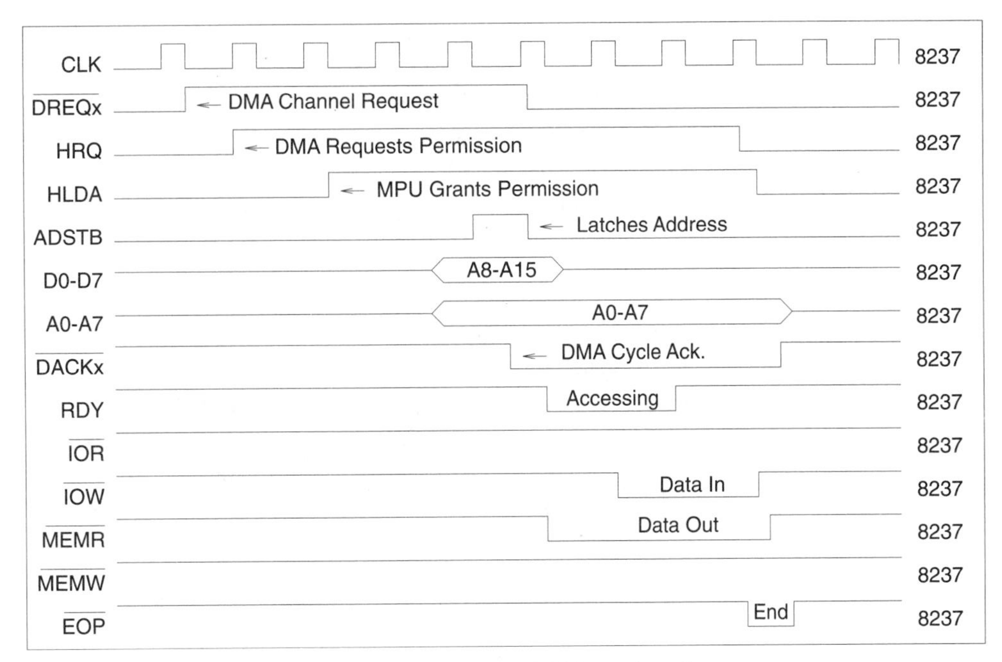

Figure 3-32 8237 DMA read cycle.

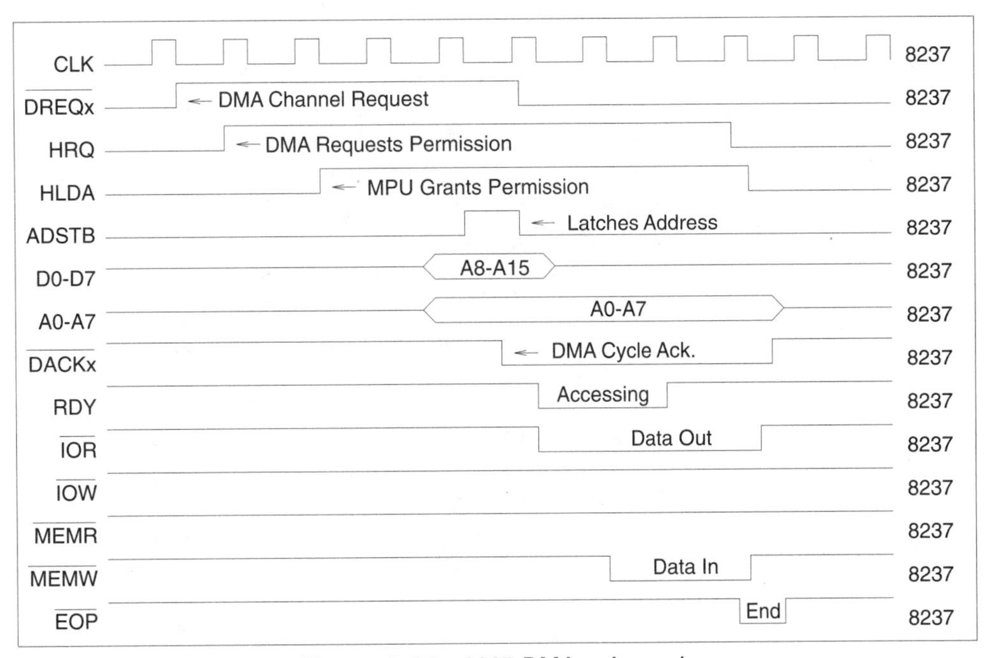

Figure 3-33 8237 DMA write cycle.

2. When one or more of the DREQx input lines go high, the DMA is informed that a DMA cycle is needed by the device connected to the channel making the request. The DMA determines which channel has the highest priority and processes that channel first. Under most transfer modes, this line stays high until the DACKx line becomes active, at which time this line normally resets. The resetting of this line is controlled by the device requesting the DMA cycle, not the DMA.
3. The DMA then requests permission to use the address bus by bringing the HRQ (Hold ReQuest) line high. The high is applied to an inverter, and the output of the inverter is applied to the external control logic of the IBM PC.

4. The external control logic of the IBM PC then causes the $\overline{\text{DMA WAIT}}$ line to go low after the 8088 finishes current instruction.
5. The low on the $\overline{\text{DMA WAIT}}$ line causes the READY output of the 8284 to go low. This causes the 8088 to go into a wait state.
6. The external control logic disables the 8288 bus controller by bringing the $\overline{\text{AEN}}$ line low. This level disables the command outputs (read and write lines) of the 8288 bus controller, causing them to go into tristate. This process allows the DMA to control the read and write lines.
7. The external control logic also disables (tristate) the outputs of the address latches of the microprocessor subsystem. This process allows the DMA to control the address bus.
8. The external control logic next causes the HLDA to go high. This high is used to grant permission to the DMA to take control of the bus; the DMA is now the bus master. At this time the external control logic has disabled the local (multiplexed) bus.
9. The DMA then outputs the address to the address bus. The DMA outputs the low byte of the address offset on address lines XA0 to XA7. At the same time, the DMA places the high byte of the address offset on the data bus.
10. The DMA pulses high the ADSTB line, which latches the levels from the data bus into the 74373. After the ADSTB goes low, the data lines of the DMA go into tri-state.
11. Near the end of the ADSTB high pulse, the DMA causes one of the $\overline{\text{DACKx}}$ lines to go low. Which $\overline{\text{DACKx}}$ lines go low is determined by which DREQx line started the DMA cycle. The $\overline{\text{DACKx}}$ line is used to enable the I/O device. The $\overline{\text{DACK2}}$ and $\overline{\text{DACK3}}$ lines are connected not only to the I/O channel bus but also to the RA and RB inputs of the 74LS670 DMA page register. The levels on these two inputs determine which 4-bit value is sent to the outputs of the 74LS670.
12. The external control logic causes the $\overline{\text{DMA AEN}}$ to go low, which performs the following:
    a. Enables the outputs of the 74LS244; this produces the low byte of the address offset on the address bus.
    b. Enables the outputs of the 74LS373; this produces the high byte of the address offset on the address bus.
    c. Enables the outputs of 74LS670; this produces the page value of the address (A16 to A19) bus.

DMA Read Cycle (Fig. 3-32)

13R. The DMA causes the $\overline{\text{MEMR}}$ (memory read) line to go low; this starts the process of accessing the memory location.

14R. Some time after the $\overline{\text{MEMR}}$ line goes low, the external control logic causes the RDY TO DMA line to go low. This informs the DMA that the memory location is in the process of being accessed. The RDY TO DMA line reacts the same as the READY line to the 8088. When high, the bus is ready, and when low, the bus is not ready (in the process of accessing the selected device).

Note: The external control logic responds differently to a memory refresh cycle, which may cause a delay in the RDY TO DMA line going low. Also, when a memory refresh cycle is in progress, the RDY TO DMA line does not have to go high before the cycle can end.

15R. Shortly after the $\overline{\text{MEMR}}$ and RDY TO DMA lines go low, the DMA causes the $\overline{\text{IOW}}$ line to go low, triggering the I/O device that is being accessed to begin accepting (writing) data.

(*Note:* If the terminal count is reached, the $\overline{\text{EOP}}$ line goes low and remains low until the $\overline{\text{IOW}}$ line goes high again at the end of the DMA cycle.)

16R. After the standard 250 to 300 ns (150 to 250 ns in an xxxPC) from the point where the RDY TO DMA line has gone low (longer if the I/O device is slow in responding), the RDY TO DMA line goes high. It may take longer if a memory refresh cycle is in the DMA cycle.

17R. After about two clock cycles from the point where the $\overline{IOW}$ line went low, the DMA causes the $\overline{IOW}$ line to go high, provided that the RDY TO DMA line is high (unless a memory refresh cycle is in progress). On the transition from low to high, whatever levels are on the data bus are latched (stored) into the memory location. Once high, the memory location no longer accepts any changes.

18R. The DMA causes the $\overline{MEMR}$ line to go high, which causes the memory location to stop producing data for the data bus.

19R. At this time, the DMA causes the HRQ line to go low, which informs the external control logic that the DMA no longer needs the bus. This causes the external control logic to begin switching control of the bus from the DMA to the MPU.

20R. Shortly after the HRQ line goes high and the extended control logic begins giving back control of the bus to the MPU (and bus controller), the HLDA input to the DMA goes low. The extended control logic also causes the $\overline{DMA\ AEN}$ line to go high, disabling the 74373, 74244, and 74670 ICs, which produce the address for the address bus in the DMA section. At this time, the DMA section has no control over the address bus.

21R. Shortly after the HLDA changes, the DMA causes the $\overline{DACKx}$ line to go high, disabling the I/O device used during the DMA cycle; the DMA then tristates its A4 to A7 lines and makes A0 to A3 inputs. The DMA cycle is now complete.

DMA Write Cycle (Fig. 3-33)

Note: The first 12 steps in the DMA cycle are the same for write and read operations.

13W. The DMA causes the $\overline{IOR}$ (I/O read) line to go low; this starts the process of accessing the I/O device selected by the $\overline{DACKx}$ line.

14W. Some time after the $\overline{IOR}$ line goes low, the external control logic causes the RDY TO DMA line to go low. This informs the DMA that the I/O device is in the process of being accessed. The RDY TO DMA line reacts the same as the READY line to the 8088. When high, the bus is ready, and when low, the bus is not ready (in the process of accessing the selected device).

15W. Shortly after the $\overline{IOR}$ and the RDY TO DMA lines go low, the DMA causes the $\overline{MEMW}$ line to go low. When the $\overline{MEMW}$ line goes low, the memory device accessed by the address on the address bus begins accepting (writing) data. (*Note:* If the terminal count is reached, the $\overline{EOP}$ line goes low and remains low until the $\overline{MEMW}$ line goes high again at the end of the DMA cycle.)

16W. After the standard 250 to 300 ns (150 to 250 ns in a xxxPC) from the point where the RDY TO DMA line has gone low (longer if the I/O device is slow in responding), the RDY TO DMA line goes high.

17W. After about two clock cycles from the point where the $\overline{MEMW}$ line went low, the DMA causes the $\overline{MEMW}$ line to go high, provided that RDY TO DMA line is high. On the transition from low to high, whatever levels are on the data bus are latched (stored) into the memory location. Once high, the memory location no longer accepts any changes.

18W. The DMA causes the $\overline{IOR}$ line to go high, which causes the I/O device to stop producing data for the data bus.

19W. At this time, the DMA causes the HRQ line to go low, which informs the external control logic that the DMA no longer needs the bus. This causes the external control logic to begin switching control of the bus from the DMA to the MPU.

20W. Shortly after the HRQ line goes high and the extended control logic begins giving back control of the bus to the MPU (and bus controller), the HOLDA input to the DMA goes low. The extended control logic also causes the $\overline{\text{DMA AEN}}$ line to go high, disabling the 74373, 74244, and 74670 ICs, which produce the address for the address bus in the DMA section. At this time, the DMA section has no control over the address bus.

21W. Shortly after the HOLDA change, the DMA causes the $\overline{\text{DACKx}}$ line to go high, disabling the I/O device used during the DMA cycle; the DMA then tristates its A4 to A7 lines and makes A0 to A3 inputs. The DMA cycle is now complete.

3.4.7 Troubleshooting the DMA Section

Condition: The $\overline{\text{DACK0}}$ line does not go low or does not go low at the proper times (about once every 16 μs in an IBM PC or once every 15 to 26 μs in an xxxPC).

Result: The data stored in memory is lost or changes. Parity errors may occur and the computer may halt.

Cause: Check the DREQ0 line; this line should go high for about 840 ns (about 400 ns in an xxxPC), once every 16 μs. If a high pulse is not present, troubleshoot the PIT section of the computer.

Cause: If floppy and hard disk operations function, most likely the DMA is programmed correctly; therefore, if the DREQ0 line goes high and the $\overline{\text{DACK0}}$ line does not go low once every 16 μs, the DMA itself could be bad; change the DMA.

Condition: The $\overline{\text{DACK2}}$ line never goes low during a floppy disk operation.

Result: The floppy disk operation does not occur or the data transferred is not correct.

Cause: Check the DREQ2 line; this line should go high for about 840 ns once a floppy disk operation is issued. If a high pulse is not present, go to the I/O channel bus pin B6 and verify the pulse; if present, there is a broken connection somewhere between the I/O channel bus socket and the DREQ2 pin. Repair as necessary. If the pulse is not present, check for the pulse on pin B6 on the floppy disk adapter; if present, there is a bad connection on the contacts between the adapter in the I/O channel bus socket.

Cause: If memory refresh and hard disk operations function, most likely the DMA is programmed correctly; therefore, if the DREQ2 line goes high and the $\overline{\text{DACK2}}$ line does not go low once a floppy disk operation is issued, the DMA itself could be bad; change the DMA.

Condition: The $\overline{\text{DACK3}}$ line never goes low during a hard disk operation.

Result: The hard disk operation does not occur or the data transferred is not correct.

Cause: Check the DREQ3 line; this line should go high for about 840 ns once a hard disk operation is issued. If a high pulse is not present, go to the I/O channel bus pin B16 and verify the pulse; if present, there is a broken connection somewhere between the I/O channel bus socket and the DREQ3 pin. Repair as necessary. If the pulse is not present, check for the pulse on pin B16 on the hard disk adapter; if present, there is a bad connection on the contacts between the adapter in the I/O channel bus socket.

Cause: If memory refresh and floppy disk operations function, most likely the DMA is programmed correctly; therefore, if the DREQ3 line goes high and the $\overline{\text{DACK3}}$ line does not go low once a hard disk operation is issued, the DMA itself could be bad; change the DMA.

Condition: The HRQ line never goes high.

Result: The computer never boots up; computer halts because there is no memory refresh or disk operations.

Cause: Check the $\overline{CS}$ line of the DMA; this line should pulse low while the computer is trying to boot itself. If this line never goes low, the DMA is not being programmed; check the address decoding section or BIOS.

Cause: If the $\overline{CS}$ pulse is present, check for a high on the DREQ0 input (memory refresh occurs automatically). If this line goes high and there is no high on the HRQ line, the problem is either the DMA itself or the programming of the DMA; check BIOS. If the DREQ0 line does not go high, either the PIT section is not programmed or the PIT section is bad.

Condition: The HLDA line never goes high.

Result: The computer never boots up; the computer halts because there is no memory refresh or disk operations.

Cause: Check the HRQ line; this line should go high at some time. If the HRQ line goes high, troubleshoot the extended control logic section. If the HRQ does not go high, troubleshoot the HRQ line.

Condition: The ADSTB line never pulses high or pulses high at the wrong time during a DMA cycle.

Result: The computer may boot up (in BASIC for an IBM PC), but floppy or hard disk operations may not occur correctly. The data from any transfer is not in the correct memory location.

Cause: Either the DMA is bad, or the 74373 is loading down the ADSTB line.

DMA SECTION SUMMARY

1. The DMA is responsible for the transfer of data on the external bus only.
2. The DMA can only control the bus after the MPU has given up control over the bus via the extended control logic.
3. The following are the important lines of the DMA.
   a. The DREQx lines are used to request a DMA cycle from a hardware device.
   b. The HRQ (Hold ReQuest) line is used to ask permission to use the bus.
   c. The HLDA (HLD Acknowledge) input is used to give permission to use the bus.
   d. A0 to A3 act as address inputs when the MPU has control over the bus. These lines act as outputs during a DMA cycle.
   e. A4 to A7 act as outputs during a DMA cycle and are disabled during an MPU cycle.
   f. D0 to D7 are used as bidirectional data lines when the MPU has control over the bus. These lines supply address bits A8 to A15 during part of the DMA cycle; while these lines are valid, the ADSTB line of the DMA latches the levels into an address latch in the DMA section.
   g. The $\overline{MEMR}$ and $\overline{MEMW}$ lines act as outputs to control the memory operation during a DMA cycle; during an MPU cycle, these lines are in tristate.
   h. The $\overline{IOR}$ and $\overline{IOW}$ lines act as inputs during an MPU cycle and act as outputs during a DMA cycle.
   i. The $\overline{DACKx}$ (Dma ACKnowledge) lines are used to inform the I/O device requesting the DMA cycle that the DMA cycle is in progress.
4. In the PC, the demand transfer mode is the most used transfer mode.
5. During a DMA read operation, data is transferred from a memory location into the selected I/O device.
6. During a DMA write operation, data is transferred from the selected I/O device into a memory location.

DMA SECTION REVIEW QUESTIONS

1. Which of the following DMA lines must go high before the HRQ line goes high?
   - **a.** ADSTB
   - **b.** HLDA
   - **c.** DREQx
   - **d.** A0 to A7
   - **e.** None of the above answers is correct.
2. Which one of the following lines is used to cause the selected memory location to output its data?
   - **a.** $\overline{\text{MEMW}}$
   - **b.** $\overline{\text{MEMR}}$
   - **c.** There is no such line.
3. Which one of the following lines is used to latch the data on the data bus into the address latch of the DMA section during the DMA cycle?
   - **a.** HRQ
   - **b.** HLDA
   - **c.** ADSTB
   - **d.** RDY
   - **e.** $\overline{\text{EOP}}$
4. Which one of the following lines is used to grant permission to the DMA to use the bus?
   - **a.** HRQ
   - **b.** HLDA
   - **c.** ADSTB
   - **d.** RDY
   - **e.** $\overline{\text{EOP}}$
5. Which one of the following lines is used to indicate the terminal count during a DMA cycle?
   - **a.** HRQ
   - **b.** HLDA
   - **c.** ADSTB
   - **d.** RDY
   - **e.** $\overline{\text{EOP}}$
6. Which one of the following lines is used to inform an I/O device asking for a DMA cycle that a DMA cycle is in progress?
   - **a.** DREQx
   - **b.** HRQ
   - **c.** HLDA
   - **d.** $\overline{\text{DACKx}}$
   - **e.** $\overline{\text{EOP}}$
7. During a DMA read operation, the $\overline{\text{MEMR}}$ line goes low first during the DMA cycle.
   - **a.** True
   - **b.** False
8. A DMA write cycle causes data to be transferred from the selected I/O device into a memory location.
   - **a.** True
   - **b.** False
9. The DMA causes the READY line to the MPU to go low when the bus is not ready during a DMA cycle.
   - **a.** True
   - **b.** False

10. Which IC should be checked first if the page value of the address is not correct during a DMA cycle?
   a. 8237
   b. 74244
   c. 74670
   d. 74373

3.5 READ-ONLY MEMORY (ROM)

The ROM section in the IBM PC (Fig. 3-34) allows the 8088 access to the ROMs (EPROM in xxxPC). This section buffers the address and data bus for the I/O devices on the system board and, with the help of the extended control logic, buffers the read and write lines between the output of the bus controller and the DMA.

This section in the IBM PC contains five 8K × 8 ROMs (one 8K × 8 EPROM in xxxPC), two 74244 octal buffers/line drivers, and two 74245 octal data bus transceivers.

3.5.1 8364 ROM

The 8364 is an 8K × 8 (8192 by 8 bit) masked programmable ROM integrated circuit. The IBM PC contains five ROMs; one contains BIOS and the remaining four contain the cassette version of BASIC (in IBM PCs only). Because this is masked programmable, the data stored in the ROM is programmed at the factory where the ROM is made. The user sends a truth table to the manufacturer; the truth table indicates the data for each of the addresses in the ROM. The manufacturer, when making the ROM, alters its mask (the blueprint that is used to create the circuitry of the ROM) to place the correct data in the correct address locations indicated by the user's truth table. Because the data is placed in the ROM at the factory, it is permanent (nonvolatile) and does not require power to retain the information.

ROM Block Diagram (Fig. 3-35). The row decoder has 8 buffered address inputs and 256 row outputs. The 8 address inputs are used to select 1 of the 256 row outputs. The 256 row outputs are connected to the ROM matrix (256 × 256), only 1 row output will ever be selected at any one time.

The column decoder has 5 buffered address inputs, 256 memory matrix inputs and 8 column outputs. The 5 address inputs are used to select which one set of 8 columns of 32 sets of columns (32 sets of 8 columns each) will be applied to the output circuitry. The outputs from the selected set of columns are applied to 8 sense amplifiers. The sense amplifiers are used to buffer the voltage levels from the selected set of columns.

The memory matrix is made up of 256 row and 256 column locations. If there is a MOSFET connected between the selected row and column, the data in the location is a logic zero. If there is no MOSFET connected between the selected row and column, the data in the location is a logic one. The number and locations of the MOSFET connections are determined by the truth table. Because there are 256 row and 256 column locations, the total memory capacity of this ROM is 65,536 (64K for short) data bits.

The output circuitry has a chip select ($\overline{CS}$) input, 8 column inputs and 8 tri-state data outputs. If the $\overline{CS}$ input is low, the data from the select columns is supplied to the data outputs (D0–D7). If the $\overline{CS}$ input is high, the data outputs will be in tri-state.

3.5.2 ROM Pin Configurations (Fig. 3-36)

A0–A12 These 13 address inputs are used to select one of the memory locations inside the ROM; there are 8K memory locations inside this IC.

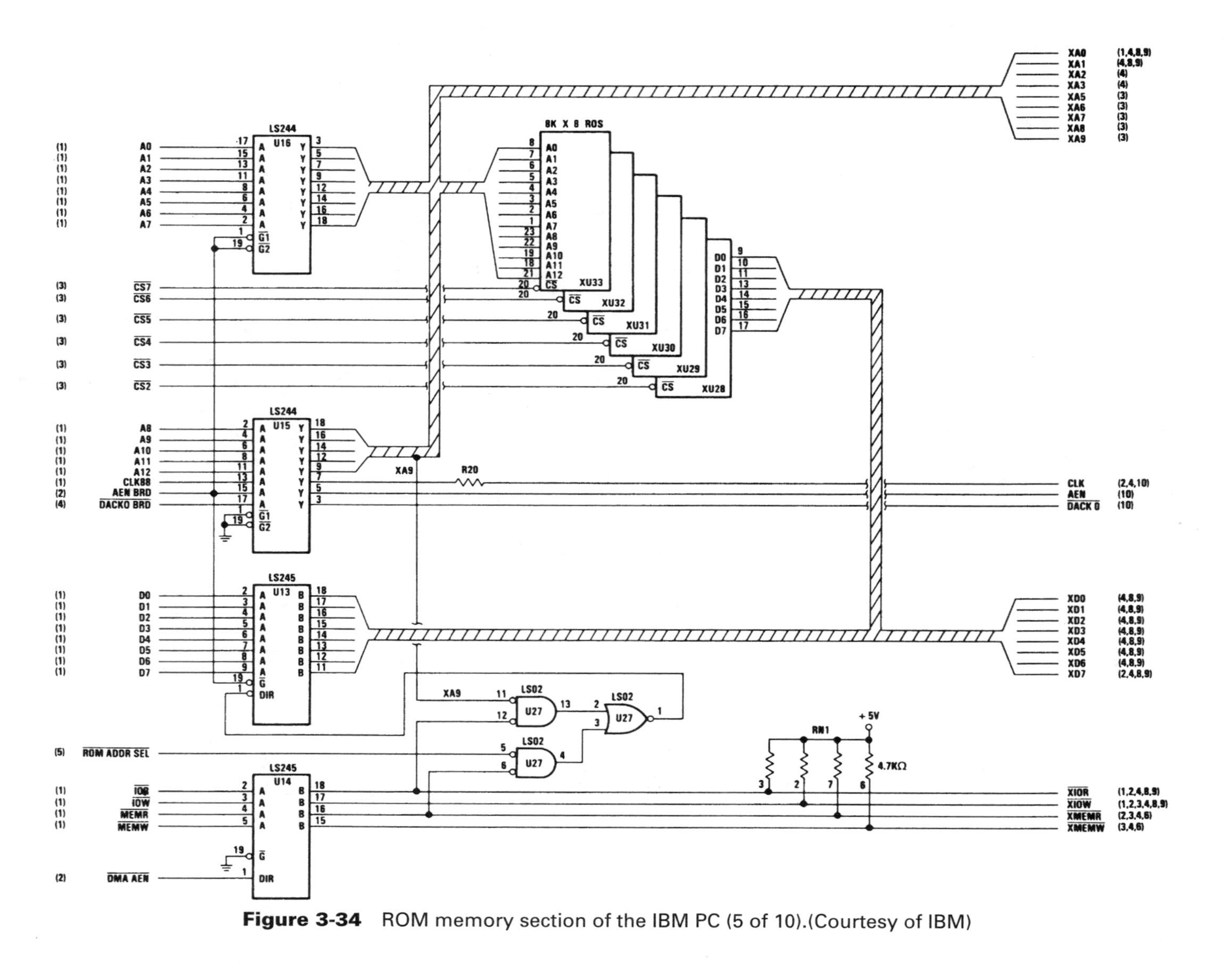

Figure 3-34 ROM memory section of the IBM PC (5 of 10).(Courtesy of IBM)

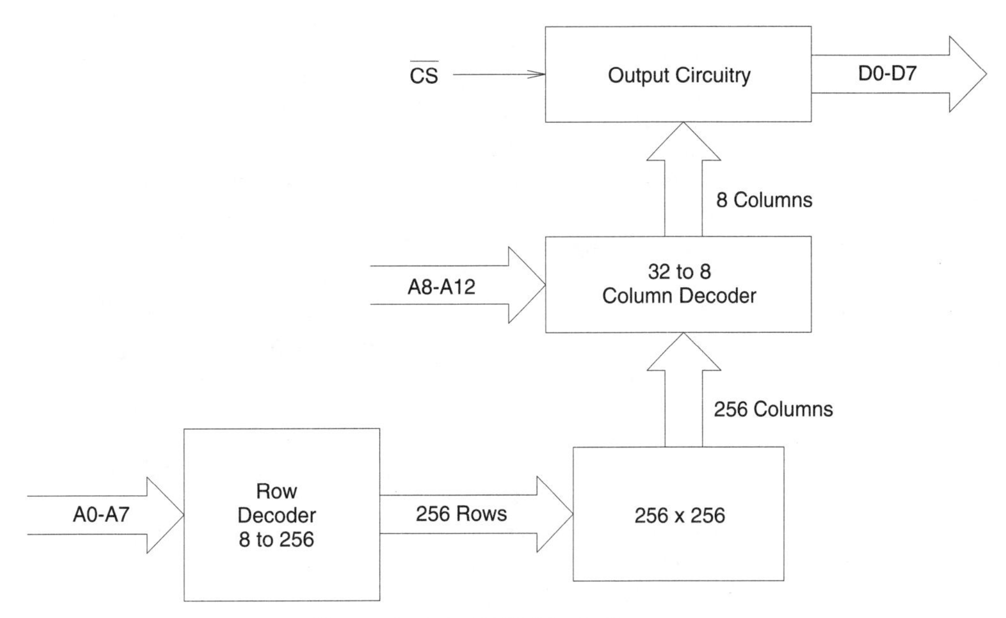

Figure 3-35 ROM block diagram.

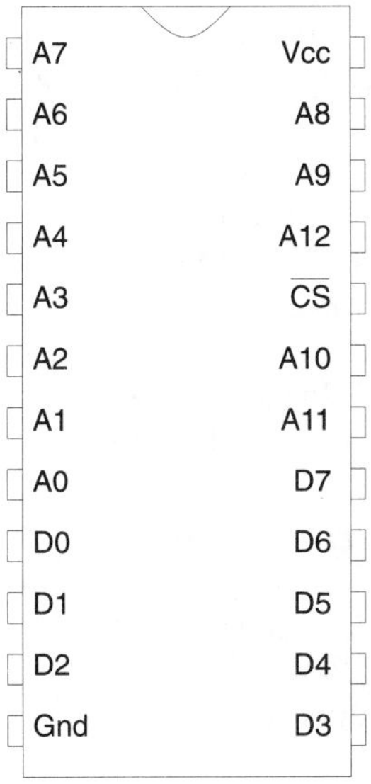

A7	Vcc
A6	A8
A5	A9
A4	A12
A3	$\overline{CS}$
A2	A10
A1	A11
A0	D7
D0	D6
D1	D5
D2	D4
Gnd	D3

Figure 3-36 ROM pin configuration.

$\overline{CS}$ The chip select is an active-low input that is used to enable the outputs of the IC when selected.

D0–D7 These are the eight output lines for the ROM; when $\overline{CS}$ is low, the data in the selected memory location is placed on the data bus. When $\overline{CS}$ is high, these outputs are placed in tristate, making them transparent to the data bus.

3.5.3 2764 EPROM

Most compatible PCs use EPROM instead of ROM; this is because of the number of changes in BIOS that most compatible PC manufacturers must perform to make their PC more compatible with the IBM PC. EPROM (Erasable Programmable Read-Only Memory) functions as a masked programmable ROM except that it can be programmed after the IC is manufactured. The 2764 is an 8K × 8 EPROM; this means that there are 8000 memory locations, with each memory location supplying 8 bits of data. Therefore, the 2764 has a memory capacity of 65,536 (64K) bits, just like the ROM.

The EPROM is made differently from the ROM in the sense that the memory matrix has a special ultraviolet light–sensitive MOSFET connected between each row and column in the matrix. When the memory matrix is made, the special MOSFETs all have open (floating) gates; therefore, when a row and column are selected, there is no conduction, and the data at all memory locations is a logic 1. Because of the special material used in making EPROMs, they tend to operate more slowly than ROMs. This condition also exists after the EPROM has been erased properly.

Programming data in the EPROM requires a special circuit to supply a high voltage to the programming voltage pin, along with the selected address and the data that is to be programmed on the data lines; the $\overline{CE}$ line must be low, and the $\overline{PGM}$ line must be pulsed low for a certain time period. What happens to the MOSFET during the programming of the EPROM is that when this high voltage is applied to the programming voltage pin with the logic level on the selected column low, the high potential forces electrons into the gate region. When enough time has passed, enough electrons accumulate in the gate region to allow the gate to conduct. Once the programming voltage is removed, these electrons remain trapped and allow the MOSFET to conduct when selected under normal operating voltages. If a mistake is made when programming the EPROM, the logic level at the selected memory location cannot be switched; it must be erased and reprogrammed. Once programmed, the user must cover the erase window with some type of nontransparent tape or sticker; this prevents the electrons from bleeding out of the gate region.

The only way to remove these extra electrons from the gate region is to shine a high-intensity ultraviolet light through the glass or quartz erase window. The ultraviolet light causes the material in the gate region of the MOSFET to lower its resistance. Once enough electrons have bled off, the gate region no longer conducts when the row and column is selected; the data reverts back to a logic 1. The process of erasing the EPROM is performed outside the PC. The time it takes to erase the EPROM depends on the intensity of the light and the size of the EPROM and can vary from 1 to 10 hours. The major problem with the EPROM is that all memory locations are erased; the user cannot selectively erase any one memory location once programmed. Basically, the EPROM can be programmed and erased as many times as necessary.

EPROM Block Diagram (Fig. 3-37). The row decoder has 8 buffered address inputs and 256 row outputs. The 8 address inputs are used to select 1 of the 256 row outputs. The 256 row outputs are connected to the EPROM matrix (256 × 256), only 1 row output will ever be selected at any one time.

The column decoder has 5 buffered address inputs, 256 memory matrix inputs and 8 bidirectional column inputs/outputs. The 5 address inputs are used to select which 1 set of 8 bidirectional columns of 32 sets of columns (32 sets of 8 columns each) will be applied to the I/O circuitry. The selected set of bidirectional columns are connected to 8 bidirectional sense amplifiers. The sense amplifiers are used to buffer the voltage levels between the EPROM matrix and I/O circuitry for the selected set of columns.

The EPROM memory matrix contains 256 row and 256 column locations. There is a special MOSFET connected between each row and column. Before programming, all MOSFETs have open gate regions that do not allow them to conduct when selected, which represents a logic one. After programming, some of the MOSFETs will conduct when selected, only locations that have or contain a logic zero will conduct.

The I/O circuitry has 3 control inputs ($\overline{OE}$, $\overline{CE}$, and $\overline{PGM}$), 8 bidirectional columns and 8 bidirectional data (O0–O7) lines. During a read operation, the chip enable and output enable lines go low, which allows the data in the selected 8 columns to be applied to the data bus as outputs. During programming, the chip enable and programming inputs must go low causing the data on the data bus to enter the I/O circuitry and then be applied to the selected 8 columns, which then writes the data to the EPROM matrix.

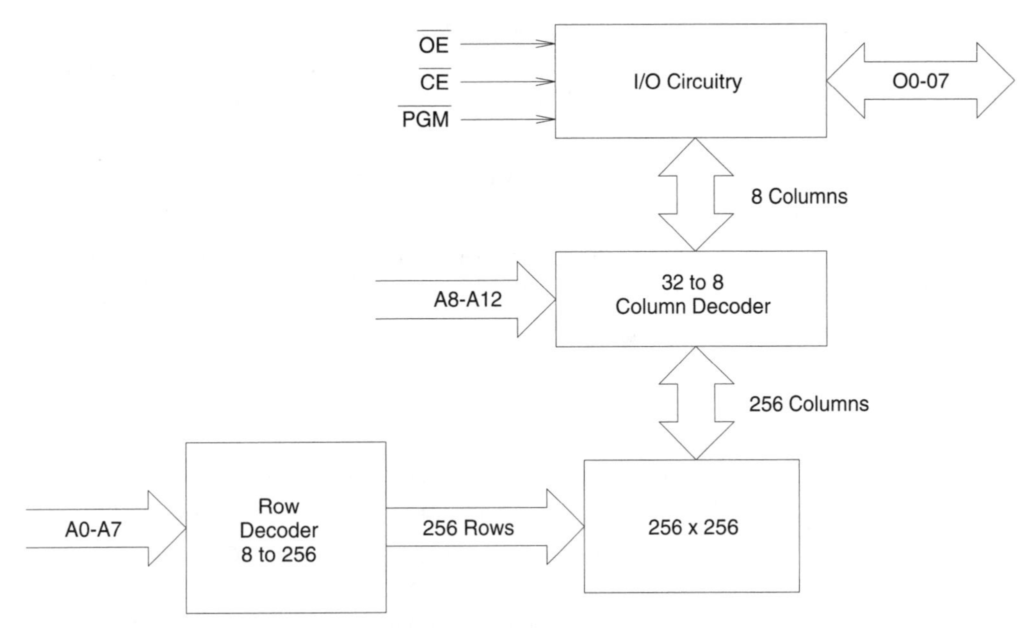

Figure 3-37 EPROM block diagram

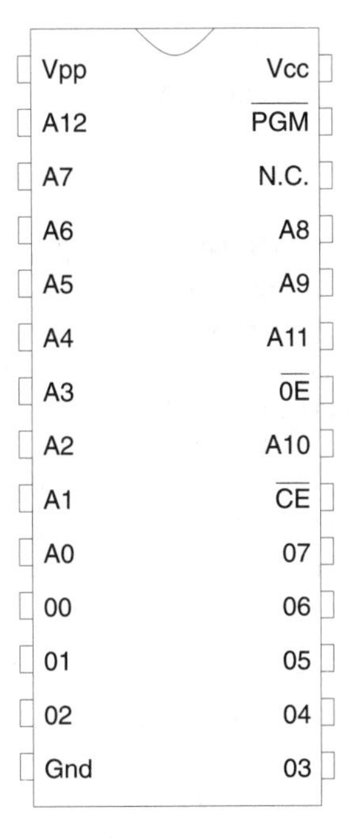

Figure 3-38 2764 EPROM pin configuration.

3.5.4 EPROM Pin Configuration (Fig. 3-38)

A0–A12	Perform the same function as the address lines in the ROM.
O0–O7	Perform the same function as the data lines (D0 to D7) in the ROM during a read operation. During programming, these lines input the data into the selected memory location.
$\overline{\text{CE}}$	The Chip Enable line is an active-low line. When low, this line helps to enable the EPROM during programming and during a read operation.

$\overline{OE}$ — The Output Enable line is an active-low line. When low, along with the $\overline{CE}$ line being low, the output buffers of the EPROM are enabled and allow the data to be output to the data bus.

Vpp — The programming voltage is used to supply a 21-V voltage to the EPROM, which is needed during the programming of the EPROM.

$\overline{PGM}$ — The ProGraMming input causes the EPROM to program the data in the selected memory location, when pulsed low, with the proper inputs on the Vpp, $\overline{OE}$, and $\overline{CE}$ lines.

3.5.5 ROMs Versus EPROMs

1. ROMs in large quantities (1000+) cost much less than EPROMs in the same quantities.
2. ROMs are usually faster than EPROMs.
3. ROMs usually have higher bit density than EPROMs.
4. ROMs can be programmed only once; EPROMs can be programmed as often as necessary.
5. ROMs cost much more in small quantities than do EPROMs.

3.5.6 ICs of the ROM Section

The ROM section consists of two address buffers, two bus transceivers, some simple gates, and five ROMs. One of the ROMs contains BIOS, and the other four ROMs contain BASIC.

74244 Octal Buffer and Line Driver. There are two 74244 octal buffers in the ROM memory section of the IBM PC. The 74244 has eight inputs and outputs, which are enabled when $\overline{G1}$ and $\overline{G2}$ are both low. If $\overline{G1}$ and/or $\overline{G2}$ is high, the outputs of this device are in tristate, making the outputs transparent to the bus.

U16 (IBM PC) buffers the address bus which can be controlled by either the MPU or DMA (XA0 to XA7) from the outputs of the address latches in the microprocessor subsystem. When AEN BRD is low, the MPU controls the XA0 to XA7 address line. When AEN BRD is high, the DMA controls the XA0 to XA7 address lines.

U15 (IBM PC) is always enabled and provides address buffering for address lines A8 to A12, which are applied to the ROM section. The other three outputs of U16 buffer the CLK88, AEN BRD, and $\overline{DACK0}$ control lines. The buffering keeps the local address latches from overloading during addressing.

74245 Octal Data Bus Transceiver. There are two 74245 octal data bus transceivers in the ROM memory section of the IBM PC. The 74245 is enabled only when $\overline{G}$ is low; otherwise, the chip is in tristate and the inputs and outputs are transparent. The level on DIR determines the direction of data flow. When low, data flows from side B to side A and, when high, data flows from A to B, once the $\overline{G}$ line is low.

U13 in the IBM PC buffers the MPU data bus (D0 to D7) from the DMA data bus (XD0 to XD7). If the gating input (G) is low, the MPU has control over the data bus and the levels on D0 to D7 controls the levels on XD0 to XD7. If the gating input (G) is high, the DMA has control and U13 will be disabled, the levels on XDO to XD7 will have no effect on D0–D7. If the DIR input is high, data will flow from D0 to D7 into XD0 to XD7, when the MPU has control. If the DIR input is low, data will flow from XD0 to XD7 to D0 to D7, when the MPU has control.

U14 in the IBM PC determines whether the DMA or bus controller controls the read/write command lines. The enable pin is grounded, enabling the transceiver at all times; the only thing that changes is the direction of data flow. When the DIR pin on U14 is low, the DMA controls the read/write command lines. When the DIR pin on U14 is high, the bus controller controls the read/write lines.

3.5.7 ROM/EPROM Timing Diagram

Figure 3-39 is a typical timing diagram for a ROM or EPROM. Remember that a ROM chip select goes low only when the addressing of page F of memory (F0000 to FFFFF hex) and a memory read are being performed.

T1 ROM/EPROM Read Cycle

1. Near the end of T1 of the bus cycle, the address from the microprocessor subsystem becomes valid on the address bus in the ROM section of the computer. The address selects the memory location that is to output its data, but data is not available on the output at this time.

T2 ROM/EPROM Read Cycle

1. Some time in T2, the address-decoding section of the PC causes the $\overline{\text{CS}}$ to the ROM or EPROM to go low. Once low, the $\overline{\text{CS}}$ allows the ROM/EPROM to output its data on the data bus; the data becomes valid. This line must stay low throughout the remainder of the bus cycle.

T3 ROM/EPROM Read Cycle

1. No changes take place.

T4 ROM/EPROM Read Cycle

1. When the bus controller causes the $\overline{\text{MEMR}}$ line to go low, the address decoding section causes the $\overline{\text{CS}}$ to the ROM/EPROM to go high. This ends the ROM/EPROM read cycle and completes the data transfer.

3.5.8 Troubleshooting the ROM Section

Very little can go wrong with the ROM/EPROM section of the computer. Most problems can be traced to one of the sections of the computer that control the ROM/EPROM section.

Condition: The $\overline{\text{CS7}}$ goes low during the boot-up of the computer, but the computer halts before it is fully booted or does not boot at all.

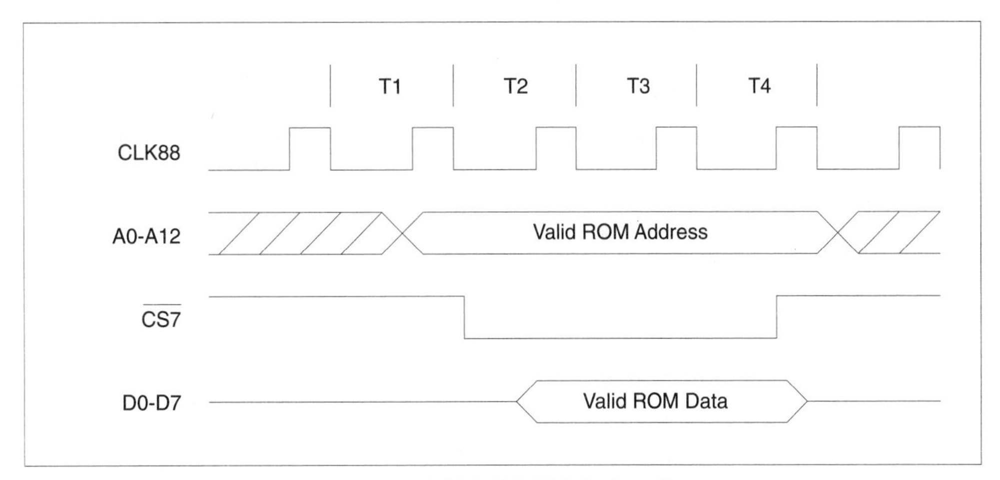

Figure 3-39 ROM/EPROM timing diagram.

Result: Computer halts.
Cause: The ROM could have gone bad (very rare).
Cause: The EPROM is electronically bad (very rare) or the EPROM has lost some of its programming. Check to make sure the erase window of the EPROM is still covered. Replace or erase and reprogram if necessary.

Condition: The $\overline{CS7}$ does not go low during the boot-up of the computer.
Result: Computer fails to boot and halts.
Cause: Troubleshoot the address-decoding section if the computer is producing an address in page F of memory, with the $\overline{MEMR}$ line going low at the same time.

ROM/EPROM SUMMARY

1. The ROM/EPROM section of the PC contains the system BIOS and in the IBM PC the cassette version of the language of BASIC.
2. BIOS can be accessed only during a memory read operation in the correct memory area of page F of memory.
3. This section buffers A0 to A12 of the address bus.
4. This section is used to buffer the data bus between the microprocessor subsystem and the data bus that can be controlled by either the MPU or DMA.
5. This section buffers the CLK, AEN, and $\overline{DACK0}$ lines on the external bus.
6. This section also helps determine whether the MPU (via the bus controller) or the DMA controls the read ($\overline{IOR}$ and $\overline{MEMR}$) and write ($\overline{IOW}$ and $\overline{MEMW}$) control lines.

ROM/EPROM REVIEW QUESTIONS

These questions may require the reader to use not only the schematics from the ROM/EPROM section, but also all the schematics provided up to this point.

1. From the ROM section schematic for the IBM PC (Fig. 3-34), which $\overline{CS}$ goes low when addressing BIOS?
   a. $\overline{CS7}$
   b. $\overline{CS6}$
   c. $\overline{CS5}$
   d. $\overline{CS4}$
   e. $\overline{CS3}$

2. Using Figure 3-34, if the DMA has control over the bus, U16 will be
   a. enabled
   b. disabled (tristate)

3. Using Figure 3-34, if the MPU has control over the bus and a ROM read operation is being performed, in which direction does the data flow through U13?
   a. From side A to side B
   b. From side B to side A

4. Using Figure 3-34, which bus master has control over the read and write lines when the DIR input of U14 is low?
   a. MPU (8088)
   b. DMA (8237)

5. If an EPROM is used, what must be the levels on $\overline{CE}$ and $\overline{OE}$ to produce an output?
   a. $\overline{CE} = 0, \overline{OE} = 0$
   b. $\overline{CE} = 0, \overline{OE} = 1$
   c. $\overline{CE} = 1, \overline{OE} = 0$
   d. $\overline{CE} = 1, \overline{OE} = 1$

3.6 RANDOM ACCESS MEMORY (RAM)

The RAM section for the IBM PC system board (Figs. 3-40 and 3-41) may contain up to a maximum of 256K of DRAM (Dynamic Random-Access Memory). The memory is set up in four banks of 64K (65,356 memory locations) each, giving a total of 256K of memory on the system board. Each bank of memory contains nine 64K × 1 DRAMS (64K memory locations by 1 bit wide; MT4264 or 4164), with their data lines connected in parallel. Eight of the DRAMs are used to hold the data byte, and the remaining DRAM holds the parity bit. The parity bit is created and used only by the RAM section to determine if a data error (a change in the data) has occurred.

The RAM section of a clonePC system board may contain a maximum of 640K of DRAM. Two banks of memory contain nine 256K × 1 (MT1259 or 41256) DRAMs, providing 256K of memory for each bank. Two banks of memory contain two 64K × 4 (MT4067 or 4464) plus one 64K × 1 DRAMs, providing 64K of memory for each bank. All four banks together provide 640K of total memory.

Besides the memory ICs, the RAM section also contains a parity generator/checker, address multiplexers, a data bus transceiver, and support circuitry.

3.6.1 DRAM Review

DRAM differs from static RAM in the fact that, in DRAM, data is stored as electrical charges in the small capacitance of the MOSFETs that make up the memory matrix. In static RAM, the data stored in a memory location is determined by which set of MOSFETs are conducting

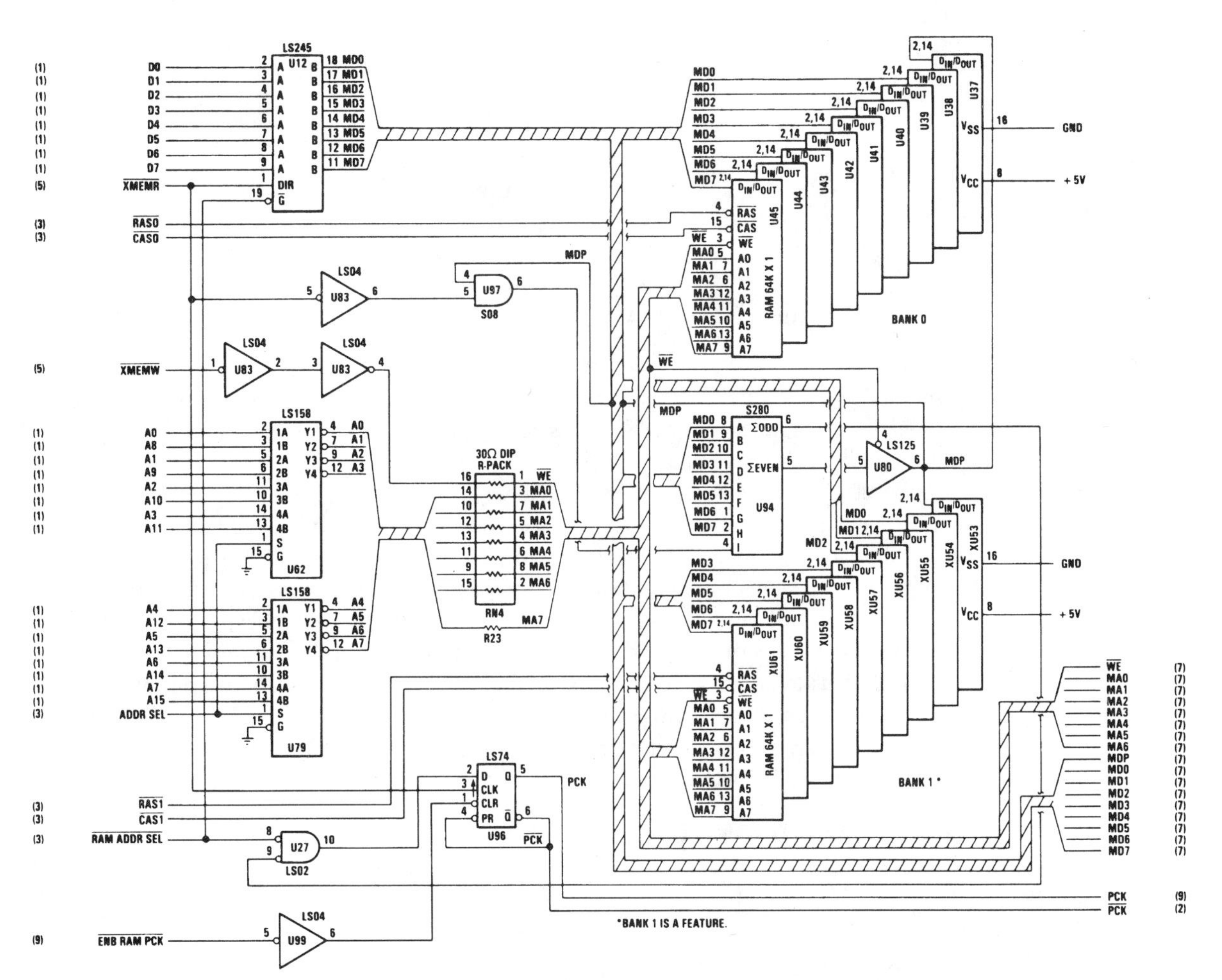

Figure 3-40 DRAM section part 1 (6 of 10). (Courtesy of IBM)

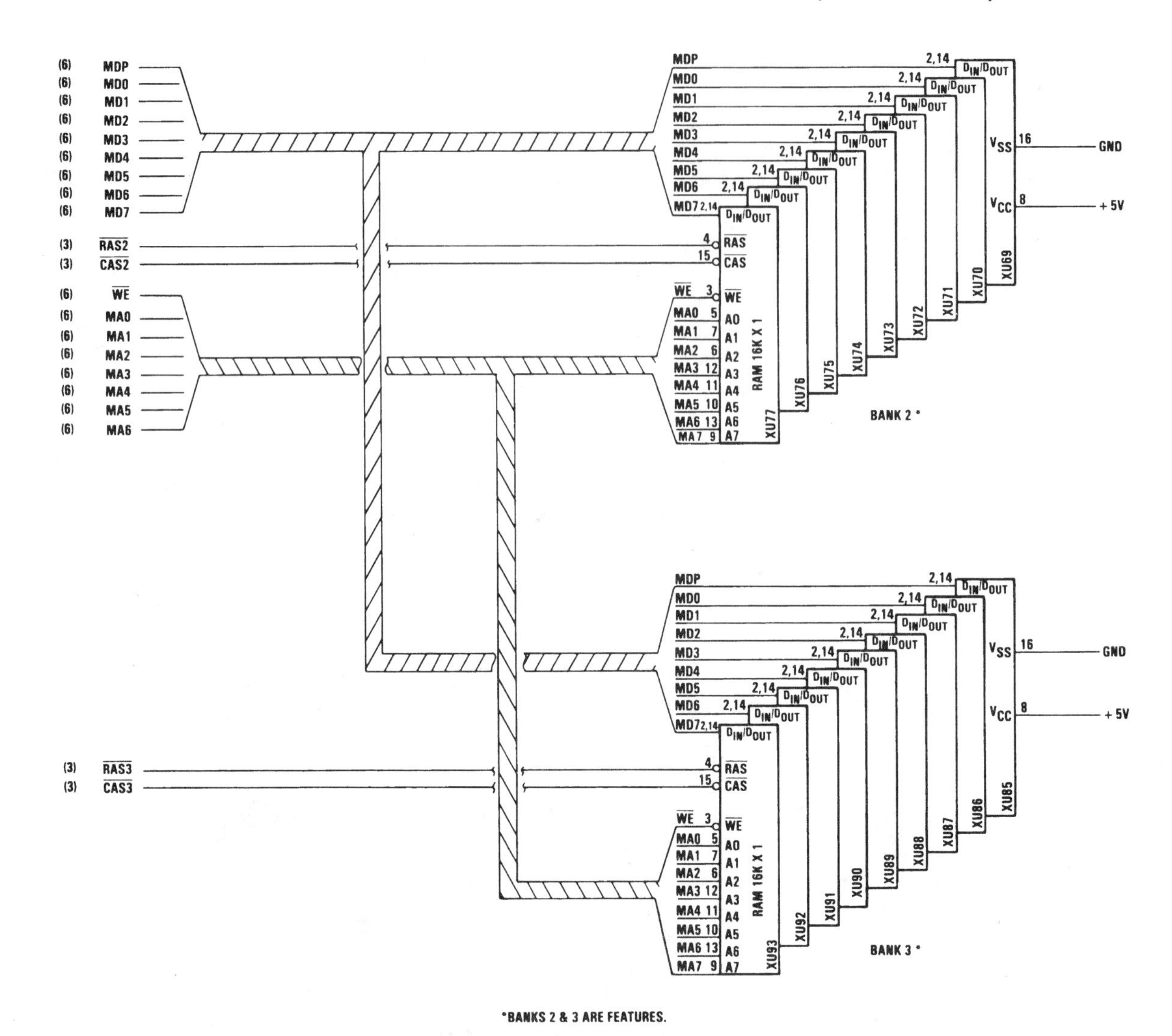

Figure 3-41 DRAM section part 2 (7 of 10). (Courtesy of IBM)

at any one time. When one set of MOSFETs is conducting, that data in the memory location is a logic 0; if the other set of MOSFETs is conducting, the data is a logic 1. The two sets of MOSFETs in the same memory location never conduct at the same time.

Because DRAM uses charge levels on the capacitance of the MOSFETs and the capacitance is very small, the charge levels must be refreshed (re-established) within a certain time period; otherwise, the data may change as a result of leakage of the capacitance. During refresh, the DRAM cannot be used to store or retrieve data; therefore, during this time period, the bus cannot be used.

PCs use DRAMs because they dissipate very little power (because data is stored as charge levels), have very high bit density (because of low power dissipation), and require fewer pin connections to address the memory (multiplexed address lines). These advantages mean that the cost of the PC is much lower, and it is easier to design. These advantages come at a price: DRAMs are somewhat slower than static RAMs. This slow speed does not have much effect on PC and XT class computers, but it does slow down AT class computers.

3.6.2 MT4264 DRAM Functional Block Diagram

Figure 3-42 is the functional block diagram for the MT4264 64K × 1 DRAM. This block diagram and the functions of each of the blocks are the same for all sizes of DRAMs; the only difference is in the size of the blocks.

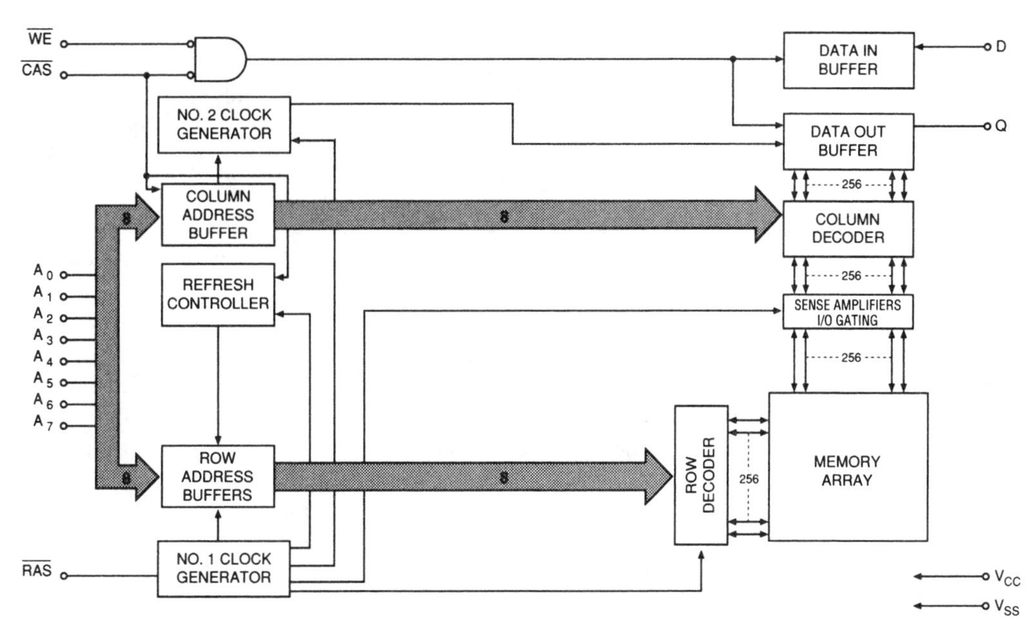

Figure 3-42 DRAM block diagram. (Reprinted by permission of Micron Technology, Inc.)

The row address buffer is used to hold the row part of the address during the DRAM cycle, once latched. The row part of the address is the lower 8 bits of the address that the MPU produces on the address bus. Once the address is latched into the row address buffer, its outputs are applied to the row decoder, which selects one of the rows in the memory array.

Clock generator 1 is used to latch the logic levels on the A0 to A7 address inputs of the DRAM IC into the row address buffer. The latching takes place when the $\overline{RAS}$ line goes low. The address decoding section of the PC causes the $\overline{RAS}$ line to go low at the proper time, and it stays low throughout the rest of the DRAM bus cycle. The $\overline{RAS}$ line, when low, also enables the sense amplifier I/O gating block.

The column address buffer is used to hold the column part of the address during the DRAM cycle, once latched. The column part of the address is the upper 8 bits of the address offset that the MPU produces on the address bus. Once the address is latched into the column address buffer, its outputs are applied to the column decoder, which selects one of the columns in the memory array.

Clock generator 2 is used to latch the logic levels on the A0 to A7 address inputs of the DRAM IC into the column address buffer. The latching takes place when the $\overline{CAS}$ line goes low. The address decoding section of the PC causes the $\overline{CAS}$ line to go low at the proper time. Once low, the $\overline{CAS}$ line remains low until the end of the DRAM bus cycle. The $\overline{CAS}$ line also completes the address of the DRAM and helps enable the data out and data in buffers.

The refresh controller circuitry is used for a hidden refresh cycle. Because the PC uses the DMA to perform refresh, this block is not used in the PC.

The row decoder gets its eight input levels from the row address buffer, which enables one of the 256 outputs. Each output is connected to one of the rows in the memory array. Once a row is selected, all MOSFETs connected to the row become enabled, which applies the charge levels from each MOSFET connected to the selected row to the columns to which the MOSFETs are connected. The row decoder has its outputs in tristate while the $\overline{RAS}$ line is high; this reduces the power dissipation of the DRAM.

The memory array is a series of 256 rows and 256 columns with a MOSFET connected between each row and column. The capacitance of this MOSFET holds the data in the form of a charge level. Because there are 256 rows and 256 columns, there is a total of 65,536 MOSFETs in the DRAM; therefore, this IC contains 65,536 bits of data. Each row has 256 columns connected to it; once a row is selected, the charge levels (data) from all 256 MOSFETs apply their levels to the columns to which they are connected.

The sense amplifiers I/O gating block is used to re-establish the charge levels on the columns once a row is selected and during a read operation. During a write cycle, the sense amplifiers overwrite the charge level on the selected column while re-establishing the levels on the unselected columns. Once the $\overline{RAS}$ line goes low, the sense amplifiers I\O gating becomes enabled. There is one sense amplifier for each column.

The column decoder contains eight address select inputs that are derived from the outputs of the column address buffer. These inputs determine which one of the 256 columns becomes enabled. The one enabled column allows data to pass between the data out buffer and sense amplifier I/O gating.

The output Q of the data out buffer becomes enabled only after the $\overline{RAS}$ has gone low, followed by the $\overline{CAS}$ line going low, when $\overline{WE}$ is high. While the IC is performing a read operation, the data from the selected column is transferred into the data out buffer and placed on its output Q. During a write operation, the output of the data out buffer is disabled, and the data from the data in buffer is passed through the data out buffer and then to the selected column.

The data in buffer is enabled only during a write cycle. This means that the $\overline{WE}$ line must be low and the $\overline{RAS}$ line must go low, followed by the $\overline{CAS}$ line going low. At this point, the data on the D input is placed in the data in buffer, which transfers the data to the data out buffer, which passes the logic level through to the selected column.

3.6.3 DRAM Internal Operation

Write Operation

1. The $\overline{RAS}$ line goes low and stays low throughout the remaining bus cycle. The high to low on the $\overline{RAS}$ line latches the logic levels on the A0 to A7 inputs into the row address buffers. The outputs from the row address buffers are applied to the row decoder. The row decoder selects one of the 256 rows in the memory array. All MOSFETs connected to the selected row become enabled and the charge levels on the MOSFETs are applied to the 256 columns in the memory array. When $\overline{RAS}$ goes low, the sense amplifiers I/O gating becomes enabled. The sense amplifiers begin to re-establish the charge levels on all 256 columns.
2. The $\overline{WE}$ line goes low, indicating a write operation, which enables the data in buffer and disables the output of the data out buffer as soon as the $\overline{CAS}$ line goes low. Data flows from outside the DRAM into the selected memory location.
3. The $\overline{CAS}$ line goes low and stays low throughout the remaining bus cycle. When the $\overline{CAS}$ line goes low, the logic levels on the A0 to A7 inputs are latched into the column address buffer. The outputs from the column address buffer are applied to the column decoder, which enables the data flow for one of the columns in the column decoder. The data from the data in buffer is applied to the data out buffer, which is passed through to the select column. The data causes the sense amplifier to overwrite the data (charge level) on the selected column. The sense amplifiers continue to re-establish the charge levels on the unselected columns of the memory array. The bus cycle begins to end when the $\overline{CAS}$ goes high, which disables the data out buffer, which turns off the overwriting of the selected column. This is followed by the $\overline{RAS}$ line going high, which deselects all rows in the memory array and disables the sense amplifiers. The memory array is now in tristate, which reduces power dissipation.

Read Operation

1. The $\overline{RAS}$ line goes low and stays low throughout the remaining bus cycle. The high to low on the $\overline{RAS}$ line latches the logic levels on the A0 to A7 inputs into the row address buffers. The outputs from the row address buffers are applied to the row decoder, which selects one of the 256 rows in the memory array. All MOSFETs connected to the selected row are enabled and the charge levels on the MOSFETs are

applied to the 256 columns in the memory array. By $\overline{\text{RAS}}$ going low, the sense amplifiers I/O gating becomes enabled. The sense amplifiers begin to re-establish the charge levels on all 256 columns.

2. The $\overline{\text{WE}}$ line goes high, indicating a read operation, which disables the data in buffer and enables the output of the data out buffer as soon as the $\overline{\text{CAS}}$ line goes low. Data flows from inside the DRAM to the outside world.
3. The $\overline{\text{CAS}}$ line goes low and stays low throughout the remaining bus cycle. When the $\overline{\text{CAS}}$ line goes low, the logic levels on the A0 to A7 inputs are latched into the column address buffer. The outputs from the column address buffer are applied to the column decoder, which enables the data flow for one of the columns in the column decoder. The data from the select column of the sense amplifiers I/O gating block is applied to the input of the data out buffers and then to the Q output of the DRAM. During this time, the sense amplifiers continue to re-establish the charge levels on all the columns of the memory array. The bus cycle begins to end when the $\overline{\text{CAS}}$ goes high, which disables the data out buffer, which in turn disables the output Q and places the output in tristate. This is followed by the $\overline{\text{RAS}}$ line going high, which deselects all rows in the memory array and disables the sense amplifiers. The memory array is now in tristate, which reduces power dissipation.

3.6.4 DRAM Pin Assignments

The following are the pin assignments for the MT4264 64K × 1 DRAM; these pins perform the same functions for all sizes of DRAMs. The big difference is the number of pins for each size of DRAM. Refer to Figures 3-43, 3-44, and 3-45.

Pin Assignments

A0–A7 — These are the address lines for the DRAM. When $\overline{\text{RAS}}$ goes low, the levels on these inputs are latched into the DRAM and used to select the row part of the address. When $\overline{\text{CAS}}$ goes low, the levels on these inputs are latched into the DRAM and used to select the column part of the address. It is the responsibility of the address-decoding section of the PC to make sure that the proper levels are on these inputs at the proper times.

$\overline{\text{RAS}}$ — The $\overline{\text{Row Address Strobe}}$ line is used to latch the row part of the address into the DRAM when the line goes low. The address decoding section of the PC determines when this line goes low and keeps it low until the end of bus cycle.

$\overline{\text{CAS}}$ — The $\overline{\text{Column Address Strobe}}$ line is used to latch the column part of the address into the DRAM when the line goes low. This line also enables the data in buffer or data out buffer and their outputs, depending on the operation. The address-decoding section of the PC determines when this line goes low and keeps it low until the end of the bus cycle.

$\overline{\text{WE}}$ — The $\overline{\text{Write Enable}}$ is used to determine where the DRAM accepts the data from the data bus (write operation) or places the DRAM's data on the data bus (read operation).This line must be low before the $\overline{\text{CAS}}$ line goes low, and it must stay low throughout the remaining write cycle. This line must be high before the $\overline{\text{CAS}}$ line goes low, and it must stay high throughout the remaining read cycle.

D — The Data input is the line that allows the data on the data bus to be transferred into the DRAM during a write operation.

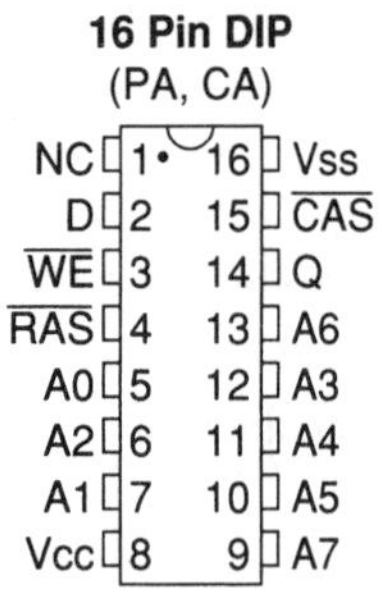

16 Pin DIP
(PA, CA)

Pin	Signal	Pin	Signal
1	NC	16	Vss
2	D	15	$\overline{CAS}$
3	$\overline{WE}$	14	Q
4	$\overline{RAS}$	13	A6
5	A0	12	A3
6	A2	11	A4
7	A1	10	A5
8	Vcc	9	A7

Figure 3-43 MT4264 64K × 1 DRAM pin assignment. (Reprinted by permission of Micron Technology, Inc.)

18 Pin DIP
(PB, CB)

Pin	Signal	Pin	Signal
1	$\overline{OE}$	18	Vss
2	DQ1	17	DQ4
3	DQ2	16	$\overline{CAS}$
4	$\overline{WE}$	15	DQ3
5	$\overline{RAS}$	14	A0
6	A6	13	A1
7	A5	12	A2
8	A4	11	A3
9	Vcc	10	A7

20 Pin ZIP
(ZB)

Pin	Signal	Pin	Signal
1	DQ3	2	$\overline{CAS}$
3	DQ4	4	Vss
5	$\overline{OE}$	6	DQ1
7	DQ2	8	$\overline{WE}$
9	$\overline{RAS}$	10	NC
11	NC	12	A6
13	A5	14	A4
15	Vcc	16	A7
17	A3	18	A2
19	A1	20	A0

18 Pin PLCC
(EJA)

Pin	Signal	Pin	Signal
1	$\overline{OE}$	10	A7
2	DQ1	11	A3
3	DQ2	12	A2
4	$\overline{WE}$	13	A1
5	$\overline{RAS}$	14	A0
6	A6	15	DQ3
7	A5	16	$\overline{CAS}$
8	A4	17	DQ4
9	Vcc	18	Vss

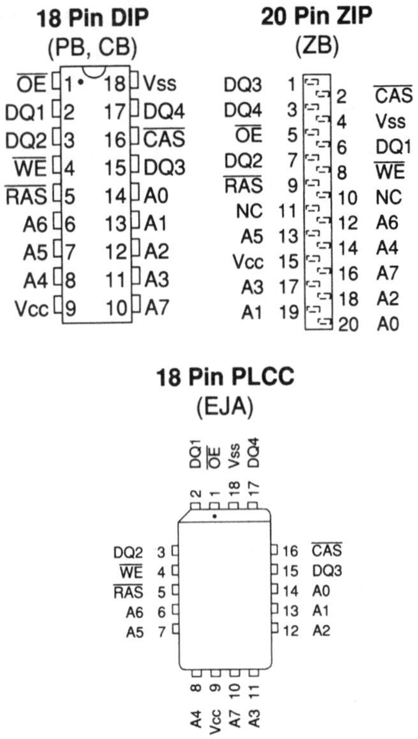

Figure 3-44 MT4067 64K × 4 DRAM pin assignment. (Reprinted by permission of Micron Technology, Inc.)

16 Pin DIP
(PA, CA)

Pin	Signal	Pin	Signal
1	A8*	16	Vss
2	D	15	$\overline{CAS}$
3	$\overline{WE}$	14	Q
4	$\overline{RAS}$	13	A6
5	A0	12	A3
6	A2	11	A4
7	A1	10	A5
8	Vcc	9	A7

16 Pin ZIP
(ZA)

Pin	Signal	Pin	Signal
1	A6	2	Q
3	$\overline{CAS}$	4	Vss
5	A8*	6	D
7	$\overline{WE}$	8	$\overline{RAS}$
9	A0	10	A2
11	A1	12	Vcc
13	A7	14	A5
15	A4	16	A3

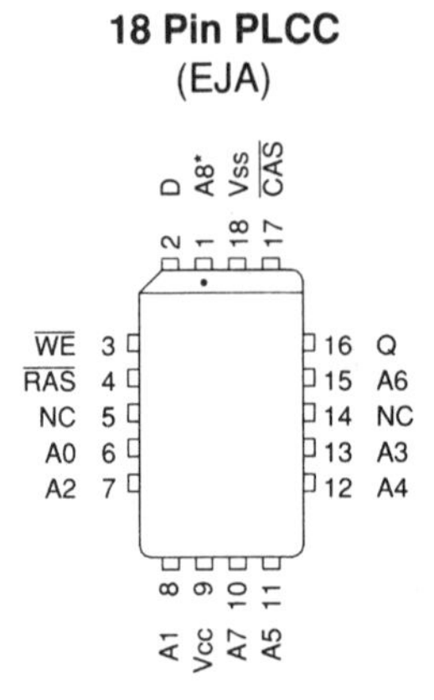

Figure 3-45 MT1259 256K × 1 DRAM pin assignment. (Reprinted by permission of Micron Technology, Inc.)

* ADDRESS NOT USED FOR $\overline{RAS}$ ONLY REFRESH

Q The Q output is the line that allows the data in the DRAM to be placed on the data bus during a read operation. Because today's PCs contain a bidirectional bus, the D and Q pins of the DRAM are tied together.

3.6.5 The ICs in the RAM Section of the PC

The RAM section of the IBM PC (Figs. 3-40 and 3-41) and xxxPC contain not only the DRAMs for the computer, but many support ICs. The support ICs perform a number of different functions in the RAM section. The following describes the functions of these support ICs.

74245. The data bus transceiver in the RAM section of the PC is used to allow data to pass between the data bus from the microprocessor subsystem and the data bus of the RAM section. The data is allowed through the data bus transceiver only when a valid RAM address is on the data bus and a memory read or write is being performed. The control signal that enables the data to flow is produced by the address-decoding section of the PC ($\overline{\text{RAM ADDR SEL}}$) and is an active low. The direction of data flow is determined by the level on the $\overline{\text{XMEMR}}$ line; when low, data flows from the RAM section to the microprocessor subsystem data bus. When high, data flows from the microprocessor data bus into the RAM section of the PC.

74158. Two Quad 2-to-1 Multiplexer with inverted outputs are used to multiplex the address from the microprocessor subsystem address bus for the RAM address bus. When the ADDR SEL line (which is produced by the address-decoding section) is low (beginning of the bus cycle), the levels on address lines A0 to A7 of the address bus from the microprocessor subsystem are applied to Memory Address lines MA0 to MA7 (where MA0 = A0, MA1 = A1, MA2 = A2, MA3 = A3, MA4 = A4, MA5 = A5, MA6 = A6, and MA7 = A7) on the RAM section. The levels on each MAx line are inverted from the levels on the A0 to A7 inputs of the address multiplexers. The MA0 to MA7 lines are connected to the address inputs A0 to A7 of each of the DRAMs in the PC. Seventy-five nano seconds after the memory read or write line has gone low, the ADDR SEL line goes high. A high on the S (select) input causes the levels on address lines A8 to A15 to be applied to MA0 to MA7 (where MA0 = A8, MA1 = A9, MA2 = A10, MA3 = A11, MA4 =A12, MA5 = A13, MA6 = A14, MA7 = A15). The inversion of the address levels has no adverse effect on the RAM section, except that if you try to view the address from the address lines of DRAMs with a logic analyzer or probe, each address bit is the opposite of the levels from the address latches of the microprocessor subsystem.

74280. The 9-bit parity generator/checker is used to produce a parity bit and to determine if a parity error has occured. This IC has two outputs, labeled summation ODD and summation EVEN, and nine inputs labeled A to I. If the total number of logical 1s is even (0, 2, 4, 6, 8), the EVEN output is high and the ODD output is low. If the total number of logical 0s is odd (1, 3, 5, 7, 9), the ODD output is high and the EVEN output is low. How the PC uses this IC is determined by the supporting IC in the RAM section.

The IBM PC and xxxPCs use odd parity; this means that every memory location has an odd number of logical 1s. Each memory location contains 8 data bits and 1 parity bit. The 8 data bits are the data the user places in memory. The parity bit is an additional bit used only by the parity generator/checker. The parity bit is generated during the write cycle; the bit is either high or low, depending on the number of logical 1s in the data byte. The following example assumes that the computer is set for odd parity. If the number of logical 1s in the data byte is odd, the parity generator produces a logic-low parity bit. If the number of logical 1s in the data byte is even, the parity generator produces a logic-high parity bit. Therefore, an odd number of logical 1s is always stored in every memory location. At the end of the read cycle, the parity checker in the PC reads the data byte and parity bit; if the total number of logical 1s is odd, no parity error has occurred. If the total number of logical 1s is even, a parity error has occurred, which produces an NMI in the 8088 MPU.

Note: The circuitry in Figure 3-41 shows two additional banks of DRAMs for the IBM PC; these two banks of DRAMs use the same signals and lines as in Figure 3-40.

3.6.6 DRAM Timing Diagrams

The following describes the DRAM write operation, using the timing diagram in Figure 3-46.

1. The 8088 places the address on the local address bus (AD0 to AD7, A8 to A15, and A16 to A19).
2. The 8088 sends a write memory status code to the 8288 bus controller ($\overline{S0}$, $\overline{S1}$, and $\overline{S2}$).
3. The bus controller latches the address on the local bus to the address latches on the microprocessor subsystem using the ALE line. The address becomes valid on the external address bus about 140 ns into the bus cycle.
4. While the bus controller is latching an address into the address latches, the DT/$\overline{R}$ line goes high. The high selects the direction of data flow through the 74245 data bus transceiver in the microprocessor subsystem. Once enabled, the data flows from the local to the external data bus.
5. The bus controller causes the $\overline{MEMW}$ line to go low, which causes the $\overline{XMEMW}$ line to go low.
6. The low on the $\overline{XMEMW}$ line causes the ADDR SEL line in the address-decoding section to stay low.
7. The ADDR SEL line is applied to the S select inputs of the 74158 RAM address bus multiplexers. This causes the logic levels on the A0 to A7 (row address) address

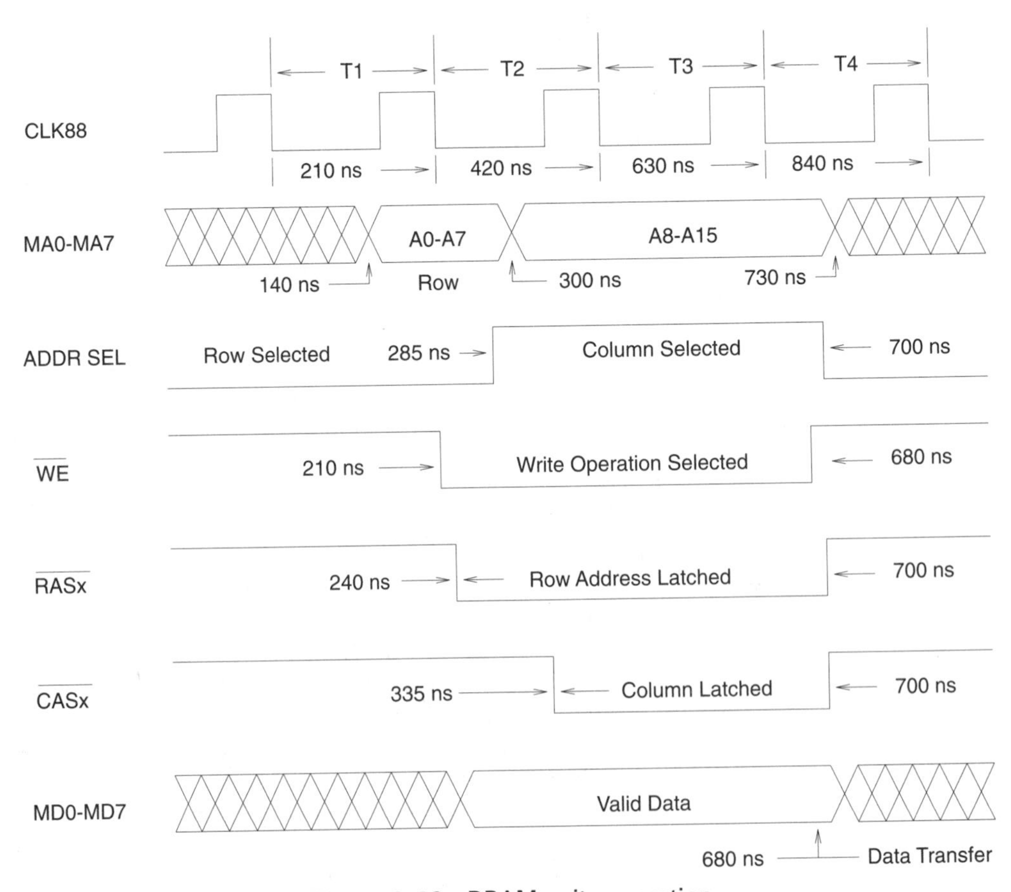

Figure 3-46 DRAM write operation.

lines from the external address bus to apply their levels to the MA0 to MA7 RAM memory address bus. The MA0 to MA7 lines are connected to the A0 to A7 address inputs of all DRAMs in the RAM section.

8. With the $\overline{\text{XMEMW}}$ low, the $\overline{\text{XMEMR}}$ line is high. The high on the $\overline{\text{XMEMR}}$ sets the direction of the 74245 data bus transceiver in the RAM section. Once enabled, the data flows from the external data (D0 to D7) bus into the RAM data (MD0 to MD7) bus.
9. The low on the $\overline{\text{XMEMW}}$ line also applies a low to the $\overline{\text{WE}}$ on all the DRAMs in the RAM section. The $\overline{\text{WE}}$ line enables the 74125 tristate buffer connected to the EVEN output of the 74280 parity generator/checker. The output of the tristate buffer is connected to the data line of all parity DRAMs.
10. The high on the $\overline{\text{XMEMR}}$ line is inverted by U83, and its output is used to place a low on the input of a two-input AND gate (U97). The output from the AND gate is low, which is applied to the I input of the 74280. The I input is always low during a write operation.
11. The address decoding section causes the $\overline{\text{RAM ADDR SEL}}$ to go low. This low causes the 74245 data bus transceiver in the RAM section to become enabled.
12. The $\overline{\text{RAM ADDR SEL}}$ line, along with the address lines connected to the address decoding section, causes one of the $\overline{\text{RASx}}$ lines to go low (240 ns into the bus cycle). The value of "x" is the value of the page of the current address. The high-to-low transition on the $\overline{\text{RASx}}$ line latches the logic levels on the MA0 to MA7 lines into the DRAMs that are connected to the $\overline{\text{RASx}}$ line that went low. The address-decoding section keeps the $\overline{\text{RASx}}$ line low until the end of the bus cycle. Only one bank of DRAMs is selected at any one time. At this time, the MA0 to MA7 memory address bus represents the row part of the address (A0 to A7).
13. The 8088 MPU places its data on the data bus.
14. The bus controller causes the DEN line to go high, which enables the data bus transceiver in the microprocessor subsystem. The data begins to flow from the local data (AD0 to AD7) bus through the data bus transceiver to the external data (D0 to D7) bus, and then through the data bus transceiver in the RAM section to the RAM data (MD0 to MD7) bus.
15. At 285 ns into the bus cycle, the address-decoding section causes the ADDR SEL line to go high, which causes the logic levels on the A8 to A15 address (column address) lines from the microprocessor subsystem to be applied to the RAM memory address bus MA0 to MA7 (where A8 = MA0,. . ., A15 = MA7). The logic levels on the MA0 to MA7 address lines are applied to the A0 to A7 address inputs for all DRAMs.
16. At this time, the data from the local bus is now valid on the RAM memory data bus, which is connected to all DRAMs, and the A to H (MD0 = A, . . ., MD7 = H) inputs of the 74280 parity generator/checker. (Example data = 3F hex.) Because in this example there are seven logical 1s in the data byte and the I input to the 74280 is 0 (from step 10), the EVEN output goes low and the ODD output goes high. The low on the EVEN output is applied to the data I/O line of the parity DRAMs of the RAM section.
17. The extended control logic causes the RDY/$\overline{\text{WAIT}}$ line to go high. This high causes the READY output from the 8284 to go low, which is applied to the 8088.
18. At 335 ns into the bus cycle, the address decoding section causes the $\overline{\text{CASx}}$ (the value of "x" is the same value used with the $\overline{\text{RASx}}$ line) to go low. The low on the $\overline{\text{CASx}}$ line latches the logic levels on the MA0 to MA7 (A8 to A15 column address) into the selected bank of DRAMs. The address-decoding section keeps this line low until the end of the bus cycle. The $\overline{\text{CASx}}$ line going low completes the address for the DRAM and causes the DRAMs in the selected bank of memory to begin writing the data into their selected memory location (MD0 to MD7 in the data-byte DRAMs and the output from the EVEN output of the 74280 into the parity DRAM).

If the data stored in the memory location is 3F hex in the data-byte DRAMs and a low parity bit in the parity DRAM, the total number of logical 1s is odd. (*Note*: The logic levels on all lines should remain the same until the next step.)

19. At about 510 ns into the bus cycle, the extended control logic causes the $\overline{\text{RDY}}$/WAIT line to go low. This low causes the READY output of the 8284 to go high about 40 ns before the falling edge of CLK88.
20. At 600 ns into the bus cycle (near the end of T3), the 8088 reads the READY line. Because the READY line is high, the 8088 informs the bus controller to end the bus cycle within one more clock cycle. The status lines to the bus controller go passive.
21. About 650 ns into the bus cycle, the bus controller causes the $\overline{\text{MEMW}}$ line to go high. This high is applied to the $\overline{\text{XMEMW}}$ line, which causes the address-decoding section to bring the $\overline{\text{CASx}}$ and $\overline{\text{RASx}}$ lines high. (Whatever logic levels are on the data inputs to the DRAMs in the selected bank of DRAMs are latched into the memory locations in the DRAMs.) The $\overline{\text{WE}}$ goes high and ADDR SEL line goes low.
22. The bus control disables the data bus transceiver and sets the DT/$\overline{\text{R}}$ line in the microprocessor subsystem. The 8088 stops producing data for the data bus. The cycle is now complete.

DRAM Read Operation. The following describes the DRAM read operation using Figure 3-46 as a reference point and the timing diagram in Figure 3-47.

1. The 8088 places the address on the local address bus (AD0 to AD7, A8 to A15, and A16 to A19).

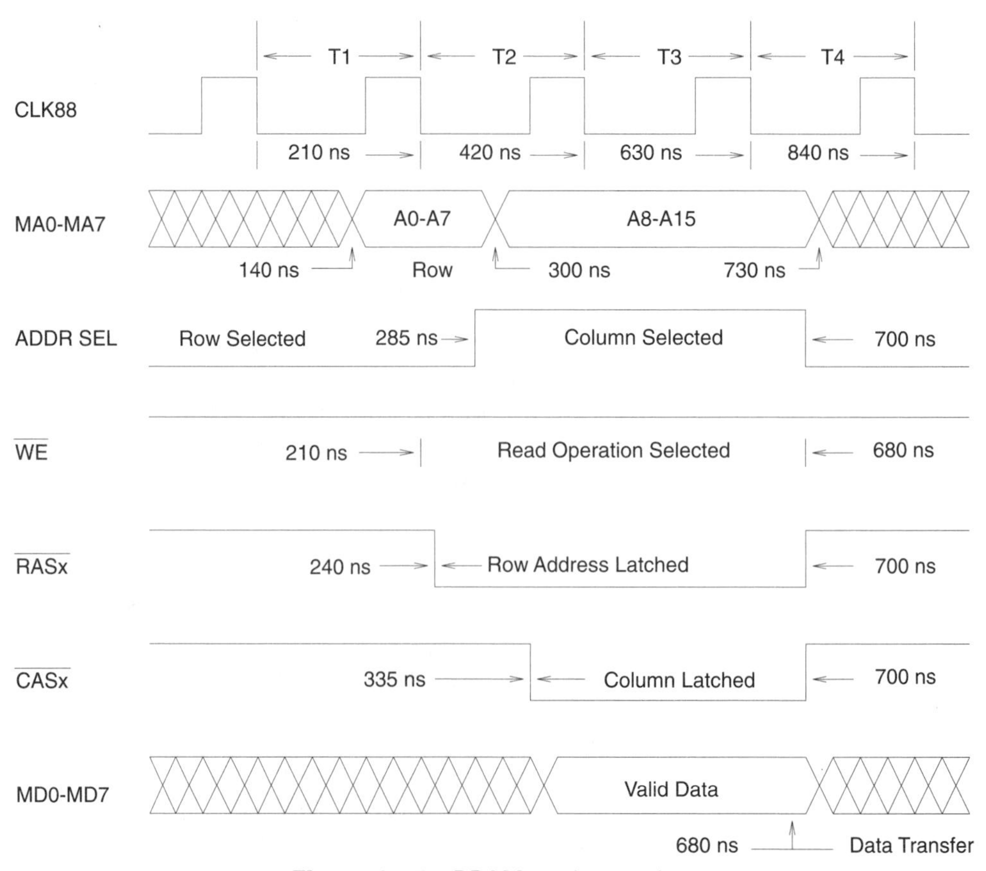

Figure 3-47 DRAM read operation.

2. The 8088 sends a read memory status code to the 8288 bus controller ($\overline{\text{S0}}$, $\overline{\text{S1}}$, and $\overline{\text{S2}}$).
3. The bus controller latches the address on the local bus to the address latches on the microprocessor subsystem using the ALE line. The address becomes valid on the external address bus about 140 ns into the bus cycle.
4. While the bus controller is latching an address into the address latches, the DT/$\overline{\text{R}}$ line goes low. The low selects the direction of data flow through the 74245 data bus transceiver in the microprocessor subsystem. Once enabled, the data flows from the external to the local data bus.
5. The bus controller causes the $\overline{\text{MEMR}}$ line to go low, which causes the $\overline{\text{XMEMR}}$ line to go low.
6. The low on the $\overline{\text{XMEMR}}$ line causes the ADDR SEL line in the address-decoding section to stay low.
7. The ADDR SEL line is applied to the S select inputs of the 74158 RAM address bus multiplexers. This causes the logic levels on the A0 to A7 (row address) address lines from the external address bus to apply their levels to the MA0 to MA7 RAM memory address bus. The MA0 to MA7 lines are connected to the A0 to A7 address inputs of all DRAMs in the RAM section.
8. With the $\overline{\text{XMEMR}}$ low, the $\overline{\text{XMEMW}}$ line is high. The low on the $\overline{\text{XMEMR}}$ sets the direction of the 74245 data bus transceiver in the RAM section. Once enabled, the data flows from the RAM data (MD0 to MD7) bus into the external data (D0 to D7) bus.
9. Because the $\overline{\text{XMEMW}}$ line is high, the $\overline{\text{WE}}$ on all the DRAMs in the RAM section is high. The $\overline{\text{WE}}$ line disables the 74125 tristate buffer connected to the EVEN output of the 74280 parity generator/checker. This allows the parity bit that will be developed by the selected parity DRAM to be supplied to one of the inputs of the two-input AND gate (U97).
10. The low on the $\overline{\text{XMEMR}}$ line is inverted by U83, and its output is used to place a high on one of the inputs of a two-input AND gate (U97). The other input is the parity bit input, developed by a read operation on the parity DRAM. The output of the AND gate is at the same level as the parity bit because a high is placed on the other input of the inverter (U83). The output of the AND gate is applied to the I input of the 74280 parity generator/checker. During a read operation, the I input always represents the value of the parity bit.
11. The address-decoding section causes the $\overline{\text{RAM ADDR SEL}}$ to go low. This low causes the 74245 data bus transceiver in the RAM section to become enabled.
12. The $\overline{\text{RAM ADDR SEL}}$ line, along with the address lines connected to the address-decoding section, causes one of the $\overline{\text{RASx}}$ lines to go low (240 ns into the bus cycle). The value of "x" is the value of the page of the current address. The high-to-low transition on the $\overline{\text{RASx}}$ line latches the logic levels on the MA0 to MA7 lines into the DRAMs connected to the $\overline{\text{RASx}}$ line that went low. The address-decoding section keeps the $\overline{\text{RASx}}$ line low until the end of the bus cycle. Only one bank of DRAMs is selected at any one time. At this time, the MA0 to MA7 memory address bus represents the row part of the address (A0 to A7).
13. At 285 ns into the bus cycle, the address-decoding section causes the ADDR SEL line to go high, which causes the logic levels on the A8 to A15 address (column address) lines from the microprocessor subsystem to be applied to the RAM memory address bus MA0 to MA7 (where A8 = MA0, . . ., A15 = MA7). The logic levels on the MA0 to MA7 address lines are applied to the A0 to A7 address inputs to all DRAMs.
14. The extended control logic causes the $\overline{\text{RDY}}$/WAIT line to go high. This high causes the READY output from the 8284 to go low, which is applied to the 8088.
15. At 335 ns into the bus cycle, the address-decoding section causes the $\overline{\text{CASx}}$ (the value of "x" is the same value used with the $\overline{\text{RASx}}$ line) to go low. The low on the $\overline{\text{CASx}}$

line latches the logic levels on the MA0 to MA7 (A8 to A15 column address) into the selected bank of DRAMs. The address-decoding section keeps this line low until the end of the bus cycle. The $\overline{\text{CASx}}$ line going low completes the address for the DRAM and causes the DRAMs in the selected bank of memory to begin outputting the data from their selected memory location (MD0 to MD7 for the data-byte DRAMs and the parity DRAM output applied to the U97 AND gate). The output from the AND gate (which represents the parity bit) is applied to the I input of the 74280 parity generator/checker. If the data from the data byte is 3F hex from the previous example, then the parity bit is low. The total number of logical 1s is odd in the IBM PC, and compatibles should have a odd number of logical 1s in all memory locations.

16. The bus controller causes the DEN line to go high, which enables the data bus transceiver in the microprocessor subsystem. The data begins to flow from the RAM section data bus (MD0 to MD7) through the RAM data bus transceiver to the external data bus (D0 to D7) and through the data bus transceiver in the microprocessor subsystem to the local data (AD0 to AD7) bus.
17. At this time, the data from the RAM data bus is now valid. The RAM memory data bus is connected to all DRAMs and the A to H (MD0 = A, . . ., MD7 = H) inputs of the 74280 parity generator/checker. (Example data = 3F hex.) The value on the memory data bus is 3F hex and is applied to the A to H inputs of the 74280, and the parity bit (low in this example) is applied to input I of the 74280. The EVEN output is high but is not allowed to exit the tristate buffer (U80); therefore, this output has no effect on the PC. The ODD output of the 74280 is high, and this high is applied to an inverted input to the U27 two-input AND gate. Because the input to the inverted input of the AND gate is high, the AND gate has a logic low on one of its inputs. This produces a logic low on the output of the AND (U27) gate, even though the $\overline{\text{RAM ADDR SEL}}$ line is low, which is inverted and places a logic high on the other input of the AND gate. The low from the AND gate is applied to the D data input of the 7474 (U96). The data is not latched into the data latch until the end of the read operation, when the $\overline{\text{XMEMR}}$ lines go high. The transition causes the logic level on the data input to be latched into the data latch. (*Note:* The logic levels on all lines should remain the same until the next step.)
18. The 8088 begins reading the data from the data bus.
19. At about 510 ns into the bus cycle, the extended control logic causes the $\overline{\text{RDY/WAIT}}$ line to go low. This low causes the READY output of the 8284 to go high about 40 ns before the falling edge of CLK88.
20. At 600 ns into the bus cycle (near the end of T3), the 8088 reads the READY line. Because the READY line is high, the 8088 informs the bus controller to end the bus cycle within one more clock cycle. The status lines to the bus controller go passive.
21. About 650 ns into the bus cycle, the bus controller causes the $\overline{\text{MEMR}}$ line to go high. This high is applied to the $\overline{\text{XMEMR}}$ line, which causes the address-decoding section to bring the $\overline{\text{CASx}}$ and $\overline{\text{RASx}}$ lines high. (Whatever logic levels are on the data inputs to the 8088 MPU are latched into the 8088.) The ADDR SEL line goes low. The DRAMs stop producing data for the data bus.
22. On the low-to-high transition of the $\overline{\text{XMEMR}}$ line, the logic level on the data input of the data latch is latched into the data latch (U96). In our example, the Q output of the data latch goes low, which is applied to the PPI section of the PC for the I/O channel bus. The $\overline{\text{Q}}$ output is high, which is applied to the extended control logic. A high on this input indicates no parity error. If the $\overline{\text{Q}}$ output is low, the extended control logic causes an NMI output to go high, which is applied to the NMI input of the 8088. A high on the NMI input of the 8088 causes a parity error sequence to begin.
23. The bus control disables the data bus transceiver and sets the DT/$\overline{\text{R}}$ line in the microprocessor subsystem. The cycle is now complete.

DRAM Refresh Cycle. The IBM PC and compatibles use RAS-only distributed refreshing to maintain valid data in DRAM memory. In RAS-only refreshing the DMA places an address on the address bus. The address-decoding section, along with the help of the control lines from the DMA ($\overline{\text{DACK0}}$), causes the address on the address bus to be latched into all $\overline{\text{RASx}}$ lines at the same time (row address). During this process, none of the $\overline{\text{CASx}}$ lines goes low; therefore, none of the DRAMs becomes enabled enough to place data on the data bus. The selection of the row address is enough internally to re-establish the charge levels on the memory cells connected to the selected row. Hence the name RAS-only refresh. The distributed part of the refresh cycle comes from the fact that one row of memory is refreshed every 72 clock cycles (about 15 μs). Because there are 256 rows in the DRAMs and one refresh cycle takes about 1 μs, once every 15 μs (16 μs total) all memory is refreshed every 4 ms. This means about 7% of the bus cycle is used for refresh and cannot be used to process data transfers. In a PC compatible running at turbo speeds about 3.5% of the bus cycle is used for refresh. Figure 3-48 shows the DRAM cycle, and the following description describes how the refresh cycle works. It will be useful to refer to Figures 3-1, 3-26, 3-27, 3-31, 3-40 and 3-49 when reading the following description.

1. Every 15μs (72 clocks) the programmable interval time (PIT) causes the DREQ0 line to go high.
2. The high on the DREQ0 line begins a DMA cycle on channel 0 of the DMA.
3. The DMA brings the HRQ line high. This informs the external control logic that the DMA wants access to the bus.
4. The extended control logic causes the $\overline{\text{DMA WAIT}}$ line to go low.
5. The $\overline{\text{DMA WAIT}}$ causes the 8088 to go into a wait state after completing the current 8088 instruction.

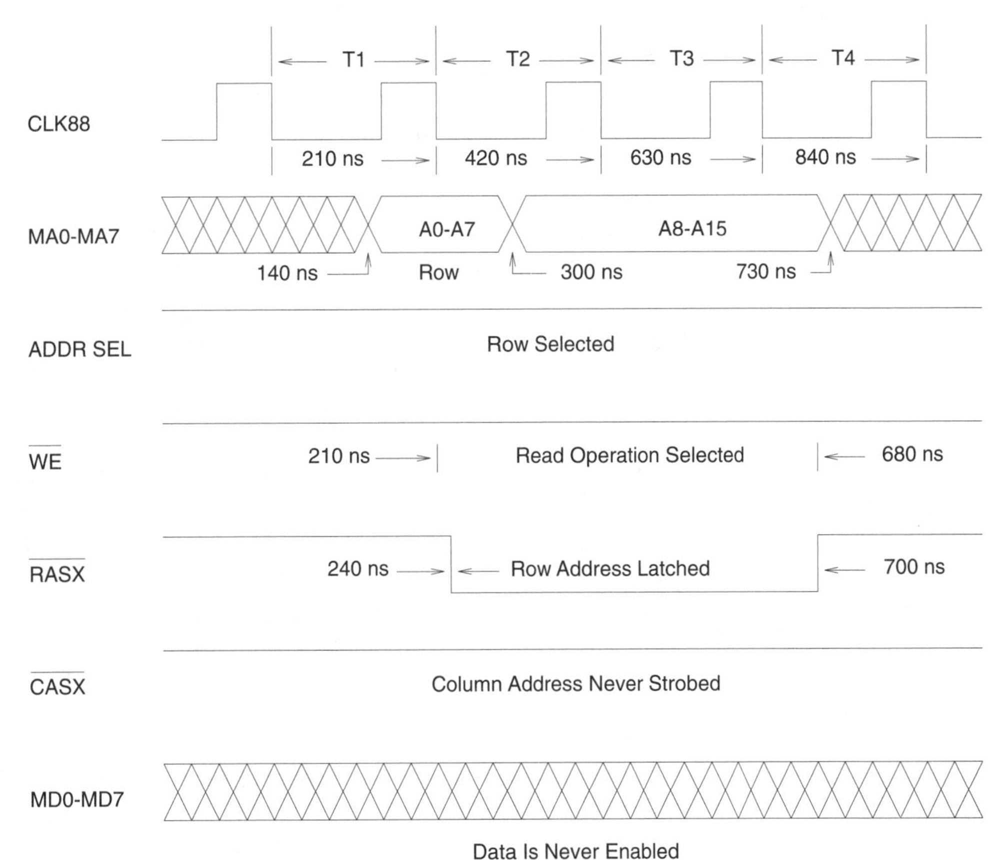

Figure 3-48 DRAM refresh operation.

6. The extended control logic causes the outputs of the bus controller to go into tristate by bringing the $\overline{\text{AEN}}$ line low.
7. The extended control logic causes the outputs of the address latches in the microprocessor subsystem to go into tristate by bringing the AEN BRD line high. At this time, the local bus is disabled from the external bus.
8. The extended control logic brings the HLDA line high, which tells the DMA to take control of the external bus.
9. The extended control logic enables the address latches in the DMA section; $\overline{\text{DMA AEN}}$ is low.
10. The DMA places the low byte of the address offset on the address bus (this stays valid for the rest of the DMA cycle).
11. The DMA places the high byte of the address offset on the data bus D0 to D7.
12. While the data bus represents the high byte of the address offset, the DMA pulses the ADSTB line high. This latches the high byte of the address offset on the data bus D0 to D7 into the address latches of the DMA section. The high byte of the address offset becomes valid.
13. The DMA causes the $\overline{\text{MEMR}}$ line to go low.
14. The low on the $\overline{\text{MEMR}}$ line causes the address decoding section to produce a low on the ADDR SEL line.
15. The low on the $\overline{\text{MEMR}}$ line causes the $\overline{\text{WE}}$ line in the DRAMs to go high, which places all DRAMs in the read mode. The tristate buffer on the output from the EVEN output of the 74280 parity generator/checker becomes disabled.
16. Data flows through the RAM data bus transceiver from the RAM data bus to the external data bus.
17. The low on the ADDR SEL line causes the address multiplexers in the RAM section to place the row part of the address (A0 to A7) on the external address bus to the RAM address bus (MA0 to MA7).
18. The DMA causes the $\overline{\text{DACK0}}$ output line to go low, this line is renamed $\overline{\text{DACK 0 BRD}}$, which is applied to the extended control logic section, the address decoding section, the ROM section, and the PIT section. The $\overline{\text{DACK0}}$ output from the DMA is also inverted to produce the DACK 0 line for the address decoding section.
19. The address-decoding section causes the $\overline{\text{RAM ADDR SEL}}$ line to go low, which enables the data bus transceiver in the RAM section.
20. The high on the DACK 0 line applied to the address decoding section causes the refresh gate (U81) to go low. This causes all $\overline{\text{RASx}}$ lines in the computer to go low, which latches the address on the RAM address into all DRAMs in the computer at the same time. This is the only time more than one $\overline{\text{RASx}}$ ever goes low at the same time.
21. The low on the $\overline{\text{DACK 0 BRD}}$ line applied to the extended control logic causes the extended control logic to produce a low on the RDY TO DMA line.
22. The DMA causes the $\overline{\text{IOW}}$ line to go low, but because no I/O device was ever enabled by any $\overline{\text{CS}}$ line, this line's going low has no effect on the PC.
23. The low on the $\overline{\text{DACK 0 BRD}}$ causes the data latch (U67) in the PIT section to reset. This causes the DREQ0 line to go low. (*Note*: Because no $\overline{\text{CASx}}$ line is selected by the address-decoding section because the DMA caused all the $\overline{\text{RASx}}$ lines to go low, the DRAMs do not put any of their data on the data bus. The DRAMs are still in tristate.)
24. After about 250 ns, the extended control logic causes the RDY TO DMA line to go high.
25. The DMA causes the $\overline{\text{IOW}}$ line to go high, followed by the $\overline{\text{MEMR}}$ lines, followed by the $\overline{\text{DACK0}}$ line.

26. The high on the $\overline{\text{DACK0}}$ line causes the $\overline{\text{DACK 0 BRD}}$ line to go high, which causes the extended control logic to cause the HLDA line to go low. This disables the DMA and allows the 8088 to regain control of the bus. The $\overline{\text{DMAWAIT}}$ line goes high, and the READY line to the 8088 goes high. The $\overline{\text{AEN}}$ line goes high to enable the control outputs of the bus controller. AEN BRD goes low to enable the address latches in the microprocessor subsystem.

3.6.7 DRAM Troubleshooting

Condition: None of the $\overline{\text{RASx}}$ lines goes low, even when no program is running.
Result: Memory errors occur, computer halts.
Cause: Check the DREQ0 line in the DMA section; this line should go high once every 16 μs. If it does not, troubleshoot the PIT section. If it does, check to make sure the HRQ line goes high. If HRQ goes high, troubleshoot the extended control logic section. If HRQ does not go high, replace the DMA.
Cause: If $\overline{\text{DACK0}}$ goes low once every 16 μs, troubleshoot the address-decoding section. If the $\overline{\text{DACK0}}$ line does not go low, troubleshoot the DMA or extended control logic section.

Condition: The $\overline{\text{CASx}}$ line goes low before the $\overline{\text{RASx}}$ line goes low during any RAM operation.
Condition: The $\overline{\text{CASx}}$ line goes low, but the $\overline{\text{RASx}}$ line never goes low during any RAM operation.
Condition: The values of "x" in the $\overline{\text{RASx}}$ and $\overline{\text{CASx}}$ lines are not the same during any RAM operation.
Condition: The $\overline{\text{RASx}}$ line goes low, followed by the $\overline{\text{CASx}}$ line going low, during a valid RAM operation.
Condition: The ADDR SEL line does not switch from a low to a high during a valid RAM operation.
Result: Computer halts.
Cause: Troubleshoot the address-decoding section.

Condition: The levels on the RAM memory address bus (MA0 to MA7) do not change during most RAM addresses (the only exception is when the low byte of the address is the same value as the high byte of the address) in a valid RAM operation.
Result: Computer halts.
Cause: Check for a change in ADDR SEL line. This line should be low at the beginning of the RAM operation and should go high partway through the RAM operation. If the ADDR SEL line does not change levels or starts at a low and then goes high, troubleshoot the address-decoding section. If the ADDR SEL line changes correctly, go to the next cause.
Cause: Troubleshoot the RAM address multiplexers in the 74158s in the IBM PC.

Condition: The ODD output of the parity generator/checker is low at the end of a RAM memory read.
Result: Computer generates parity errors; computer may halt.
Cause: Check power supply for noise on the + 5 V applied to the system board.
Cause: Check for a low on any $\overline{\text{RASx}}$ line on any DRAM once every 16 μs. If not present, go to the first troubleshooting condition in this section.
Cause: If the above two causes are not the problem, run a memory diagnostic test program (like the IBM PC Advance Diagnostic or CHECK-IT Diagnostic programs). These programs perform a sequence of read and write operations and produce a code that indicates the DRAM or DRAMs that may be bad. These programs should at least indicate the bad bank of memory. Replace the DRAM or DRAMs in question.

DRAM SUMMARY

1. The RAM section has its own data bus transceiver, which allows data to flow between the RAM data bus and the external data bus only when a valid RAM address is selected. This reduces the loading of the external data bus. The direction of data flow is determined by the $\overline{\text{XMEMR}}$ line.
2. The RAM section has a multiplexed address bus called the memory address bus (MA0 to MA7). The address-decoding section causes the address multiplexers in the RAM section to place the row part of the address (low byte of the address offset, A0 to A7) and the column part of the address (high byte of the address offset, A8 to A15) on the memory address bus at the proper times.
3. The address decoding section causes one of the $\overline{\text{RASx}}$ lines to go from low while the row part of the address is on the memory address bus. This is the first part of the address of the DRAMs. The value of "x" indicates the page value of the address on the external address bus. When the $\overline{\text{RASx}}$ line goes low, the DRAMs that use that $\overline{\text{RASx}}$ line latch the row part of the address into the DRAM. This address causes the DRAM to select one of the rows in the memory array. Once a row is selected, all memory locations connected to the selected row have their charge levels (data) re-established. The $\overline{\text{RASx}}$ line goes high either at the same time as or shortly after the $\overline{\text{CASx}}$ line.
4. The address-decoding section causes one of the $\overline{\text{CASx}}$ lines to go low while the column part of the address is on the memory address bus. Once the $\overline{\text{CASx}}$ goes low, the DRAMs connected to the selected $\overline{\text{CASx}}$ become enabled. If a read operation is selected, the DRAMs start producing their data on the RAM data bus, as indicated by the address selected. If a write operation is selected, the DRAMs start accepting data from the RAM data bus. The value of "x" for the $\overline{\text{CASx}}$ line must be the same as the "x" value selected by the $\overline{\text{RASx}}$ line. The cycle ends when the $\overline{\text{CASx}}$ line goes high.
5. Each bank of DRAMs in the RAM section contains nine DRAMs (64K $\times$ 1 in the IBM PC). Eight of the DRAMs are connected to the memory data bus (MD0 to MD7); one DRAM is used for each bit of data. These DRAMs hold the data byte for the RAM section. The ninth DRAM is used to hold the parity bit, which is only used in the RAM section and is not available on the data bus.
6. The 74280 is a parity generator/checker that is used to produce a parity bit during a write cycle and to check for proper parity during a read cycle. There are nine inputs, A to I; eight of these inputs are connected to the memory data bus (MD0 = A, . . ., MD7 = H). The ninth input, the I input, is always high during a write operation. During a read operation, the I input is the value of the parity bit. There are two outputs: The EVEN output is used to produce the parity bit during a write operation. The ODD output is used to indicate a parity error during a read operation; if low, a parity error has occurred; if high, no parity error has occurred.
7. The parity error data from the RAM section is used by the extended control logic section to produce an NMI (parity error) signal for the 8088.

DRAM REVIEW QUESTIONS

1. If the address on the external address bus is 20012 hex, which $\overline{\text{RAS}}$ goes low during a memory read operation?

a. $\overline{\text{RAS0}}$

b. $\overline{\text{RAS1}}$

c. $\overline{\text{RAS2}}$

d. $\overline{\text{RAS3}}$

2. During the time the column part of the address is on the RAM address bus, the ADDR SEL line will be
   a. a logic low
   b. a logic high
3. The data bus transceiver in the RAM section is enabled when the $\overline{\text{RAM ADDR SEL}}$ line is
   a. a logic low
   b. a logic high
4. The data on the RAM data bus is valid when the $\overline{\text{CASx}}$ line is ________ after the $\overline{\text{RASx}}$ line has gone low during a memory write line.
   a. low
   b. high
   c. tristate
5. During a memory write operation, the I input to the parity generator/checker is
   a. low
   b. high
   c. tristate
6. If the data on the RAM data bus is 1E hex, the parity bit produced by the parity generator/checker during a write operation is
   a. low
   b. high
   c. tristate
7. Using Figure 3-41, which DRAM number is used to hold the parity bit?
   a. XU90
   b. XU77
   c. XU69
   d. XU86
8. If the ODD output is low during a write operation, a parity error has occurred.
   a. True
   b. False
9. U90 is enabled during what type of memory cycle?
   a. Read
   b. Write
   c. Refresh
10. During a refresh cycle, which bank(s) of DRAMs places its (their) data on the RAM data bus?
    a. Bank 0
    b. Banks 0 and 1
    c. Banks 2 and 3
    d. All banks of DRAMs
    e. None of the banks of DRAMs

3.7 PROGRAMMABLE INTERVAL TIMER (PIT)

The PIT section (Fig. 3-49) of the system board supplies low frequencies to produce sounds (cassette data interface signals on the old IBM PC only) and precise time delays for the computer. Because the cassette interface has not been used to store data since 1982, this part of the PIT section is not discussed, although the circuitry may still be found in older PCs. The cassette interface was never incorporated into compatibles.

The main controlling element in the PIT section is the 8253 PIT (Fig. 3-50). The 8253/8254 from Intel is a three-channel counter-timer; in the IBM PC this device is addressed

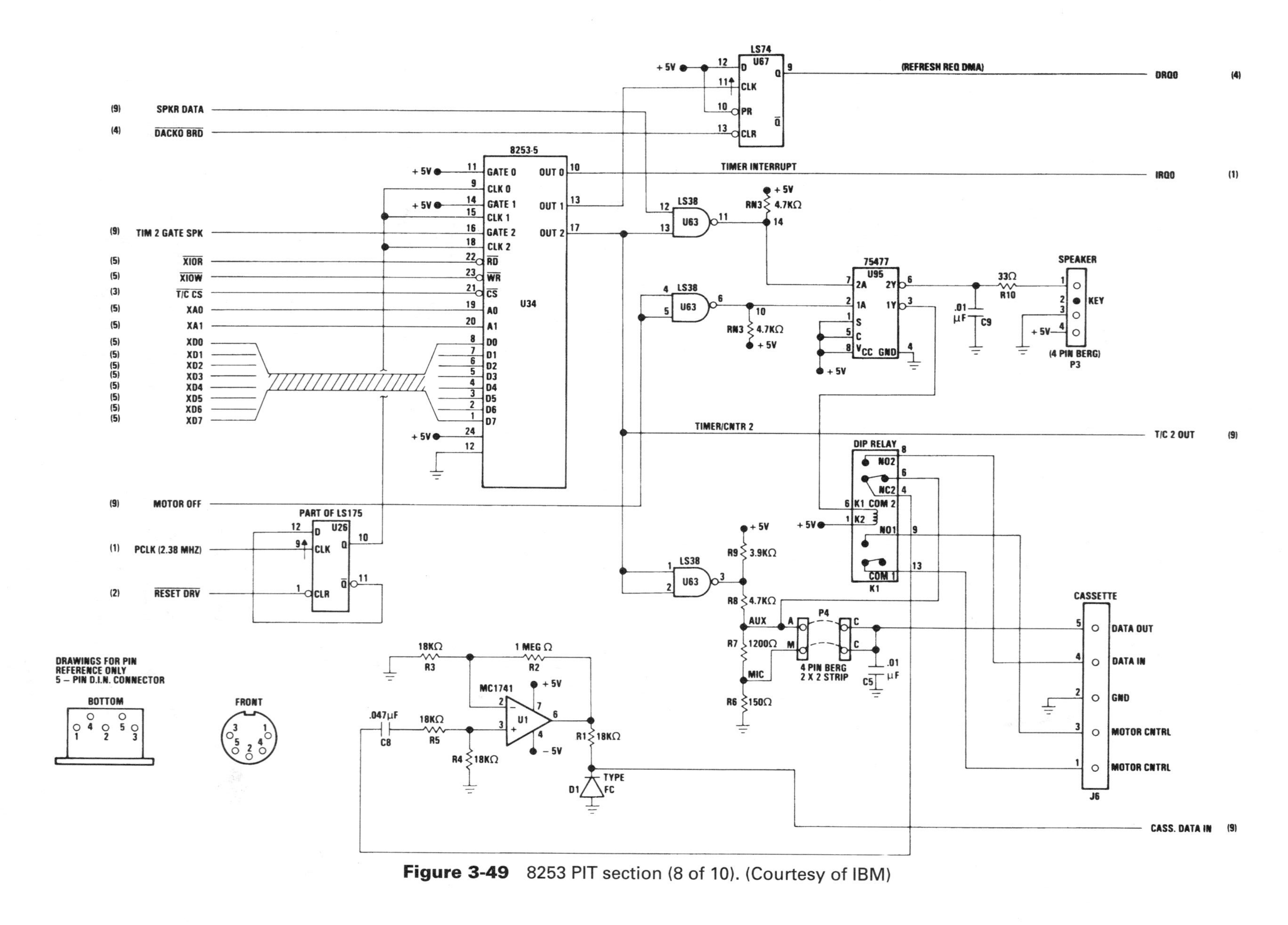

Figure 3-49 8253 PIT section (8 of 10). (Courtesy of IBM)

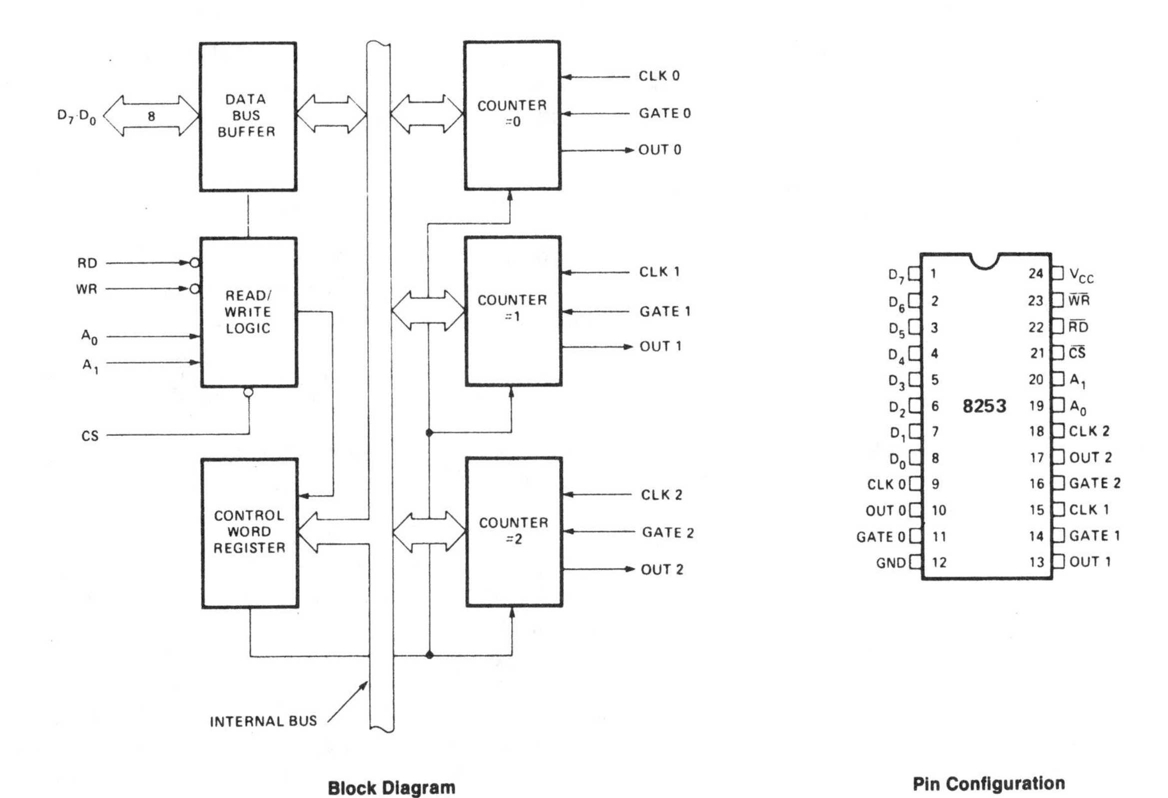

Figure 3-50 8253 PIT block diagram and pin configuration. (Courtesy of Intel)

as an I/O device. It contains three 16-bit counters that can count in binary or BCD (binary coded decimal), each with their own clock inputs. The maximum input frequency is 5 MHz for the 8253 and 10 MHz for the 8254. The input frequency to the timer is derived from a data flip-flop. The PCLK is applied to a data flip-flop, which divides the frequency in half and applies a 1.19-MHz clock to the timer. The same frequency is applied to each clock of each channel in the IBM PC and compatibles.

3.7.1 8253 Block Diagram Functions (Fig. 3-50)

DBB — The Data Bus Buffer is used to control the direction of data flow to and from the 8253 PIT. This section also temporarily holds data that is used by the 8253.

R/WL — The Read/Write Logic controls the data bus buffer and control word register. This logic circuit determines whether a read or write operation is performed, which one of the internal registers is programmed, and when the chip is enabled.

CWR — The Control Word Register holds information about the operation of each of the three counters. This is a write-only register, used to program each of the counters in the chip.

C1,2,3 — There are three programmable counters in the 8253, all identical in construction and operation but each independent of the others. Each counter is a programmable (pre-settable) 16-bit down counter capable of counting in binary or BCD. The counters also contain separate clock and gating inputs with separate outputs. In the IBM the same clock frequency is applied to all three counters.

3.7.2 8253 Pin Configuration (Fig. 3-50)

D0–D7

D0 to D7 are bidirectional data lines used to supply data for the chip during programming. The lines also allow data to be supplied to the data bus during a read status operation.

$\overline{RD}$

The $\overline{ReaD}$ data input enables a read operation when it is low and the $\overline{CS}$ line is low. The $\overline{RD}$ line is controlled by the $\overline{XIOR}$ line.

$\overline{WR}$

The $\overline{WRite}$ data input enables a write operation when it is low and the $\overline{CS}$ line is low. The $\overline{WR}$ line is controlled by the $\overline{XIOW}$ line.

A0–A1

The two Address lines select the internal register to be accessed during programming or a read status operation. For the levels on the address lines to have any effect on this IC, either the $\overline{RD}$ or $\overline{WR}$ line must be low, along with a low on the $\overline{CS}$ line.

$\overline{CS}$

The $\overline{Chip Select}$ input, when low, enables the 8253 during programming or reading status. The $\overline{CS}$ line is controlled by the $\overline{T/C CS}$ (Timer/Counter) line and is developed by the address-decoding section of the PC. This line does not have to be low for the counters to perform their programmed functions.

CLK 0–2

There are three CLocK inputs in the 8253, one for each counter. The counters use the clock input to determine their timing.

GATE 0–2

There are three GATE inputs, one for each counter. The level on the gate input is used by some of the operational modes to trigger the counter when high.

OUT 0–2

There are three counter OUTputs, one for each counter. The outputs change their levels according to the programming of the counters.

3.7.3 Operational Modes of the 8253/8254

ITC

The Interrupt on Terminal Count mode (mode 0) causes the output of the counter to go high as soon as the count is completed, provided that the gate input is high. The counter is disabled when the gate is low or going low. In order to reset the counter and its output, the counter must be reloaded. The IBM uses this mode for channel 0, to produce the timing signals so that DOS can keep time.

POS

The Programmable One Shot mode (mode 1) causes the output of the counter to go high for the time period specified by the value of the counter. The count starts on the rising edge of the gate input. Once the sequence begins, the counter can be reprogrammed without affecting the operation of the counter until the next cycle.

RG

The Rate Generator mode (mode 2) causes the output of the counter to go low for one clock time period each time the count of N is reached. If the value of N is 11 hex (17 decimal), the output goes low for one clock cycle on the 18th decimal (12 hex) clock cycle. A high on the gate input enables the counting, while a low disables the count and sets the output. The rising edge of the gate input reloads the counter and restarts the count. The IBM uses this mode for counter 1 to produce the refresh pulse for the DMA.

SWRG — The Square Wave Rate Generator mode (mode 3) causes the output of the counter to go high for half of the counts and low for the other counts. The gate input in this mode functions the same as in mode 2. This mode is used to create the frequency produced by counter 2.

STS — The Software Triggered Strobe mode (mode 4) causes the output of the counter to go high until the final clock, at which time the output goes low for the final count. A low on the gate disables the counting, while a high enables the counting.

HTS — The Hardware Triggered Strobe mode (mode 5) causes the output of the counter to go low for one clock period on the final cycle of the count. The count is retriggered by a low-to-high transition on the gate input.

3.7.4 Functions of the 8253 in the IBM PC

Counter OUT 0 is used by the IBM to produce the timing pulse signal necessary to update the time of day for DOS. A high pulse is produced 18.2 (Hz) times per second (once every 55 ms for about 840 ns). OUT 0 of the 8253 is applied to IRQ0 (Time of Day) input of the 8259 PIC. This causes an INT 8 hex to be performed, which updates the memory location that DOS uses to keep track of the time of day.

Counter OUT 1 is used by the IBM to initiate a DRAM refresh cycle. Counter 1 is programmed as a mod 17 counter, and produces a high for about 840 ns on the 18th clock. Because the clock for this counter is derived from U26 data latch that divides the PCLK frequency by one-half, the input frequency to the clock is 1.19 MHz, or one-fourth the CLK88 frequency. Therefore, the output frequency of counter 1 is 66.1 KHz, which produces an 840-ns pulse once every 15 μs (72 clock cycles). On the low-to-high transition of OUT 1, the refresh data latch (U67) sets its output, which is applied to the DREQ0 input of the DMA. This causes a DMA cycle on channel 0. After the DMA takes control of the bus and places the address on the address bus, the $\overline{\text{DACK0}}$ line goes low, causing the $\overline{\text{DACK0 BRD}}$ line to go low and thus resetting the refresh data latch.

Counter OUT 2 is used to produce a programmable pulse to create frequencies needed for the speaker to create sound. BIOS programs this counter to produce a 400-Hz frequency, which is produced once the gate enables counter 2. This counter can be reprogrammed by software to create almost any frequency. During start-up BIOS programs this counter to produce a 400-Hz frequency once Gate 2 goes high. As POST completes the memory check, the computer produces one short beep. This beep is created by the computer to bring the Gate 2 input high, which produces a 400-Hz frequency on OUT 2. The Gate 2 line is controlled by the TIME 2 GATE SPK (PB0) from the PPI section. The frequency of OUT 2 is applied to a two-input NAND (U63) gate, which is used to determine length of time the frequency is available for the speaker. The NAND gate also allows the computer to frequency modulate the OUT 2 frequency. The computer brings the SPKR DATA (PB1) line from the PPI section high for 250 ms (one-fourth of a second); this allows the OUT 2 frequency to the 75477 (amplifier) to produce the beep. The 75477 is used to increase the current capability of the frequency; this is applied through a low-pass filter to an 8-Ω speaker. In a compatible PC, the 75477 is often replaced by a common emitter transistor circuit.

The frequency from OUT 2 is also used to create the frequency for the T/C 2 OUT input to the PPI section (PC5), which can be applied to the TIME 2 GATE SPK to modulate counter 2.

3.7.5 8253 Troubleshooting

Condition: No OUT from any counter is producing a frequency once enabled.
Result: Computer fails to boot completely, POST fails, and no sound comes from the speaker.
Cause: Check the $\overline{CS}$ line to the 8253; this line should be pulsed low during boot-up. If the $\overline{CS}$ line does not pulse low, troubleshoot the address-decoding section. If $\overline{CS}$ line pulses low, troubleshoot ROM BIOS or change the 8253 PIT.

Condition: The OUT 0 line does not produce a high pulse once every 55 ms.
Result: The computer either does not update the time of day, or the time of day is either too fast or too slow.
Cause: A common cause of this type of problem is software that disables this counter to reduce the number of interrupts in the PC. This increases the speed of the computer. This normally happens with older types of graphic programs, tape back-up software, and some games. The result is that the clock loses time, proportional to the amount of time the software is operating.
Cause: Make sure GATE 0 is tied high; correct if necessary.
Cause: Use diagnostic software to determine if the PIC is bad or the programming of counter 1 is bad. If the programming of counter 1 is bad, the problem lies with the contents of BIOS (usually not the program).

Condition: OUT 1 does not produce an output pulse.
Condition: OUT 1 produces an pulse but not at the proper times.
Result: Computer does not boot up completely and may produce parity errors because memory is not being refreshed at all or often enough.
Cause: See the second cause of the previous condition.
Cause: GATE 1 should be tied high; correct if necessary.

Condition: OUT 1 produces an output, but the DREQ0 line does not go high.
Result: Computer does not boot up completely and may produce parity errors because memory is not being refreshed.
Cause: Check the "D" and "PR" inputs of the refresh data latch; these lines should be tied to Vcc. Correct if necessary.
Cause: Check the $\overline{CLR}$ input to the refresh data latch; this line should normally be high and should be pulsed low once every 15 μs. If this input is stuck low, troubleshoot the DMA section ($\overline{DACK0}$).
Cause: Check for proper logic level changes on the CLK input of the refresh data latch. If levels are OK, change 7474; otherwise, change PIT.

Condition: OUT 1 produces the correct output pulse, but the DREQ0 line never goes low.
Result: Computer may halt because the DMA is continuously producing refresh cycles.
Cause: Check the $\overline{CLR}$ input to the refresh data latch; this line should normally be high and should be pulsed low once every 15 μs. If the correct low pulse is produced, change the 7474 refresh data latch. If the input is always low, troubleshoot the DMA section ($\overline{DACK0}$).

Condition: OUT 2 never produces an output frequency.
Result: The computer speaker does not make any sounds.
Cause: Make sure the TIME 2 GATE SPK goes high during the time the OUT 2 should be producing a frequency.

Cause: Use diagnostic software to determine if the PIC is bad or the programming of counter 2 is bad. If the programming of counter 2 is bad, the problem lies with the contents of BIOS (usually not the program) or the software.

8253 SUMMARY

1. Counter 0 is used to produce the Time of Day interrupt for the PC. This interrupt is produced 18 times a second.
2. Counter 1 is used to initiate a DRAM refresh cycle. A refresh cycle occurs 66,100 times each second. This allows the computer to refresh all DRAM in the computer 258 times each second.
3. Counter 2 is used to produce the frequencies used by the speaker of the computer. In only the old orginal IBM PC, this frequency is used to help store data on a cassette tape.

8253 REVIEW QUESTIONS

1. In order to program the PIT, the $\overline{CS}$ lines must go
   a. low
   b. high
   c. tristate

2. The NAND gate U63 becomes enabled when the SPKR DATA line goes
   a. low
   b. high

3. Memory parity errors may indicate which PIT output is bad?
   a. OUT 0
   b. OUT 1
   c. OUT 2

4. If the computer loses time on the DOS time of day, which PIT output is bad?
   a. OUT 0
   b. OUT 1
   c. OUT 2

5. If the computer does not produce any sound from the speaker, which PIT output is bad?
   a. OUT 0
   b. OUT 1
   c. OUT 2

3.8 PROGRAMMABLE PERIPHERAL INTERFACE (PPI)

The PPI section (Fig. 3-51) controls peripheral signals for the keyboard, modifies the sound created by the 8253 PIT, enables counter 2 of the PIT, and reads the equipment switches in the IBM PC to determine the equipment in the computer. The operation of this section is basically controlled by the 8255 PPI chip. The other major chip in this section is the 74322 8-bit shift register with extension sign. The shift register and keyboard interface sections are covered in Chapter 7, Keyboards and Power Supplies.

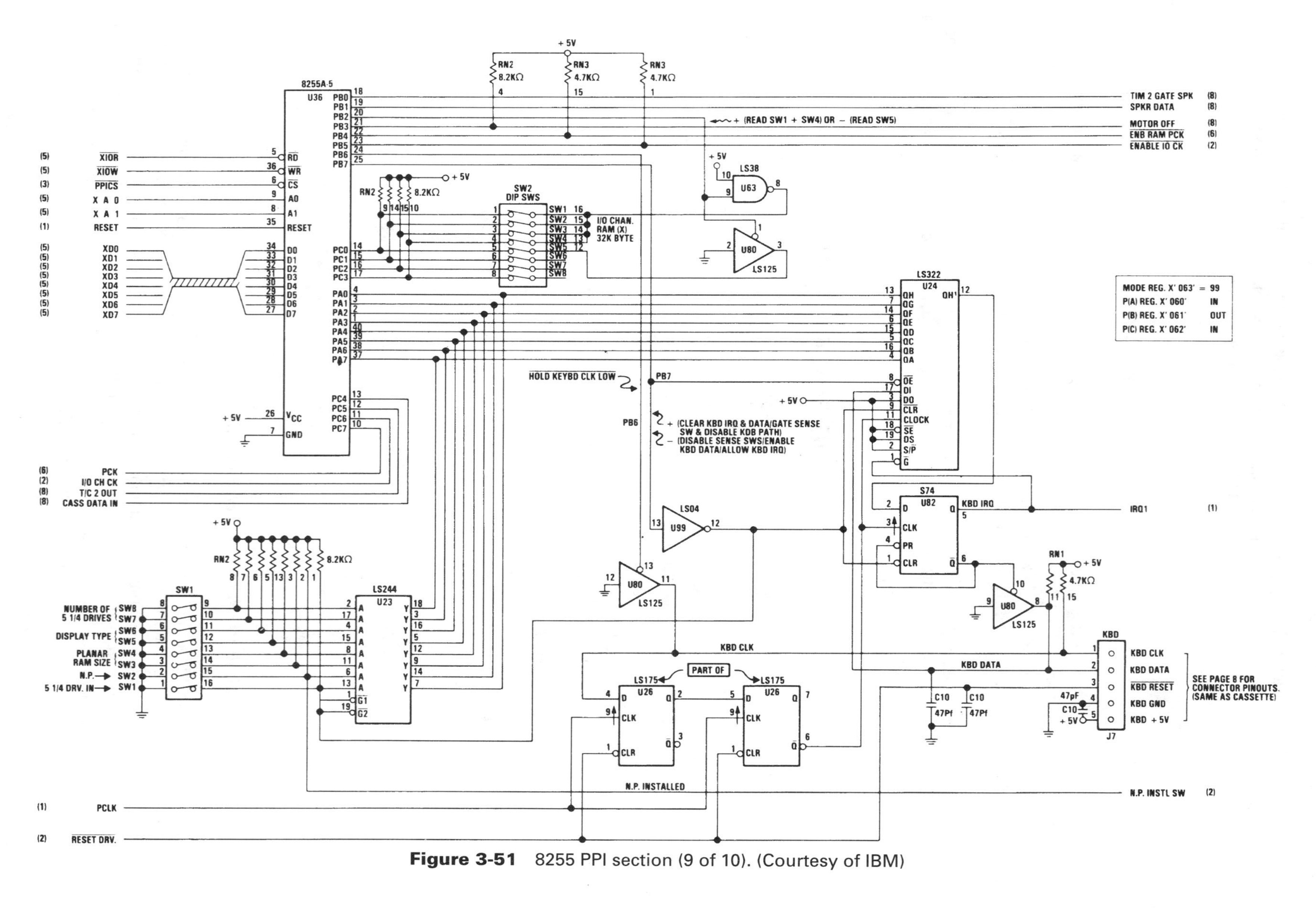

Figure 3-51 8255 PPI section (9 of 10). (Courtesy of IBM)

3.8.1 8255 Programmable Peripheral Interface

The IBM uses the Intel 8255, 24-line PPI chip to perform the following tasks.

1. Sensing of the two DIP switches used to determine the type of equipment in the IBM.
2. Input for the data keyboard codes from the keyboard.
3. Control of a variety of peripheral units on the system board:
   a. Enables the frequency applied to the speaker
   b. Enables counter 2 in the PIT
   c. Controls cassette motor
   d. Enables above-board RAM
   e. Enables I/O channel check (for testing)
   f. Disables keyboard clock when keyboard is full.
   g. Enables keyboard or clears keyboard
   h. Enables sense switches (for testing or dedicated setup)
   i. Enables PCK (Parity ChecK) line

3.8.2 8255 Block Diagram Functions (Fig. 3-52)

DBB — The Data Bus Buffer block controls the direction of data flow to and from the chip. This section also enables or disables the internal data bus of the device from the external data bus.

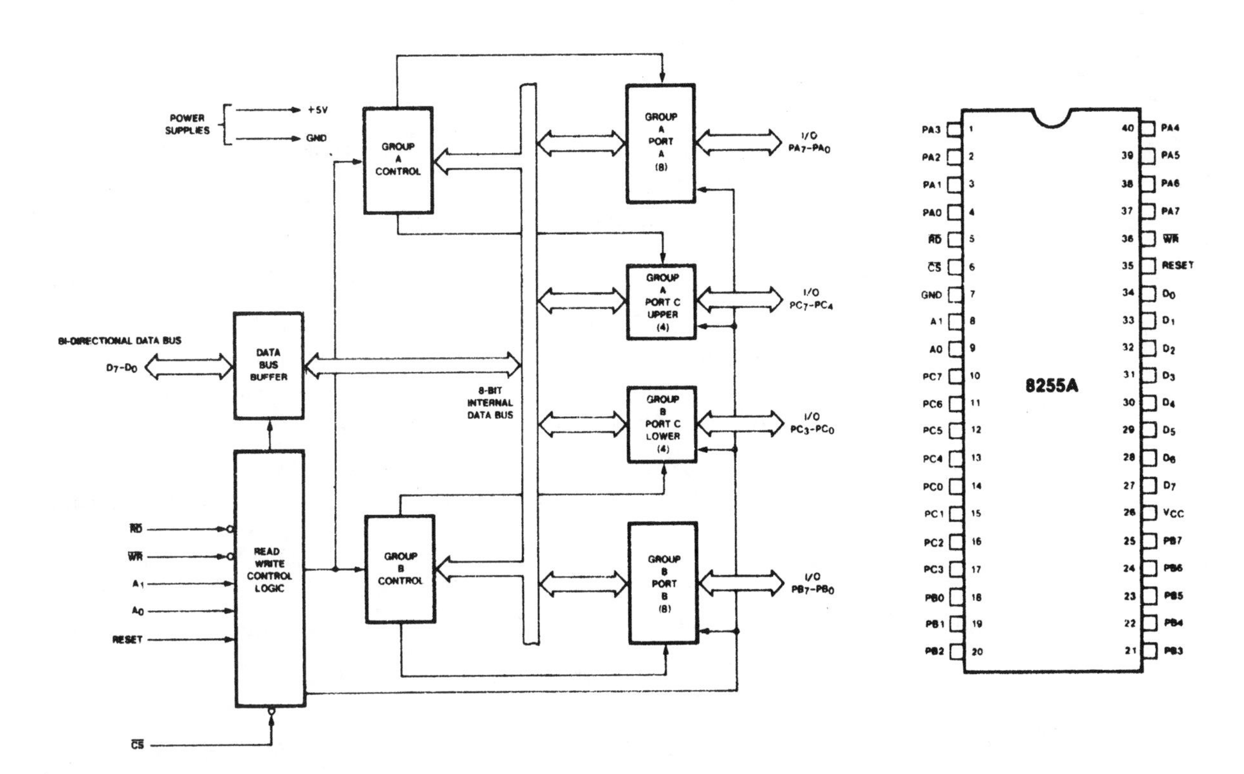

8255A Block Diagram **Pin Configuration**

Figure 3-52 8255 PPI block diagram and pin configuration. (Courtesy of Intel)

RWCL

The Read/Write Control Logic section controls the type of data transfer inside the chip. This section enables the selected registers for programming or for reading status. It also controls the direction of data flow in the bus buffer block.

CGAB

The Control Groups A-B write-only registers are used to format the three ports (channels). Each control group can operate independently or in combination with others. The command words loaded inside these registers determine the operation of the I/O lines. Group A control register formats the 8 bits of port A (PA0 to PA7) and the highest 4 bits of port C (PC4 to PC7). Group B control register formats the 8 bits of port B (PB0 to PB7) and the lowest 4 bits of port C (PC0 to PC7).

PORTS

Each of the three 8-bit ports (A, B, C) contain eight I/O lines. The following are the types of configurations possible for each port.

PORT A 8-BIT output latch-buffer or
8-bit input latch

PORT B 8-bit input/output latch-buffer or
8-bit input buffer

PORT C 8-bit latch-buffer or
8-bit input buffer or
two 4-bit latches used for control outputs or status inputs

3.8.3 8255 Pin Configuration (Fig. 3-52)

D0–D7

D0 to D7 are bidirectional data lines used to transfer data to or from this device.

A0–A1

These two address lines are used to access one of the internal registers.

$\overline{RD}$

The $\overline{ReaD}$ input enables a read operation when low while $\overline{CS}$ is low. The $\overline{RD}$ input is controlled by the $\overline{XIOR}$ line.

$\overline{WR}$

The $\overline{WRite}$ input enables a write operation when low while $\overline{CS}$ is low. The $\overline{WR}$ input is controlled by the $\overline{XIOW}$ line.

$\overline{CS}$

The $\overline{\text{Chip Select}}$ enables the transfer of data to or from this chip when low, depending on whether the $\overline{RD}$ or $\overline{WR}$ is low. The $\overline{CS}$ line does not have to go low for the PPI 8255 to operate.

RESET

The RESET input line resets the 8255 when high. The reset function makes all ports act as inputs so no damage can occur from output being connected together.

PA0–PA7

These eight lines are I/O lines controlled by port A command formats.

PB0–PB7

These eight lines are I/O lines controlled by port B command formats.

PC0–PC7

These eight lines are I/O lines controlled by port C command formats.

3.8.4 8255 Operational Modes

BIO

The Basic Input/Output mode (mode 0) can control any of the three ports in the chip. This mode requires no hardware control signals (handshaking signals) for the operation of the ports. This is the mode that the IBM PC and compatibles use. The parameters for this mode are as follows.

1. Sixteen different combinations of I/O line formats.
2. Ports A and B are 8-bit–only ports.
3. Port C operates as two 4-bit ports.
4. All outputs are latched.
5. All inputs are unlatched.

SIO

The Strobe Input/Output mode (mode 1) allows data transfer through the use of hardware handshaking signals. The parameters for this mode are as follows.

1. Port A acts as an 8-bit–only latched I/O port. The high nibble of port C acts as a control (status/data) port for port A.
2. Port B acts as an 8-bit–only latched I/O port. The low nibble of port C acts as a control (status/data) port for port B.

SBB

The Strobe Bidirectional Bus I/O mode (mode 2) allows bidirectional communications with peripheral devices and requires some type of handshaking. The parameters are as follows.

1. Control of port in control group A only. This allows port A to act as a bidirectional latched port, with the five most significant bits of port C acting as the control port.

3.8.5 8255 Operation Inside the IBM PC

Port A

PA0–PA7

The Port A I/O lines all act as inputs. The first function of this port is to read the keycodes from the keyboard shift register. The following is the sequence.

1. PB7 of the PPI is normally low; this enables the output of the keyboard shift register, which applies the key code to Port A. This low also disables the equipment dip switch buffer, which is also connected to Port A.
2. Once the keyboard shift register circuitry has converted the serial data from the keyboard into a parallel key code, a keyboard interrupt (IRQ1) is generated in the 8259 PIC.
3. Once the keyboard interrupt routine begins, the computer reads the key code from Port A and transfers the key code into the keyboard buffer (16-byte memory location in RAM).
4. If the keyboard buffer is full, the computer causes the PB6 line to go low, which disables the keyboard clock. If the keyboard buffer is not full, the PB6 line stays high.
5. After the transfer of the key code into the keyboard buffer, PPI causes PB7 to go high. This resets the data latch that produces the keyboard interrupt (IRQ1) and clears the keyboard shift register. The keyboard circuitry is now ready to accept another key code.
6. The computer, via the keyboard interrupt routine, then processes the key codes.

The second function of Port A is to read the equipment DIP switches, the levels on which are used by BIOS to configure the basic equipment in the computer. This information is used only when the computer is first booted. Whenever PB7 is high, the equipment DIP switch buffer is enabled. With the equipment DIP switch buffer enabled, Port A represents the basic equipment in the computer. The following explains what each of the DIP switches means.

IBM PC

SW1-1	Boot from drive A diskette enable
SW1-2	Reserved
SW1-3/4	Amount of motherboard RAM (64K, 128K, 192K, or 256K)
SW1-5/6	Type of video controller (MDA, CGA 40 column, CGA 80 column, or no video)
SW1-7/8	Number of floppy drives in the system (1, 2, 3, or 4)

Clone PC

SW1-1	Normal operation
SW1-2	8087 math coprocessor present
SW1-3/4	Amount of motherboard RAM (In some clones these switches have no function.)
SW1-5/6	Same as for the IBM PC
SW1-7/8	Same as for the IBM PC

Port B

PB0–PB7

These eight lines of port B act as all outputs; their functions are as follows:

PB0 [Timer 2 gate speaker] Enables counter 2 of the 8253, which generates square waves for the speaker or cassette.

PB1 [Speaker data] Allows the frequency from the output of counter 2 to be applied to the speaker. This output is also used to modulate the frequency for the speaker.

PB2 [Read read/write memory size] Enables part of port C to read the off-board memory size. In most compatibles there is not a second DIP switch to determine off-board memory, because they have 640K of address decoding on the motherboard. (In some compatibles this output line is used to switch the speed of a computer between normal (4.772727 MHz) and turbo (8 MHz or 10 MHz) speed under keyboard control.)

PB3 [Cassette motor off] Turns off motor. (In some compatibles this output performs no function.)

PB4 [Enable read/write memory] Enables RAM parity checking.

PB5 [Enable I/O channel check] Enables I/O channel parity checking.

PB6 [Hold keyboard clock low] Disables clock to the keyboard by bringing the clock line low.

PB7 [Enable keyboard/equipment switch] Causes port A to read either the keyboard key code or the equipment DIP switch.

Port C

PC0–PC7 These eight lines of port C all act as inputs, which perform the following functions:

PC0–PC3 [Off-board read/write memory] Determine the size of off-board memory devices (either RAM or I/O). The values are determined by the values of equipment switch 2. (In most compatibles these outputs do not serve any function.)

PC4 [Cassette data in] Reads serial data from the data cassette. (In most compatibles this output does not serve any function.)

PC5 [Timer channel 2 out] Allows the PPI to monitor and use the frequency produced by the counter 2 outputs.

PC6 [I/O channel check] Allows I/O adapter to provide a READY signal to the extended control logic section of the PC.

PC7 [Read/write memory parity check] Allows an I/O adapter to produce a parity error status.

3.8.6 8255 Troubleshooting

Other than the keyboard interface in the PPI, not much can go wrong in this section.

Condition: DIP switch settings are incorrect, or the 74244 equipment DIP switch buffer is not functional.
Result: Computer may halt, some of the eqiupment in the computer may not function, or POST produces some type of error.
Cause: Equipment DIP switches have mechanically failed.
Cause: PB7 of the PPI is not producing a logic high.
Cause: The 74244 buffer is bad.
Cause: The 7404 (U99) inverter is bad.

Condition: PPI is not configured correctly or does not operate properly.
Result: Computer halts, keyboard no longer functions, or POST produces some type of error.
Cause: $\overline{CS}$ does not go low during POST.
Cause: $\overline{WR}$ does not go low during POST.
Cause: $\overline{CS}$ and $\overline{WR}$ do not go low during POST.
Cause: If $\overline{CS}$ and $\overline{WR}$ go low during POST, PPI is most likely bad.

8255 SUMMARY

1. Port A funtions as inputs. They are used to read the information from the equipment DIP switches or the key code from the keyboard shift register.
2. Port B functions as outputs. Not all of the outputs from this port are used by compatibles.
3. Port C functions as inputs. Not all of the inputs are used by compatibles.
4. The equipment DIP switches are used to inform BIOS about the basic equipment in the PC. These switches indicate whether a math coprocessor is present, the amount of motherboard RAM, the type of video present, and the number of floppy drives in the PC.

8255 REVIEW QUESTIONS

1. If POST, when displaying the size of RAM being tested, does not test all the memory in the computer but does not produce any error codes, the problem is most likely
   a. bad equipment DIP switch setting
   b. bad 74LS244 buffer
   c. bad ports A of the PPI
   d. bad PPI
2. If during POST no video display is seen on the screen, but the computer produces one long and two short beeps continuously, the problem is most likely an incorrect setting on the equipment DIP switch.
   a. True
   b. False
3. If PB4 is stuck high, the computer does not report any parity errors.
   a. True
   b. False
4. If the computer produces memory errors in memory above 256K in the IBM PC, the most likely cause is a problem with
   a. PC0 to PC3
   b. PA2 to PA3
   c. PB0 to PB2
5. Which one of the following ports of the PPI may cause the sound out of the PIT section to be incorrect or missing?
   a. PC4
   b. PC5
   c. PB1
   d. PB3

3.9 I/O INTERFACE BUS CHANNEL

The IBM PC and clones use a 62-pin I/O expansion channel bus to connect the other necessary I/O and memory devices (Fig. 3-53). Typical expansion cards are the video controller card (color graphics or monochrome), disk controller, and multifunction I/O cards. The early versions of the IBM PC contained only five I/O interface bus slots, which were increased to eight slots in later versions. All clones use eight interface slots.

3.9.1 I/O Interface Bus Channel Functions

D0–D7 These eight lines act as the bidirectional Data bus.

A0–A19 These 20 lines are the unidirectional Address bus.

IRQ2–IRQ7 These six Interrupt ReQuest lines are the inputs for the PIC in the microprocessor subsystem.

$\overline{\text{DACK0}}$–$\overline{\text{DACK3}}$ These four lines are the ACKnowledge output lines from the DMA. When low, the DMA request has been acknowledged and will be processed.

DRQ1–DRQ3 These three lines are the ReQuest input lines for the DMA. When high, a device is requesting a DMA cycle for the channel.

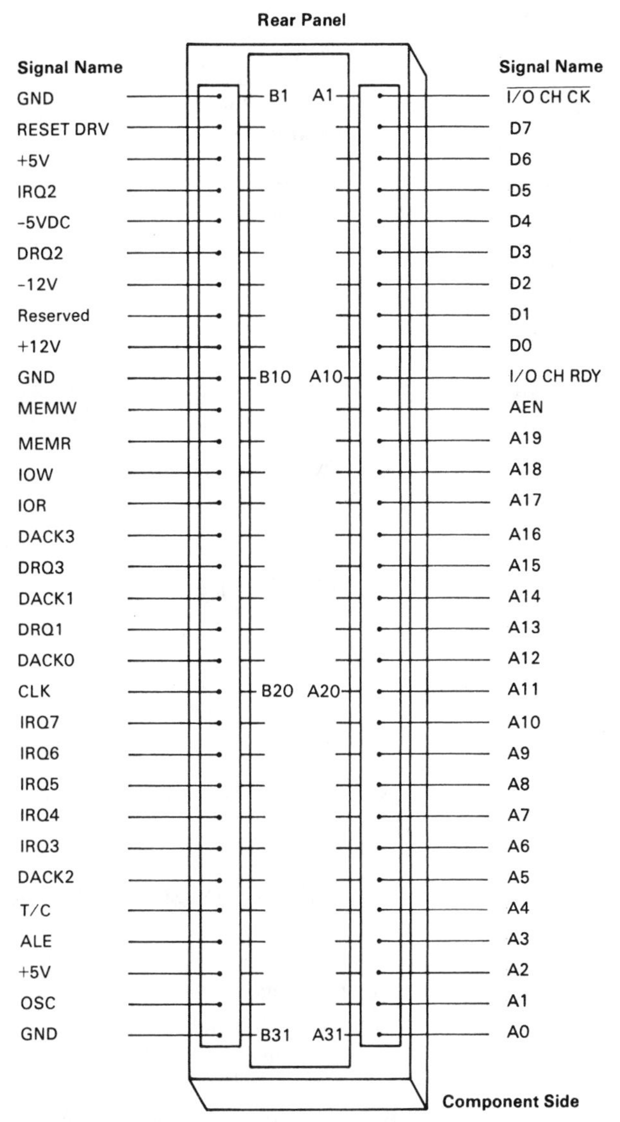

Figure 3-53 I/O channel socket (10 of 10). (Courtesy of IBM)

ALE	The Address Latch Enable latches the address on the address bus for any nonsystem I/O or memory boards.
AEN	The Address ENable enables the address latches on any nonsystem boards.
T/C	The Terminal Count (EOP) line is an output from the 8237 DMA and can be used by any nonsystem boards for memory.
OSC	This is the undivided crystal OSCillator frequency from the 8284 clock drive chip (14.383 MHz).
RESET DRV	The RESET DRiVer line, when high, causes all smart ICs in the IBM PC to reset to their original values.
$\overline{\text{I/O CH CK}}$	The $\overline{\text{I/O CHannel CheckK}}$ indicates to the PPI that one of the I/O channels has found an error.

I/O CH RDY	The I/O CHannel ReaDY line indicates that an I/O channel is ready to perform the specified operation.
$\overline{\text{MEMW}}$	The $\overline{\text{MEMory Write}}$ line, when low, enables a write operation on a memory device.
$\overline{\text{MEMR}}$	The $\overline{\text{MEMory Read}}$ line, when low, enables a read operation on a memory device.
$\overline{\text{IOW}}$	The $\overline{\text{I/O Write}}$ line, when low, enables a write operation on an I/O device.
$\overline{\text{IOR}}$	The $\overline{\text{I/O Read}}$ line, when low, enables a read operation on an I/O device.
GND	There are three GrouND lines on the I/O channel bus. These lines allow current to be supplied to any nonsystem boards.
+5V	Two +5-V lines supply the necessary positive voltage potential for any nonsystem boards.
−5V	One −5-V line supplies the necessary negative voltage potential for any nonsystem boards.
+12V	One +12-V line supplies the necessary positive voltage potential for any nonsystem boards.
−12V	One −12-V line supplies the necessary negative voltage potential for any nonsystem boards.
CLK	The CLocK line supplies the 4.772727 MHz-MPU clock to any nonsystem boards.

3.9.2 I/O Interface Bus Channel Troubleshooting

The I/O channel bus allows the technician to verify all signals that enter or leave the system board. With the proper card adapter, all signals can be monitored, allowing the technician to locate missing or wrong I/O signals between the system board and the different interfaces used.

The two major problems are as follows:

1. Bad interface slot connections between the interface slot and the edge connector on the adapter itself. Use compressed air to blow out any dirt in the slots, and clean the edge connectors on the adapter with alcohol. If the dirt on the edge connector is hard, use an ink eraser to clean the edge connector.
2. Broken solder connections on the interface slot soldered to the system board. Reheat the solder connections on the system board.

I/O INTERFACE BUS CHANNEL REVIEW QUESTIONS

1. Which line is used to initiate a memory refresh cycle on an external memory board?

a. ALE
b. T/C
c. OSC
d. $\overline{\text{DACK0}}$
e. GND

2. Which I/O channel bus line is used to cause an I/O adapter to output its data to the data bus?
   a. ALE
   b. I/O CH RDY
   c. $\overline{IOR}$
   d. $\overline{IOW}$
3. Address line 19 is on which side of the I/O channel interface slot?
   a. Side A
   b. Side B
4. DRQ2 is an active-low line.
   a. True
   b. False
5. Which line is used to supply the CLK88 signal to an I/O adapter?
   a. OSC
   b. CLK
   c. T/C
   d. ALE

chapter 4

80286 Microcomputer System Board

OBJECTIVES

Upon completion of this chapter, you should be able to meet the following objectives.

1. Define the purpose and function of the major ICs on a typical 286 system board—80286 MPU, 80287 NP, 82284 CGD, 82288 BC, 8259 PIC, 82237 DMA, 8042 MC, 27256 EPROM, DRAM, and 6818 RTC.
2. List and define the purpose of each section of a typical 286 AT block diagram—microprocessor subsystem, programmable interrupt controller section, direct memory access controller section, programmable interval timer section, real time clock, universal peripheral interface microcontroller, ROM section, and DRAM section.
3. List and define the purpose of the following control sections of the 80286; BU, IU, EU, and AU.
4. Define the purpose of each pin of the 80286.
5. Define the purpose of each pin of the 80287.
6. List and define the function of each section of the block diagrams of the Intel 82230/82231 AT chip set.
7. Define the function of each pin of the 82230 and 82231.
8. Using the timing of a typical bus cycle, identify each processor state and phase.
9. Using the timing of a typical bus cycle, determine if the sequence is correct.
10. Define the function of the blocks in the block diagram of an AT system board using the Intel AT chip set.

4.1 AT SYSTEMS

The first major step beyond the 8088 XT system board came with the development of the 80286 MPU. System boards using the 80286 were called ATs (Advanced Technology). All motherboards now made use AT-type technology even though the 80286 is for the most part no longer used. Most systems are designated by the chip number; for example, the 286 uses the 80286 MPU, the 386SX uses the 80386SX MPU, the 386DX uses the 80386DX MPU, the SX or 486SX uses the 80486SX MPU, the DX or 486DX uses the 80486DX MPU, the DX2 uses the 80486DX2, the DX4 uses the 80486DX4, and the Pentium uses the Pentium. All of these system boards use AT-type technology and are in many ways very similar to each other. The differences lie in the number of address, data, and control lines and their speed. Therefore, many areas of the system board discussed in this chapter are also used in system boards found in the following chapters covering more advanced Intel MPUs.

Many of the ICs and circuitry used on the AT system board are also found in XT system boards. In fact, many of the ICs in the AT system boards were part of the XT system boards. Because the MPUs in the AT system boards have more address, data, and control lines, many more IC support chips are needed, almost doubling the number of ICs. The size of the system boards is increased by more than 50%; this also increases the amount of power needed by the system boards. To reduce the disadvantages of the first AT-type system boards, Intel and other IC manufacturers started integrating many support chip functions in large-scale and very large-scale integrated chip sets, which reduced the number of ICs on a system board. This has kept the price of system boards to a reasonable level.

In this chapter we first discuss the block diagram of an earlier version of the AT system that did not use chip sets. The chapter concludes with a more detailed discussion of a 286 AT system board that uses the Intel 82230/82231 AT-compatible chip set.

4.1.1 Typical 286 AT Block Diagram

Just as with the XT system board, the AT system board (Fig. 4–1) may be divided into smaller sections to allow us to understand their function more easily.

4.1.2 Microprocessor Subsystem

The microprocessor subsystem, just as in the XT system board, contains the MPU, math coprocessor, clock generator, bus controller, address latches, and data bus transceiver. The only thing missing is the programmable interrupt controller, which in the AT system board is located in its own section. Unlike the 8088 and 8086, the 80286 MPU does not multiplex its address and data lines; therefore the 80286 has more than 40 pins and does not come in a DIP-type package. The 286 contains 24 address lines, 16 data lines, and 28 assorted control/status and power lines. The 80286 comes in a 68-pin package, which is discussed later in this chapter. The 286 supports all software instructions found on the 8088; this feature allows older software to operate on a system that uses the 286 MPU. The 286 also allows 16 bits of data to be transferred during any one bus cycle and allows addressing above 1MB. The 286 also operates at higher frequencies than the 8088; speeds range from 6 MHz through 12 MHz.

A math coprocessor (also known as a numeric processor) socket is also available on the system board. The numeric processor is labeled the 80287 and operates at the same frequency as the 286.

The clock generator used in a 286 system is the Intel 82284 clock generator. This IC basically operates the same as the 8284 used in the XT except for its speed. This IC supplies the system clock for the system board, provides reset circuitry, and is used to indicate the end of the bus cycle.

The bus controller used in a 286 system is the Intel 82288 bus controller. This IC basically operates the same as the 8288 used in the XT except for its speed. This IC controls that flow of data between the local and system data buses, controls the address latches in the microprocessor subsystem, and generates the read and write signals for the system board.

Because there are more address and data lines in the 286, more address latches and data bus transceivers are used in AT system boards. Many 286 AT system boards, however, still use the 74373 octal latches (address latches) and 74245 octal bus transceivers (data bus transceiver).

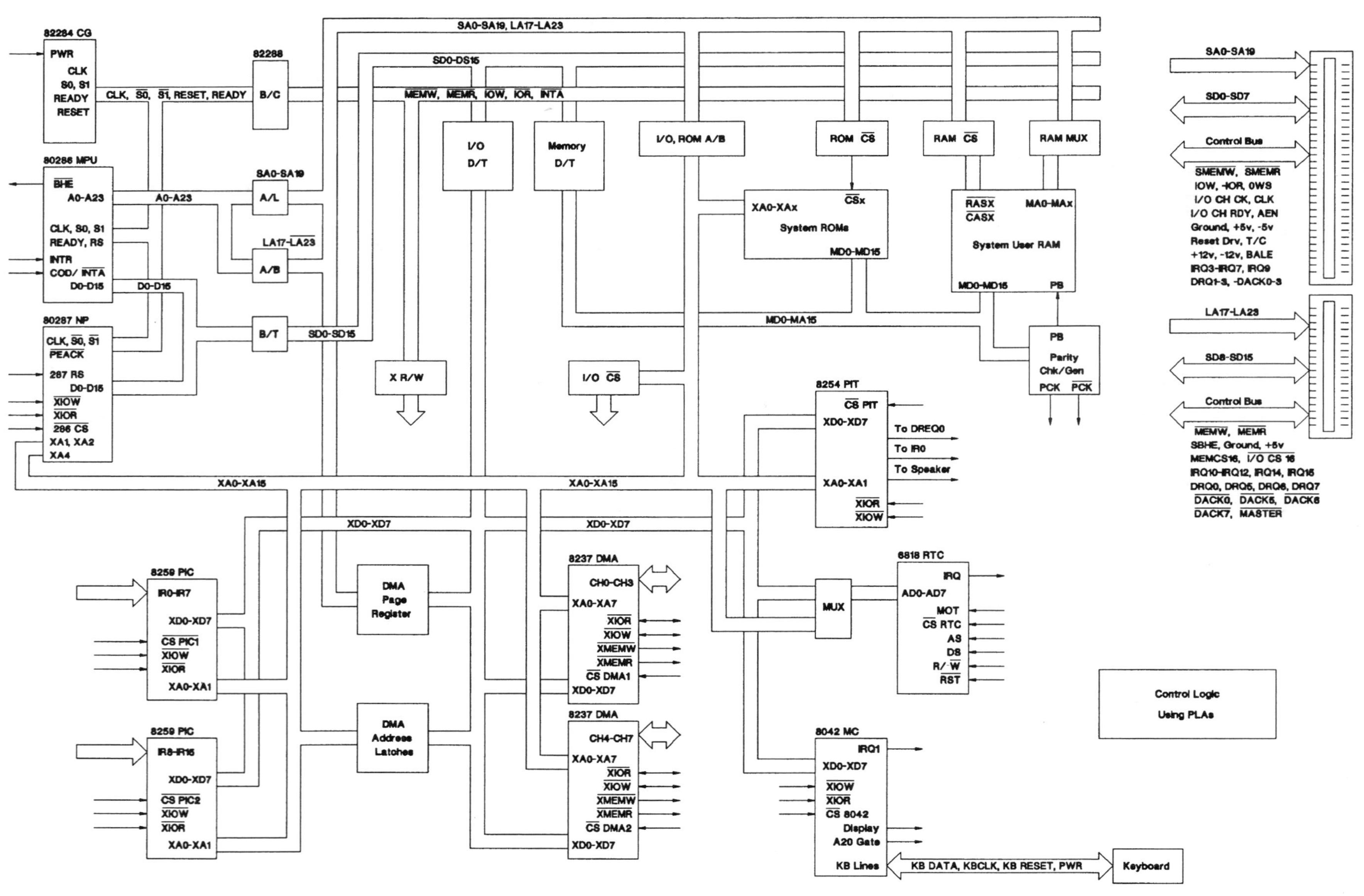

Figure 4-1 80286 system block diagram.

Additional circuitry is also needed to control which of the two data bus transceivers is enabled during byte or word data bus cycles.

4.1.3 8259 Programmable Interrupt Controller (PIC)

In the XT system board, the 8259 PIC is contained within the microprocessor subsystem, but in the AT system board the PIC is found within its own section and is not connected to the local bus. The 286 system board uses two 8259 PICs, as opposed to one in the XT. The AT system board contains 16 interrupt lines, as opposed to 8 in the XT. The second PIC was added to allow more hardware devices to generate hardware interrupts, which can be accessed faster than software interrupts.

The master (first) 8259 contains IRQ0 (system timer, highest priority [priority 0]), IRQ1 (keyboard, priority 1), IRQ2 (cascaded 8259, priority 2), IRQ3 (serial port 1 [COMM1], priority 12), IRQ4 (serial port 2 [COMM2], priority 13), IRQ5 (parallel port 2 [LPT2], priority 13), IRQ6 (floppy disk controller, priority 14), IRQ7 (parallel port 1 [LPT1 or PRN], priority 15). The slave (second) 8259 contains IRQ8 (real time clock, priority 4), IRQ9 (software redirected to INT 0Ah, priority 5), IRQ10 (reserved, priority 6), IRQ11 (reserved, priority 7), IRQ12 (reserved, priority 8), IRQ13 (20287, priority 9), IRQ14 (fixed disk controller, priority 10), and IRQ15 (reserved, priority 11).

4.1.4 8237 Direct Memory Access (DMA) Controller

AT systems use two DMAs to allow both 8-bit and 16-bit data transfers. The other reason for two DMAs is to allow more I/O devices to directly access DMA cycles. The first DMA is used to control DMA channels 0 to 3, and the second DMA controls DMA channels 4 to 7. DMA channel 4 is used to cascade the second DMA to the first DMA to allow for 16-bit data transfers. The DMAs used are the same as in the IBM XT.

4.1.5 8254 Programmable Interval Timer (PIT)

AT systems use a PIT just like the one in the IBM XT, and the timer performs the same functions as in the XT. Channel 0 is used to generate IRQ0, the Time of Day pulse. Channel 1 is used to generate DRQ0, which causes a refresh cycle to occur in DRAM. Channel 2 is used to produce sound from the speaker.

4.1.6 6818 Real Time Clock (RTC)

An AT system incorporates an RTC IC that keeps time when power is removed from the system. The RTC is also used to provide information for BIOS during boot up indicating the type and amount of equipment in the computer (number and types of floppy drives, hard drive number and parameters, type of video, amount of RAM in the system) and special information about how the computer should configure itself. Connected to the RTC is a battery that allows the RTC to maintain its data and time when the power is off.

4.1.7 8042 Universal Peripheral Interface Microcontroller

AT systems incorporate a microcontroller to control the interfacing between the system board and the keyboard. The 8042 is a single-chip computer with ROM, RAM, and programmable I/O lines. By using this, IC keyboard operations can be transparent to the 286 except when the 286 must read the keyboard buffer, which reduces the amount of housekeeping that the 286 must perform. The 8042 is also used to generate a signal that allows the 286 to switch from the real to the protected mode of operation. This controller is also used to provide some hardware boot-up information to the system.

4.1.8 ROM

The system ROMs in an AT system are much larger than those used in the basic XT system. Because the AT system contains more hardware to initialize, more memory is required. Most AT systems also contain extra firmware located in ROM, which may contain the following: hard drive parameter look-up tables, password protection, hardware diagnostics, hard drive testing, and set-up programs. The ROM section contains two ROMs; one supplies the low byte and the second the high byte for the data bus. AT ROMs are normally 128K in a 16K × 8 format, compared with the ROM of an XT, which is 64K in an 8K × 8 format.

4.1.9 DRAM

The 286 AT system contains more address lines than the 20 available in the XT system. Besides containing 24 address lines, data can be transferred one or two bytes at a time. The DRAM section in most 286 AT systems allows the user to put up to 4MB of memory on the motherboard; additional memory may be added to the system by using a memory card on the I/O channel bus. One or more banks of memory are used for odd addresses, with other banks of memory used for even addresses. Special gating must be used to access the correct memory location so that the data from the memory location is applied to either the upper or lower bytes of the data bus.

4.2 80286 MICROPROCESSOR

The 80286 is a 16-bit high-performance microprocessor made by Intel and is compatible with 8088 and 8086 software. Unlike the 8088/86 MPUs, which use a multiplexed address/data bus, the 286 uses a separate 24-bit address bus and a 16-bit data bus. The 286 has four different clock frequencies: 80286-6 (6-MHz clock), 80286-8 (8-MHz clock), 80286-10 (10-MHz clock), and 80286-12 (12.5-MHz clock). As opposed to the 40-pin DIP package with the 8088/86, the 80286 MPU is available in three types of 68-pin packages (Fig. 4-2)—ceramic LCC (Leadless Chip Carrier), PGA (Pin Grid Array), and PLCC (Plastic Leaded Chip Carrier).

The 80286 has two operating modes, both of which support all 8088/86 software. The 8086 real address mode, which makes the 80286 use the address bus like an 8086/88, limits addresses to 1 MB. The second operating mode, called the protected virtual address mode, allows the 80286 to use all of its address lines to access 16 MB of physical memory with 1 GB of virtual memory.

The 80286 also comes with built-in memory management and protection, which allows the MPU to be used in a multiple-user and multitasking system. The built-in memory management system internally realigns data bytes during a fetch (read) operation if the data is stored at odd address locations. The memory management system also realigns data words during a fetch bus cycle stored at odd address locations. During a fetch bus cycle of an odd addressed word, the memory management system automatically transfers the data one byte at a time and realigns each byte as it enters the MPU. An odd address memory fetch cycle requires two bus cycles.

4.2.1 80286 Base Architecture

The 80286 (Fig. 4-3) contains four controlling units. Although all the controlling units work together to perform the bus cycle, each unit performs its task independently.

Bus Unit (BU). The BU is responsible for interfacing the internal architecture of the 80286 with the outside world. The BU applies the 24-bit address to the address bus and controls the flow of data to and from the 80286. The data bus is the path by which the data enters or exits the MPU, while the bus control/status lines control the direction and timing of data flow.

Component Pad Views—As viewed from underside of component when mounted on the board.

P.C. Board Views—As viewed from the component side of the P.C. board.

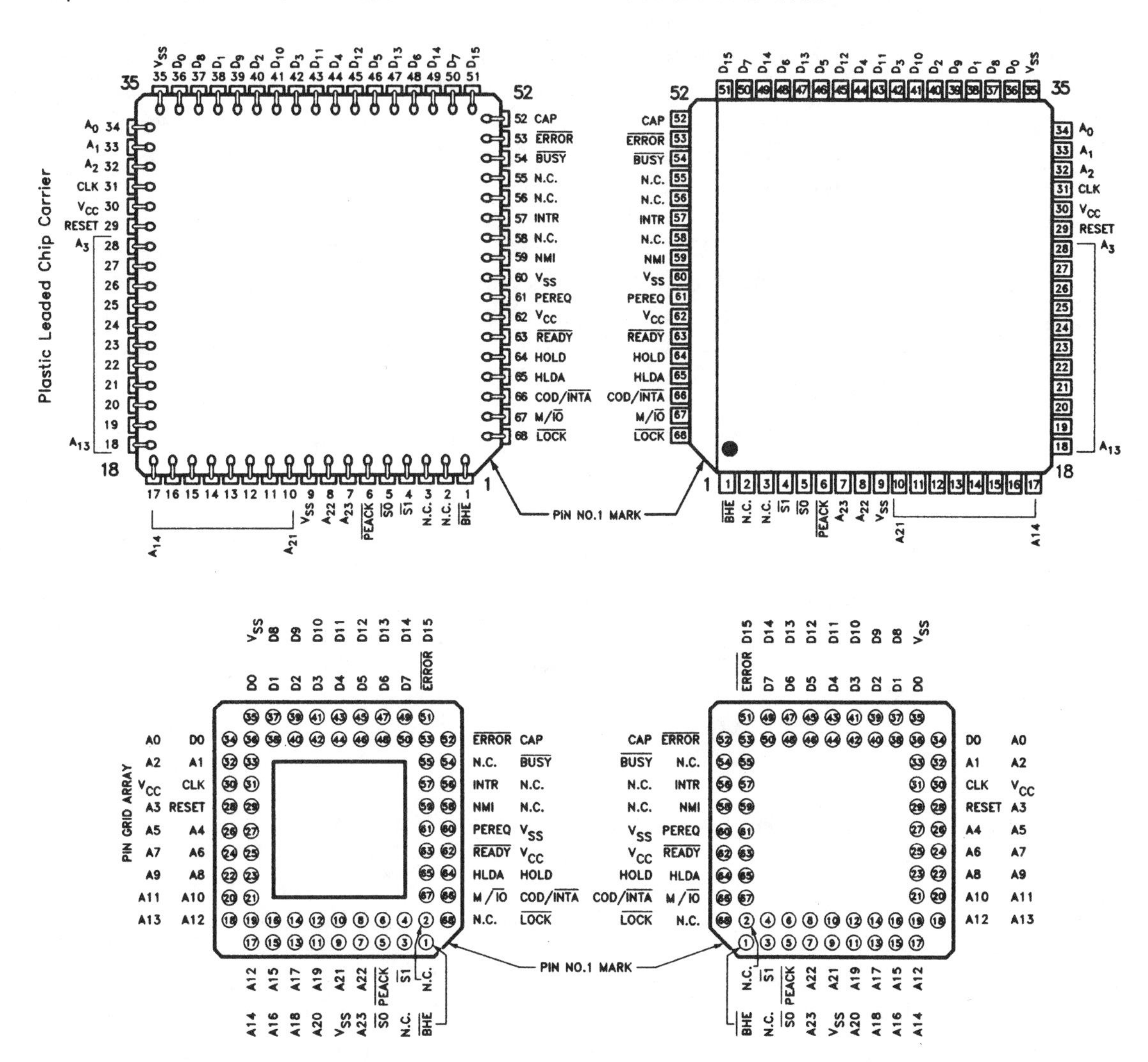

Figure 4-2 80286 pin configuration. (Courtesy of Intel)

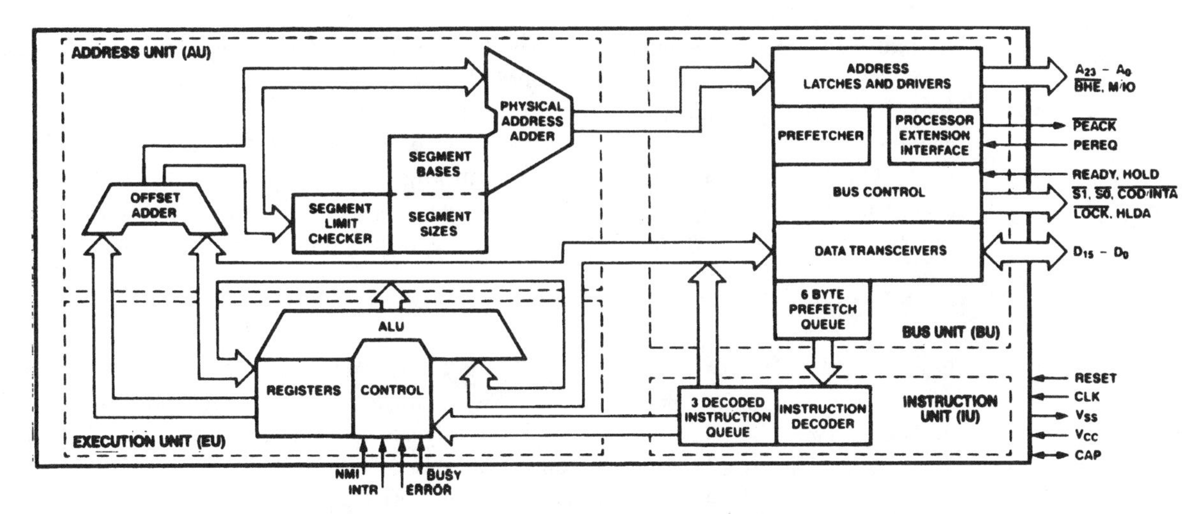

Figure 4-3 80286 internal block diagram. (Courtesy of Intel)

The following output lines are controlled by the BU: A23 to A0 (address bus), $\overline{\text{BHE}}$, M/$\overline{\text{IO}}$, $\overline{\text{PEACK}}$, $\overline{\text{S1}}$, $\overline{\text{S0}}$, COD/ $\overline{\text{INTA}}$, $\overline{\text{LOCK}}$, and HLDA. The following input lines are used to supply status to the BU: PEREQ, READY, and HOLD. Bidirectional data lines D15 to D0 are controlled by the BU. The BU can hold up to 6 bytes of data in a 6-byte prefetch queue (buffer) until the instruction unit needs the data. This buffer acts the same as in the 8088/86 by allowing the BU to keep working while the instruction unit is busy processing an instruction. There must be at least 2 bytes free in the prefetch queue before the BU prefetches the next byte. The 6-byte queue allows the BU and instruction unit to function simultaneously during most operations.

Instruction Unit (IU). The IU is responsible for decoding the instruction from the BU. Once decoded, the instruction goes into a decoded instruction queue, where up to three decoded instructions wait until they can be processed by the MPU. This unit gets status from the RESET and CLK inputs. The IU outputs the decoded instruction to either the BU or the execution unit, depending on the type of instruction.

Execution Unit (EU). The EU contains the ALU and the general purpose registers. This unit is responsible for processing data and executing instructions, once the data is inside the MPU. The EU contains most of the processing power of the MPU.

Address Unit (AU). The AU is used to develop the physical address for the MPU. How the address is developed depends on the operating mode of the 286. The AU gets its data from the EU and outputs its data to the BU. While the 286 is operating in the real mode, only the 20 least significant bits of the address bus are valid. When the 286 is in the protected mode, all 24 address bits are valid.

4.2.2 Memory Management

The 80286 contains a built-in memory management system; this circuitry is used to realign data that has been stored at odd address locations. The memory in an 80286 system is divided into two areas; half of the memory is located at even address locations and the other half at odd address locations. The even address locations use data lines D0 to D7, while odd address locations used data lines D8 to D15.

If the MPU needs to fetch a byte of data located at an even memory location, the data is transferred directly from the D0 to D7 data lines to the low byte of the specified register or location inside the MPU. If the MPU needs to fetch a byte of data located at an odd memory location, the data is transferred directly from the D8 to D15 data lines into the BU section of the MPU and then into the low byte of the specified register or location inside the MPU.

If the MPU needs to fetch a word of data located at an even memory location, the data is transferred directly from the D0 to D15 data lines to the specified register or location inside the MPU. If the MPU needs to fetch a word of data located at an odd memory location, the MPU performs a byte fetch, the data enters the bus unit of the MPU using data lines D8 to D15, and the bus unit transfers the byte into the low byte of the specified register or location. Next the MPU increments the address by one and performs another byte fetch; now the address is even. The data from the even memory location enters the BU of the MPU using data lines D0 to D7, and the BU transfers the byte into the high byte of the specified register or location.

4.2.3 Modes of the 80286

Upon reset, the 286 defaults to the real mode of operation. While the 286 is in the real mode, it acts like a very fast 8086. Addresses while in the real mode of operation function the same as in the 8086. Address lines A20 to A24 are disabled; only address lines A0 to A19 operate correctly.

Once booted up, the computer can be placed into the protected mode. While the 286 is in the protected mode, it can access memory above 1MB. The 286 automatically maps 1GB of virtual memory. Because the 286 has only 24 address lines, real address space is limited to 16MB. Also, when the 286 is in the protected mode, the computer can perform multitasking, allowing each task to occupy its own area of memory.

4.2.4 80286 Registers

General Registers (Real or Protected Mode). The purpose of the eight 16-bit general registers (Figs. 4-4 and 4-5) inside the 286 is the same as for those in the 8088 (see chapter 3); Accumulator (AX, AH, and AL), Date register (DX, DH, and DL), Count register (CX, CH, and CL), Base register (BX, BH, and BL), Base Pointer (BP), Source Index (SI), Destination Index (DI), and Stack Pointer (SP).

Flags Register (Real or Protected Mode). The flags register performs the same function as the flags register in the 8088 (see chapter 3). The only difference is that the flags register contains two additional flags found in three bits. The 2-bit IOPL (I/O Privilege Level) flag is used to determine the privilege level during multitasking in the protected mode. The NT (Nested Task) flag determines which task is in control during multitasking.

Machine Status Word Register (Real or Protected Mode). While in the real mode, the MSWR (Machine Status Word Register) is only monitored. The 16-bit MSWR contains data that controls and provides status for the current operating mode of the 286 MPU. Only 4 of the 16 bits are used; the following 3 bits are control bits: PE (Protect Enable), MP

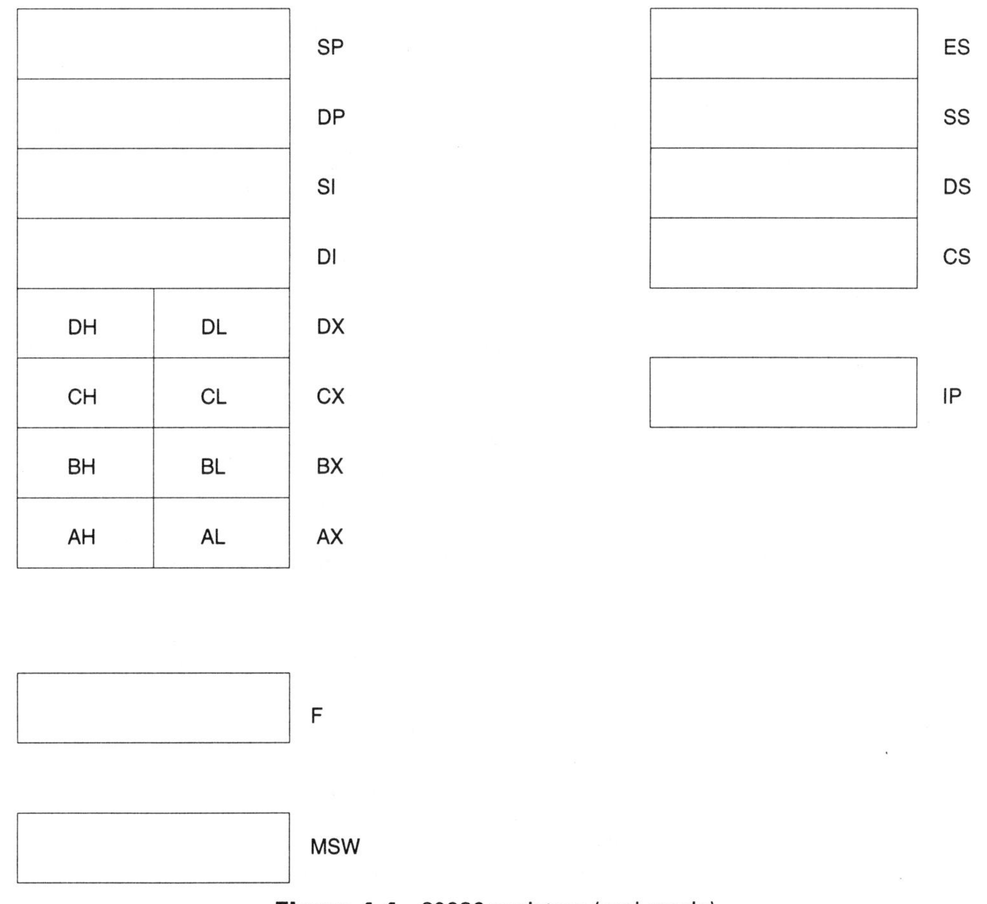

Figure 4-4 80286 registers (real mode).

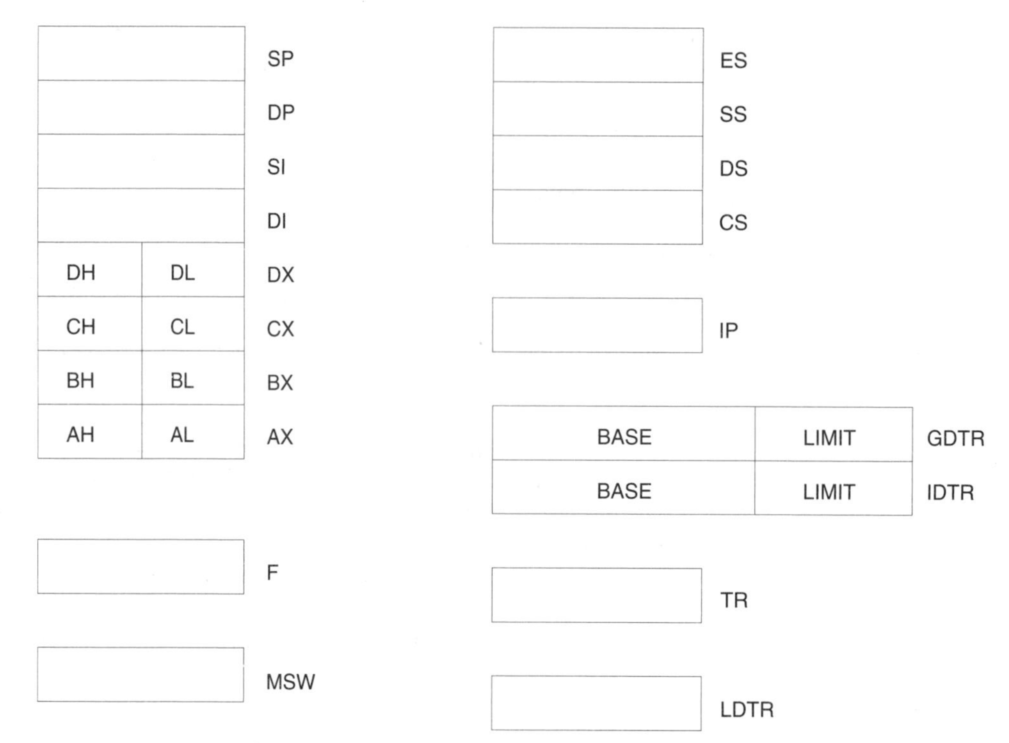

Figure 4-5 80286 registers (protected mode).

(Monitor Processor extension), EM (EMulate processor extension); the last bit is status bit TS (Task Switched).

The reset default level for the PE bit is 0, which means that the 286 is operating in the real mode. When the PE bit sets, the computer enters the protected mode; once in the protected mode, the MPU cannot return to the real mode without a hardware reset. If the MP bit is set, it indicates the presence of an 80287 numeric coprocessor; otherwise this bit is reset. If the EM bit is set, the 286 software emulates the 80287. *Note:* MP and EM can never be set at the same time. TS automatically sets whenever the MPU switches tasks using a processor extension and must be reset through software.

Addressing Registers (Real Mode). The four 16-bit segment registers (CS, DS, SS, and ES) and the 16-bit IP (Instruction Pointer) function the same as in the 8088 while the 286 is in the real operating mode. To develop a physical address (also known as an effective address), the value of the selected 16-bit segment register (known as the segment base) is transferred into the physical address adder, where four binary 0s are shifted into the least significant location, which creates a shifted segment value of 20 bits. The value of the IP or offset value is added to the shifted segment value. The 20 least significant bits of the result are used as the physical address, which is then applied to the address bus through the address latches and drivers.

Addressing Registers (Protected Mode). The Global Descriptor Table Register (GDTR) is a 40-bit register used to control the contents of the global descriptor table, which is located in physical memory. The contents of these memory locations are used to help produce addresses in physical memory above 1MB. The 16 least significant bits of the GDTR define the size of the memory block (ranges from 0 to 65,535). The 24 most significant bits represent the starting point of the memory block. A software driver controls the bidirectional movement of data between the GDTR and the global descriptor table. The size of the table is one plus the limit size. Each global descriptor entry in physical memory is 8 bytes in size, which specifies size, starting point, and access rights of the global memory segment. A maximum of 8192 descriptors can be used.

The Interrupt Descriptor Table Register (IDTR) is also a 40-bit register that is used like the GDTR except that it is used when processing an interrupt. The descriptor during an interrupt sequence is called the interrupt gate and specifies the starting address and attributes of the service routine.

When the MPU goes into the protected mode, the purpose of the segment registers changes from pointing to a segment value to providing a selector value. Of the 16 bits, bits 0 and 1 show the privilege level. Bit 2 is used to determine whether the global or local descriptor table is used, while bits 3 through 15 are used to select which descriptor entry is used.

Multitasking. Multitasking is the process by which the MPU allows two or more tasks to be done in the MPU at the same time. In reality the MPU time-shares each task. This means that each task is given a certain amount of the processor's time. When the time period of each task ends, all the data and the contents of the internal environment are stored. The MPU then recalls all the data for the next task, so that the new task can take control over the MPU.

The Task Register (TR) is a 16-bit register (protected mode only) that holds a value called the task state segment descriptor. During task switch, the value of the TR (one value for each task running) is used with the base value stored in the global descriptor table to determine the memory locations for the next selected task.

The Local Descriptor Table Register (LDTR) is a 16-bit register (protected mode only) that holds a value called the local descriptor. This value is used to produce the physical address for the current task.

Therefore, the TR is used to determine the address for the next task, while the LDTR defines the current task running. The memory management drivers must keep track of and control these values.

4.2.5 80286 Instructions

The 286 supports all 8088/8086 instructions, addressing modes, and data types. The 286 also has added more instructions and is capable of handling more types of data than the 8088/8086. Tables 4-1 and 4-2 briefly list the number of instructions and types of data.

4.2.6 80286 Pin Configurations

CLK — The CLocK input (Fig. 4-6) is used to provide the basic timing for the 286 MPU and any device that must be synchronized with bus operations. The system clock frequency is divided by 2 internally to produce a processor (bus state) clock; the first system clock is called phase 1 and the second phase 2. All bus cycles require a minimum of two processor clocks (bus states).

Memory — *Note:* In the 80286 system memory is divided into two blocks. The low block of memory (even memory locations, when A0 = 0) is used to

TABLE 4-1
286 Instruction Types

Type of Instruction	*Number of Instructions*
Data transfer	17
Arithmetic	20
String	10
Shift/rotate logical	12
Program transfer	28
Processor control	14
High level	3

TABLE 4-2
286 Data Types

Type of Data	*Description*
Integer	Signed 8-bit and 16-bit binary values
Ordinal	Unsigned 8-bit or 16-bit binary values
Pointer	32-bit value (segment selector and offset)
String	1 byte to 64K of contiguous data
ASCII	Alphanumeric code using ASCII format
BCD	A byte of data used to represent 0 to 9
Packed BCD	A byte of data, where each nibble is a value between 0 and 9
Floating point	Signed 32- to 80-bit value supported by the 287

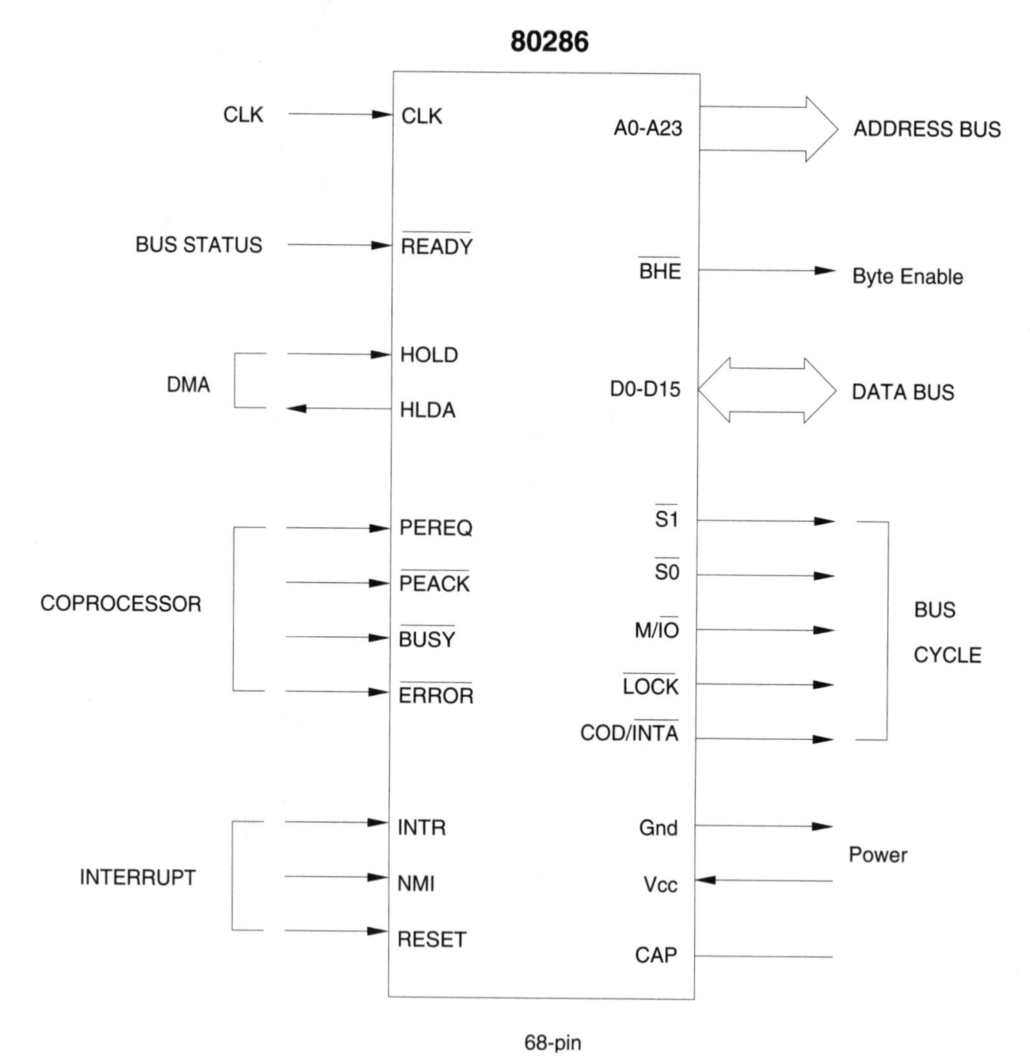

Figure 4-6 80286 pin groupings.

store data transferred by the lower 8 bits of the data bus. The high block of memory (odd memory locations) is used to store data transferred by the upper 8 bits of the data bus.

A0–A23

As opposed to the 8088/86, which uses a multiplexed address/data bus, the 80286 uses a dedicated 24-bit address bus. These 24 output lines are used to address the devices in the 286 system, depending on the operating

TABLE 4-3
$\overline{\text{BHE}}$/A0 Operations

$\overline{BHE}$	*A0*	*Operation*
0	0	Word transfer (D0 to D15)
0	1	High byte transfer (D8 to D15)
1	0	Low byte transfer (D0 to D7)
1	1	Disallowed state

TABLE 4-4
286 Bus Cycle

COD/$\overline{INTA}$	*M/$\overline{IO}$*	*$\overline{S1}$*	*$\overline{S0}$*	*Bus Cycle*
0	0	0	0	Interrupt acknowledge
0	1	0	0	A1 = 1 halt, else shutdown
0	1	0	1	Memory read
0	1	1	0	Memory write
0	1	1	1	Idle (passive)
1	0	0	1	I/O read
1	0	1	0	I/O write
1	1	0	1	Memory instruction read
1	1	1	1	Idle (passive)

mode. While the 80286 is in the real mode, it develops the effective address in the same way as the 8088; only address lines A0 to A19 represent the address. While in the protected mode, address lines A0 to A23 represent the valid address. A0 is low whenever the lower byte of the data bus is used. During I/O transfers, A16 to A23 are low. During a bus hold acknowledge (DMA cycle), these lines go into tristate.

$\overline{\text{BHE}}$ The $\overline{\text{Bus High Enable}}$ is used when the upper data byte is accessed in memory. The level on this output is used with the value of A0 to determine which memory block or blocks receive the data from the transfer (Table 4-3).

The $\overline{\text{BHE}}$ line is used with the chip selects to control which block of memory is used during the transfer. The $\overline{\text{BHE}}$ line goes into tristate during a bus hold acknowledge.

D0–D15 D0 to D15 represent the 16-bit bidirectional data bus. During a memory or I/O write operation, these lines act as outputs. During a memory, I/O read, or interrupt acknowledge operation, these lines act as inputs. During a byte operation D8 to D15 (high byte) or D0 to D7 (low byte) lines are used. During a word operation all data lines (D0 to D15) are used.

M/$\overline{\text{IO}}$ The Memory/$\overline{\text{Input/Output}}$ select line is used in conjunction with the bus cycle status lines to inform the bus controller what type of operation is to take place. When high, a memory operation takes place. When low, an I/O operation takes place.

$\overline{\text{S0}}$, $\overline{\text{S1}}$ The bus cycle Status lines are used in conjunction with the M/$\overline{\text{IO}}$ line to determine which bus operation takes place. This information is supplied to the 82288 bus controller to perform the selected bus operation (Table 4-4).

Any combination of signals not listed in Table 4-4 is either a disallowed state or not a status cycle. Bus cycle status lines go into a tristate condition during a bus hold acknowledge.

COD/$\overline{\text{INTA}}$

The CODe/$\overline{\text{INTerrupt Acknowledge}}$ line is used to determine when an instruction fetch or a memory data read cycle occurs. This line is also used to indicate to the bus controller that an interrupt acknowledge cycle is in progress as opposed to an I/O cycle. When high, an instruction is being transferred or an I/O operation is taking place. When low, an interrupt acknowledge is taking place. This line goes into tristate during a bus hold acknowledge.

$\overline{\text{READY}}$

The bus $\overline{\text{READY}}$ is an active-low input that is used to terminate the current bus cycle when low (the bus is ready to complete its transfer). When high, the bus is still in the process of accessing the selected device and is not ready to complete the cycle. During a bus hold acknowledge the $\overline{\text{READY}}$ line is ignored.

INTR

The INTerrupt Request (maskable interrupt) input informs the MPU that the PIT has requested the servicing of an interrupt, when this line goes high. The MPU samples this input at the beginning of each bus cycle. This line must be high for two processor cycles for the interrupt sequence to start. Once the interrupt sequence starts, this line must remain high until the end of the first of the two interrupt acknowledge cycles. The MPU sends out two interrupt acknowledge cycles during an interrupt sequence.

NMI

The NonMaskable Interrupt input, when high, produces an interrupt vector of 2 internally. This line must be high for at least four system clock cycles and must remain high for another four system clock cycles.

$\overline{\text{LOCK}}$

The bus $\overline{\text{LOCK}}$ line is used to lock out any other bus master (DMA) when low. When high, another bus master can ask for control of the bus.

HOLD

The bus HOLD request is an input used by the bus master to request control of the local bus. When the DMA wants to use the bus, it causes this line to go high. A high requests permission to use the bus. The 286 must complete its current cycle before it can give up control of the bus. This input must remain valid throughout the bus master cycle; otherwise permission is taken away.

HLDA

The bus HoLD Acknowledge is the output that the 286 uses to grant permission to the DMA (bus master) to use the bus. When the 286 is ready to give up control of the bus, it causes this line to go high.

PEREQ

Processor Extension operand REQuest input informs the 286 data operand transfer. This signal is used for memory management and protection for the 286. This input is controlled by the 80287 math coprocessor.

$\overline{\text{PEACK}}$

The $\overline{\text{Processor Extension operand ACKnowledge}}$ output indicates that a processor extension operand request is being transferred. This output is applied to the 80287 math coprocessor.

$\overline{\text{BUSY}}$

The $\overline{\text{BUSY}}$ input is controlled by the 80287 math coprocessor. When low, this line makes the 286 wait until the 80287 is ready for more data.

$\overline{\text{ERROR}}$	The $\overline{\text{ERROR}}$ input is controlled by the 80287 math coprocessor. When low, the 286 performs an internal processor extension interrupt.
RESET	The RESET input is used to reset the 286 to its starting values when this line goes low for at least 16 system clock cycles.
Vss	Vss is the ground for the 286; it provides a path for current and acts as a reference point for the 286.
Vcc	Vcc supplies the voltage potential for the 286.
CAP	The substrate filter CAPacitor is connected to a capacitor to ground. It reduces any internal substrate bias generated by the switching of the gates in the 286. The size is 0.047 μF at 12 volts.

4.3 80287 80-BIT NUMERIC PROCESSOR EXTENSION

The 80287 numeric processor (also known as a math coprocessor) (Figs. 4-7 and 4-8) is a device that perfroms math operations at least 10 times faster than the 286. The 287 can handle 32-, 64-, and 80-bit floating point values and 32- and 64-bit integers. The 287 is also able to process 18-digit BCD values. The disadvantage of the 287 is that if the software does not use the 287 math instruction coding, the 287 does not process the data. Not all software uses the 287 even if the math coprocessor is present.

4.3.1 80287 Pin Configuration and Functions

CLK	The CLocK input allows the 287 to synchronize its operations with the 286 when the 287 is enabled. The CLK input is the same as that supplied to the 286.
CKM	The ClocK Mode input is used to select whether the 287 uses the CLK input frequency directly or divides it by 3. This line is normally tied high so that the 287 uses the CLK frequency directly.

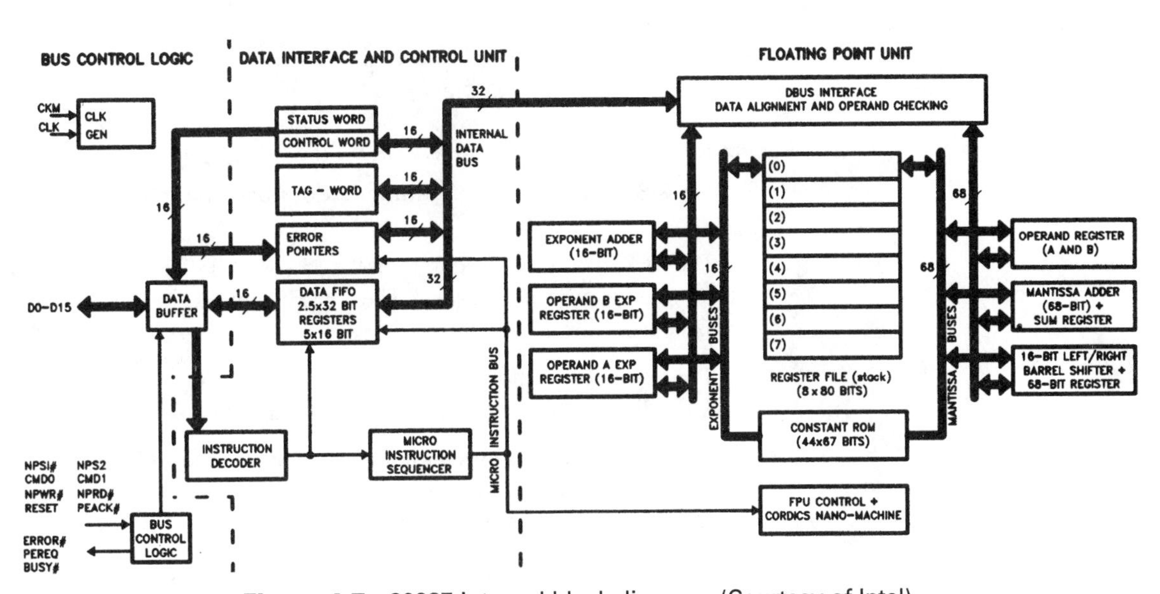

Figure 4-7 80287 internal block diagram. (Courtesy of Intel)

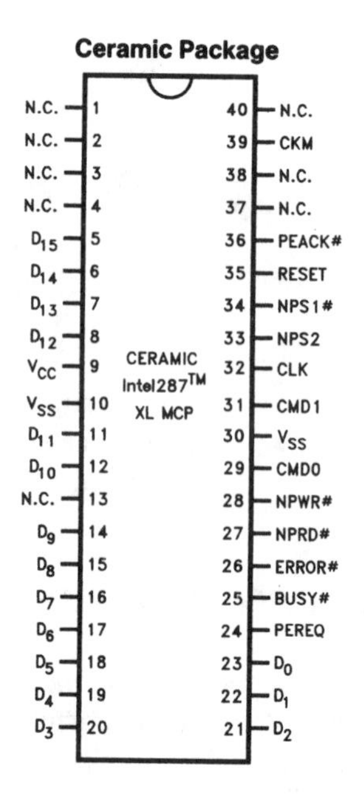

NOTE:
The Intel287 XL will operate in any coprocessor socket designed for an NMOS 80287 or the CMOS 80C287A.

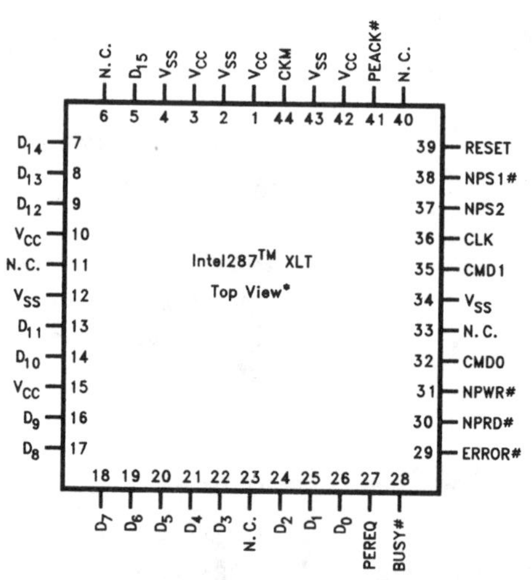

*"Top View" means as the package is seen from the component side of the board.

Figure 4-8 80287 pin configuration. (Courtesy of Intel)

RESET	The RESET input is used to reset the 287 to its initial values when high for four or more clock cycles. The RESET input is connected to the same reset line as the 286.
D0–D15	The Data inputs/outputs are used to allow data to enter and exit the 287, depending on the operation and the communications with the 286.
$\overline{\text{BUSY}}$	The $\overline{\text{BUSY}}$ output is used to inform the 286 that the 287 is busy processing information, when low. The 286 does not send any data to the 287 until the busy line goes high again.
$\overline{\text{ERROR}}$	The $\overline{\text{ERROR}}$ output line indicates that an unmasked error has occurred during the processing of the instruction.

TABLE 4-5
287 Bus Cycle

NPS1	*NPS2*	*CMD0*	*CMS1*	*NPRD*	*NPWR*	*Bus Cycle*
0	1	0	0	1	0	Opcode write 287
0	1	0	0	0	1	CW or SW read 287
0	1	1	0	0	1	Read data 287
0	1	1	0	1	0	Write data 287
0	1	0	1	1	0	Write exception pointer

PEREQ — The Processor Extension data channel operand transfer REQuest output is used to signal the 286 that the 287 is ready to transfer its data, when high. Once the data is transferred, this line resets. A low on the PEACK input line also resets this line.

PEACK — The Processor Extension data channel operand transfer ACKnowledge input is used by the 286 to acknowledge the data transfer, when low. The 286 sets the PEACK line once the transfer is complete.

NPRD — The Numeric Processor ReaD input causes the data from the 286 to enter the 287 using the data bus, when low.

NPWR — The Numeric Processor WRite input causes the data to exit the 287 and be applied to the data bus, where it is transferred to the 286.

NPS1/NPS2 — The Numeric Processor Select inputs (NPS1 active high and NPS2 active low) are used to indicate that the 286 is in the process of performing an Escape instruction. If both lines are inactive, no data transfers using the 287 occur. If both of these lines are active concurrently, the 287 is allowed to perform floating point instructions.

CMD1/CMD0 — The CoMmanD lines 0 and 1 inputs are used by the 286 to control the type of operation that takes place in the 287. These lines are used in conjunction with the NPS1 and NPS2 lines (Table 4-5).

4.3.2 80287 and 80286 Operational Sequence

If the 286 performs an escape instruction during the execution of a program, it checks the BUSY line from the 287. If the BUSY line is low, the 286 writes the opcode (operation code) into the 287, followed by any data necessary for the instruction to be performed. Upon completion of the instruction, the 287 informs the 286 that the instruction is complete, at which time the 286 reads the data that the 287 placed on the data bus. A typical operation is as follows:

1. The 286 processes an escape code instruction.
2. The 286 checks the BUSY line; if high, the process continues. If low, the 286 waits.
3. The 286 brings the NPS1, CMD0, and CMD1 lines low while bringing NPS2 and NPRD lines high, enabling the 287.
4. The 287 causes the PEREQ line to go high (indicating that the 287 is ready).
5. The 286 places the instruction onto the data bus.
6. The 286 causes the NPWR line to go low, which causes the 287 to perform an opcode write (data from the data bus enter the 287).
7. The 286 then acknowledges the transfer by causing the PEACK line to go low.
8. Once the 287 accepts the data on the data bus, it causes the PEREQ line to go low.

9. Once the 286 sees the PEREQ line go low, it sends the data for the math instruction to the 287.
10. The 286 causes the NPWR line to go high, ending the write operation.
11. The 287 causes the BUSY line to go high. *Note:* If an error occurs at any time during the 287 bus cycle, the 287 causes the ERROR output to go low, which in turn generates an interrupt.
12. Steps 1 through 11 are executed a number of times to write the data for the instruction into the 287. The difference is that levels on the NPS1, CMD0, CMD1, NPS2, NPRD, and NPWR lines change with each type of bus operation. If the data is located in memory locations, the 286 addresses the memory locations so that the data is available on the data bus for the 287. Upon completion of the math instruction, the 286 either processes the results of the 287 operation or stores those results in memory. The instruction may require the 286 to read the status word to determine if the operation was successful.

4.4 INTEL 82230/82231 AT CHIP SET

Not long after the 80286 MPU became available, Intel and other IC manufacturers began thinking of ways to reduce the number of support ICs for this new type of MPU. Their approach to reducing the number of ICs was to integrate many of the support IC functions in large scale and very large scale ICs. The following was one solution by Intel to reduce the number of ICs necessary to build an AT-type microcomputer. Other manufacturers also make AT chip sets.

The 82230 is a 84-pin IC (Fig. 4-9) that contains the following:

1. 82284 clock generator/driver
2. Two 8259A PITs
3. 6818 real time clock (battery backed-up)

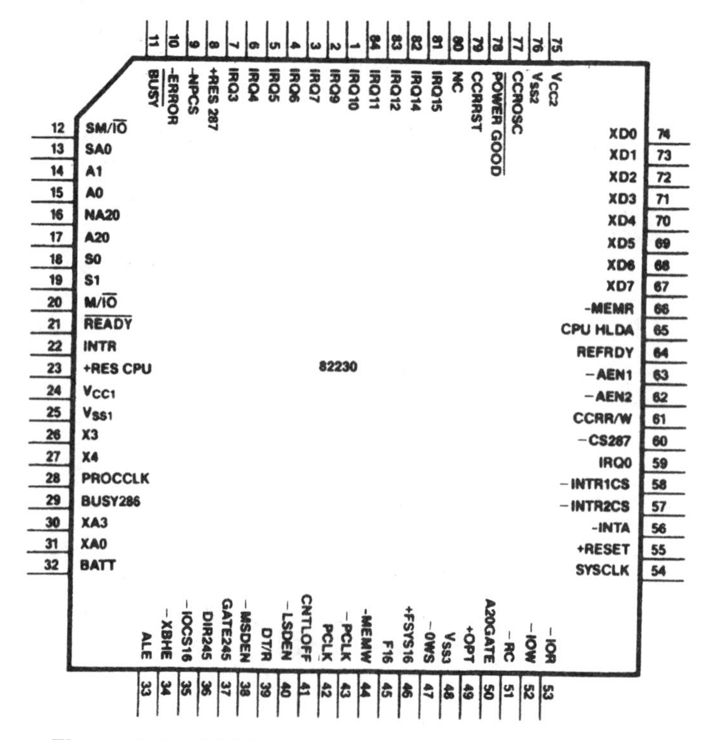

Figure 4-9 82230 pin configuration. (Courtesy of Intel)

4. 82288 bus controller
5. Address and data bus control logic
6. Bus control logic
7. CPU shutdown logic
8. Coprocessor interface logic

The 82231 is a 84-pin IC (Fig. 4-10) that contains the following:

1. R/W logic
2. I/O chip select
2. Support logic for the PIT and I/O devices
3. 8254 PIT
4. Memory mapper (74612)
5. Two 8237 DMAs
6. DRAM refresh counter, latch, and logic
7. Parity check logic

4.4.1 82230 (Fig. 4-11)

82284 Clock Generator/Driver

X3–X4 X3 (input) and X4 (output) together are used to provide the crystal frequencies for the PROCCLK and SYSCLK signals. The crystal connected to these pins must operate at twice the frequency of the processor clock.

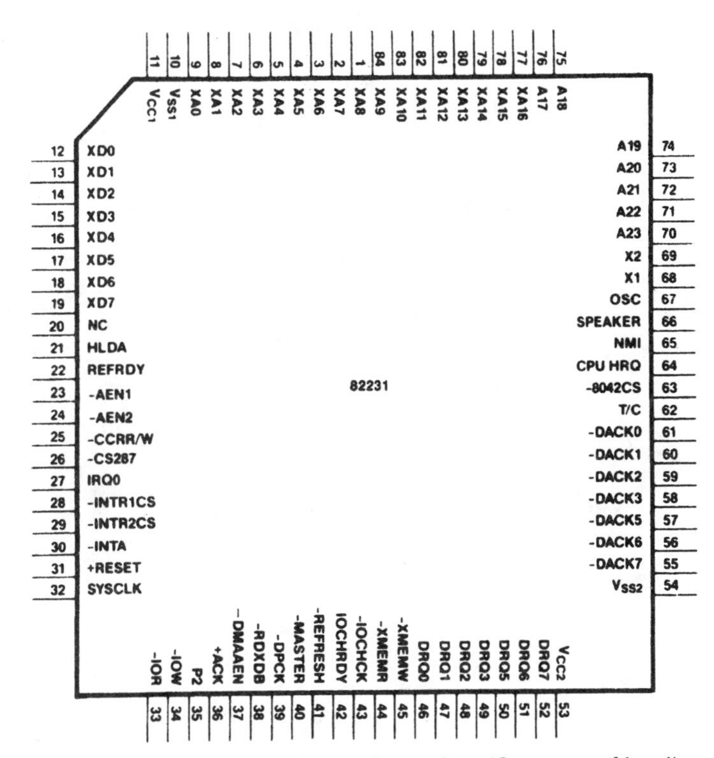

Figure 4-10 82231 pin configuration. (Courtesy of Intel)

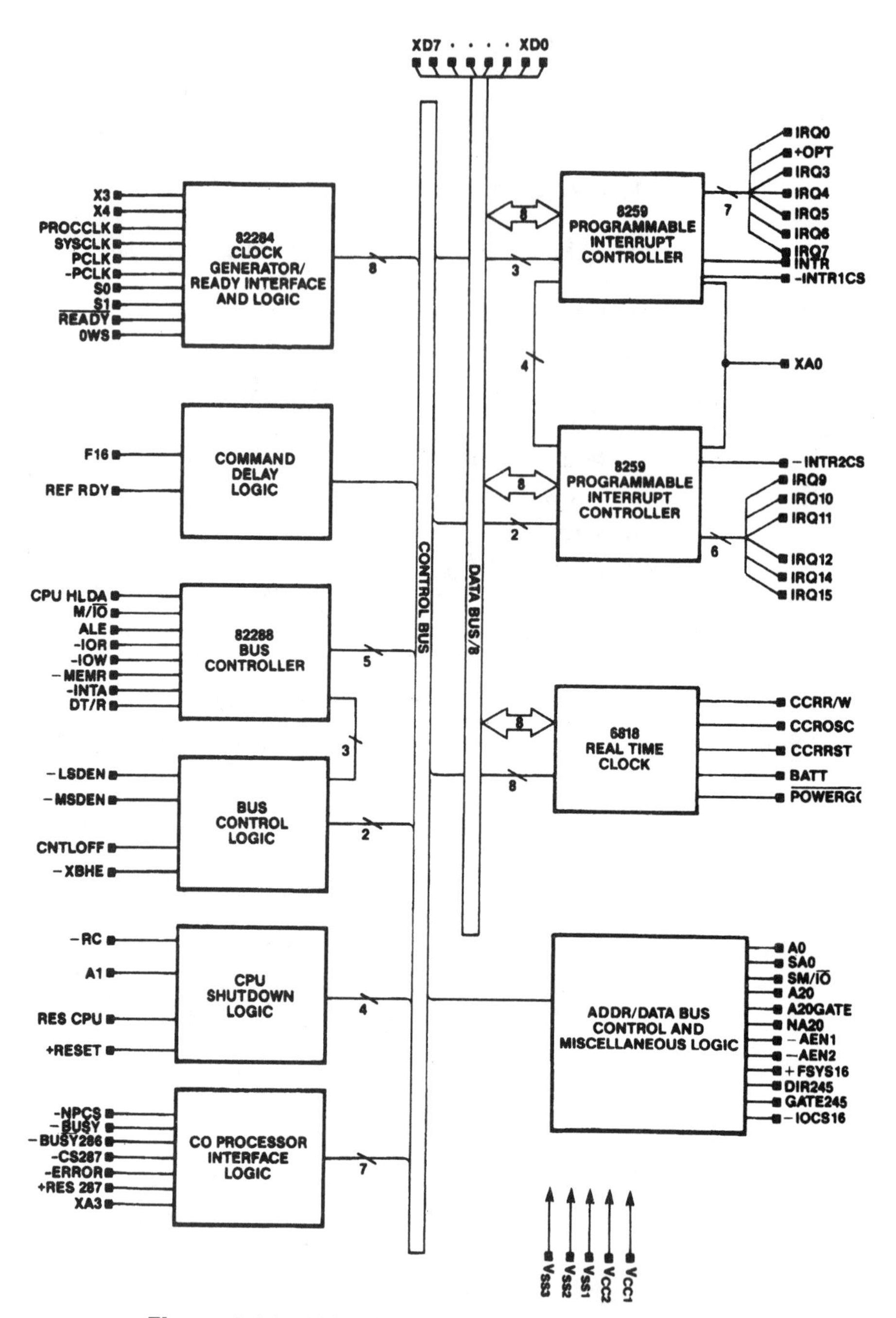

Figure 4-11 82230 block diagram. (Courtesy of Intel)

PROCCLK — The PROCessor CLocK output is used to supply the clock signal for the MPU and math coprocessor. This line is connected directly to the clock input of the 286 and 287.

SYSCLK — The SYStem CLocK output operates at one-half of the processor clock (PROCCLK). Because the system is synchronized with the T-state's processor clock, it can be used for peripherals that must be synchronized with the MPU. The duty cycle of the system clock is 50%.

PCLK, $\overline{\text{PCLK}}$ — The Peripheral CLocK and inverted peripheral clock outputs are used to provide clock signals that are one-half the processor clock to peripheral devices such as the crystal inputs for the 8042 keyboard controller (PCLK to XTAL1 and $\overline{\text{PCLK}}$ to XTAL2).

S0, S1 — The MPU uses the Status inputs (S0 and S1) to provide the current status of the instruction being processed to the 82230. The levels on these lines are generated by the S0 and S1 outputs from the 286.

$\overline{\text{READY}}$ — The $\overline{\text{READY}}$ output is used to indicate to the MPU that the current bus cycle is ready to be completed, when low. The level on the $\overline{\text{READY}}$ line is determined by the levels on $\overline{\text{0WS}}$, $\overline{\text{POWER-GOOD}}$, S0, and S1. This output is connected to the $\overline{\text{READY}}$ input of the 286.

$\overline{\text{0WS}}$ — The zero Wait State input is connected to the keyboard controller and, when pulled low during a bus cycle, terminates the current processor instruction by causing the $\overline{\text{READY}}$ line to go low. This line can be tied high, and the number of wait states can be programmed into the 82230. The default number of wait states is 1 for memory operations and 4 for I/O operations.

Command Delay Logic

F16 — The F16 input line, when high, is used to indicate a 16-bit memory access. The level on this line is generated by the address-decoding section, which can be used to reduce command delays for memory bus cycles.

REFRDY — The REFresh I/O channel ReadDY input is used to insert wait states into the bus cycle when high. This signal is developed by the REFRDY output of the 82231. When high, this line indicates that the device on the I/O channel is not ready to complete the bus cycle. This line causes the $\overline{\text{READY}}$ output of the 82230 to stay high until the device on the I/O channel is ready.

82288 Bus Controller

CPU HLDA — The CPU HoLD Acknowledge input from the 286 (HLDA) is used to indicate that the MPU has given up control of the bus to a different bus master, when high. In the case of the IBM AT, the other bus master is the DMAC (Direct Memory Access Controller).

M/$\overline{\text{IO}}$ — The Memory/$\overline{\text{Input/Output}}$ from the 286 is used to determine the type of bus cycle. When high, a memory bus cycle is indicated; a low indicates an I/O bus cycle.

ALE — The Address Latch Enable output is used to latch address lines A0 through A19 into the system the address latches for the bus cycle. This output is active high, which allows the address to enter the address latches. Address lines A20 to A23 are not latched for the system address bus; instead, they are latched by the address-decoding section. ALE is low during a HALT cycle.

$\overline{\text{IOR}}$ — The $\overline{\text{I/O Read}}$ output is used to instruct the I/O device being accessed to output its data to the data bus as long as this line is low. During a DMA cycle this line is in tristate, and during CNTLOFF this line is held low.

$\overline{\text{IOW}}$ — The $\overline{\text{I/O Write}}$ output is used to instruct the I/O device being accessed to accept the data on the data bus as input as long as this line is low,

During a DMA cycle this line is in tristate, and during a CNTRLOFF this line is held low.

$\overline{\text{MEMR}}$ — The $\overline{\text{MEMory Read}}$ input/output is used to instruct the memory device being accessed to output its data to the data bus as long as this line is low. During a DMA cycle this line is in tristate, and during CNTLOFF this line is held low.

$\overline{\text{MEMW}}$ — The $\overline{\text{MEMory Write}}$ input/output is used to instruct the memory device being accessed to accept the data on the data bus as input as long as this line is low. During a DMA cycle this line is in tristate, and during a CNTRLOFF this line is held low.

$\overline{\text{INTA}}$ — The $\overline{\text{INTerrupt Acknowledge}}$ output signal is used to signal the acknowledgment of an interrupt sequence to the 82231 when pulsed low. During a DMA cycle this line is in tristate, and during a CNTRLOFF this line is held low.

DT/$\overline{\text{R}}$ — The Data Transmit/$\overline{\text{Receive}}$ output line sets the direction that data flow in the 74245 data bus transceivers once they are enabled. The level of this line is determined by the two status (S0 and S1) lines. During a write (transmit) cycle this line is high, and data flows from the MPU to the device being accessed. During a read (receive) cycle this line is low, and data flows from the device being accessed to the MPU. When the bus is in an inactive state, this line is high.

Bus Control Logic

$\overline{\text{LSDEN}}$ — The $\overline{\text{Least Significant Data ENable}}$ output line is used to enable the flow of data in the 74245 local data bus transceiver, which controls the flow of data in the lower byte of the data bus when low.

$\overline{\text{MSDEN}}$ — The $\overline{\text{Most Significant Data ENable}}$ output line is used to enable the flow of data in the 74245 local data bus transceiver, which controls the flow of data in the upper byte of the data bus when low.

CNTLOFF — The CoNTroL OFF output is used to enable the low byte of the data bus during a byte operation, when high.

$\overline{\text{XBHE}}$ — The $\overline{\text{eXternal Bus High Enable}}$ input/output line is used by the 82230 to generate the $\overline{\text{MSDEN}}$ signal. When low, the $\overline{\text{MSDEN}}$ line is low, which allows the data to flow in the upper byte of the data bus through the 74245 local data bus transceiver.

CPU Shutdown Logic

$\overline{\text{RC}}$ — The $\overline{\text{Reset CPU}}$ input is developed by pin 21 of the 8042 keyboard controller. When this pin goes low, the 82284 section in the 82230 causes the RES CPU line to go high, thus resetting the MPU.

A1 — Address line 1 developed by the MPU is used with the levels on M/$\overline{\text{IO}}$, S0, and S1 to determine if the MPU is in a shutdown condition (HALT). If M/$\overline{\text{IO}}$ is forced high while S0, S1, and A1 are forced low, the HALT condition occurs.

RES CPU — The RESet CPU output is used as the line that resets the MPU when high. This signal is developed by the 82284 section of the 82230 and is connected to the reset input of the MPU. If the $\overline{\text{Power-Good}}$ or $\overline{\text{RC}}$ lines go active or if a HALT bus cycle is processed, this line goes high for 16 PROCCLK cycles before it can go inactive.

RESET — The system RESET output is used to reset the 82231 (pin 55) and any other smart IC in the computer system when high.

Coprocessor Interface Logic

$\overline{\text{NPCS}}$ — 80287 Chip Select ($\overline{\text{NPS1}}$ pin).

$\overline{\text{BUSY}}$ — Input from the 287 to indicate the status of the 287; this line is used to generate the $\overline{\text{BUSY286}}$ line.

$\overline{\text{BUSY286}}$ — The $\overline{\text{BUSY286}}$ output is sent to the 286 and informs the MPU of the readiness of the 287. This line is tied to the $\overline{\text{BUSY}}$ line of the 286.

$\overline{\text{CS287}}$ — Math coprocessor $\overline{\text{Chip Select}}$ controls the $\overline{\text{NPCS}}$ from the $\overline{\text{CS287}}$ pin of the 82231. This line, when low, enables the 287 to operate.

$\overline{\text{ERROR}}$ — This negative-edge triggered input from the numeric processor indicates that an unmarked error condition exists: $\overline{\text{Error (287)}}$ to $\overline{\text{Error (82230)}}$.

RES 287 — RESet signal for the 287.

XA3 — Address line 3 of the I/O bus is used to help generate the reset and chip select signals for the 287.

8259 PIC

IRQ0 — Interrupt ReQuest line 0 input is used to request an interrupt sequence for the Time of Day in PIT 1 of the 82230. This signal is developed by the IRQ0 output line of the 82231, which is internally generated by channel 0 of the 8254. IRQ0 has the highest priority of all interrupt requests.

OPT — OPT [interrupt request line 1] input is used to request an interrupt sequence for the keyboard in PIT 1 of the 82230. This signal is developed by pin 24 of the 8042 keyboard controller. When high, it indicates that the keyboard buffer is full. This input has the second highest priority of all interrupt requests.

IRQ3–IRQ7 — Interrupt ReQuest inputs 3 through 7 are used to generate interrupt sequences in interrupt controller 1. These interrupt request lines have the lowest priorities of all interrupt request inputs, with IRQ7 having the very lowest priority. In the AT system, IRQ3 is used for COMM2 (serial port 2), IRQ4 is used for COMM1 (serial port 1), IRQ5 is used for LPT2 (parallel printer port 2), IRQ6 is used for the floppy disk adapter, and IRQ7 is used for LPT1 (parallel port 1).

Interrupt request lines IRQ8 and IRQ13 are generated internally in the 82230 and 82231. IRQ8 generates an interrupt sequence for the real time clock in the 82231, and IRQ13 generates an interrupt sequence for the 80287 math coprocessor.

IRQ9–IRQ15 — Interrupt ReQuest inputs 9 through 12, 14, and 15 are used to generate interrupt sequences in interrupt controller 2. These interrupt request lines have the highest priority following IRQ0 and IRQ1, with IRQ9 having the highest priority of the set. In the AT system, IRQ9 is used for software redirected to interrupt 0A (hex); IRQ10, IRQ11, and IRQ15 are not reserved for any particular device and may be used by special I/O adapters when needed. IRQ14 is used for the fixed disk controller.

INTR

The INTerrupt Request output is used to inform the MPU that an interrupt needs to be serviced. A high on this line causes the MPU, after completing its current instruction, to generate an interrupt acknowledge bus cycle.

$\overline{\mathrm{INTR1CS}}$

The $\overline{\text{INTeRrupt 1 Chip Select}}$ input is used to select PIC 1 in the 82230 when low. This signal is generated by the 82231. PIC 1 uses IRQ0, OPT, and IRQ3 to IRQ7 interrupt request inputs. This input allows PIC 1 to be programmed or to have its status read.

$\overline{\mathrm{INTR2CS}}$

The $\overline{\text{INTeRrupt 2 Chip Select}}$ input is used to select PIC 1 in the 82230 when low. This signal is generated by the 82231. PIC 2 uses IRQ9 to IRQ15 interrupt request inputs. This input allows PIC 2 to be programmed or to have its status read.

XA0

Address line 0 input from the X bus (peripheral) is used with the $\overline{\mathrm{INTR1CS}}$, $\overline{\mathrm{INTR2CS}}$, $\overline{\mathrm{XIOR}}$, and $\overline{\mathrm{XIOW}}$ lines to determine which of the registers of the selected PICs is accessed during a read or write operation.

6818 Real Time Clock

CCRR/W

The Clock CalendaR Read/Write line input is used as a chip select for the 6818 real time clock. When high, data can be read from or written into the real time clock. The level on this line is developed by the CCRR/W output of the 82231.

CCROSC

The Clock/CalendaR OSCillator input is used to apply the 32.768-KHz clock to the 82230. This frequency is necessary for the real time clock to maintain proper timing.

CCRRST

The Clock/CalendaR ReSeT input is used to reset the values in the real time clock and its RAM data, when low. This line is usually developed by a header and jumper on the system board. This line is brought low only if the user wishes to erase all the data in the real time clock and its RAM.

BATT

The BATTery input is used to supply a +5-V voltage from a battery to maintain the operation of the real time clock and its RAM information when the power is turned off. The battery may be located on the system board, or an external battery pack may be used.

$\overline{\mathrm{POWER\text{-}GOOD}}$

The $\overline{\mathrm{POWER\text{-}GOOD}}$ input line is generated by the power-good output line from the power supply. If the power of any one of the outputs from the power supply goes out of tolerance, this line goes low, which forces a reset condition to occur.

Address/Data Bus Control and Miscellaneous Logic

A0

Address line 0 input is developed by local address line 0 from the MPU. The level on this input is used to develop the system address line for the system board. This line is needed to help determine the size of the data transfer to be performed.

SA0 The System Address line 0 output is applied to the system address line 0 of the system board. It is developed by local address line 0. Just like the other system address lines, this line goes high during a DMA cycle and is forced low during an interrupt acknowledge.

SM/$\overline{\text{IO}}$ The System Memory $\overline{\text{Input/Output}}$ line input is used to help if the bus cycle is dealing with memory or I/O devices. When high a memory bus cycle is indicated, and when low an I/O bus cycle is indicated. This signal is used with the S0 and S1 status lines to determine the bus cycle.

A20 Address line 20 output is used to provide the level for the system address line 20. This line is developed by the 82230 from the levels on A20GATE and NA20. A20 goes high during a hold bus cycle.

A20GATE The Address line 20 GATE input is generated by the keyboard controller. This line is used to switch the computer between the read and protected modes of the 286. When this line is high, the computer is in the protected mode and A20 output from the 82230 follows the NA20 input from the A20 line of the 286. When low, the A20 output line is forced low, and the 286 is in the read mode.

NA20 Normal Address line 20 input is generated by A20 from the 80286. The level on address line 20 is processed by the 82230.

$\overline{\text{AEN1}}$ The $\overline{\text{Address Enable 1}}$ input is generated by the DMA section of the 82231. When low, the address latches in the DMA section are enabled for a byte operation.

$\overline{\text{AEN2}}$ The $\overline{\text{Address Enable 2}}$ input is generated by the DMA section of the 82231. When low, the address latches in the DMA section are enabled for a word operation.

FSYS16 The Fetch System 16-bit signal is the latched signal that, when high, indicates a 16-bit word memory access.

DIR245 The DIRection 245 output line is used to control data multiplexed in the extended data bus transceiver section. When performing an 8-bit I/O operation on the extended data bus, this line is used to transfer the high byte from the system bus to the low byte on the extended bus at the proper times. This line also converts a low byte on the extended data bus to the high byte on the system data bus at the proper times during an 8-bit I/O operation.

$\overline{\text{GATE245}}$ The $\overline{\text{GATE 245}}$ enable output line is used to enable the transfer.

$\overline{\text{IOCS16}}$ The $\overline{\text{Input/Output Chip Select}}$ 16-bit input line is controlled by the address-decoding section to indicate that a 16-bit I/O cycle is in progress. One wait state is added to the bus cycle.

Extended Data Bus

XD0–XD7 The bidirectional data lines are used to allow data to program information into the 82230. During a write operation, data enters the 82230. During a read operation these lines act as outputs, so that the status of the 82230 can be read. Data enters and exits this IC from and to the extended data bus. Because this IC is connected to the extended data bus, it is considered a peripheral device.

4.4.2 82231 (Fig. 4-12)

Read/Write Logic and Peripheral Select Decode

$\overline{\text{IOR}}$ The $\overline{\text{Input\Output Read}}$ line of the 82231 can act as an input or output, depending on the device controlling the bus. During a DMA cycle this line acts as an output and, when low, enables the selected I/O device to output its data onto the data bus. During an MPU cycle, this line acts as an input to access the different sections of the 82231.

$\overline{\text{IOW}}$ The $\overline{\text{Input/Output Write}}$ line of the 82231 can act as an input or output, depending on the device controlling the bus. During a DMA cycle this line acts as an output and, when low, enables the selected I/O device to

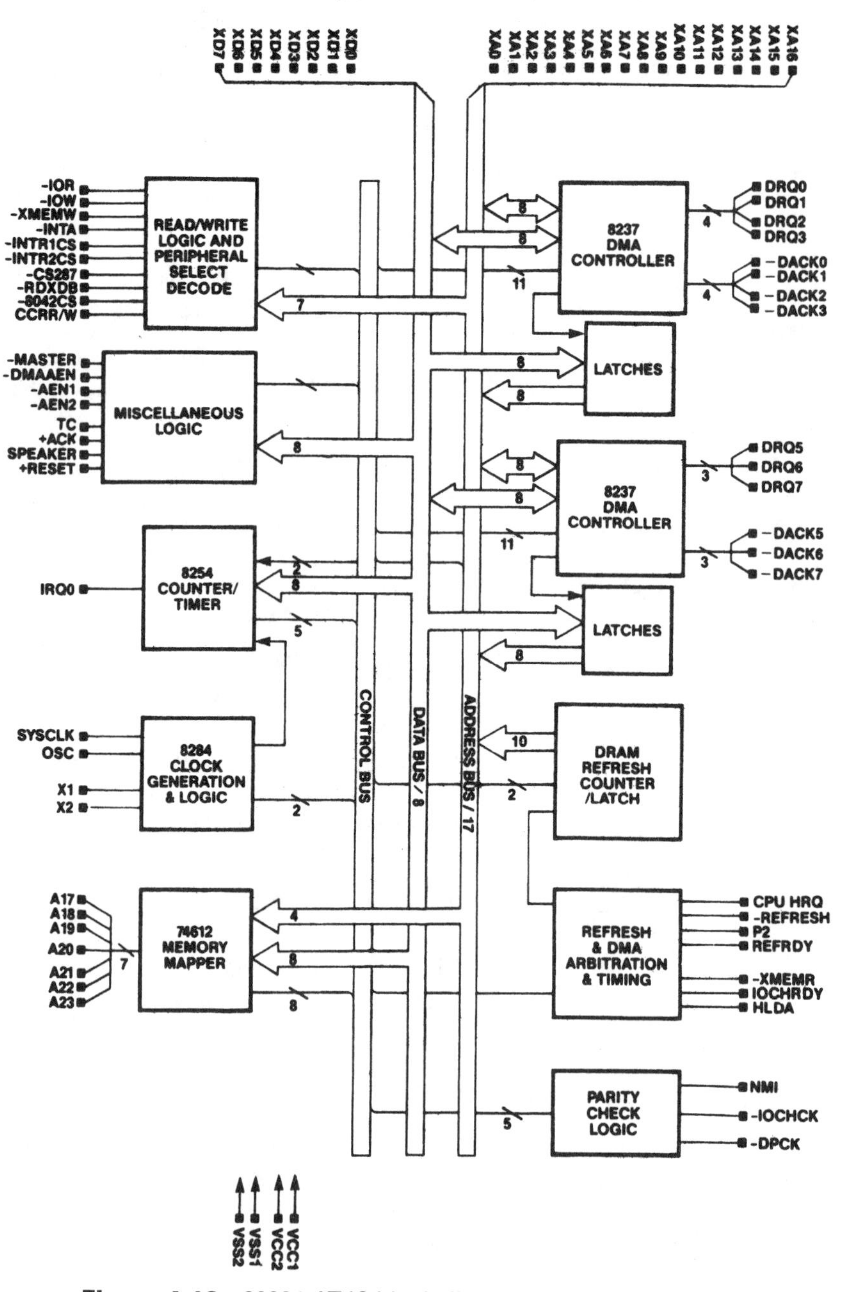

Figure 4-12 82231 AT IC block diagram. (Courtesy of Intel)

accept the data on the data bus. During an MPU cycle, this line acts as an input to access the different sections of the 82231.

XMEMW — The peripheral MEMory Write line of the 82231 acts as an output only during a DMA cycle. This line causes the selected memory device to accept the data on the data bus. During an MPU cycle this line is in tristate.

INTA — The INTerrupt Acknowledge input is generated by the INTA output line from the 82230. When low, this line informs the device requesting the interrupt to release its vectors code onto the data bus. In the 82231, this line causes the RDXDB to control the flow of the low byte of the data bus between the peripheral data bus (XD0 to XD15) and the system bus (SD0 to SD15).

INTR1CS — The INTeRrupt controller 1 (master) Chip Select output is used to control the INTR1CS input of the 82230. A low on this line allows the MPU to select the master interrupt controller to program or to read its status, using the peripheral data bus for the data transfers.

INTR2CS — The INTeRrupt controller 2 (slave) Chip Select output is used to control the INTR2CS input of the 82230. A low on this line allows the MPU to select the slave interrupt controller to program or to read its status, using the peripheral data bus for the data transfers.

CS287 — The Chip Select 287 output is used to control the CS287 input of the 82230. A low on this line causes the math coprocessor chip select (NPCS) to go low, allowing the math coprocessor to perform a math operation on the data.

RDXDB — The ReaD X Data Bus output is used to control the flow of the lower byte of data between the system and peripheral (X) data buses (Table 4-6).

8042CS — The 8042 (keyboard controller) Chip Select output is used to select the keyboard controller to program or to read its status, when low.

CCRR/W — The Clock/CalendaR Read/Write output signal controls the CCRR/W input lines of the 82230. When low, the real time clock is selected so that its data can be read or it can be programmed.

Miscellaneous Logic

MASTER — The MASTER input, along with a DRQ line, is used to gain control of the system bus. When low, this line causes the DMA to take control of the system bus. After the DMA cycle is complete, the MASTER line goes high, which gives control back to the MPU.

DMAAEN — The DMA Address ENable output line is used to enable the address latches during a DMA cycle. This line is low during read and write DMA cycles and during a refresh cycle.

TABLE 4-6
RDXDB Operation

ACK	*RDXDB*	*Operation*
0	0	Peripheral read X-bus to S-bus
0	1	Peripheral write S-bus to X-bus

$\overline{\text{AEN1/2}}$ The $\overline{\text{Address ENables 1 and 2}}$ outputs are used to control the $\overline{\text{AEN1}}$ and $\overline{\text{AEN2}}$ lines of the 82230. The output level on these two lines is generated by NANDing the DMA address-enable lines to the $\overline{\text{MASTER}}$ line. The $\overline{\text{AEN1}}$ line goes low during 8-bit DMA operations, while $\overline{\text{AEN2}}$ goes low during 16-bit DMA operations.

TC The Terminal Count output line pulses high at the terminal count of a DMA cycle. This signal is provided to the I/O interface bus slots of any device that requires the pulse.

ACK The ACKnowledge output line is used to enable the flow of data between the X-data and S-data buses. This line controls the gating input of the data bus transceiver to determine when data flows. The $\overline{\text{RDXDB}}$ line determines the direction of data flow.

SPEAKER The SPEAKER output is controlled by the 8254 OUT2 output of the 82231. This output is normally connected to either a transistor circuit or an op-amp circuit to provide more drive current for the speaker. The circuitry connected to this output also buffers the output from the speaker.

RESET The RESET input is controlled by the reset output of the 82230. When high, the system resets to its initial state.

8254 Counter Timer

IRQ0 The Interrupt ReQuest 0 output controls the IRQ0 input of the 82230. This line goes high every 55 ms to provide the Time of Day update pulse to keep the real time clock functioning properly.

8284 Clock Generation and Logic

SYSCLK The SYStem CLocK input gets its signal from the SYSCLK output from the 82230. This clock signal is used to synchronize the 82231 with bus operations and the 80286.

OSC The OSCillator output is supplied to the I/O interface bus slots. The frequency of this output is determined by the frequency of the crystal applied to the X1 and X2 crystal inputs. The duty cycle of this output is 50%.

X1/X2 The crystal inputs X1 and X2 are used to generate the clock frequencies for I/O devices (mainly the PIT section of the 82231). The crystal used must have 12 times the frequency needed by the clock inputs of the internal 8254. A crystal of 14.318 MHz in series with two capacitors is used for the input frequency. The duty cycle of this output is 50%.

74612 Memory Mapper

A17–A23 Address lines 17 through 23 are outputs that supply page addresses for memory locations to the address bus only during a DMA cycle. During an MPU cycle, these lines are in tristate.

8237 DMA Controller

DRQ0–3 DMA ReQuest lines 0 through 3 inputs are used to allow a hardware device to request a DMA cycle. These four lines are used to initiate a DMA cycle in the master DMA, which is normally used for 8-bit data transfers.

DRQ5–7 — DMA ReQuest lines 5 through 7 inputs are used to allow a hardware device to request a DMA cycle. These three lines are used to initiate a DMA cycle in the slave DMA. DRQ4 internally is used to cascade the slave to the master DMA to provide 16-bit data transfers. Table 4-7 is used for the DMA inputs.

$\overline{\text{DACK0–3}}$ — DMA ACKnowledge lines 0 through 3 outputs are used to inform the I/O device that the DMA cycle is in progress. These lines, when low, enable the I/O device requesting the DMA cycle.

$\overline{\text{DACK5–7}}$ — DMA ACKnowledge lines 5 through 7 outputs are used to inform the I/O device that the DMA cycle is in progress. These lines, when low, enable the I/O device requesting the DMA cycle.

Refresh and DMA Arbitration and Timing

CPU HRQ — The CPU Hold ReQuest output line is used to control the HRQ input on the 286. When this line goes high, the DMA is requesting permission to take control of the bus. The pin also goes high during a refresh cycle.

$\overline{\text{REFRESH}}$ — The $\overline{\text{REFRESH}}$ output pin goes low to cause a refresh cycle in the system DRAMs. In the XT computer, refresh was accomplished by using the DACK0 line with additional gating.

P2 — The P2 line indicates that a valid address for the refresh cycle is on the peripheral address bus. This line is gated with the $\overline{\text{REFRESH}}$ line to cause a RAS-only refresh in the DRAM.

REFRDY — The REFresh I/O channel ReaDY output is used to indicate that the device on the I/O channel is ready to complete the bus cycle. This line is developed by ORing the REFRESH and IOCHRDY signals, and this is used to enable the $\overline{\text{READY}}$ line of the 82230. When high, the bus is ready to complete the bus cycle; when low, the bus is not ready.

$\overline{\text{XMEMR}}$ — The eXtended MEMory Read line output is used by the DMA during a DMA cycle to control a memory read operation, when low. During an MPU cycle, this line acts as an input.

IOCHRDY — The I/O CHannel ReaDY input is generated by the selected I/O device. When low, the I/O device is not ready, which causes the $\overline{\text{READY}}$ line to stay high and inserts wait states into the bus cycle. When high, the I/O device is ready and the $\overline{\text{READY}}$ line goes low, allowing the bus cycle to be completed.

HLDA — The HoLD Acknowledge input line is generated by the 286 HLDA output. When high, this line indicates that the 286 has given up control of the bus to the DMA.

TABLE 4-7
DMA Request Table

DMA Number	*DRQ*	*Type of Hardware*	*Size*
1	DRQ0	Reserved	8-bit
1	DRQ1	SDLC	8-bit
1	DRQ2	Floppy disk controller	8-bit
1	DRQ3	Reserved	8-bit
2	DRQ5	Reserved	16-bit
2	DRQ6	Reserved	16-bit
2	DRQ7	Reserved	16-bit

Parity Check Logic

NMI	The NonMaskable Interrupt output is connected to the NMI input of the 286. This output, when high, indicates that a parity error has been detected. The detection of the parity error occurs only after a memory read.
$\overline{\text{IOCHCK}}$	The $\overline{\text{I/O CHannel ChecK}}$ input line is controlled by an I/O device or a memory device connected to the I/O bus channel. When low, a parity error has occurred in an I/O or memory device connected to the I/O channel.
$\overline{\text{DPCK}}$	The $\overline{\text{Data Parity ChecK}}$ line is controlled by the DRAM section of the system board. During a memory read, if this line goes low, it indicates that a parity error has occurred, which generates a high on the NMI line.

Extended Data Bus

XD0–XD7	The bidirectional data lines are used to allow data to program information into the 82231. During a write operation, data enters the 82231. During a read operation, these lines act as outputs, so that the status of the 82231 can be read. Data enters and exits this IC from and to the extended data bus. Because this IC is connected to the extended data bus, it is considered a peripheral device.

4.5 80286 AT SYSTEM BOARD USING THE 82230/82231 AT CHIP SET

The remainder of this chapter deals with a typical 286 AT system board that uses the Intel 82230/82231 AT chip set (Fig. 4-13). As each section is discussed, examples are given of typical problems that may occur and how to check for them. Because most system board manufacturers do not provide their schematics or allow them to be used by anyone other than factory representatives, our discussion deals with items that can be tested to determine proper operation.

4.5.1 80286 MPU

The 286 is the main controlling element on the system board. This IC is responsible for fetching, decoding, and executing a software program. Software is a list of instructions that cause the MPU to perform some specific task. The 286 is usually in control of the address, data, and status buses. The only other device that can control the buses is the DMA. The DMA controls the buses primarily during I/O-to-memory and memory-to-I/O data block (more than one byte per block, 64K maximum) transfers. If a problem develops in the 286, the computer normally does not function or execute any type of program, depending on the exact problem. The main causes of a 286 failure are heat, electrical spikes, and modification of equipment by untrained personnel.

All measurements should be taken from the pin of the IC in question and a good ground point on the system board. If the pin of the IC is not accessible, use the corresponding pin on the IC socket. The most common checks are listed; these checks can be made in the field where a logic analyzer may not be available. A good DVM or DMM and a dual-trace oscilloscope with a band width of 50 MHz are sufficient. Logic levels for a logic 0 should be no more than 0.8 V and a logic 1 should be no less than 2 V, which follows standard TTL threshold levels, unless otherwise noted.

Condition: Computer fails to boot.
Computer fails to perform POST.
No step of the boot-up sequence is performed other than turning on the fan and the power LED.

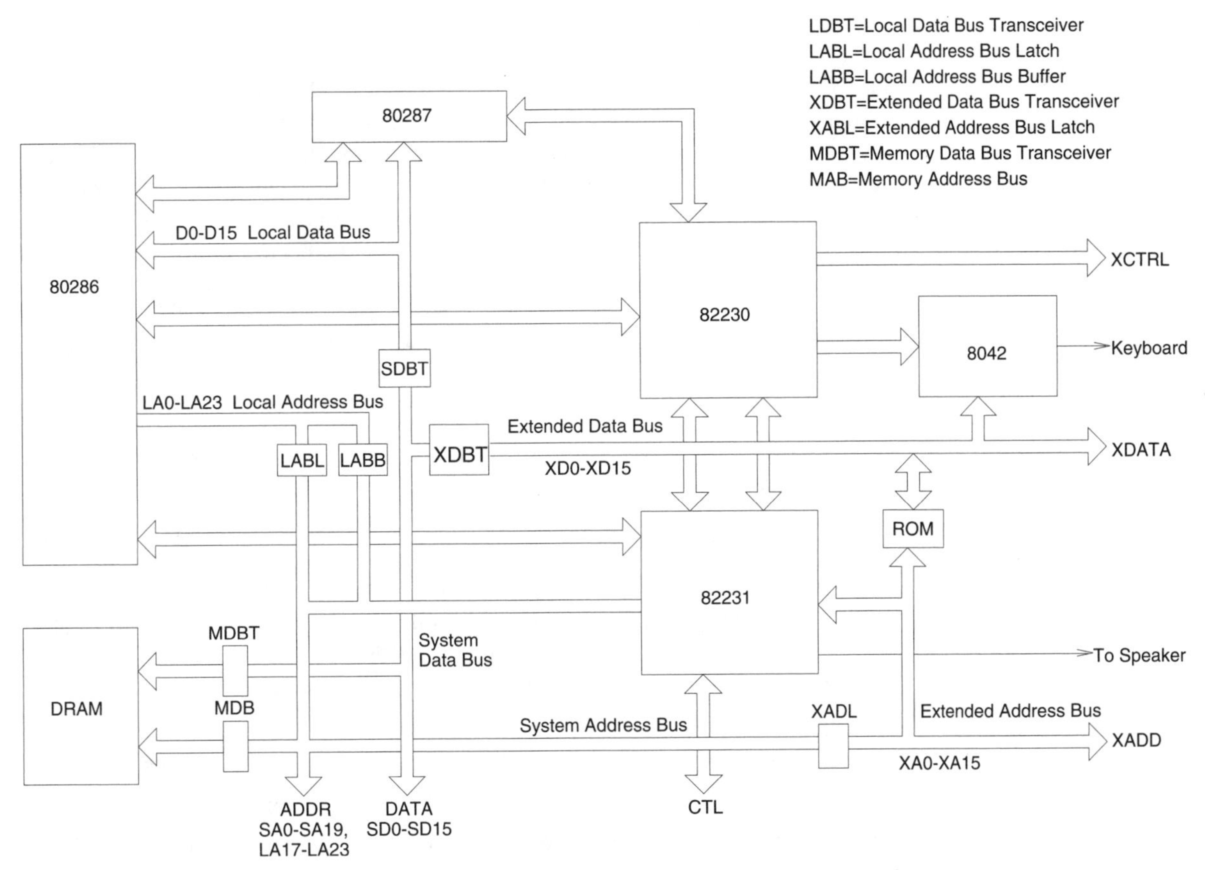

Figure 4-13 80286 system board using Intel AT Chip Set.

Check 1: Measure the voltage of Vcc to any ground point on the system board. The voltage should be between +4.5 V and +5.5 V. If the voltage is lower than +4.5 V, there may be a loading problem with the power supply. If the voltage is higher than +5.5 V, the power supply is not regulating the voltage correctly.

Check 2: Measure the voltage of ground on the 286 to any other ground point on the system board. The voltage should be 0 V. If more than a few millivolts is measured, a ground loop is possible.

Check 3: Measure the frequency, duty cycle, and peak levels of the CLK input of the 286. The frequency is not critical; it should be 6 MHz, 8 MHz, 10 MHz, or 12 to 12.5 MHz. Check the clock speed on the 286 (80286-6 is a 6-MHz MPU, 80286-8 is an 8-MHz MPU, 80286-10 is a 10-MHz MPU, and 80286-12 is a 12-MHz MPU); if the frequency is more than 10% higher, there may not be anything wrong with the 286. Simply reduce the frequency. If the frequency is 10% slower than the standard frequency, normally it is not a problem. The duty cycle should be about 50% ± 10%. The logic 0 point of the CLK signal should be not higher than 0.6 V, and a logic 1 needs to be a minimum of 3.8V. If any of the signals is not correct, check the CLK output of the 82284.

Check 4: Measure the voltage level on the $\overline{\text{READY}}$ line. Use an oscilloscope to verify the logic highs and lows. The $\overline{\text{READY}}$ line should normally be high except during a bus cycle, when it is pulsed high–low–high. If voltages are not correct, check the 82284 $\overline{\text{READY}}$ output.

Check 5: Measure the voltage levels on the RESET line; this line should normally be low.

4.5.2 80287 Math Coprocessor

The math coprocessor is used in conjunction with the 286 and is preceded by an escape instruction to perform mathematical operations at least 10 times faster than the 286. If the software does not support the 287 instructions, the 287 is not used. In some software the 287 is required; if the computer tries to operate this software, it either locks up or displays an error message. The 287 does have the ability to control the address bus; therefore, the 286 controls the address bus for the 287. During a 287 cycle, the 287 and 286 are in constant contact with each other. Every time the 287 or 286 performs a step, it sends a signal to the other IC to verify or acknowledge the step before the next step is taken. Most 286 systems do not use the 287 because of its cost, especially in the earlier days of the 286 AT systems. By the time the price of the 287s dropped to a level that most users could afford, Intel came out with more advanced MPUs and math coprocessors.

Condition: Computer operates correctly except when running software that requires a math coprocessor.

Check 1: Use the diagnostic software program that comes with most 287s to verify that the IC is responding. If the 287 does not respond, verify that the 287 is in the 287 socket (normally located near the 286 IC) and is a 40-pin DIP. If the 287 is in the socket, make the following checks. If the diagnostic software verifies the 287, the problem is with the software or the configuration.

Check 2: Verify Vcc; see check 1 of the 286.

Check 3: Verify proper ground; see check 2 of the 286.

Check 4: Verify CLK frequency, duty cycle, and logic levels; see check 3 of the 286.

Check 5: Verify the voltage level on the RESET line; it should normally be low.

Check 6: Check the voltage on the $\overline{\text{BUSY}}$ line; this line should normally be high.

Check 7: Check the voltage on the $\overline{\text{ERROR}}$ line; this line should normally be high.

4.5.3 The 82230 IC

The 82230 is half of the Intel integrated AT chip set. The 82230 contains the 82284 clock generator/driver, two 8259 PICs, the 6818 real time clock (battery backed-up), the 82288 bus controller, address and data bus logic, bus control logic, CPU shutdown logic, and coprocessor interface logic. The function of each section is discussed here. *All conditions and checks assume that all the Vcc and ground lines are valid.*

82284 Clock Generator/Driver. The 82284 section of the 82230 is responsible for producing the bus timing for the system board and the ready signal necessary to conclude the bus operation. The processor clock output is used as the main clocking signal for all bus operations. It is directly connected to the 286 and 287. The system clock output is one-half the frequency of the processor clock but is synchronized with the T-states of the processor clock; this output is not used in many system boards. The peripheral clock and its inverted counter are used to supply a frequency to the 8042 keyboard controller. The ready line is used to inform the 286 that the bus is ready and the bus cycle needs to end. The $\overline{\text{0WS}}$ is used to modify the number of wait states on the system. This function is normally not used in most of today's systems.

Condition: Computer fails to boot and does not perform POST.

Check 1: Check the frequency of PROCCLK.

Check 2: Check the duty cycle of PROCCLK.

Check 3: Check the voltage levels of PROCCLK.

Check 4: Check the logic level for $\overline{\text{READY}}$. This line should normally be high except during a bus cycle, where it is pulsed high–low–high.

Condition: Computer fails to fully boot but runs through POST; a 3XX error code or keyboard error is displayed. Computer boots but the key pressed does not perform the proper function.

Check 1: Make sure that the keyboard is connected to the back of the computer.

Check 2: If the keyboard has a XT/AT switch, make sure that the switch is set to the AT setting.

Check 3: Make sure that the keyboard is not locked. The lock on many computers locks out keyboard operations.

Check 4: Check the frequency of PCLK.

Check 5: Check the duty cycle of PCLK.

Check 6: Check the voltage levels for PCLK.

Check 7: Check the frequency of $\overline{\text{PCLK}}$.

Check 8: Check the duty cycle of $\overline{\text{PCLK}}$.

Check 9: Check the voltage levels for $\overline{\text{PCLK}}$.

Check 10: Check the phase relationship between PCLK and $\overline{\text{PCLK}}$; they must be 180 degrees out of phase.

Command Delay Logic. The command delay logic section is used to increase or decrease the number of wait states in the computer system. The F16 input is used to reduce the command delays for 16-bit operations. The REFRDY input is used to insert wait states during I/O operations. This section is the most likely to go bad, and troubleshooting this section requires specialized software and a logic analyzer.

8288 Bus Controller. The 82288 bus controller section is the part of the computer system that controls bus sequencing for the 286. This section gets its signals from the 286 (M/$\overline{\text{IO}}$, CPU HLDA, S0, and S1 from the 82284 section) to cause the data to flow in the proper directions, at the proper times, and in the proper areas of the computer. If any of the following outputs (ALE, $\overline{\text{IOR}}$, $\overline{\text{IOW}}$, $\overline{\text{MEMR}}$, $\overline{\text{MEMW}}$, $\overline{\text{INRTA}}$, DT/$\overline{\text{R}}$) fail, the computer fails to boot. A logic analyzer is needed to verify the proper operational sequence of these outputs.

The $\overline{\text{IOR}}$, $\overline{\text{IOW}}$, $\overline{\text{MEMR}}$, and $\overline{\text{MEMW}}$ output lines control all read and write operations when the MPU is in control of the bus. These lines go into tristate during a DMA cycle. The ALE line is used to latch address lines A0 to A19 from the local bus into the local address bus latches (LABL); the outputs from the LABL become the system address bus.

Bus Control Logic. The output lines of the bus control logic section are used to enable the size of the data being transferred during a bus cycle. The least significant and most significant data enables, control off, and external bus high enable outputs are used to control which byte of data is enabled during a byte or word operation. To troubleshoot these outputs, a logic analyzer is needed to verify the proper operational sequence.

CPU Shutdown Logic. The CPU shutdown logic section is used to allow the keyboard ($\overline{\text{RC}}$ input) to generate a reset of the 286 (RES CPU output) and any other device in the system using the RESET output. The A1 input from the 286 is used to verify a reset condition rather than a CPU HALT at the proper times.

Condition: Computer fails to boot or execute POST.

Check 1: Verify that AC power is supplied to the computer.

Check 2: Verify that the fan on the power supply is operational.

Check 3: Check the voltage on the power-good line from the power supply. The voltage should not be less than 2.6 V; if less, change the power supply.

Check 4: Verify Vcc to all Vcc inputs of the 82230.

Check 5: Verify a good ground.

Check 6: Check the voltage level on $\overline{RC}$; it should be a logic high. If low, check the keyboard.

Check 7: Check the voltage level on RES CPU; it should be a logic low.

Check 8: Check the voltage level on RESET; it should be a logic low.

Coprocessor Interface Logic. This section of the 82230 is used to assist the 286 in working with the 287 to perform math instructions requiring a math coprocessor. The best way to troubleshoot this section is to use the diagnostic program supplied with the 287 or some other type of diagnostic program. Troubleshooting this section requires diagnostic software and possibly a logic analyzer.

8259 PIC. The PIC section is used to allow hardware devices to generate maskable interrupts. Hardware interrupts are processed faster than software interrupts. This section contains two 8259s cascaded together, allowing the system board to handle 16 interrupts. The circuitry in the 8259s determines the priority of each input and outputs a vector code onto the data bus during an interrupt acknowledge from the 286. Table 4-8 lists the interrupts in order of priority and defines which PIC is controlling the interrupt, its interrupt request number, the vector code placed on the data bus, and the type of device connected to the IRQ input.

Condition: Computer fails to boot or run POST.
I/O device does not respond properly.
Computer loses time.
Keyboard does not respond when a key is pressed.

Check 1: Use a scope to check IRQ0 for a high 40-ns pulse once every 55 ms.

Check 2: Use a scope to check IRQ1 for a high pulse every time a key is pressed on the keyboard.

Check 3: Use a scope to check IRQ3 for a high pulse if software is operating COMM2.

Check 4: Use a scope to check IRQ4 for a high pulse if software is operating COMM1.

Check 5: Use a scope to check IRQ6 for a high pulse at the beginning of a floppy disk operation.

TABLE 4-8
PIC Interrupt Table

PIC No.	*IRQ No.*	*Vector*	*Type*
1	0	08h	Time of day
1	1	09h	Keyboard
1	2	0Ah	Second 8259
2	8	10h	Real time clock
2	9	11h	Software INTA
2	10	12h	Reserved
2	11	13h	Reserved
2	12	14h	Reserved
2	13	15h	287 coprocessor
2	14	16h	Hard drive
2	15	17h	Reserved
1	3	0Bh	COMM2
1	4	0Ch	COMM1
1	5	0Dh	LPT2
1	6	0Eh	Floppy disk
1	7	0Fh	LPT1

Check 6: Use a scope to check IRQ14 for a high pulse at the beginning of a hard disk operation.

Check 7: Use a scope to check the INTR line for a high level after one or more of the IRQ lines go high.

6818 Real Time Clock. The real time clock section is used to keep time and the proper date in the computer when the power is removed. An external connection for a battery is used to supply power to keep the time of the clock and date operating. The POWER-GOOD line is used to switch between the computer's power supply and the battery input voltage. The POWER-GOOD line is also used to help control the CPU shutdown logic. This section also contains 54 memory locations that keep data for the CMOS set-up and POST, which is used to configure the computer upon boot-up. If battery power is lost, the time, date, and contents of the memory locations are lost when the power is turned off on the system board. The computer tells the user that an error has occurred in the CMOS and prompts the user to reset the time and date and re-enter the CMOS data.

If the voltage on the battery input falls below 2.8 V, replace the battery. Another problem that sometimes occurs is that if the peripherals are turned on after the computer or the peripherals are turned off before the computer, a noise spike may cause the CMOS data to change, producing the same type of problem as a bad battery. After 1 year the battery may start causing problems.

Address and Data Bus Control and Miscellaneous Logic. This section of the 82230 is used to control the majority of the address latches and data bus transceivers on the system board. These lines control the direction and timing of data flow during different types of data transfers. The control lines in this section are also used to control address latching and enabling of buffers. Some of the lines in this section are also used to control and get status from the 82231 IC. To troubleshoot these inputs and outputs, a logic analyzer and software routine are needed to verify the proper operational sequence.

Extended Data Bus. The XD0 to XD7 lines allow the 82230 to be programmed or to have its status read. During a read operation data exits the IC, and during a write cycle data enters the IC. To troubleshoot these lines, a logic analyzer and software are needed to verify the proper operational sequence.

4.5.4 The 82231 IC

The second part of Intel's AT chip set controls the read and write lines during a DMA cycle, produces I/O chip selects (address decoding of I/O devices), and contains a PIT, memory map page information for DMA operations, two 8237 DMAs, and parity check logic. Troubleshooting most of this IC requires a logic analyzer and a software routine to determine proper operational sequence.

Read/Write Logic and Peripheral Select Decode. The read and write lines in this section are used as outputs during DMA cycles. The rest of this section is used to enable I/O devices on the system board when the proper address is placed on the address bus. Table 4-9 lists the I/O devices on the system board and the addresses associated with each device. Whenever the address on the address bus matches the address listed in Table 4-9 and an I/O read or write operation is performed, either one of the devices in this table is programmed or the status is read. To troubleshoot these lines, a logic analyzer and software are needed to verify the proper operational sequence.

Miscellaneous Logic. The miscellaneous logic section is used by the DMA to enable the address bus during a DMA cycle and to take control of the address, data, and control

TABLE 4-9
System I/O Address Table

Address	*Function*
000000–00000F	8237 DMA 1
000020–000021	8259 PIC 1
000040–000043	8254 PIT
000060–000063	8042 keyboard
000070–00007F	Real time clock
000080–000087	DMA page register
0000A0–0000A1	8259 PIC 2
0000C0–0000CF	8237 DMA 2
0000F0	Clear 287 busy
0000F1	Reset 287

buses. This section also is used to output a speaker frequency to the drive circuitry for the speaker. Although many of the signals in this section require a logic analyzer to troubleshoot, some checks can be made to verify most of the operations of this section.

Condition: Computer boots but produces parity errors; software fails to load. Computer passes POST, but floppy and hard drive do not load software into the computer.

Check 1: Verify Vcc.

Check 2: Verify a good ground.

Check 3: Scope the $\overline{\text{MASTER}}$ line; this line should pulse low for 200 ns once every 15 μs. If the pulse does not occur, the DMA is not working properly or the refresh circuitry is not functioning. If the pulse is present, the problem lies with memory for the parity error problem, and possibly with the floppy and hard drive controllers.

Check 4: Scope the TC output line, which should pulse high at the same time the $\overline{\text{MASTER}}$ lines goes high.
Condition: Computer boots but does not beep during the successful completion of POST.

Check 1: Verify that the speaker is connected to the system board.

Check 2: Scope the speaker output of the 82231 while holding down a key on the keyboard. Once the buffer fills up, a 400-Hz frequency should appear on the speaker output. If no frequency is seen on the scope, the problem may be the 82231. If a frequency is produced, check the drive circuitry between the speaker output of the 82231 and the speaker.

8254 Counter Timer. The IRQ0 output from the PIT outputs a pulse every 55 ms. This signal is used to update the time of day while the computer is on. If this output does not pulse every 55 ms, the computer loses time. If the output pulses faster, the computer gains time. This section can be checked by scoping the IRQ0 output; all levels should be TTL compatible.

8284 Clock Generation and Logic. The only output of this section is the OSC output. This output should operate at the frequency of the crystal attached to the X1 and X2 inputs. The duty cycle should be 50%. If the OSC output fails, the user will probably not detect the problem because only a few special devices use the frequency from the I/O interface slots.

74612 Memory Mapper. The memory mapper section is used to control the page address lines during a DMA cycle. The address lines A17 to A23 on the 82231 are enabled to output the address to the A17 to A23 address lines only during a DMA cycle. During an MPU cycle these lines are in tristate. To troubleshoot these lines, a logic analyzer and software are needed to verify the proper operational sequence.

8237 DMA Controller. The 82231 contains two 8237 DMAs, which allow the DMA to transfer 8-bit and 16-bit data packages. During 8-bit data transfers, only one DMA is used; during 16-bit data transfers, one channel from each DMA is used. The DMA handles only data transfers from I/O to memory and from memory to I/O devices. Although some of the operations of this section can be verified using a scope on the specific DREQ and $\overline{\text{DACK}}$ lines, the entire sequence cannot be verified without a logic analyzer and a software routine to work the DMA channels.

Refresh and DMA Arbitration and Timing. This section of the 82231 is used to provide supporting signals for DMA operations and to indicate a refresh cycle for DRAM. Although a logic analyzer and software are needed to thoroughly troubleshoot this section, a few checks can verify whether some of the circuitry is functioning.

Condition: Computer boots but produces parity errors.
Computer fails to boot or run POST.
Check 1: Scope the CPU HRQ line for a high pulse during a DMA cycle.
Check 2: Scope the $\overline{\text{REFRESH}}$ line for a low pulse once every 15 μs.
Check 3: Scope the HLDA line for a high pulse once every 15 μs or at the end of a DMA cycle.

Parity Check Logic. The parity check logic is used to hold the NMI parity bit from the DRAM section following a DRAM memory read cycle. The NMI output is generated by either the $\overline{\text{IOCHCK}}$ line or $\overline{\text{DPCK}}$ line going low. The NMI output line resets after executing the NMI interrupt sequence.

Extended Data Bus. The XD0 to XD7 lines allow the 82231 to be programmed or to have its status read. During a read operation data exits the IC, and during a write cycle data enters the IC. To troubleshoot these lines, a logic analyzer and software are needed to verify the proper operational sequence.

4.5.5 286 System Board Timing

The timing diagrams in Figures 4-14 through 4-17 demonstrate different types of bus cycles and data transfer sizes. These timing diagrams use the block diagram from Figure 4-13 as a reference. The 286 initiates the cycle, and the bus controller section of the 82230 completes the process.

During any data transfer the MPU requires two processor states, and each processor state requires two CLK pulses. The first clock is labeled phase 1 and the second phase 2. The first processor state is called the status state (Ts). During this part of the bus cycle the operation is determined (uses M/$\overline{\text{IO}}$, COD/$\overline{\text{INTA}}$, $\overline{\text{S0}}$, and $\overline{\text{S1}}$), the address is selected and latched into the address latches, direction of data flow is selected, and transceivers are enabled. The status state sets up the hardware for the data transfer. The second processor state is called the command state. During the command state, the logic levels on the selected equipment are allowed to stabilize and the data becomes ready to be transferred. The data is transferred at the end of the command state if the bus is ready. If the bus is not ready by the end of the command state, the command state repeats itself, at which time the old command state is called a wait state. Once the bus becomes ready at the end of either the command state or one of the wait states, the operation is concluded.

Memory Read Bus Cycle (Low Byte). In the timing diagram in Figure 4-14, the lower 8 bits (lower byte) of the data bus are transferred from a memory location into the MPU.

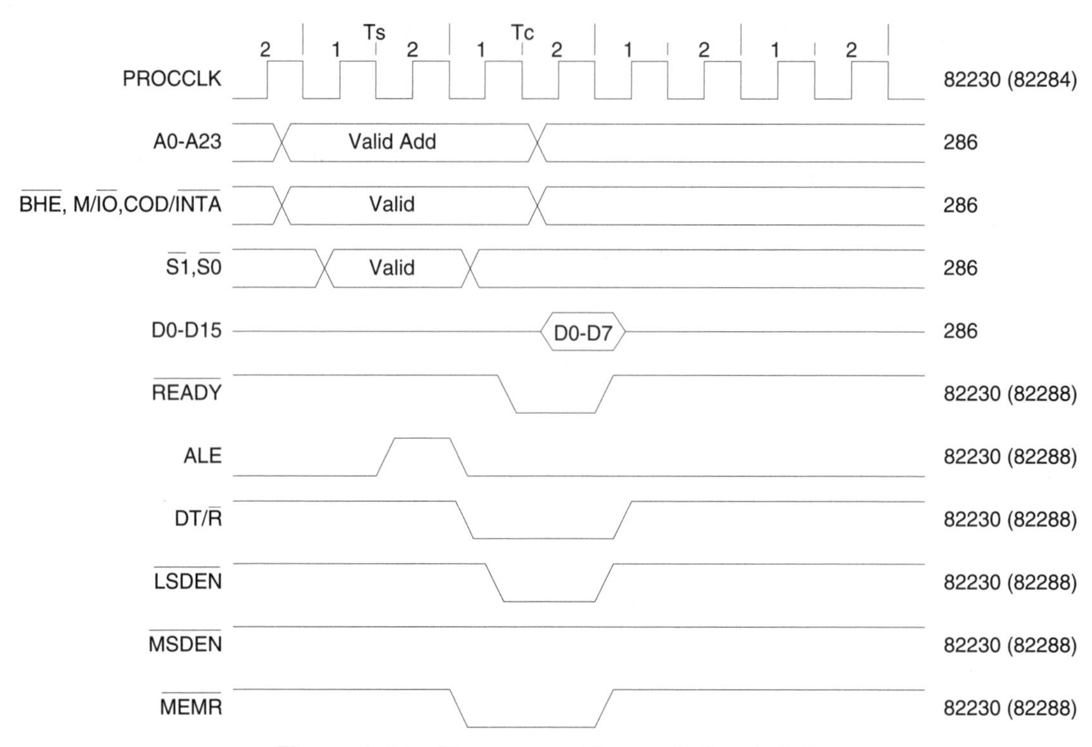

Figure 4-14 Memory read bus cycle (low byte).

This example shows no wait states. Only two processor clocks are needed to transfer the data. The following lists the names and defines the purpose of each trace line.

PROCCLK — The PROCessor CLocK is an output from the 82284 section of the 82230. This frequency is applied to the CLK input of the 286 and is the operational frequency of the 286.

A0–A23 — The Address lines are outputs from the 286. The levels on these lines are used to address the different memory and I/O devices in the computer system. Address lines A0 to A19 are latched in the local address bus latches during the status state of the bus cycle. Address lines 17 to 19 are also buffered with address lines A20 to 23 for the system bus.

M/$\overline{\text{IO}}$, COD/$\overline{\text{INTA}}$, and $\overline{\text{BHE}}$ — These lines are outputs from the 286. The M/$\overline{\text{IO}}$ line is used to determine if the operation is to occur on a memory location or in a selected I/O device (high = memory and low = I/O). The COD/$\overline{\text{INTA}}$ is used to help determine if the data transfer is an instruction or part of an interrupt sequence. The $\overline{\text{BHE}}$ line, when low, enables the data transfer of the high byte of the data bus. This line is used with the A0 line to determine the size of the data transfer.

$\overline{\text{S0}}$, $\overline{\text{S1}}$ — The two MPU bus status lines are used to initialize the bus cycle. These lines, along with the M/$\overline{\text{IO}}$ and COD/$\overline{\text{INTA}}$ lines, are used to inform the 82288 bus controller section of the 82230 which bus cycle should take place.

D0–D15 — These bidirectional data lines are controlled by the 286. Depending on the operation selected, data may enter or exit the MPU. Not all lines have to be used during a data bus transfer. The MPU can, through software, select data to be transferred in 8-bit packages called bytes (low byte uses D0 to D7 and high byte uses D8 to D15), or a word can be transferred, in which case all data lines are used.

$\overline{READY}$ — The $\overline{READY}$ line is controlled by the clock generator driver (82284) section of the 82230. The $\overline{READY}$ line indicates to the bus controller and the 286 that the device being addressed is ready for the data transfer. This line is checked at the end of each command state to determine if the bus is ready to make the data transfer and complete the operation. If low, the bus controller causes the command state to conclude by transferring the data on the next phase of the clock. If high, the bus controller causes another command state to be generated by not making any changes in the levels of any of its control lines. The old command state is then labeled a wait state because the operation was not completed with that state.

ALE — The Address Latch Enable line is controlled by the bus controller (82288) in the 82230. This line, when it goes high, causes the local address bus inputs of the local address bus latches to become latched. The outputs from these latches are used to feed the levels for the system address bus. Only address lines A0 to A19 are latched for the system bus. A copy of address lines A17 to A19 are buffered, along with address lines A20 to A23.

$DT/\overline{R}$ — The Data Transmit/$\overline{\text{Receive}}$ line is used to control the direction of data flow through the system data bus transceiver (SDBT). The line is controlled by the bus controller section of the 82230. During a read operation, the data flows from the local to the system data bus. During a write operation, the data flows from the system to the local data bus.

$\overline{LSDEN}$ — The $\overline{\text{Least Significant Data ENable}}$ line, when low, enables data to flow through the data bus transceiver that controls data lines D0 to D7. The bus controller section of the 82230 controls the levels on this line. The 82230 uses the $\overline{BHE}$ and A0 lines to determine if this line goes low at some point during the bus cycle. This line, along with the $\overline{MSDEN}$ line, goes low during a word transfer.

$\overline{MSDEN}$ — The $\overline{\text{Most Significant Data ENable}}$ line, when low, enables data to flow through the data bus transceiver that controls data lines D8 to D15. The bus controller section of the 82230 controls the levels on this line. The 82230 uses the $\overline{BHE}$ and A0 lines to determine if this line goes low at some point during the bus cycle. This line, along with the $\overline{LSDEN}$ line, goes low during a word transfer.

$\overline{MEMR}$ — The $\overline{\text{MEMory Read}}$ line goes low during a memory read operation. This line is controlled by the bus controller section of the 82230. When low, the memory device that is being addressed outputs its data to the data bus. Once this line goes high, the data stops exiting the memory device.

The memory read bus cycle (low byte)(Fig. 4-14) consists of the following steps:

Step	**Cycle State Phase**	**Description**
1	Previous bus cycle Tc (command state) Phase 2 clock	The 286 begins the current bus cycle at the beginning of the phase 2 clock of the previous bus cycle or idle state. The process begins when the 286 places an address on the

Step	Cycle State Phase	Description
		address bus. At the same time the 286 sets or resets the levels on the $\overline{BHE}$, M/$\overline{IO}$, and COD/$\overline{INTA}$ lines, depending on the instruction. The 286 keeps all the lines in this step constant until another bus cycle is ready to be started or this new bus cycle has transferred its data, at which time a change in levels would not affect the bus cycle.
2	Current bus cycle Ts (status state) Phase 1 clock	1. In the status state, the 286 causes the status lines ($\overline{S0}$ and $\overline{S1}$) to go to their proper levels shortly after the beginning of the phase 1 clock of the current bus cycle. 2. On the falling edge of the phase 1 clock of the Ts state, the bus controller section of the 82230 checks the status lines. The levels of the status lines, along with the levels on the M/$\overline{IO}$ and COD/$\overline{INTA}$ lines, determine which bus operation takes place. The $\overline{BHE}$ and A0 lines are used to determine the size of the data being transferred. Once the bus controller of the 82230 checks the status lines and determines the operation and the size of the data transfer, the bus controller takes control of the bus operation. At this point the 286 does not control the bus operation.
3	Current bus cycle Ts Phase 2 clock	1. During the phase 2 clock of the status state, the levels on the lines that have been discussed to this point should remain at their logic levels. 2. At the beginning of the phase 2 clock of the status state, the bus controller section of the 82230 causes the ALE line to go high, and this line stays high throughout the rest of the phase 2 clock. A high on this line allows the address levels on the local bus to enter the address latches. This concludes the status state (Ts) of the current bus cycle.
4	Current bus cycle Tc (command state) Phase 1 clock	During the phase 1 clock of the command state of the current bus cycle, the data starts to be positioned for the data transfer. 1. The ALE line goes low. The logic levels on the inputs of the address latches are locked (latched) into the address latches on the falling edge of ALE. The address is now available on the system address bus and does not change until the end of the current bus cycle or the beginning of a new bus cycle. 2. At the same time the ALE line goes low, the bus controller causes whichever read/write line is selected by the command levels to go low. In this example the read/write line that changed is the memory read $\overline{MEMR}$ line. A low on this line causes the memory location selected by the address to start outputting its data onto the data bus. There is a time delay, usually about one phase clock, before the data actually appears on the data bus. 3. The next thing to change is the DT/$\overline{R}$ line, which is controlled by the bus controller. Because this is a read

Step	Cycle State Phase	Description
		cycle, the DT/$\overline{R}$ line goes low and stays low until the end of the data transfer. 4. The bus controller enables the data bus. In this example, the $\overline{LSDEN}$ line goes low while the $\overline{MSDEN}$ line stays high. Only the low byte of the data bus is enabled (D0 to D7); the upper byte of the data bus is disabled. At this point, the data bus is now waiting for the data to exit the memory device selected by the address. 5. On the falling edge of the phase 1 clock of the Tc, the 286 and bus controller checks the $\overline{READY}$ line to see if the bus is ready. If low, the bus is ready and the current bus cycle ends shortly after the end of the phase 2 clock. If the $\overline{READY}$ line is still high, no changes will take place until the end of the phase 2 clock where the $\overline{READY}$ line is checked again. Because in this example the $\overline{READY}$ line is low, the current bus cycle ends following the falling edge of the phase 2 clock of the Tc. If the 286 had another bus cycle waiting, the 286 would get the next cycle ready to start at the beginning of the phase 2 clock.
5	Current bus cycle Tc Phase 2 clock	1. If the $\overline{READY}$ line goes low by the end of this clock or the phase 1 clock, the bus cycle ends, beginning with the falling edge of the phase 2 clock. 2. If the $\overline{READY}$ line does not go low by the falling edge of the phase 2 clock, another command state is issued by the bus controller and the 286 does not issue a new bus cycle until the current bus cycle is ready to be completed. If another command state is issued, the current command state is called a wait state and the next new command state acts as a command state for the current bus cycle. This sequence may be repeated until the $\overline{READY}$ line goes low. 3. Because in this example of the timing diagram the $\overline{READY}$ line is low during the phase 1 clock, no wait state is generated. The data transfer begins on the falling edge of the phase 2 clock.
6	Next bus cycle Ts Phase 1 clock	1. The previous bus cycle ends at the beginning of the phase 1 clock by causing the $\overline{MEMR}$ line to go high. During this transition the data on the low byte of the data bus is transferred into the MPU. 2. Once the read cycle ends, the $\overline{READY}$ line goes high, indicating that the bus is ready now, so that it can be used for the next bus cycle. 3. The bus controller then disables the lower byte of the data bus transceiver by making the $\overline{LSDEN}$ line go high. 4. The last item to change is the bus controller, which causes the DT/$\overline{R}$ line to go high (its normal state). 5. While these changes are happening, if another bus cycle was pending in the phase 2 clock of the previous bus cycle, the new bus cycle is already in progress.

Memory Read Bus Cycle (Word). In the timing diagram in Figure 4-15, a word is transferred from a memory location into the MPU. This example shows no wait states; only two processor clocks are needed to transfer the data.

Step	Cycle State Phase	Description
1	Previous bus cycle Tc (command state) Phase 2 clock	This part of the bus cycle is the same as in the previous example (Fig. 4-14) except that the $\overline{\text{BHE}}$ line is low during this bus cycle.
2	Current bus cycle Ts (status state) Phase 1 clock	This part of the bus cycle is the same as in the previous example (Fig. 4-14) except that the size of the bus cycle is 16 bits (one word).
3	Current bus cycle Ts Phase 2 clock	This part of the bus cycle is the same as in the previous example (Fig. 4-14).
4	Current bus cycle Tc Phase 1 clock	This part of the bus cycle is the same as in the previous example (Fig. 4-14) with the following change: Both the $\overline{\text{LSDEN}}$ and $\overline{\text{MSDEN}}$ lines go low to enable both the low and high byte bus transceivers. Data lines D0 to D15 are used during this bus cycle.
5	Current bus cycle Tc Phase 2 clock	This part of the bus cycle is the same as in the previous example (Fig. 4-14).
6	Next bus cycle Ts Phase 1 clock	This part of the bus cycle is the same as in the previous example (Fig. 4-14), except for the following changes: 1. The data from all 16 data lines is transferred into the MPU. 2. Both the $\overline{\text{LSDEN}}$ and $\overline{\text{MSDEN}}$ lines go high.

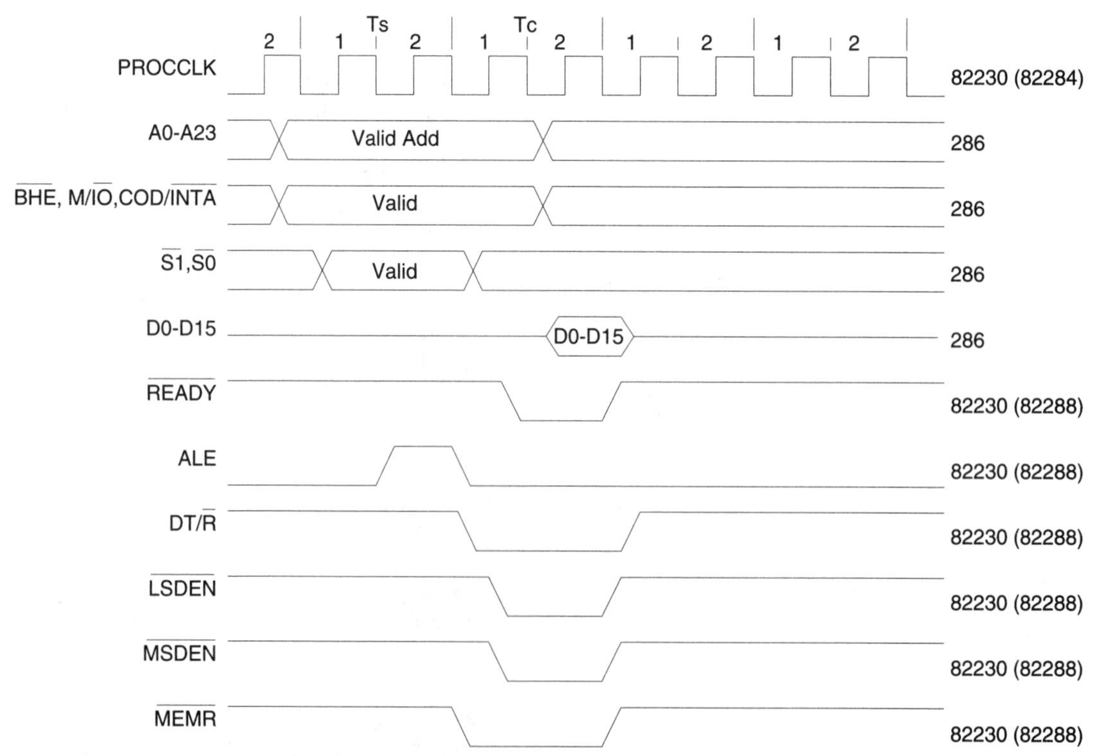

Figure 4-15 Memory read bus cycle (word).

Memory Write Bus Cycle (High Byte). In the timing diagram in Figure 4-16, the high byte is transferred from the MPU into a memory location. This example shows no wait states; only two processor clocks are needed to transfer the data.

Step	Cycle State Phase	Description
1	Previous bus cycle Tc (command state) Phase 2 clock	This part of the bus cycle is the same as in the first bus cycle example (Fig. 4-14) except that the level on the $\overline{BHE}$ line is different.
2	Current bus cycle Ts (status state) Phase 1 clock	This part of the bus cycle is the same as in the first bus cycle example (Fig. 4-14) except that the levels on the $\overline{S0}$ and $\overline{S1}$ lines are different.
3	Current bus cycle Ts Phase 2 clock	This part of the bus cycle is the same as in the first bus cycle example (Fig. 4-14) except for the following addition: The $\overline{MSDEN}$ line goes low at the beginning of the clock and the data is already valid because it is produced by the MPU and transferred into the selected memory location, so the data bus is enabled sooner. This sequence occurs during all write operations for both memory and I/O.
4	Current bus cycle Tc Phase 1 clock	This part of the bus cycle is the same as in the first bus cycle example (Fig. 4-14) except for the following changes: 1. The $\overline{MEMW}$ line goes low instead of the $\overline{MEMR}$ line. This causes the memory location selected by the address on the address bus to begin accepting data from the data bus; the data bus transceivers have already

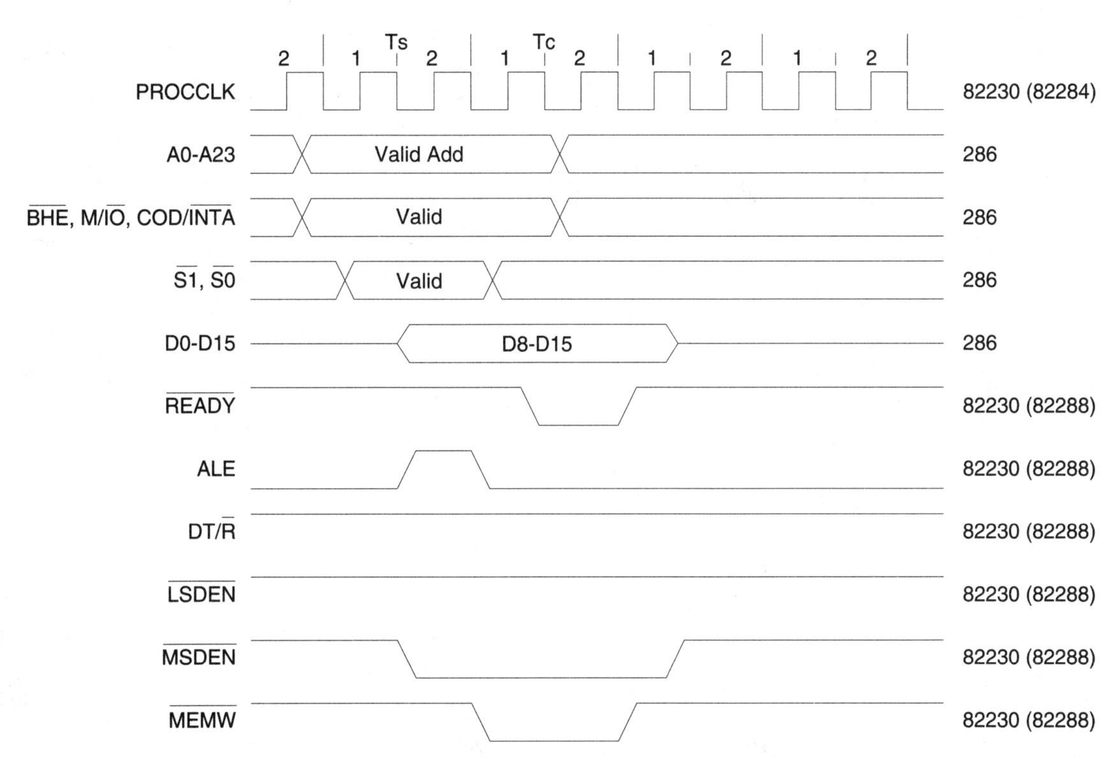

Figure 4-16 Memory write bus cycle (high byte).

Step	Cycle State Phase	Description
		been enabled, so the data is available. A write operation takes longer to perform than a read operation because the write operation may have to cause the data stored in the memory location to change, which is a longer process. 2. Because the DT/$\overline{\text{R}}$ line is normally high, this line remains high. 3. The $\overline{\text{LSDEN}}$ line remains high because the high byte of the data is being transferred. Also the data enabling was performed during the phase 1 clock of the status state.
5	Current bus cycle Tc Phase 2 clock	This part of the bus cycle is the same as in the first bus cycle (Fig. 4-14).
6	Next bus cycle Ts Phase 1 clock	This part of the bus cycle is the same as in the first bus cycle (Fig. 4-14) except for the following changes: 1. The $\overline{\text{MEMW}}$ line goes high, which causes the data on the high byte of the data bus to be transferred into the memory location. 2. The $\overline{\text{MSDEN}}$ line goes high instead of the $\overline{\text{LSDEN}}$ line. 3. The DT/$\overline{\text{R}}$ line remains high.

I/O Write Bus Cycle (Word). In the timing diagram in Figure 4-17, a word is transferred from the MPU into an I/O device. This example shows one wait state, so three processor clocks are needed to transfer the data.

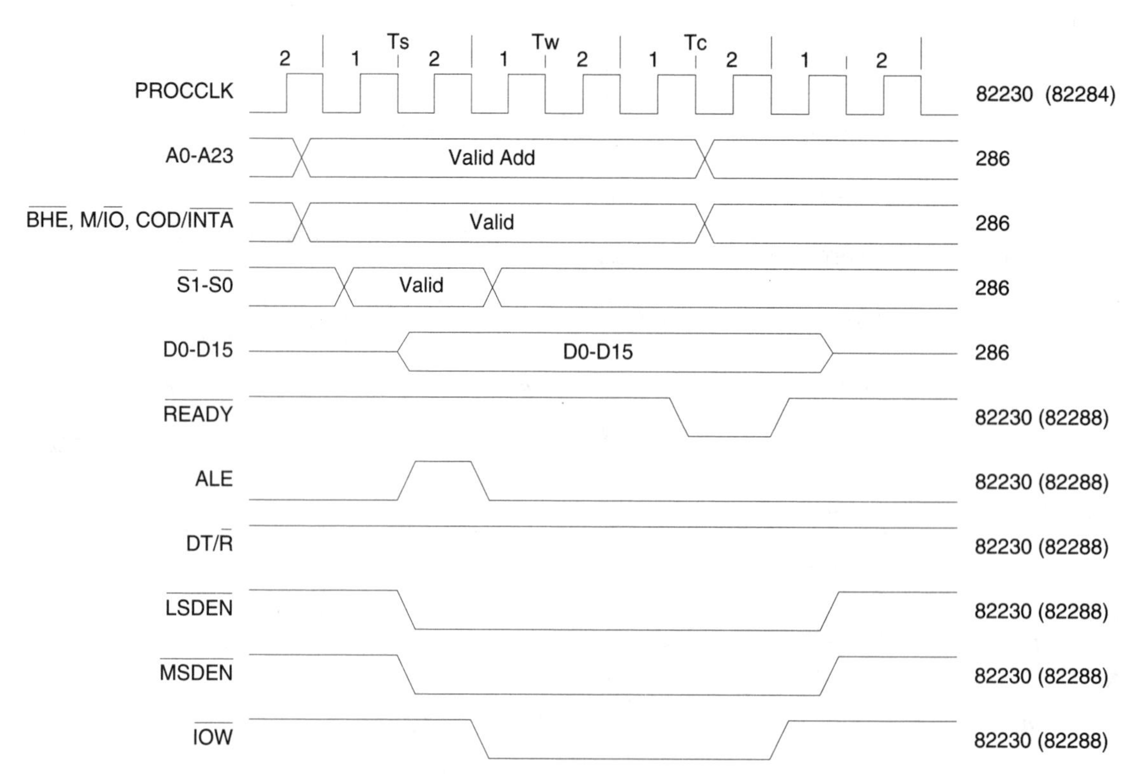

Figure 4-17 I/O write bus cycle with one wait state (word).

Step	Cycle State Phase	Description
1	Previous bus cycle Tc (command state) Phase 2 clock	This part of the bus cycle is the same as in the first bus cycle example (Fig. 4-14) except for the following changes: 1. The address on the address bus selects an I/O device. 2. The level on the $\overline{BHE}$ line is different. 3. The level on the M/$\overline{IO}$ line is different. 4. The level on the COD/$\overline{INTA}$ line is different.
2	Current bus cycle Ts (status state) Phase 1 clock	This part of the bus cycle is the same as in the first bus cycle example (Fig. 4-14) except that the levels on the $\overline{S0}$ and $\overline{S1}$ lines are different.
3	Current bus cycle Ts Phase 2 clock	This part of the bus cycle is the same as in the first bus cycle example (Fig. 4-14) except for the following addition: The $\overline{MSDEN}$ and $\overline{LSDEN}$ lines go low at the beginning of the clock and the data is already valid because it is being produced by the MPU and transferred into the selected I/O device, so the data bus is enabled sooner.
4	Current bus cycle Tc (command state) Phase 1 clock	This part of the bus cycle is the same as in the first bus cycle example (Fig. 4-14) except for the following changes: 1. The $\overline{IOW}$ line goes low instead of the $\overline{MEMR}$ line. This causes the I/O device selected by the address on the address bus to begin accepting the data from the data bus. The data bus tranceivers have already been enabled, so the data is available. 2. The $\overline{LSDEN}$ and $\overline{MSDEN}$ lines remain low from the previous clock. 3. On the falling edge of the phase 1 clock the MPU and bus controller check the $\overline{READY}$ line. This line is high; therefore a wait state may be produced if the $\overline{READY}$ line does not go low before the falling edge of the phase 2 clock. The bus is currently not ready to complete the transfer.
5	Current bus cycle Tc Phase 2 clock Tc is now called the first Tw (wait state)	This part of the bus cycle is the same as in the first bus cycle (Fig. 4-14) except that on the falling edge of the phase 2 clock, the $\overline{READY}$ line is still not low, so the bus controller and the MPU add another command state.This first command state is now called the first wait state of the bus cycle.
6	Current bus cycle Second Tc Phase 1 clock	On the falling edge of the phase 1 clock of the second command state, the MPU and bus controller check the $\overline{READY}$ line. The $\overline{READY}$ line is now low; therefore, after the phase 2 clock of this second command state the bus cycle comes to an end.
7	Current bus cycle Second Tc Phase 2 clock	On the falling edge of the phase 2 clock of the second command state, the $\overline{READY}$ line is still low; therefore at the beginning of the next bus cycle the current bus cycle comes to an end.

Step	Cycle State Phase	Description
8	Next bus cycle Ts Phase 1 clock	This part of the bus cycle is the same as Step 6 in the first bus cycle (Fig. 4-14) except for the following changes: 1. The $\overline{\text{IOW}}$ line goes high, which causes the data on D0 to D15 of the data bus to be transferred into the selected I/O device. 2. The $\overline{\text{MSDEN}}$ and $\overline{\text{LSDEN}}$ lines go high. 3. The DT/$\overline{\text{R}}$ line remains high.

4.5.6 ROM

The ROM section (Fig. 4-18) of the system board of Figure 4-13 contains three basic ICs—one PAL16L2 (Programmable Array Logic) IC used for address decoding for the ROM section and two EPROMs 27256 that contain the BIOS monitor program. One EPROM is used for even addresses and the other for odd addresses.

The PAL is a chip that can simulate many discrete gates. The type of simulation and how the outputs respond to the inputs are programmable into the IC. The system board manufacturer can program the outputs to respond to the levels on the inputs within the parameters of the PAL used. In the ROM section the PAL works to perform address decoding for the ROM section. The O1 output of the PAL goes low only when the address on the address bus is between 0F7FFF and 0FFFFF while a memory write command is performed. The output is an active low, which enables the outputs of the EPROMs.

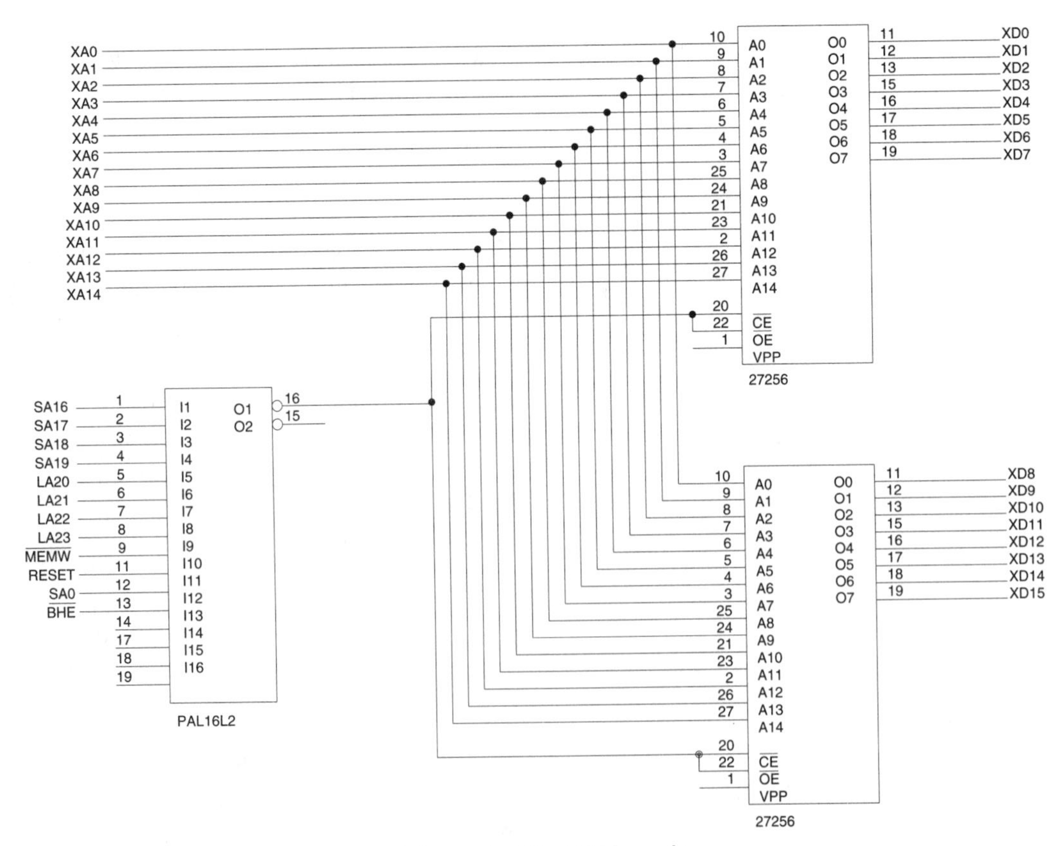

Figure 4-18 ROM section.

The two 27256 EPROMs are used to hold the BIOS (Basic Input Output System) program in nonvolatile memory. BIOS tells the computer how to act like a computer, configures the hardware during boot-up, provides CMOS configuration video screens, provides a password protection program, provides a look-up table for configuring a hard drive, a hard drive diagnostic and set-up program, and most importantly, POST.

Table 4-10 is a memory map from a typical AT system. The PAL sets up the memory map to be enabled only when a read memory operation is performed at the ROM BIOS locations; at all other times this ROM section is disabled. The other memory locations in Table 4-10 are decoded by the other sections or hardware adapters, which contain their own address-decoding sections.

To troubleshoot this section, the same procedures as used in the ROM section of the XT computer can be used. If this section goes bad, the computer most likely does not boot, in which case a POST card can be used to verify at which point during POST the system failed. A logic analyzer can be used, but is not of much help unless BIOS information is available.

4.5.7 DRAM

The DRAM section of the block diagram is contained in two figures. The first (Fig. 4-19) is the address-decoding and timing section; the second (Fig. 4-20) contains the actual DRAM, address multiplexing circuitry, data bus transceivers, and parity generator/check IC.

Address Decoding and Timing. The address decoding and timing section (Fig. 4-19) is used to provide the proper $\overline{\text{RASx}}$, $\overline{\text{CASx}}$, and ADDR SEL signals at the proper time. This section contains one time-delay IC support logic and a PAL. The PAL that controls the $\overline{\text{RASx}}$ lines is used to provide the row address strobe signal for the DRAMs at the proper times. This PAL is also responsible for enabling all banks of DRAMs for a RAS-only refresh cycle. Table 4-11 shows the address range for each $\overline{\text{RAS}}$ line. During a normal read or write cycle, only one of these lines ever goes low at the proper time. During a memory refresh cycle, all $\overline{\text{RASx}}$ lines go low at the same time, so the address on the address bus can refresh all memory cells in the selected row in all banks of DRAMs. In this example of DRAM address decoding and timing, the first 8MB of DRAM are mapped.

The PAL is also used to generate the $\overline{\text{CASx}}$ lines, two for each bank of memory. The $\overline{\text{CASx}}$ lines followed by (LO) in Table 4-12 are enabled when the LSB of the address is even during a byte operation or a word operation. The $\overline{\text{CASx}}$ lines followed by (HI) are enabled when the LSB of the address is odd during a byte operation or a word operation. The A0 and $\overline{\text{BHE}}$ lines from the 286 control the type of operation. The same $\overline{\text{CASx}}$ bank selects to match the $\overline{\text{RASx}}$ bank.

Table 4-12 shows the address ranges for each $\overline{\text{CASx}}$ line.

DRAM Memory and Parity. The second figure (Fig. 4-20) for the DRAM section contains the DRAMs, address multiplexers, DRAM data bus transceivers, and parity generator/checkers.

Two 74245 data bus transceivers are used in the DRAM section of the computer. The 74245 that has the $\overline{\text{LSDEN}}$ line connected to it allows data to flow only when the low byte

TABLE 4-10
286 Memory Map

Size	*Address*	*Description*
0K–640K	000000–09FFFF	Conventional user RAM
640K–768K	0A0000–0BFFFF	128K video RAM
768K–960K	0C0000–0EFFFF	192K I/O expansion ROM
960K–1MB	0F0000–0FFFFF	64K system BIOS ROM
1MB—16MB	100000–FFFFFF	15MB extended/expanded RAM

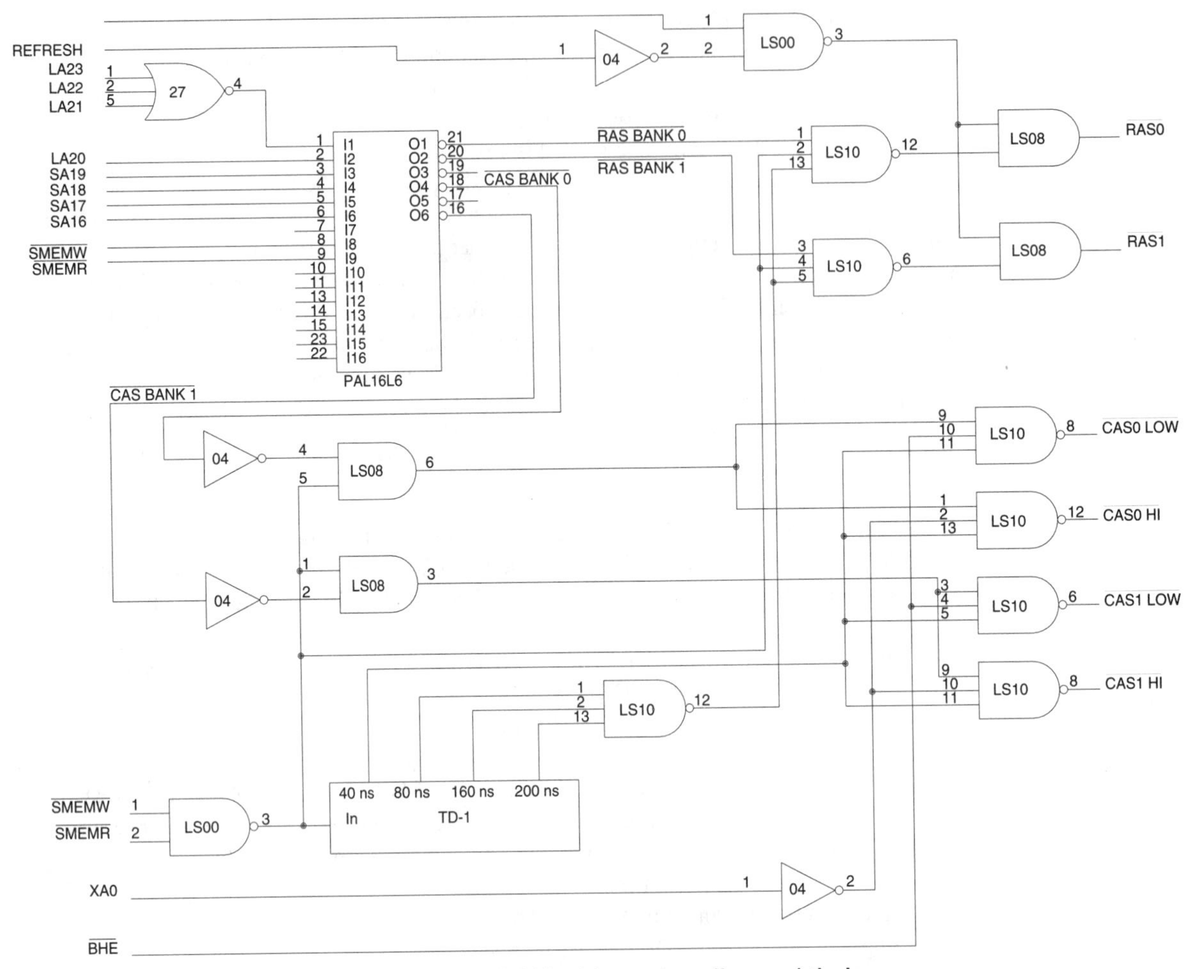

Figure 4-19 DRAM address decoding and timing.

TABLE 4-11
RAS Memory Map

Memory Address Range	*RASx*
000000–1FFFFF	RAS0
200000–3FFFFF	RAS1

TABLE 4-12
CAS Memory Map

Address Type	*Memory Address Range*	*CASx*
even	000000–1FFFFF	CAS0 (LO)
odd	000000–1FFFFF	CAS0 (HI)
even	200000–3FFFFF	CAS1 (LO)
odd	200000–3FFFFF	CAS1 (HI)

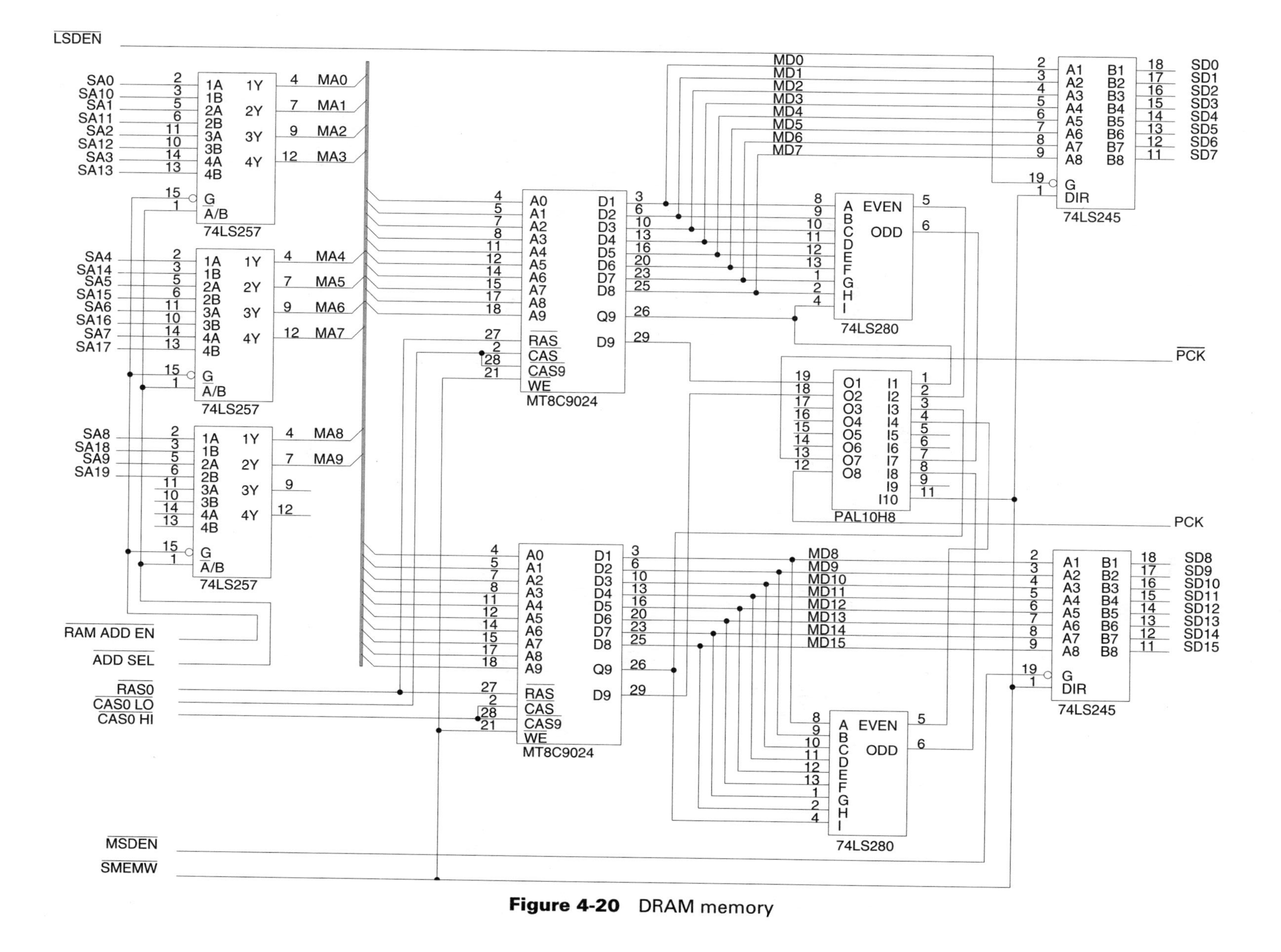

Figure 4-20 DRAM memory

(even addresses) of the data bus is enabled, which occurs during a low byte operation or a word operation. The second 74245 enables the flow of data when the high byte of the data bus is enabled using the $\overline{\text{MSDEN}}$ line (odd addresses), which occurs during a high byte operation or a word operation. The direction of data flow is selected by the $\overline{\text{MEMW}}$ line.

Three 74257 quad 2 to 1 multiplexers are used to multiplex the address from the system address bus into a multiplexed DRAM address bus. During the row part of the address the levels on address lines A1 to A10 are supplied to the MA0 to MA9 address lines, and during the column part of the address lines A11 to A20 are supplied to the MA0 to MA9 address lines. This provides 1 million memory locations for each DRAM. System address line A0 is used to help determine which bank—odd or even—is enabled. The address select line (ADDR SEL) is used to multiplex the addresses.

There are two 74280 9-input parity generator/checkers, one to check for parity for the even address bank and the other for the odd address bank. If either of the parity checkers detects a parity error, an NMI error is produced. The PAL provides the Logic gating to produce the PCK and $\overline{\text{PCK}}$ signals to indicate parity errors, just like in the XT system.

The memory used in this section of the computer uses DRAM modules in the form of a SIMM (Single-In-line Memory Module). This example shows only one bank of memory; most computers contain two or more banks of memory. Each bank of memory requires two SIMMs. One SIMM supplies the low byte of data (even addresses) while the other supplies the high byte of data (odd addresses). The address and size of the operation determines which SIMM is used during the bus cycle. Each SIMM (Figure 4-21) used in this example contains 1 million memory locations with each memory location controlling 9 bits of data (1MB × 9). The data inputs/outputs D1 to D8 are used to input or output the data byte, Q9 outputs the parity bit (only during a read cycle), and D9 inputs the parity bit (only during a write cycle). One SIMM has its D1 to D8 lines connected to the MD0 to MD7 (lower byte) memory data lines. The other SIMM connects its D1 to D8 lines to MD8 to MD15 (high byte). The SIMMs used here are made up of 9 low-power surface-mount DRAMs placed on a small printed circuit card that fits a special edge card–type socket. SIMM modules are normally about 0.75 by 3.5 inches with 30 edge connection points. The number of ICs on the SIMM depends on the surface-mount DRAMs used. Some contain nine DRAMs and some contain three DRAMs, which are configured differently but provide the same amount of memory. SIMMs are now used in most computer systems because of their small size, low power dissipation, lower cost than DRAMs using DIPs, small amount of board space needed, and ease of upgrading. SIMMs come in three common sizes—256 K (256K × 9), 1MB (1MB × 9), and 4MB (4MB × 9).

The timing sequence for this section is the same as for the XT computer except that the time periods are much shorter. The best way of troubleshooting this section is to use a diagnostic program if the computer can boot up. If the computer does not boot, a POST card can be used to find the failure. Unless the problem is within the first 256,000 memory locations, the computer should boot. The troubleshooting section on DRAM in the XT computer can be used for this section.

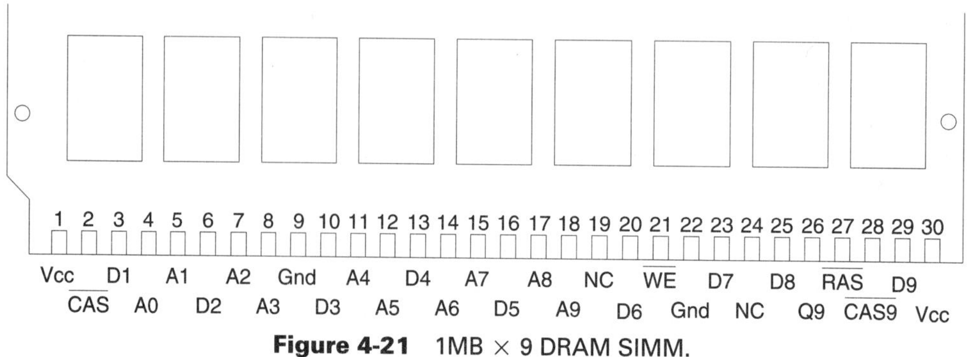

Figure 4-21 1MB × 9 DRAM SIMM.

4.5.8 8042 Keyboard Microcontroller

AT system boards do not use the 8255 PPI and support circuitry to decode the equipment DIP switches or convert the serial data from the keybaord into parallel data for the computer. Because there are no DIP switches for equipment, the AT uses data stored in the battery–backed-up CMOS section of the RTC for equipment information. Instead AT system boards use the Intel 8042 universal peripheral interface 8-bit slave microcontroller. The 8042 section (Fig. 4-22) on the AT system board is responsible for converting serial keyboard data into parallel data for the data bus. It generates the IRQ1 keyboard interrupt request, and it is also responsible for controlling the A20GATE line, which is used to switch the 286 MPU between the real and protected operating modes. Depending on the system board, the 8042 is also responsible for locking out the keyboard using a key lock on the front of the computer, switching the speed of the 286 from normal to turbo speed, and in some cases detecting the default video monitor type (color or monochrome).

The 8042 is basically an 8-bit single-chip computer system that contains the following features: 256 bytes of RAM, 2K of ROM, two programmable 8-bit I/O ports, and over 90 instructions in a 40-pin DIP package. The best way to troubleshoot this section is to use a diagnostic program that tests all the functions of the section. If the computer does not boot or the keyboard does not function correctly, a POST card provides the best results.

4.5.9 I/O Interface Bus Slots

The I/O interface bus slots are the edge card adapter near the back of the system board. When the user places an interface card in one of these slots, the card has access to most of the bus signals to control the card. If the card does not need a particular signal, a connection is not made. Most AT system boards contain eight interface slots; normally two are 8-bit interface slots (single 62-pin edge connector) and six are 16-bit interface slots, which are identified by an additional 36-pin edge connector in line with the larger connector. The signals for each of the slots are connected in parallel; therefore, a 16-bit I/O adapter can be placed in any 16-bit I/O slot and any 8-bit adapter can be placed in any 8-bit slot or any 16-bit slot (but only the 8-bit signals are used).

Each interface adapter contains its own address-decoding circuitry, address latches if required by the adapter, and data bus transceivers. Table 4-13 displays the memory map for AT systems and their I/O interface adapters.

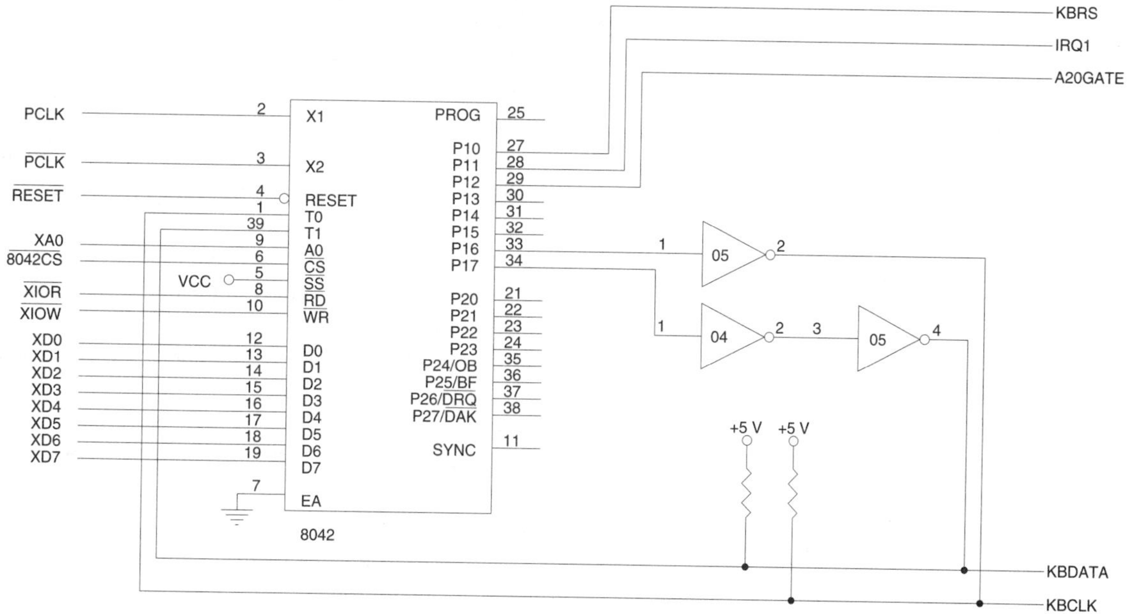

Figure 4-22 8042 keyboard microcontroller.

TABLE 4-13
AT I/O Adapter Memory Map

Address	*Description*
0001F0–0001F8	Floppy disk controller
000200–000207	Game port
000278–00027F	Parallel printer port 2 (LPT2)
0002F8–0002FF	Serial port 2 (COMM2)
000300–00031F	Prototype card
000360–00036F	Reserved
000378–0003FF	Parallel printer port 1 (LPT1)
000380–00038F	SDLC 2
0003A0–0003AF	SDLC 1
0003B0–0003BF	MDA video card (PRN)
0003C0–0003CF	Reserved
0003D0–0003DF	CGA video card
0003F0–0003F7	Floppy disk controller
0003F8–0003FF	Serial port 1 (COMM1)

Figure 4-23 diagrams the location and function of each pin on an ISA (Industry Standard Architecture) AT I/O interface slot. Usually the only things that go bad with the interface slots are dirt or corrosion on the contact pins, or broken solder connections during the insertion or removal when the user moves the edge connectors from side to side, which should never be done.

CHAPTER SUMMARY

1. AT system boards contain any type of the following MPUs: 80286, 80386SX, 80386DX, 80486SX, 80486DX, 80486DX2, 80486DX4, and the Pentium.
2. The 286 in an AT system board is the main controlling element of the system. It is responsible for the execution of instructions and controls support circuitry to control the operation of the system board. The 80286 MPU has a 16-bit data bus, 24-bit address bus, two operational modes (real mode with an address limit of 1MB and protected mode with a real address limit of 16MB), and built-in memory management; is capable of multitasking software; and is compatible with the 8088/8086 instruction set. The speeds of the 286 MPUs are 6 MHz, 8 MHz, 10 MHz, and 12 MHz.
3. The 82284 is the AT version of the 8284 clock generator/driver found on the XT system board. It is responsible for producing the main timing signals for the system board, helps determine if the bus is ready to complete a bus cycle, and contains reset circuitry to reset the computer if the power supply voltages are bad.
4. The 82288 is the AT version of the 8288 bus controller found on the XT system board. It is responsible for producing the control line sequence for the operation of the address, data, and control buses on the system board. The 286 tells the bus controller which type of bus operation is to take place and when the bus controller is to take over the workload of the 286.
5. The AT system board contains 74373 address latches and 74245 data bus transceivers, just like the ones used on the XT system board. Because of the number of address and data lines, the AT system boards contain more address latches and data bus transceivers.
6. The AT system board contains two 8259 programmable interrupt controllers (PICs), which allow the AT system to accept up to 16 different interrupt requests from different hardware devices in the system; the XT contains only one 8259.

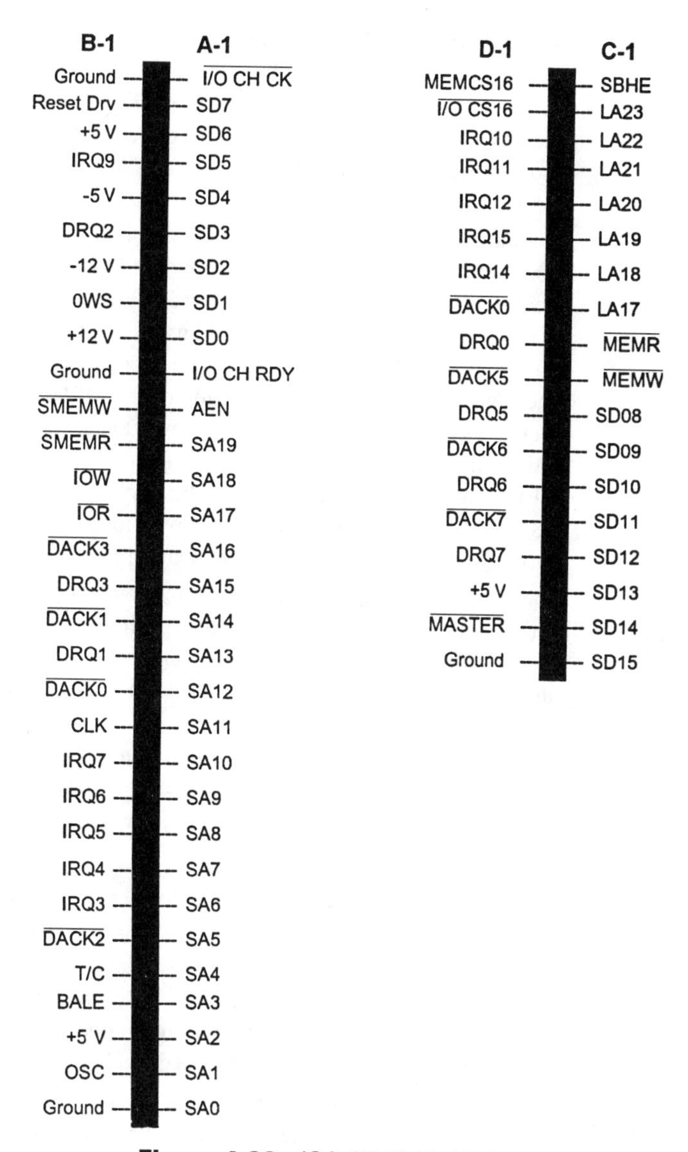

Figure 4-23 ISA AT 16-bit I/O bus.

7. The AT system board contains two 8237 DMAs, providing eight channels for hardware devices to transfer data from I/O devices to memory and from memory to I/O devices. The XT contains only one DMA.
8. The AT system board, like the XT board, contains only one 8254 programmable interval timer (PIT). The 8254 is used to generate the Time of Day pulse and refresh cycle and produces the frequencies for the speaker in the computer.
9. The AT system board contains a real time clock (RTC), which is used to keep the correct time and date while the computer is turned off. The RTC also contains RAM memory locations, which hold data about the set-up and standard equipment in the computer.
10. The AT system board also contains an 8042 microcontroller that controls the keyboard signals.
11. The ROM section in the AT is much like the ROM section of the XT except that the ROM that contains BIOS is much larger in size (memory).
12. The DRAM section in the AT system is much like the DRAM section of the XT except that more memory can be placed on the system board, the memory is faster, and the memory data bus is 16-bits wide.

13. Internally the 80286 has four control units. The bus unit (BU) is responsible for the transfer of data to and from the 286. The execution unit (EU) performs manipulation of data once it is inside the 286. The instruction unit (IU) is responsible for decoding instructions and has a buffer that helps keep all the control units inside the 286 running simultaneously under most conditions. The address unit (AU) is responsible for developing the address being used for the instruction.
14. The memory management unit in the 286 is responsible for realigning data that has been stored at odd address locations. If the MPU needs to fetch a byte of data from an even memory location, the data is transferred directly from D0 to D7 data lines to the low byte of the specified register or location inside the MPU. If the MPU needs to fetch a byte of data from an odd memory location, the data is transferred directly from the D8 to D15 data lines and then, in the bus unit section of the MPU, into the low byte of the specified register or location inside the MPU. If a word needs to be fetched from an even memory location, the data is directly transferred from D0 to D15 to the specified location inside the MPU or register. If a word needs to be fetched from an odd memory location, the data on D8 to D15 is transferred into the low byte location and the data on D0 to D7 is transferred into the high byte of the specified location inside the MPU or register.
15. The 286 has two operational modes. In the real mode the 286, just like an 8088, limits the address memory in the computer to 1 million locations (only address lines A0 to A19 are used). In the protected mode the 286 automatically maps 1 GB of virtual memory, but the number of actual memory locations is limited to 16MB (all address lines are used, A0 to A23).
16. The 286 boots up in the real mode and must be programmed to enter the protected mode.
17. There are eight 16-bit general registers, as in the 8088; they may be used as 8-bit or 16-bit registers. The general purpose registers can be used in both the real and protected modes.
18. The flags register is a 16-bit register that performs basically the same function as the flags register inside the 8088 except that three additional bits are used. The flags register is used in both modes of the MPU.
19. The machine status word register is a 16-bit register that contains data that controls and provides status for the current operating mode. Although this register is used in the protected mode, it is monitored in the real mode.
20. The four 16-bit segment registers and the 16-bit instruction pointer in the MPU perform the same function as in the 8088 while the MPU is in the real mode.
21. In the protected mode the address is developed by the global descriptor table register, which defines a memory location above 1MB.
22. The interrupt descriptor table register performs the same function as the global descriptor table register when addressing occurs during an interrupt.
23. Multitasking in the 286 uses the information found in the task register and local descriptor table register to allow more than one program to be executed in the computer at any one time.
24. The 286 supports all 8088/8086 instructions, addressing modes, and data types, with an increase in the number and size of the data that can be transferred at any one time, the types of data, and the number of instructions.
25. The following is a brief description of the function of the pins of the 286.
    - CLK is the clock input for the 286 MPU; internally this frequency is divided into two processor cycles. All data transfers require a minimum of two processor cycles.
    - A0 to A23 are unidirectional address output lines. Only A0 to A19 are used while in the real mode; all address lines may be used during the protected mode. Address line A0 is used to help determine if the low byte of data is to be transferred. When A0 is low, the low byte is transferred, indicating an even address.

- $\overline{\text{BHE}}$, the byte high enable line, when low, is used to enable the high byte, which indicates that the data is located in an odd address.
- D0 to D15, the 16 bidirectional data lines, are used to input or output data to and from the MPU.
- M/$\overline{\text{IO}}$, the memory and I/O line, is used with other lines to determine the type of operation. A high indicates a memory operation, while a low indicates an I/O operation.
- $\overline{\text{S0}}$ and $\overline{\text{S1}}$ (status) lines are used with other lines to determine which operation is to be performed. This information is supplied to the bus controller to select the proper bus operation.
- COD/$\overline{\text{INTA}}$ is used with other lines to determine which operation is to be performed.
- $\overline{\text{READY}}$, the ready line, is used to inform the MPU to terminate the current bus cycle, when low.
- INTR, the interrupt request line, is used to inform the MPU that an interrupt needs to be serviced.
- NMI, the nonmaskable interrupt input line, is used to inform the MPU that a parity error has occurred, when high.
- $\overline{\text{LOCK}}$, the lock line, is used to lock out any other bus master from controlling the bus.
- HOLD, the hold request line, is used by the DMA to request control of the bus, when high.
- HLDA, the hold acknowledge, is used to inform the DMA that the MPU is giving up control of the bus to the DMA.
- PEREQ, the processor extension operand request, informs the MPU of a data operand transfer from the math coprocessor.
- $\overline{\text{PEACK}}$, the processor extension operand acknowledge output, indicates a processor extension operand request is being transferred to the 287, when low.
- $\overline{\text{BUSY}}$, the busy input, informs the MPU that the math coprocessor is busy processing data and that the 286 needs to wait, when low.
- $\overline{\text{ERROR}}$, the error input, informs the 286 that the 287 math coprocessor has generated an error in its operation, when low.
- RESET is used to reset the MPU when high for the proper time period.
- Vss supplies the current for the MPU and is the ground reference.
- Vcc supplies the voltage potential for the MPU.
- CAP, the capacitor line, is used to reduce any internal substrate bias generated by the switching of the gates inside the MPU. A 0.047 μF at 12 V capacitor is used between this line and ground.

26. The 287 is a math coprocessor that performs only math operations a minimum of 10 times faster than the 286. The 287 operates only if the software uses the math instructions of the 287. The 286 controls the data transfer for the 287; the 287 just performs the math operation.
27. Intel's 82230 IC is one-half of the AT chip set. This IC contains much of the support circuitry for the 286. The 82230 contains the 82284 clock generator/driver, two 8259 PICs, a 6818 time clock with CMOS memory, an 82288 bus controller, address and data bus control logic, bus control logic, CPU shutdown logic, and coprocessor interface logic.
28. The 82231 is the other half of Intel's AT chip set. This IC contains read and write logic, I/O chip selects, support logic for the PIT and I/O devices, an 8254 PIT, a 74612 memory mapper, two 8237 DMAs, DRAM refresh counter, latch, and logic, and parity check logic.
29. Each bus cycle must contain a minimum of two processor states. During the status state, the command is issued and support circuitry becomes enabled. The status state contains two clocks, labeled phase 1 and phase 2 clocks. During the second processor state (com-

mand state) the data transfer occurs and ends. This state also contains two clocks, labeled phase 1 and phase 2. A slow memory or I/O device may extend the command state, at which point the first command state is renamed the first wait state. This process continues until the memory or I/O device is ready to terminate the bus cycle.

30. The ROM section of the AT contains address decoding for the ROM and the ROM ICs themselves. There are normally two ROMs, one supplying the low byte and the second the high byte. The ROMs or EPROMs contain BIOS in the AT system.
31. The DRAM section of the AT contains address decoding for the DRAM section, the DRAM ICs, and a parity generator/checker. The amount of memory that can be placed on the system board is much larger than on an XT: 4MB is normal and 16MB is the maximum. There are two sets of DRAM per bank of memory. One set of DRAMs supplies the low byte for the data bus and lies on even addresses, while the other set of DRAMs supplies the high byte for the data bus and lies on odd addresses.
32. The main purpose of the 8042 keyboard microcontroller is to process the keycode data received by the system board from the keyboard. This IC converts the serial data from the keyboard into parallel data for the data bus and also is used to generate the keyboard interrupt request (IRQ1).
33. The I/O interface slots in the 286 use two edge connector sockets. The larger connector is basically the same as the XT I/O interface slot. The smaller connector is used to supply the additional address, data, and control signals found in the AT system. Because all slots are in parallel, any I/O adapter can be placed in any slot. An 8-bit I/O adapter can be placed in a 16-bit slot, but only the larger connector and signals are available for the adapter.

REVIEW QUESTIONS

1. Which of the following Intel MPUs is not found in an AT system?

a. 80286

b. 386SX

c. 8088

d. None of the given answers

2. What is the slowest operating speed of the 286?

a. 8 MHz

b. 10 MHz

c. 12 MHz

d. None of the given answers

3. Which of the following ICs is responsible for generating the ready signal in an AT system?

a. 82284

b. 82288

c. 82231

d. None of the given answers

4. In a standard AT system, which IC is responsible for directly controlling the levels on the memory and I/O read/write lines?

a. 82284

b. 82288

c. 80286

d. None of the given answers

5. Which IC controls the flow of data on the data bus?

a. 74245

b. 74157

c. 74373
d. 74244
e. None of the given answers

6. Which interrupt is used to generate a keyboard interrupt service routine?
   a. IRQ0
   b. IRQ1
   c. IRQ3
   d. IRQ7
   e. None of the given answers
7. Which one of the following ICs can act as a bus master?
   a. 8237
   b. 8259
   c. 80287
   d. 8254
   e. None of the given answers
8. Which one of the following ICs supplies the frequency for the speaker?
   a. 8237
   b. 8259
   c. 80287
   d. 8254
   e. None of the given answers
9. Which part of the AT system board holds the CMOS data for the system?
   a. 8042
   b. ROM
   c. DRAM
   d. RTC
   e. None of the given answers
10. Which part of the AT system board converts serial keyboard data into parallel keyboard data for the data bus?
    a. 8042
    b. ROM
    c. DRAM
    d. RTC
    e. None of the given answers
11. How many ROM or EPROM ICs are normally found in an AT system?
    a. 0
    b. 1
    c. 2
    d. 4
    e. None of the given answers
12. The DRAM section of the 286 AT system board is _____ bits wide.
    a. 4
    b. 8
    c. 16
    d. 32
    e. 64
13. Which control section of the 286 is used to generate the address for the current instruction?
    a. BU
    b. EU

c. IU
d. AU
e. None of the given answers

14. Which part of the 286 is used to realign data during a bus cycle?
a. BU
b. EU
c. MMU
d. IU
e. None of the given answers

15. Which operational mode in the 286 limits the amount of addressable memory space to 1 million memory locations?
a. real
b. protected
c. None of the given answers

16. Which operational mode is the default mode for the 286 upon reset?
a. real
b. protected
c. None of the given answers

17. AX is used to indicate a ____-bit register.
a. 4
b. 8
c. 16
d. 32
e. 64

18. Which register is used to indicate special conditions that have occurred inside the 286?
a. AX
b. Flags
c. SI
d. CS
e. ES

19. Which register is used to provide control data and status for the current operating mode?
a. TR
b. IDTR
c. SS
d. MSWR
e. None of the given answers

20. The segment registers in the 286 function the same as the segment registers in the 8088 in the protected mode.
a. True
b. False

21. Which register in the 286 is used to define a memory location above 1 MB?
a. TR
b. MSWR
c. Flags
d. GDTR
e. None of the given answers

22. Which register in the 286 is used to define a memory location above 1 MB during an interrupt?
a. TR
b. MSWR

c. Flags
d. GDTR
e. None of the given answers

23. Which register is not used to control multitasking in the 286?
a. TR
b. LDTR
c. DX
d. None of the given answers

24. One advantage of the 286 over the 8088 is that the 286 has a 32-bit data bus.
a. True
b. False

25. Which 286 pin is used to enable the high byte for a bus cycle?
a. D0–D15
b. $M/\overline{IO}$
c. CLK
d. $\overline{BHE}$
e. None of the given answers

26. Which 286 pin is used to terminate a bus cycle?
a. $\overline{S0}$
b. $\overline{S1}$
c. $\overline{READY}$
d. $\overline{BUSY}$
e. None of the given answers

27. Which 286 pin is used to grant permission to the DMA to use the bus?
a. INTR
b. NMI
c. $\overline{HOLD}$
d. $\overline{PEACK}$
e. HLDA

28. The math coprocessor controls the data on the bus during a math operation calling for a 287.
a. True
b. False

29. Which IC of Intel's AT chip set contains the real time clock?
a. 82230
b. 82231
c. Neither of the given answers

30. Which IC of Intel's AT chip set contains the DMA?
a. 82230
b. 82231
c. Neither of the given answers

31. All I/O adapters can be placed in any I/O interface slot.
a. True
b. False

chapter 5

80386 Microcomputer System Board

OBJECTIVES

Upon completion of this chapter, you should be able to perform the following tasks:

1. List and define the purpose of the following control sections of the 80386SX and DX MPUs: bus interface, central processing unit, and memory management unit.
2. List and define the real and protected addressing modes of the 386.
3. Define the different ways of specifying the size of the registers of the 386.
4. List and define the differences, advantages, and disadvantages of the 386SX and DX.
5. Define the purpose of each pin of the 386DX and SX.
6. Define the purpose of each pin of the 387DX and SX math coprocessors.
7. Define the purpose of the 82335 interface in a 386SX system using the Intel 82230/82231 AT chip set.
8. Define the main purpose of each block inside the block diagram of the 82335 interface IC.
9. List and define the purpose of each block inside the block diagram of a typical 386SX system board using the Intel AT chip set.
10. Define the difference between non–page mode, page mode, and page mode interleaving addressing of DRAM.
11. Define the purpose and operation of the 82385DX cache controller and cache static RAM.
12. List and define the purpose of each block inside the block diagram of a typical 386DX system board using the Intel AT chip set.

Not long after the 286 AT became popular in computer systems, Intel released the 80386DX 32-bit high-performance microprocessor. This MPU was developed to allow a larger address bus, larger packages of data to be transferred during a bus cycle, and a more effective operational mode. The 80386SX 32-bit high-performance microprocessor was developed after the 386DX hit the market. The 386SX is basically the same internally as the 386DX except that it has only a 16-bit data bus and a 24-bit address bus. The SX, with its lower power dissipation rating, smaller size, and lower price than the DX, was developed for use in portable or notebook AT computers. The 386SX, because of the size of its data and address buses, can be designed into a 286 system board with few changes to the board.

5.1 80386 DX/SX INTERNAL BLOCK DIAGRAMS

The internal architecture of the DX and SX is basically the same. Both MPUs contain the same size registers, instructions, and operating modes. The only differences are in the size of the address and data buses, power dissipation, and size of the unit.

The 386 (Figs. 5-1 and 5-2) contains three main controlling units; each unit is divided into smaller groups that perform specific tasks for the controlling unit.

5.1.1 Bus Interface

The bus interface contains the request prioritizer, address driver, pipeline bus size control unit, and multiplexer transceivers. The request prioritizer is used to control handshaking between the MPU and other devices that request interrupts and the coprocessor. The address driver outputs the address to the address bus and the levels on the byte enable lines at the proper times and selects the device being accessed. The 386DX has 4 byte enable lines ($\overline{BE3}$, $\overline{BE2}$, $\overline{BE1}$, and $\overline{BE0}$) and 30 address lines (A2 to A31), while the 386SX has 2 byte enable lines ($\overline{BHE}$ and $\overline{BLE}$) and 23 address lines (A1 to A23). The pipeline bus size control unit supplies information to the bus controller and support logic to indicate the type and size of the bus cycle. This section allows the current bus cycle in process to end, while the next bus cycle is placed on the control lines so that the next bus cycle can begin without any delay. The multiplexer transceiver controls the size and direction of data transfers between the MPU and the outside world. This section is also used to convert data from byte and word locations as the data enters the MPU. The 386DX contains 32 data lines (D0 to D31) while the 386SX contains 16 data lines (D0 to D15).

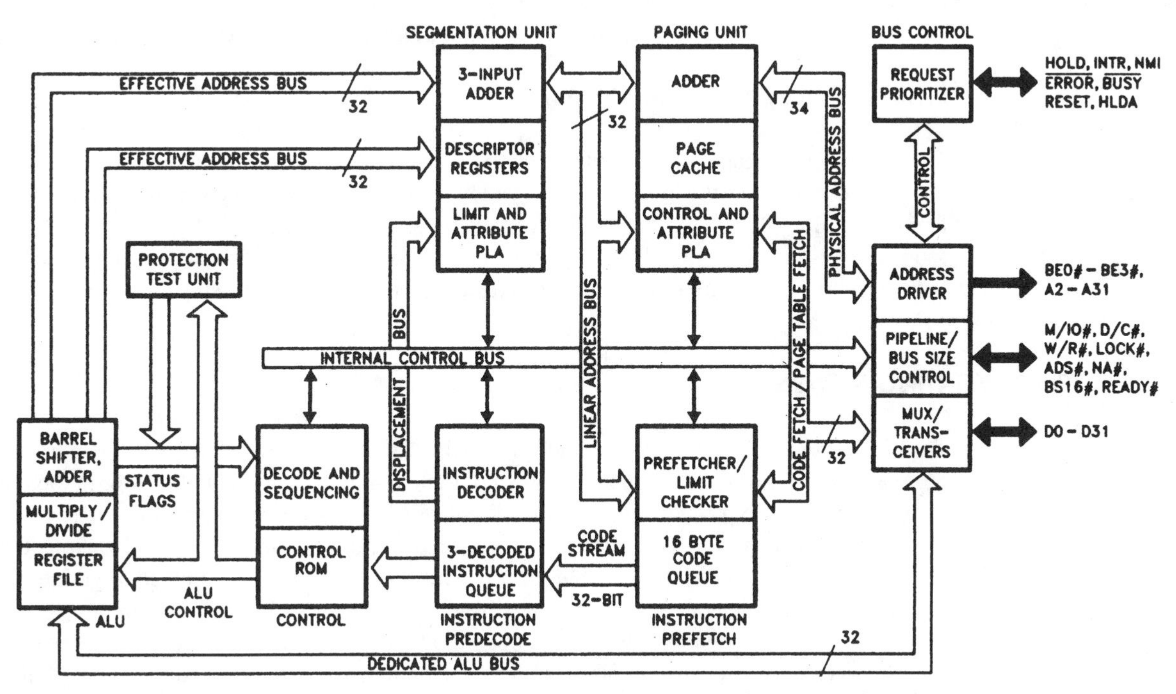

Figure 5-1 386DX internal block diagram. (Courtesy of Intel)

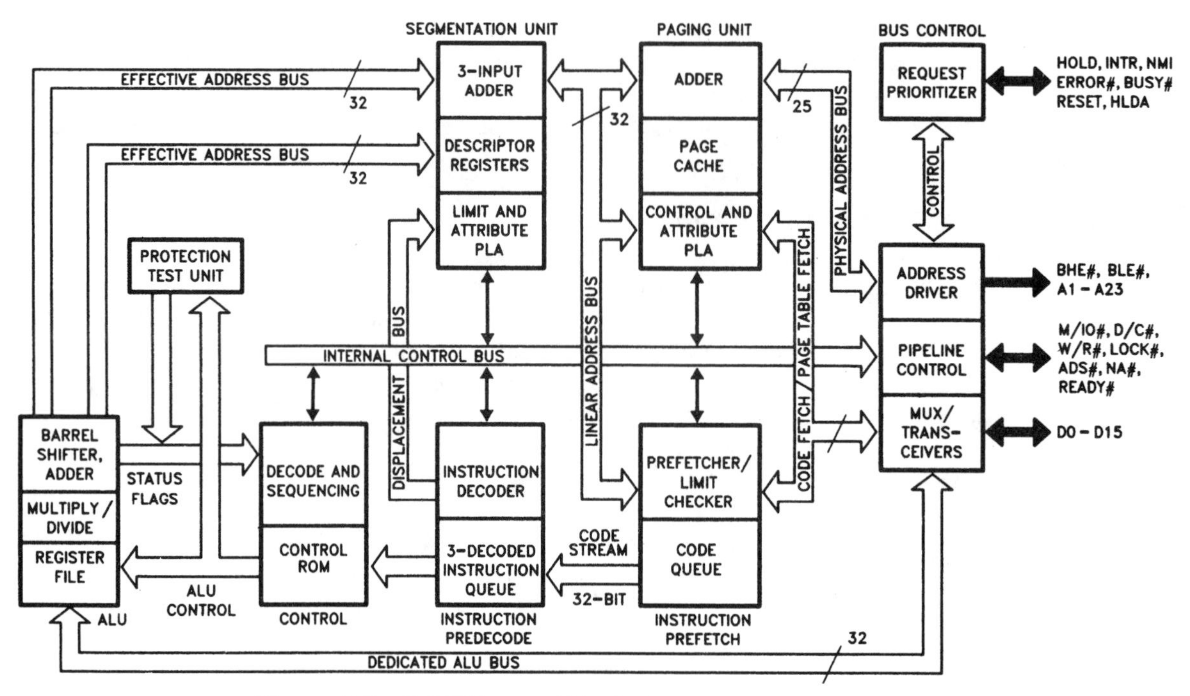

Figure 5-2 386SX internal block diagram. (Courtesy of Intel)

5.1.2 Central Processing Unit

The central processing unit (also known as the execution unit) contains the general purpose registers, the ALU, and a 64-bit barrel shifter. The general purpose registers are basically the same as those in the 286 except that they are 32 bits wide and they can be specified as two 8-bit registers, a 16-bit register, or one 32-bit register. This section also contains a flags register. The ALU performs most math and logical operations on the data as specified by the instruction. The 64-bit barrel shifter is used to perform complex math operations that would normally take many clock cycles for the ALU to perform; these are division, multiplication, and rotate and shift instructions.

5.1.3 Memory Management Unit

The memory management unit contains the segmentation and paging units. The segmentation unit is used to generate the logical address, which provides protection and isolation between different programs while operating in the protected mode (virtual 86 mode). The paging unit is responsible for producing the physical address that is placed on the address bus and byte enable lines. The operation of these two units changes with the different addressing modes of the 386.

5.2 80386 OPERATING MODES

The 386 has two different operating modes, which define how the MPU accesses memory and processes instructions. The computer boots up in the real mode, just as in the 286. In the real mode the 386 develops addresses like the 8088/8086 MPUs; this limits the number of memory locations to 1 million. The difference between the real mode in an 8086 and a 386 is that the 386 has a 32-bit data bus available for use. The MPU must be programmed to go into the protected mode, which operates somewhat differently than the protected mode in the 286 and for that reason is referred to as the virtual 8086 mode. In the protected mode, each task can be set up to operate as a separate 8086 MPU. These virtual MPUs are protected and isolated from the 386 itself and from each other.

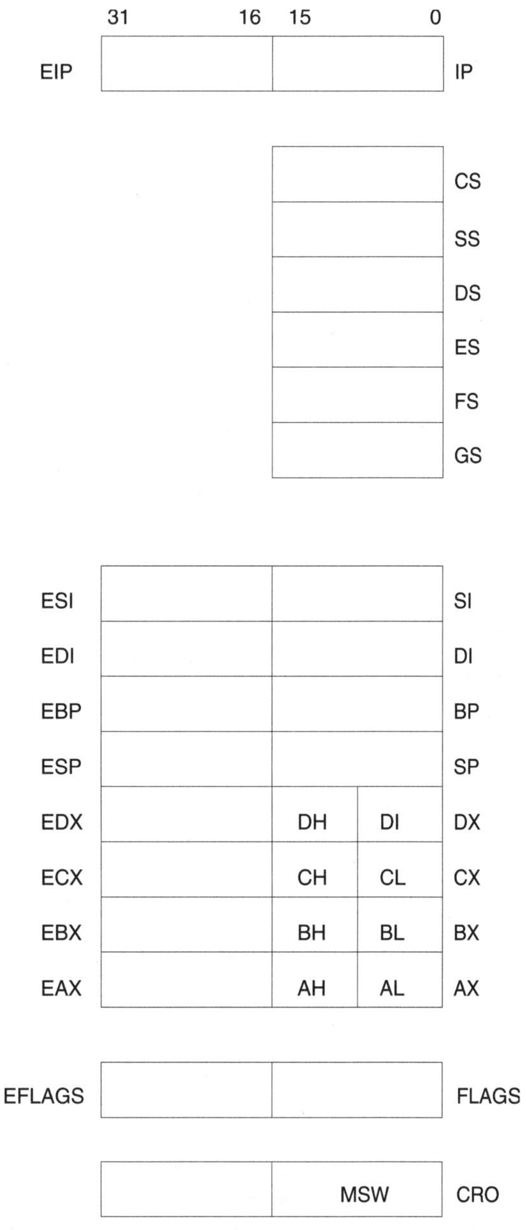

Figure 5-3 386DX/SX registers (real mode).

5.2.1 80386 Registers

General Registers (Real or Protected). The eight 32-bit general registers (Figs. 5-3 and 5-4) inside the 386 have the same purposes as those in the 8088 (see Chapter 3): Accumulator (EAX = 32 bits, XA = 16 bits, AH = 8 bits high byte and AL = 8 bits low byte), Data register (EDX, DX, DH, and DL), Count register (ECX, CX, CH, and CL), Base register (EBX, BX, BH, and BL), Base pointer (EBP = 32 bits and BP = 16 bits), Source index (ESI and SI), Destination index (EDI and DI), and Stack pointer (ESP and SP).

Flags Register (Real and Protected Modes). The flags register (EFLAGS) has the same purpose as the flags register in the 8088 (see Chapter 3) except that in the 386 this is a 32-bit register. In the real mode only the 16 least significant bits (LSB) are used, the same as in the 286. In the protected mode all 32 bits are used. The lower 16 bits are used in the same

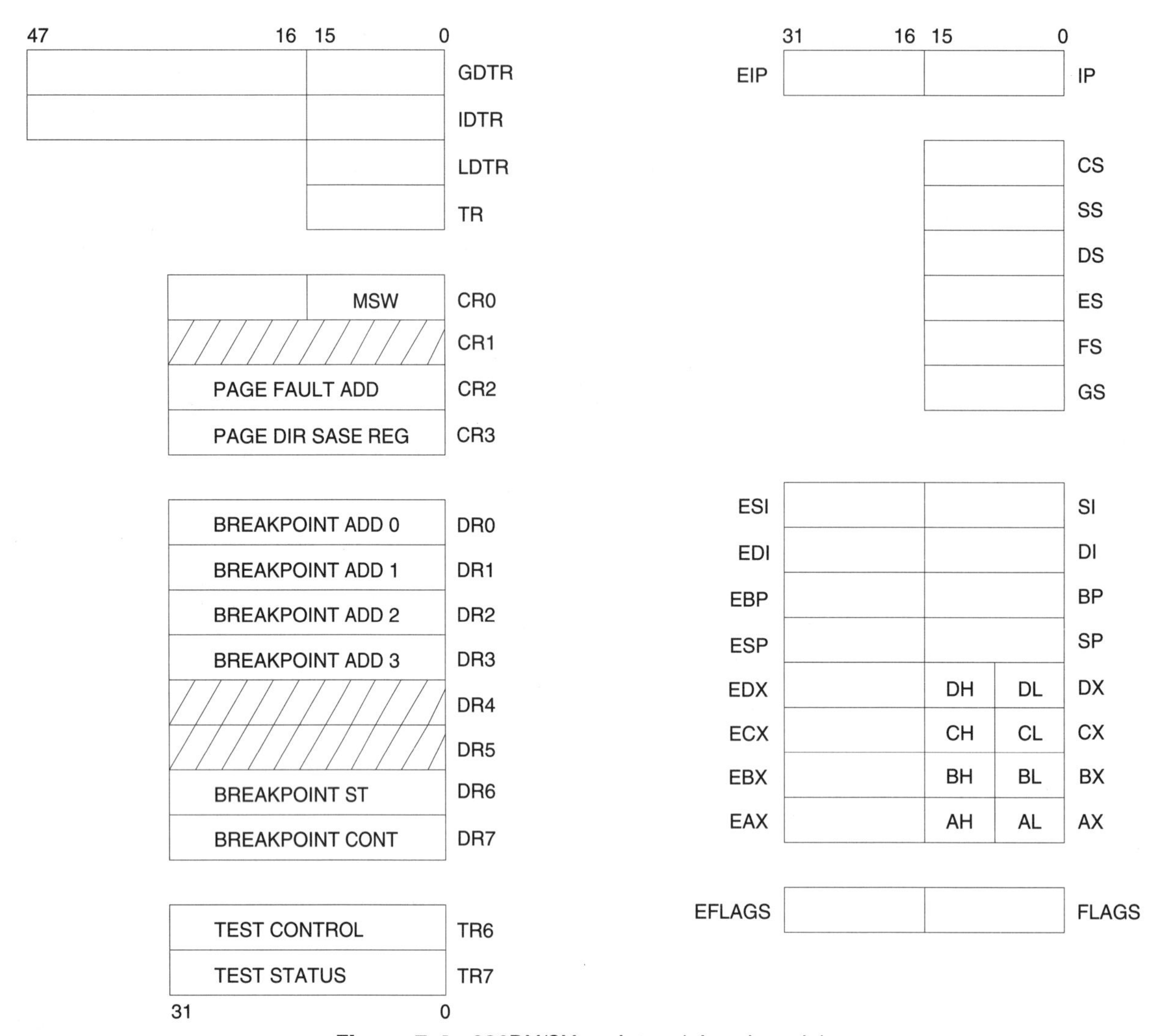

Figure 5-4 386DX/SX registers (virtual mode).

way as in the real mode and 2 of the 16 most significant bits (MSB) are used to supply information during the protected mode. Bit 16, the resume flag, is used with the debug registers to determine if a breakpoint will be processed; while bit 17, the virtual mode flag, is used to identify that the 386 is in the virtual address mode.

5.2.2 Addressing (Real Mode)

There are six 16-bit only segment registers (CS, DS, SS, ES, FS, and GS) and a 16-bit instruction pointer. The registers in this section function the same as in the 8088 and 286, except that there are two additional segment registers (F and G data segment registers) that can be used to support data segments. To develop a physical address (also known as an effective address), the value of the selected 16-bit segment register is transferred into the physical address adder, where four binary 0s are shifted into the least significant location. This creates a 20-bit shifted segment value. The value of the instruction pointer or offset value is added to the shifted segment value. The 20 LSBs of the result are used as the physical address, which is then applied to the address bus through the address drivers.

5.2.3 Addressing (Protected Mode)

When the 386 is in the protected mode, registers CR0 to 3, LDTR, GDTR, IDTR, TR, DR0 to 7, and TR6 to 7 are used to develop the address for each task.

Word Control Registers. There are four 32-bit word control registers (CR0, CR1, CR2, and CR3) in the 386. These registers contain data that control and provide status information for the current operating mode while providing information for the paging unit. Only part of CR0 is used during the real mode, while all word control registers are used during the protected mode.

In the real mode only the lower 16 bits of CR0 are used; of the 16 bits, the five LSBs are used to perform the same function as the machine status word register in the 286. Of those five bits, only bit 0—the protection enable bit—is used, which indicates the addressing mode of the 386.

Word control register zero (CR0) is 32 bits in size, only 6 bits of which are predefined when the 386 is in the protected mode. The 16 LSBs function as the machine status word register to maintain compatibility with the 286. Of these 16 bits, only the 5 LSBs are used. Bit 0 is the protection enable bit; when it is reset the 386 is in the real mode, and when it is set the 386 is in the protected mode. Bit 1 is the monitor coprocessor (MP) bit; when set it indicates the presence of a math coprocessor. Bit 2 is the emulate coprocessor (EM) bit; when set, it causes the 386 to perform math operations. If the MP is set, EM must be reset; and if MP is reset, EM normally is set. Bit 3, the task switched (TS) bit, sets whenever the MPU switches tasks using a processor extension and must be reset through software. Bit 4 is the extension type (R) bit; when set, it indicates that an 80387 NP is being used, and when low it indicates that an 80287 NP is used. The sixth bit of word control register 0 is bit 31, the paging enable (PG) bit. When set, the PG bit enables the paging unit in the memory management unit.

Word control register 1 (CR1) is 32 bits in size and is reserved by Intel. Word control registers 2 (CR2) and 3 (CR3) are 32-bit registers used during the protected addressing mode to provide information for the paging unit of the memory management unit. CR2 holds the page fault linear address, and CR3 holds the page directory base address.

Global Descriptor Table Register. The global descriptor table register (GDTR) is a 48-bit register that contains the address for the global descriptor table located in physical memory. The contents of these memory locations are used to produce the address in physical memory above 1MB. The 16 lowest bits of this register define the size of the memory block (ranges from 0 to 65,535). The 32 most significant bits represent the starting point of the memory block. A software driver controls the transfer between the GDRT and the global descriptor table. The size of this table is one plus the limit size. Each global descriptor entry in physical memory is 8 bytes, which specifies the size, starting point, and access rights of the global memory segment. A maximum of 8,192 descriptors can be used if the maximum size is selected.

Interrupt Descriptor Table Register. The interrupt descriptor table register is also a 48-bit register that is used in the same way as the GDTR except that it is used when processing an interrupt. Because the 386 is limited to 256 interrupts, the number of interrupt descriptor tables in memory is also limited to 256. During an interrupt the interrupt descriptor table specifies the starting address and the attributes of the interrupt service routine.

Task Register. Multitasking is the process that allows the MPU to execute two or more programs at the same time. In reality the MPU executes only one program at a time, splitting its processing time between the programs being executed. This means that each task is given a certain amount of the processor time, as defined by the software. When the time period for each task ends, all the data and the contents of the internal environment are stored. The MPU then recalls all the data for the next task so that the new task can take control of the MPU.

The Task Register is a 16-bit register that holds a value called the task state segment descriptor. During task switching the value of the task register (one value for each task running) is used with the base value stored in the global descriptor table to determine the memory locations for the next selected task.

Local Descriptor Table Register. The local descriptor table register is a 16-bit register that points to a local address space for the current task. Each task has its own local address space that defines its own local descriptor table space; therefore, the value in the local descriptor table register changes as the current task changes. The number of local descriptor tables located in memory is determined by the number of tasks configured in the system.

Segment Registers. When the 386 is in the protected mode, 16-bit segment registers are renamed segment selector registers. The purpose of these registers changes from pointing to a segment value (real mode) to providing a selector value. Of the 16 bits in the segment selector register, bits 0 and 1 define the privilege level. Bit 2 indicates whether the global or local descriptor table is used for the task. Bits 3 to 15 are used to select which descriptor entry is used for the task. The select segment descriptor defines the segment base address (32 bits), segment limit (32 bits), and attribute bits.

Debug Registers (Protected Mode). The 386 contains internal hardware support of debugging. Debugging of the 386 occurs when breakpoints are set within the program and when single-stepping the program. During these operations the 386 uses its debug (DR0 to DR7) registers and test registers (TR6 to 7).

Of the eight 32-bit debug registers, DR4 and DR5 are reserved by Intel. Debug registers 0 to 3 are used to hold the 32-bit linear address for one of the breakpoint sets. If paging is disabled, the 32-bit linear address is equal to the physical address. DR7 provides control information for debugging, and DR6 provides the status for the debug operation.

Test Registers (Protected Mode). There are two 32-bit test registers. The purpose of the test registers is to test internal operations of the MPU. TR6 provides the control bits that determine the types of tests and how they are performed. TR7 provides the status of the test performed.

5.3 80386 INSTRUCTIONS

The 386 supports all 8088/8086 and 80286 instructions, addressing modes, and data types. The 386 also has added more instructions and is capable of handling larger packages of data and more tasks during multitasking in a more efficient manner. Tables 5-1 and 5-2 briefly list the types of instructions and data.

TABLE 5-1
80386 Instruction Types

Type of Instruction	*Number of Instructions*
Data transfer	21
Arithmetic	18
String	10
Shift/rotate logical	10
Program transfer	35
Processor control	17
High level	17

TABLE 5-2
80386 Data Types

Type of Data	*Description*
Bit	Single bit of data
Bit field	Contiguous bits used as a group (32 bits maximum)
Integer	Signed 8-bit (byte), 16-bit (word), 32-bit (double word), and 64-bit (quad word) binary values
Unsigned	Unsigned 8-bit, 16-bit, 32-bit, and 64-bit binary values
String	1 byte to 4GB of contiguous bytes, words, or double words
ASCII (character)	Alphanumeric code using ASCII format
BCD	A byte of data used to represent a decimal value from 0 to 9
Packed BCD	A byte of data that represents a decimal value from 00 to 99. Each nibble represents one value from 0 to 9.
Floating point	Signed 32- to 80-bit values when using the 387 for 386DX systems and 287 or 387SX for 386SX systems

TABLE 5-3
80386 Features

Description	*386DX*	*386SX*
Address lines	32	24
Physical address space	4 GB	16 MB
Date lines	32	16
Clock speeds	16 MHz	16 MHz
	20 MHz	20 MHz
	25 MHz	25 MHz
	33 MHz	33 MHz
Number of pins	132	100

5.4 80386 MICROPROCESSOR

The 80386 is a 32-bit high-performance microprocessor made by Intel and is compatible with 8088, 8086, and 80286 software. Like the 80286, the data and address buses are not multiplexed. The following pages describe the pin configurations of both the 386DX and 386SX MPUs (Table 5-3).

The 80386 has two operating modes, both of which support all 8088/86 software. The 8086 real address mode, which makes the 80386 use the address bus like an 8086/88, limits address to 1 MB but has the ability to transfer up to 32 bits of data during a cycle. The second operating mode is called the protected mode, known as the virtual 8086 mode, and differs from the 286 protected mode in that the 386 sets up each task as a separate 8086 that can be isolated and protected from the others and from and the host 386 microprocessor operating system by the use of paging and the I/O permission bitmap.

5.4.1 80386 Pin Configurations

CLK2 (DX/SX) The CLocK 2 input (Figs. 5-5 to 5-8) is used to provide the basic timing for the 386 MPU and any device that must be synchronized with bus operations. The clock input (also known as the system clock) is divided by 2 internally to produce a processor (bus state) clock. The first bus state is called the status state and is made up of two clocks, called phase 1 and phase 2. The second bus state is called the command state and is also made up of two clocks. All bus cycles require a minimum of two processor clocks (bus states).

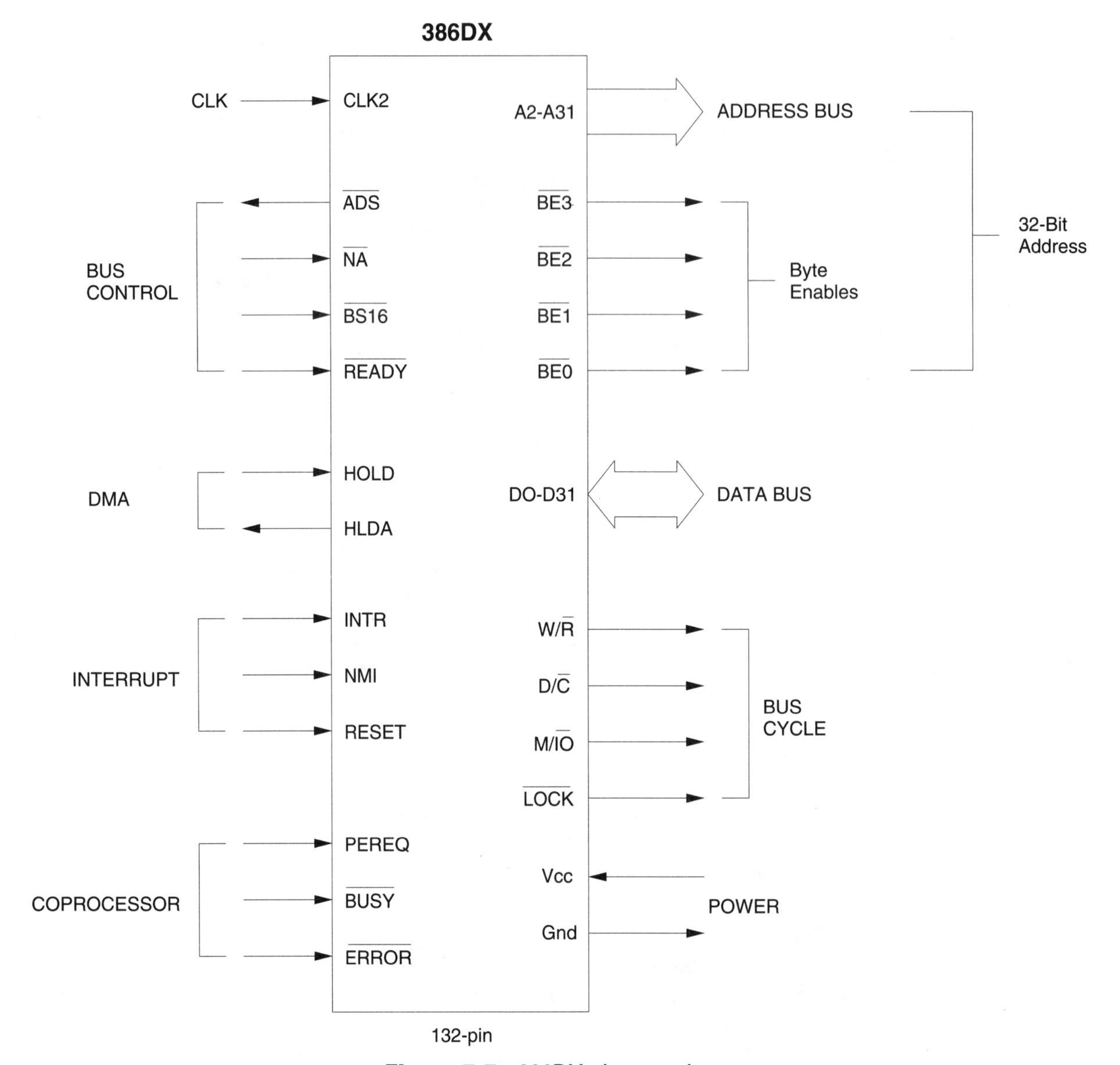

Figure 5-5 386DX pin groupings.

5.4.2 Address Bus

Memory in an 80386SX system is divided into two memory blocks. One block of memory is used to transfer data on data lines D0 to D7, the low byte, and the other block of memory is used to transfer data on data lines D8 to D15, the high byte.

Memory in an 80386DX system is divided into four memory blocks. One block of memory is used to transfer data on data lines D0 to D7, another on lines D8 to D15, still another on lines D16 to D23, and the fourth on lines D24 to D31.

A1–A23(SX) There are 23 actual address output lines for the 386SX. No address line A0 is developed directly by the 386SX because one address line can be at only one logic level during any one bus cycle. Therefore, if a word transfer is needed, two bus cycles are required, which would slow down the operation. If all memory locations were 16 bits wide, when a byte of data was transferred, two memory locations would be used when only one was needed. For these reasons the 386SX uses two byte

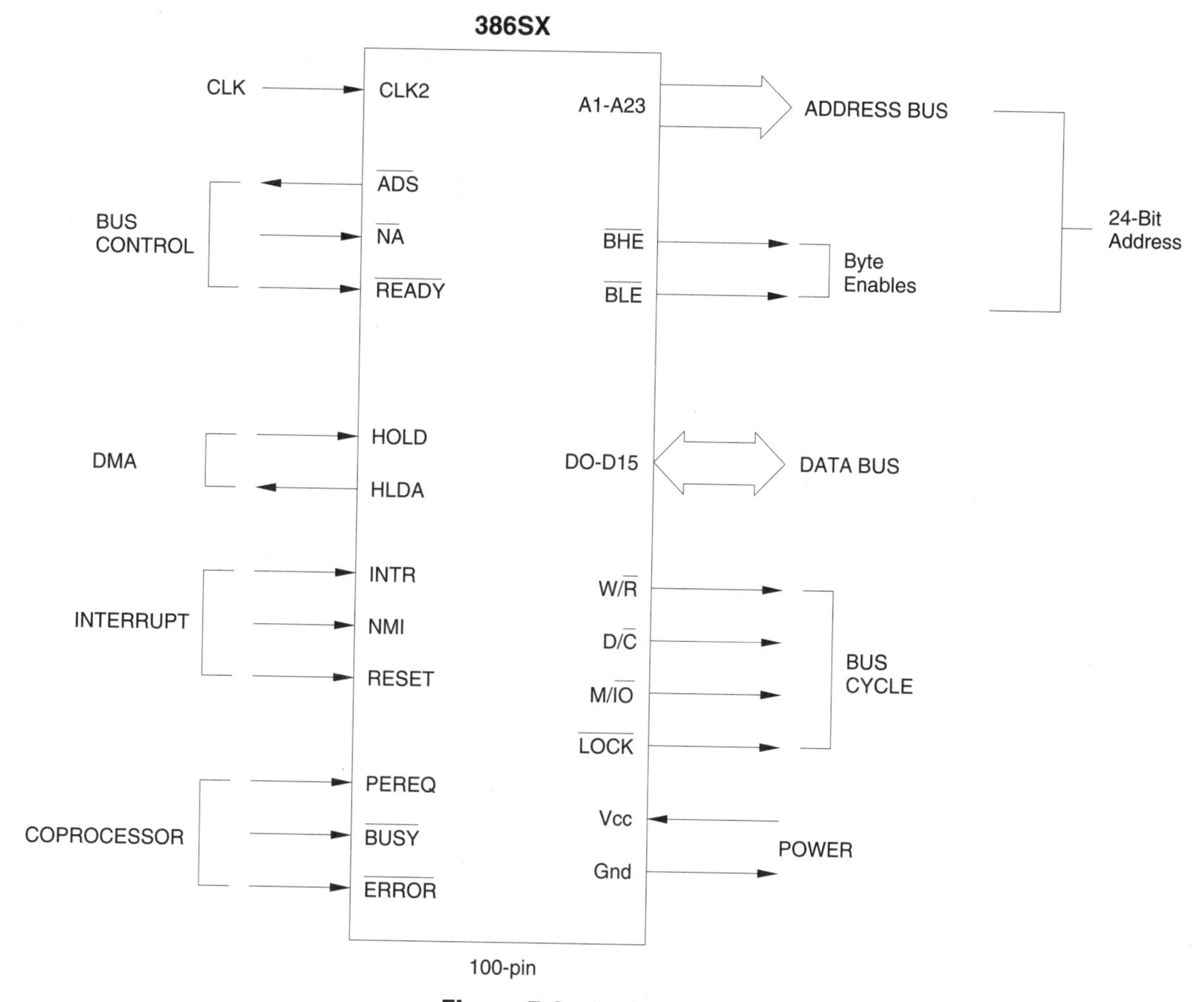

Figure 5-6 386SX pin groupings.

enable lines to enable odd and even memory to compensate for the problems that would occur with address line A0.

$\overline{BHE}/\overline{BLE}$ — The $\overline{\text{Byte High}}$ and $\overline{\text{Low Enable}}$ lines are both active lows. The levels on these lines are used to determine which block of memory is accessed. The low byte memory block represents even memory locations and the high byte memory block represents odd memory locations. When $\overline{BLE}$ is low, the memory at even addresses is enabled and the data flow on D0 to D7. When $\overline{BHE}$ is low, the memory at odd addresses is enabled and the data flow on D8 to D15. When both $\overline{BHE}$ and $\overline{BLE}$ are low, both odd and even addresses are accessed at the same time, allowing data to flow on all data lines (D0 to D15) (Table 5-4).

A2–A31(DX) — There are 30 actual address output lines for the 386DX. Address lines A0 and A1 are not developed by the 386DX directly. Instead the levels on A0 and A1 are developed by the levels on $\overline{BE3}$, $\overline{BE2}$, $\overline{BE1}$, and $\overline{BE0}$ lines, which are all active lows. The levels on the byte enable lines determine which block of memory is enabled. By using the byte enable lines, the 386DX can transfer one byte on any one of its four sets of data lines. The byte enable lines also allow the 386DX to transfer data

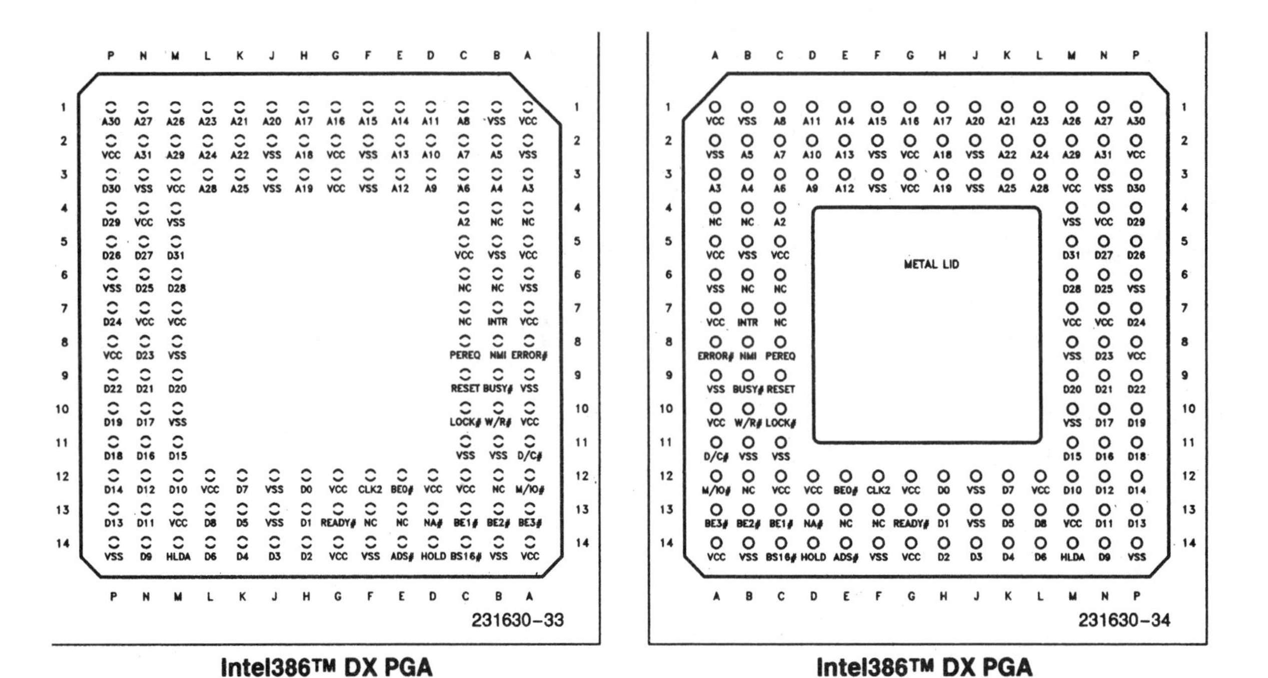

Intel386™ DX PGA Pinout—View from Top Side

Intel386™ DX PGA Pinout—View from Pin Side

Intel386™ DX PGA Pinout—Functional Grouping

Signal	Pin	Signal	Pin	Signal	Pin	Signal	Pin	Signal	Pin	Signal	Pin
A2	C4	A24	L2	D6	L14	D28	M6	V_{CC}	C12	V_{SS}	F2
A3	A3	A25	K3	D7	K12	D29	P4		D12		F3
A4	B3	A26	M1	D8	L13	D30	P3		G2		F14
A5	B2	A27	N1	D9	N14	D31	M5		G3		J2
A6	C3	A28	L3	D10	M12	D/C#	A11		G12		J3
A7	C2	A29	M2	D11	N13	ERROR#	A8		G14		J12
A8	C1	A30	P1	D12	N12	HLDA	M14		L12		J13
A9	D3	A31	N2	D13	P13	HOLD	D14		M3		M4
A10	D2	ADS#	E14	D14	P12	INTR	B7		M7		M8
A11	D1	BE0#	E12	D15	M11	LOCK#	C10		M13		M10
A12	E3	BE1#	C13	D16	N11	M/IO#	A12		N4		N3
A13	E2	BE2#	B13	D17	N10	NA#	D13		N7		P6
A14	E1	BE3#	A13	D18	P11	NMI	B8		P2		P14
A15	F1	BS16#	C14	D19	P10	PEREQ	C8		P8	W/R#	B10
A16	G1	BUSY#	B9	D20	M9	READY#	G13	V_{SS}	A2	N.C.	A4
A17	H1	CLK2	F12	D21	N9	RESET	C9		A6		B4
A18	H2	D0	H12	D22	P9	V_{CC}	A1		A9		B6
A19	H3	D1	H13	D23	N8		A5		B1		B12
A20	J1	D2	H14	D24	P7		A7		B5		C6
A21	K1	D3	J14	D25	N6		A10		B11		C7
A22	K2	D4	K14	D26	P5		A14		B14		E13
A23	L1	D5	K13	D27	N5		C5		C11		F13

Figure 5-7 386DX MPU PGA pinout and table. (Courtesy of Intel)

in a word (16-bit) format. Either the low or high word can be transferred at one time, and the byte enable lines can allow a long word (32 bits) of data to be transferred at one time. The 386DX allows these different types of transfers to take place without wasting memory locations.

$\overline{BE0}$–$\overline{BE3}$ (DX) There are four byte enable lines for the 386DX, all of which are active lows. The levels on these lines are used to determine which block of memory is accessed. The first block of memory (block 1) is enabled if the address levels on A0 = 0 and A1 = 0, if they are used. Block 2 is enabled if the address levels on A0 = 1 and A1 = 0, if they are used. Block 3 is enabled if the address levels on A0 = 0 and A1 = 1, if they are used. Block 4 is enabled if the address levels on A0 = 1 and A1 = 1,

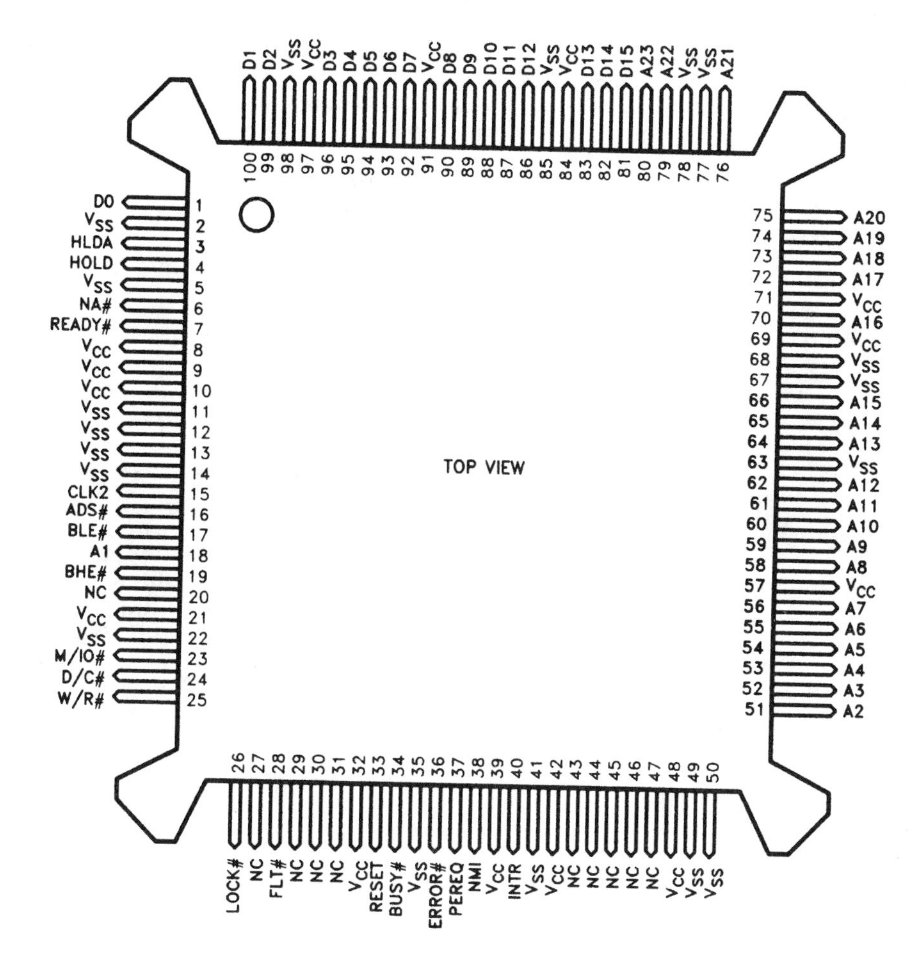

NOTE:
NC = No Connect

Intel386™ SX Microprocessor Pin out Top View

Alphabetical Pin Assignments

Address		Data		Control		N/C	V_{CC}	V_{SS}
A_1	18	D_0	1	ADS#	16	20	8	2
A_2	51	D_1	100	BHE#	19	27	9	5
A_3	52	D_2	99	BLE#	17	29	10	11
A_4	53	D_3	96	BUSY#	34	30	21	12
A_5	54	D_4	95	CLK2	15	31	32	13
A_6	55	D_5	94	D/C#	24	43	39	14
A_7	56	D_6	93	ERROR#	36	44	42	22
A_8	58	D_7	92	FLT#	28	45	48	35
A_9	59	D_8	90	HLDA	3	46	57	41
A_{10}	60	D_9	89	HOLD	4	47	69	49
A_{11}	61	D_{10}	88	INTR	40		71	50
A_{12}	62	D_{11}	87	LOCK#	26		84	63
A_{13}	64	D_{12}	86	M/IO#	23		91	67
A_{14}	65	D_{13}	83	NA#	6		97	68
A_{15}	66	D_{14}	82	NMI	38			77
A_{16}	70	D_{15}	81	PEREQ	37			78
A_{17}	72			READY#	7			85
A_{18}	73			RESET	33			98
A_{19}	74			W/R#	25			
A_{20}	75							
A_{21}	76							
A_{22}	79							
A_{23}	80							

Figure 5-8 386SX MPU PGA pinout view and table. (Courtesy of Intel)

TABLE 5-4
80386 SX Byte Enable Table

$\overline{BHE}$	$\overline{BHL}$	*Size of Transfer*	*Memory Block*
0	0	D0 to D15 word	Odd and even
0	1	D8 to D15 high byte	Odd
1	0	D0 to D7 low byte	Even

TABLE 5-5
80386 DX Byte Enable Table

$\overline{BE3}$	$\overline{BE2}$	$\overline{BE1}$	$\overline{BE0}$	*Size of Transfer*	*Memory Block*
1	1	1	0	D0 to D7 byte (1) (LSB)	1
1	1	0	1	D8 to D15 byte (2)	2
1	0	1	1	D16 to D23 byte (3)	3
0	1	1	1	D24 to D31 byte (4) (MSB)	4
1	1	0	0	D0 to D15 word (low)	1 and 2
0	0	1	1	D16 to D31 word (high)	3 and 4
0	0	0	0	D0 to D31 long word	1, 2, 3, and 4

if they are used. Table 5-5 lists all the valid levels on the byte enable line and how they affect memory access.

5.4.3 Data Bus

D0–D15 (SX) The 386SX contains 16 bidirectional data lines. These lines act as outputs during a write operation and as inputs during a read operation. During a byte operation either the D0 to D7 or D8 to D15 lines are enabled. During a word operation all lines are enabled.

D0 to D31 (DX) The 386DX contains 32 bidirectional data lines. These lines act as outputs during a write operation and as inputs during a read operation. During a byte operation either the D0 to D7, D8 to D15, D16 to D23, or D24 to D31 lines are enabled. During a word operation either the D0 to D15 or D16 to D31 lines are enabled. During a long word (also called a double word) all data lines D0 to D31 are enabled.

5.4.4 Bus Cycle (in the XT known as the status lines)

W/$\overline{R}$ (DX/SX) The Write and $\overline{\text{Read}}$ line is used to determine if some type of write or read bus cycle is to be performed. When high, a write operation is indicated, which means the data lines enabled are used as outputs. When low, a read operation is indicated, and the enabled data lines act as inputs. This output is used with other outputs to inform the bus controller which type of bus cycle is to be performed.

D/$\overline{C}$ (DX/SX) The Data/$\overline{\text{Control}}$ line is used to determine if a data transfer or some type of control cycle is to be performed. When high the transfer contains data, and when low some form of control bus cycle is performed. This output is used with other outputs to inform the bus controller which type of bus cycle is to be performed.

TABLE 5-6
80386 Bus Cycle Table

M/$\overline{IO}$	*D/$\overline{C}$*	*W/$\overline{R}$*	*Bus Cycle*
0	0	0	Interrupt acknowledge
0	1	0	I/O read
0	1	1	I/O write
1	0	0	Memory code read
1	0	1	Halt/shutdown
1	1	0	Memory read
1	1	1	Memory write

M/$\overline{\text{IO}}$ (DX/SX) The memory/$\overline{\text{I/O}}$ line is used to determine if the bus operation uses a memory location or an I/O device. When high a memory operation is indicated, and when low an I/O operation is indicated. This output is used with other outputs to inform the bus controller which type of bus cycle is to be performed. Table 5-6 lists valid bus cycles available with the 386DX and SX MPUs.

$\overline{\text{LOCK}}$ (DX/SX) The $\overline{\text{LOCK}}$ output line is used to lock out the use of the bus by the DMA during an interrupt sequence or the time the lock instruction is used in a program.

5.4.5 Bus Control

$\overline{\text{READY}}$ (DX/SX) The $\overline{\text{READY}}$ line is an active-low input that is used to terminate the current bus cycle in the MPU when it goes low. The $\overline{\text{READY}}$ line goes high at the beginning of a bus cycle and goes low when the cycle is ready to end. During a DMA cycle the $\overline{\text{READY}}$ line is ignored. This line is used to insert wait states in a bus cycle when slow devices are being accessed.

$\overline{\text{ADS}}$ (DX/SX) The $\overline{\text{ADdress Status}}$ line goes low to indicate that a valid address is on the byte enable lines, address bus, and W/$\overline{\text{R}}$, D/$\overline{\text{C}}$, and M/$\overline{\text{IO}}$ output lines during the current bus cycle.

$\overline{\text{NA}}$ (DX/SX) The $\overline{\text{Next Address}}$ line goes low to indicate that the system is prepared to accept new values for the byte enable lines, address bus, and W/$\overline{\text{R}}$, D/$\overline{\text{C}}$, and M/$\overline{\text{IO}}$ output lines, even if the current bus cycle is not complete. This line, along with the address status line, is used to allow pipelining of instructions on the bus. This means that the control signals for the next bus cycle become valid before the current bus cycle ends.

$\overline{\text{BS16}}$ (DX) The $\overline{\text{Bus Size}}$ input line is used to allow the 386DX to run multiple bus cycles to complete a request from devices that cannot provide or accept 32-bit data in a single bus cycle.

5.4.6 Interrupt Bus

INTR (DX/SX) The INTerrupt Request (maskable interrupt) input informs the MPU that the programmable interrupt controller has requested the servicing of an interrupt, when this line goes high. The MPU samples this input at the beginning of each bus cycle. This line must be high for two processor cycles for the interrupt sequence to start. This must remain high until the end of the first of the two interrupt acknowledge signals issued by the MPU.

NMI (DX/SX) — The NonMaskable Interrupt input, when high, indicates a parity error. This line causes an internal interrupt vector code of 2. This line must be valid for two processor cycles before the interrupt is acknowledged and must remain high for two additional processor cycles once acknowledged.

RESET (DX/SX) — The RESET input is used to reset the MPU to its starting value when this goes low for at least 16 system clock cycles. This input is controlled by the clock generator and normally indicates a problem with the power supply.

5.4.7 DMA Bus

HOLD (DX/SX) — The bus HOLD request is used by the DMA and its support circuitry to request control of the bus. When the DMA wants to use the bus, this line goes high. The high requests permission to use the bus. Before the MPU gives control of the bus to the DMA, the current bus cycle must be completed. This input must remain valid throughout the DMA bus cycle; otherwise permission is taken away by the MPU.

HLDA (DX/SX) — The bus HoLD Acknowledge is the output that the MPU uses to grant permission to the DMA to control the bus, once the MPU has disabled its control of the bus. The MPU, through the bus controller and control logic, disables the address latches and the read/write lines controlled by the bus controller. Once high, the DMA controls the address, data, and control buses.

5.4.8 Coprocessor

PEREQ (DX/SX) — The Processor Extension operand REQuest input line informs the MPU of a data operand transfer during a math coprocessor operation. When high, the MPU transfers data from memory to the coprocessor or from the coprocessor into memory. The MPU controls the flow of data based on the math instruction.

$\overline{\text{BUSY}}$ (DX/SX) — The $\overline{\text{BUSY}}$ input is controlled by the coprocessor (80387). When low, this line causes the MPU to wait until the coprocessor is ready for more data. While the coprocessor is busy, the MPU goes into a wait state.

$\overline{\text{ERROR}}$ (DX/SX) — The $\overline{\text{ERROR}}$ input is also controlled by the coprocessor. When low, this line causes the MPU to perform an internal processor extension interrupt when an error occurs during a math operation being performed by the coprocessor.

5.4.9 Power Bus

Vss (DX/SX) — The Vss is connected to ground. This line provides a path current to flow into the MPU and also acts as the reference point of the MPU.

Vcc (DX/SX) — The Vcc line is connected to a +5-V supply. This line provides the voltage potential for the MPU.

5.5 80387DX/SX NUMERIC COPROCESSOR

The 80386SX system can use either the 80287 or the 80387SX 80-bit numeric coprocessor, whereas the 80386DX uses the 80387DX. The advantages and disadvantages of these coprocessors are the same as those for the 286 system. The operation of the 387DX (Fig. 5-9) and 387SX (Fig. 5-10) is basically the same except that the DX version has a 32-bit data bus while the SX version has a 16-bit address bus to match the 386SX.

CPUCLK2 (DX/SX) — The CPU CLocK 2 input allows the 387 to synchronize its operations with the 386 when the 387 is enabled in the synchronous mode. When the 387 uses this input, the data and handshaking lines between the 387 and 386 operate asynchronously, but bus operations still occur synchronously. This clock input is the same as that supplied to the 386. Internally this signal is divided by 2 to produce the processor clock for the 387.

NUMCLK2 (DX/SX) — The NUMeric CLocK 2 input allows the 387 to synchronize its operations with the 386 when the 387 is enabled in the asynchronous mode. When the 387 uses this input, the data and handshaking interface lines between the 387 and 386 operate asynchronously, but bus operations still occur synchronously. The frequency of this input must be between 62.5% and 71.4% slower than the CPUCLK2 input. When the 387 is in the synchronous mode, the effect that this input has on the 387 is disabled. Internally this signal is divided by 2 to produce the processor clock for the 387 when used.

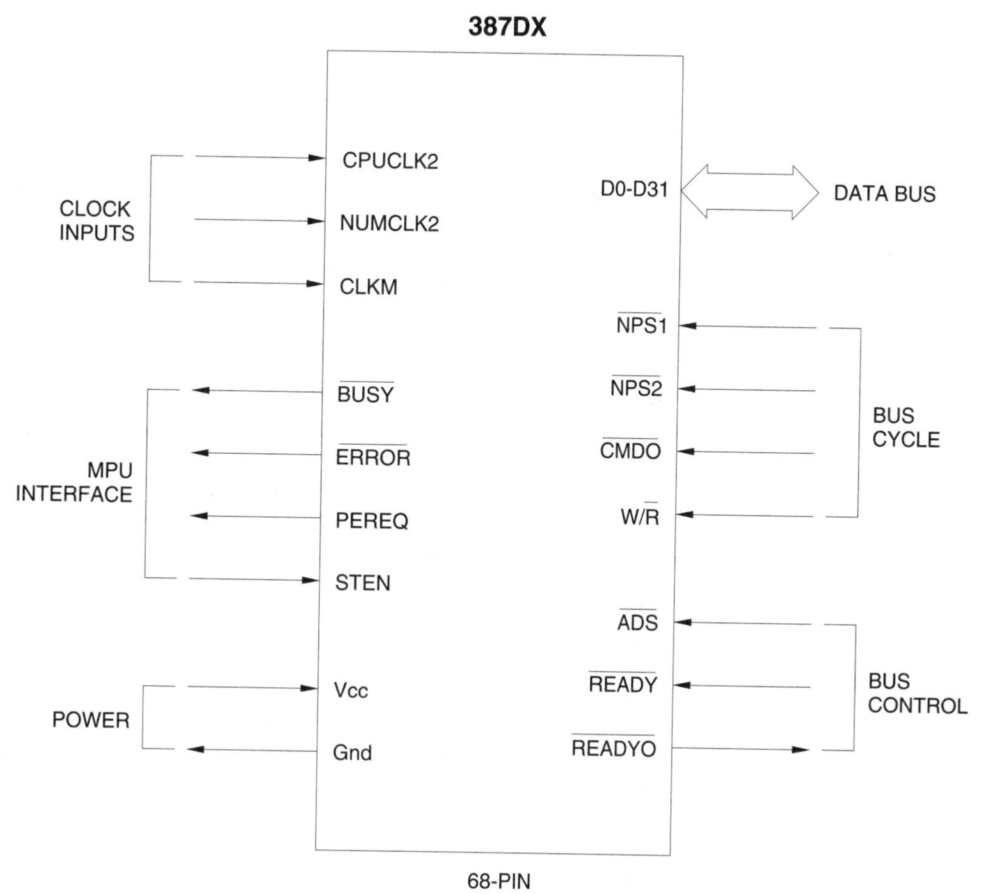

Figure 5-9 387DX pin groupings.

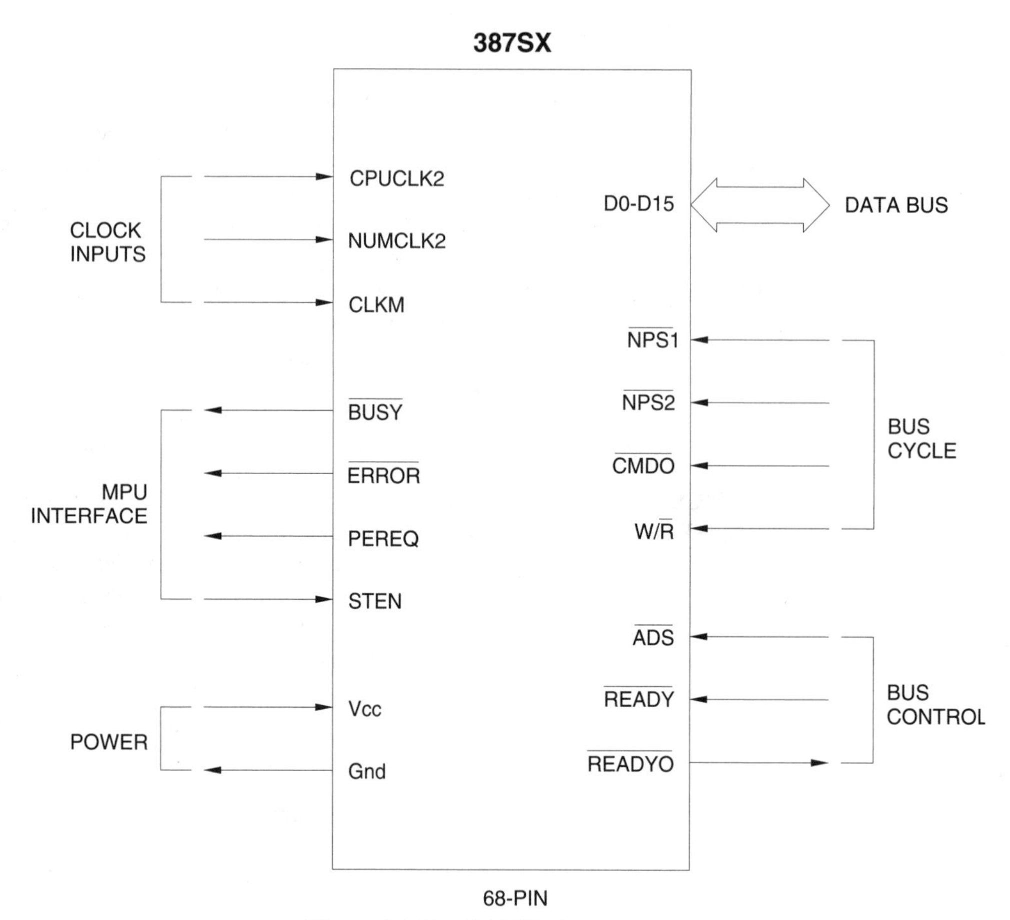

Figure 5-10 387SX pin groupings.

CLKM (DX/SX) — The CLocK Mode input is used to select whether the 387 uses the CPUCLK2 (synchronous clock) or the NUMCLK2 (asynchronous clock) input to control the data and handshaking interface lines between the 387 and the 386. When high the 387 uses the synchronous clock input, and when low the asynchronous clock input is used.

RESETIN (DX/SX) — The RESET INput is an active-high input. On the rising edge of this pin, the 387 terminates activity and goes into a dormant (wait) state. A valid reset does not occur until this pin stays high for a minimum of 40 NUMCLK2 (asynchronous) clocks. The operation of the 387 does not go out of reset until it is low for a minimum of 50 NUMCLK2 clocks.

D0–D31 (DX) — The Data inputs/outputs are used to allow data to enter and exit the 387, depending on the operation and the communications with the 386DX.

D0–D15 (SX) — The Data inputs/outputs are used to allow data to enter and exit the 387, depending on the operation and the communications with the 386SX.

$\overline{\text{BUSY}}$ (DX/SX) — The $\overline{\text{BUSY}}$ output, when low, is used to inform the 386 that the 387 is busy processing information. The 386 does not send any data to the 387 until the busy line goes high again.

$\overline{\text{ERROR}}$ (DX/SX) — The $\overline{\text{ERROR}}$ output line, when low, indicates that an unmasked error has occurred during the processing of the instruction.

TABLE 5-7
80387 Enable Table

STEN	*$\overline{\text{NPS1}}$*	*NPS2*	*387*
1	0	1	Enabled
0	X	X	Disabled
X	1	X	Disabled
X	X	0	Disabled

PEREQ (DX/SX) The Processor Extension data channel operand transfer REQuest output, when high, is used to signal the 386 that the 387 is ready to transfer its data. Once the data is transferred, this line resets.

W/$\overline{\text{R}}$ (DX/SX) The Write/Read input is used to indicate to the 387 that a read or write cycle is to be processed. This line has an effect on the 387 only when it is enabled by the proper lines. When the 387 is enabled, a high on this line indicates a write operation and a low indicates a read. This line basically takes the place of the $\overline{\text{NPRD}}$ and $\overline{\text{NPWR}}$ lines in the 287. This line is connected to the W/$\overline{\text{R}}$ line of the 386.

$\overline{\text{ADS}}$ (DX/SX) The ADdress Strobe line, when low, is used to help determine when a bus cycle begins and when the 387 should check the chip select and W/$\overline{\text{R}}$ lines. This line is connected to the $\overline{\text{ADS}}$ ouput of the 386.

$\overline{\text{READY}}$ (DX/SX) The bus READY line performs the same function as the ready input of the 386. When low, the bus cycle is ready to end. This line is used to insert wait states into the bus cycle. The $\overline{\text{READY}}$ line is connected to the same $\overline{\text{READY}}$ line as the 386.

$\overline{\text{READYO}}$ (DX/SX) The Optional bus READY line performs the same function as the ready input of the 386. If no additional wait clocks are to be applied to the 387, this line is connected to the $\overline{\text{READY}}$ line. If additional wait clocks are to be applied to the 387 during a 387 operation, external circuitry is required to generate the level on this line. This line is an active low and has an effect on the 387 only during a 387 operation.

STEN (DX/SX) The STatus ENable line, when high, is used along with other inputs to help enable the 387. When low, the handshaking interface ($\overline{\text{BUSY}}$, $\overline{\text{ERROR}}$, and PEREQ), $\overline{\text{READYO}}$, and data lines go into tristate. This line may either be tied high through a pull-up resistor or be tied to a line that is used to test for the 387 (Table 5-7).

$\overline{\text{NPS1}}$ (DX/SX) The Numeric Processor Select line 1 is used with the levels on STEN and NPS2 to enable the 387. This line is an active low that is connected to the M/$\overline{\text{IO}}$ line of the 386 so that it is low only during an I/O operation.

NPS2 (DX/SX) The Numeric Processor Select line 2 is used with the levels on STEN and $\overline{\text{NPS1}}$ to enable the 387. This line is an active high that is connected to address line A31(DX) or A15(SX) so that it is high during I/O addresses for the 387.

$\overline{\text{CMD0}}$ (DX/SX) The CoMmand Data line 0 is used to determine if the 387 operation is a command or some form of data. If low during a read the control register outputs its data; if high the data register outputs its data. If low during a write cycle the 387 accepts the data on the data bus as the instruction (opcode); if high the 387 accepts the data as the data for the instruction. This line is connected to address line A2 of the 386.

5.6 386 SYSTEM BOARDS

Since the development of the 386, many companies have started manufacturing AT chip sets to reduce the number of ICs needed to support AT-level MPUs. Intel is one of the biggest manufacturers, and Chips and Technologies, Opti Inc., and VLSI are the most popular. There are more than a dozen other companies, along with motherboard manufacturers themselves. Because there are so many different companies and AT chip sets on the market and the number changes rapidly, the remaining part of this book on AT system boards on focuses on block diagrams and troubleshooting from a field view.

In most cases a technician needs to use the basic concepts presented in Chapters 3 and 4 and apply them to more advanced systems. Because schematics and pinouts are very difficult to find, we focus on some basic checks with test equipment, diagnostic software, and POST card testing.

In Chapter 6 I/O bus structures other than the AT ISA are discussed. When the 486 hit the market, the AT ISA bus was found to be too slow and cumbersome, although some of the advanced bus structures do use 286 and 386 MPUs.

5.6.1 386SX System Board

82335 Interface Device for 386SX Microprocessor–Based AT System. The 82335 is used in conjunction with the 82230/82231 AT chip set to complete an AT microprocessor system (Fig. 5-11). The main purpose of this IC is to convert the signals to and from the 386SX into signals needed by the 82230/82231 AT chip set. This IC provides DRAM address decoding, parity checking and generation, memory timing, modifications of coprocessor handshaking interface signals between the 387SX and 386SX, and the bus ready signal. By using the 82335 the number of support ICs other than the AT chip set is reduced, making the system board easier to design, more reliable, and more cost effective. The following describes the function of each section block in the block diagram of the 82335 (Fig. 5-12).

82335 Clock Generator and Reset Synchronizer. This section creates the system clocks (CLK2 and $\overline{\text{PCLK}}$) for the 386SX, 387SX, and AT chip set so that all operations are synchronized with the 82335. The 32-MHz frequency is supplied to the EFI input from a TTL-compatible oscillator. The CLK2 line supplies the clock frequency to the MPU and coprocessor, while the $\overline{\text{PCLK}}$ (one-half of the EFI frequency) is supplied to the AT chip set for synchronization. This section also modifies the reset (SYSRESET and CPURESET) output signals from the 82330 to create the reset for the MPU (RESETSX) and math coprocessor (RESETNPX).

82335 Parity Generator and Checker. The parity generator and checker section of the 82335 is responsible for generating high and low parity bits during a write operation. The parity bits are placed directly into the selected DRAM bank and do not pass through the memory data bus transceiver. During a read operation the parity is checked just as in the XT system except that two parity bits are used (high bit for the high byte and low bit for the low byte of data). Parity checking can be disabled through programming. If a parity error occurs at the end of a read operation, a nonmaskable interrupt is generated.

82335 Address Mapper and Decoder. The address mapper and decoder section is programmable. This section allows the BIOS program stored in either ROM or EPROM to be transferred into faster DRAM, which increases the speed of the computer because the interrupt service routines can be accessed faster. The 82335 also allows the contents of video and other adapter ROMs to be placed into DRAM, which makes those operations run much faster. This section also produces the address decoding for the ROM section.

Also included in this section is circuitry responsible for roll address mapping. Roll address mapping allows the 82335 to remap the memory locations in the upper 512K (four 128K)

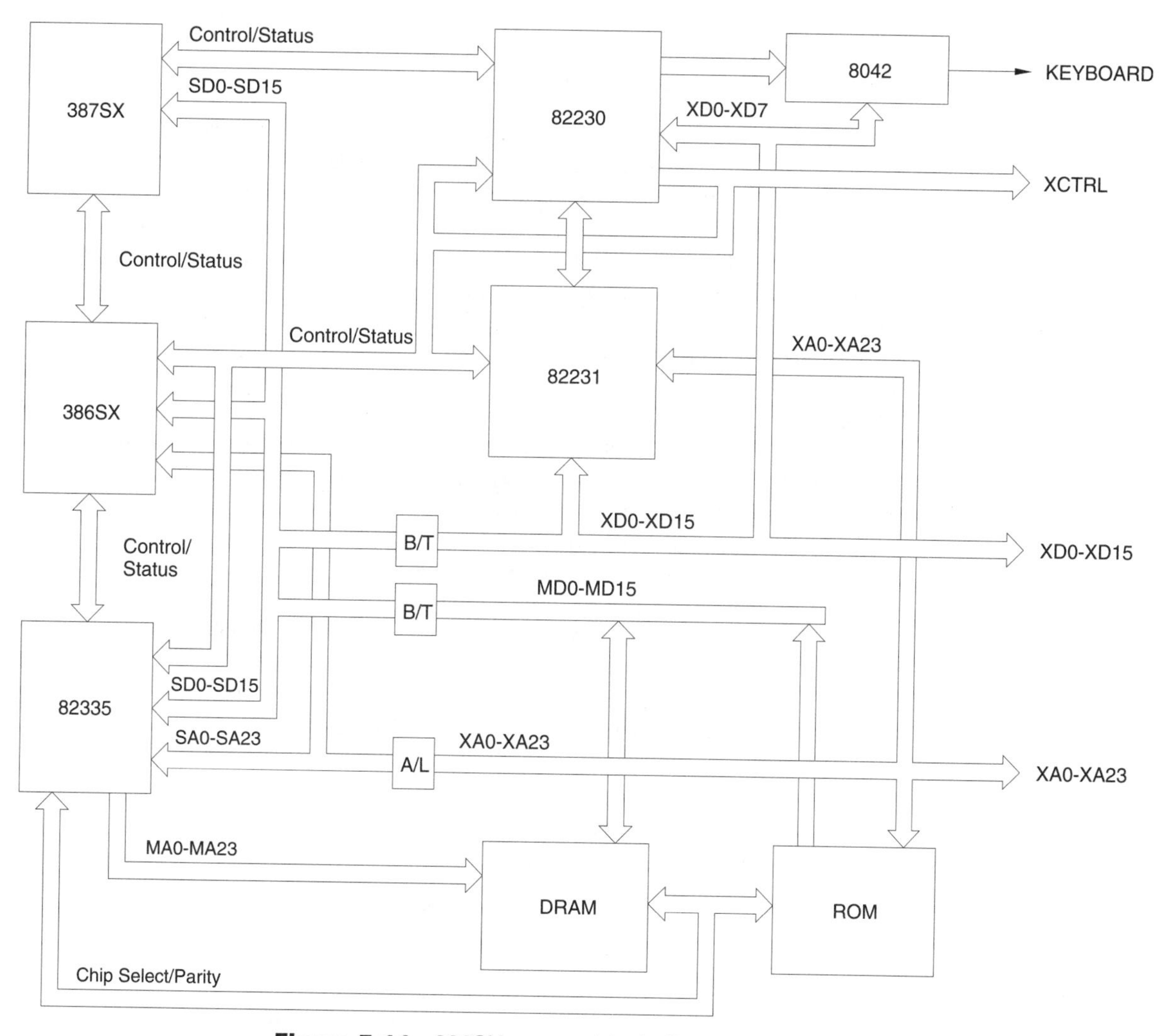

Figure 5-11 386SX system block diagram with 82335.

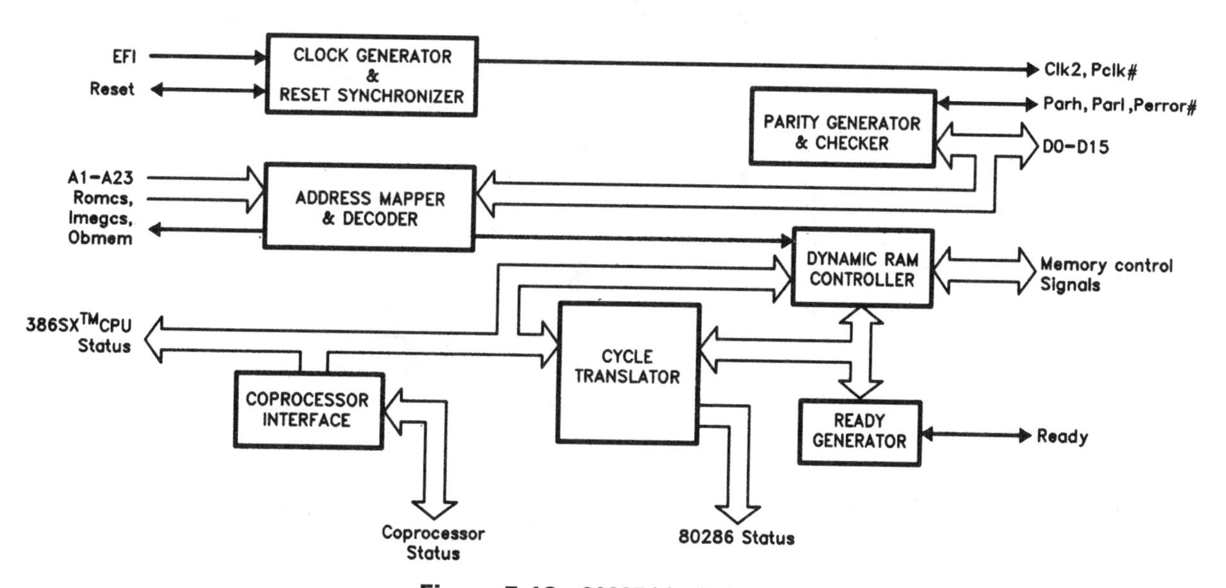

Figure 5-12 82335 block diagram. (Courtesy of Intel)

of the first 1MB of memory and locate at the top of physical memory, thus allowing the memory to be used by the system.

82335 DRAM Controller Section. The DRAM controller section of the 82335 controls the access of up to four 16-bit banks of DRAM. Each 16-bit bank is divided into two 8-bit banks. One 8-bit bank supplies the low byte for the data bus and the other the high byte. Besides the 8 bits per bank, each bank contains one additional bit, called the parity bit, which is not data but is used to determine if the data stored is correct. The size of the DRAMs can be either 256K or 1MB, allowing up to 16MB of addressable DRAM space on the system board. The amount of memory is determined by an autoscan routine located in BIOS, which programs the 82335 during boot-up. Memory must be in groups of two 9-bit DRAMs to provide the 16 bits necessary for the data bus.

The DRAM (Fig. 5-13) controller controls the row and column address lines, address multiplexing for the DRAM section, the write enable level, parity generation and checking, and the direction and enabling signals for the data bus transceiver in the DRAM section. This controller also supports page mode and page mode interleaving addressing modes, along with four different DRAM modes for different access times. The addressing and operational modes of the 82335 are all programmable and are programmed during boot-up, depending on the information supplied by BIOS and CMOS set-up data.

Regardless of which addressing and operational mode the 82335 is using, the 82335 performs the DRAM address decoding, which generates the $\overline{\text{RASx}}$ (0 to 3), $\overline{\text{CASHx}}$ (high byte 0 to 3), and $\overline{\text{CASLx}}$ (low byte 0 to 3) lines. The 82335 also supplies the level to the $\overline{\text{WE}}$ line, multiplexes the A1 to A23 system address bus, and creates the levels on the MA0 to MA9 memory address bus.

Addressing DRAM Non-Page Mode. Normally during the addressing of DRAM, the system address bus is multiplexed into two groups of bits. If a 1MB DRAM is used, the DRAM address decoding circuitry places logic levels on the system address lines A1 to A10 onto the DRAM memory address lines MA0 to MA9. While the levels are constant, the address-decoding section causes the proper row address strobe line to go low. The low on the

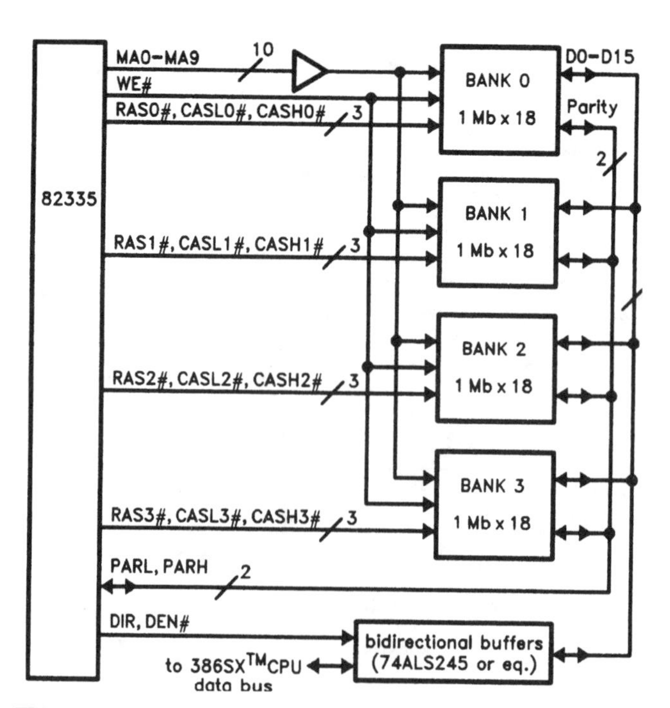

Figure 5-13 82335 to DRAM interface. (Courtesy of Intel)

$\overline{\text{RASx}}$ line causes the logic levels on MA0 to MA9 to be stored in the selected bank of DRAM (the 18 DRAMs that make up the bank); this selects the row in the selected bank of DRAM, referred to as the row part of the address. The small letter x specifies the value for the selected bank of memory. The $\overline{\text{BLE}}$ and $\overline{\text{BHE}}$ lines are used to enable the byte of data to be used. When the $\overline{\text{BLE}}$ line is low, the address is even, and when the $\overline{\text{BHE}}$ line is low the address is odd. When both inputs are low, two bytes of data are transferred (Table 5-8).

Once the row part of the address is strobed into the DRAMs, the address-decoding section causes the address multiplexers in the DRAM section to place the logic levels on system address lines A11 to A21 onto the DRAM memory address lines MA0 to MA9. While the levels are constant, the address-decoding section causes the proper column address strobe line to go low. The low on the $\overline{\text{CASx}}$ line enables the DRAMs in the selected bank or banks (depending on the size of the data transfer), and the transfer takes place. If the write enable line is low, the DRAMs accept the data on the data bus. If the write enable line is high, the DRAMs output their data onto the data bus.

Once the transfer has taken place, the operation ends when the $\overline{\text{RASx}}$ line goes high. In non–page mode addressing the $\overline{\text{CASx}}$ line also goes high. This cycle repeats for each new address. The disadvantage of non–page mode addressing is that both the row and column address strobe lines must cycle each time a new word address is selected. In faster MPUs—386SX (Fig. 5-14) and higher—this causes a minimum of two wait states during a normal memory bus cycle.

Page Mode Addressing. In page mode addressing (Fig. 5-14) a new row address is supplied to the DRAMs only when the row address is different from the last row address latched into the DRAMs. The column address strobed into the DRAMs is the same. This reduces the number of wait states introduced into the bus cycle, thus speeding up memory transfers (Table 5-9).

During page mode addressing, the 82335 places the logic levels on the system address lines A11 to A21 onto the DRAM memory address lines MA0 to MA9. *Note:* In non–page mode addressing these levels would represent the column part of the address; but because the levels on these lines change the least, they are used as the row part of the address during page mode addressing. While these levels are constant, the 82335 causes the proper row address strobe line to go low. The low on the $\overline{\text{RASx}}$ line causes the logic levels on MA0 to MA9 to be stored in the selected bank of DRAMs. This process selects the row in the selected bank of DRAMs and is referred to as the row part of the address. The $\overline{\text{BLE}}$ and $\overline{\text{BHE}}$ lines are used to specify the byte of data to be used. When the $\overline{\text{BLE}}$ line is low the address is even, and when the $\overline{\text{BHE}}$ line is low the address is odd. When both inputs are low, two bytes of data are transferred. Each row address is valid for about 1024 word address locations.

TABLE 5-8
DRAM Non-Page Mode Addressing

	Bank 0		*Bank 1*	
Row Add	*HI Byte*	*LO Byte*	*HI Byte*	*LO Byte*
000h	000001h	000000h	100001h	100000h
001h	000003h	000002h	100003h	100001h
002h	000005h	000004h	100005h	100002h
003h	000007h	000006h	100007h	100003h
004h	000009h	000008h	100009h	100004h
005h	00000Bh	00000Ah	10000Bh	100005h
006h	00000Dh	00000Ch	10000Dh	100006h
007h	00000Fh	00000Eh	10000Fh	100007h
008h	000011h	000010h	100011h	100010h
009h	000013h	000012h	100013h	100012h

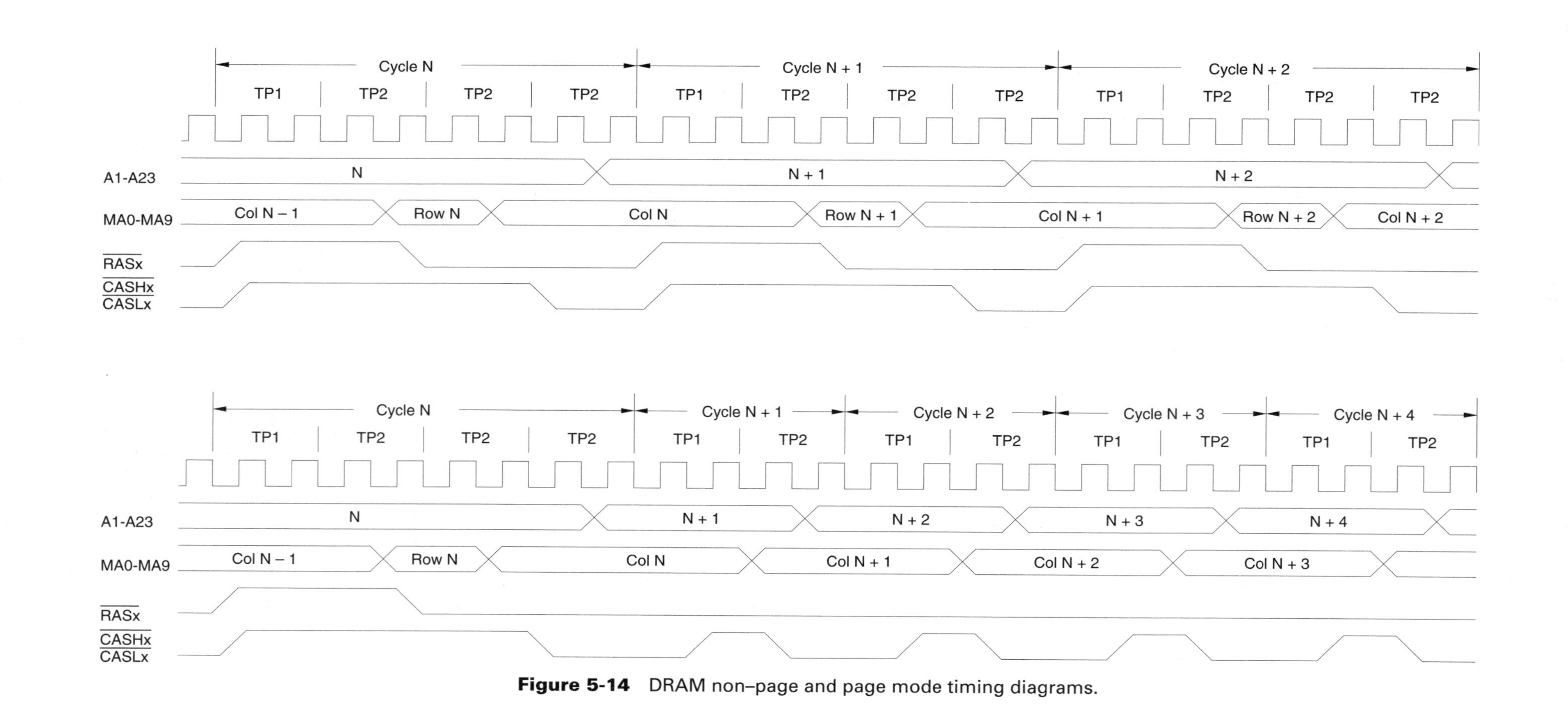

Figure 5-14 DRAM non–page and page mode timing diagrams.

TABLE 5-9
DRAM Page Mode Addressing

	Bank 0		*Bank 1*	
Row Add	*HI Byte*	*LO Byte*	*HI Byte*	*LO Byte*
000h	000001h	000000h	100001h	100000h
000h	000003h	000002h	100003h	100001h
000h	000005h	000004h	100005h	100002h
000h	000007h	000006h	100007h	100003h
000h	000009h	000008h	100009h	100004h
000h	00000Bh	00000Ah	10000Bh	100005h
000h	00000Dh	00000Ch	10000Dh	100006h
000h	00000Fh	00000Eh	10000Fh	100007h
000h	000011h	000010h	100011h	100010h
000h	000013h	000012h	100013h	100012h

Once the row part of the address is strobed into the DRAM, the 82335 places the logic levels on system address lines A1 to A11 onto the DRAM memory address lines MA0 to MA9. While the levels are constant, the address decoding section causes the proper column address strobe line to go low. The low on the $\overline{\text{CASx}}$ line enables the DRAMs in the selected bank or banks (depending on the size of the data transfer), and the transfer takes place. If the write enable line is low, the DRAMs accept the data on the data bus. If the write enable line is high, the DRAMs output their data onto the data bus.

If the 82335 determines that another memory transfer is to take place and the next address uses the same row as the last memory transfer, the $\overline{\text{CASx}}$ line goes high to end the first transfer. Next the 82335 places the levels on the A1 to A11 system address lines onto the MA0 to MA9 memory address lines and brings the column address strobe line low again, thus selecting a new address. The $\overline{\text{RASx}}$ line remains low. This process continues until a new row address is needed, at which time the $\overline{\text{RASx}}$ line goes high, which ends the page mode address cycle for the row address stored in the banks of DRAM. The $\overline{\text{CASx}}$ also goes high and ends the page mode addressing cycle. The advantage of page mode addressing is that the 82335 has to pulse the $\overline{\text{RASx}}$ line only when a new row address is required. Because each memory cycle does not require both a $\overline{\text{RASx}}$ and $\overline{\text{CASx}}$, no wait clocks are needed except during the first row address cycle, which speeds up memory cycle transfers.

Page Mode Bank Interleave Addressing. Page mode bank interleave addressing is similar to page mode addressing except that instead of the row strobe address line being enabled for only one bank of memory, two or more banks of memory are enabled at one time. If two banks are enabled, the row address is good for twice as many addresses as in page mode addressing. If three banks are enabled, the row address is good for three times as many addresses as in page mode addressing. If all four banks for memory are enabled, the row address is good for four times as many addresses as in page mode addressing. In page mode bank interleave addressing, all DRAMs in all banks must have the same access time.

DRAM Refresh. The DRAM controller section is also responsible for generating and controlling the refresh cycle for the DRAM section. The internal 10-bit address refresh counter increments by 4 after each refresh cycle. The $\overline{\text{RASx}}$ lines do not all go low at the same time; instead they are staggered to lessen the effect of surges during the refresh cycle.

82335 Bus Cycle Translator. This section of the 82335 is responsible for translating 386SX control signals into 286 control signals for the 82330/82331 AT chip set. This section also converts the signals from the 82330/82331 AT chip set into 386SX control signals for I/O access, memory access, interrupt, halt, and shutdown cycles.

82335 Ready Generator. The ready generator modifies the ready signal for the system. The modified ready signal is used for the DRAMs and 387SX math coprocessor.

82335 Math Coprocessor Interface. The math coprocessor interface section is responsible for detecting the presence of the 387SX math coprocessor and for modifying the synchronous signals between the 386SX, 387SX, and 82330.

Intel 82330/82331 AT Chip Set. The 82330 and 82331 ICs that make up the Intel AT chip set still perform most of the MPU support functions for a 386SX (Fig. 5-11) system. Some of the signals are modified from the original source through the 82335 to make them compatible with the signals and timing needed for the 386SX MPU.

The 82230 contains the following circuitry: 82284 clock generator/driver, two 8259A PICs, 6818 real time clock (battery backed-up), 82288 bus controller, address and data bus control logic, bus control logic, CPU shutdown logic, and coprocessor interface logic. See Chapter 4 for a more detailed description of the operation of this IC.

The 82231 contains the following circuitry: read/write logic, I/O chip select, support logic for the PIT and I/O devices, 8254 PIT, DMA memory mapper (74612), two 8237 DMAs, DRAM refresh counter, latch and logic (not used in the SX system), and parity check logic.

DRAM Section. The DRAM section of the computer is actually a DRAM module in the form of a SIMM (Single In-line Memory Module). SIMMs are used in most of today's computer systems because of their small size, low power dissipation, lower cost than DRAMs using DIPs, small amount of board space needed, and ease of upgrading. SIMMs come in three common sizes—256K (256K × 9), 1MB (1MB × 9), and 4MB (4MB × 9). The address decoding and parity generation and checking are performed by the 82335 DRAM controller section.

ROM Section. The 82335 performs the address decoding for the EPROMs in the system and allows remapping of the contents of BIOS into DRAM. The EPROM used for most AT system boards is the 27256 (32K × 8) or 27512 (64K × 8), depending on the size of the BIOS program itself and any utility programs. As in the 286 AT system, BIOS tells the computer how to act like a computer, configures the hardware during boot-up, provides CMOS configuration video screens, a password protection program, a hard drive look-up table for configuring a hard drive, and a hard drive diagnostic and set-up program, and contains POST.

8042 Keyboard Microcontroller. AT system boards do not use the 8255 PPI and support circuitry to decode the equipment DIP switches or convert the serial data from the keyboard into parallel data for the computer. Because there are no DIP switches for equipment, the AT uses data stored in the battery–backed-up CMOS section of the RTC for equipment information. Instead AT system boards use the Intel 8042 universal peripheral interface 8-bit slave microcontroller. The 8042 section on the AT system board is responsible for converting serial keyboard data into parallel data for the data bus. It generates the IRQ1 keyboard interrupt request and is also responsible for controlling the A20GATE line, which is used to switch the 386 between the real and protected operating modes. Depending on the system board, the 8042 is also responsible for locking out the keyboard using a key lock on the front of the computer, switching the speed of the 386 from normal to turbo speed, and sometimes detecting the type of default video monitor (color or mono).

I/O Interface Bus Slots. The I/O interface bus slots are the edge card adapter near the back of the system board. An interface card placed in one of these slots has access to most of the bus signals to control the card. If the card does not need a particular signal, a connection is not made. Most AT system boards contain eight interface slots, normally two 8-bit interface slots (single 62-pin edge connector) and six 16-bit interface slots, which are identified by an additional 36-pin edge connector in line with the larger connector. The signals for each of the slots are connected in parallel; therefore, a 16-bit I/O adapter can be placed in any 16-bit I/O slot and any 8-bit adapter can be placed in any 8-bit or 16-bit slot (but only the 8-bit signals are used).

Troubleshooting the 386SX System. Use the same troubleshooting sequence as in Chapter 4. The frequencies are higher but for the most part the sequences are the same. DRAM operations that use page mode addressing do not produce the $\overline{\text{RASx}}$ pulse for each address. The $\overline{\text{RASx}}$ pulse is produced only when a different row address is required.

5.6.2 386DX System Board

The 386DX system board in Figure 5-15 shows a typical 386DX system. As mentioned before, Intel is one of about a half a dozen manufacturers that make AT chip sets. Most chip sets use the same labeling of signals and timing, but the signal may be located at different pins. The number of chips in the chip set and the amount of external support circuitry that is required may also differ. To troubleshoot a system board to chip set level is nearly impossible without the pinout for the chip. Most troubleshooting on system boards with chip sets involves running diagnostic software or using an emulation board. (An emulation board is a computer board that allows the user to simulate any signal or group of signals in the computer. These signals are applied to the computer being tested to try to identify the problem area.) The other problem often encountered is that most chips in the chip sets are not socked but soldered onto the board. Because these chips are normally surface mounted, most companies do not have the equipment or trained personnel to perform this type of repair. Such a repair is made only when the cost of the board is far greater than the cost in time and equipment to replace the component or components. In general, if the system board costs $200.00 or less, repair may not be cost effective unless this is the main business of the company or the company is an authorized factory service center.

The following pages describe the functions of each chip in the block diagram in Figure 5-15. The example uses the 16-bit AT ISA bus standard. Currently there is no standard 32-bit AT ISA bus, although some system manufacturers include a proprietary 32-bit slot used for one of their proprietary 32-bit memory cards. Proprietary slots work only with the special proprietary adapters supplied by the manufacturer of the system board. In Chapter 6, larger 32-bit bus structures are discussed.

82385 32-Bit Cache Controller. The Intel 82385 (Fig. 5-16) is a 32-bit cache controller that is designed to work with the 80386DX to provide near zero wait memory operations. The main purpose of the chip is to keep a copy of frequently used codes and data in fast cache memory. The cache controller is programmable and allows cache to function with both memory read and write cycles. The 82385 also supports direct and two-way set associative cache mapping. In most cases the 82385 generates 386DX-like signals or controls the choice and timing of 386DX signals to the system bus.

82385 Internal Block Diagram. The 82385 contains four controlling blocks, although each block has its own separate inputs and outputs, the blocks are tied together to control cache memory.

The 82385 local bus interface block is used to control the flow of data, addresses, and cycle definitions for the 82385 local bus. The 82385 local bus separates the 386DX local bus from the system bus.

The processor interface block is used to monitor and control signals between the 386DX and 82385 cache controller. The 82835 monitors some signals while intercepting others. The 82385 produces signals that simulate 386DX signals so that the system bus can operate properly.

The cache directory block contains the directory listings identifying the address locations that have been cached. This block also provides control over bus snooping of other bus

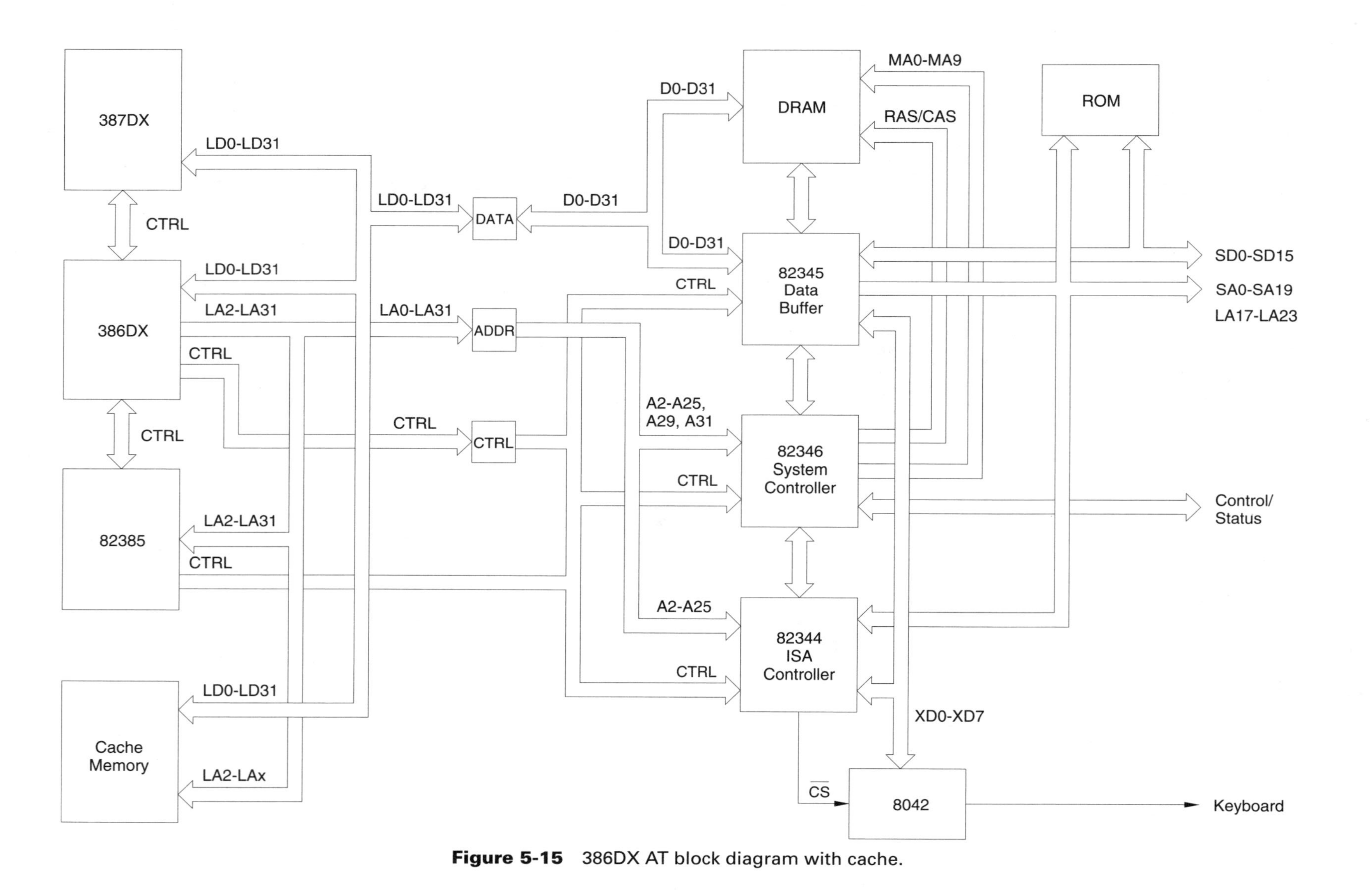

Figure 5-15 386DX AT block diagram with cache.

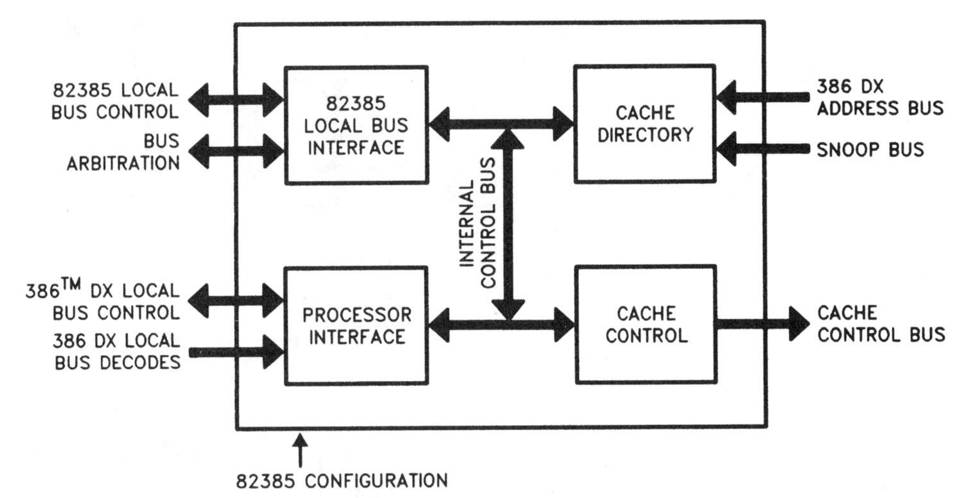

Figure 5-16 82385 internal block diagram. (Courtesy of Intel)

masters. The snooping option allows the 82385 to monitor bus cycles on the system bus that is controlled by a different bus master (DMA). If the other bus master changes the contents of a memory location that has been cached, the 82385 will update the data stored in the cache memory that corresponds to the memory location changed.

The Cache Controller. Before the operation of cache memory is discussed, a brief discussion on programming is needed. A program is a list of instructions that tell the computer how to perform a selected task. To reduce the size of the program and increase its speed, a well-written program often repeats a set of instructions a number of times. These often-used sets of instructions are referred to as routines or subroutines. Most software programs repetitively cycle through a number of routines, depending on input data and conditions that occur during the execution of the program. The fact that these routines reside at the same memory locations during the execution of the program allows the program to operate much faster. During the normal execution of a program, the computer must repeatedly access the same memory locations in order to execute the program routines. If the information that is used repeatedly were made available to the computer faster, the computer could execute the program faster. This is the reason behind cache memory.

The block diagram of the system board example uses a cache controller and static cache RAM, as do most of today's computer systems. The reason for the cache controller and cache memory is that computer systems operating at or above 25 MHz are so fast that DRAMs do not allow the computer to access the information without causing wait states in the bus cycle. The additional wait states increase the bus cycle time, which reduces the number of bus cycles that can be performed in any given time period. This additional time delay in the transfer of data is called bottlenecking, which becomes greater as the speed of the MPU increases. The job of the cache controller is to maintain a copy of frequently used codes and data in cache static RAM (32K to 256K in size with 15 to 25 ns access time) so that the MPU does not always have to access the slower DRAM.

The cache controller, located on the local bus of the MPU (Fig. 5-17), intercepts all command signals between the MPU and bus controller. Depending on the command, the cache controller either passes the command signals to the bus controller or processes the memory request itself. Commands passed to the bus controller deal with interrupts, I/O operations, and noncached memory reads. During memory writes, either the main memory is updated (noncache write) or both main and cache memory is updated (cache write). In either case, the cache controller allows most write operations to occur with no wait states generated. The cache controller monitors the local address bus of the MPU and controls the cache address bus when necessary. The cache controller also controls the transfer of data and addresses between the MPU local bus and the system bus.

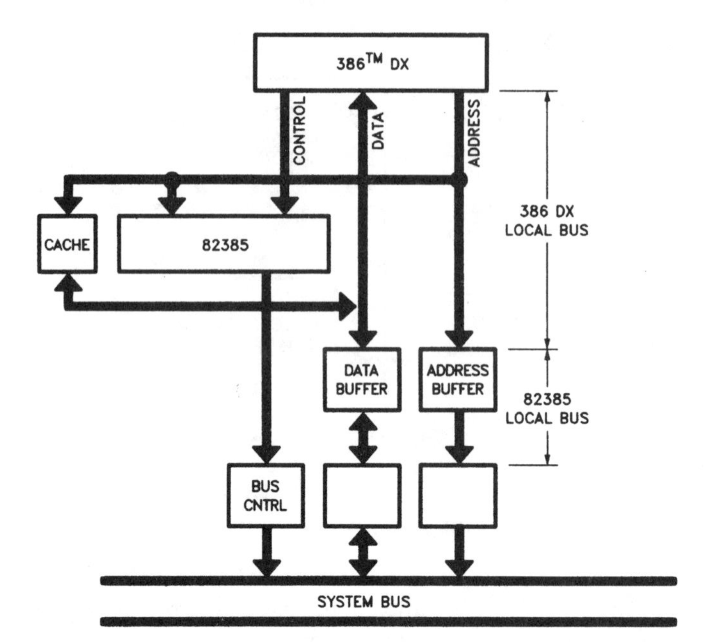

Figure 5-17 82385 system bus structure. (Courtesy of Intel)

Direct Mapped Cache Memory. In direct mapped cache memory, all the cache memory is used as one large block of memory. The 82385 allows up to 32K of cache to be controlled. Because the data bus is 32 bits in size, the memory is broken up into four 8K × 8 memory blocks. Each block of memory provides eight different bits for the 32-bit data bus. When the 82385 is set for direct cache mapping, 1024 internal directory entries are available. Each directory entry represents eight double word (32-bit) addresses. The directory entry points to the starting address in main memory and provides other information needed by the cache controller to maintain proper cache control. When the cache controller detects that the address of the data needed by the MPU (read cycle) matches one of its directory entries (listing of addresses stored in cache memory), the cache controller blocks the MPU request from reaching the bus controller. The cache controller instead supplies the data directly from the cache memory to the local bus for the MPU by enabling the proper cache memory locations. This process is called a cache hit. Because cache memory is on the local bus, no wait states are required when retrieving data from cache memory. If the address does not correspond to a directory entry, the cache controller passes the MPU command request to the system bus, where the bus controller and other chips perform the bus cycle. As the data is transferred to the MPU, the cache controller places a copy of the data into cache memory. The cache controller then updates its directory.

During a write cycle the 82385 intercepts the control signals from the MPU and examines the address bus. The 82385 stores the data in the data buffer and address latches located between the local and system buses and causes the operation to be completed so that the MPU does not have to wait for main memory. If the address matches one of the directory entries, the data is also written into cache memory. This process is called write through cache and allows most write operations to take place with no wait states.

The 82385 blocks the MPU cycle definition signals from passing through to the bus controller and system bus only during either a memory or memory code read cache hit. Otherwise the command is passed through to the bus controller and system bus so that it can be executed.

When another bus master gains control of the bus, the 82385 monitors the address, data, and control status. If the other bus master causes a change in a memory that is cached, the 82385 updates the proper cache memory location. This process is called snooping, and it occurs in both direct and two-way set associative mapping.

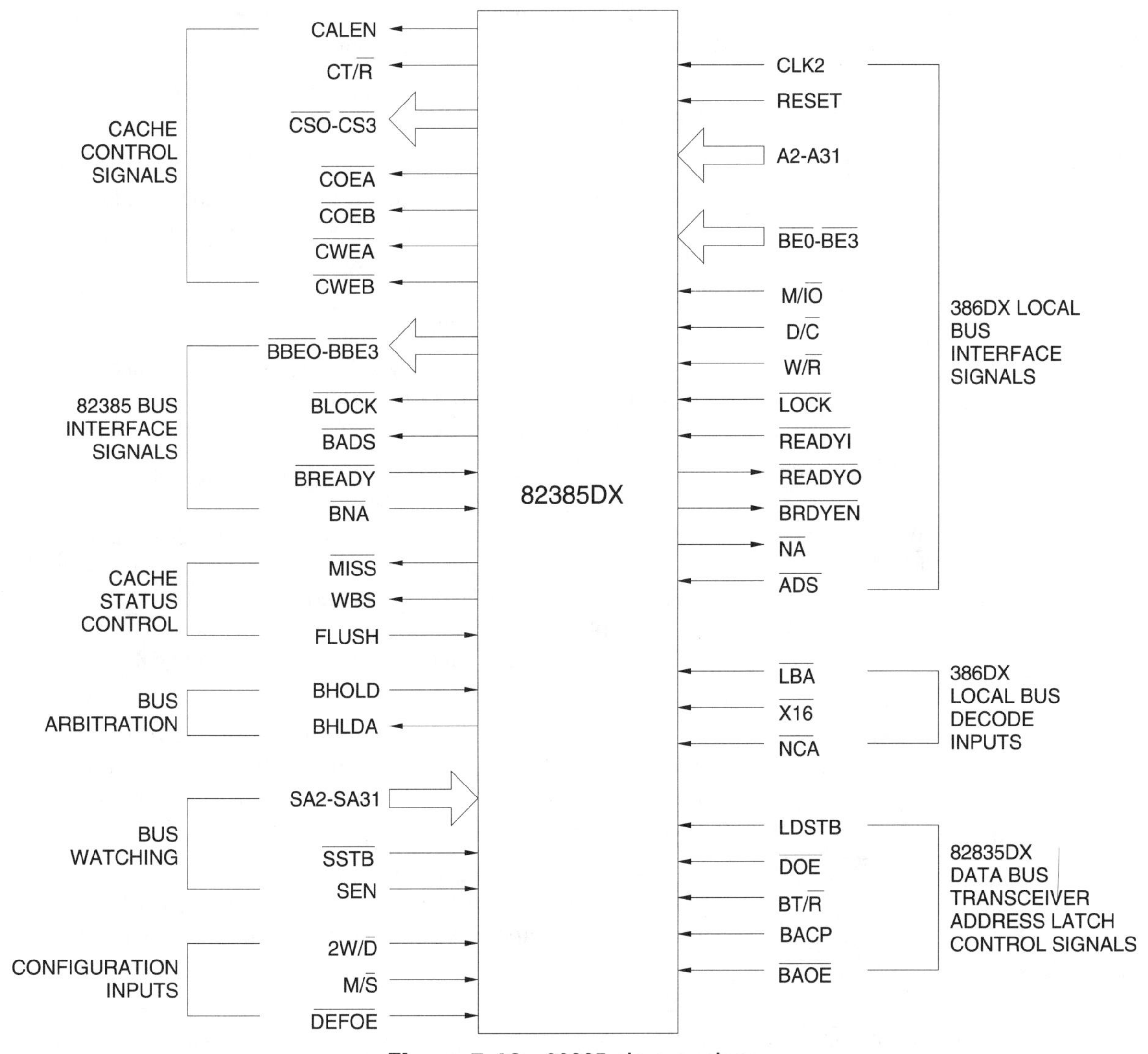

Figure 5-18 82385 pin groupings.

Two-Way Set Associative Cache Memory. In two-way set associative cache memory, the one block of 8K double words is divided into two smaller blocks of 4K double words. One is called block A and the other block B. Each block contains 512 double word directory entries. The 82385 checks both blocks of directories each time a memory operation is performed and, just as in direct mapped cache, controls the flow of data. The 82385 marks one of the directory blocks as being least recently used, depending on how often it is used. The least recently used block most likely contains data not needed by the MPU; therefore, that is the block of memory replaced first by the 82385 controller. Although the 82385 does not support four-way set associative mapped cache, other cache controllers do allow this type of mapping to occur. Two-way associative mapping normally produces a 99% hit rate.

5.6.3 82385 Pin Configurations

386DX MPU Interface Signals. The signals in this section share the local bus with the 386DX. Most of these signals are either monitored or intercepted by the 82385 (Figure 5-18). The 82385 supplies simulated 386DX signals that have been intercepted to provide proper control of the system bus.

CLK2	The CLK2 input is derived from the same source as the CLK2 input for the 386DX and performs the same function.
RESET	The RESET input is derived from the same source as the RESET input for the 386DX and performs the same function.
A2–A31	The levels on address lines A2 to A31 are derived from the A2 to A31 outputs of the 386DX. These lines are also applied to the address latches that supply the address to the system bus. These lines are used by the 82385 to determine the DRAM memory location to be accessed by the MPU.
$\overline{\text{BE0–3}}$	The 82385 intercepts the $\overline{\text{BE0 to 3}}$ outputs from the 386DX. The levels on these inputs indicate which bytes are enabled on the data bus. These signals are not directly supplied to the bus controller section in the AT chip set (see $\overline{\text{BBE0–3}}$ outputs).
M/$\overline{\text{IO}}$ D/$\overline{\text{C}}$ W/$\overline{\text{R}}$ $\overline{\text{LOCK}}$	The M/$\overline{\text{IO}}$, D/$\overline{\text{C}}$, W/$\overline{\text{R}}$, and $\overline{\text{LOCK}}$ bus definition inputs are derived from the 386DX. The M/$\overline{\text{IO}}$, D/$\overline{\text{C}}$, and W/$\overline{\text{R}}$ outputs are also supplied to a system control latch. The output of the latch supplies these signals to the bus controller section in the AT chip set only during noncached operations. During a cache operation the output of the control latch is disabled while the 82385 performs the bus cycle. The $\overline{\text{LOCK}}$ input is intercepted by the 82385 and supplies the proper level on its $\overline{\text{BLOCK}}$ output line when necessary.
$\overline{\text{READYI}}$	The $\overline{\text{READY Input}}$ line gets its signal from the same device that supplies the $\overline{\text{READY}}$ signal for the 386DX. This signal is used with the $\overline{\text{ADS}}$ input to determine when the bus is ready during a 386DX bus cycle.
$\overline{\text{READYO}}$	The $\overline{\text{READY Output}}$ is gated with other ready signals from the ready control circuitry in the AT chip set to indicate when the bus is ready to terminate the current bus cycle. The output basically tells the 386DX when the cache memory is ready to terminate a memory bus cycle that involves cache memory.
$\overline{\text{BRDYEN}}$	The $\overline{\text{Board ReaDY ENable}}$ output line is used to enable or disable the board ready signal from the AT chip set from being applied to the gating circuitry to create the ready input signal for the 386DX.
$\overline{\text{NA}}$	The $\overline{\text{Next Address}}$ output is connected to the $\overline{\text{NA}}$ input of the 386DX. When low, this signal informs the 386DX that the 82385 is ready for the next address.
$\overline{\text{ADS}}$	The $\overline{\text{ADdress Status}}$ line is derived from the $\overline{\text{ADS}}$ output of the 386DX. This signal, along with the 82385 $\overline{\text{READYI}}$ input, allows the 82385 to keep track of when the bus is ready for a 386DX bus cycle.

386DX Local Bus Decode Inputs

$\overline{\text{LBA}}$	The $\overline{\text{Local Bus Access}}$ input, when low, is used to indicate the 386DX access of a local bus device other than cache.
$\overline{\text{NCA}}$	The $\overline{\text{NonCache Access}}$ input, when low, indicates a noncache 386DX bus cycle.
$\overline{\text{X16}}$	The 16-bit access input, when low, indicates a 16-bit noncache 386DX bus cycle.

82385 Bus Data Transceiver and Address Latch Control Signals

LDSTB
: The Local Data STroBe line latches the data latches that control the flow of data between the system and local data buses. The latching occurs on the rising edge of this output.

$\overline{\text{DOE}}$
: The $\overline{\text{Data Output Enable}}$ line, when low, enables outputs of the data latches, allowing data that was latched into the address latches to flow between the local and system data buses.

BT/$\overline{\text{R}}$
: The system Bus Transmit/$\overline{\text{Receive}}$ output controls the direction of data flow in the address latches between the local and system data buses.

BACP
: The system Bus Address Clock Pulse is used to clock in the levels on the address and cycle definition latches from the local bus.

$\overline{\text{BAOE}}$
: The system $\overline{\text{Bus Address Output Enable}}$ line is used to supply the address and cycle definition levels to the system bus.

Cache Control Signals

CALEN
: The Cache Address Latch ENable output is used to enable the lower address lines to the address latches for cache memory, beginning on the falling edge and as long as the line stays low. When high, this line causes the address latches for cache memory to become transparent.

CT/$\overline{\text{R}}$
: The Cache Transmit/$\overline{\text{Receive}}$ output line is used to control the direction of data flow in the data transceivers in the cache memory section.

$\overline{\text{CS0–3}}$
: The cache $\overline{\text{Chip Selects}}$ are used like the byte enable lines ($\overline{\text{BER0}}$ through $\overline{\text{BE3}}$) in the 386DX. During a cache write operation, $\overline{\text{CS0}}$ enables the static RAM chip that supplies D0 to D7 to the data bus. $\overline{\text{CS1}}$ enables the data for D8 to D15, $\overline{\text{CS2}}$ for D16 to D23, and $\overline{\text{CS3}}$ for D24 to D31. During a cache read hit or miss, all four cache chip selects are enabled regardless of whether the MPU has requested that bytes be enabled.

$\overline{\text{COEA–B}}$
: The $\overline{\text{Cache Output Enable (bank A and B)}}$ lines are used to help enable the cache RAM chips or the bus transceivers in either bank A or B for two-way set associative cache configuration. In direct cache configuration, both outputs are enabled at the same time.

$\overline{\text{CWEA–B}}$
: The $\overline{\text{Cache Write Enable (bank A and B)}}$ lines are used to enable a write operation in cache RAM chips in either bank A or B for two-way set associative cache configuration. In direct cache configuration, both outputs are enabled at the same time.

82385 Bus Interface Signals

$\overline{\text{BBEO–3}}$
: The system $\overline{\text{Bus Byte Enable Output lines (0 through 3)}}$ are used to supply the byte enable signals from the 386DX that was intercepted by the 82385 to the system bus. During any cache read, all four lines are enabled at the same time; otherwise these lines emulate the $\overline{\text{BE0}}$ through $\overline{\text{BE3}}$ lines of the 386DX.

$\overline{\text{BLOCK}}$	The system $\overline{\text{Bus LOCK}}$ output line is used to supply the lock signal from the 386DX to the system bus.
$\overline{\text{BADS}}$	The system $\overline{\text{Bus ADdress Status}}$ output line is used to supply the address status from the 386DX to the system bus. When low, it indicates when the address on the system bus and bus definition are valid.
$\overline{\text{BREADY}}$	The system $\overline{\text{Bus READY}}$ line is used to supply the ready from the AT chip set before it is gated with any cache ready signals. This signal, along with the $\overline{\text{READYI}}$ input signal, is used to terminate bus cycles for both the 82385 and 386DX MPU.
$\overline{\text{BNA}}$	The system $\overline{\text{Bus Next Address}}$ request input line is used to inform the 82385 that the system bus is ready for the next address and bus cycle definition.

Status Control Signals

$\overline{\text{MISS}}$	The cache $\overline{\text{MISS}}$ indication output goes low after a memory cache read or write cycle is missed.
WBS	The Write Buffer Status output is used to indicate that the data in the data latches has not been written into cache even though the next memory bus cycle has started.
FLUSH	The flush input causes the 82385 to clear all directory tag bits, when high for eight CKL2 clocks.

Bus Arbitration Signals

BHOLD	The system Bus HOLD input performs the same function as the HOLD signal of the 386DX for the 82385.
BHLDA	The system Bus HoLD Acknowledge output performs the same function as the HLDA signal of the 386DX for the 82385.

Coherency and Support Signals

SA2–SA31	The Snoop Address lines (SA2 to SA31) are used to allow the 82385 to monitor the address bus when a different bus master (DMA) is using the bus. The lines allow the 82385 to determine if a current cache address has been modified.
$\overline{\text{SSTB}}$	The $\overline{\text{Snoop STroBe}}$ input line goes low to indicate that a valid address is on the system bus when another bus master has control over the bus during a write cycle.
SEN	The Snoop ENable line goes high during a write cycle when the 82385 is configured with snooping.

Configuration Inputs

$2W/\overline{D}$	The 2-Way set associative/$\overline{\text{Direct}}$ line controls whether the 82385 is configured for in a two-way set associative or direct cache memory configuration. When tied high, two way is selected; when low, direct cache memory configuration is selected.

M/$\overline{\text{S}}$ — The Master/$\overline{\text{Slave}}$ line controls whether the 82385 operates in the master or slave mode. When this line is high, the 82385 operates in the master mode; when low, it operates in the slave mode. The master and slave modes cause the BHOLD and BHLDA lines to process the operation differently.

$\overline{\text{DEFOE}}$ — The $\overline{\text{DEFine Output Enable}}$ is used to control how the cache output enables function. When low, the cache outputs control cache memory directly. When high, data bus transceivers are needed to control cache memory.

5.6.4 Intel 82340 AT Chip Set

The Intel 82330/82331 AT chip set was designed to provide support circuitry for the 286 and 386SX systems. The 82330/82331 AT chip set cannot be used with the 386DX MPU because this chip set supports only a 16-bit data bus and a 24-bit address bus, whereas the 386DX has a 32-bit data and address bus. Therefore Intel developed the 82340 AT chip set for the 386DX MPU. In Figure 5-15, three chips from the 82340 chip set are used. The 82345 data buffer is used to reduce the number of ICs needed to control the different data buses and parity generation and checking for DRAM. The 82346 system controller is used to produce control and status signals, the multiplexed DRAM address bus, and row and column address strobe signals. The 82344 ISA controller controls interrupts, DMA channels, timer signals, and assorted other signals. These three ICs, along with some external support circuitry, make up the remaining controlling elements of the 386DX system.

82345 Data Buffer. The 82345 data buffer is a one-chip solution to the control of data between the different data buses in the computer system. This IC contains data bus transceivers to control the direction of data flow on the different data buses. It also determines the choice and timing of data flow in each of the data buses. The main input for this IC is the D0 to D31 data bus, which is controlled by the 82385 data latches in the local bus of the 82385. Other inputs include the chip selects developed by the other ICs in the AT chip set, the byte enable lines from the local bus to determine which bytes are enabled on the data bus, and assorted other control lines. Based on these inputs, the 82345 directs the data to and from the proper data buses. The memory data bus MD0 to MD31 is used to transfer data between the DRAM and ROM sections of the computer and one of the other data buses of the system. The system data bus SD0 to SD15 is used for I/O interface slots on the system board. The peripheral data bus is an 8-bit bus (XD0 to XD7) that is also used for the transfer of data between peripherals. The 82345 only enables the data bus or buses and the size of the data bus that is required by the current bus cycle.

Besides controlling the data buses of the computer system, the 82345 also contains parity circuitry. During a DRAM write operation, this IC produces the parity bits (one parity bit for each byte of data) stored in DRAM; this is referred to as parity generation. During a DRAM read operation the data bus and parity bits are applied to a parity checker circuit. If the parity is not correct (the number of logical 1s has changed), this IC produces a parity error signal.

82346 System Controller. The Intel 82346 system controller is responsible for the timing, directing, and modifying of most of the control signals of the system. This IC gets signals from the local bus, the data buffer, and the ISA controller. The following paragraphs list the functions of this IC in the 386DX system.

The clock generator section produces the CLK2, which is used to provide the timing for the 386DX, math coprocessor, and other devices that must be synchronized with bus operations. This section also provides a programmable BUSCLK output, which is applied to the interface slots.

The bus arbitration logic is used to help process the request and system acknowledgments for DMA cycles between the DMA and the 82385 cache controller. This section also has an effect on the DRAM refresh control for the system board.

The bus control logic is basically the bus controller of the system. In this example the signals that control the bus controller circuitry are derived from the 82385 cache controller under the control of the 386DX. The control signals produced by this section are used to enable the data, direction of data flow (type of bus cycle), and timing of latching into address latches on the system bus (operates like the 82288 bus controller).

The address-decoding section, along with the DRAM control section and the remapping logic, is used to multiplex the address bus for the DRAM section (MA0 to MA9), produce row ($\overline{RAS0}$ through $\overline{RAS3}$) and column address strobe signals ($\overline{CAS0}$ through $\overline{CAS3}$), provide the control for the refresh cycle, allow the memory to be remapped in the address map of the system, and provide the chip select for ROM BIOS. This section is also responsible for producing some of the bus-watching signals needed by the 82385.

The ready control circuitry produces the ready signal for the system. When this line becomes active, it indicates that the bus is ready to complete the bus cycle, which terminates the cycle.

The math coprocessor interface circuitry is used to supply status and control signals for the 387DX math coprocessor. This includes the $\overline{ERROR}$, $\overline{BUSY}$, $\overline{RESET}$, IRQ13 (387 interrupt), and coprocessor extension request signals.

The Port A section is used to process the A20GATE signal from the keyboard controller, which allows the 386DX to switch from the protected mode to the real mode.

The last section produces the reset signal for the MPU.

82344 ISA Controller. The last IC in the chip set is used to supply most of the ISA signals for the system. This IC contains data conversion and wait state control logic, I/O bus ready and control, peripheral signals (8259, 8237, 8254, RTC, and DMA page register), chip select for the keyboard controller, and address bus controls.

The data conversion and wait state control performs two functions: (1) It produces even and odd byte addresses for 8-bit and 16-bit addresses. This processes and translates the byte enable signals for the 386DX. (2) It inserts wait states during I/O operations.

The 264 and 266 ready and bus control circuitry modifies and produces the necessary interface ready and bus control signals that are supplied to the chips in the 82340 AT chip set.

The peripheral control section contains system board I/O devices, which perform the same functions as in the XT and other AT system boards. The peripheral control section contains two 8259 PICs and an 8254 used to generate the refresh cycle (via the refresh counter) and produce the frequency for the speaker. The Port B logic produces the NMI signal for the 386DX, which indicates a parity error. The section also contains two 8237 DMAs, which are used to provide high-speed data transfers between memory and I/O devices, and a DMA page register used to supply the page values of the address bus during a DMA cycle. The real time clock is used to provide a battery–backed-up clock source for the system, and the battery–backed-up CMOS allows the user to store set-up data about the type of hardware and configuration needed by BIOS during boot-up.

The last section provides address bus latching and conversion for the local and system buses. The address-decoding section provides an external chip select for the 8042 keyboard controller as well as the internal chip selects for the peripheral control section.

ROM Section. The ROM section contains either one or two ROMs/EPROMs. Because the ROM address can be remapped most of the time, the contents on ROM BIOS are relocated into faster DRAM so that BIOS routines operate much faster. In older systems two 1-byte ROMs were used in parallel to supply 16 bits of data, but in most newer systems one 8-bit ROM is used and the contents are relocated into 32-bit DRAM. If two ROMs are used, each ROM is 8K $\times$ 8, while in a one-ROM configuration a 16K $\times$ 8 ROM is used. No other external support circuitry is needed for the ROM section.

8042 Keyboard Controller. The 8042 keyboard controller section contains the same circuitry as the keyboard controller section used in the 286 and 386SX system.

DRAM Section. Because the data buffer and system controller supply all the necessary signals for the DRAM section (multiplexed memory address bus, row and column address strobes, parity generator/checker, and refresh gating), very little external circuitry is needed. This section contains four banks of four SIMM modules each. Because this board is set up for 1MB SIMMs, this DRAM section can contain up to 16MB.

I/O Interface Bus Slots. The I/O interface bus slots are the edge card adapter near the back of the system board. The I/O interface bus slots in a 386DX ISA system are the same as in a 386SX system because the ISA standard supports only 16-bit adapters. Some system board manufacturers supply a special 32-bit slot, which can be used only by an interface card (usually a memory card) made by the manufacturer. This slot is not standardized and may differ from manufacturer to manufacturer. An interface card placed in one of these slots has access to most of the bus signals to control the card. If the card does not need a particular signal, a connection is not made. Most AT system boards contain eight interface slots, normally two 8-bit slots (single 62-pin edge connector) and six 16-bit slots, which are identified by an additional 36-pin edge connector in line with the larger connector. The signals for each of the slots are connected in parallel; therefore, a 16-bit I/O adapter can be placed in any 16-bit I/O slot and any 8-bit adapter can be placed in any 8-bit or 16-bit slot (but only the 8-bit signals are used).

CHAPTER SUMMARY

1. The bus interface block is used to provide the address for the address bus outputs, determine the size of the data bus to be used, control the data bus multiplexer transceivers, and control the handshaking between the MPU and other devices needed to perform the bus cycle. The address bus in the SX is 24 bits and that in the DX is 32 bits. The SX has a 16-bit data bus while the DX has a 32-bit data bus.
2. The central processing unit (execution unit) contains the general purpose registers, ALU, and a 64-bit barrel shifter. This section processes only data inside the MPU.
3. The memory management unit contains the segmentation and paging units. This section of the MPU creates the address for the bus interface unit.
4. The real operating mode causes the MPU to develop addresses just like the 8086/8088. Addressing is limited in this mode to 1 million address locations.
5. In the protected operating mode (virtual 8086 mode), all memory can be accessed. The protected mode also allows the MPU to create separate 8086 MPU multiprocessors that are protected and isolated from the 386 itself and from each other. The protected address mode allows multitasking to be performed.
6. Register names followed by the letter L or H indicate that the register is used as an 8-bit register (high or low byte). Register names followed by the letter X mean that the register is used as a 16-bit register. If a register is followed by the letter X and preceded by the letter E, the register is used as a 32-bit register.
7. The advantages of the 386SX over the 386DX are that it consumes less power, costs less, can be integrated into already designed 286AT system boards without many modifications, and uses all 386DX instructions and operating modes. The disadvantages of the 386SX are that it has only a 16-bit data bus and 24-bit address bus and is slower in executing software written to use 32-bit instructions (because of the 16-bit data bus).
8. The advantages of the 386DX over the 386SX are that it has a 32-bit data bus and 32-bit address bus (access to more memory locations) and can perform 32-bit operations faster

(because 32 bits of data can be transferred during one bus cycle). The disadvantages of the 386DX are that it requires more support circuitry and different AT chip sets, consumes more power, and costs more because more support ICs are required.

9. CLK2 is the clock input for the MPU, which provides synchronization for bus cycles.

   A1 to A23 and A2 to A31 are the address outputs for the SX and DX, respectively. The address bus is used to select the device used during the bus cycle.

   $\overline{BHE}$/$\overline{BLE}$ are outputs for the SX that are used to determine which byte or bytes are transferred during the bus cycle. When $\overline{BHE}$ is low the high byte is enabled, and when $\overline{BLE}$ is low the low byte is enabled. If both lines are low, a 16-bit operation is performed.

   $\overline{BE0}$ to $\overline{BE3}$ are outputs for the DX that are used to determine which byte or bytes are transferred during the bus cycle. All of these outputs are active low. These lines allow the low byte, high byte, low word, high word, or a long word (32 bits) to be transferred during the bus cycle.

   D0 to D15 and D0 to D31 are bidirectional data lines used to input or output data to and from the MPU.

   W/$\overline{R}$ is the write and read output used by the MPU to indicate whether a read or write operation is performed during the bus cycle.

   D/$\overline{C}$ is the data and control output used by the MPU to indicate if the bus cycle contains data or control codes.

   M/$\overline{IO}$ is the memory and I/O output used by the MPU to indicate if the bus cycle occurs for memory or an I/O device.

   $\overline{LOCK}$ is used to lock output from any bus master asking permission to control the bus.

   $\overline{READY}$ input informs the MPU when to terminate the bus cycle.

   $\overline{ADS}$, the address status line, indicates that a valid address is on the byte enable lines, address bus, and W/$\overline{R}$, D/$\overline{C}$, and M/$\overline{IO}$ lines.

   $\overline{NA}$, the next address input, informs the MPU that the system is prepared to accept a new bus cycle.

   $\overline{BS16}$, the bus size input (DX only), allows the 386DX to run multiple bus cycles to complete a request from devices that cannot provide or accept 32-bit data in a signal bus cycle.

   INTR, the interrupt request input, informs the MPU that the programmable interrupt controller has requested an interrupt.

   NMI, the nonmaskable interrupt input, informs the MPU that a parity error has occurred.

   RESET is used to reset the MPU.

   HOLD, the bus hold request input, is used by the DMA to request a DMA cycle from the MPU.

   HLDA, the bus hold acknowledge output, is used by the MPU to grant permission for the DMA to use the bus.

   PEREQ, the processor extension operand request input, informs the MPU of a data operand transfer during a math coprocessor operation.

   $\overline{BUSY}$ informs the MPU that the math coprocessor is busy.

   $\overline{ERROR}$ informs the MPU that an error has occurred during a math coprocessor bus cycle.

10. CPUCLK2 is the synchronizing clock input for the math coprocessor operating in the synchronous mode.

    NUMCLK2 is the synchronizing clock input for the math coprocessor operating in the asynchronous mode.

CLKM, the clock mode input, is used to select which clock is used.

RESETIN, the reset input, is used to reset the math coprocessor.

$\overline{\text{READYO}}$ supplies ready information to external circuitry to indicate when the math coprocessor is ready to complete its cycle.

STEN, the status enable input, is used with other inputs to enable the 387.

$\overline{\text{NPS1}}$, the numeric processor select line 1, is used with the levels on STEN and NPS2 to enable the 387.

NPS2, the numeric processor select line 2, is used with the levels on STEN and $\overline{\text{NPS1}}$ to enable the 387.

$\overline{\text{CMD0}}$, the command data line 0, is used to determine if the 387 operation is a command or some form of data.

11. The main purpose of the 82335 IC is to convert the signals to and from the 386SX into signals needed by the 82230/82231 AT chip set from Intel. This IC also provides DRAM address decoding, parity checking and generation, memory timing, and modifications of coprocessor handshaking interface signals between the 387SX and the 386SX.
12. The 82335 clock generator and reset synchronizer create the system clocks for the 386SX and 387SX and the AT chip set so that all operations are synchronized with the 82335. This section also modifies the reset output signals from the 82230 to create the reset for the MPU and math coprocessor.
13. The 82335 parity generator and checker section is responsible for generating a high and low parity bit during a write operation; during a read operation parity checking is performed.
14. The 82335 address mapper and decoder allow the BIOS program to be relocated into DRAM. This section also allows the contents of the video adapter to be relocated into DRAM and provides ROM chip selects. This section is responsible for remapping high memory.
15. The 82335 DRAM controller section controls the access of up to four 16-bit banks of DRAM. This section allows non-page, page, and page mode bank interleave addressing modes. This section is also responsible for DRAM refresh.
16. The 82335 bus cycle translator is responsible for translating 386SX control signals into 286 control signals for the 82230/82231 AT chip set.
17. The 82335 ready generator section modifies the ready signals for the system.
18. The 82335 math coprocessor interface section is responsible for detecting the 387SX math coprocessor and for modifying the synchronous signals between the 386SX, 387SX, and 82230.
19. The 82230 contains the following circuitry: 82284 clock generator/driver, two 8259A PICs, 6818 real time clock, 82288 bus controller, address and data bus control logic, bus control logic, CPU shutdown logic, and coprocessor interface logic.
20. The 82231 contains the following circuitry: read and write logic, I/O chip select, support logic for the PIT and I/O devices, 8254 PIT, DMA memory mapper, two 8237 DMAs, DRAM refresh counter, latch and logic, and parity check logic.
21. The DRAM section contains the actual DRAM modules in the form of SIMMs. This section contains four banks of 16 bits each.
22. The ROM section contains BIOS and set-up routines that are stored in either ROMs or EPROMs.
23. The 8042 keyboard microcontroller section is responsible for converting serial keyboard data into parallel data for the data bus. This section also generates the IRQ1 keyboard interrupt request.
24. The I/O interface bus slots use 8-bit (62-pin XT slots) and 16-bit (62-pin AT slots with another connector that supplies an additional 32 pins). A 16-bit slot can be used by an 8-bit I/O adapter.

25. In non–page mode addressing, each address requires a new row and column address. In page mode and page mode interleave addressing, a row and column address is needed only when a new row address is needed. In page and page mode interleave addressing, once a row is defined the DRAM section needs only to supply the column address; the previous row address is used until a new row is selected. Essentially in page and page mode interleave addressing, the row address is represented by the upper bits of the address bus and the column address by the low bits of the address bus. Therefore the row address does not have to change as often.
26. The Intel 82385 is a 32-bit cache controller designed to work with the 80386DX to provide near zero wait memory operations. The main purpose of the chip is to keep a copy of frequently used codes and data in fast cache memory. The cache controller is programmable and allows cache to function with both memory read and write cycles. The 82385 also supports direct and two-way set associative cache mapping. In most cases the 82385 generates 386DX-like signals or controls the selection and timing of 386DX signals to the system bus.
27. The 82345 data buffer in the 386DX system controls the data between the different data buses in the computer system. This IC contains data bus transceivers to control the direction of data flow on the different data buses. It also determines the selection and timing of data flow in each of the data buses. The main input for this IC is the D0 to D31 data bus, which is controlled by the 82385 data latches in the local bus of the 82385.
28. The Intel 82346 system controller in the 386DX system is responsible for timing, directing, and modifying most of the control signals of the system. This IC gets signals from the local bus, the data buffer, and the ISA controller.
29. The 82344 ISA controller in the 386DX system is used to supply most of the ISA signals for the system. This IC contains data conversion and wait state control logic, I/O bus ready and control, peripheral signals (8259, 8237, 8254, RTC, and DMA page register), chip select for the keyboard controller, and address bus controls.
30. The ROM section of the 386DX system contains BIOS and set-up routines stored in either ROMs or EPROMs.
31. The DRAM section of the 386DX system contains up to four banks of 32 bits each of SIMM memory modules.
32. The keyboard controller section of the 386DX system performs the same functions as in the 286 and 386SX systems.
33. The I/O interface slots in a 386DX system contain both 8-bit XT-type and 16-bit ISA-type interface slots. These slots are used to allow the computer to access different types of interface adapters (floppy disk/hard disk adapters usually 16 bits, multi-I/O interface adapters usually 8 bits, and video adapters usually 8 or 16 bits).

REVIEW QUESTIONS

1. Which control section inside the 386DX or 386SX is responsible for supplying data for the data bus?

   a. Bus interface
   b. Central processing unit
   c. Memory management unit

2. Which 386 operating mode allows the 386 to perform multitasking operations?

   a. Real mode
   b. Protected mode

3. AX is used as a ________ bit register.

a. 4

b. 8

c. 16

d. 32

e. None of the given answers

4. ECX is used as a ________ bit register.

a. 4

b. 8

c. 16

d. 32

e. None of the given answers

5. The 386SX uses more power than a 386DX.

a True

b. False

6. The 386SX has a larger data bus than the 386DX.

a. True

b. False

7. What must be the levels on the $\overline{\text{BHE}}$ and $\overline{\text{BLE}}$ inputs for the 386SX for a word to be transferred?

a. $\overline{\text{BHE}} = 0$, $\overline{\text{BLE}} = 0$

b. $\overline{\text{BHE}} = 1$, $\overline{\text{BLE}} = 0$

c. $\overline{\text{BHE}} = 0$, $\overline{\text{BLE}} = 1$

d. $\overline{\text{BHE}} = 1$, $\overline{\text{BLE}} = 1$

8. A2 to A31 on the 386DX always act as outputs.

a. True

b. False

9. If all of the byte enable lines on the 386DX are low, a 32-bit data transfer occurs during the bus cycle.

a. True

b. False

10. During a bus cycle if the D/$\overline{\text{C}}$ line is high, the data on the data bus is data and not an opcode.

a. True

b. False

11. The $\overline{\text{ADS}}$ line indicates that the address on the address bus is valid for the next bus cycle.

a. True

b. False

12. Which 386 signal is used by the DMA to ask for permission to use the bus?

a. NMI

b. HOLD

c. INTR

d. HLDA

e. RESET

13. When the error output of the 387SX is high, an error has occurred in the 387.

a. True

b. False

14. If the $\overline{\text{READY}}$ line to the 387 is stuck high, the computer system halts.

a. True

b. False

15. The 82335 is responsible for creating the CLK2 signals for the 386SX system.
    a. True
    b. False
16. The 82335 contains the DRAMs for the 386SX system.
    a. True
    b. False
17. Which IC in the 386SX system is responsible for controlling the DRAM refresh cycle?
    a. 82230
    b. 82231
    c. 82335
    d. 82385SX
    e. 80387SX
18. Which type of addressing mode provides the fastest access to memory for a large group of memory locations?
    a. Non-page mode
    b. Page mode
    c. Page mode interlace
19. The main purpose of the cache controller is to keep a copy of seldom-used codes and data in fast cache memory.
    a. True
    b. False
20. Which type of memory is the fastest?
    a. DRAM
    b. ROM
    c. EPROM
    d. SRAM
21. Which type of cache memory mapping uses all the memory in cache as one large block of memory?
    a. Direct
    b. Two-way set associative
22. The 82385 cache controller controls the data and address bus during any type of bus cycle.
    a. True
    b. False
23. The data on the data bus goes into the 82385 cache controller and the controller sends it to the proper location.
    a. True
    b. False
24. The 386DX uses the address on its address bus to select which cache address the data are placed in.
    a. True
    b. False
25. Which 386DX chip may be bad if the CMOS set-up data is lost every time the computer is turned off?
    a. Cache controller
    b. Data buffer
    c. System controller
    d. ISA controller
    e. ROM section

chapter 6

80486/Pentium and Advanced AT Systems

OBJECTIVES

Upon completion of this chapter, you should be able to meet the following objectives:

1. List the features of the 80486 and the Pentium.
2. List and define the purpose of each of the controlling sections of the 80486 and the Pentium.
3. Define the purpose of each pin of the 80486 and the Pentium.
4. List the features, advantages, and disadvantages of the MCA, EISA/VL, and PCI advanced AT buses.
5. List and define the purpose of each IC in the block diagram of an MCA, EISA/VL, and PCI microprocessor system board.

6.1 80486DX/DX2/DX4/SX/SX2 MPU

The 486DX from Intel is actually an 80386 MPU, 80387 math coprocessor, cache controller (82385), 8K of cache RAM, and additional control circuitry all integrated into one IC. The only difference between the 486DX and 486SX is that the 486SX does not contain the built-in math coprocessor. The DX2 operates internally at twice the speed of the DX, and the DX4 operates internally at three times the speed of the DX. The SX2 operates internally at twice the speed of the SX.

6.2 80486 INTERNAL ARCHITECTURE

The block diagrams in Figures 6-1 to 6-3 show the 32-bit microarchitecture of the 486DX, 486DX2, and 486SX. The basic difference is the lack of the floating point unit and floating point register file on the 486SX.

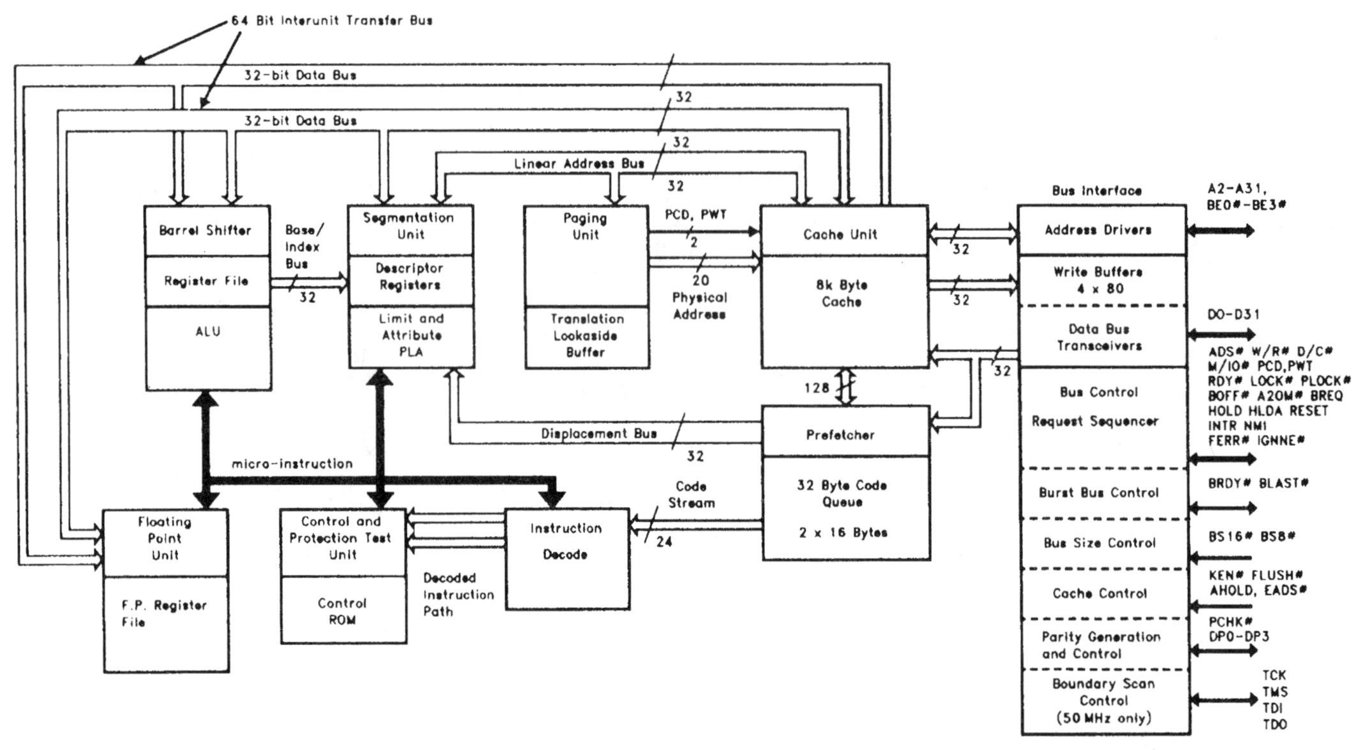

Figure 6-1 Intel 486DX microprocessor pipelined 32-bit microarchitecture. (Courtesy of Intel)

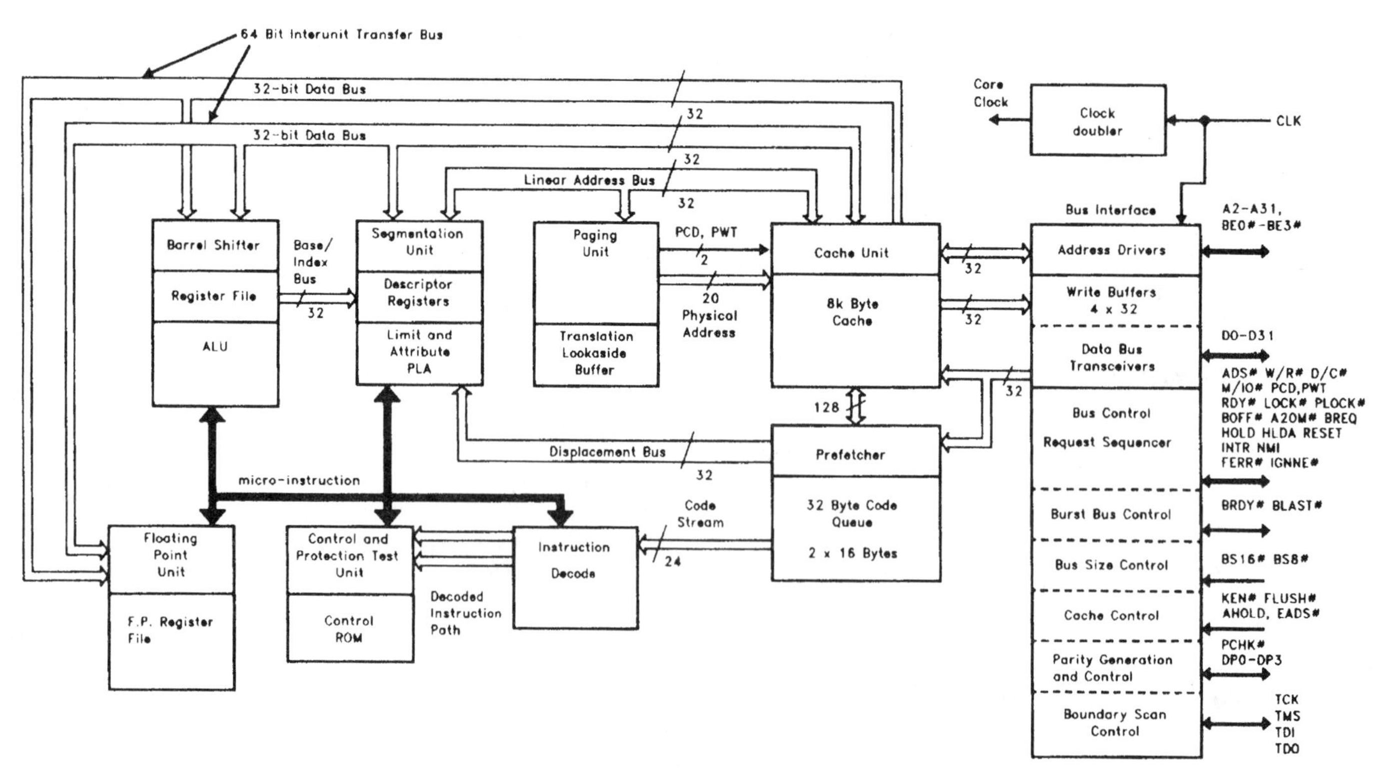

Figure 6-2 Intel 486DX2 microprocessor pipelined 32-bit microarchitecture. (Courtesy of Intel)

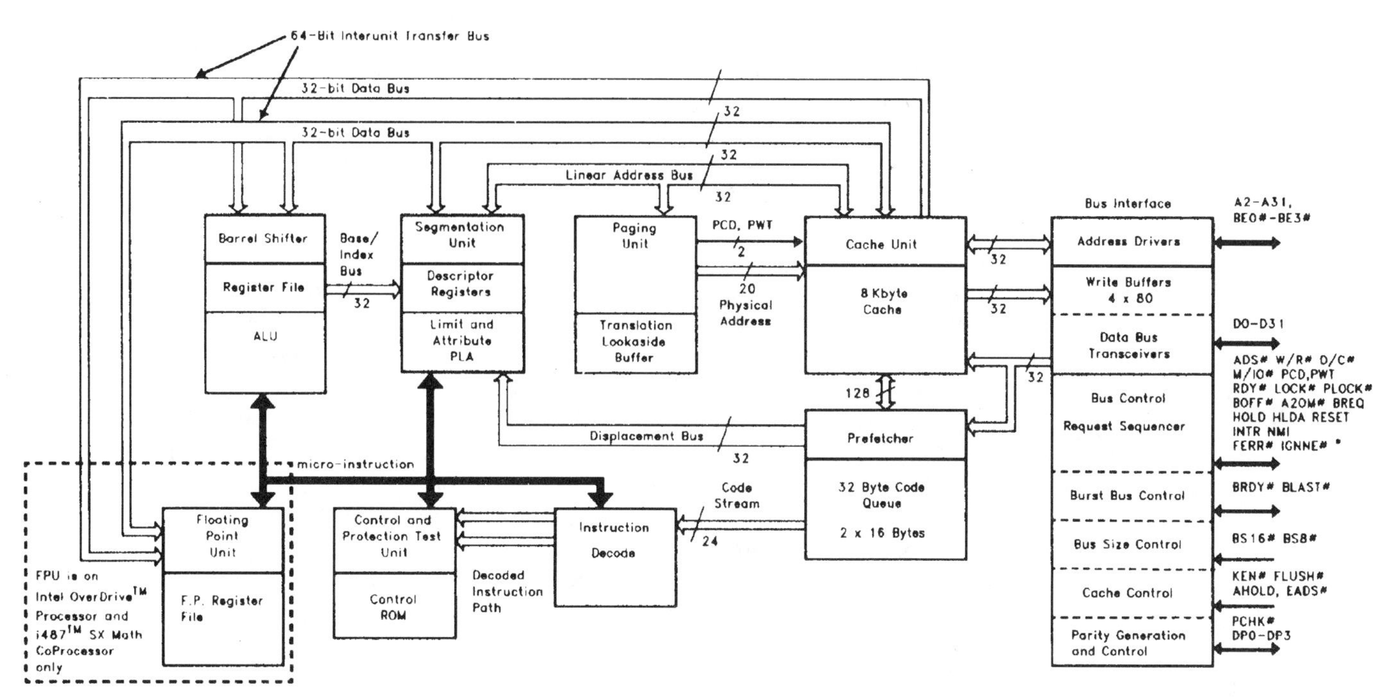

*FERR# and IGNNE# are present on the Intel OverDrive Processor/Intel487 SX Math CoProcessor only.

Figure 6-3 Intel 486SX microprocessor/Intel overdrive processor/Intel 487SX math coprocessor pipelined 32-bit microarchitecture. (Courtesy of Intel)

6.2.1 Bus Interface Unit (80486DX/DX2/DX4/SX/SX2)

The bus interface unit contains the following sections and controls the following pin functions. The address drivers are used to place the address on the address bus. The address bus is made up of address lines A2 to A31 and byte enable lines BE0# through BE3#. Note the latest change in nomenclature from chip manufacturers in designating the active state of a pin: Pins followed by the pound sign (#) are active low. This nomenclature replaces the negation line above the pin or the minus sign preceding the pin. Some of the address lines in this section are bidirectional rather than just outputs, as with most addresses produced by the MPU. The reason for this is discussed in the pin configuration section.

The data control section is made up of two parts. The write buffer section is used to temporarily hold data for the data bus during a write operation so that a wait state is not needed. The cache memory feeds the data to this section as the data bus becomes available to transfer the data. The data bus transceiver section is used to control the flow of data into and out of the MPU on the data bus. During a write operation the data is received from the write buffer section and sent out onto the data bus at the proper times. During a read operation the data on the data bus is placed into both the internal 8K cache memory and the prefetcher. The bus control request sequencer controls and receives the status for all control signals needed to complete the bus cycle (ADS#, W/R#, D/C#, M/IO#, PCD, PWT, RDY#, LOCK#, A20M#, BREQ, HOLD, HDLA, RESET, INTR, NMI, FERR#, and IGNNE#). The burst control section uses the BRDY# and BLAST# lines to control a cache burst control bus cycle. The bus size control uses the BS16# and BS8# lines to determine the size of the bus cycle other than a 32-bit bus cycle. The cache control KEN#, FLUSH#, AHOLD, and EADS# input lines are used to help the cache controller in the 486 determine how to control the cache cycle. The parity generation and control section is used to generate a parity bit for each byte during a write cycle on DP0 to DP3. During a read operation the DP0 to DP3 lines are used to input the parity bit for each byte, and this section checks the parity of the operation. The boundary scan control section and its control and status lines are used only in the 50-MHz operation of the 486DX and are not discussed here.

The clock doubler block in the 486DX2 is used to double the internal clock frequency of the 486. The doubler is used in the DX2 and SX2 486 MPUs. In the DX4 a clock tripler block is incorporated to triple the internal clock speed of the 486. In any case, the internal speed is increased while the bus cycles externally use the standard clock speed.

6.2.2 Instruction Unit (80486DX/DX2/DX4/SX/SX2)

The instruction control section contains the prefetcher, which receives data either from the data bus transceivers if the data is coming from outside the MPU or from the 8K cache memory block if the data is cached. The displacement part of an instruction can be sent directly from the prefetcher to the memory management unit when necessary, thus increasing the speed of the MPU. The prefetcher feeds the 32-byte code queue (buffer), which holds two 16-byte lines of opcodes for the instruction decode section. This queue fills up if the time it takes to execute the instruction is longer than the time it takes to fetch the instruction and empties out if it takes less time to execute the instruction than to fetch it. The queue ensures that an instruction is always ready for the instruction decode section when it is ready to accept the new instruction. The instruction code section breaks down the instruction so that it can be transferred into either the ALU or floating point unit (math coprocessor) or the control and protection test unit (control logic). The control ROM is used to supply the microcode sequence to the control and protection test unit.

6.2.3 Floating Point Unit (80486DX/DX2/DX4)

The floating point unit and floating point register file make up the math coprocessor section (387DX). This section receives its instruction in parallel with the ALU in the central processing unit of the 486, which increases the speed of processing 387-type math instructions. The floating point unit and the execution unit are connected to the internal 64-bit interunit transfer bus, which is made up of two 32-bit data buses. The bidirectional 32-bit data transfer bus is used to supply data to the floating point unit and to transfer the results of the math coprocessor out of the floating point unit to wherever it is needed. The unidirectional 32-bit data bus is used to supply one 32-bit value into the floating point unit. The operation speed of the 486 has been increased dramatically because of the 64-bit interunit transfer bus, as opposed to a separate 386 MPU and 387 coprocessor.

6.2.4 Central Processing Unit (80486DX/DX2/DX4/SX/SX2)

The central processing unit (also known as the execution unit) contains the barrel shifter, register file (general purpose registers), and ALU. The speed of the section is increased because it is connected to the 64-bit interunit transfer bus, which means that two 32-bit data values can be transferred into the central processing unit at one time. The one 32-bit result value can then be transferred out of the central processing unit once the instruction is complete. The purposes of all of the parts in the central processing unit are the same as for those found in the central processing unit of the 386DX (see Chapter 5).

6.2.5 Memory Management Unit (80486DX/DX2/DX4/SX/SX2)

The memory management unit is made up of the segmentation section and the paging section. The segmentation section contains the segmentation unit, descriptor registers, and limit and attribute PLA. The purpose of the segmentation section is to create the linear address, which is applied to the cache controller (cache unit) and the paging unit. While the processor is operating in the protected mode the linear address becomes the physical address, which is used as the effective address for the address bus via the cache controller. When the MPU is in the real mode, the paging section (paging unit and translation lookaside buffer) becomes enabled and takes the linear address and creates the physical address, which is used as the effective address for the address bus via the cache controller.

6.2.6 Cache Section (80486DX/DX2/DX4/SX/SX2)

The internal 486 cache section (82385 compatible), which contains the cache unit (cache controller) and an 8K cache memory block, is one of the important sections of the 486 MPU. This section is set up in a four-way set associative mode with write-through policy enabled. This section is used to intercept data and keep that data stored in the cache memory block. The frequently used codes and data that are stored in cache memory can be processed without the need for a bus cycle. A factor that increases the speed of the 486 is the fact that the cache unit controls the 64-bit interunit transfer bus, allowing twice as much data to be transferred internally at one time. During non-MPU write bus cycles, the 486 monitors the address bus. If a write occurs on a main memory location stored in cache, the cache entry for that location is invalidated. If the main memory address being written to is not in cache, no action is taken by the MPU. In the case of cache invalidations, the data in cache is invalid therefore, the MPU must update the data before the data in cache can be used again. The cache invalidation function of the 486 ensures that the data stored in cache is the same as the data stored in main memory. Cache memory can be increased by adding the 82485 second level cache controller and external cache memory, which works with the internal cache unit to increase cache memory from 64K to 128K.

6.3 80486 REGISTERS

6.3.1 Base Architecture Registers

General Purpose Registers (Real or Protected). The general purpose registers (Fig. 6-4) of the 486 are the same as those used in the 386DX. Table 6-1 lists the size and nomenclature for each of these registers. These registers have the same purpose as the general purpose registers of the 386DX.

Flags Register (Real or Protected). The flags register (EFLAGS) performs the same purpose as the flags register in the 386 (see Chapter 5). The lower 16 bits of this register are flags associated with 16-bit operations. In the real mode only the 16 least significant bits (LSBs) are used in the same way as in the 386. In the protected mode all 32 bits are used in the same way as in the 386DX, with the addition of one bit. Bit 18 is an alignment check bit found only in the 486 MPUs; its purpose is to enable the generation of memory reference misalignment address faults, when set. A misaligned address occurs when a word access starts on an odd address, a double word is not within a double word boundary, or an 8-byte reference is not within a 64-bit boundary.

Segment Registers. The function of the segment registers differs between the two operating modes of the 486. There are six 16-bit only segment registers that perform the same function as they do in the 386. The code segment (CS) is used to help define code addresses (opcodes). The stack segment (SS) is used to help define stack pointer addresses. The remaining four segment registers—data segment (DS), extra segment (ES), F data segment (FS), and G data segment (GS)—are used to help define data addresses. The segment registers of the 486 operate in the same way as in the 386.

TABLE 6-1
80486 Register Size and Nomenclature

Register Name	*Lo-Byte*	*Hi-Byte*	*Word*	*Double Word*
Accumulator	AL	AH	AX	EAX
Base	BL	BH	BX	EBX
Count	CL	CH	CX	ECX
Data	DL	DH	DX	EDX
Source index	—	—	SI	ESI
Destination index	—	—	DI	EDI
Base pointer	—	—	BP	EBP
Stack pointer	—	—	SP	ESP

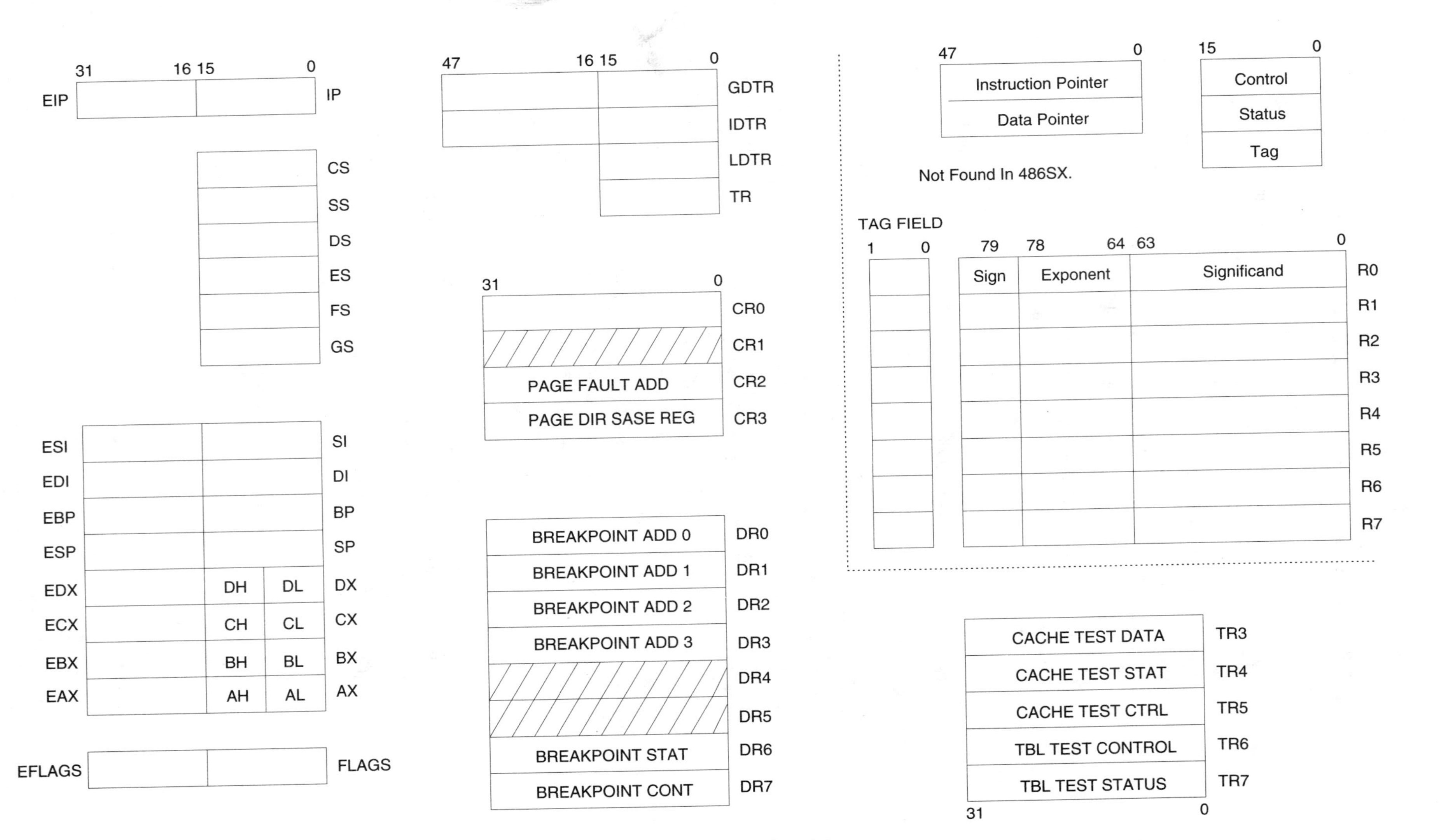

Figure 6-4 486 internal registers.

Instruction Pointer. This is a 32-bit register called EIP; of the 32 bits, the lower 16 bits (IP) are used in the real mode to provide the offset of the next instruction to be executed. The upper 16 bits are not currently defined in the 486. In the protected mode this register is not used. The function of this register is the same as in the 386.

Addressing (Real). The 486 real mode addressing functions the same as in the 386DX. The difference is that instead of the physical address being applied directly to the address drivers, the effective address (same as the physical address in the real mode) is applied to the cache unit, which controls the address drivers and cache memory. To develop a physical address (also known as an effective address) in the real mode, the value of the selected 16-bit segment register is transferred into the paging unit, where four binary 0s are shifted into the last significant location. This creates a 20-bit shifted segment value. The value of the instruction pointer (lower 16 bits of EIP) or offset is added to the shifted segment value. The 20 LSBs of the result are used as the physical address. The physical address is then applied to the cache unit.

6.3.2 System Level Registers

System level registers are used in the protected mode of the 486; the contents of these registers are used to control the cache unit and the math coprocessor. System level registers include the following: CR0 to 3 and the system address registers (GDT, IDT, LDT, and TSS).

Word Control Registers. The four 32-bit word control registers (CR0, CR1, CR2, and CR3) in the 486 operate basically the same as they do in the 386DX. The only difference between these registers and the ones found in the 386DX is that a few more bits are defined by Intel and used to help the cache unit control the address. Machine control register 0 (CR0) is used to define how the address is developed in the protected mode. Machine control register 1 (CR1) is reserved by Intel, just as in the 386DX. Word control register 2 (CR2) holds the page fault linear address. Word control register 3 (CR3) holds the page directory base address.

System Address Registers. The system address registers are used to develop the effective address while the 486 is in the protected mode. The registers in this section are the global descriptor table register, interrupt descriptor table register, local descriptor table register, and task state segment register.

Global Descriptor Table Register: The global descriptor table register (GDTR) is a 48-bit register that contains the address for the global descriptor table located in physical memory. The contents of these memory locations are used to produce the address in physical memory above 1MB. The function and contents of this register are the same as in the 386DX.

Interrupt Descriptor Table Register: The interrupt descriptor table register (IDTR) is also a 48-bit register that is used in the same way as the GDTR except that it is used when processing an interrupt. The function and contents of this register are the same as in the 386DX.

Local Descriptor Table Register: The local descriptor table register (LDTR) is a 16-bit register that points to a local address space for the current task. The function and contents of this register are the same as in the 386DX.

Task State Segment: The task state segment (TSS) register is the same as the task register in the 386DX. The function and contents of this register are the same as in the 386DX.

6.3.3 Floating Point Registers

Because the 486DX contains a math coprocessor, the floating point registers are found inside the 486DX. These registers are the same as those found in the 387DX math coprocessor. The registers in this section are the data registers, tag word register, control register, status register, instruction pointer, and data pointer.

There are eight 80-bit data registers in the floating point section of the 486DX. Each data register is divided into three fields that represent the type of data the math coprocessor can work with. The lower 64 bits represent the significant value, the next 16 bits represent the exponent value, and the most significant bit represents the sign bit.

The 16-bit tag word register is divided into eight 2-bit values, one 2-bit tag for each data register. The purpose of the tags is to increase the speed of the floating point unit by specifying the status of each of the data registers.

The 16-bit status word register is used to provide that status of the floating point unit, which is similar to the flags register.

The 16-bit control word register is used to configure how the floating point unit processes the math instruction.

There are two 48-bit registers in the floating point unit, one called the instruction pointer and the other the data pointer. When an error occurs in the floating point unit, these registers point to the address of the instruction and data causing the error. These two registers are needed because the floating point unit operates in parallel with the ALU section of central processing unit.

6.3.4 Debug and Test Registers

The eight 32-bit debug registers of the 486 perform the same functions as they do in the 386. In the 486 there are five 32-bit test registers, compared with only two in the 386. The three additional test registers are used to supply information about cache test data, cache test status, and cache test control.

6.4 80486 INTERNAL CACHE

The internal cache memory (Fig. 6-5) in the 486 is organized using four-way set associative mapping. The 8K cache in the 486 is divided into four blocks of 2K each. Each block uses 128 directory entries (tags) of 16 bytes each. There are 128 21-bit tags in each 2K block; each tag defines the physical address locations of the data stored in cache memory for the tag selected.

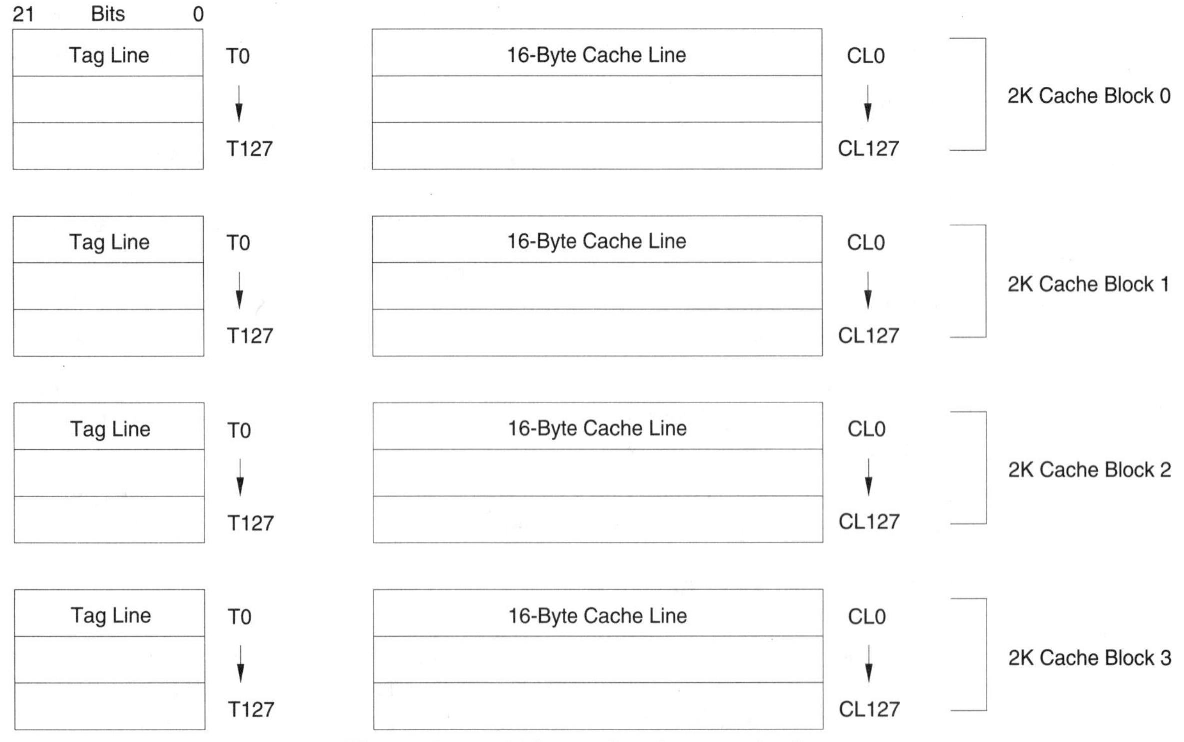

Figure 6-5 486 internal cache organization.

During an MPU read operation, the cache unit determines if the data required is cache memory. If present, the data is transferred immediately to the proper locations. If the data is in a memory location that is not cached, the cache unit fetches the data from the outside world and transfers it into the proper location and into cache memory, at which time the cache tag is updated. During an MPU write operation the data is written into both cache memory and the selected memory location. The transfer to cache is immediate while the transfer to the outside world is buffered so that the MPU is not slowed down. During a non-MPU bus cycle, the 486 monitors the bus and if the transfer changes a memory location that is currently in cache, the cache unit invalidates the cache entry. An invalidated cache entry means that when that cache location is needed, the cache unit must update the data in cache before they can be used.

6.5 80486 INSTRUCTIONS

The 486 supports all 8088/8086/80286/80386 instructions, data types, and addressing modes. The 486 also has added more instructions and is capable of handling larger packages of data and more tasks during multitasking in a more efficient manner. Tables 6-2 and 6-3 are brief listings of the types of instructions and data. The 486DX series contains floating point instructions while the 486SX series does not.

TABLE 6-2
80486 Instruction Types

Type of Instruction	*Number of Instructions*
Date transfer	11
Arithmetic	17
Logical	20
Control transfer	31
String	8
High level language	2
Flag control	9
Floating point	59
Floating point control	15
Segment register	7
Miscellaneous	6

TABLE 6-3
80486 Data Types

Type of Data	*CPU DX/SX*	*FPU DX*	*Description*
Unsigned	X		8 bits (byte), 16 bits (word), and 32 bits (double word)
Signed (integer)	X		8 bits, 16 bits, and 32 bits
		X	16 bits, 32 bits, and 64 bits
Floating point		X	Single precision (32 bits), double precision (64 bits), and extended precision (80 bits)
BCD unpacked	X		8 bits
BCD packed	X		8 bits
		X	80 bits
String and ASCII	X		Byte, word, and double word
Pointer	X		48 bits (16-bit selector and 32-bit offset)
			32 bits (32-bit offset)

6.6 80486 MICROPROCESSOR

The 80486 is a 32-bit high-performance microprocessor made by Intel and is compatible with 8088/86, 286, and 386 software. As in 286 and 386, the data and address buses are not multiplexed. The following pages describe the pin configurations of both the DX and SX series of the 486 MPU.

The major difference between the DX and SX series (Table 6-4) is that the DX series contains a floating point unit (math coprocessor) while the SX series does not. High-speed complex math operations require a 487SX math coprocessor. Because the SX series does not contain the built-in floating point unit, the power dissipation and cost are much less than for the DX series of the same speed.

The 486 has two operating modes, both of which support all 8088/86 software. The 8086 real addressing mode makes the 486 use the address bus like the 8088/86, which limits address space to 1MB, with the ability to transfer up to 32 bits of data during each bus cycle. The second address mode, called the protected mode (virtual 8086 mode), operates like the 386 virtual 8086 mode. In this mode the 486 sets up each task as a separate 8086 MPU that can be isolated and protected from one another and from the host 486 microprocessor operating system by using a paging and I/O permission bitmap.

6.6.1 486 Pin Configurations (Fig. 6-6)

Address Bus. The 486 address bus is made up of address lines A2 to A31 and byte enable lines BE0# through BE3#. The address bus and byte enable lines function the same in both the DX and SX series.

A2–A31 — The Address lines perform two functions. During an MPU bus cycle, address lines A2 to A31 are used to select the upper 30 bits of the address. During an MPU cycle, these lines act as outputs. During a non-MPU bus cycle, address lines A4 to A31 act as inputs. These lines are used by the cache unit for cache invalidations. During a non-MPU bus cycle, the cache unit monitors these lines, and if the other bus master writes to a memory location in DRAM that is currently stored in the internal cache of the 486, a copy of the data is stored in the associated cache location.

BE0#–BE3# — The Byte Enable output lines (active low) are used to enable the bytes of data to be transferred on the data bus during the current bus cycle. Address lines A2 to A31 select the address, and the byte enable lines determine the size of the transfer. Table 6-5 specifies which bytes are transferred.

TABLE 6-4
80486 Specifications

Description	*486SX*	*486SX2*	*486DX*	*486DX2*	*486DX4*
Address lines	32	32	32	32	32
Physical address space	4 GB	4 GB	4 GB	4 GB	4 GB
Data lines	32	32	32	32	32
External clock speed	20 MHz	—	20 MHz	—	—
	25 MHz	25 MHz	25 MHz	25 MHz	—
	33 MHz	33 MHz	33 MHz	33 MHz	33 MHz
	—	—	50 MHz	—	—
Internal clock speed	—	50 MHz	—	50 MHz	—
	—	66 MHz	—	66 MHz	99 MHz
Number of pins	168	168	168	168	168
Math coprocessor (internal)	no	no	yes	yes	yes

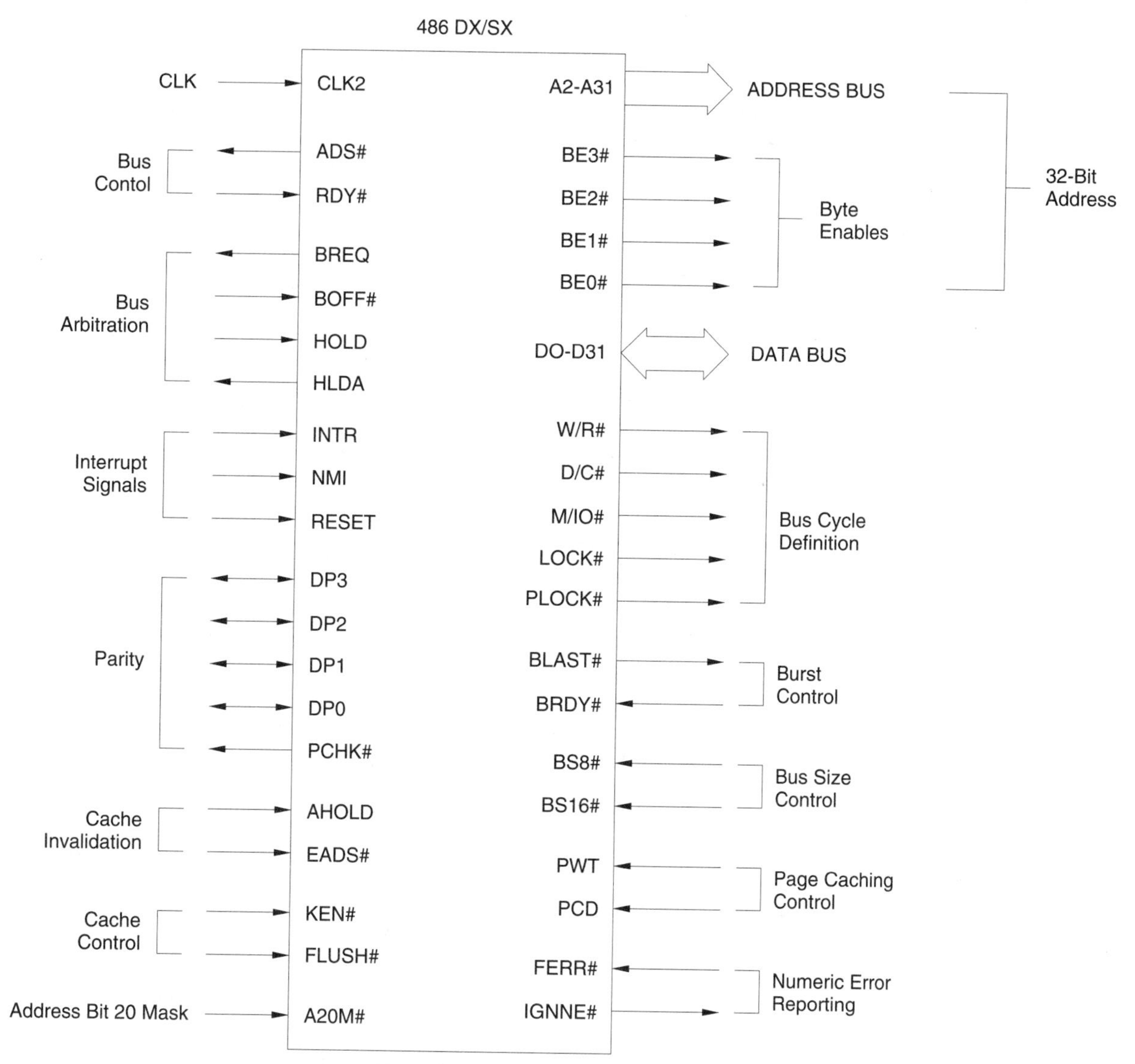

Figure 6-6 486DX/SX pin groupings.

TABLE 6-5
80486 Byte Enable

BE3#	*BE2#*	*BE1#*	*BE0#*	*Size of Transfer*	*Memory Block*
1	1	1	0	D0–D7 byte (1) (LSB)	1
1	1	0	1	D8–D15 byte (2)	2
1	0	1	1	D16–D23 byte (3)	3
0	1	1	1	D24–D31 byte (4) (MSB)	4
1	1	0	0	D0–D15 word (low)	1 and 2
0	0	1	1	D16–D31 word (high)	3 and 4
0	0	0	0	D0–D31 double word	1, 2, 3, and 4

TABLE 6-6
80486 Bus Cycle Definition

M/IO#	*D/C#*	*W/R#*	*Bus Cycle*
0	0	0	Interrupt acknowledge
0	0	1	Halt/special cycle
0	1	0	I/O read
0	1	1	I/O write
1	0	0	Memory code read
1	0	1	Reserved
1	1	0	Memory read
1	1	1	Memory write

Bus Cycle Definition (Table 6-6)

M/IO# The Memory/I/O line is used to determine if the bus operation uses a memory location or an I/O device. When high a memory operation is indicated, and when low an I/O operation is indicated. This output is used with other outputs to inform the bus controller which type of bus cycle is to be performed.

D/C# The Data/Control line is used to determine if a data transfer or some type of control cycle is to be performed. When high, the transfer contains data, and when low, some form of control bus cycle is performed. This output is used with other outputs to inform the bus controller which type of bus cycle is to be performed.

W/R# The Write/Read line is used to determine if some type of write or read bus cycle is to be performed. When high, a write operation is indicated, which means that the data lines enabled are used as outputs. When low, a read operation is indicated, and the enabled data lines act as inputs. This output is used with other outputs to inform the bus controller which type of bus cycle is to be performed. This ADS# line initiates the bus cycle.

LOCK# The bus LOCK output line is used to lock out the use of the bus by any non-MPU bus master during an interrupt sequence or a cache read-modify-write cycle. LOCK becomes active during the first clock of the bus cycle and remains active until the ready line goes active for the last locked bus cycle. This action is somewhat different from the 386 because the 486 contains a built-in cache controller.

PLOCK# The Pseudo LOCK output line is used to lock out the use of the bus by any non-MPU bus master during read or write cycles that are larger than the size of the data bus (32 bits). This line becomes active during 64-bit floating point transfers, 64-bit segment table descriptor reads, and 128-bit cache line fills. These types of transfers require more than one bus cycle, and the PLOCK# remains active until the transfers are complete.

Bus Arbitration

HOLD The bus HOLD request is an input used by the DMA or local bus master to request control of the bus. When high the local bus master wants to use the bus. Just before the MPU gives permission to use the bus it tristates most of its lines (high impedance). The MPU keeps the BREQ, HLDA, FERR#, and PCHK# lines active during the hold request.

HLDA — The bus HoLD Acknowledge output is used by the MPU to grant permission to the DMA or local bus master to use the bus. Just before this line becomes active, the MPU tristates most of its lines.

BOFF# — The BackOFF input is used to cause the same effect in the MPU as the HOLD input. The difference is that when this line goes low, the MPU floats (tristates) its lines on the next clock cycle and the HLDA line does not go active in response to the BOFF# signal. If this line goes active during a bus cycle, the bus cycle is aborted, as apposed to waiting until the bus cycle is complete when using the HOLD signal. This signal has a higher priority than HOLD. It is used with an advanced AT bus system that has local bus masters.

BREQ — The Bus REQuest output is used by the MPU to indicate that a bus cycle is pending internally. The line becomes active during the first clock of an MPU bus cycle, along with the ADS# line, and during the first clock of a non-MPU bus cycle. This line never floats, regardless of the device in control of the bus cycle.

Burst Control. The burst control lines are used with the cache unit of the 486 MPU. These lines are active only during a cache burst cycle. During a burst cycle multiple data transfers occur, with the first cycle requiring two clocks and all remaining transfers during the burst cycle requiring only one clock each.

BRDY# — The Burst ReaDY input line performs the same function as the RDY# line during a burst cycle. This signal informs the MPU that the data bus is ready to complete the bus cycle. This line informs the MPU that valid data is on the data bus during a read operation or that the device being addressed has accepted the data during a write operation. During a multiple burst transfer address lines A2 and A3 change their values to indicate the new address. After each burst transfer, the ADS# goes active to indicate the valid address change.

BLAST# — The Burst LAST output is used to indicate that the next time the BRDY# line goes active it will be treated as a non-burst RDY# signal. When this line goes active, the next BRDY# signal is the final burst transfer.

Bus Size Control

BS8# — The Bus Size 8-bit input is used to inform the MPU to use only the lower 8 bits of the data bus when this line is low. This line causes the MPU to translate large data transfers (larger than 8 bits) into multiple smaller data transfers (8 bits each).

BS16# — The Bus Size 16-bit input is used to inform the MPU to use only the lower 16 bits of the data bus when this line is low. This line causes the MPU to translate large data transfers (larger than 16 bits) into multiple smaller data transfers (16 bits each).

Parity

DP0–DP3 — There are four Data Parity input/output lines, one for each byte of data on the data bus. During a memory write, the 486 generates a parity bit and outputs the parity bit to the bus. During a read operation (code access, memory, and I/O read) the 486 uses the lines as inputs and uses the parity bit along with the data bits to check for proper parity. DP0 is used with data bits D0 to D7, DP1 with bits D8 to D15, DP2 with bits D16 to D23, and DP3 with bits D24 to D31. The 486 generates and checks for even parity. Even parity means that the total number of logical 1s (data bits plus the parity bit) must be an even value (0, 2, 4, etc.).

PCHK# The Parity CHecK status output line is used to indicate that a parity error has occurred, when low. This line is valid only following a read operation.

Clock

CLK The CLocK input provides the basic timing for the internal circuitry in the 486. The clock input is not divided by 2 as in the 386. A non-burst bus cycle requires a minimum of two clocks (labeled T1 and T2). Burst bus cycles require a minimum of two clocks for the first transfer and one clock for each additional data transfer.

Data Bus

D0–D31 The 486 contains 32 bidirectional data lines. These lines act as outputs during a write operation and as inputs during a read operation. During a byte operation either the D0 to D7, D8 to D15, D16 to D23, or D24 to D31 lines are enabled. During a word operation either the D0 to D15 or D16 to D31 lines are enabled. During a long word (also called a double word) all data lines (D0 to D31) are enabled.

Bus Control

ADS# The ADdress Status output is used to indicate that the address and bus cycle definition signals are valid for the current bus cycle. This line goes low during the first clock of the bus cycle and goes high in the second clock of the bus cycle.

RDY# The ReaDY input is used to indicate that the data on the data bus is valid and the transfer should be terminated. This line is used to indicate the readiness of the bus during a non-burst bus cycle. In a read cycle this line indicates that the MPU should read the data bus. During a write cycle this line indicates that the external device has accepted the data on the data bus.

Interrupt Signals

INTR The INTerrupt Request (maskable interrupt) input informs the MPU that the PIC has requested the servicing of an interrupt, when this line goes high. The MPU samples this input at the beginning of each bus cycle. This line must be high for two processor cycles for the interrupt sequence to start and must remain high until the end of the first of the two interrupt acknowledge signals issued by the MPU. Once acknowledged, the MPU sends out two interrupt acknowledge signals on the bus definition lines. At the end of the second interrupt acknowledge, the MPU reads the lower 8 bits for the data bus to determine the vector code.

RESET The RESET input is used to reset the MPU to its starting value when this goes low for at least 15 clock cycles during a warm boot.

input is controlled by the clock generator or AT chip set and normally indicates a problem with the power supply.

NMI — The NonMaskable Interrupt input indicates a parity error when high for an external interrupt. If this line goes active, the internal NMI signal is disabled. The NMI signal does not cause the MPU to generate the interrupt acknowledge signals because the NMI generates an internal interrupt vector code of 02. This line must be valid for two processor cycles before the interrupt is acknowledged and must remain high for two additional processor cycles once acknowledged.

Cache Invalidation. The cache invalidation section of the 486 is used to ensure that the data stored in cache memory is the same as the data stored in main memory. During a non-MPU write cycle, the 486 monitors address lines A4 to A31. If the write cycle affects an address in main memory that is currently stored in cache memory, the cache entry of that memory location is invalidated. This means that the next time the MPU needs to use the contents of that cache location, the cache unit inside the 486 must update the cache entry from main memory before the operation can take place.

AHOLD — The Address HOLD request input goes high when another bus master is ready to take control of the bus. When this line goes active, the address bus floats.

EADS# — The External ADdresS valid input is used with an external device to indicate to the MPU that the address on the address bus is valid. When this line becomes active, the MPU causes address lines A4 to A31 to act as inputs so that the cache unit can monitor the address bus. The function of this pin is valid only during a non-MPU write operation.

Cache Control

KEN# — The cache ENable input is used to inform the 486 that the data being read for the current bus cycle is cacheable. The current bus cycle (usually memory read) is converted into a cache line fill cycle. Each cache line fill cycle requires 16 bytes of data; therefore the bus cycle requires 4 (32-bit), 8 (16-bit), or 16 (8-bit) transfers to complete the cycle. The BS8# and BS16# lines are used to specify whether the transfer is less than 32 bits. During a cache line fill cycle, the byte enable lines are ignored.

FLUSH# — The cache FLUSH input is used to force a flush of data from the internal cache memory in the 486. When active for one clock, the contents of cache memory are flushed.

Page Caching Control

PWT — The Page Write-Through output signal is used for the secondary cache system. The level on this output corresponds to the level of the PWT bit in the page table entry in the 486. When high the write-through policy is enabled for the current page. When low the write-through policy is disabled for the current page.

PCD — The Page Cache Disable output signal is used for the secondary cache system. The level on this corresponds to the level on the PCD bit in the page table entry in the 486. When this line is low, the on-chip cache is enabled. When high the on-chip cache is disabled.

Numeric Error Reporting

FERR# — The Floating point ERRor output is used to indicate that an error has occurred during the operation of a floating point instruction. This output can be applied to external control logic for error reporting.

IGNNE# — When active, the IGNore Numeric Error input causes the 486 to ignore any floating point error while allowing the execution of noncontrol floating point instructions. If this line is inactive during a floating point error, the 486 freezes when trying to execute a noncontrol floating point instruction.

Address Bit 20 Mask

A20M# — The Address bit 20 Mask bit input is used to help the 486 switch between the real and protected operating modes of the 486. While in the real mode this input must be low. In the real mode only the 20 LSBs of the address bus are available. In the protected mode all address bits of the address bus are available.

6.6.2 80486 Bus Cycles

The following are some of the most common bus cycles of the 486.

Basic 2-2 Bus Cycle. Figure 6-7 shows the shortest basic bus cycle, referred to as a 2-2 bus cycle. The following describes what happens during each bus cycle.

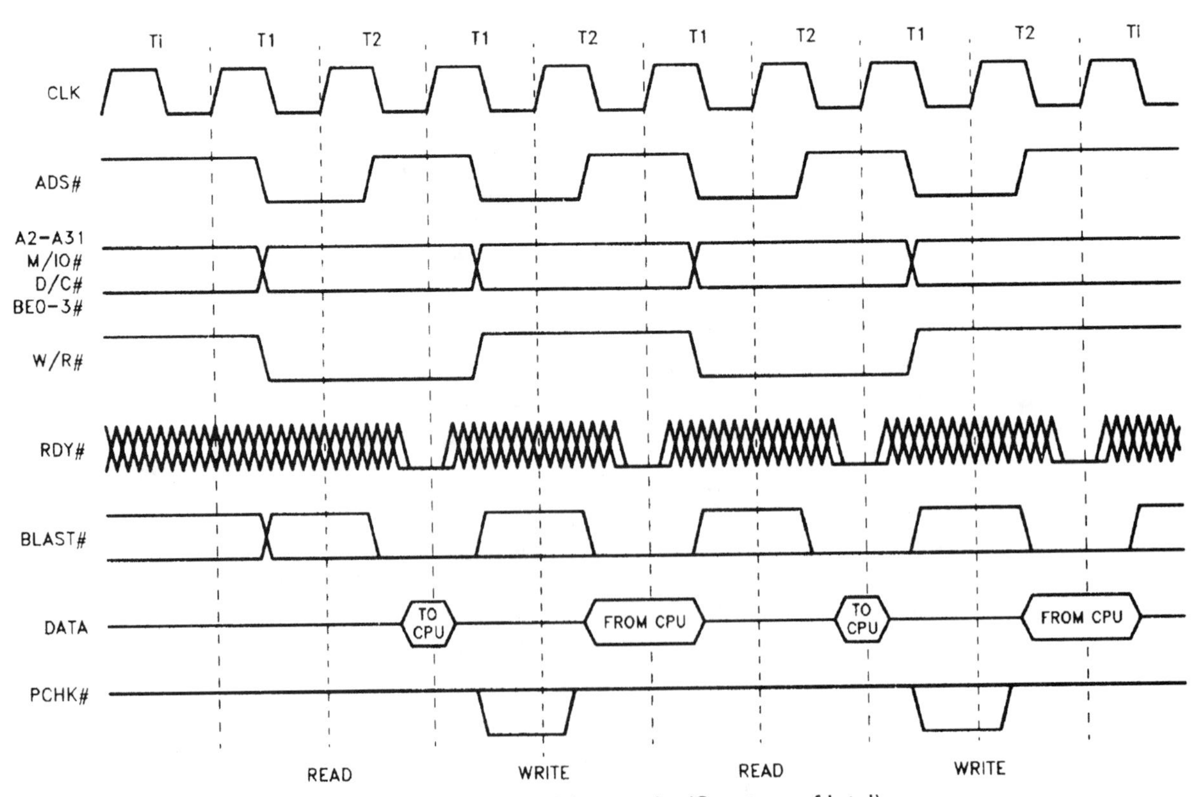

Figure 6-7 Basic 2-2 bus cycle. (Courtesy of Intel)

2-2 Read Bus Cycle

Clock 1 (T1)

1. The MPU places the address on the addresss bus (A2 to A31).
2. The MPU places the proper logic levels on the byte enable lines (BE0# to BE3#).
3. The M/IO# line goes to its proper logic state (depends on the operation).
4. The D/C# line goes to its proper logic state (depends on the operation).
5. The W/R# line goes low.
6. The ADS# line goes low (indicating a valid address and bus definition).
7. The levels on the BLAST# and PCHK# line have no bearing on the operation during this clock.

Clock 2 (T2)

1. The BLAST# line goes low.
2. The ADS# line goes high about halfway through T2.
3. At the end of T2 the MPU checks the levels on the non-burst ready input (RDY#)line. If the RDY# line is low (data is valid), the MPU transfers the data on the data bus into the MPU and the levels on the parity data lines (PD0 to PD3, depending on the size of the transfer), and the bus cycle ends. If the RDY# line is high at the end of T2, the MPU produces another T2 clock cycle. The levels on the address bus, bus definition, W/R#, BLAST#, and PCHK# lines do not change until the end of the new T2, at which time the RDY# is checked again.
4. As the data is being transferred into the MPU from the data bus and parity data line, the MPU checks for even parity. If the PCHK# goes high following a read operation, no parity error was generated; if low, a parity error was generated.

2-2 Write Bus Cycle

Clock 1 (T1)

1. The MPU places the address on the address bus (A2 to A31).
2. The MPU places the proper logic levels on the byte enable lines (BE0# to BE3#).
3. The M/IO# line goes to its proper logic state (depends on the operation).
4. The D/C# line goes to its proper logic state (depends on the operation).
5. The W/R# stays high throughout the rest of the bus cycle.
6. The ADS# line goes low (indicating a valid address and bus definition).
7. The levels on the BLAST# and PCHK# lines have no bearing on the operation during this clock.

Clock 2 (T2)

1. The BLAST# line goes low.
2. The ADS# line goes high about halfway through T2.
3. The MPU places the data on the data bus.
4. At the end of T2 the MPU checks the levels on the non-burst ready input (RDY#)line. If the RDY# is low, the device that has been addressed accepts the data on the data bus and the levels on the parity data lines (PD0 to PD3 depending on the size of the transfer) if a memory write is being performed, and the bus cycle ends. If the RDY# line is high at the end of T2, the MPU produces another T2 clock cycle. The levels on the address bus, bus definition, W/R#, BLAST#, and PCHK# lines do not change until the end of the new T2, at which time the RDY# is checked again.

5. Because during a write operation the MPU does not check parity, the level on the PCHK# line has no effect on the bus cycle.

Basic 3-3 Bus Cycle. Figure 6-8 shows the nonminimum basic bus cycle, referred to as a 3-3 bus cycle. The following describes what happens during each bus cycle.

The only difference between the 2-2 and 3-3 basic bus cycles is that at the end of clock 2 (T2) the RDY# line is high, which causes the MPU to produce a second clock 2 (T2), at which time the RDY# line is low, ending the bus cycle. Therefore the bus cycle takes 50% more time to complete.

Non-Cacheable Burst Cycle. Figure 6-9 shows a typical non-cacheable read burst cycle, during which time two transfers take place. A burst cycle takes place when the data to be transmitted contains more bytes than the data bus can handle at one time. The first transfer takes a minimum of two clocks, and each transfer that follows requires a minimum of one clock. The BRDY# line is used in conjunction with the RDY# and BLAST# lines to control the transfers.

Clock 1 (T1) (First Transfer Cycle)

1. The MPU places the address on the address bus (A2 to A31).
2. The MPU places the proper logic levels on the byte enable lines (BE0# to BE3#).
3. The bus definition lines (M/IO#, D/C#, and W/R#) go to their proper levels.
4. The ADS# line goes low.
5. The KEN# line stays high because the transfer is not cacheable.
6. The level on the BLAST# line during this clock has no effect.
7. The data bus is in tristate.

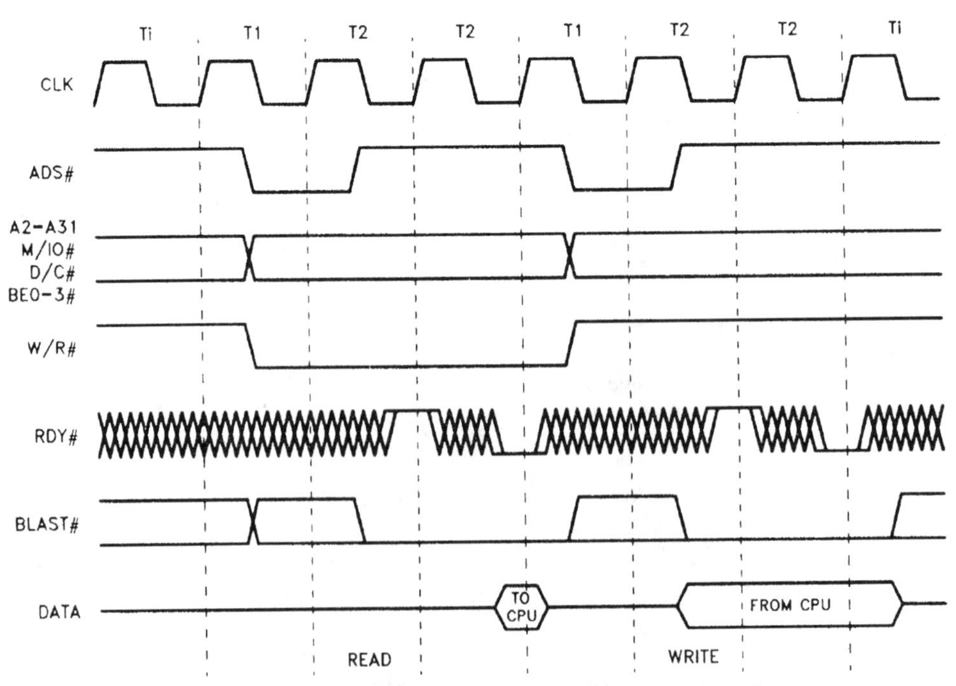

Figure 6-8 Basic 3-3 bus cycle. (Courtesy of Intel)

Clock 2 (T2) (First Transfer Cycle)

1. The ADS# line goes low about halfway through T2.
2. At the end of T2, RDY# goes high, BRDY# goes low, and BLAST# goes high. This indicates that a burst cycle is in progress (BRDY# low and RDY# high), and because the BLAST# line is high this is not the last transfer in the burst cycle.
3. At the end of T2, the data is transferred into the MPU.

Clock 2 (T2) (Second Transfer Cycle)

1. The MPU places the address on the address bus (A2 to A31) based on the burst bus cycle information inside the MPU.
2. The MPU places the proper logic levels on the byte enable lines (BE0# to BE3#) based on the burst bus cycle information inside the MPU.
3. Because this is a burst bus cycle the ADS# line does not have to change.
4. Near the end of the second T2 clock RDY# goes high, BRDY# goes low, and BLAST# goes low. The RDY# and BRDY# lines indicate that the bus is ready. The BLAST# line going low indicates that this is the last transfer in this burst bus cycle.
5. At the end of the second T2 the data is transferred into the MPU, which ends the burst bus cycle.

Non-Burst Cacheable Cycles. Figure 6-10 shows a typical non-burst cacheable read cycle with a cache line fill. Each cache entry is 16 bytes wide. Because the data bus is only 32 bits (4 bytes) wide, whenever a cacheable read is performed, the MPU performs four 32-bit transfers (or eight 16-bit transfers or 16 8-bit transfers) depending on the device being accessed. Figure 6-10 shows four 32-bit data transfers. The KEN# line is used to indicate the beginning of the cacheable bus cycle. The KEN# and BLAST# lines indicate the end of the cacheable bus cycle.

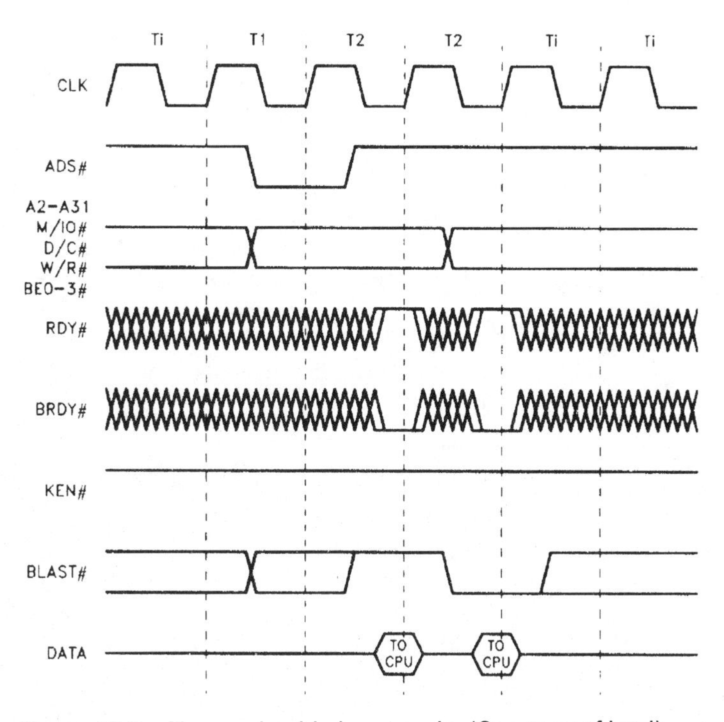

Figure 6-9 Non-cacheable burst cycle. (Courtesy of Intel)

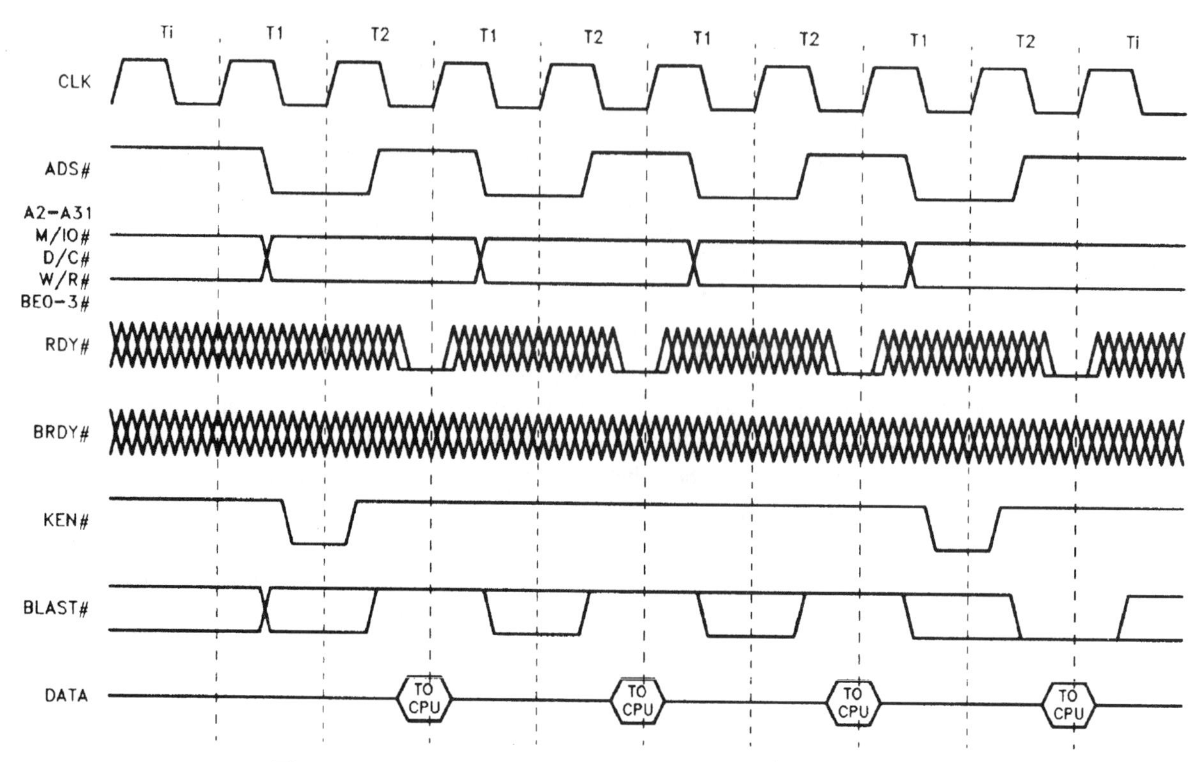

Figure 6-10 Non-burst cacheable cycle. (Courtesy of Intel)

Clock 1 (T1) (First 4-Byte Transfer Cycle)

1. The MPU places the address on the address bus (A2 to A31).
2. The MPU enables all byte enable lines (BE0# to BE3#) in this example.
3. The bus definition lines (M/IO#, D/C#, and W/R#) go to their proper levels.
4. The ADS# line goes low.
5. The KEN# line goes low (this indicates a cacheable operation).
6. The level on the BLAST# line during this clock has no effect.
7. The data bus is in tristate.

Clock 2 (T2) (First 4-Byte Transfer Cycle)

1. The KEN# line goes high about one-third of the way through T2.
2. The ADS# line goes low about halfway through T2.
3. The BLAST# line goes high (inactive, which indicates that this is not the end of the cacheable cycle).
4. The RDY# line goes high by the end of T2.
5. The BRDY# line does not affect the operation because this is not a burst cycle.
6. The data is transferred into the MPU at the end of T2. *Note:* Because this is not a burst cycle, the ADS# line is processed on each transfer.

Clock 1 (T1) (Second 4-Byte Transfer Cycle)

1. The MPU places the next address on the address bus (A2 to A31).
2. The MPU enables all byte enable lines (BE0# to BE3#) in this example.

3. The bus definition lines (M/IO#, D/C#, and W/R#) go to their proper levels.
4. The ADS# line goes low.
5. The KEN# line stays high (the cache unit is already on).
6. The level on the BLAST# line during this clock has no effect.
7. The data bus is in tristate.

Clock 2 (T2) (Second 4-Byte Transfer Cycle)

1. The ADS# line goes low about halfway through T2.
2. The BLAST# line goes or stays high (inactive, which indicates that this is not the end of the cacheable cycle).
3. The RDY# line goes high by the end of T2.
4. The BRDY# line does not affect the operation because this is not a burst cycle.
5. The data is transferred into the MPU at the end of T2.

Third 4-Byte Transfer Cycle

The third 4-byte transfer cycle uses the same sequence as the second 4-byte transfer cycle except that a new address is used.

Clock 1 (T1) (Fourth 4-Byte Transfer Cycle)

1. The MPU places the next address on the address bus (A2 to A31).
2. The MPU enables all byte enable lines (BE0# to BE3#) in this example.
3. The bus definition lines (M/IO#, D/C#, and W/R#) go to their proper levels.
4. The ADS# line goes low.
5. The KEN# line goes low (the cache unit becomes disabled at the end of the next data transfer).
6. The level on the BLAST# line during this clock has no effect.
7. The data bus is in tristate.

Clock 2 (T2) (Fourth 4-Byte Transfer Cycle)

1 The KEN# line goes high about one-third of the way through T2.
2. The ADS# line goes low about halfway through T2.
3. The BLAST# line goes low (this indicates that this is the last cacheable data transfer).
4. The RDY# line goes high by the end of T2.
5. The BRDY# line does not affect the operation because this is not a burst cycle.
6. The data is transferred into the MPU at the end of T2.

Burst Cacheable Cycle. Figure 6-11 shows a typical burst cacheable read bus cycle with a cache line fill. Each cache entry is 16 bytes wide. Because the data bus is only 32 bits (4 bytes) wide, whenever a cacheable read occurs the MPU performs four 32-bit transfers (or eight 16-bit transfers or 16 8-bit transfers), depending on the device being accessed. Figure 6-11 shows four 32-bit data transfers. The KEN# line is used to indicate the beginning of the cacheable bus cycle. The KEN# and BLAST# lines are used to indicate the end of the cacheable bus cycle. Because this shows a burst cycle, the ADS# line goes active only once during the entire bus cycle. Because this is a burst bus cycle, only the first transfer requires two clocks; all remaining transfers require one clock each.

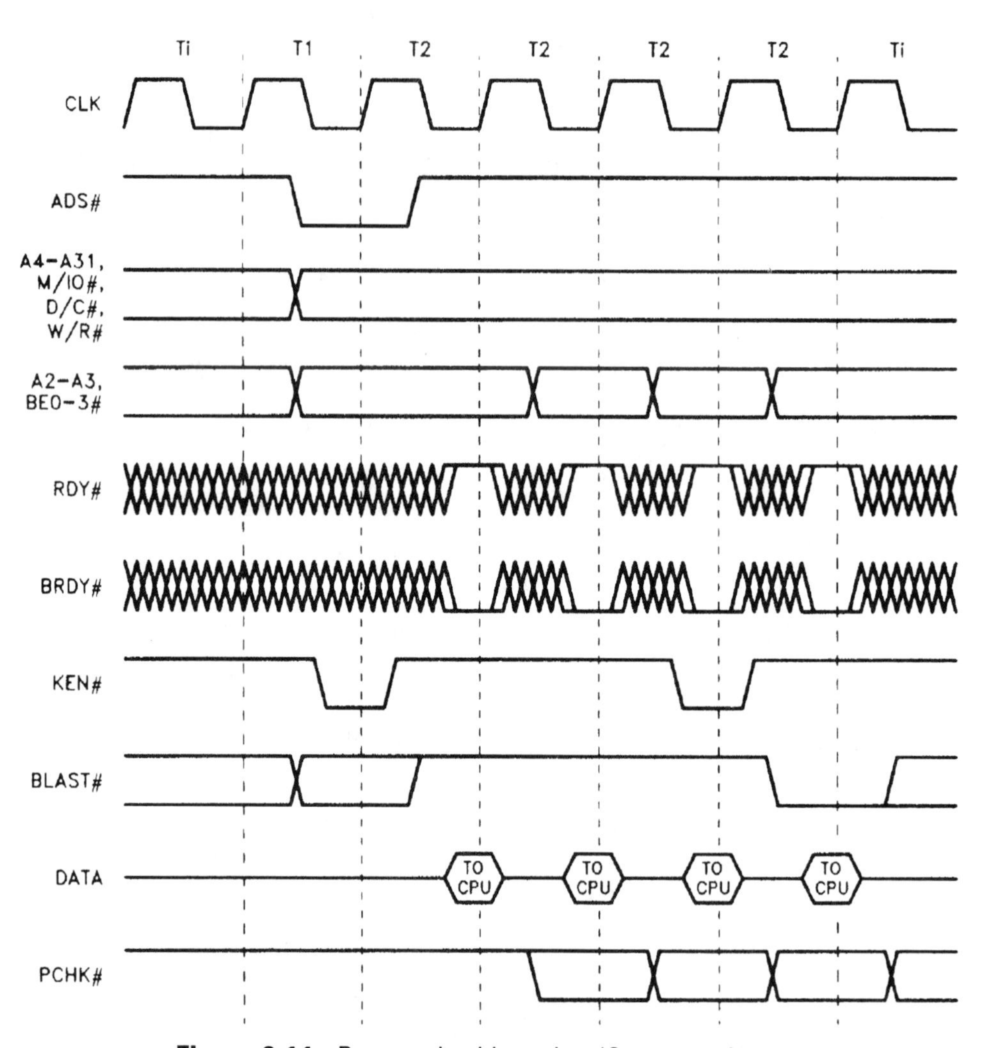

Figure 6-11 Burst cacheable cycles. (Courtesy of Intel)

Clock 1 (T1) (First 4-Byte Transfer Cycle)

1. The MPU places the address on the address bus (A2 to A31).
2. The MPU enables all byte enable lines (BE0# to BE3#) in this example.
3. The bus definition lines (M/IO#, D/C#, and W/R#) go to their proper levels.
4. The ADS# line goes low.
5. The KEN# line goes low (this indicates a cacheable operation).
6. The level on the BLAST# line during this clock has no effect.
7. The data bus is in tristate.

Clock 2 (T2) (First 4-Byte Transfer Cycle)

1. The KEN# line goes high about one-third of the way through T2.
2. The ADS# line goes low about halfway through T2.
3. The BLAST# line goes or stays high (inactive, which indicates that this is not the end of the cacheable cycle).
4. The RDY# line goes high by the end of T2.
5. The BRDY# line goes low at the same time RDY# goes high.
6. The data is transferred into the MPU at the end of T2. *Note:* Because this is a burst cycle, the ADS# line is not needed for the remaining part of the bus cycle.

Clock 2 (T2) (Second 4-Byte Transfer Cycle)

1. The MPU places the next address on the address bus (A2 to A31).
2. The MPU enables all byte enable lines (BE0# to BE3#) in this example.
3. The bus definition lines (M/IO#, D/C#, and W/R#) go to their proper levels.
4. The ADS# line stays high.
5. The KEN# line stays high (the cache unit is already on).
6. The level on the BLAST# line is high (inactive).
7. The RDY# line goes low before the end of T2.
8. The BRDY# line goes high at the same time RDY# goes low.
9. The MPU accepts the data on the data bus.

Clock 2 (T2) (Third 4-Byte Transfer Cycle)

The sequence of T2 for the third 4-byte transfer cycle is the same as T2 for the second 4-byte transfer cycle.

Clock 2 (T2) (Fourth 4-Byte Transfer Cycle)

1. The MPU places the next address on the address bus (A2 to A31).
2. The MPU enables all byte enable lines (BE0# to BE3#) in this example.
3. The bus definition lines (M/IO#, D/C#, and W/R#) go to their proper levels.
4. The ADS# line stays high.
5. The KEN# line goes low just before the beginning of T2 (the cache unit is disabled at the end of the bus cycle). This line remains low for about one fourth of the way through T2.
6. The BLAST# line goes low about halfway through T2 (indicating that this is the last transfer of the bus cycle).
7. The RDY# line goes low before the end of T2.
8. The BRDY# line goes high at the same time the RDY# line goes low.
9. The MPU accepts the data on the data bus, and the bus cycle ends.

6.7 PENTIUM PROCESSOR

The Pentium processor is Intel's newest line of microprocessors. The Pentium is compatible with all other Intel xxx86 microprocessors. The technical innovations of the Pentium processor include superscalar architecture, separate 8K code and data caches with MESI (modified, exlusive, shared, invalid) protocol writeback data cache, high-performance pipelined floating point unit, 64-bit data bus, dynamic branch prediction, data integrity and error detection, multiprocessing support, memory page sizing, and performance monitoring. Figure 6-12 shows the internal processor block diagram of the Pentium.

Superscalar Architecture. The Pentium uses superscalar architecture, which means that the processor contains more than one execution unit. The Pentium contains two separate execution units (general purpose registers, flags registers, and ALUs) and address generation logic for each of the two pipelines, which allows the execution of more than one instruction per clock cycle. Each execution unit has its own pipeline, known as the U and V pipelines. The pipelines share the other resource blocks of the Pentium processor to achieve maximum processing speed. The U pipeline executes any type of integer and floating point instructions, while the V pipeline executes FXCH floating point and simple integer instructions. The Pentium has increased the number of hardwired instructions, which can be processed much faster than instructions requiring microcode. The Pentium allows

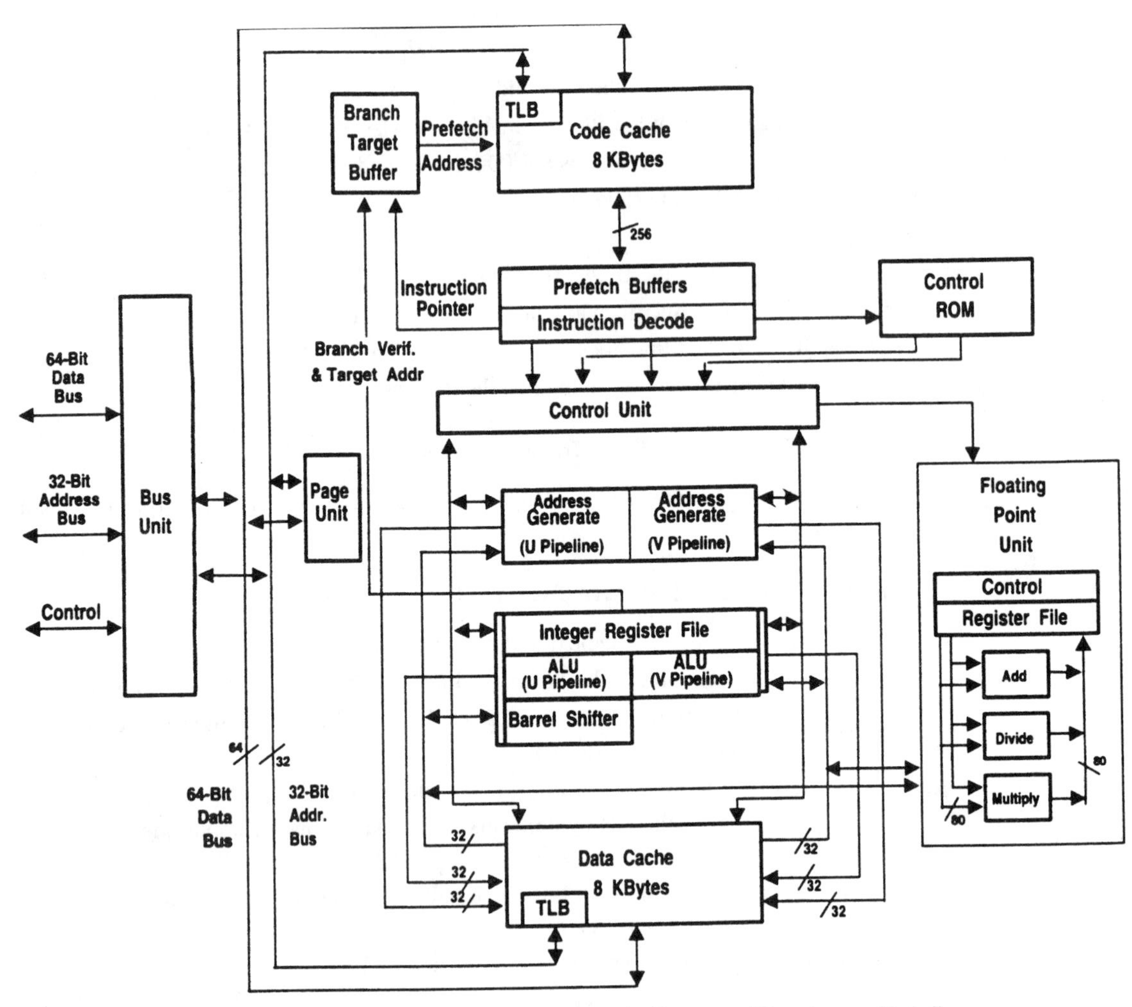

Figure 6-12 Pentium processor block diagram. (Courtesy of Intel)

the execution of integer instruction in stages, so that several instructions can be at different stages of execution.

Separate Code and Data Caches. The Pentium contains two separate 8K code and data caches. Each of these caches is configured with a 32-byte cache line as a two-way set associative cache; this reduces the time it takes to search for a cache line. Each cache unit has its own translation lookaside buffer (TLB), which translates linear addresses into physical addresses. The 64-bit internal data bus is applied to both of these caches to allow full bandwidth transfers to take place. The separate code and data caches reduce bus conflicts because the caches are available for most of the bus cycles, which increases the performance of the processor and bus cycle.

The data cache has dual ports (one port per pipeline), while tags have triple ports. The data cache uses the MESI protocol algorithm, which ensures the consistency of the data in both cache and main memory. The data cache also employs a write-back caching configuration. Write-back caching allows writes to update cache without the need to access main memory, as with older write-through configurations.

The code cache holds the instructions that have been fetched from the external bus. The branch target and prefetch instruction buffers are used to supply instructions to the execution units. The code cache also uses its own TLB to convert linear addresses into physical addresses.

High-Performance Floating Point Unit. The Pentium's floating point unit performs much faster than previous floating point units because many of the instructions are now hardwired. The floating point unit appends a three-stage instruction pipeline to the integer pipelines to speed up integer instructions and employs a seven-stage pipeline for other data types.

64-Bit Data Bus. The 64-bit data bus doubles the bandwidth of the 486 data bus. The 64-bit internal data bus is also pipelined, which allows a new bus cycle to start before the previous bus cycle is complete. The pipelining reduces bottlenecking in the external system by providing more time for external circuitry to process the bus cycle.

Branch Prediction. Branch prediction is a technique normally found in mainframe computers. The branch prediction section is responsible for predetermining the most likely group of instructions needed for a branch routine. The reliability of branch prediction is increased because instructions are usually already in the code cache.

Data Integrity and Error Detection. The Pentium contains parity generation and checking for both addresses and data. The Pentium also performs internal parity checks to determine if any parity errors have been generated. A functional redundancy check function is also available in multiple Pentium systems, which ensures data integrity for mission-critical applications.

Multiprocessing Support. The Pentium supports multiple Pentium systems because of its superscalar architecture and separate code and data caches. Additional control and status lines, along with a functional redundancy checking mode, allow for more stable interfacing.

Memory Page Sizing. The memory page size is variable from 4K to 4MB. Larger page sizes reduce the number of times page swapping is needed in different operating systems, graphics, and frame buffers, which increases the speed of data access.

Performance Monitoring. Performance monitoring provides special signals to indicate performance bottlenecking in software coding and hardware, which allows designers and software developers to optimize Pentium performance.

6.7.1 The Pentium Processor Versus the 80486

The Pentium processor was built to function in much the same way as the 486 during many bus cycles. Table 6-7 lists the characteristics of the Pentium. The reason for this similarity was to maintain compatibility with current hardware and software. The major differences are the

TABLE 6-7
Pentium Specifications

Description	*Pentium*
Address lines	32
Physical address space	4 GB
Data lines	64
Clock speeds	60 MHz
	66 MHz
	75 MHz
	90 MHz
	100 MHz
	120 MHz
Number of pins	273
Type of package	GAP

size of the data bus, operational speed, superscalar architecture, dual pipelining, how the Pentium processes information internally, additional error detection, performance monitoring, and testing. On average, the Pentium has a minimum of twice the throughput of a 486DX2, and in certain operations more than five times the throughput. The Pentium supports all previous instruction sets of the Intel xxx86 family of processors.

Many of the bus cycle sequences are the same in the Pentium as in the 486, with the exception of a larger data bus, greater bus speed, and additional signals.

6.7.2 Pentium Pin Configuration

32-Bit Address Bus. The Pentium address bus is made up of address lines A3 to A31 and byte enable lines BE0# to BE7#.

A3–A31 — During a normal bus cycle, address lines A3 to A31 act as outputs to supply the address to the address bus. During a non-MPU bus cycle (such as a DMA cycle), address lines A5 to A31 act as inputs, which allows the MPU to monitor the address being accessed. If an address is accessed that is currently stored in cache, the cache for the address is invalidated.

BE0#–BE7 — The byte enable lines (BE0# to BE7#) are used to determine which bytes on the data bus are enabled. Because the data bus is 64 bits in size, eight byte enable lines are needed to control the transfer. The values on the internal address bits A0 to A2 and the size of the transfer determine which of the byte enable lines is enabled and when. The size of the transfer and the internal levels on address lines A0 to A2 determine if one or more than one transfer is needed. Table 6-8 shows the relationship between the byte enable lines, the internal address lines A0 to A2, and the size of the data transfer.

The byte enable lines are applied in different forms to the different sizes of memory and I/O devices. Figure 6-13 shows how the byte enable signals are converted into the proper signals for different sizes of memory and I/O devices. When addressing 64-bit memory, address lines A3 to A31 and byte enable lines BE0# to BE7# lines are used directly. When addressing 32-bit memory, the byte select logic creates address line A2 and BE0# through BE3#, which are used to control the 32-bit wide memory addresses. When addressing 16-bit memory, the byte select logic creates address lines A2 and A1 and BHE# and BLE# lines, which are used to control the 16-bit memory addresses. When addressing 8-bit memory, the byte select logic creates address lines A0, A1, and A2, which are used to control the 8-bit wide memory addresses. Table 6-9 demonstrates how data transfers are developed for byte, word, double word, and quad word. In some transfers more than one bus cycle are required. A transfer requires more than one bus cycle if it requires data from two different address boundaries.

TABLE 6-8
Pentium Address, Byte Enables, and Data Bus Chart

A2	*A1*	*A0*	*Byte Enable*	*Data Bus*
0	0	0	BEO#	Byte 0 (D0–D7)
0	0	1	BE1#	Byte 1 (D8–D15)
0	1	0	BE2#	Byte 2 (D16–D23)
0	1	1	BE3#	Byte 3 (D24–D31)
1	0	0	BE4#	Byte 4 (D32–D39)
1	0	1	BE5#	Byte 5 (D40–D47)
1	1	0	BE6#	Byte 6 (D48–D55)
1	1	1	BE7#	Byte 7 (D56–D63)

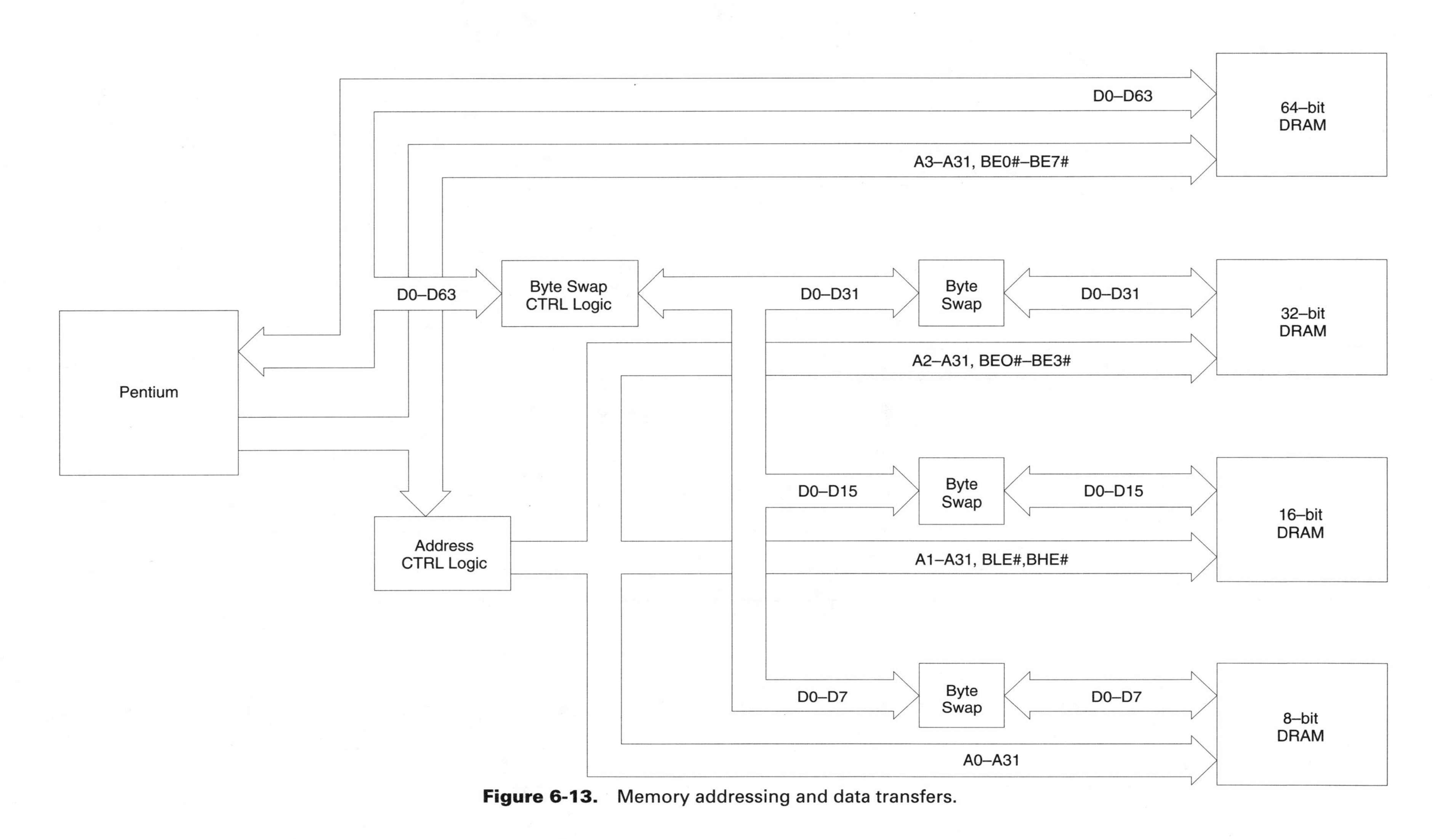

Figure 6-13. Memory addressing and data transfers.

TABLE 6-9

Transfer Bus Cycles for Bytes, Words, Double Words, and Quad Words. (Courtesy of Intel)

Length of Transfer	*1 Byte*	*2 Bytes*							
Low Order Address	*xxx*	*000*	*001*	*010*	*011*	*100*	*101*	*110*	*111*
First Transfer	b	w	w	w	hb	w	w	w	hb
Byte enables driven	0	BE0# to BE1#	BE1# to BE2#	BE2# to BE3#	BE4#	BE4# to BE5#	BE5# to BE6#	BE6# to BE7#	BE0#
Value driven on A3		0	0	0	0	0	0	0	1
Second transfer (if needed)					lb				lb
Byte enables driven					BE3#				BE7#
Value driven on A3					0				0
Length of Transfer		*4 Bytes*							
Low Order Address		*000*	*001*	*010*	*011*	*100*	*101*	*110*	*111*
First transfer		d	hb	hw	h3	d	hb	hw	h3
Byte enables driven		BE0# to BE3#	BE4#	BE4# to BE5#	BE4# to BE6#	BE4# to BE7#	BE0#	BE0# to BE1#	BE0# to BE2#
Low order address		0	0	0	0	0	1	1	1
Second transfer (if needed)			l3	lw	lb		l3	lw	lb
Byte enables driven			BE1# to BE3#	BE2# to BE3#	BE3#		BE5# to BE7#	BE6# to BE7#	BE7#
Value driven on A3			0	0	0		0	0	0
Length of Transfer		*8 Bytes*							
Low Order Address		*000*	*001*	*010*	*011*	*100*	*101*	*110*	*111*
First Transfer		q	hb	hw	h3	hd	h5	h6	h7
Byte enables driven		BE0# to BE7#	BE0#	BE0# to BE1#	BE0# to BE2#	BE0# to BE3#	BE0# to BE4#	BE0# to BE5#	BE0# to BE6#
Value driven on A3		0	1	1	1	1	1	1	1
Second transfer (if needed)			l7	l6	l5	ld	l3	lw	lb
Byte enables driven			BE1# to BE7#	BE2# to BE7#	BE3# to BE7#	BE4# to BE7#	BE5# to BE7#	BE6# to BE7#	BE7#
Value driven on A3			0	0	0	0	0	0	0

Key: b= byte transfer w=2-byte transfer 3=3-byte transfer d=4-byte transfer h=high order
5=5-byte transfer 6=6-byte transfer 7=7-byte transfer q=8-byte transfer l=low order

8-byte operand:

high order byte	byte 7	byte 6	byte 5	byte 4	byte 3	byte 2	low order byte

↑ byte with highest address ↑ byte with lowest address

Data Bus

D0–D63 The data bus contains 64 bidirectional data lines. These lines are used as outputs during a write operation and as inputs during a read operation. Because not all bus cycles use all the data bus lines, the level on these lines during certain bus cycles is not valid. Because the data bus is connected to 64-bit, 32-bit, 16-bit, and 8-bit devices, additional external logic, called byte swap logic, is used to make sure the correct data is supplied to the Pentium during the bus cycle. The byte swap logic is used to translate the data between the device being accessed and the proper data lines of the Pentium. Address lines A3 to A31 and byte enable lines BE0# to BD7# are used to control how the byte swap logic translates the data to and from the memory blocks.

Bus Cycle Definition. The bus cycle definition lines are used to define the type of cycle to be performed. These lines perform the same function as in the 486 MPUs.

M/IO# The Memory and I/O output is used with the other bus cycle definition lines to define the type of bus cycle to be performed. When high, memory is accessed. When low, an I/O device is accessed.

D/C# The Data and Code output is used with the other bus cycle definition lines to define which type of bus cycle is performed. When high, the information on the data bus is data. When low, the information on the data bus is some type of code.

W/R# The Write and Read output is used with the other bus cycle definition lines to define the type of bus cycle to be performed. When high, a write operation takes place (data is transferred from the MPU into the device being accessed). When low, a read operation takes place (data is transferred from the device being accessed into the MPU).

LOCK# The LOCK bus output line, when low, is used to lock out any device from asking permission to control the bus.

Tables 6-10 and 6-11 show the type of bus cycles selected by the bus cycle definition lines. Because the Pentium also contains a built-in internal cache unit and cache memory, Table 6-10 also specifies the condition of the CACHE# and KEN# lines to indicate if the operation is to be cached. Table 6-11 specifies special bus cycle encoding.

Bus Arbitration

HOLD The bus HOLD request is an input used by the DMA or local bus master to request control of the bus. When high, the local bus master wants to use the bus. Just before the MPU gives permission to use the bus, it tristates most of its lines (high impedance).

HLDA The bus HoLD Acknowledge output is used by the MPU to grant permission to the DMA or local bus master to use the bus. Just before this line becomes active, the MPU tristates most of its lines.

BOFF# The BackOFF input is used to cause the same effect in the MPU as the HOLD input. The difference is that when this line goes low, the MPU floats (tristates) its lines on the next clock cycle and the HLDA line does not go active in response to the BOFF# signal. If this line goes active during a bus cycle, the bus cycle is aborted, as opposed to

TABLE 6-10
Pentium Processor Initialed Bus Cycle. (Courtesy of Intel)

M/IO#	*D/C#*	*W/R#*	*CACHE#**	*KEN#*	*Cycle Description*	*Number of Transfers*
0	0	0	1	x	Interrupt acknowledge (2 locked cycles)	1 transfer each cycle
0	0	1	1	x	Special cycle	1
0	1	0	1	x	I/O read, 32-bits or less, non-cacheable	1
0	1	1	1	x	I/O write, 32-bits or less, non-cacheable	1
1	0	0	1	x	Code read, 64-bits, non-cacheable	1
1	0	0	x	1	Code read, 64-bits, non-cacheable	1
1	0	0	0	0	Code read, 256-bit burst line fill	4
1	0	1	x	x	Intel reserved (will not be driven by the Pentium processor)	n/a
1	1	0	1	x	Memory read, 64-bits or less, non-cacheable	1
1	1	0	x	1	Memory read, 64-bits or less, non-cacheable	1
1	1	0	0	0	Memory read, 256-bit burst line fill	4
1	1	1	1	x	Memory write, 64-bits or less, non-cacheable	1
1	1	1	0	x	256-bit burst write-back	4

*CACHE# will not be asserted for any cycle in which M/IO# is driven low, or for any cycle in which PCD is driven high.

TABLE 6-11
Special Bus Cycle Encoding. (Courtesy of Intel)

BE7#	*BE6#*	*BE5#*	*BE4#*	*BE3#*	*BE2#*	*BE1#*	*BE0#*	*Special Bus Cycle*
1	1	1	1	1	1	1	0	Shutdown
1	1	1	1	1	1	0	1	Flush (INVD, WBINVD instr)
1	1	1	1	1	0	1	1	Halt
1	1	1	1	0	1	1	1	Write-back (WBINVD instruction)
1	1	1	0	1	1	1	1	Flush acknowledge (FLUSH# assertion)
1	1	0	1	1	1	1	1	Branch trace message

waiting until the bus cycle is complete when using the HOLD signal. This signal has a higher priority than HOLD. It is used with advanced AT bus systems that have local bus masters. Once the bus hold is complete using the BOFF# line, the MPU restarts the aborted bus cycle.

BREQ

The Bus REQuest output is used by the MPU to indicate that a bus cycle is pending internally. The line becomes active during the first clock of an MPU bus cycle along with the ADS# line and during the first clock of a non-MPU bus cycle. This line never floats, regardless of the device in control of the bus cycle.

Burst Control

BRDY# — The Burst ReaDY input line informs the Pentium that the data bus is ready to complete the bus cycle. This signal informs the Pentium that valid data is on the data bus during a read operation or that the device being addressed has accepted the data during a write operation. During a single transfer only one BRDY# signal is returned. During a cache line fill, the cycle ends after the fourth BRDY# is returned.

Error Detection

DP0–DP7 — There are eight Data Parity input/output lines, one for each byte of data on the data bus. During a memory write, the Pentium generates a parity bit to provide even parity and outputs the parity bit to the bus. During a read operation (code access, memory, and I/O read), the Pentium checks for even parity. DP0 is used with data bits D0 to D7, DP1 with bits D8 to D15, DP2 with bits D16 to D23, DP3 with bits D24 to D31, DP4 with bits D32 to D39, DP5 with bits D40 to D48, DP6 with bits D48 to D55, and DP7 with bits D56 to D63.

PCHK# — The data Parity CHecK status output line is used to indicate that a parity error has occurred on the data bus, when low following a read operation. The parity checker section of the Pentium checks for even parity on the bits specified in the read operation. This signal becomes valid two clocks after the BRDY# signal and stays valid for one clock for each parity error detected. System circuitry is responsible for processing and determining what action should be taken.

PEN# — The Parity ENable input indicates whether a machine check interrupt is enabled as a result of the detection of a data parity error. When PEN# is active during the reception of a data parity error, the cycle information and physical address are saved in the machine check type and machine check address registers. The MEC bit of CR4 determines how the Pentium processes this information.

AP — The Address Parity line is bidirectional. When the Pentium controls the bus cycle, a parity bit is generated to produce even parity for the address bus. During a bus inquire cycle (address bus not controlled by the Pentium), this line acts as an input and checks for even parity for the address bus.

APCHK# — The Address Parity CHecK status output line indicates that a parity error has occurred during a bus inquire cycle (address bus not controlled by the Pentium). This line becomes valid two clocks after EADS# is sampled when active. This line remains active for one clock for each address parity error detected.

BUSCHK# — The BUS CHecK input is used to inform the Pentium that the system was unsuccessful in completing a bus cycle, when low. The BUSCHK# line is sampled during read and write cycles on any clock edge that BRDY# is sampled. This line indicates that an error has occurred in a synchronous bus cycle. If this line becomes active, the cycle type and address are latched into the machine check type and machine check address registers. The MCE bit of CR4 determines how the Pentium processes this information.

IERR# — The Internal ERRor output is used to indicate one of two types of internal errors in the Pentium: (1) A parity error during the reading of any of the internal arrays of the Pentium (data or instruction cache storage or tag arrays, data or instruction TLB storage and tag arrays, and the microcode ROM). This output goes high for one clock and the Pentium shuts down. (2) When the Pentium is configured as a checker in a dual Pentium processor system. In this type of configuration one Pentium (checker) performs functional redundancy checking (FRC) for the other Pentium (master) to detect 99% of errors and preserve system integrity.

Clock

CLK — The CLocK input provides the basic timing for the internal circuitry in the Pentium.

Bus Control

ADS# — The ADdress Status output is used to indicate that the new address and bus cycle definition signals are valid for the current bus cycle.

NA# — The Next Address input informs the Pentium that the system is ready to accept a new bus cycle, even though the current bus cycle transfers are not complete. The Pentium can support two outstanding bus cycles.

Interrupt Signals

INIT — The INITialization input line forces the Pentium to begin execution of a known state. Initialization causes a reset sequence to begin, except that the contents of the internal caches, write buffers, and floating point registers prior to the INIT becoming active are used instead of being reset.

INTR — The INTerrupt Request (maskable interrupt) input informs the Pentium that the PIC has requested the servicing of an interrupt, when this line goes high. This input is processed only if the interrupt flag of the flags register is set. The Pentium samples this input at the beginning of each bus cycle. This line must be high for two processor cycles for the interrupt sequence to start. This must remain high until the end of the first of the two interrupt acknowledge signals issued by the Pentium. Once acknowledged, the Pentium sends out two locked interrupt acknowledge signals on the bus definition lines. At the end of the second interrupt acknowledge, the Pentium reads the lower 8 bits of the data bus to determine the vector code.

RESET — The RESET input is used to reset the Pentium to its starting value when this goes low. During reset the internal caches in the Pentium are invalidated.

NMI — The NonMaskable Interrupt input, when high, indicates that an external interrupt has been generated.

Cache Invalidation. The cache invalidation section of the Pentium is used to ensure that the data stored in cache memory is the same as the data stored in main memory. During a non-MPU write cycle the Pentium monitors address lines A5 to A31. If the write cycle affects an address in main memory that is currently stored in cache memory, the cache entry of that memory location is invalidated. This means that the next time the Pentium needs to use the contents of that cache location, the cache unit inside the Pentium must update the cache entry from main memory before the operation can take place.

AHOLD — The Address HOLD request input goes high when another bus master is ready to take control of the bus. When this line goes active, the Pentium stops driving the address bus (A3 to A31) and the AP line.

EADS# — The External ADdreSs valid input is used with an external device to indicate to the Pentium that the address on the address bus is valid. When this line becomes active, the Pentium causes address lines A5 to A31 to act as inputs so that the cache unit can monitor the address bus. The function of this pin is valid only during a non-MPU write operation.

INV — The INValidation input indicates the type of invalidation state to the Pentium during an inquire cycle. If this line is active when EADS# is sampled, the cache line is invalidated. If this line is inactive when EADS# is sampled, the cache line is marked shared. This line has no effect during a cache line miss.

Cache Control

CACHE# — The CACHE output is used by the Pentium to indicate a cacheable cycle. If this line is active along with KEN#, the cycle is converted into a line fill. If the CACHE# line is inactive, the data is not cached regardless of the level on the KEN# line. If this line is active during a write cycle, the Pentium performs a burst write-back cycle.

FLUSH# — The cache FLUSH input forces the Pentium to invalidate both internal caches and to write all modified lines in the data cache.

HIT# — The HIT output is used by the Pentium to indicate a cache hit when high during an inquire cycle. A low during an inquire cycle indicates a cache miss.

HITM# — The HIT Modified line output is used by the Pentium to indicate a hit in a modified line when low. This causes the Pentium to schedule a write-back cycle on the next bus cycle. If this HITM# line stays high, a miss has occurred on a modified cache line. This line is used to stop another bus master from accessing the data until the line is written back.

KEN# — The cache (K) ENable input is used to inform the Pentium that the data being read for the current bus cycle is cacheable. The current bus cycle (usually memory read) is converted into a cache line fill cycle. Each cache line fill cycle requires 32 bytes of data; therefore, the bus cycle requires four (64-bit) transfers to complete the cache line fill cycle. During a cache line fill cycle, the byte enable lines are ignored.

WB/WT# — The Write-Back and Write-Through input, along with the PWT pin, determines how the Pentium defines the MESI state of the cache line.

Page Caching Control

PCD — The Page Cache Disable output signal is used by the Pentium to indicate the page cacheability attribute for a secondary cache system.

PWT — The Page Write-Through output signal is used by the Pentium to indicate page write-through for a secondary cache system.

Numeric Error Reporting

FERR# — The Floating point ERRor output is used to indicate that an error has occurred during the execution of a floating point instruction.

IGNNE# When active, the IGNore Numeric Error input causes the Pentium to ignore any floating point error while allowing the execution of noncontrol floating point instructions.

Address Bit 20 Mask

A20M# The Address bit 20 Mask bit input is used to help the Pentium switch between the real and protected operating modes. While in the real mode, this input must be low and only the 20 LSBs of the address bus are available. In the protected mode all address bits of the address bus are available.

Operational Control

FRCM/C# The Functional Redundancy Checking Master/Checker mode input is used to determine the operational mode of the Pentium in a dual processor system. When high, the Pentium is in the master mode. When low, the Pentium is in the checker mode.

System Management

SCYC The Split CYCle output is used by the Pentium to indicate a misaligned locked transfer, in which two or more transfers are locked together. This keeps any other bus master from interrupting the transfer until it is complete.

SMI# The System Management Interrupt input is used to inform the Pentium that a management interrupt has occurred and is internally latched. This causes the Pentium to enter the system management mode.

SMIACT# The System Management Interrupt ACTive output is used by the Pentium to indicate that it is in the system management mode.

External Bus Write Status

EWBE# The External Write Buffer Empty input is used to inform the Pentium that there is a pending write-through cycle on the external system, when high. The Pentium does not process writes to any E or M state lines until the external write buffer is empty (indicated by a low).

Testing Bus. The testing bus lines are used in the testing and debugging of the Pentium processor. These lines are not used during normal operation.

BP0–3 Two of these four outputs are multiplexed and serve two functions in the Pentium. When BP0 and 1 lines are configured as BreakPoint outputs, they are combined with the levels on BP2 and 3 (which always act as breakpoint lines) to indicate a breakpoint match when using the debug registers of the Pentium. When BP0 and 1 are configured as performance monitoring outputs (renamed PM0 and 1 lines), their purpose in this configuration is confidential and proprietary to Intel.

BT0–3 The Branch Trace outputs are used to provide the three LSBs of the branch target linear address and the operand size during one of the Pentium's special cycles (branch trace message).

IBT The Instruction Branch Taken output indicates that a branch was taken, when high. When low, the instruction did not take the branch.

IU	The Instruction U pipeline complete output is used to indicate that the instruction in the U pipeline is complete.
IV	The Instruction V pipeline complete output is used to indicate that the instruction in the V pipeline is complete.
PRDY	The Probe mode ReaDY output indicates that the Pentium has responded to an R/S# signal and has stopped normal execution or that the Pentium has entered the probe mode.
R/S#	The Reset and Set edge sensitive input causes the Pentium to stop normal execution of the processor and places it in an idle state.
TCK	The Testability ClocK input supplies the clock for the Pentium while using the boundary scan interface.
TDI	The Test Data Input is used for entering serial data for the test logic in the Pentium.
TDO	The Test Data Output is used for outputting serial data from test logic in the Pentium.
TMS	The Test Mode Select input is used to help control the sequence of the controller state changes of the TAP (test access port) controller inside the Pentium.
TRST#	The Test ReSeT input is used to asynchronously initialize the TAP controller of the Pentium.

6.8 ADVANCED AT I/O BUSES

Although advanced AT buses were developed before the advent of the Intel 486 microprocessor, these buses were not very popular and most did not become widely available until the 486 became the microprocessor of choice. The first advanced AT bus was the IBM microchannel bus found in most PS/2 models. The microchannel AT bus was followed by EISA, VLB, and PCI. The purpose of each of these advanced AT buses was to increase the transfer rate of data on the data bus.

6.8.1 Microchannel Architecture (MCA)

In early 1987, IBM introduced its first MCA computer system—PS/2 model 50. It contained a 286 MPU with a 16-bit MCA bus. The 16-bit MCA bus was continued through a number of PS/2 models operating the 286 (modules 50, 50Z, and 60) and 386SX (models 55SX, 55LS, 56SX, 56SLC, 56LS, 56SLCLS, 57SX, 57SLC, M57SLC, and 65SX) MPUs. In 1988, with the introduction of the PS/2 model 70-386, the MCA bus was increased to 32 bits. The 32-bit MCA bus was incorporated with PS/2 using the 386DX (models 70-386, P70-386, and 80-386), 486DX (models 70-486 and P75-486), and 486SX or DX (models 90-XP-486 and 95-XP-486) MPUs. Besides the different MPUs and sizes of the MCA bus that were used in the different PS/2 models, the other differences were the type of built-in video standard ranging from VGA to XGA, the amount of memory available on the system board, the size and type of hard drive, the floppies available, and the on-board I/O interfaces (serial, parallel, and mouse ports).

The IBM MCA bus is not compatible with any other advanced AT bus. Many of the ideas used in the MCA bus relate to mainframe buses. The edge connector is much smaller (2.8 inches, compared with 5.29 inches) than the standard ISA edge connector and also contains 89 contacts on each side, compared with 31 contacts (31 contacts on the 8-bit bus plus an additional 18 contacts for the 16-bit extension) on the ISA bus. In the attempt to make the MCA bus the next standard, IBM licensed its MCA bus to other computer companies, but only a few companies took the offer. Today basically only IBM still uses the MCA bus.

6.8.2 Intel 82311 High-Integration Microchannel Compatible Peripheral Chip Set

The Intel MCA chip set is used to support the 386DX/SX MPUs used in the IBM PS/2 models. This chip set supports 386DX processor speeds up to 25 MHz and 386SX processor speeds up to 16 MHz. This chip set supplies most of the circuitry needed to implement MCA specifications with a minimum number of ICs. There are basically seven ICs in this chip set—82303 local I/O support chip, 82304 local I/O support chip, 82307 DMA/CACP controller, 82308 microchannel bus controller, 82309 address bus controller, 82706 VGA graphics controller, and 82077 floppy disk controller. Figure 6-14 shows a basic block diagram for the implementation of the 82311 MCA chip set on a system board. The blocks in the diagram that are not listed above are not discussed here because they have been described elsewhere in this book.

82303 Local I/O Support Chip. The Intel 82303 MCA local I/O support chip performs four basic functions in the MCA. First, it provides a parallel port interface (usually connected to a printer) that can be configured as a standard (unidirectional) or an extended (bidirectional) parallel port. This IC contains the card set-up port and provides system board set-up local that allows MCA I/O adapters to be configured to the MCA system. This IC also provides address latches for the peripheral address bus.

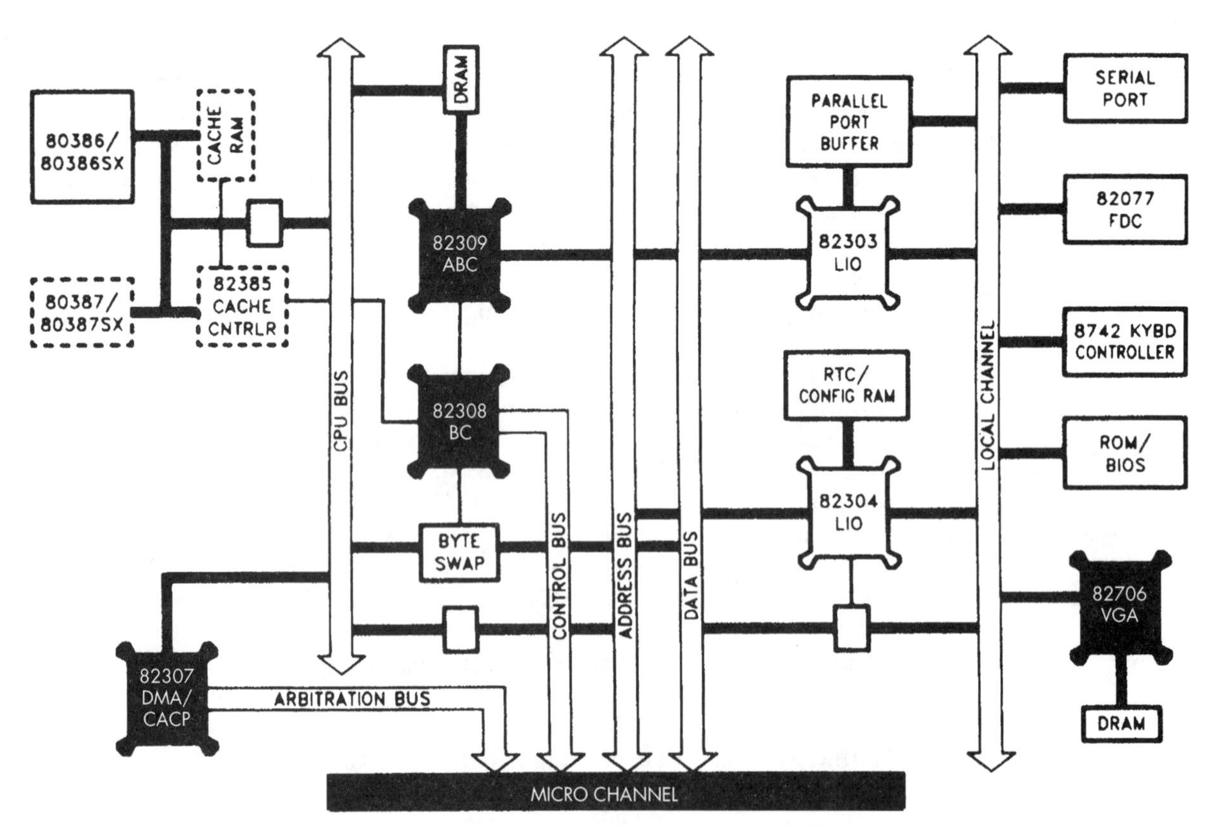

Figure 6-14 Intel 82311 MCA system block diagram. (Courtesy of Intel)

82304 Local I/O Support Chip. The Intel 82304 local I/O support chip performs several functions. The bus interface and control block are used to convert MCA signals into signals for the peripheral buses, provide external latching of the data bus transceiver that separates that MCA data bus from the peripheral bus, and contain circuitry that can extend the bus cycle for slow peripherals. The interrupt control section integrates two 8259 MCA-compatible PICs and provides the interrupt signals for the MCA bus and peripheral bus. The timer/counter block supplies three outputs; output 0 is used for multitasking time slice interrupt, timer 2 is used as the audio tone generator, and timer 3 provides the watchdog function of the MCA bus. This IC provides the interface to access and configure the following peripheral devices on the MCA system board; real time clock, battery–backed-up CMOS RAM, serial ports, parallel port, card set port, keyboard, mouse, numeric coprocessor, floppy disk controller, VGA graphics controller, and system status and control ports (system control ports A and B, card selected feedback register, system board set-up port, system board POS (power on setup) port, and the NMI enable port).

82307 DMA/CACP Controller. The Intel 82307 DMA/MCA arbitration controller performs the following functions: The numeric coprocessor interface section is used to provide the chip select for the math coprocessor and coprcoessor error processing. The address decoder section provides an address decoding chip selected for the 8259 PIC and generates POS address space output for the card set-up expansion slots. The MCA refresh address generation cycling is used to allow the DMA to perform a refresh cycle on main memory. The MCA arbitration section is responsibile for producing the signals for bus arbitration (control and status signals necessary to perform a bus cycle) for the MCA bus. The DMA section contains eight 8/16-bit wide channels and controls the 24-bit address bus for the MCA.

82308 Microchannel Bus Controller. The Intel 82308 microchannel bus controller performs the following functions: This IC is responsible for controlling the transfer of data between the MPU, DMA, main memory, and MCA bus, as well as providing the signals required for external data conversion (size swapping) and alignment. The 82308 provides signals necessary to interface the 82385 cache controller with the MCA bus. This IC also provides state machine logic, which is used to control the basic MCA signals. The 82308 also provides reset detection logic for the different types of reset cycles. The hardware-enforced I/O recovery section is used to provide delays and control I/O recovery signals for different MCA I/O devices.

82309 Address Bus Controller. The Intel 82309 address bus controller performs the following functions: It provides address decoding for DRAM (produces the RASx and CASx signals, multiplexes the address into the memory address bus, produces the write enable signal, produces the channel ready signal, and provides all the timing necessary to access DRAM). It contains the circuitry necessary to perform memory refresh, and it provides up to 16MB of page interleave RAM memory and shadowed RAM addressing of BIOS.

82706 VGA Graphics Controller. The Intel 82706 VGA graphics controller performs the following functions: It produces the necessary interface signals to allow the video system access to the MCA bus. The CRT controller section is used to access video memory during screen refresh, generates the sync timing signals (horizontal and vertical), and controls the cursor (type and size). When the video is in the alphanumeric mode, the 82706 accesses two bytes of video data; one byte is sent to the character generator, where the ASCII code is converted to row scan code to produce the dots of light for each row that makes up the character on the screen. The second byte is sent to the attribute controller, where the data is converted into the necessary signals to define how the character (background and foreground colors, intensity, blinking) is displayed on the screen. When the video is in the graphics mode (all points addressable), the data from video memory is serialized and transferred to the attribute controller. The attribute controller then uses this information to control which of the three color signals are enabled and their intensity during the horizontal and vertical scanning times.

82077 Floppy Disk Controller. The Intel 82077 floppy disk controller converts the interrupts from the operating system into the signals needed to control the hardware on the floppy disk drive unit. It converts and conditions parallel data from the data bus into serial data for the read and write heads on the floppy disk drive unit during a write cycle and serial data from the read and write heads from the floppy disk drive unit into parallel data during a read cycle. During and following a disk operation, it determines if the disk cycle has been performed without errors and informs the MPU of the status.

MCA Bus Slots and Edge Connectors. The IBM PS/2 MCA bus slots (Figure 6-15) and edge connectors are divided into five different sections. The 8-bit section contains the basic signals needed by all boards (therefore all slots have this section). This section contains 90 connections (45 on each side, where side A is the foil side and side B is the component side). The 16-bit section is an extension used with the 8-bit slot to extend the signals necessary to access the 16-bit data bus. The 16-bit extension contains 32 connections (16 on each side). The 32-bit slot is used with the 16-bit slot to extend the signals necessary to access the 32-bit data bus. The 32-bit extension contains 60 connections (30 on each side). Some of the slots may also contain the auxiliary video extension slots, which supply the additional signals for an external video adapter. The video extension contains 22 connectors (11 on each side). The MCA bus also has a matched memory extension slot that is normally used with the 32-bit slot for special high-speed memory boards. The matched memory extension contains 8 connections (4 on each side).

6.9 EXTENDED INDUSTRY STANDARD ARCHITECTURE (EISA)

The EISA bus is a 32-bit bus that was developed for 386 and 486 MPUs. The developers of the bus architecture used the 16-bit ISA bus as a base and then added signals to extend the communications so that data transfers and bus operations could occur at higher speeds (33 MHz) than with the standard 8-MHz bandwidth of the 16-bit ISA bus. In order to maintain compatibility

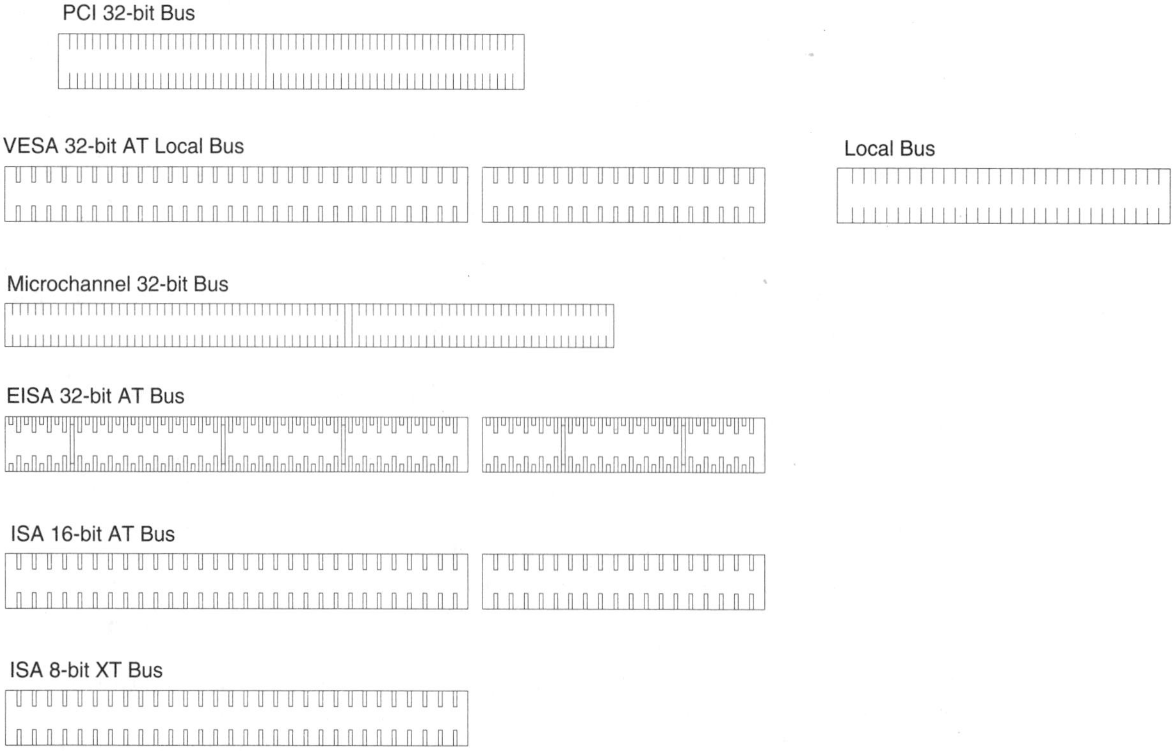

PCI 32-bit Bus

VESA 32-bit AT Local Bus

Local Bus

Microchannel 32-bit Bus

EISA 32-bit AT Bus

ISA 16-bit AT Bus

ISA 8-bit XT Bus

Figure 6-15 I/O bus slots.

with current ISA interface adapters, the EISA slots allow the insertion of both ISA and EISA adapters. The EISA slots have a key that allows the ISA adapters to fit only halfway down the slot, so only the ISA connections make contact with the ISA adapter. The EISA adapters have deeper edge connectors that are notched, which allow the extra EISA connections to make contact with the EISA edge connectors. An EISA adapter does not fit into an ISA slot because of the design of the edge connectors.

The EISA bus data transfer speed from the chip set allows each EISA adapter to access the high-speed EISA bus (33 MHz) as a bus master. Because each EISA adapter acts as a bus master, the host MPU (386 or 486) is not needed to control the transfer. Each EISA adapter contains a bus master interface controller (BMIC), which controls the interface signals between the MPU on the EISA adapter and the EISA bus on the system board. The host MPU needs to interact with ISA interface adapters used in an EISA system, because ISA adapters are not allowed to be bus masters and usually do not have their own MPU or smart ICs. EISA adapters are normally used with interfaces that require high-speed 32-bit data transfers (floppy/hard drive controllers, video systems, LAN adapters, and high-speed synchronous communication interfaces). Typical 8/16-bit ISA adapters used in an EISA are modems, multi-I/O adapters, scanner adapters, and sound cards.

6.9.1 Intel 82350 EISA Chip Set

The Intel 82350 EISA chip set is composed of the 82357 integrated system peripheral (ISP), 82358 EISA bus controller (EBC), and optional 82352 EISA bus buffers (EBB). Figure 6-16 shows a typical EISA chip set implementation using a 486 as the host MPU.

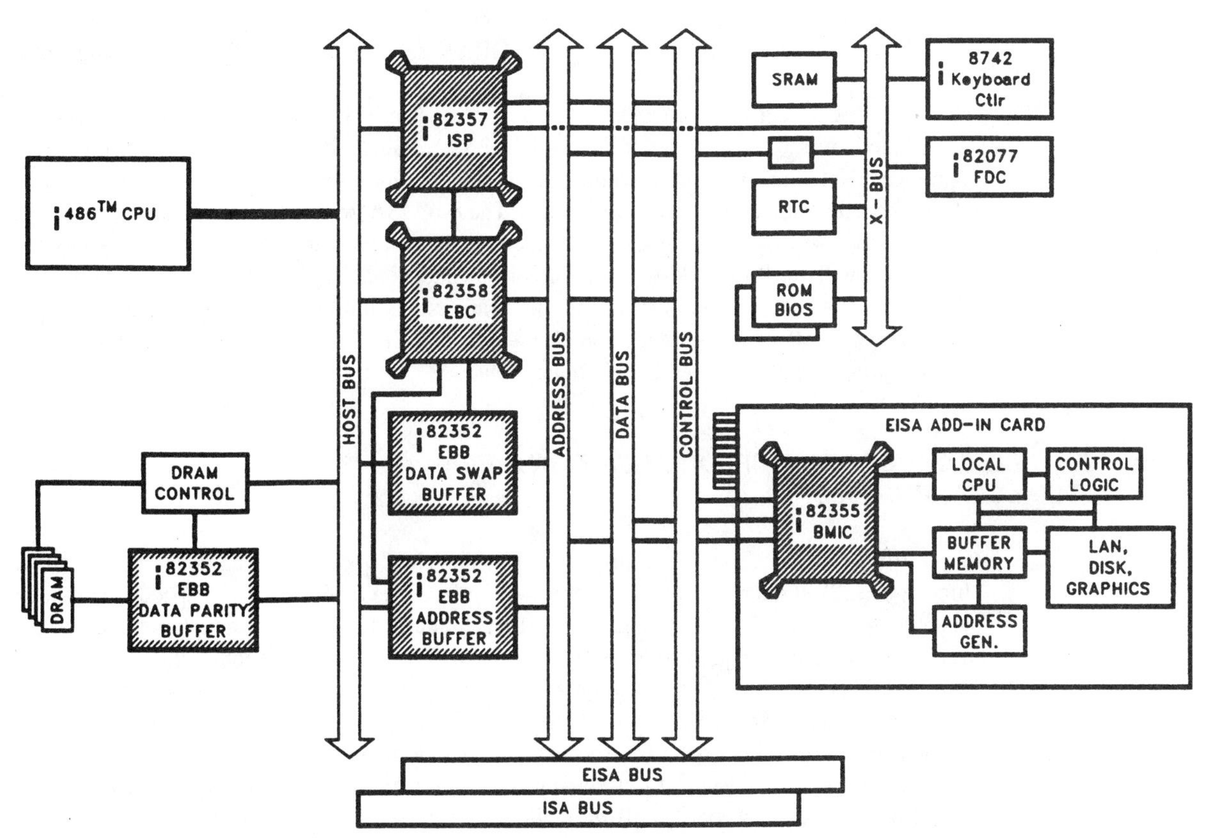

Figure 6–16 Intel's i486 CPU system with 82350 EISA chip set. (Courtesy of Intel)

82350 Chip Set Block Diagram. The block diagram in Figure 6-16 contains four buses. The host bus allows the MPU to communicate with the EISA chip set directly. The EISA chip set controls and generates the signals necessary based on the information provided by the MPU. The speed of this bus is the bus speed of the MPU. The EISA and ISA buses are the EISA bus slots that allow interface adapters access to the system board. The necessary EISA signals are developed by the EISA chip set. The X-bus in the system board peripheral bus contains devices discussed in earlier chapters.

82358 EISA Bus Controller. The 82358 EBC is an intelligent bus controller that develops controls for most of the communications between devices on the EISA bus. This IC controls and generates all the state machine interface signals between the host MPU, EISA, and ISA buses and the other chips of the chip set. The EBC also controls the communications between the bus masters and slaves on the 8-bit, 16-bit, and 32-bit buses. The EBC is also responsible for controlling all of the 82352 EBB ICs in the block diagram and the 82357 integrated system peripheral IC.

82357 Integrated System Peripheral. The 82357 ISP contains the support circuitry for the MPU and EISA and ISA buses. The ISP contains the following functions: two compatible 8259 PICs; bus arbitration for the EISA and ISA buses, bus masters, host MPU, DRAM refresh, and the DMA channels; a five output 16-bit counter/timer section; and control logic signals (NMI parity system and I/O memory, EISA fail-safe timer, and bus time-out functions).

82352 EISA Bus Buffers. The configurable 82352 EBBs in the block diagram are used to replace discrete logic that is necessary in performing three different tasks. One of the 82352 bus buffers is configured to perform data swapping, which is needed to translate different size data transfers between the host and EISA/ISA buses. Another 82352 is configured to buffer the address from the host address bus to the EISA/ISA buses. The third 82352 is configured to control the parity bus between the DRAM section and the MPU using the host bus.

EISA Bus Slots and Edge Connectors. The EISA bus slots and edge connectors are divided into two groups of connections. The ISA connections are located at the top of the interface adapter, and the EISA connections at the bottom of the edge connector and slot. The ISA connections are divided into two groups. The 8-bit ISA group contains 62 connections (31 on each side—side A is the foil side and side B is the component side). The 16-bit ISA extension works with the 8-bit ISA signals to provide an additional 32 connections (16 on each side). The EISA slots are 32 bits is size, and 100 connections minus 10 connections are used for EISA notches (50 minus 5 notches on each side). The EISA notches prevent the ISA adapter from making contact with the EISA connections.

6.10 VIDEO ELECTRONIC STANDARD ASSOCIATION (VESA) VIDEO LOCAL BUS (VLB)

VESA VLB. The VLB, also known as the video local bus, was developed to allow high-speed 32-bit microcomputer systems to interface effectively with high-speed smart video adapters without as much overhead as the EISA standard. The local bus adapters are under the control of the local bus controller to access the MPU's local bus without using the standard ISA bus. Therefore the local bus adapter can access data at the same speed as the MPU. The VLB incorporates the standards set by the Video Electronics Standards Association (VESA). In the early 1990s VLB systems boards became one of the most used buses because the additional circuitry needed to meet this standard was easily to incorporated into current ISA chip sets. Local bus system boards cost only about 10% more than standard ISA boards and provide a two to four-fold increase in local bus performance using local bus interface cards. The circuitry necessary to incorporate local bus interfacing increases the cost of the adapter by only about 20%. Because of the acceptance of the local bus standard, computer manufacturers have incorporated VESA specifications to provide 32-bit access to hard drive and LAN adapters. Most of the current AT chip sets now contain the additional circuitry for the local bus. The only disadvantage of the local bus is that only two or three local bus adapters can be connected to the bus because of the loading of the bus.

Video Local Bus Slots and Edge Connectors. The local bus slot is an additional slot placed inline with a standard 16-bit ISA slot. This slot contains 116 connections (58 on each side), which supply the necessary signals for the 32-bit access and control of the local bus. Local bus adapters, in addition to the standard AT connections, have 116 edge connections that fit into the local bus slot. The extra edge connections require that the local bus adapter not be shorter than two-thirds of a full adapter card, which sometimes makes it difficult to use in small computer cases. In some of the newer small footprint computer cases, the local bus adapters are integrated into the system board.

6.11 PERIPHERAL COMPONENT INTERCONNECT (PCI) BUS

The latest type of bus used in microcomputer systems is Intel's PCI bus. The PCI bus is the only bus standard that is currently giving the local bus serious competition. The PCI bus was developed to support Intel's Pentium processor and is currently the only bus that supports a 64-bit data bus and allows the Pentium to operate at maximum efficiency. The PCI data bus and PCI adapters use 32 bits of the data bus at any one time. The biggest advantage the PCI bus has over all other advanced AT buses is how it controls main memory. Figures 6-17 and 6-18 show typical implementations of the PCI bus with ISA and EISA secondary buses.

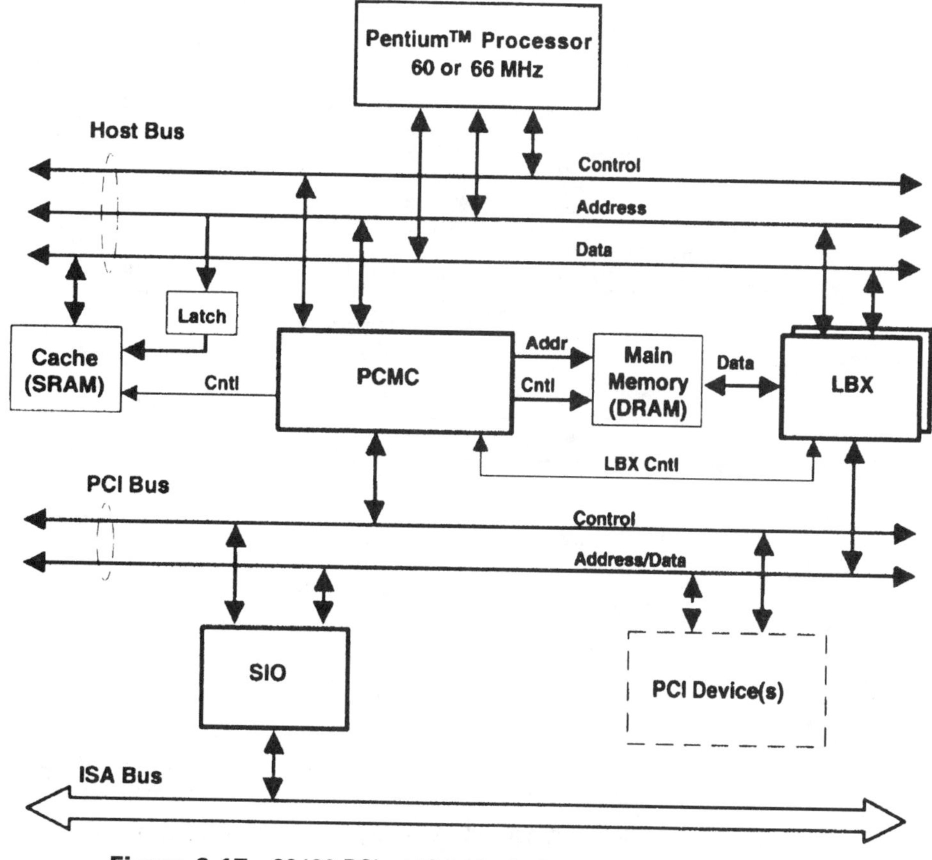

Figure 6-17 82430 PCIset ISA block diagram. (Courtesy of Intel)

6.11.1 82430 PCIset

The Intel 82430 PCIset is used to provide PCI signals for the Pentium processor. The PCIset is made up of two high-integration ICs with an option of either an ISA bridge controller or EISA bridge set. The 82434LX PCI/cache/memory controller (PCMC) and the 82433LX local accelerator (LXB) make up the controlling elements of the PCI bus. To make a PCI bus system board, a secondary bus is normally used to provide interfacing for slow interface adapters. The secondary bus can be either ISA or EISA or both and requires one or more additional support ICs. The 82378 system I/O (SIO) ISA bridge controller provides interface signals to a secondary ISA bus. The 82374EB EISA system component (ESC) and 82375EB PCI/EISA (PCEB) provide the interface signals to a secondary EISA bus.

82434LX PCI/Cache/Memory Controller. The 82434LX PCMC chip is the heart of the PCI system and provides three basic functions. The host/PCI bridge is responsible for translating CPU cycles into PCI bus cycles and maintains concurrence between the CPU and main memory, CPU and second-level cache, PCI and PCI, and PCI and main memory transactions. This section also translates memory writes into PCI burst cycles and burst writes to PCI with zero PCI wait states. The second-level cache controller supports between 256K and 512K of cache memory, supports the write-back and write-through modes of the Pentium, and has the ability to control both standard and burst SRAM. It produces cache hit cycles of 3-1-1-1 using burst SRAM and 3-2-2-2 (read)/4-2-2-2 (write) using standard SRAM. The DRAM

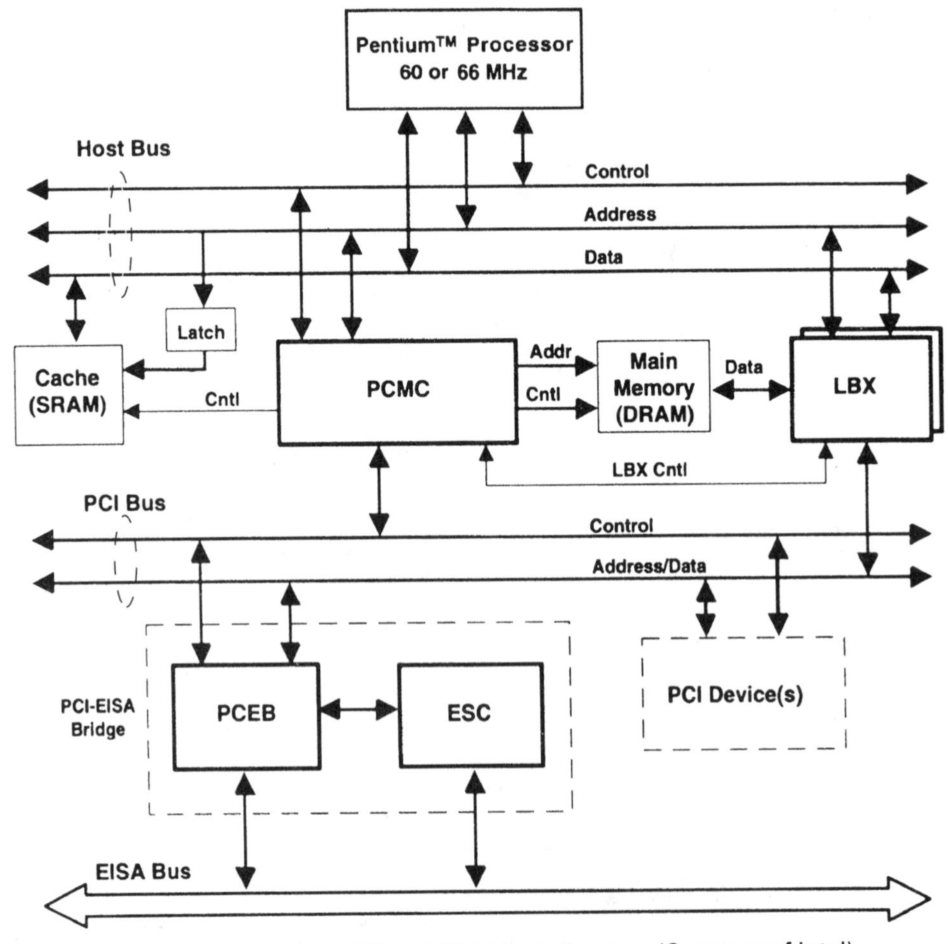

Figure 6-18 82430 PCIset EISA block diagram. (Courtesy of Intel)

controller supports cacheable main memory ranging from 2BM to 192MB using 60-ns or 70-ns low-cost DRAMs, provides single or burst RAS only or CAS before RAS DRAM refresh, and decouples refresh cycles from ISA refresh.

82433LX Local Bus Accelerator. The 82433 LBX uses dual-port architecture that allows concurrent bus cycles on the host and PCI buses, interfaces the 64-bit DRAM data bus to the 32-bit PCI data bus, provides parity generation and checking support for both main and host memory buses, and provides data bus buffering between the CPU, the main memory, and the PCI bus.

82378 System I/O. The 82378 system I/O functions as a bridge that translates PCI bus signals into ISA signals, and produces all peripheral support signals (32-bit DMA channels, interrupts, counters/timers, NMI support, generation of chip selects, and data bus buffering) necessary for ISA adapters.

82375EB PCI/EISA Bridge/82374EB EISA System Controller. The 82375EB PCEB functions as a bridge that allows the PCI bus to interface with the EISA system controller (ESC). The 82374EB ESC generates all the control signals necessary to control EISA bus adapters. This chip also translates bus cycles between the EISA and ISA buses and produces all ISA support signals needed to operate ISA adapters.

CHAPTER SUMMARY

1. Features of the 486 MPU:

Description	*486SX*	*486SX2*	*486DX*	*486DX2*	*486DX4*
Address lines	32	32	32	32	32
Physical address space	4 GB	4 GB	4 GB	4 GB	4 GB
Data lines	32	32	32	32	32
External clock speed	20 MHz	—	20 MHz	—	—
	25 MHz	25 MHz	25 MHz	25 MHz	—
	33 MHz	33 MHz	33 MHz	33 MHz	33 MHz
	—	—	50 MHz	—	—
Internal clock speed	—	50 MHz	—	50 MHz	—
	—	66 MHz	—	66 MHz	99 MHz
Number of pins	168	168	168	168	168
Math coprocessor (internal)	no	no	yes	yes	yes

2. Features of the Pentium processor:

Description	*Pentium*
Address lines	32
Physical address space	4 GB
Data lines	64
Clock speeds	60 MHz
	66 MHz
	75 MHz
	90 MHz
	100 MHz
	120 MHz
Number of pins	273
Type of package	GAP

3. Controlling elements of the 486:

   - Bus interface unit—interfaces the internal address/data/control buses with the outside world.
   - Instruction unit—decodes and buffers instructions until they can be used by one of the control units of the MPU.
   - Floating point unit (DX only)—Performs high-speed math operations.
   - Central processing unit—performs all ALU operations; also contains the general purpose registers and flags register.
   - Memory management unit—develops and controls logical and physical addresses.
   - Cache section—contains the cache controller and 8K cache memory. Its purpose is to maintain commonly used codes and data within the 486 to increase the speed of the MPU.

4. Controlling elements of the Pentium processor:

   - Superscalar architecture allows more than one execution unit so that more than one instruction can be executed at the same time.
   - Separate code and data caches allow faster access to either instruction code or data currently in cache.
   - High performance floating point unit processes complex math functions quickly; in the Pentium many of the instructions are hardwired.
   - 64-bit data bus allows 64 bits of data to be transferred during any one bus cycle.
   - Branch prediction allows the MPU to prepare for a possible branch instruction.
   - Data integrity and error detection are used to verify that the data and register data are valid.
   - Multiprocessing support allows two or more MPUs to operate with each other to increase speed.
   - Memory page sizing allows programmers to develop programs that use memory in the most efficient way by changing the size of memory pages.
   - Performance monitoring allows software and hardware developers to maximize MPU performance.

5. 486 pin configurations:

   - A2–A31—Bidirectional address bus. During an MPU bus cycle A2 to A31 act as outputs to produce the address for the address bus. During a non-MPU bus cycle A4 to A31 act as inputs that allow the 486 to monitor the address for the cache section of the MPU.
   - BE0#–BE3#—The byte enable outputs are used to determine which bytes are transferred on the data bus during a bus cycle.
   - M/IO#—The memory and I/O output is used to determine if the bus cycle uses memory or an I/O device.
   - D/C#—The data and code output is used to determine if the bus cycle deals with data or codes (instructions).
   - W/R#—The write and read output is used to indicate a read or write bus cycle.
   - LOCK#—The lock output is used to lock out other bus masters from using the bus during non-burst bus cycles.
   - PLOCK#—The pseudo lock output is used to lock out other bus masters from using the bus during burst bus cycles.
   - HOLD—The hold input is used by a bus master to request permission to use the bus.
   - HLDA—The hold acknowledge output is used to grant permission to a bus master to use the bus.

- BOFF#—The back-off input performs the same function as the HOLD input under special bus cycles.
- BREQ—The bus request output indicates that a bus cycle is pending internally.
- BRDY#—The burst ready input is used to indicate the readiness of the bus during a burst bus cycle.
- BLAST#—The burst last output is used to indicate the end of a burst bus cycle.
- BS8#—The bus size 8-bit input is used to instruct the MPU to use only the lower 8 bits of the data bus.
- BS16#—The bus size 16-bit input is used to instruct the MPU to use only the lower 16 bits of the data bus.
- DP0–DP3—The data parity bit bidirectional lines. During a write operation the MPU generates and then outputs one parity bit for each byte of data. During a read operation the MPU reads the levels on these lines and checks for parity errors on the data received.
- PCHK#—The parity check output is used to indicate that a parity error has occurred following a read operation.
- CLK—The clock input provides the basic timing for the 486.
- D0–D31—The 32 bidirectional data lines are used as inputs during a read cycle and as outputs during a write cycle.
- ADS#—The address status output is used to indicate that the address and bus cycle definition signals are valid.
- RDY#—The ready input is used to indicate that the bus is ready to complete the current bus cycle.
- INTR—The interrupt request input informs the MPU that a device is requesting an interrupt sequence.
- RESET—The reset input is used to reset the MPU.
- NMI—The nonmaskable interrupt input is used to indicate a parity error.
- AHOLD—The address hold request indicates when another bus master is ready to take control of the bus.
- EADS#—The external address valid input is issued to indicate to the MPU that the address on the address bus is valid.
- KEN#—The cache enable input is used to inform the MPU of a cacheable cycle.
- FLUSH#—The cache flush input is used to force the MPU to flush the data from the internal cache memory.
- PWT—The page write-through output is used to provide information for the secondary cache controller.
- PCD—The page cache disable output is used to provide information for the secondary cache controller.
- FERR#—The floating point unit error output is used to indicate that an error has occurred in the FPU.
- IGNNE#—The ignore numeric error input causes the MPU to ignore any FPU error.
- A20M#—The address bit 20 mask is used to help the MPU to switch operational modes.

6. Pentium processor pin configurations:

   The following pins perform the same functions as in the 486, although the number of lines may be different: A3–A31, BE0#–BE7#, D0–D63, M/IO#, D/C#,W/R#, LOCK#, HOLD, HLDA, BOFF#, BREQ, BRDY#, DP0–DP7, PCHK#, CLK, ADS#, INTR, RESET, NMI, AHOLD, EADS#, FLUSH#, KEN#, PCD, PWT, FERR#, IGNNE#, and A20M#.

   - PEN#—The parity enable input is used to enable parity checking.

- AP—The address parity line is bidirectional. During an MPU cycle the MPU generates and outputs an address parity bit. During a non-MPU cycle the MPU reads the AP line and determines if a parity error has occurred.
- APCHK#—The address parity check output is used to indicate that a parity error has occurred during a non-MPU cycle.
- BUSCHK#—The bus check input is used to inform the Pentium that the system was unsuccessful in completing the bus cycle.
- IERR#—The internal error output indicates some type of internal error.
- NA#—The next address input is used to inform the Pentium that the system is ready to accept a new bus cycle.
- INIT—The initialization input forces the Pentium to begin execution of a known state.
- INV—The invalidation input indicates the type of invalidation state to the Pentium during an inquire cycle.
- CACHE#—The cache output is used to indicate a cacheable bus cycle.
- HIT#—The hit output is used by the Pentium to indicate a cache hit.
- HITM#—The hit modified line output is used by the Pentium to indicate a hit in a modified line.
- WB/WT#—The write-back and write-through inputs determine how the Pentium defines the MESI state of the cache line.
- FRCMC#—The functional redundancy checking master/checker mode input is used to determine the operation mode of the Pentium in a dual processor system.
- SCYC—The split cycle output is used to indicate a misaligned locked transfer.
- SMI#—The system management interrupt input is used to inform the Pentium that a management interrupt has occurred.
- SMIACT#—The system management interrupt active output is used by the Pentium to indicate that it is in the system management mode.
- EWBE#—The external write buffer empty input is used to inform the Pentium that there is a pending write-through cycle on the external system.
- BP0–3—These pins are used for breakpoint monitoring or performance monitoring.
- BT0–3—The branch trace outputs are used to provide address information for branch instructions.
- IBT—The instruction branch taken output indicates that a branch was taken.
- IU—The instruction U pipeline complete output is used to indicate that the instruction in the U pipeline is complete.
- IV—The instruction V pipeline complete output is used to indicate that the instruction in the V pipeline is complete.
- PRDY—The probe mode ready output indicates that the Pentium has responded to an R/S# signal.
- R/S#—The reset and set input is used to cause the Pentium to stop normal execution.
- TCK—The testability clock input supplies the clock for the Pentium while using the boundary scan interface.
- TDI—The test data input is used for entering serial data for the test logic in the Pentium.
- TDO—The test data output is used for outputting serial data from test logic in the Pentium.
- TMS—The test mode select input is used to help control the sequence of the controller state changes of the TAP inside the Pentium.
- TRST#—The test reset input is used to asynchronously initialize the TAP controller of the Pentium.

7. Advanced AT Buses:

MCA—Microchannel architecture
EISA—Extended industry standard architecture
VESA VLB—Video electronics standards association video local bus
PCI—Peripheral component interconnect

Description	*MCA*	*EISA*	*VESA VLB*	*PCI*
8-bit bus support	x	x	x	x
16 bit bus support	x	x	x	x
32-bit bus support	x	x	x	x
64-bit memory bus support				x
Maximum bus speed	25 MHz	33 MHz	33 MHz	33 MHz
Bus mastering	x	x	x	x

REVIEW QUESTIONS

1. Which of the following processors have a 32-bit address bus?
- **a.** 486 SX
- **b.** 486 DX
- **c.** Pentium
- **d.** a, b, and c.
- **e.** None of the above answers

2. The 486DX2-66 has an external clock speed of _____.
- **a.** 132 MHz
- **b.** 66 MHz
- **c.** 33 MHz
- **d.** None of the given answers

3. Which of the following processor(s) does not contain an internal math coprocessor?
- **a.** 486SX
- **b.** 486DX
- **c.** 486DX2
- **d.** Pentium
- **e.** None of the given answers

4. Which of the following processors does not have a 32-bit data bus?
- **a.** 486SX
- **b.** 486DX
- **c.** Pentium
- **d.** a, b, and c.
- **e.** None of the given answers

5. Which of the following processors contains superscalar architecture?
- **a.** 486SX
- **b.** 486DX
- **c.** Pentium
- **d.** a, b, and c.
- **e.** None of the given answers

6. Which section of the 486 controls the levels on the address lines?
- **a.** Bus interface unit
- **b.** Instruction unit
- **c.** Floating point unit

   d. Central processing unit
   e. Memory management unit
7. Which section of the 486 is responsible for creating the linear address?
   a. Bus interface unit
   b. Instruction unit
   c. Floating point unit
   d. Central processing unit
   e. Memory management unit
8. Which 486 pin(s) are used to act as address bits 0 to 1?
   a. D/#C
   b. BE0# to BE3#
   c. DPO to DP3
   d. None of the given answers
9. Which of the following pins act as outputs only?
   a. A2 to A31
   b. D0 to D31
   c. HOLD
   d. a. and b.
   e. None of the given answers
10. When the M/IO# line is low, a memory operation is taking place.
   a. True
   b. False
11. When the LOCK# pin is high, the DMA cannot ask permission to use the bus.
   a. True
   b. False
12. The DP0 to DP3 lines act as inputs during a memory read operation.
   a. True
   b. False
13. BE6# of the Pentium enables data bits 48 to 55 when low.
   a. True
   b. False
14. The Pentium contains how many data parity pins?
   a. 4
   b. 6
   c. 8
   d. 16
   e. None of the given answers
15. If a parity error occurs in the Pentium, the PCHK# line goes _____.
   a. Low
   b. High
   c. Tristate
16. The BUSCHK# Pentium input goes high when the system fails to complete a bus cycle.
   a. True
   b. False
17. The INTR line is used to indicate a maskable interrupt request to the Pentium.
   a. True
   b. False
18. The AHOLD# pin of the Pentium functions the same as the AHOLD pin of the 486.
   a. True
   b. False

19. The CACHE# pin of the Pentium will go high during a cacheable cycle.

a. True

b. False

20. The HIT# pin of the Pentium goes high to indicate a cache hit during an inquire cycle.

a. True

b. False

21. Which of the following advance AT I/O bus standards was developed especially for the Pentium processor?

a. EISA

b. ISA

c. MCA

d. PCI

22. Which of the following advance AT I/O bus standards was developed by IBM for their PS2 line?

a. EISA

b. VLB

c. MCA

d. PCI

23. Which Intel MCA IC is responsible for generating maskable interrupts?

a. 82303

b. 82304

c. 82307

d. 82308

e. 82309

24. Which Intel EISA IC is responsible for generating the state machine signals between the MPU, EISA, and ISA buses?

a. 82352

b. 82357

c. 82358

25. Which Intel PCI IC is responsible for translating CPU cycles into PCI bus cycles?

a. 82433LX

b. 82434LX

c. 82378

chapter 7

Keyboards and Power Supplies

OBJECTIVES

Upon completion of this chapter, you should be able to perform the following tasks:

1. Describe the function of each component in a capacitive and positive response keyboard.
2. List the advantages and disadvantages of a capacitive and positive key.
3. Determine the make and break keycodes from a key that has been depressed.
4. List the sequence of events that occur on the system board when the serial keyboard data is converted into parallel data for the data bus.
5. List some common keyboard problems.
6. List the common types of AC power problems.
7. List the operational sequence of linear and switching power supplies.
8. List a sequence that can be used to troubleshoot power supply problems.
9. List the purposes of power back-up systems.

7.1 KEYBOARDS

The keyboards used in the IBM PC XT and AT systems perform basically the same functions. The differences lie in how the keys are encoded and the types of actual key codes produced. There are three basic sections of any keyboard—the single-chip microcontroller, the keyboard matrix, and support logic.

7.1.1 8048/8049

The Intel 8048 and 8049 are single-component 8-bit microcontrollers. Both of these ICs are basically single-chip microcomputers that contain ROM, RAM, and I/O interface ports. Most of the instructions are single-byte instructions. These ICs are called microcontrollers because they are usually configured to perform only a specific set of tasks. Older XT-type keyboards use the 8048 8-bit microcontroller while most keyboards now manufactured use the 8049.

Both microcontrollers have the same pin configurations; the only difference is the size of ROM and RAM contained on each microcontroller (Table 7-1). The purpose of the keyboard microcontroller is to scan the keyboard for valid key closures and for stuck keys during POST, to produce key codes for the system board once a key has been pressed or released, to transfer the key codes in a serial format to the system board, and to produce a keyboard clock signal needed to convert the serial key code into parallel data for the system board. The 8048/8049 is also responsible for providing the user with information regarding the condition of the numeric lock, caps lock, and scroll lock by driving LEDs to indicate their current condition (LED is on when the key is in the locked condition and off in the unlocked condition).

8048/8049 Pin Configurations (Figs. 7-1 and 7-2)

P10–P17 These eight quasi-bidirectional lines are referred to as Port 1. They can be programmed as either inputs, outputs, or bidirectional lines, which is why they are called quasi-bidirectional lines. When configured as inputs they are not latched. When configured as outputs they are statically latched.

P20–P27 These eight quasi-bidirectional lines are referred to as Port 2. These lines can act the same way as the lines of Port 1. P20 to P23 provide the four most significant bits (MSBs) of the program count and supply I/O signals needed to interface the 8243 4-bit I/O expander bus controller (not used in the keyboard).

DB0–DB7 The 8-bit bidirectional Data Bus is made up of DB0 to DB7.

T0 Testable input 0 can be used with clock input or with conditional transfer instructions.

T1 Testable input 1 can be used with timer/counter or with conditional transfer instructions.

INT# The INTerrupt input is used to indicate that an interrupt has occurred. This interrupt input is maskable.

RD# The ReaD output strobe line is used to indicate that a read cycle is on the bus.

RESET# The RESET input is used to reset the microcontroller to its starting values.

WR# The WRite output strobe line is used to indicate that a write cycle is on the bus.

ALE The Address Latch Enable output line is used to latch the address into program and external memory.

PSEN# The Program Store ENable output line indicates that a read has occurred from external memory.

TABLE 7-1
8048/8049 Specifications

Description	*8048AH*	*8049AH*
Amount of RAM	64x8	128x8
Amount of ROM	1Kx8	2Kx8
Number of I/O lines	27	27
Clock speed	1–11 MHz	1–11 MHz
Number of pins	40	40

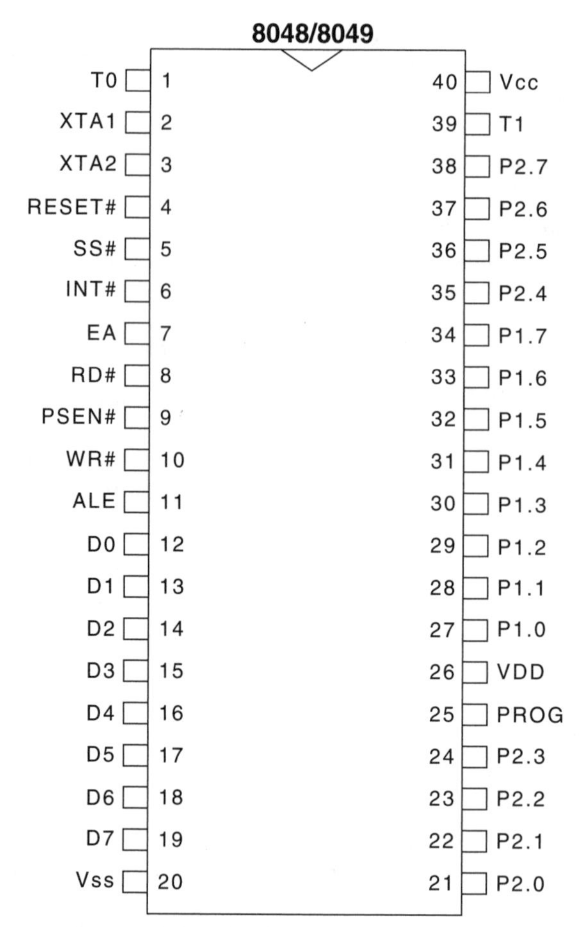

Figure 7-1 8048/8049 pin configuration.

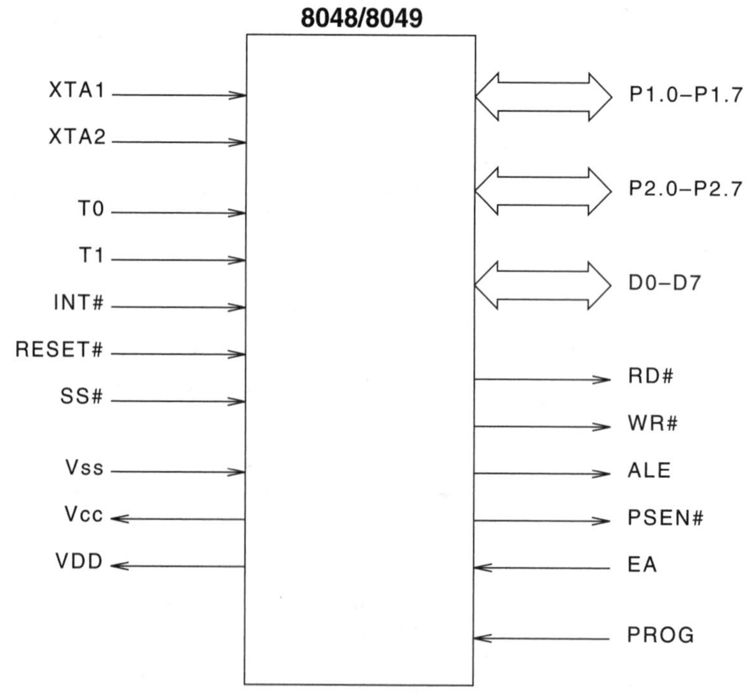

Figure 7-2 8048/8049 pin groupings.

SS#	The Single Step input allows the microcontroller to single step through instructions.
EA	The External Access input is used to cause all memory reads to occur on external memory.
XTA1/2	The crystal inputs 1 and 2 are used to provide a crystal frequency to the microcontroller. XTA1 can also be used to input an external source frequency.
Vss	The Vss input is the ground reference point for the microcontroller.
Vcc	The Vcc supplies the operating voltage for the microcontroller.
VDD	The VDD line supplies both low-power stand-by power and normal operating voltage. This pin is usually tied to the same line as the Vcc pin.

7.1.2 Key Types

Currently two different types of key closures are found in standard keyboards. The first type is called the capacitive key closure and the second a positive response (mechanical) key closure. Each type of key closure has its advantages and disadvantages, which are discussed.

Capacitive keys (Fig. 7-3), when pressed, form a small capacitor by bringing a plate made of conductive or semiconductive material in close proximity to two conductive circuit pads covered with an insulated material. When the row and column connected to the key being pressed are selected, the capacitor starts to charge, causing current to flow in the selected column. Once the key becomes unselected, the capacitor discharges because of leakage of the capacitance. If the key is not pressed, the capacitor is not charged because of the distance between the conductive plate and the two conductive circuit pads. The advantages of capacitive keys are that, because no physical contact is made when a key is pressed, they are not as susceptible to dirt or corrosion and keys do not wear out as quickly. The disadvantages of a capacitive key are that a single key cannot be replaced when it goes bad because it is actually the pad underneath the key that has gone bad, the cost is somewhat higher than for a positive response keyboard, and the feel of the key when pressed is not very positive except in the more expensive keyboards.

A positive response key (Fig. 7-4) causes a mechanical connection between a set of contacts when pressed; this connects the row and the column together. When a positive response key is not pressed, the contacts for the row and column are separated by some type of nonconductive material or air. The advantages of a positive response key are low cost, ease of design, and replacement of a single key if it goes bad. The disadvantages are susceptibility to dirt, corrosion, and mechanical stress.

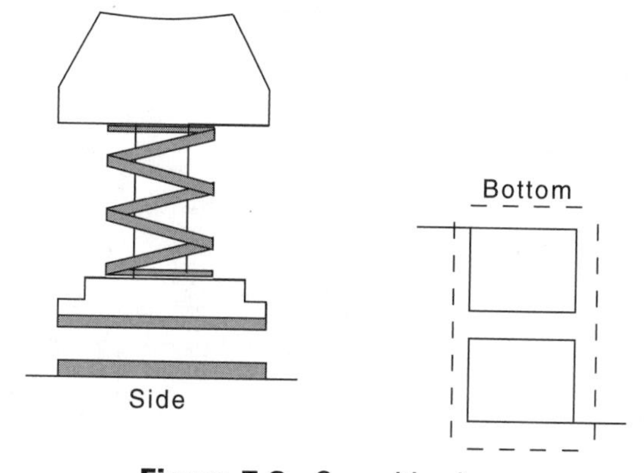

Figure 7-3 Capacitive key.

7.1.3 Keyboard Matrix

The keyboard matrix is how keys of the keyboard are connected to the microcontroller. Each key is connected between a row and a column. The microcontroller selects one of the rows of the keyboard matrix by making the selected row go low; all other rows are at a logic high. While a row is selected, the microcontroller samples each column. If the key connected to the selected row is pressed, either current (capacitive key) flows in the column or a logic low (positive response key) is seen on the column, which indicates a key closure. Absence of current flow or any other logic level indicates the absence of a key closure.

As can be seen in the schematic diagrams of an IBM PC XT keyboard (Fig. 7-5) and a clone AT-type keyboard (Fig. 7-6), the number of rows and columns that make up the keyboard can vary. The IBM PC XT (83-key) keyboard contains 12 rows and 8 columns to decode each of the keys connected to the keyboard matrix. The clone AT-type (101-key) keyboard contains 13 rows and 8 columns to decode each of the keys connected to the keyboard matrix. The greater the number of keys that must be decoded, the greater the number of rows used.

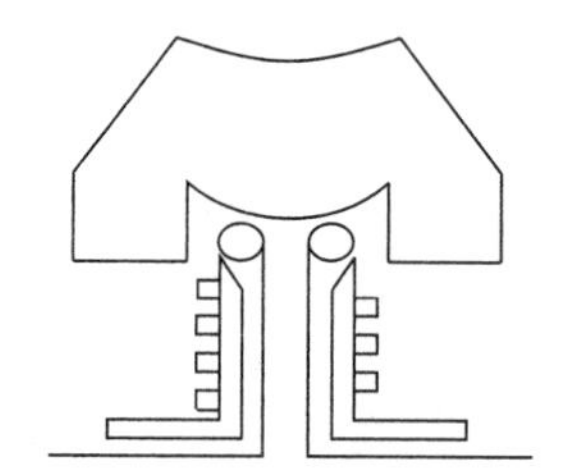

Figure 7-4 Positive response key.

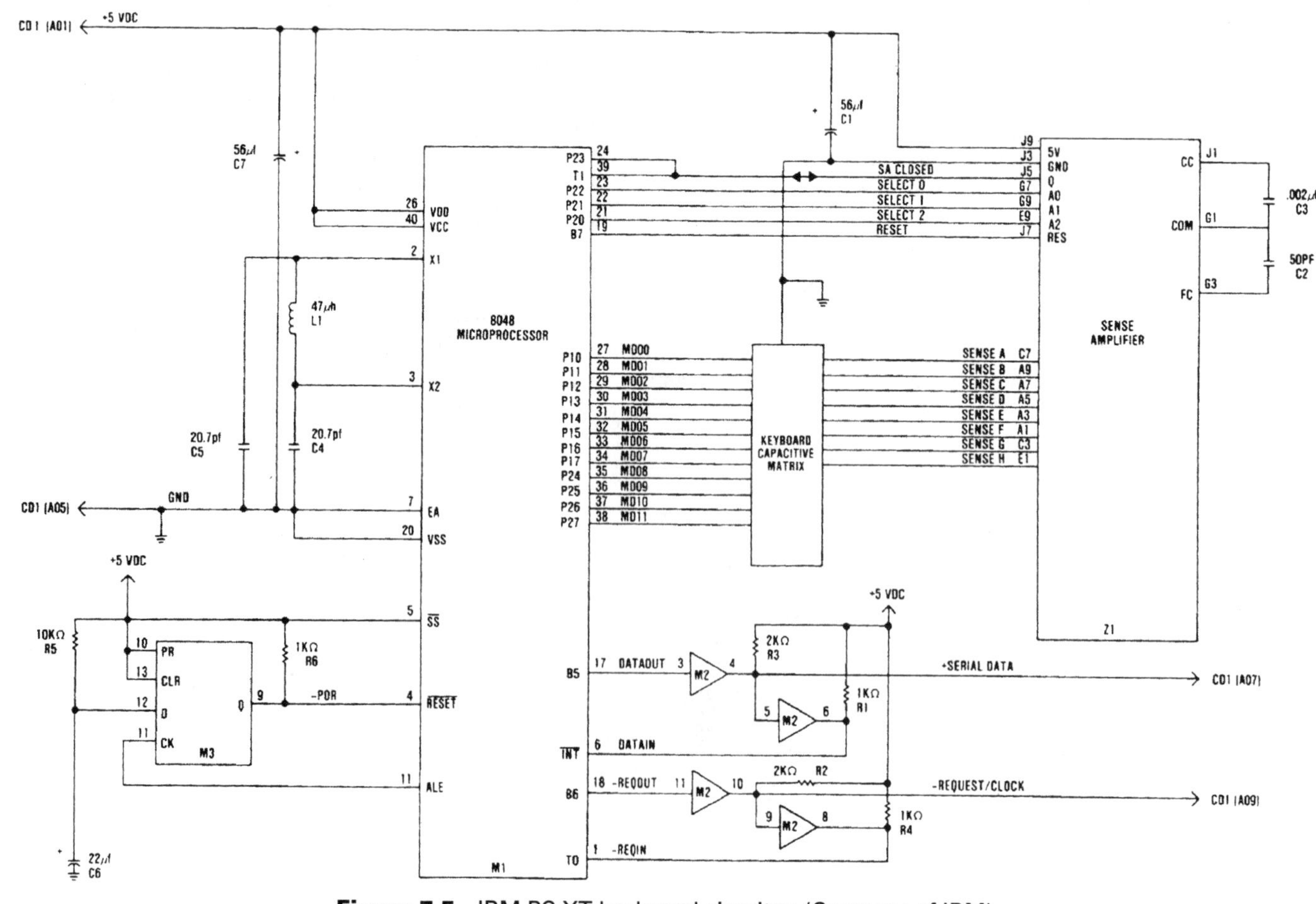

Figure 7-5 IBM PC XT keyboard circuitry. (Courtesy of IBM)

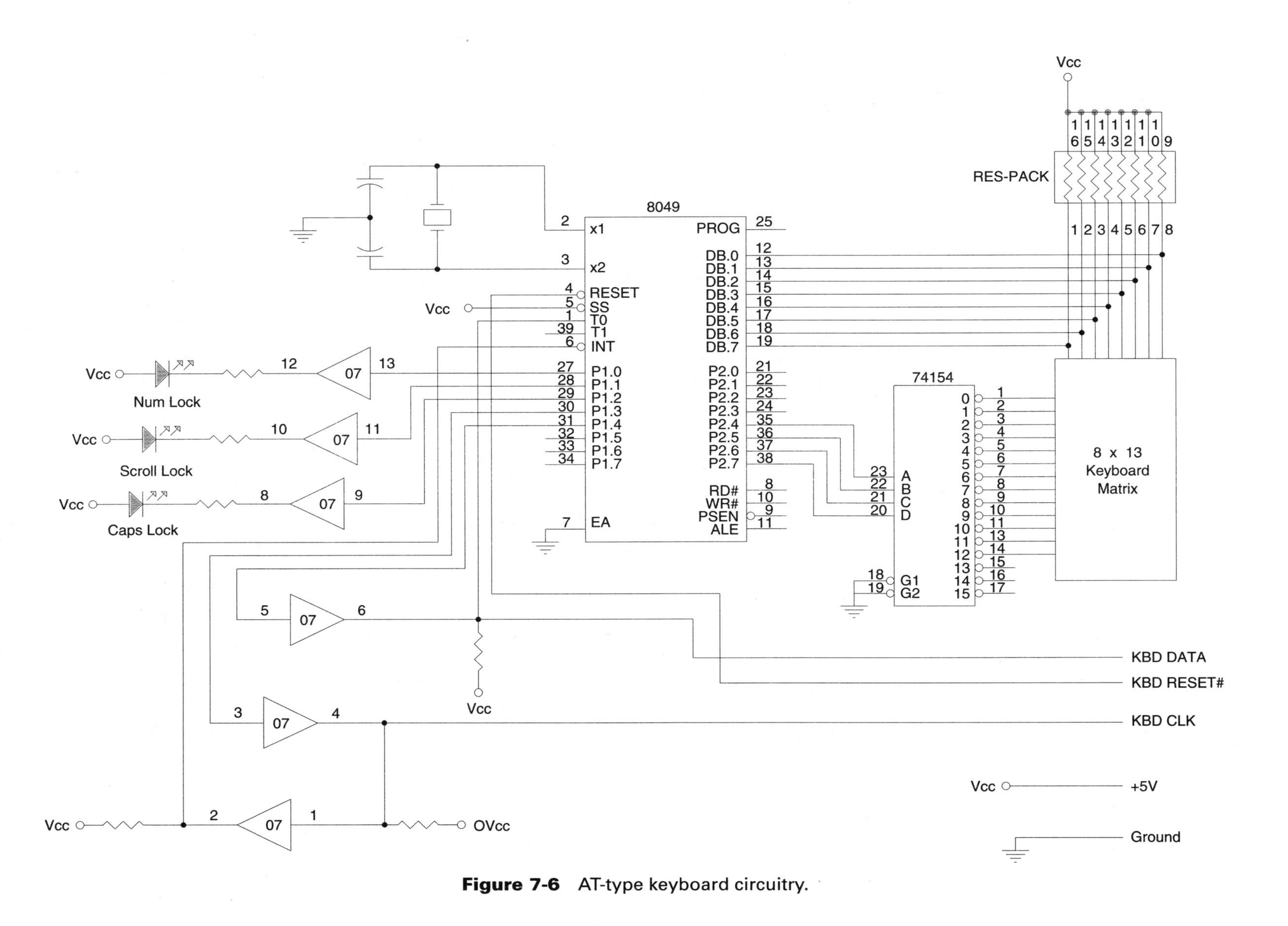

Figure 7-6 AT-type keyboard circuitry.

7.1.4 IBM PC XT Keyboard Operation

The 8048 is always working, scanning the keyboard matrix. It selects one row at a time and samples each of the eight columns connected to the row. If no key is pressed, the column sense amplifier/decoder sees no current flow. If current flows while the microcontroller is selecting one of the rows and sampling one of the columns, the column sense amplifier/decoder produces a high on its Q output. The high is applied to the testable input pin of the 8048, which causes the address of the current row and column to be stored in RAM, and the scanning continues. If, at the end of five complete keyboard scans, the key in question is still depressed, the 8048 produces a make key code (this software delay is used to verify that a key closure is valid and is used for key debounce). If another key is also pressed, the 8048 does not recognize the second key closure until the first key is released, except for three special keys (Ctrl, Alt, and Shift), which are used to modify how the key codes are decoded by the system board logic. These modifier keys must be pressed first and held down while the second key is pressed to modify how the second key is processed. In the case of modified keys, the key code for the modifier is transmitted first, followed by the key code for the second key pressed. Normal key function requires that the first key pressed be released before the second key is processed. The make key code is a value between 00 and 7F, which is sent to the IBM PC system board as a serial data package.

The serial package contains 10 bits of data (Fig. 7-7). The data package starts with two start bits (highs), which are used to initialize the circuitry on the IBM PC system board, followed by the 8-bit make or break key code. The least significant bit (LSB) of the key code is sent first following the two start bits, with the MSB of the key code being transmitted last. During the transmission of the keyboard data (key code), the 8048 also transmits a keyboard clock signal, which is used to synchronize the data reception. Each data bit in the transmission occurs on the falling edge of the keyboard clock. Each bit in the package has a time duration of 0.1 ms; thus it takes 1ms to transmit one complete key code, which yields a baud rate of 10K bps. Baud rate is the number of bits transmitted in 1 second from one device to another. The make key code in Figure 7-7 is 4C (hexadecimal) (Table 7-2), which means that the number "5" key on the numeric keypad was pressed.

If the key is released before 0.5 second passes, the 8048 produces a break code, which is transmitted to the IBM PC system board in a serial data format. The break code is the same as the make code except that the last bit transmitted (7 bit of the break code) is set. The break code (Fig. 7-7) tells the IBM PC BIOS key routine that the function of that key sequence is completed and the key has been released (the break code is needed if a program is used to change the function of the key pressed). If more than one nonshift key is pressed, the 8048 transmits the first make key code (the first key pressed) and sends the next make key code. If, however, the Ctrl, Alt, or Shift key is pressed before a second key, they produce a 00 hexadecimal make code, which allows the next key pressed to produce a special modified make key code. This, when processed by the BIOS key-encoding routine, produces an extended ASCII code.

If the key is still pressed after 0.5 second (72 keyboard scans), the 8048 produces a make key code (the same as the original make key code) and transmits it at a rate of six every second (one every 166.7 ms). This autorepeat function continues until the break code for the key pressed is sensed by the 8048. The break code is the same as the make code except that the most significant bit is always set, which gives a break code range from 80 to FF.

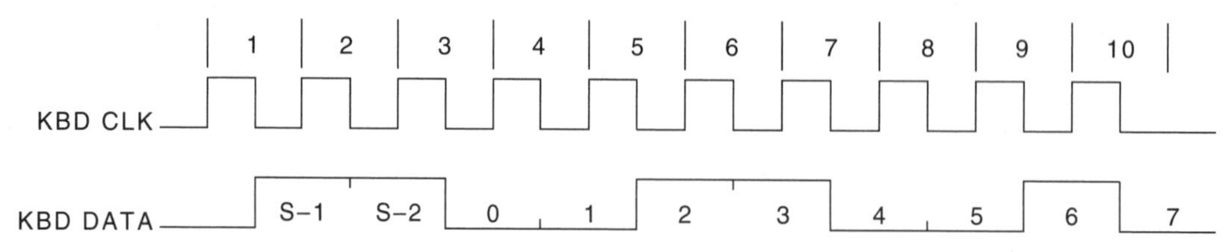

Figure 7-7 XT keyboard serial transmission.

TABLE 7-2
XT Key Codes

Key Code	*Key*	*Key Code*	*Key*	*Key Code*	*Key*
1	Esc	1D	Ctrl	39	Space Bar
2	1	1E	a	3A	Caps Lock
3	2	1F	s	3B	F1
4	3	20	d	3C	F2
5	4	21	f	3D	F3
6	5	22	g	3E	F4
7	6	23	h	3F	F5
8	7	24	j	40	F6
9	8	25	k	41	F7
A	9	26	l	42	F8
B	0	27	;	43	F9
C	-	28	'	44	F10
D	=	29	`	45	Num Lock
E	Backspace	2A	Lt. shift	46	Scroll Lock
F	Tab	2B	\	47	7/Home
10	q	2C	z	48	8/Up Arrow
11	w	2D	x	49	9/Pg Up
12	e	2E	c	4A	Keypad−
13	r	2F	v	4B	4/Lt. Arrow
14	t	30	b	4C	Keypad 5
15	y	31	n	4D	6/Rt. Arrow
16	u	32	m	4E	Keypad +
17	i	33	,	4F	1/End
18	o	34	Period	50	2/Dn Arrow
19	p	35	/	51	3/Pg Dn
1A	[	36	Rt. Shift	52	0/Ins
1B	]	37	*	53	./Del
1C	Enter	38	Alt		

The open collector buffers connected to the 8048 in Figure 7-5 are used to buffer the keyboard data and keyboard clock. The open collector buffers are needed because the system board disables these lines by grounding them while the system board reads the key code from the keyboard-encoding section on the system. If totem pole buffers were used, the outputs of the buffers would be damaged when the system board disables the keyboard lines.

7.1.5 AT Keyboard Operation

The AT keyboard in Figure 7-6 uses positive response keys. The AT keyboard uses the 8049 microcontroller and a 13-row by 8-column key matrix. The AT keyboard performs basically the same function as the XT keyboard. The 8049 scans the keyboard and selects one of the rows by bringing that row low while all other rows are high. In Figure 7-6 the 8049 selects the row by applying the binary value of the row on the select inputs of the 74LS154 4-to-16 demultiplexer/decoder, which brings the selected row low while all other rows remain high. The diodes connected between the decoder and the key matrix are used to provide a 0.6V drop when a key is pressed to reduce transient noise and provide a valid logic low. The 8049 then samples the eight columns one at a time. At the end of each column a pull-up resistor connected to +5-V supply provides a logic 1 on the column when a key is not pressed. When a key is pressed, the column being sampled produces a logic low. Once a low is detected on the sampled column, the 8049 stops scanning the keyboard and waits about 1.3 ms (time delay for key debounce) and samples the same column again. If the column is still low, the row and column address is stored in the 8049, and the 8049 prepares to transmit the key code. The 8049 scans the key matrix again until the next key is pressed or released. If the column is high, the 8049 scans the keyboard matrix again and ignores the key closure as noise. If another key is also pressed, the 8049 does not recognize the second key closure until the first key is released, except for three special keys (Ctrl, Alt, and Shift), which are used to modify how the key codes are decoded by the system board logic. These modifiers keys must be pressed first and held down while the second key is pressed to modify how the second key is processed. In the case of modified keys, the key code for the modifier is transmitted first, followed by the key code for the second key pressed. Normal key function requires that the first key pressed be released before the second key is processed. As opposed to the make and break key codes on the XT keyboard, the AT keyboard uses a make key code, and the break key code is made up of two bytes of data—the value of F0 followed by the make code. The make key code is a value between 00 and 7F, which is sent to the AT system board as a serial data package.

The serial package contains 11 bits of data (Fig. 7-8). The data package starts with one start bit (low), followed by the 8-bit key code (LSB first, MSB last), followed by one parity bit (odd), and ends with one stop bit (high). During the transmission of the keyboard data (key code), the 8049 also transmits a keyboard clock signal, which is used to synchronize the data reception. Each data bit in the transmission occurs on the falling edge of the keyboard clock and has a duration of 62.4 μs; thus, it takes 686.4 ms to transmit one complete key code, which yields a baud rate of 16K bps. The make key code in Figure 7-8 is 1C (hexadecimal) (Table 7-3), which means that the letter "a" key was pressed.

If the key is released before 0.5 second passes, the 8049 produces a break code, which is transmitted to the AT system board in a serial data format. The break code for an AT keyboard is the F0 followed by the key code. The break code (Fig. 7-8) tells the AT BIOS key routine that the function of that key sequence is completed and the key has been released (the break code is needed if a program is used to change the function of the key pressed). If more than one nonshift key is pressed, the 8049 transmits the first make key code (the first key pressed) and sends the next make key code. If the Ctrl, Alt, or Shift key is pressed before a second key, it produces a special make code that allows the next key pressed to produce a special modified make key code. This, when processed by the BIOS key-encoding routine, produces an extended ASCII code.

If the key is still pressed after 0.5 second, the 8049 produces a make key code (the same as the original make key code) and transmits it at a rate of six every second (one every 166.7 ms). This autorepeat function continues until the break code for the key pressed is sensed by the 8049. The break code is a series of two bytes transmitted one after the other; the first byte is F0 followed by the key code.

The open collector buffers connected to the 8049 in Figure 7-6 are used to buffer the keyboard data and keyboard clock. The open collector buffers are needed because the system board disables these lines by grounding them while the system board reads the key code from the keyboard-encoding section on the system. If totem pole buffers were used, the outputs of the buffers would be damaged when the system board disables the keyboard lines. Another set of buffers is connected to the Num Lock, Caps Lock, and Scroll Lock LEDs on the keyboard. These buffers are used to drive the LEDs sc that the 8049 is not loaded down.

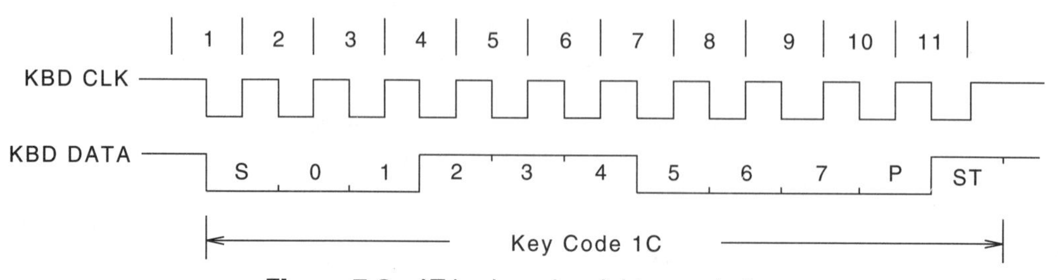

Figure 7-8 AT keyboard serial transmission.

TABLE 7-3
AT Key Codes

Key Code	*Key*	*Key Code*	*Key*	*Key Code*	*Key*
1	F9	31	n	6B	4/Lt. Arrow
3	F5	32	b	6C	7/Home
4	F3	33	h	70	0/Ins
5	F1	34	g	71	./Del
6	F2	35	y	72	2/Dn Arrow
7	F12	36	6	73	5 Keypad
9	F10	3A	m	74	6/Rt. Arrow
A	F8	3B	j	75	8/Up Arrow
B	F6	3C	u	76	Escape
C	F4	3D	7	77	Num Lock
D	Tab	3E	8	78	F11
E	~	41	,	79	+ Keypad
11	Left Alt	42	k	7A	3/Pg Dn
12	Left Shift	43	i	7B	− Keypad
14	Left Ctrl	44	o	7D	9/Pg Up
15	q	45	0	7E	Scroll Lock
16	1	46	9	83	F7
1A	z	49	Period		
1B	s	4A	/	E0 4A	/
1C	a	4B	l	E0 11	Right Alt
1D	w	4C	;	E0 14	Right Ctrl
1E	2	4D	p	E0 12	Prt Scrn
21	c	4E	-	E0 69	End
22	x	52	'	E0 6B	Left Arrow
23	d	54	[	E0 6C	Home
24	e	55	=	E0 70	Insert
25	4	57		E0 71	Delete
26	3	58	Caps Lock	E0 72	Down Arrow
29	Space Bar	59	Right Shift	E0 74	Right Arrow
2A	v	5A	Enter	E0 75	Up Arrow
2B	f	5B	]	E0 7A	Page Down
2C	t	5D	\	E0 7D	Page Up
2D	r	66	Backspace	E1 14 77	Pause
2E	5	69	1/End		

7.1.6 XT System Board Key-Encoding Circuitry

The circuitry on the XT system board that encodes the key codes (Fig. 7-9) coming from the keyboard is made up of four ICs, all located in the peripheral interface section. The key-encoding circuitry is made up of one flip-flop, two data flip-flops, and one 8-bit shift register.

The purpose of the 7474 flip-flop is to develop the IRQ1 request for the 8259 programmable interrupt controller (PIC). This interrupt occurs only after the serial key code is converted into a parallel format. The 8088 then processes the interrupt request generated by the PIC and uses the vector to the BIOS key-encoding routine. The key-encoding routine causes the 8088 to read the data on port A of the PPI. Once the data is read, encoded into ASCII, and stored in the FIFO (first in first out) RAM keyboard buffer, this flip-flop resets, ending the sequence. The RAM keyboard buffer is 8 bytes and is controlled by BIOS. If the buffer becomes full, the new key code is lost and the IBM PC beeps, indicating a loss of the key code. The size of the keyboard buffer can be increased through software.

There are two data flip-flops in the circuitry on the system board. FF-1 is used to delay the KBD CLK one clock cycle, which allows the data to stabilize on the KBD DATA line. The second flip-flop (FF-2) is used to further delay the KBD CLK by one-half cycle, which causes

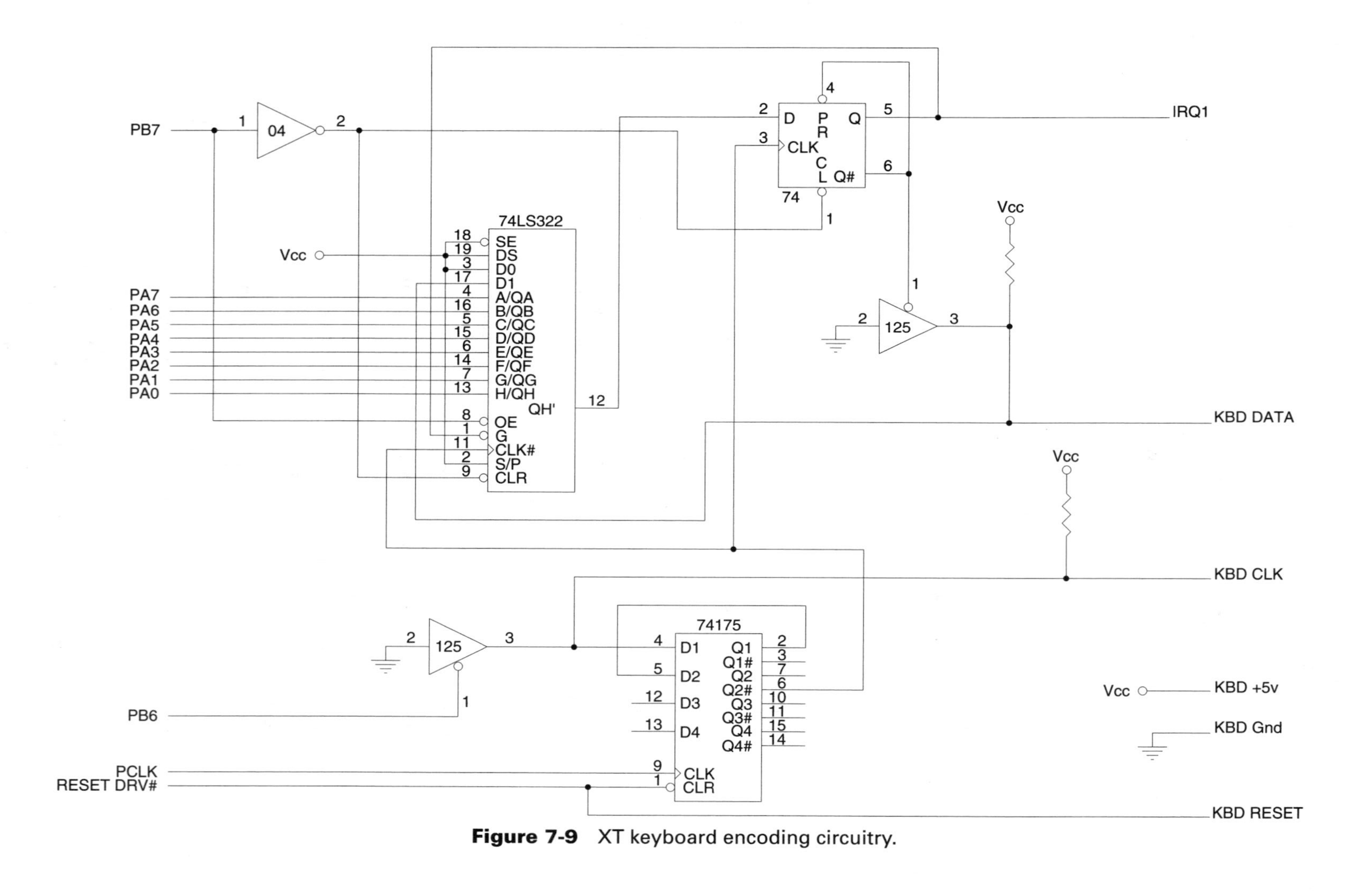

Figure 7-9 XT keyboard encoding circuitry.

the data on the KBD DATA line to be read half-way through the bit time period. The 1.5 clock time delay provides the most accurate data transfer.

The purpose of the 74322 shift register is to accept data in a serial format from the KBD DATA input and convert it into a parallel format. Once the data byte is shifted into the shift register, the second start bit flags the 7474 flip-flop, which develops the IRQ1 keyboard interrupt. This IC is configured as a serial in, parallel out, shift right shift register. Once the data has been read by port A of the PPI, this register is cleared.

7.1.7 XT Key Encoding Operation

The following sequence is what happens when key 76 (4C hex) is pressed on an XT keyboard. This is the number 5 key on the numeric keypad of the XT IBM PC keyboard; the NumLock key is locked. The timing diagram in Figure 7-10 shows the timing sequence for this key.

1. The number 5 key is pressed on the keypad (NumLock key is locked).
2. As the 8048 scans the keyboard, the testable pin goes high when key location 76 (row 9, column 4) is tested. The code 76 is placed in the RAM of the 8048 for later use.
3. If, after five complete keyboard scans, this key is still pressed, the 8048 begins transmitting the make key code data along with 10 KBD CLKs. The reason for the delay is to debounce the key and verify a proper key closure.
4. KBD DATA is sent one bit at a time, on the falling edge of each KBD CLK, and the data bit stays valid on the KBD DATA line for 100 μsc. The first two bits are high. The start bits are inserted into the data by the 8048 and are not used as part of the key code. The second start bit is used to develop IRQ1 for the PIC.

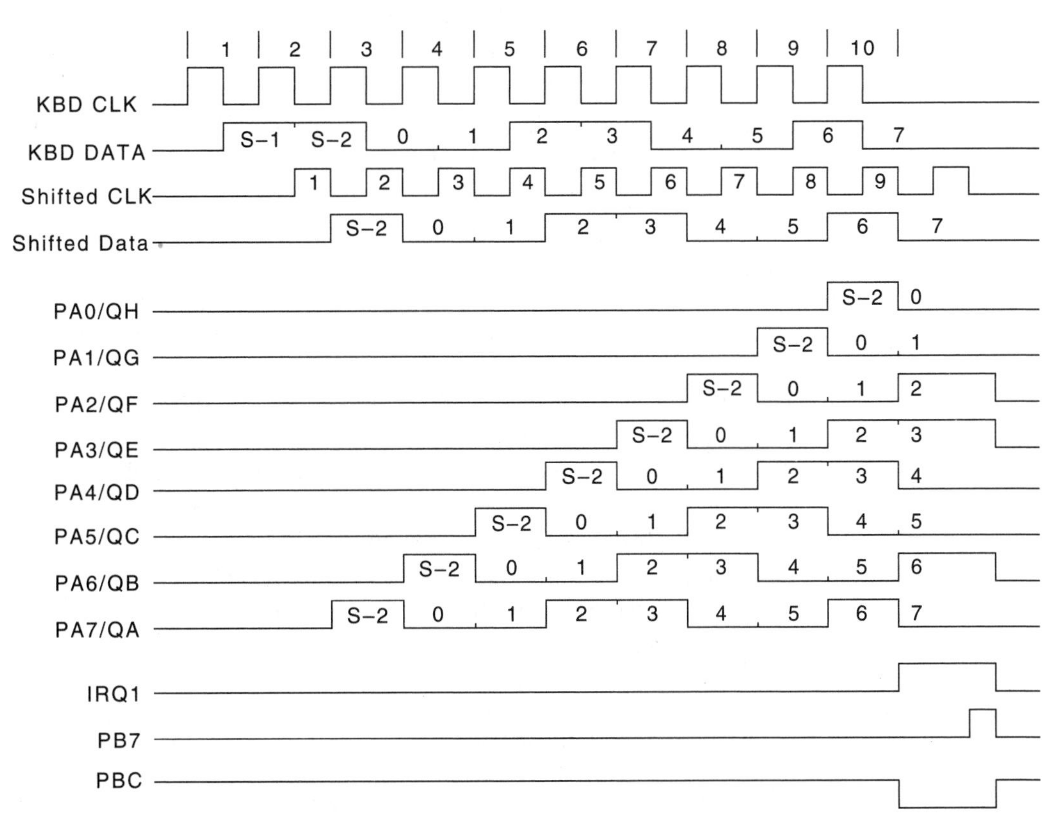

Figure 7-10 XT key encoding timing diagram.

5. The KBD CLK is first applied to data FF–1, which delays the SHIFTED CLK signal to the shift register by one cycle. The data on the data input of the shift register is not shifted into the shift register because a clock pulse was not received by the clock input of the shift register. The first start bit is lost (not shifted into the shift register).
6. The SHIFTED CLK from the fist flip-flop is applied to data flip-flop two, which delays the SHIFTED CLK by one-half cycle. This causes the data on the KBD DATA line to be shifted to the shift register halfway through each bit time.
7. The SHIFTED CLK from the second flip-flop is applied to the shift register clock input. Because of the delays, by the time the shift register receives the first clock pulse, the second start bit is on the KBD DATA line. The second start bit (first start bit is lost because of the time delay) is shifted into the LSB of the shift register. Port A now has the following values, beginning with the MSB as the left-most value: 1 0 0 0 0 0 0 0.
8. Halfway through the third KBD CLK, the data on the KBD DATA line is shifted into the shift register. The value of port A at this time is 0 1 0 0 0 0 0 0.
9. Halfway through the fourth KBD CLK, the data on the KBD DATA line is shifted into the shift register. The value of port A at this time is 0 0 1 0 0 0 0 0.
10. Halfway through the fifth KBD CLK, the data on the KBD DATA line is shifted into the shift register. The value of port A at this time is 1 0 0 1 0 0 0 0.
11. Halfway through the sixth KBD CLK, the data on the KBD DATA line is shifted into the shift register. The value of port A at this time is 1 1 0 0 1 0 0 0.
12. Halfway through the seventh KBD CLK, the data on the KBD DATA line is shifted into the shift register. The value of port A at this time is 0 1 1 0 0 1 0 0.
13. Halfway through the eighth KBD CLK, the data on the KBD DATA line is shifted into the shift register. The value of port A at this time is 0 0 1 1 0 0 1 0.
14. Halfway through the ninth KBD CLK, the data on the KBD DATA line is shifted into the shift register. The value of port A at this time is 1 0 0 1 1 0 0 1.
15. At this time, the LSB of the shift register contains the second start bit. The high causes FF–3 to produce a high on the IRQ1 line of the PIC on the next KBD CLK.
16. Halfway through the tenth KBD CLK, the data on the KBD DATA line is shifted into the shift register. The value of port A at this time is 0 1 0 0 1 1 0 0 (4C hex).
17. At this time, FF–3 sets the IRQ1 input of the PIC.
18. The IRQ1 input causes the 8088 to vector to the BIOS key-encoding routine.
19. PB6 of the PPI disables any clocks to the shift register by grounding the KBD CLK line. The KBD CLK line remains disabled until port A of the PPI is read. Because no clock signal is applied to the shift register, no new data is allowed to be shifted into the shift register.
20. The key-encoding routine causes the 8088 to read port A of the PPI. The data is then transferred into the keyboard RAM buffer.
21. The data in the keyboard buffer is converted into an extended ASCII code.
22. Once the data is converted into ASCII, the data is transferred into the locations indicated by the key-encoding routine.
23. Once port A of the PPI is read, the 8088 causes PB7 of the PPI to set, which clears the shift register so that the next key can be read.
24. PB6 of the PPI enables the KBD CLK line, allowing the keyboard to send the next key code.

7.1.8 AT Key-Encoding Circuitry

The AT system board uses the Intel 8042AH universal peripheral interface 8-bit slave microcontroller to encode key codes. This IC is also responsible for controlling the A20GATE line for AT MPUs (used to help switch between the real and protected operating modes) and in some AT systems is used to indicate equipment and default configurations of the system board. Figure 7-11 shows a typical 8042 section for an AT system board.

The 8042 from Intel is an 8-bit microcontroller that contains the following features: maximum operating frequency of 12 MHz, 2K × 8 ROM, 256 × 8 RAM, 8-bit timer/counter, and 18 programmable I/O lines. The KBD CLK is applied to the test input zero (T0), which is used to clock in the 11-bit key code package. The KBD DATA is applied to the test input one (T1) pin, which is used to input the serial key code from the keyboard. P16 of port 1 is used to generate the keyboard enable signal, which enables or disables the keybaord when high. The 7405 open collector inverter on the output of P16 is used because its output is connected to the KBD DATA output of the keyboard. P17 of port 1 is used to generate the keyboard clock enable signal, which when high disables the keyboard clock until the MPU can read the key code that has been shifted into the 8042. The open collector inverter on the output of P17 is needed because its output is connected to the KBD CLK output of the keyboard. The KBD CLK EN line is used like PB6 in the XT system except that the keyboard clock is disabled when high, not low. The idle state of the KBD CLK line from the keyboard is a logic high. Because the 8042 is used to encode the keyboard data, the data package takes on the form of a standard asynchronous data package. The 11-bit key code package begins with one start bit (logic low), followed by the 8-bit key code (LSB first and MSB last) and then a parity bit (odd parity is used), and ends with one stop bit (logic 1). Each of the 11 bits of the key code package requires the falling edge of the KBD CLK, which is produced only during the transmission of a key code. Once the key code is sent to the 8042 the KBD CLK line remains low until the key code is read from the 8042. Once the key code is shifted into the 8042, the 8042 strips away the start bit, checks for a parity error, and verifies the number of bits using the stop bit to indicate the end of the transmission. If everything is correct, the 8042 produces the IRQ1 interrupt for the PIC. If there is a problem such as an invalid start bit, parity error, or framing error (the total number of bits in the package is not correct), the 8042 flushes out the key code received and does not generate the IRQ1 interrupt.

7.1.9 AT Key-Encoding Operation

The following sequence occurs when the letter "a" key (1C hex) is pressed on an AT keyboard. The timing diagram in Figure 7-12 shows the timing sequence for this key.

1. The letter "a" key is pressed.
2. As the 8049 scans the keyboard, a low is seen on the column connected to the letter "a" key when the correct row is selected.
3. The 8049 stops scanning the keyboard, and the column and row address is saved in one of its RAM locations. After 1.3 ms the column is checked again; if still low the 8049 processes the key code value and begins scanning the keyboard again. If the column is high at the end of 1.3 ms, the RAM location is cleared and the 8049 continues scanning the keyboard as though no key were pressed.
4. Processing the key code value requires that the row and column address be converted into the key code itself (1C hex in this case).
5. The 8049 then produces a parity bit. Because the number of logical 1s in the key code is odd, the parity bit is low (odd parity).
6. Once the key code and parity bit is developed, the 8049 starts transferring the data out of the 8049.
7. The 8049 produces the first KBD CLK and sends it out on the KBD CLK line. On the falling edge of the KBD CLK, the first bit in the key code package is placed on the KBD DATA line. The first bit is the start bit, which is a low. The 8042 reads the data on KBD DATA line one-half bit time following the first KBD CLK. Because

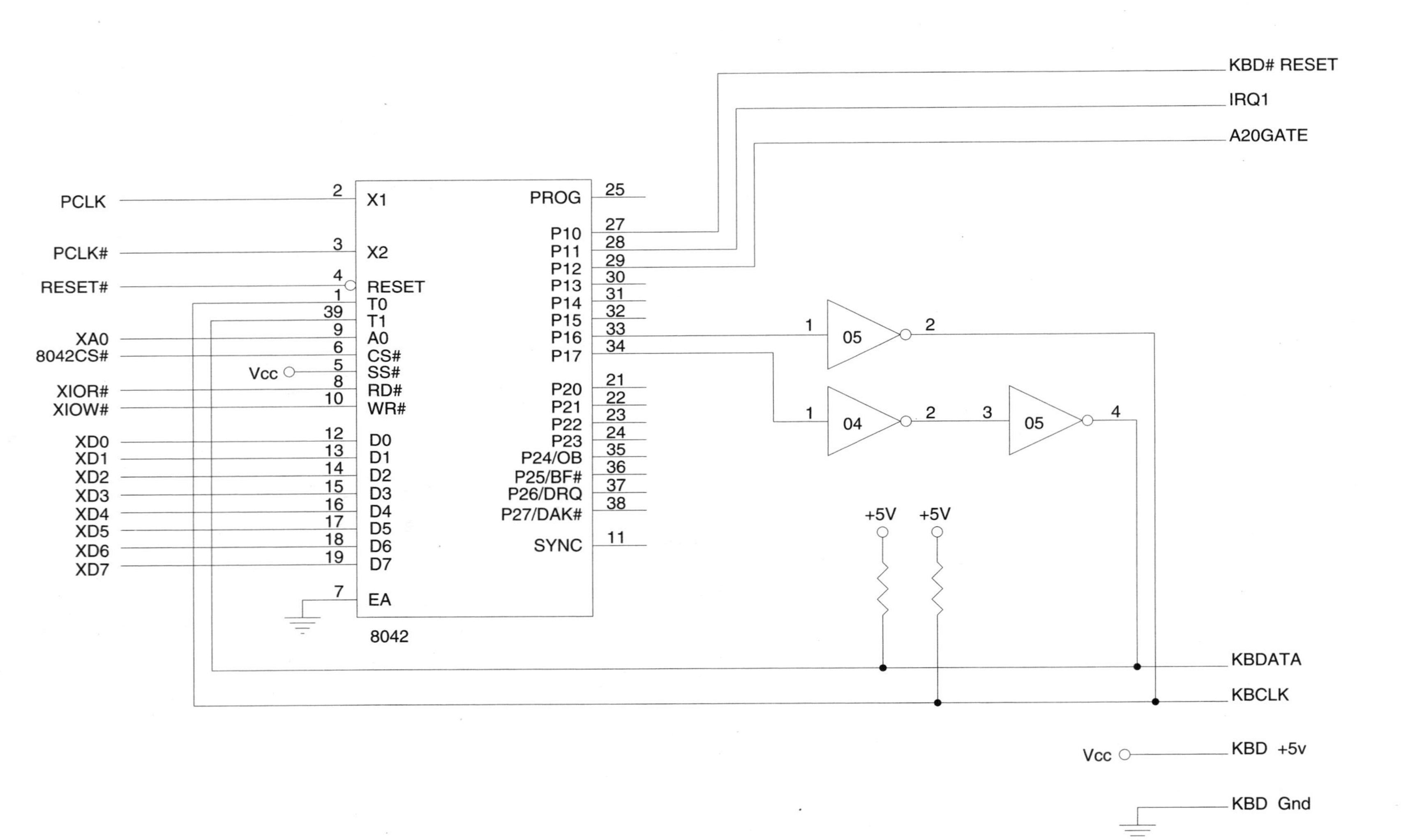

Figure 7-11 AT key encoding circuitry.

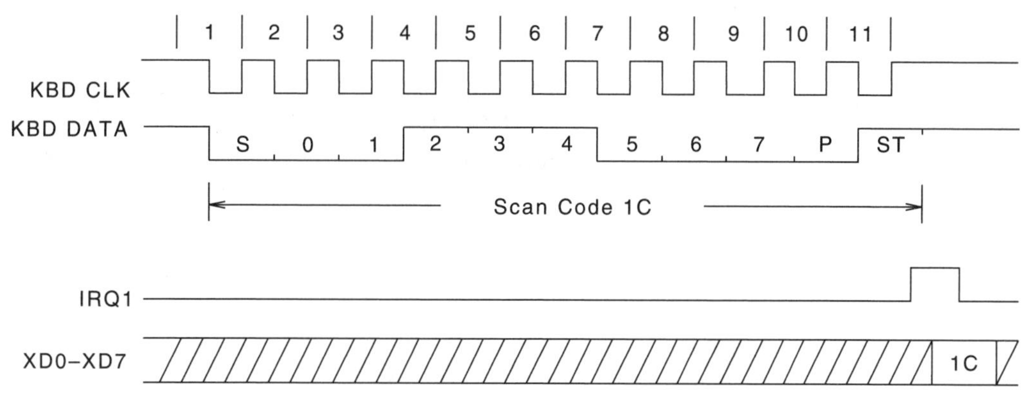

Figure 7-12 AT key encoding timing diagram.

data changes with a clock, bit time and clock time are equal. This process allows the voltage level on the KBD DATA line to settle before the data is transferred into the 8042.

8. The 8049 produces the second KBD CLK. On the falling edge of the second KBD CLK, the second bit in the key code package is placed on the KBD DATA line. The second bit is bit 0 (LSB) of the key code itself (in this example a logic 0). The 8042 reads the data on the KBD DATA line halfway through the bit time of the second bit.
9. On the falling edge of the third KBD CLK, the third bit in the key code package is placed on the KBD DATA line. The third bit is bit 1 of the key code itself (in this example a logic 0). The 8042 reads the data on the KBD DATA line halfway through the bit time of the third bit.
10. On the falling edge of the fourth KBD CLK, the fourth bit in the key code package is placed on the KBD DATA line. The fourth bit is bit 2 of the key code itself (in this example a logic 1). The 8042 reads the data on the KBD DATA line halfway through the bit time of the fourth bit.
11. On the falling edge of the fifth KBD CLK, the fifth bit in the key code package is placed on the KBD DATA line. The fifth bit is bit 3 of the key code itself (in this example a logic 1). The 8042 reads the data on the KBD DATA line halfway through the bit time of the fifth bit.
12. On the falling edge of the sixth KBD CLK, the sixth bit in the key code package is placed on the KBD DATA line. The sixth bit is bit 4 of the key code itself (in this example a logic 1). The 8042 reads the data on the KBD DATA line halfway through the bit time of the sixth bit.
13. On the falling edge of the seventh KBD CLK, the seventh bit in the key code package is placed on the KBD DATA line. The seventh bit is bit 5 of the key code itself (in this example a logic 0). The 8042 reads the data on the KBD DATA line halfway through the bit time of the seventh bit.
14. On the falling edge of the eighth KBD CLK, the eighth bit in the key code package is placed on the KBD DATA line. The eighth bit is bit 6 of the key code itself (in this example a logic 0). The 8042 reads the data on the KBD DATA line halfway through the bit time of the eighth bit.
15. On the falling edge of the ninth KBD CLK, the ninth bit in the key code package is placed on the KBD DATA line. The ninth bit is bit 7 (MSB) of the key code itself (in this example a logic 0). The 8042 reads the data on the KBD DATA line halfway through the bit time of the ninth bit.
16. On the falling edge of the tenth KBD CLK, the tenth bit in the key code package is placed on the KBD DATA line. The tenth bit is the parity bit (in this example a logic 0).

The 8042 reads the data on the KBD DATA line halfway through the bit time of the tenth bit.

17. On the falling edge of the eleventh KBD CLK, the eleventh bit in the key code package is placed on the KBD DATA line. The eleventh bit is the stop bit (always a logic 1). The 8042 reads the data on the KBD DATA line halfway through the bit time of the eleventh bit.
18. Upon shifting in the stop bit, the 8042 processes the key code package, first verifying that a total of 11 bits has been read. This is known as a framing check. If a framing error occurs, the 8042 clears the data package just received and waits for the next data package without generating a keyboard interrupt.
19. The 8042 next strips away the start and stop bits.
20. The 8042 then counts the total number of logic 1s in the 8-bit key code and the parity bit. If the total number of logic 1s is odd, no parity error is detected. If a parity error is detected, the 8042 clears the data package just received and waits for the next data package without generating a keyboard interrupt.
21. If no type of error is detected by the 8042, the 8042 disables the KBD CLK line by making P16 low, which places a logic 1 on the KBD CLK line. This keeps the keyboard from sending any more key code packages to the system board until the key code can be read from the 8042.
22. While disabling the KBD CLK line, the 8042 generates a keyboard interrupt (IRQ1) by making P11 go high.
23. The IRQ1 input causes the MPU to vector to the BIOS key-encoding routine.
24. The key-encoding routine causes the MPU to read the key code from the 8042. The data is then transferred into the keyboard RAM buffer.
25. The data in the keyboard buffer is converted into an extended ASCII code.
26. Once the data is converted into ASCII, the data is transferred into the locations indicated by the key-encoding routine.
27. Once the key code is read from the 8042, the 8042 clears the key code and re-enables the KBD CLK line. The keyboard is allowed to send the next key code package.

7.1.10 Troubleshooting the Keyboard

1. If a key or group of keys do not function:
   (a) Take apart the keyboard and check for foreign material (dirt, dust, liquids) in the keyboard on the contacts or on the pads of a capacitive keyboard. Clean any foreign material from the keyboard.
   (b) Check for missing springs or rubber grommets.
   (c) Check for missing or broken key parts.
2. The keyboard does not function, and a 301 (any type of 30x error) or no keyboard connected error message is displayed on the screen.
   (a) Verify that the keyboard cable is connected.
   (b) Verify that the keyboard connector to the system board is not loose.
   (c) Check KBD ground line.
   (d) Check KBD +5-V line.
   (e) Check KBD RESET#; it should be high for normal operation.
   (f) Press a key and check the KBD clock line for pulses.
   (g) Press a key and check the KBD data line for pulses.
3. If the keyboard checks out, make the following checks on the system board:
   (a) Check the clock input for the 74322 shift register or the clock input of the 8049.

(b) Check for a high on the IRQ1 line once a key is pressed.
(c) In an XT check PB7. It should normally be low except for a short pulse when the register is cleared.
(d) Check PB6 on the XT. It should normally be high except for a pulse when a keyboard is reading port A.

7.2 POWER SUPPLIES

As with most electronic equipment, the components of the computer operate on direct current (DC), even though most power supplies are powered from an alternating current (AC) supply source. AC is used to power the majority of electronic devices because it is easy to generate, is easy to transfer the energy from one location to another with minimal loss, is the most available power source in the world, and provides low operational cost. Batteries, which were the first form of electric power, suffer great power losses when transferring the energy from one location to another over long distances. Distribution of DC power from a centralized location is not cost effective, and DC power therefore has never been used as a major source of power.

7.2.1 AC Power

AC power sources are not without their own problems. The companies that generate AC power in the US regulate the frequency of the AC to 60 Hz. This frequency regulation is one parameter of AC that is usually maintained and very rarely presents problems with electronic equipment. On the other hand, during the transfer of energy from power stations to the end user, may things happen to the voltage level of the AC that do affect electronic equipment. Any type of AC distortion has some effect on a power supply; how much effect it has depends on the severity of the problem and the design of the power supply.

Figure 7-13 shows a typical AC sine wave with a frequency of 60 cycles per second. The voltage of a typical AC sine wave ranges from 110 to 120 volts, which means that the peak voltage is between +155 to 170 V and −155 to 170 V.

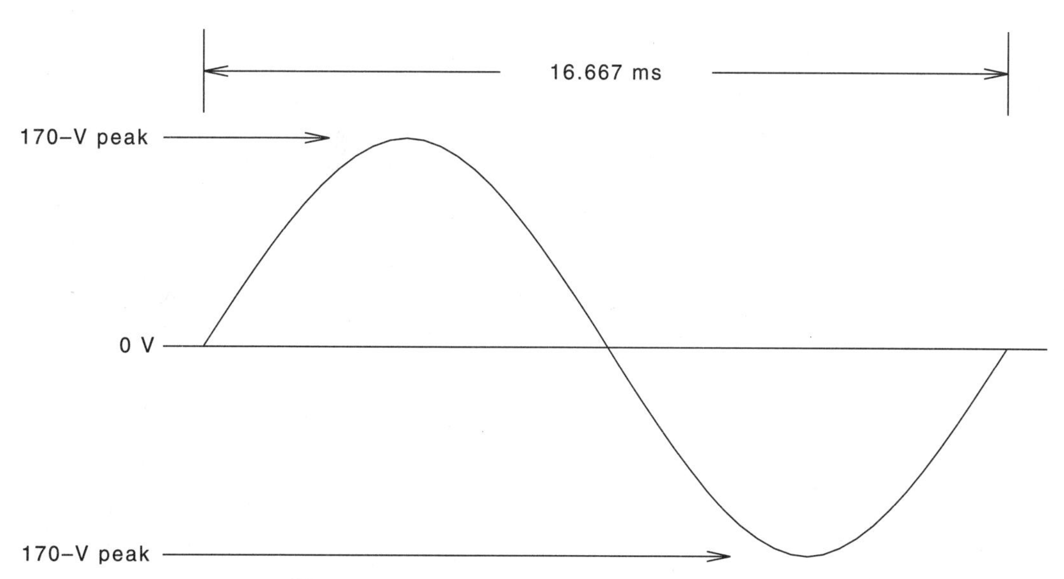

Figure 7-13 Normal AC power sine wave.

7.2.2 Types of AC Distortion

AC distortion normally occurs when the shape and/or voltage levels of the sine wave change. There are two basic forms of distortion: (1) Short-duration distortion, which normally occurs during one cycle and affects the shape of the sine wave. This includes noise, spikes (impulses), and dropouts. (2) Long-duration distortion normally affects voltage levels over a longer period of time (two or more cycles); it includes surges, sags, brownouts, overvoltage, and undervoltage.

Short-duration AC distortion (Fig. 7-14) falls into three categories—noise, spikes, and dropouts. Noise on the AC line is normally associated with high-frequency repetitive changes in the voltage level that ride on the sine wave. Noise momentarily changes the voltage level of the sine wave, therefore changing the shape of the sine wave. Noise in many cases can be caused by devices that either produce radio frequencies or control fast-changing loads. A spike (also known as impulse noise) is a positive or negative momentary (usually of less than 1 ms) change in voltage level that can occur during one cycle of a sine wave. Spikes can occur when devices on the AC line toggle from one state to another state. A dropout is a momentary loss of power of normally less than one cycle. Dropouts occur when power companies switch power transfers from one path to another path to meet power demands and reroute power during power failures in other locations on the power grid.

Long-term AC distortion (Fig. 7-15) falls into two categories—surges and sags. Surges occur when the peak voltage of the sine wave increases above the normal standard of the power company, usually for one or more cycles. (When a surge lasts more than a few seconds, it is known as an overvoltage condition.) Surges often occur during the time it takes the utility company to compensate for changes in the power grid from a heavy load to a lighter load. This condition can also occur within a building. Sags occur when the peak voltage of the sine wave decreases below the normal standard of the power company, usually for one or more cycles.

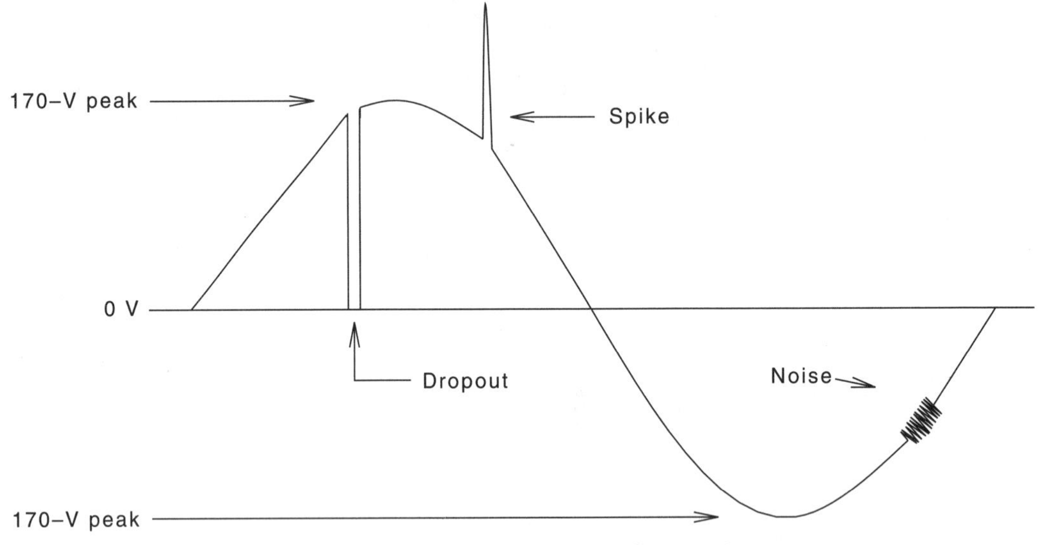

Figure 7-14 Short-duration AC distortion.

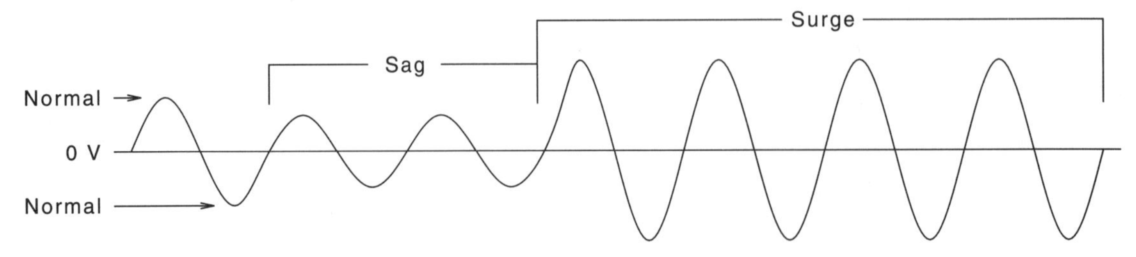

Figure 7-15 Long-duration AC distortion.

(Sags of one cycle or less are often called a flicker.) Sags that last for a few seconds are referred to as undervoltage conditions, and a sag that lasts for a minute or more is referred to as a brownout. Sags occur in the power grid or in a building when the power load changes quickly from a light load to a much heavier load. A brownout (one or more minutes) is basically a long-term sag or undervoltage (more than a few seconds) condition that lasts for long periods of time, normally during heavy-load periods.

7.2.3 AC Power Protection

The purpose of a computer power supply is to convert the AC from the wall outlet into a DC voltage that remains constant within 5% of the rating of the DC line. Better power supplies regulate this voltage under most conditions, but frequency and strong short- and long-duration AC distortion can affect the DC output voltage. The power supply must also provide enough current for its loads and changes in its loads to maintain the proper output voltage.

Today's computer power supplies include spike and noise protection and can keep its output voltage regulated during most sags. The spike protection (mistakenly referred to as surge protection) is accomplished with the use of an MOV (metal oxide varsister) to shut out high-voltage spikes between the hot and neutral AC lines (a form of lightning protection, but it cannot protect the system during a direct strike). Once the voltage on the MOV exceeds its breakdown rating, the voltage is shorted to the neutral line. MOVs go bad if the spike exceeds this maximum input rating, if the spike duration is too long, or after a number of times the MOV breaks down to protect the equipment. Noise protection in the computer power supply is provided by an EMI (elector magnetic interference) filter used to reduce the effects of radio frequency on the power supply. Computer power supplies also contain low-pass filters and sometimes notch filters to reduce the frequency created by ICs switching in the computer itself or the frequency of the switching power supply regulator from interfering with the AC input voltage before it is converted into the DC. Strong high-frequency noise may cause circuitry in the computer to oscillate, which may cause the equipment to fail. Long-duration voltage sags (undervoltage or brownouts) or surges (overvoltage) in most cases cause the voltage regulator part of the power supply to work harder and lose some efficiency.

External power protection devices can be used to help protect the computer power supply from AC distortion before the distortion gets to the computer. A low-cost surge suppressor/protector is normally nothing more than a multi-outlet strip that uses MOV to protect the equipment from high-voltage spikes (not normal surges) and normally contains a fuse or circuit breaker. Many of today's surge suppressors also allow the user to add protection to the phone lines connected to the computer. For a little more money the user can purchase a surge suppressor and EMI filter (Fig. 7-16), which is a surge suppressor with an EMI filter to reduce the effects of radio frequencies. Surge suppressors do not compensate for normal surges (overvoltage) and sags (undervoltage and brownout conditions); to correct for these types of problems a line regulator is used. Line regulators (Fig. 7-17) monitor the AC input and adjust its special transformer to maintain the AC output voltage. Line regulators cost more than $100 but maintain the AC output voltage between 110 and 130 V even though the AC input voltages may range from 80 to 150 V. Most line regulators also contain spike and EMI noise filtering, with circuit breaker protection. A line regulator must be sized to the equipment for which it regulates AC voltage. Line regulators are rated in VA (volt, amps), which must be converted into wattage ratings to determine the size needed. Line regulators are rated at 600VA and larger. To convert VA into watts, multiply the VA by 0.75; a 600VA line regulator controls about 450 watts of continuous power without overworking the regulator.

7.2.4 XT/AT Power Supply

All of the XT and AT computer systems today use switching power supplies to supply the voltages needed to operate the system units and keyboard. The following is a review of a linear power supply and an introduction to switching power supplies.

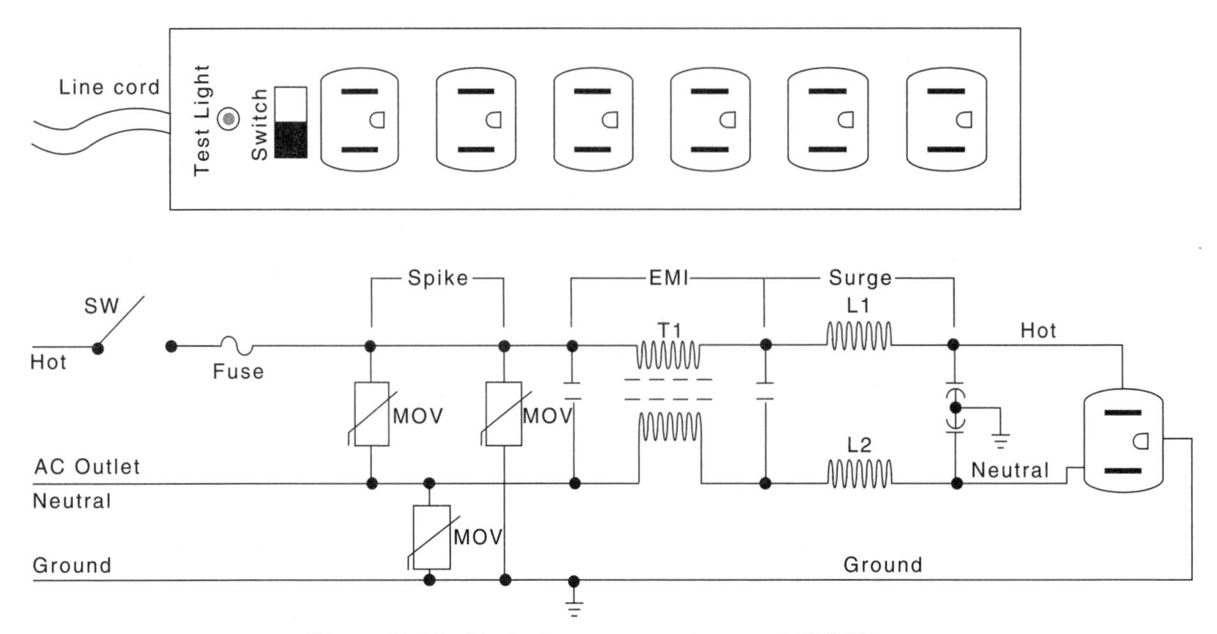

Figure 7-16 Typical surge suppressor and EMI filter.

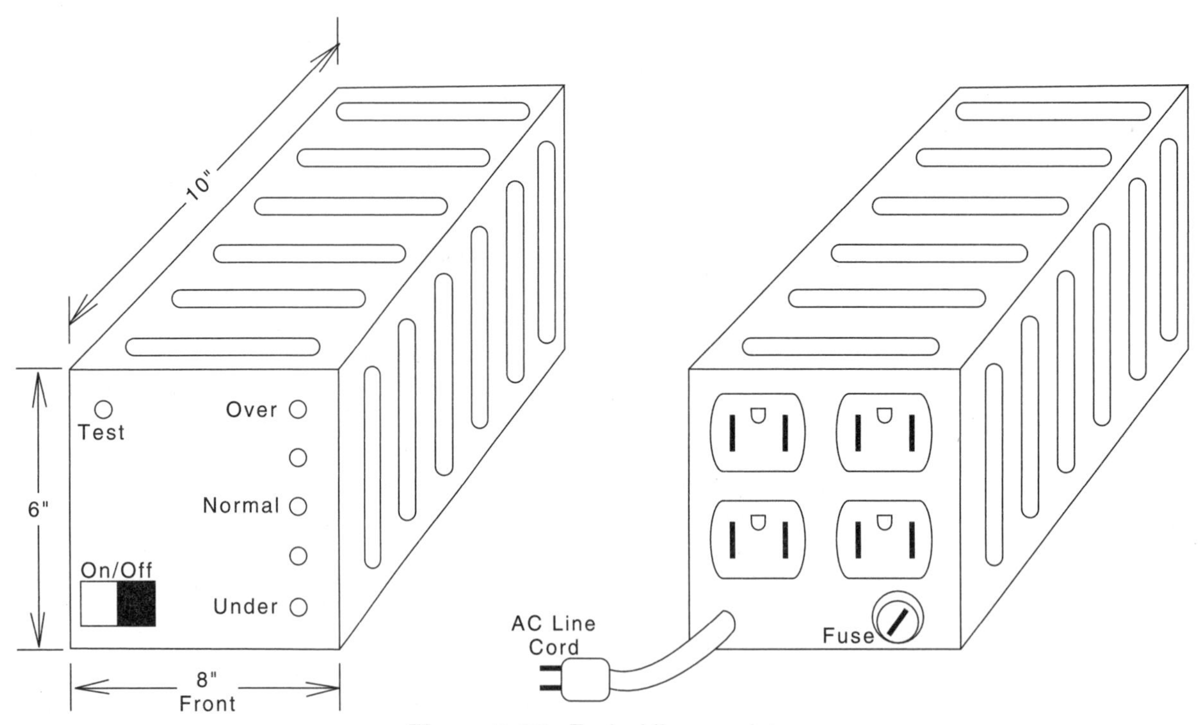

Figure 7-17 Typical line regulator.

Linear Regulated Power Supply. Figure 7-18 is a basic block diagram for a linear regulator used in low-power power supplies. Even though voltage regulators may differ in operation, this block diagram is used as the basis for all linear voltage regulators. This diagram contains four basic parts—voltage reference source, error amplifier, feedback network, and series pass transistor.

The purpose of the voltage reference source is to set up the static and dynamic output voltages from the error amplifier, which is used to control the conduction of the series pass transistor. The static output voltage keeps the transistor on when there is no load on its output.

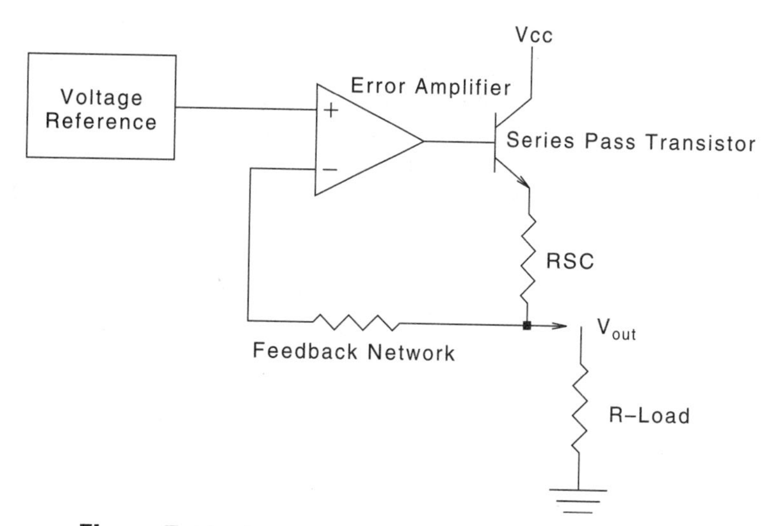

Figure 7-18 Basic linear power supply block diagram.

The dynamic output voltage level is determined by the difference between the voltage reference source and a portion of the output voltage from the regulator.

The feedback network is used to supply a portion of the voltage output of the regulator back to the error amplifier for comparison with the voltage reference source.

The error amplifier is a medium-gain amplifier circuit that compares the feedback network voltage and the voltage reference source. The difference between the two determines the output voltage. The output voltage is applied to the series pass transistor, which controls the output voltage of the regulator.

The series pass transistor is used to complete a series network so that current can pass from ground through the load and the series pass transistor to the supply source. The resistance of the series pass transistor varies to maintain the constant output voltage across the load. The conduction of the transistor is determined by the output voltage from the error amplifier. In series with the series pass transistor is an RSC (resistor short circuit) resistor which has a very small value of resistance. This resistor is used to limit the current through the series pass transistor to protect the regulator from a short circuit (its maximum current rating). Once the current through the RSC register reaches it designated value, the voltage difference between the emitter and the base of the series pass transistor is low enough to shut down the conduction of the series pass transistor.

Operation of a 7805. The 7805 is a fixed three-pin voltage regulator. Its output voltage is within 10% of its rating, with an input voltage of 7 to 35 volts. Fixed three-pin voltage regulators normally drop 2 volts internally. The error amplifier biases the series pass transistor so that the voltage across the load remains constant even though the load changes.

As the load resistance decreases, the load current increases, which tends to lower the output voltage of the regulator. The slightest decrease in the output voltage causes the feedback network voltage to decrease and the output voltage of the error amplifier to increase. The increase in error voltage causes the series pass transistor to conduct more; this lowers the voltage across it, which in turn increases the voltage at the output of the voltage regulator. The opposite occurs if the load resistance increases.

Switching Power Supply. The most-used type of switching power supply uses pulse width modulation. Pulse width modulation (PWM) switching power supplies are offshoots of the older variable-frequency and variable-duty cycle switching power supply formats.

Basically a pulse width modulation power supply regulates the output voltage of the supply by turning transistors on and off at a high frequency and then varying the duty cycle to maintain the voltage across the output controlling elements (capacitors and inductors). Because capacitors and inductors dissipate very little power in the real world, the power dissipation of the power supply is low. This differs from a linear power supply, in which the controlling element is the series pass transistor, which is always biased on, meaning that a large amount power is lost as heat from the series pass transistor (Table 7-4). The PWM switching power supply is also popular because the capacitors and inductors can be much smaller because of the high switching frequency. Because the frequency in linear power is 60 Hz, the transformers, inductors, and capacitors are much larger in both value and actual size. Because the switching power supply has a high switching frequency, the power supply can react more quickly to changes in the output voltage under varying loads. Furthermore, a switching power supply can easily be used to set the voltage up or down and still maintain good regulation. The only disadvantages of a switching power supply are that the values of the controlling elements (capacitors and inductors) must be properly chosen and have close tolerances and the switching of the transistors produces some high-frequency noise.

SG3525A PWM Switching Regulator. The SG3525A, a pulse width modulator control circuit IC, is the heart of many switching power supplies. This IC provides all of the controlling elements necessary to develop a switching power supply. Figure 7-19 shows the internal block diagram of the SG3525A.

The reference regulator is used to provide the reference voltage of the internal blocks of the IC and the reference voltage output used for the noninverting reference for the error amplifier.

The undervoltage lockout block is used to disable the A and B outputs of the SG3525A. When the voltage is less than +8 V on the Vcc pin, this block keeps the A and B outputs of the IC from switching levels. This block is also used to keep the external soft start capacitor from charging until the voltage is above +8 V and sets the RS latch.

The oscillator block is used to produce the switching frequency. RT and CT control the frequency. The discharge pin is used to discharge the capacitor quickly to produce a very short discharge time for the ramp and produce a linear charge ramp. Oscillator output and sync pins are provided to allow multiple switching regulators to operate in the same power supply without the switching frequencies of the regulators interfering with each other. The output frequency of the oscillator connected to the flip-flop is a pulse that occurs at the beginning of each cycle and is used to toggle the Q and Q# outputs. At the same time that the flip-flop toggles, the oscillator pulse resets the output of the RS latch. The output applied to the noninverting input to the pulse width modulator is a linear ramp that matches the charge time of the capacitor.

The error amplifier has two inputs. The noninverting input has the portion of the reference voltage from the voltage reference regulator. The feedback voltage is applied to the in-

TABLE 7-4
Power Supply Specifications

Parameters	*Linear*	*PWM*
Operating frequency	60 Hz	200 KHz
Duty cycle	50%	5–95%
Efficiency	35–50%	70–95%
Regulation	0.1–5%	0.001–0.5%

verting input of the error amplifier. The feedback voltage is a portion of the output voltage. The output of the error amplifier is equal to the difference between the reference and feedback voltages applied to the error amplifier. If the voltages on the two inputs are equal, the error voltage is one-half the maximum error voltage. As the feedback voltage falls below the reference voltage, the error voltage increases. As the feedback voltage goes above the reference voltage, the error voltage decreases.

The soft start pin is used to allow the switching regulator to start up slowly when power is first applied. The constant current source charges the capacitor connected to the soft start pin to provide a starting feedback voltage for the pulse width modulator until the feedback voltage has time to build up to an acceptable level.

The shutdown circuitry is used to protect the switching regulator during start up and power down by disabling the A and B outputs, undervoltage lockout, and soft start circuitry.

The pulse width modulator compares the voltage ramp from the capacitor on the oscillator with the error voltage. The pulse width modulator output is a logic 1 (which is used to set

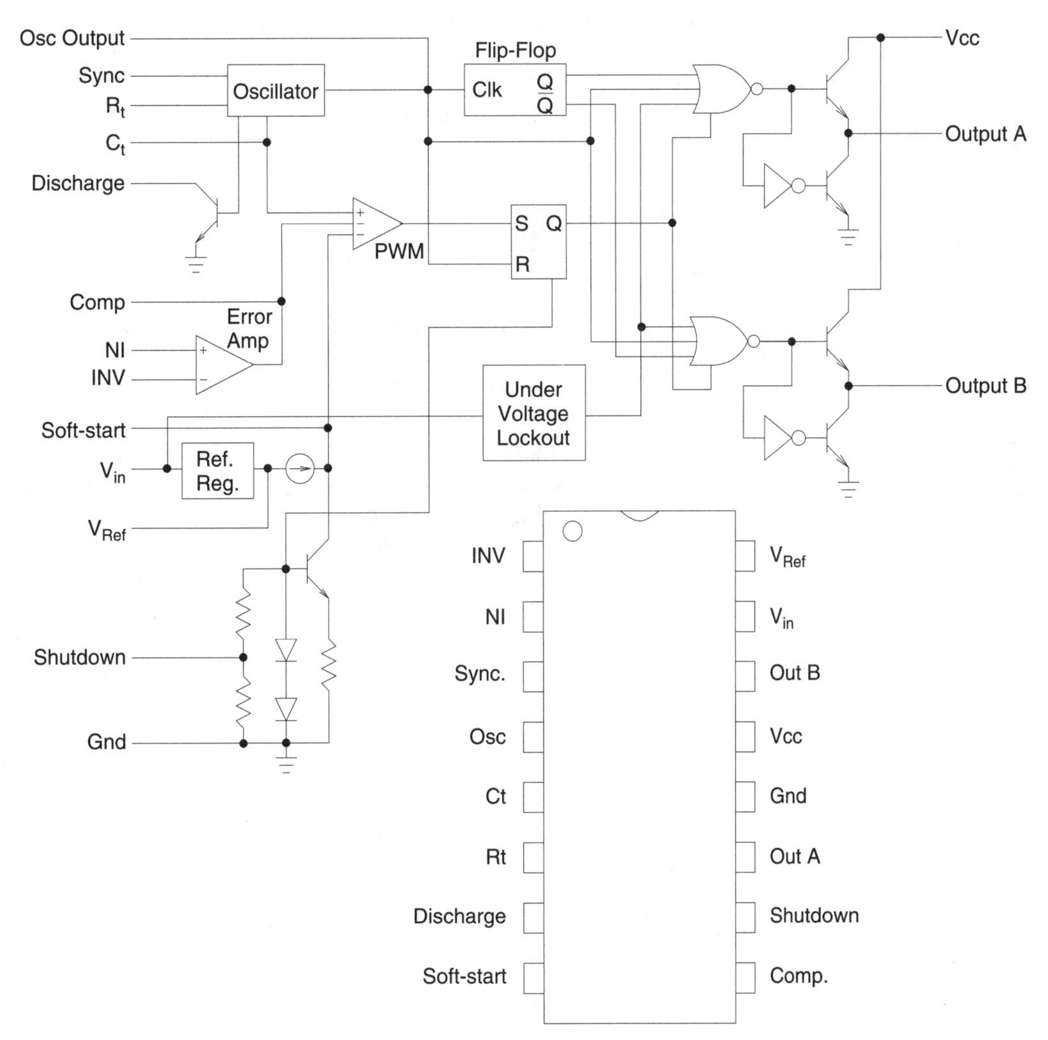

Figure 7-19 SG3525A internal block diagram and pin configuration.

the RS latch) whenever the error voltage is higher than the ramp voltage; otherwise a logic 0 (a logic low has no effect on the "S" input of the RS latch) is produced.

The flip-flop is toggled by the oscillator, the Q and Q# outputs are used with the levels from the RS latch, and shutdown circuitry determines which of the outputs goes high.

There are two NOR gates with four inputs each. The output of each NOR gate is connected to a pair of transistors configured in totem pole format. When a low is applied to each of the four inputs of the NOR gate, the transistor connected to Vcc shorts while the transistor connected to ground is turned off (open). This produces a high on the output. When any other condition occurs on the NOR gate, the transistor connected to Vcc is turned off (open) while the transistor connected to the ground shorts, producing a low level on the output. Because the Q output is connected to the top NOR gate and the Q# output is connected to the bottom NOR gate, the outputs are never turned on at the same time. The A output provides one-half the duty cycle and the B output provides the other half.

SG3525A Operation. The basic operation (Fig. 7-20) of this switching regulator is simple. When the reference voltage is the same as the feedback voltage, the duty cycle is 50%, and half of the duty cycle is provided by each output. As the load resistance decreases, the feedback decreases, which causes the duty cycle to increase and allows the switching transistors to stay on for a longer time. The higher duty cycle causes the voltage to increase. As the load resistance increases, the feedback increases, which causes the duty cycle to decrease. The decrease in duty cycle causes the switching transistor to stay off for a longer time, which lowers the output voltage.

PWM Switching Power Supply. The PWM switching power supply has basically three sections. Figures 7-21, 7-22, and 7-23 show a three-part schematic of the switching power supply.

Nonisolated AC Section: The nonisolated AC section (Fig. 7-21) of the power supply is given that name because this section is not isolated from the AC line. The AC is applied to a double-pole double-throw switch (SW1). When turned off, this switch disconnects the hot and neutral lines from the AC power supply. On the hot side of the AC line following SW1 is a fuse; this fuse is normally located on the circuit board and is not assessible from outside of the power supply. C1-2 and T1 make up the EMI filter, used to reduce radio frequency interference. L1-2 and C2-4 are used to reduce the effects of the switching frequency of the power supply on the AC line frequency. Diode bridge one (DB1) and C5-6 are used to convert the AC into DC.

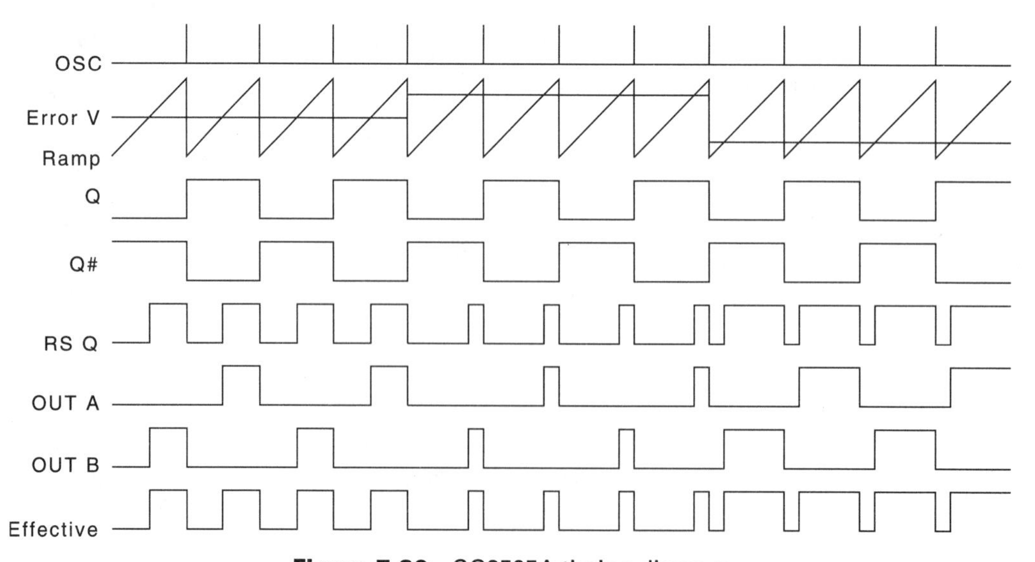

Figure 7-20 SG3525A timing diagram.

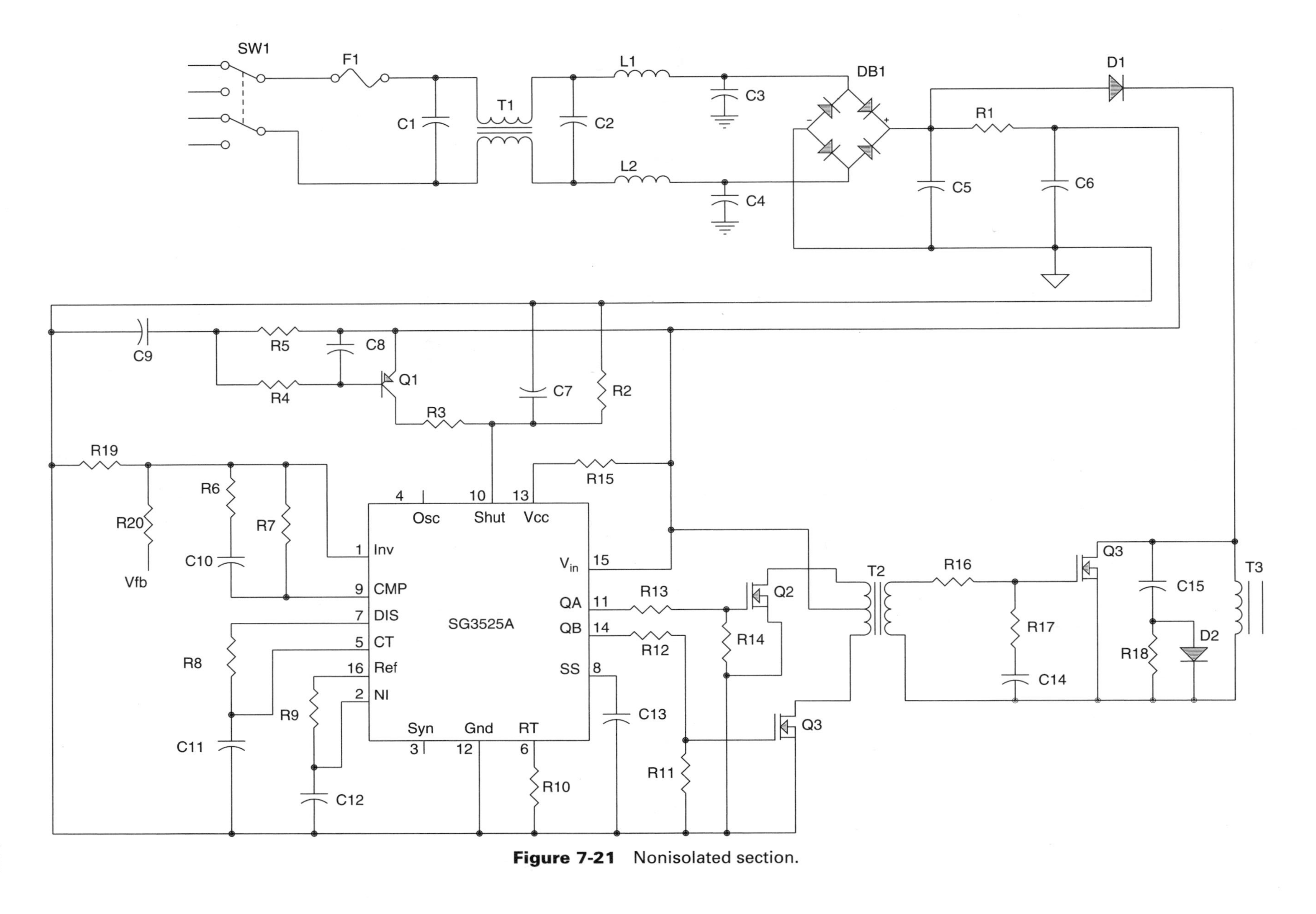

Figure 7-21 Nonisolated section.

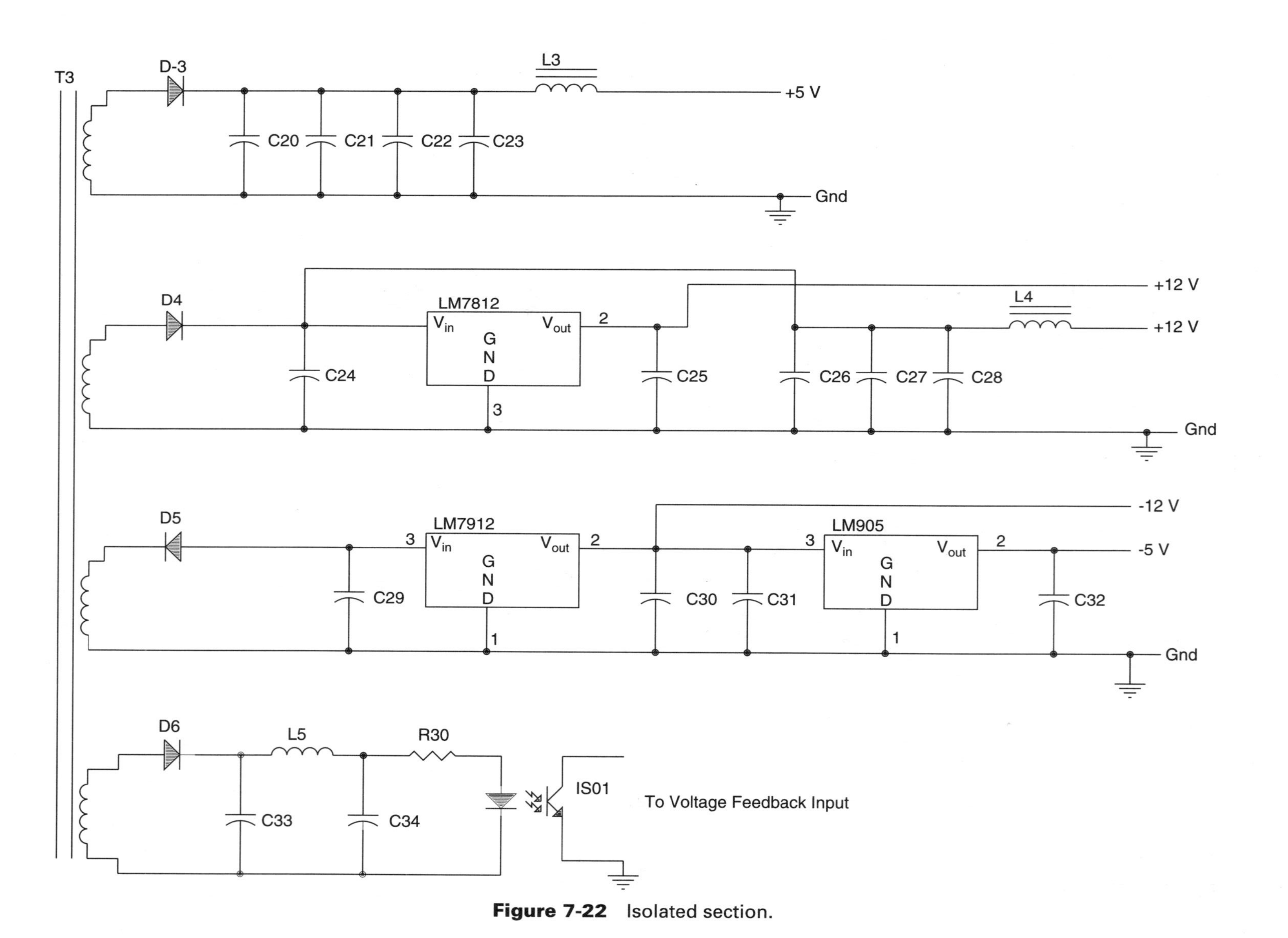

Figure 7-22 Isolated section.

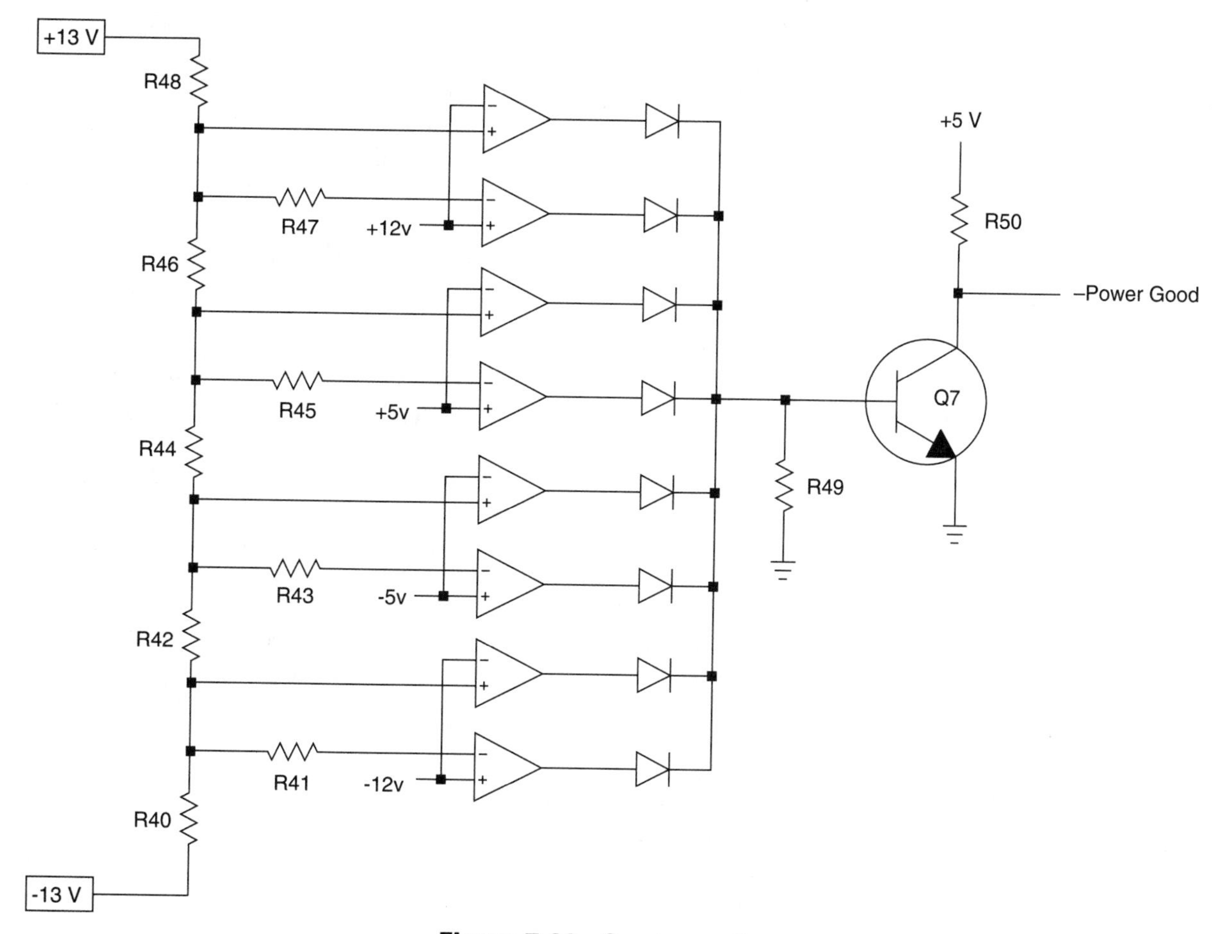

Figure 7-23 Sensing section.

R1 reduces the DC voltage to the switching regulator to about 30 V. R2-R10, R15, Q1, and C7-C13 are the biasing components for the SG3525A.

R11 and R12 set up a voltage divider for Q3 (Power MOSFET) being driven from output B, while R13 and R14 set up a voltage divider for Q2 (Power MOSFET) being driven from output A. The outputs of the MOSFETs produce pulses in the T2 primary center tapped transformer. The voltage on either primary of T2 is approximately the same as V_{in} (about 30 V) when both MOSFETs are turned off and approximately 0 V when either of the MOSFETs is on.

The secondary of T2 isolates the switching regulator from the output stage and the feedback network. D2, C15, and R18 make up the voltage feedback network, which supplies the voltage on the gate of the output stage MOSFET back to the feedback voltage input of the switching power supply. The drain of the output MOSFET is connected to the top of T3 (isolation transformer) and to the raw DC voltage across C5. The raw DC voltage across C6 provides a higher DC level on the primary of T3 because the MOSFET, when turned off, produces about 120 V DC. This connection also allows another path for current to flow, which reduces power dissipation in the power supply.

Isolated Section: The isolated section (Fig. 7-22) of the power supply is isolated from the AC line. This section starts on the secondary outputs of T3. T3 has four secondaries, one to produce the +5-V supply line, one to supply the +12-V supply line, and one to provide both the −12-V and −5-V supply lines.The fourth secondary is used to create a feedback voltage for the SG3525A.

The +5-V section of the power supply contains one diode, four capacitors, and one inductor. Because the frequency from the secondary is about 200 KHz, only a half-wave supply is needed. The four capacitors are referred to as a multistage low-pass filter. Each capacitor has a different size; the small values are used to reduce high-frequency noise while the low frequencies are handled by the larger values of capacitors. The iron core is used to maintain a constant current flow and thus helps regulate the output voltage.

The +12-V section of the power supply contains one diode, five capacitors, one inductor, and one 7812 fixed three-pin voltage regulator. The diode, C26-28, and L4 perform the same functions as these components in the +5-V section. The +12-V output that uses L4 powers the power supply fan and floppy and hard drive motors. The LM7812 is used to isolate noise caused by motors from affecting the +12-V line connected to the system board for I/O interface adapters.

The −12-V and −5-V section of the power supply contains one diode, four capacitors, and two negative linear fixed-voltage regulators. Both voltage regulators are low-power regulators that supply the −12-V and −5-V outputs to the system board for I/O interface adapters.

The fourth secondary is used to provide the optically isolated voltage feedback input to the SG3525A. D6, C33-34, and L5 are used to produce the DC voltage for the optical isolator unit. R38 is used to limit the current through the LED section of the optical isolator. As the current increases, the resistance of the output transistor decreases. As the current decreases, the resistance of the output transistor increases. The output transistor is connected in series with R19 and R20 to form a voltage divider network in the nonisolated section. Changes in resistance of the output transistor changes the voltage applied to the Inv pin of the SG3525A. The feedback voltage is used to increase or decrease the pulse width of the SG3525A and to compensate for changes in the load of the power supply.

Sensing Section. The sensing section (Fig. 7-23) of the power supply simply samples each output voltage and compares these voltages to a reference voltage. If any of the voltages are out of tolerance, the comparators cause the transistor to turn off. The output of the transistor produces the power-good signal for the system board, which is used to reset the computer system.

7.2.5 Troubleshooting the Power Supply

Most computer companies do not make their own power supplies. Figure 7-24 shows the information that most computer or power supply companies provide for technicians or users. Computer companies care only that the power supply is large enough to power the computer and is the correct size. Most power supplies are 200 watts or more and are in either an XT or AT size case. The ICs used inside the power supply vary from company to company; only the connections and voltage levels for each output are standard.

When AC power is applied to the computer, if the fan motor is not heard or no air comes out of the power supply case, there is a problem with the power supply. If the fan motor does not turn, vacuum out the power supply. Make sure that there is proper air circulation, and then check the fan motor (+12 V). In most cases the power supply is simply replaced because of its low cost ($30 to $55).

If the power supply needs repair, the following steps should be taken to determine which section of the power supply is bad.

1. Connect the power supply to an isolation transformer, making sure that the VA rating is the same as or larger than the power supply.
2. To determine if the problem is in the power supply, system board, drives, or I/O adapter, disconnect all power supply cables. Place a 47- to 51-ohm, 2-watt resistor between ground and the +5-V line. This acts as a load for the power supply; a switching power supply needs a load to verify the voltage of the supply. Check for +5-V, +12-V, −12-V, and −5-V output lines. If good, the problem is in the system board,

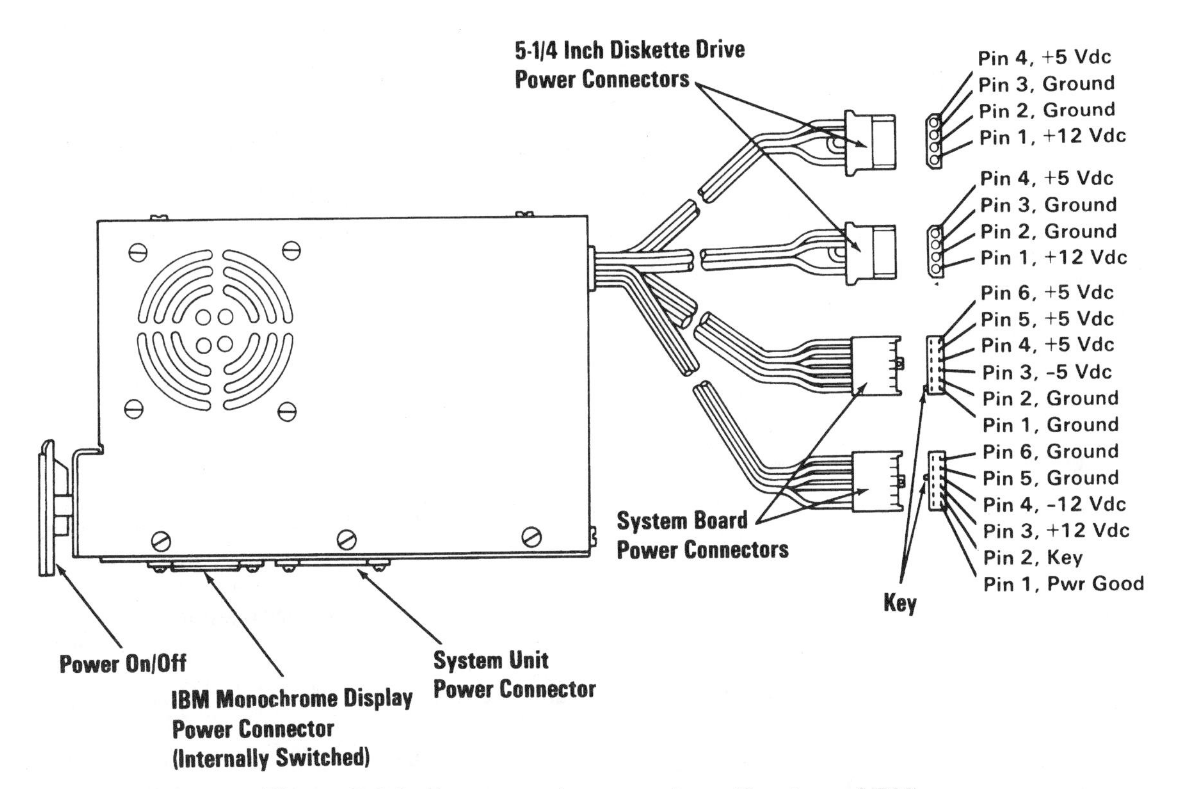

Figure 7-24 Power supply connections. (Courtesy of IBM)

drives, or I/O interface adapters. If any of the voltage outputs are missing, the problem is within the power supply itself.

3. If the correct voltages are measured in step 2, start connecting one power cable at a time until the power supply or voltage fails, indicating which area of the computer system is loading down the power supply.
4. If the voltages are missing or correct from step two, perform the following measurements. Do not use the isolated or signal grounds, which may cause damage.
   (a) Check the fuse.
   (b) Check the voltage across the MOV (should be 102 AC)
   (c) Check the output voltage of the EMI filter.
   (d) Make sure the switching transistors are switching. *Caution:* Voltages are high.
   (e) Measure the voltage across the primary of the isolation transformer. *Caution:* Voltages are high.
   (f) Measure the voltage across the secondary of the isolation transformer.
5. If not all the voltages are missing, perform the following:
   (a) Check the voltage across the proper secondary of the isolation transformer that is missing the voltage.
   (b) Verify that the diode is not bad.
   (c) Check for shorted output capacitors or open inductors.
   (d) If the voltage missing is derived from a linear voltage regulator, verify the operation of the linear voltage regulator.

7.2.6 Power Back-up Systems

There are basically two types of power back-up systems, both of which perform the same basic function—to supply AC power to the computer when the AC from the utility company fails.

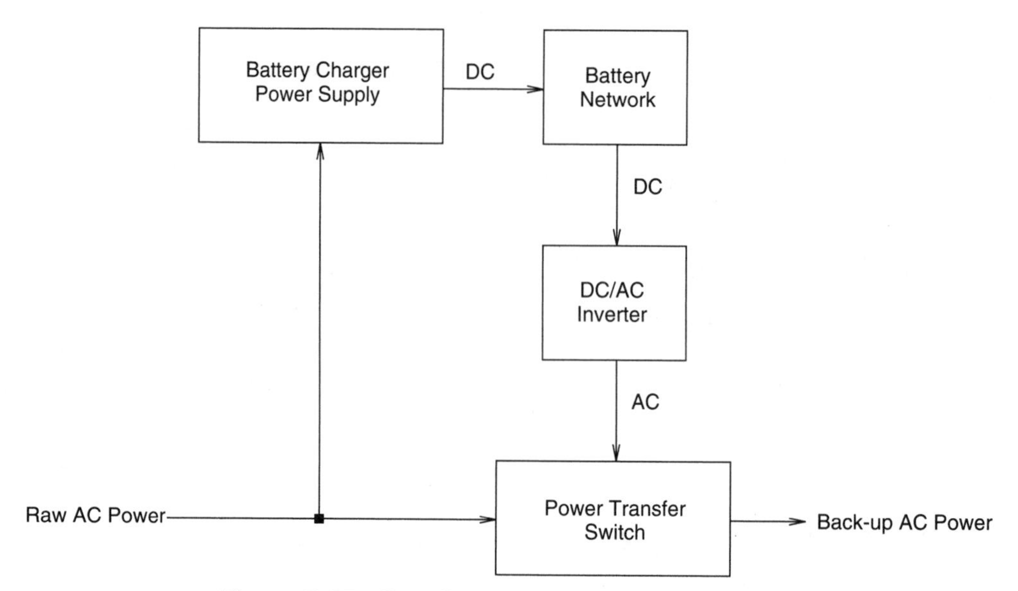

Figure 7-25 Standby power system block diagram.

Both types of systems use batteries and inverters to supply the AC signal to the computer once power fails.

A standby power system (SPS) (Fig. 7-25) takes the AC from a wall plug and directs it in two directions. The AC is applied to a DC power supply that is used to keep the batteries in the SPS charged. The output of the batteries is applied to an inverter (converts DC into AC) and keeps the inverter in a standby mode. The output of the inverter is then connected to an AC transfer switch. The original AC signal from the wall plug is also applied to the other input of the AC transfer switch. Under normal operating conditions, the AC transfer switch allows the original AC signal from the wall plug to power the computer through the AC transfer switch. But when the AC voltage falls below or rises above a certain level, the AC transfer switch causes the output of the inverter to power the computer. This condition continues until the battery voltage decreases to the point where the inverter can no longer supply the power for the AC or until the original AC signal from the wall plug returns. SPSs are rated in VA (volt-amps); the larger the rating, the longer the supply drives the inverter. SPSs are used to allow a computer system to save its data before power is lost. Because SPSs normally operate off the AC from the wall plug, they do not provide line spike, surge, and noise protection. A typical 900VA SPS supplies 7.5 amps at 110 volts for about 10 to 20 minutes, after which time power is lost. Some SPSs have a computer interface so that when power is lost, the computer automatically saves any open files and shuts down any unattained computer system. More advanced SPSs

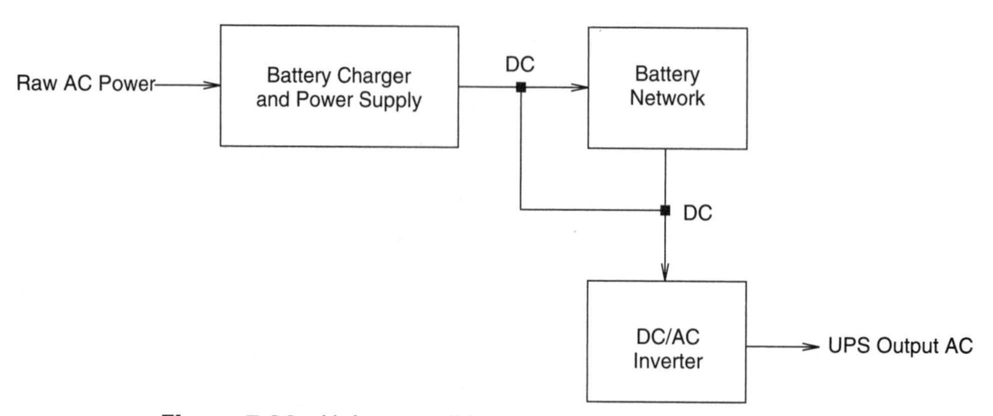

Figure 7-26 Uninterruptible power system block diagram.

also restart the computer and software once power is restored. The major problem with an SPS is that the switching transfer time may not be fast enough under heavy loads, which can cause spikes or surges during switching that cause the computer system to lock up.

The second type of power back-up system is called an uninterruptible power system (UPS). In a UPS (Fig. 7-26) the AC from the wall plug does not directly power the computer. Instead the AC is converted into DC, which is used to keep the batteries on the system charged and supplies the DC to the inverter at all times. When AC is lost, the batteries supply the voltage for the inverter. Thus there is always a DC voltage applied to the inverter used to power the computer system. Because the computer system is never powered by the AC from the wall plug directly, a UPS suppresses noise, spikes, surges, sags, and brownouts. The problem with UPS is that the batteries do not last as long, because a UPS tends to use the batteries more frequently than an SPS. The main advantages are the AC filtering and AC power regulation for the computer. UPSs are also rated in VA. A UPS supplies power for the computer for a number of minutes at a certain current.

CHAPTER SUMMARY

1. The 8048/8049 microcontrollers found in XT/AT keyboards are used to scan the keyboard and produce key codes when a key is pressed and released. These ICs also send out the key codes in a serial format over the KBD DATA line, along with a clock pulse with each bit in the serial package.
2. The keyboard matrix is a set of switches (keys); each connected between a row and column. The 8048/8049 microcontroller selects one row at a time and then scans each column to determine if the condition of a key has changed.
3. There are two types of keys—capacitive and positive response. Capacitive keys use the capacitance between the row and column when the key is pressed to determine when it has been pressed. A positive response key, when depressed, makes a physical connection between the row and column location of the key.
4. Capacitive keys are not as susceptible to dirt and corrosion, and keys do not wear out as quickly because there is no physical contact. The disadvantages of capacitive keys is that they cannot be replaced when the row and column pads go bad, they cost more, and they lack a positive response when pressed. Positive response keyboards usually cost less, have a simpler design, and a single key closure can be replaced when it goes bad. The disadvantages are that positive response keys are susceptible to dirt, corrosion, and mechanical stress.
5. In an XT keyboard the key code for pressing (make code) and releasing (break code) a key on the keyboard are the same except for the MSB bit of the key code. In a break code the MSB is set, while in the make code the MSB is reset.
6. In an AT keyboard the key code for pressing (make code) is also used when releasing (break code) the key, except that the hex code "FO" precedes the key code.
7. As the XT keyboard-encoding section receives the KBD CLK and KBD DATA, the data are shifted into the shift register in a serial format. Once the last bit of the key code is received, the second start bit is used to generate the keyboard interrupt (IRQ1), which causes the 8088 to service the key-encoding routine. PB6 disables the KBD CLK until port A of the PPI is read, and the data are transferred into the keyboard buffer. Once port A is read, PB7 clears the shift register and PB6 enables the KBD CLK.
8. As the AT keyboard-encoding section (8042) receives the KBD CLK and KBD DATA, the data is transferred into the 8042. Once the key code is received, the number of bits in the package is verified, parity is checked, and, if everything is correct, the start, parity, and stop bits are stripped. The 8042 disables the KBD CLK while generating the keyboard interrupt (IRQ1). Once the key code is read from the 8042, the 8042 resets the key code buffer and enables the KBD CLK.

9. If a key or group of keys does not function, check for mechanical problems and foreign material in the key closure area. If the keyboard does not function, check the KBD GND, KBD +5 V, KBD CLK, and KBD DATA lines for proper operation. If there is a problem in the keyboard-encoding section, check the clock and data inputs of either the shift register or the 8042, and check for the generation of the keyboard interrupt (IRQ1).
10. Short-duration AC distortion includes noise, spikes, and dropouts. Long-duration AC distortion includes surges and sags.
11. In a linear power supply, if the load increases the feedback voltage decreases, which causes the error voltage to increase. The increase in error voltage causes the series pass transistor to conduct harder to lower the voltage drop across the transistor, which increases the voltage output.
12. In a PWM switching power supply, if the load increases the feedback voltage decreases, which causes the duty cycle of the switching power to increase. The increase in the duty cycle allows an increase in the output voltage.
13. The basic power supply checks are as follows: (1) Make sure the power supply fan is operating. (2) Using a dummy load, verify the voltage on the output lines of the power supply. (3) If the power supply voltage is good, disconnect all boards and drives from the power supply, then connect one load at a time until the voltage of the power supply goes bad. This indicates which device is loading down the power supply.
14. There are two types of power back-up systems. In a standby power system the control circuitry of the system causes the battery-controlled inverter to power the computer once the AC from the wall plug goes out of a voltage range. In an uninterruptible power system, the computer is always powered by the signal from the inverter. The control circuitry causes the battery to supply the DC to the inverter when the AC from the wall plug goes out of range; otherwise the DC produced from the AC from the wall plug supplies the signal for the inverter.

REVIEW QUESTIONS

1. Which type of keyboard stops scanning the key matrix once a key has been pressed?
  a. XT keyboard
  b. AT keyboard

2. Which type of keyboard uses the MSB of the key code to indicate if a key was pressed or released?
  a. XT keyboard
  b. AT keyboard

3. Which keyboard transfers key codes at a faster rate?
  a. XT keyboard
  b. AT keyboard

4. Which interrupt is used for the keyboard interrupt?
  a. IRQ0
  b. IRQ1
  c. IRQ2
  d. IRQ5
  e. IRQ7

5. Which type of key is not susceptible to dirt or corrosion?
  a. Capacitive keys
  b. Positive response keys

6. Which type of computer system uses the 8042 to encode the key code?
   a. XT
   b. AT
7. Which type of computer does not check for the parity of the key code received?
   a. XT
   b. AT
8. If a key on the keyboard does not function, the most likely cause of the problem is ________.
   a. The 8048 or 8049 microcontroller on the keyboard
   b. The +5-V supply line to the keyboard
   c. The KBD CLK
   d. The key closure and/or contacts
   e. The ground line for the keyboard
9. Noise is typically ________-duration distortion.
   a. Short
   b. Long
10. An AC sag indicates an _____________ condition.
    a. Overvoltage
    b. Undervoltage
11. Which part of the linear power supply is the controlling element?
    a. Feedback voltage network
    b. Reference voltage
    c. Error amplifier
    d. Series pass transistor
12. Which type of power supply has the best voltage regulation?
    a. Linear power supply
    b. Switching power supply
13. If the load in a linear power supply increases, the conduction of the series pass transistor ________.
    a. Decreases
    b. Increases
    c. Stays the same
14. If the load in a linear power supply decreases, the feedback voltage ________.
    a. Decreases
    b. Increases
    c. Stays the same
15. If the load in a linear power supply decreases, the error voltage ________.
    a. Decreases
    b. Increases
    c. Stays the same
16. If the load in a switching power supply decreases, the duty cycle ________.
    a. Decreases
    b. Increases
    c. Stays the same
17. If the load in a switching power supply increases, the feedback voltage ________.
    a. Decreases
    b. Increases
    c. Stays the same
18. If the load in a switching power supply increases, the switching frequency ________.
    a. Decreases

b. Increases

c. Stays the same

19. If the feedback voltage equals the reference voltage in a switching power supply, the duty cycle equals ________.

a. 0%

b. 25%

c. 50%

d. 75%

e. 100%

20. If the fan of the power supply is not working but the computer boots, which supply is bad?

a. +5 V

b. −5 V

c. +12 V

d. −12 V

e. 120 V AC

chapter 8

Video Systems

OBJECTIVES

Upon completion of this chapter, you should be able to perform the following tasks:

1. List and define the purpose of the two parts a computer video system.
2. Define the purpose of the parts of a CRT.
3. Define the purpose of horizontal and vertical scanning.
4. Define retrace and trace times.
5. Explain the abbreviations MDA, CGA, RGB, EGA, VGA, and SVGA.
6. Define the difference between digital and analog monitors.
7. Define the terms resolution and dot pitch.
8. List the advantages and disadvantages of the MDA, CGA, and VGA video monitors.
9. Define the purpose of the following blocks in the block diagram of a video monitor: video driver-output, vertical driver-output, horizontal driver-output, vertical driver/oscillator, horizontal driver/oscillator, flyback section, CRT, color driver-outputs.
10. List the purpose of the 6845 CRTC.
11. Define how a character is created on the video screen, using a character matrix.
12. Define the purpose of character code and attribute video memory, character generator, attribute decoder, and shift register in the MDA video monitor.
13. Demonstrate how to determine the character from an extended ASCII code or the extended ASCII code from a given character.
14. List the differences between the alphanumeric and all-point-addressing modes of a video adapter.
15. Define the purpose of each block in the block diagram of a VGA adapter.

All video systems used in computer systems are made up of two parts—the video adapter and the video display. Regardless of the type of video system, these two components perform basically the same functions. Video systems differ in how fast they work, the amount of resolution (amount of screen control), the number of colors they can produce, the display type, and the cost.

The video adapter is the interface between the computer and the video monitor. The information from the computer is converted into signals that the video monitor needs to display an image on the video screen. The information on the video screen is stored in video memory, which is located on the video adapter. The video adapter contains a video controller IC, which is used to control the flow and convert the data from video memory to create the signals necessary to display the information on the screen. The amount of video memory, screen resolution, number of colors, and speed of the video system are determined by the type of video adapter.

The video display is the part of the video system that produces the video image on some sort of video screen. A video monitor is one type of video display device. It is much like a TV using a CRT (cathode ray tube) as the video screen. A video monitor is the most popular and least expensive type of video display and produces the best results for the money. Video monitors are active devices, which means that they produce their own light. Video monitors, because of their size, weight, and power dissipation, are not suited for portable computer systems. Portable systems (laptops and notebooks) use LCDs (liquid crystal displays), GPDs (gas plasma displays), or ELDs (electroluminescent displays). Each type of portable video display has its own advantages and disadvantages, but all share the same basic features: small size, light weight, and low power dissipation compared with a video monitor. These types of video displays also share some of the same disadvantages of high cost, slow speed, low resolution, and limited color availability. Today active color displays are becoming more affordable and commonplace in portable computers. As their popularity increases, their cost will decrease.

8.1 THE VIDEO DISPLAY SCREEN

In explaining how a video display works, we look at a typical video monitor that uses a CRT. Just like a TV, the image is produced on the screen by turning the electron beam on and off while deflection coils use a magnetic field to position the electron beam at different locations on the screen.

The CRT (cathode ray tube) (Fig. 8-1) is a funnel-shaped glass tube evacuated of all air. The back of the CRT is cylinder shaped and contains the electron gun (one for monochrome and three for color CRTs). The front of the tube is normally rectangular or square; this part of the tube is called the face or video screen. On the inside face of the CRT are many phosphor-coated dots that produce photons (our eyes see photons as light) when struck by the electron beam produced by the electron gun. In a monochrome CRT only one type of phosphor is used to produce one color of light. A color CRT has three types of phosphor-coated dots, one that produces red light, one green, and one blue. The three different colors of dots are arranged in a triangle format, called a triad. Because of the closeness of the three color phosphor dots, our eyes see each triad as a single spot of light on the screen. The color of the light is the combination and intensity of the three colors. In a color CRT, just behind the phosphor coating is a piece of metal stretched across the screen, called the shadow mask. The mask has many little holes in it, one for each triad (or screen location) on the video screen. These holes may be rectangular or square but are usually round. The shadow mask helps keep the red, green, and blue electron guns from hitting the wrong color phosphor dots on the face of the CRT. Most monochrome CRTs do not contain a shadow mask because only one type of phosphor is used.

Moving toward the back of the CRT, the next part is the second anode. A very high voltage is applied to the second anode, about 1 Kv per 1-inch diagonal in a monochrome CRT and about 1.5 Kv per inch in a color CRT. The purpose is to provide a high voltage potential, which causes the electrons from the electron gun to move toward the front of the CRT and strike the phosphor coating on the inside face to produce photons, which we see as light.

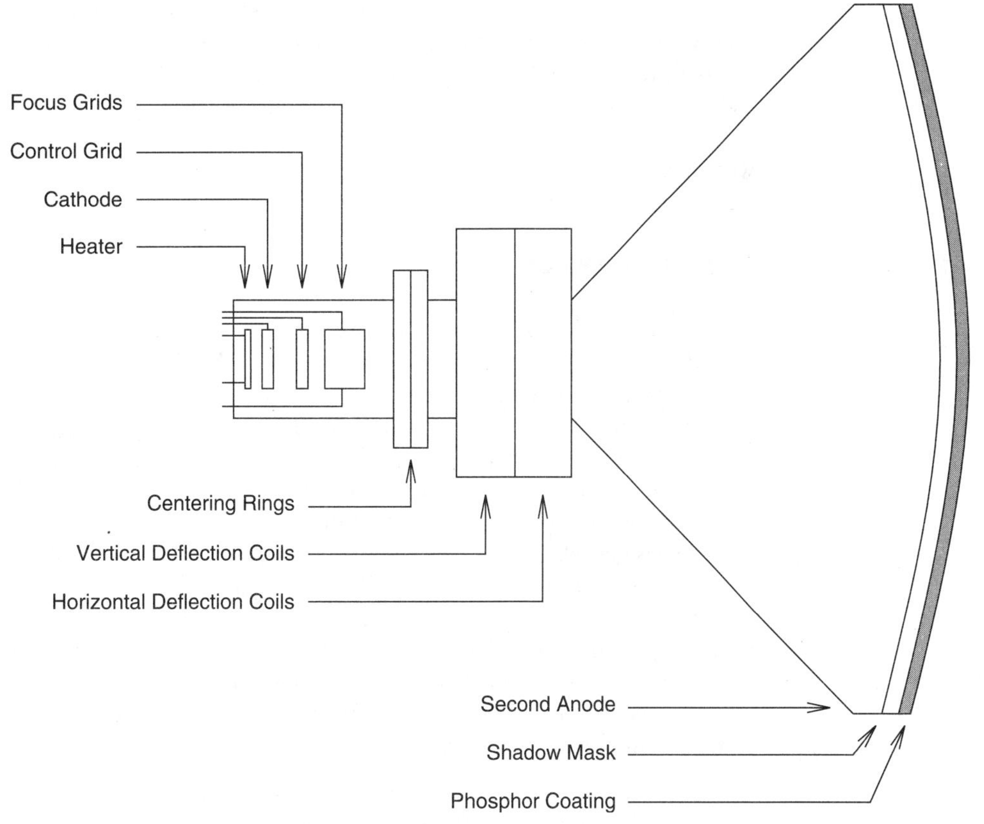

Figure 8-1 Cathode ray tube.

The deflection coils are located on the widest part of the neck of the CRT. They are constructed on a plastic collar that forms the yoke for the CRT and is located on the outside of the neck between the electron guns and the face of the CRT. The yoke contains the horizontal and vertical deflection coils, which are used to position the electron beam at different points on the video screen. The horizontal deflection coil is responsible for moving the electron beam from left to right and right to left. The vertical deflection coil is responsible for moving the electron beam from top to bottom and back to the top. The electromagnetic field set up by these coils is used to position the electron beam. The deflection coils of the CRT are used to scan the electron beam to produce a matrix of rows and columns. The horizontal deflection coil produces the columns and the vertical deflection coil produces the rows. As the deflection coils cause the electron beam to scan the video screen while the electron beam turns on and off, an image appears on the video screen.

The next part of the CRT is the electron gun, which consists of the focus grid, control grid, cathode, and heater. Located at the very back of the CRT is the heater, the element used to heat the cathode to a constant level. Just above the heater is the cathode, which is coated with a material called rare earth that emits free electrons when heated. As the voltage applied to the cathode decreases, the number of free electrons emitted by the cathode increases because the potential between the cathode and second anode has increased. The electrons emitted by the cathode are pulled toward the face of the CRT by the second anode voltage. As the free electrons head toward the face of the CRT, they pass a piece of metal called the control grid. The voltage on the control grid controls the number of electrons that reach the second anode. Because the distance between the control grid and cathode is small compared with the distance to the second anode, only a small voltage is needed to control the electron flow. In some video monitors the voltage potential on the control grid is at or near ground. The last part of the electron gun, called the focus grid, is used to condense into a narrow beam the electrons heading toward the face of the CRT. A small part of the second anode voltage is applied to the focus grid, which creates an electrostatic field to condense the electron beam. Note that a monochrome CRT has only one electron gun, while a color CRT has three electron guns (red, green, and blue). In the color CRT the red gun is aligned to hit only those phosphor dots that generate photons that produce red light; the green and blue electron guns function in the same way with their own color phosphor dots.

8.2 PRODUCING THE VIDEO IMAGE

A video image is created on the video screen by turning the electron beam on and off as it scans the video screen. The scanning sequence begins with the electron beam positioned in the upper left corner of the video screen. The horizontal deflection coil causes the electron beam to move across the video screen, which is called the horizontal trace. During the horizontal trace (Fig. 8-2), the electron beam is toggled on and off to produce a pattern of lighted dots on the video screen. Once the electron beam is at the right side of the screen, the beam is turned off. The beam heads toward the left side of the screen, a process called retrace. During the trace time, data is placed on the video screen; during retrace the video display is turned off. Because it takes time for the phosphor-coated dot to dissipate the energy of the electron beam, the dot gradually dims to the point where no light is produced. Therefore the image on the video screen must be updated before the light (image) dissipates. As the electron beam moves horizontally, the vertical deflection coil also causes it to move down the video screen.

The vertical trace occurs during the time the electron beam moves from the top to the bottom of the screen, which allows the horizontal trace to produce the dots of light on the

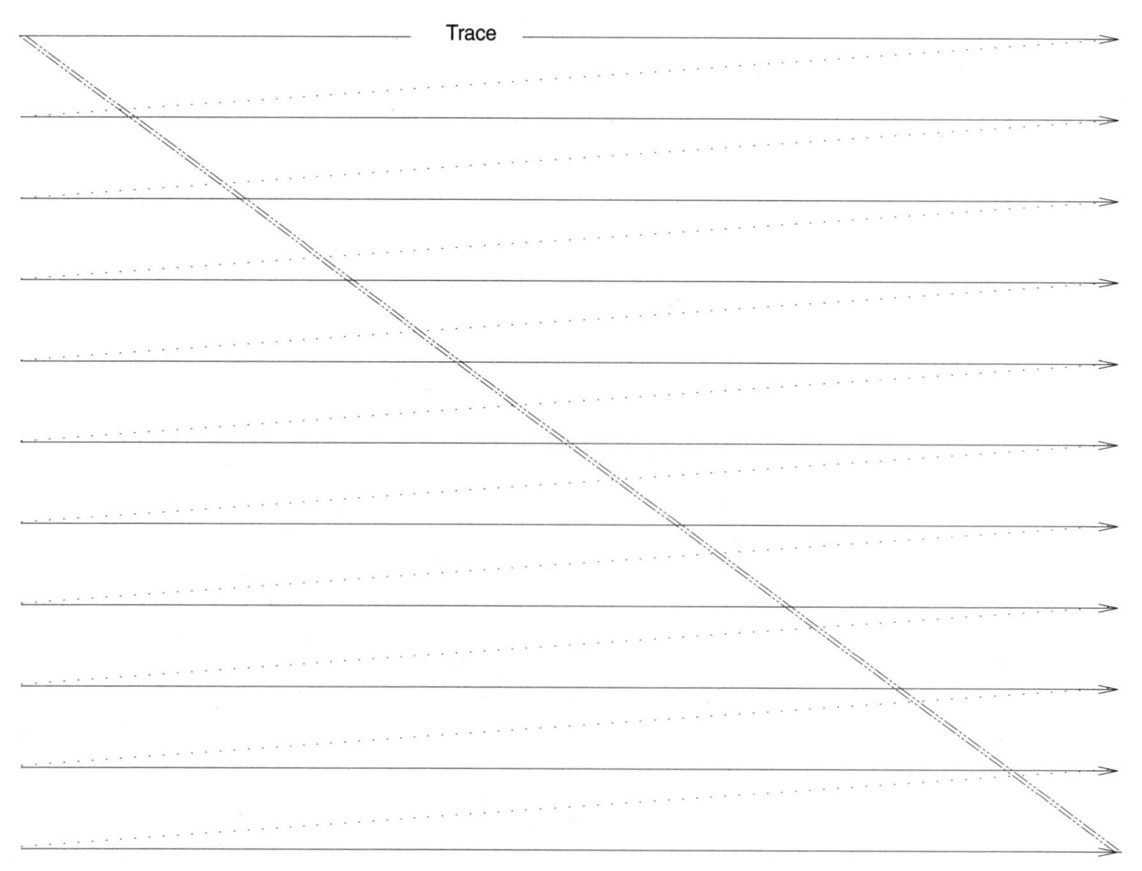

Figure 8-2 Video screen scanning.

screen. Once the electron beam reaches the bottom of the screen a vertical retrace occurs, which keeps the beam turned off until it reaches the top of the screen. During vertical retrace the electron beam is turned off regardless of whether a horizontal trace or retrace is occurring. The vertical frequency determines how many times a video image is painted on the screen and how many horizontal scans make up the image. Each image update on the screen is known as a frame, and updating occurs from 50 to 72 times a second depending on the system. The slower the refresh (frame) rate of the video screen, the more likely the image is to flicker. Most video systems used with IBM-compatible systems refresh the entire screen during one vertical scan period, which is referred to as raster or non-interlace scanning and is somewhat different from that in a TV. In a TV only every other horizontal scan line of the image is painted on the screen during one vertical scan period; during the second vertical scan period the missing horizontal scan lines are filled in. This method of scanning is called interlace scanning. Interlace scanning allows video images that change rapidly to blend into each other to provide more smoothly changing video images than does raster scanning. The disadvantage of interlace scanning is that it does not produce as sharp a stationary image, which is why raster scanning is used with most computer systems.

Because the vertical frequency is slower than the horizontal frequency of the video screen, many horizontal scans occur for each vertical scan. The speed with which the electron beam turns on and off and the number of rows per video screen specify the screen resolution. The following example shows the relationship between the vertical and horizontal scan frequencies. If a video screen has a resolution of 640×480, during each horizontal trace the electron beam can be turned on and off to control 640 points of light. The number 480 indicates that there are 480 horizontal scans (rows) on one video screen. In this example, the horizontal frequency is operating 480 times faster than the vertical frequency. There are 640 columns of dots and 480 rows of dots, a total of 307,200 dots of light possible on the screen. As the resolution increases, the number of dots per row and the number of rows increase. This means more control over the video screen, which increases both the scanning frequencies and the cost of the video monitor.

8.3 VIDEO MONITORS

The block diagrams of three video monitors are discussed here, along with their differences and common features. Two of the monitors are digital monitors—the TTL monochrome monitor, which is used in conjunction with the MDA adapter, and the RGB color monitor, which is used in conjunction with the CGA adapter. The third monitor is an analog monitor—the VGA color monitor used in conjunction with the VGA adapter, currently the most popular type of monitor. There are basically two types of video monitors; those that use TTL levels for the data inputs of the monitor are referred to as digital monitors, and those that use analog voltages (non-TTL, more than two voltage levels) are referred to as analog monitors. In general, analog monitors allow the video system to display more colors than digital monitors because they can respond to more than two voltage levels on the data inputs. Basically all monitors, whether digital or analog, work the same way. The main difference is that digital monitors have circuitry that responds to only two voltage levels (on and off) while analog monitors have circuitry that responds to multiple voltage levels. Digital monitors can be identified by a 9-pin, 2-row male connector to the video adapter. Analog monitors are identified by a 15-pin, 3-row male connector to the video adapter.

As with most electronics, early versions of the monitors were built using discrete transistors, but as computer monitors became more popular IC manufacturers started integrating many of the blocks in ICs, reducing the number of components necessary to build the monitors and increasing the reliability while reducing the cost and design time. Today most monitors sold are analog VGA or SVGA monitors. The big determination on cost today is the resolution, dot pitch, and size. The more expensive monitors have their own microcontrollers, which allow the monitor to respond to various resolution settings without having the user adjust the video screen.

8.3.1 TTL Monochrome Monitor

One of the first types of video monitors used for personal computers was the TTL monochrome monitor, which was the video display for the MDA (monochrome display adapter) video adapter. As its name implies, the image on the video screen is only one color, and the background color is black. The image is normally either green or amber. This is the simplest type of monitor and requires the least complex circuitry and CRT. Normally only two types of user adjustments are accessible in this type of monitor—the brightness control, which controls the overall brightness of the monitor, and the contrast control, which controls the difference between the brightness of the background and foreground by varing the amount of foreground brightness (the amount of light produced when the electron beam is on). Table 8-1 shows the function of each pin for both digital and analog monitors, and Figure 8-3 shows the position of each pin in both types of connectors.

Figure 8-4 shows the video driver circuit used to amplify and buffer the TTL input video signals from the display adapter for the video output circuit. The bandwidth of this circuitry must be able to handle the TTL video signal, which changes a maximum of 16.3 million times every second (Table 8-2). The intensity signal is also applied to this circuit, which can also change at the same frequency (also a TTL level). The intensity signal changes the biasing of the video driver to produce a higher output voltage when the intensity and the video signal are high. The frequency values in Table 8-2 are general frequencies and may vary from monitor to monitor.

The video output is the final amplification stage for the video circuit. The purpose of this stage is to increase the voltage and supply enough current drive for the CRT. The output voltage is normally between 15 and 50 volts. The input for this stage is from the output of the video driver circuit. Two outputs from this circuit are shown in the block diagram (may differ in different monitors); one output is fed into the cathode of the CRT. The biasing of the cathode is derived from the output signal and a voltage divider network called the contrast control, which

TABLE 8-1
Video Monitor Pin Configuration

Pin	*MDA*	*CGA*	*EGA*	*VGA*
1	Ground	Ground	Ground	Red signal
2	Ground	Ground	Secondary red	Green signal
3	No pin	Red signal	Red signal	Blue signal
4	No pin	Green signal	Green signal	No pin
5	No pin	Blue signal	Blue signal	Self test
6	Intensity	Intensity	Secondary green	Red return
7	Video	No pin	Secondary blue/ monochrome	Green return
8	Horizontal synchronization	Horizontal synchronization	Horizontal synchronization	Blue return
9	Vertical synchronization	Vertical synchronization	Vertical synchronization	No pin
10				Digital ground
11				Digital ground
12				No pin
13				Horizontal synchronization
14				Vertical synchronization
15				No pin

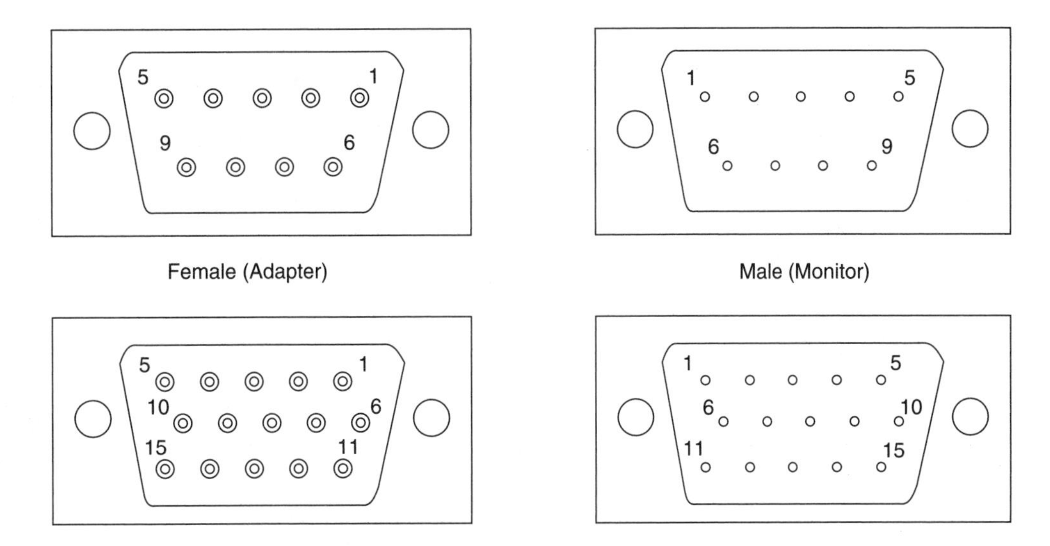

Figure 8-3 Video adapter/monitor pin configuration layout.

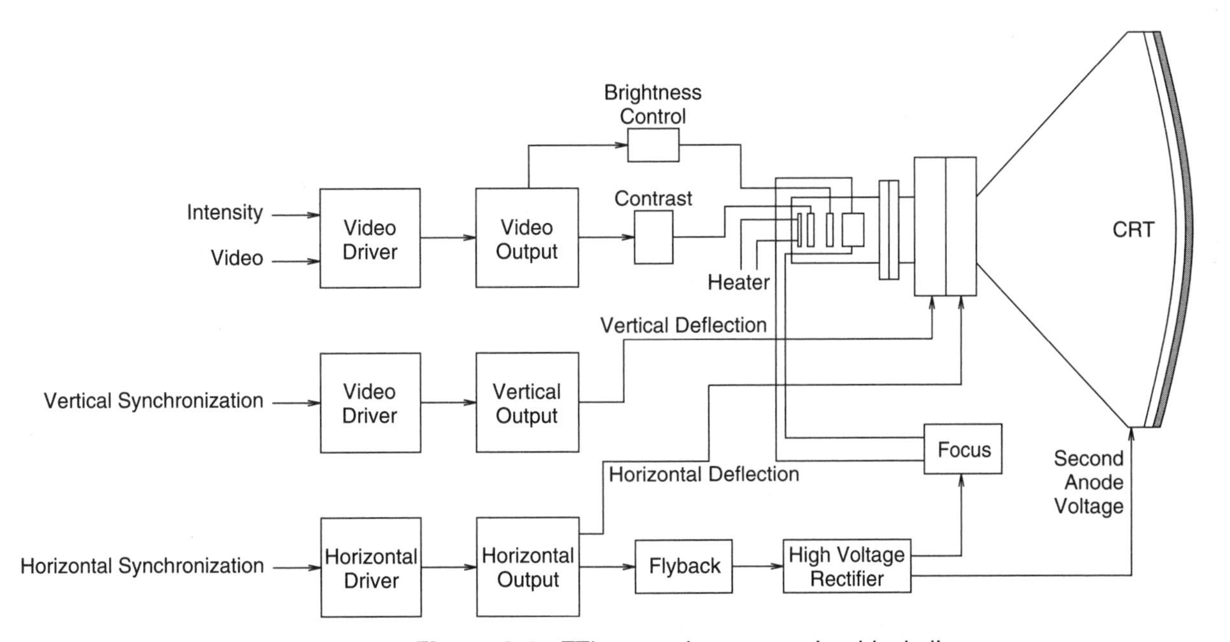

Figure 8-4 TTL monochrome monitor block diagram.

controls the brightness of the foreground (amount of voltage applied to the cathode when the electron beam is turned on). The second output is applied to another voltage divider network called the brightness control. The output of the brightness control is applied to the first control grid of the CRT. The brightness control determines the overall brightness (when the electron beam is turning on and off) of the screen.

The vertical driver circuit amplifies and buffers the TTL vertical synchronization pulse (50 Hz) from the display adapter for the vertical output circuit. The purpose of this stage is to increase the voltage and current of the vertical synchronization signal for the vertical output circuit.

TABLE 8-2
Video Monitor Frequencies

Monitor Type	*Resolution*	*Dot Clock*	*Horizontal*	*Vertical*
MDA	720 × 350	16.3 MHz	18.4 KHz	50 Hz
CGA	640 × 200	14.3 MHz	15.7 KHz	60 Hz
EGA	640 × 350	16.3 MHz	21.9 KHz	60 Hz
VGA	640 × 480	31.2 MHz	37.5 KHz	72 Hz
SVGA	800 × 600	40.0 MHz	37.9 KHz	60 Hz
8514A	1024 × 768	64.0 MHz	48.4 KHz	70 Hz

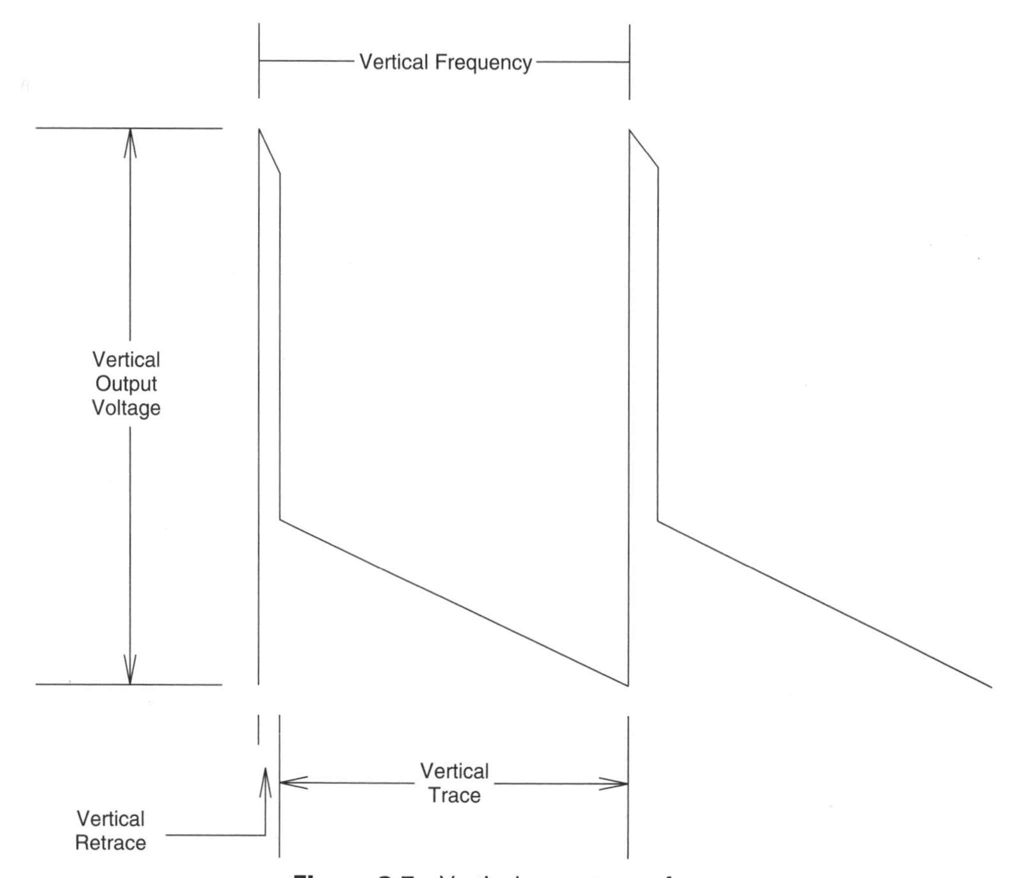

Figure 8-5 Vertical output waveform.

The vertical output circuit increases the votlage and current drive capability of the vertical synchronization signal from the vertical driver in order to drive the vertical deflection coil. Figure 8-5 shows a typical output waveform from the vertical output circuit. The peak voltage can range from 20 to 70 volts. The narrow peak of the waveform is the retrace time (about 5% of the vertical frequency), and the ramp is the trace time, the remaining time left in the vertical frequency.

The vertical deflection coil moves the electron beam from the top to the bottom of the screen during trace time, during which time the electron beam is enabled. During retrace time, the deflection coil moves the electron beam from the bottom to the top of the screen; during this time the electron beam is disabled regardless of whether the horizontal scan is in the trace or retrace mode.

The horizontal drive circuit amplifies and buffers the TTL horizontal synchronization pulse (18.4 KHz) from the display adapter for the horizontal output circuit. The purpose of this stage is to increase the voltage and current of the horizontal synchronization signal.

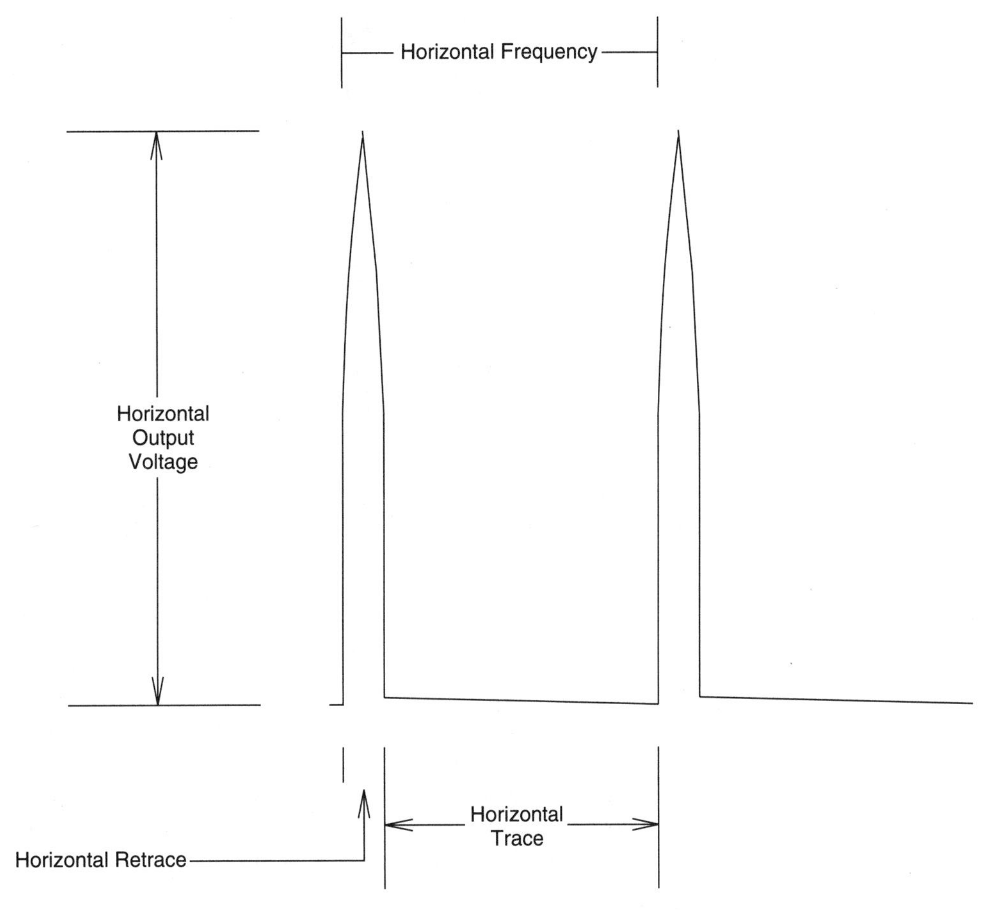

Figure 8-6 Horizontal output waveform.

The horizontal output circuit increases the voltage and current drive capability of the vertical synchronization signal from the horizontal driver. This signal is applied to the horizontal deflection coil and the primary side of the flyback transformer. Figure 8-6 shows a typical output waveform from the horizontal output circuit; the peak voltage can range from 55 to 130 volts. The pulse of the waveform is the retrace time (about 20% of the horizontal frequency), and the lower part of the waveform is the trace time, the remaining time left in the horizontal frequency. When the pulse is applied to the horizontal deflection coil, the coil modifies the waveform to produce a slight ramp.

The flyback transformer is a step-up transformer that increases the input voltage applied to the primary winding (normally between 55 and 130 volts). One secondary winding produces a 10-Kv to 25-Kv pulse that is applied to the high-voltage rectifier circuit, which converts the pulse into a DC voltage that is applied to the second anode. The second anode voltage pulls the electrons from the electron beam to the front of the screen. The amount of voltage varies with the size of the screen (about 1 Kv per 1-inch diagonal measurement). In some monitors part of the second anode voltage is applied to a voltage divider network, and the voltage from the network is applied to the focus grids (about 100 to 500 volts). In other monitors a secondary winding is used to produce a pulse that is rectified into DC to create the focus voltage. The focus voltage is applied to a voltage divider network, which is then applied to the focus grids.

8.3.2 RGB Color Monitor

The second of the first generation of video monitors used for PCs was the RGB color monitor (Fig. 8-7), which was the video display for the CGA (color graphics adapter) video adapter. Although the CGA adapter also contained a color composite output, the RGB output provides

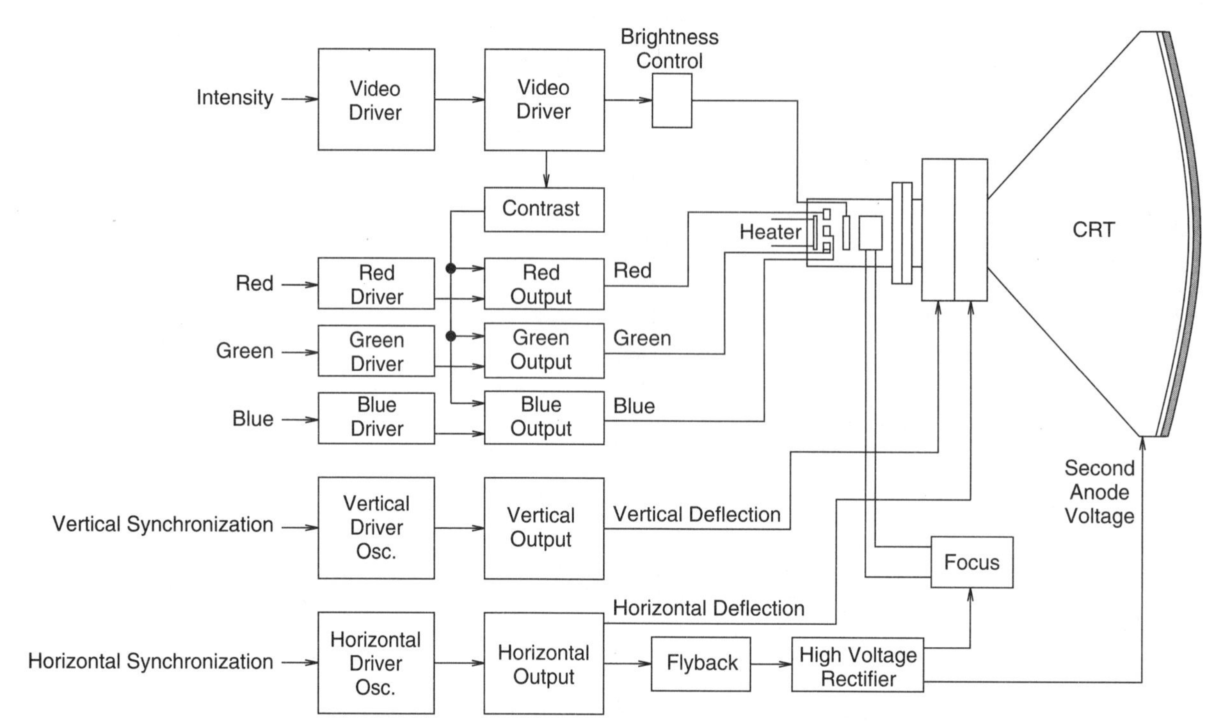

Figure 8-7 RGB color monitor block diagram.

a high-resolution display because the information for each color is fed into the monitor on its own line without mixing the color information along with the synchronization signals. As its name implies, each of the colors has its own input signal and support circuitry. Color monitors have more circuitry than monochrome monitors because each color must have its own circuitry. The CRT also contains three electron guns located in the neck of the CRT and three different phosphor coatings on the inside face of the video screen. The additional circuitry and the more complex CRT makes a color monitor cost at least twice as much as a monochrome monitor. The brightness and contrast controls are still user accessible, and some color monitors also provide color controls on the back of the monitor to adjust color levels. On a color monitor, the background is normally a color, whereas in a monochrome display the background is simply an unlit area of the screen. Table 8-1 shows the function of each pin for both digital and analog monitors, and Figure 8-3 shows the position of each pin in both types of connectors. Most of the circuitry in the RGB color monitor is the same as in the TTL monochrome monitor. The main difference is that the color circuitry operates at different frequencies and produces different voltage levels.

The video driver circuitry is used to control the intensity of the screen. The intensity input is applied to the input that amplifies and buffers the TTL level signal from the CGA adapter. The output of this signal is applied to the video output stage. In some RGB color monitors, the signals from video driver and output stages are combined with the color drivers and output signals to control the cathodes while the control grid is kept at about ground potential.

The video output circuitry amplifies the intensity signal and applies it to two different parts of the monitor. Part of the signal is applied to a voltage divider network called the brightness control, which controls the amount of voltage applied to the control grid of the CRT, which then controls the overall brightness of the CRT (number of electrons allowed to reach the video screen). The other part of the video output signal is applied to a voltage divider network called the contrast control. The voltage from the contrast control changes the biasing of the three color output circuits. The contrast control controls the voltage of each color output when the foreground color is being displayed and thereby affects the foreground color brightness.

There are three inputs to the monitor, labeled R (red), G (green), and B (blue), from the RGB output connector of the CGA adapter. These inputs are applied to three different color drivers (one for each color), the purpose of which is to amplify and buffer the TTL color signals from the adapter. The output of each driver is applied to its corresponding color output circuits.

The color output circuits further amplify the red, green, blue signals from the three color driver circuits for the three color cathodes of the CRT. Depending on the color being displayed and the contrast setting, the voltage levels on these lines vary from 0 to 50 volts. The rate at which they change levels can be as great as 14.3 million times per second.

The vertical driver/oscillator circuit amplifies and buffers the TTL vertical synchronization pulse (60 Hz) from the display adapter for the vertical output circuit. If a vertical synchronization pulse is not available, the oscillator section produces the vertical frequency for the monitor until a vertical synchronization pulse is received by the monitor. When the vertical synchronization pulse is received, the vertical driver locks onto the vertical synchronization pulse from the video adapter and the monitor uses that pulse to control the vertical scanning of the video screen. The purpose of this stage is to increase the voltage and the current of the vertical synchronization pulse from the adapter and, if necessary, produce the vertical synchronization signal for the vertical output circuit.

The horizontal drive/oscillator circuit amplifies and buffers the TTL horizontal synchronization pulse (15.7 KHz) from the display adapter for the horizontal output circuit. If a horizontal synchronization pulse is not available, the oscillator section produces the horizontal frequency for the monitor until a horizontal synchronization pulse is received by the monitor. When the horizontal synchronization pulse is received, the horizontal driver locks onto the horizontal synchronization pulse from the video adapter and the monitor uses the pulse to control the horizontal scanning of the video screen. The purpose of this stage is to increase the voltage and current of the horizontal synchronization pulse from the adapter and if necessary produce the horizontal synchronization signal for the horizontal output circuit.

The vertical output, horizontal output, vertical-horizontal deflection coils, flyback transformer, high voltage rectifier, and focus sections in the RGB monitor perform their functions much as they do in the TTL monochrome monitor. The differences are as follows: The second anode voltage is about 20% to 50% higher, the horizontal frequency is 15.7 KHz and retrace takes 30% of the time period, and the vertical frequency is 60 Hz and retrace takes 23% of the time period.

8.3.3 VGA Monitor

The VGA color monitor is the most popular type of monitor for computers today because of its reasonable cost, high resolution, and number of colors that can be displayed on it. The cost of a VGA monitor depends on the size of the CRT, the maximum resolution of the monitor, and the dot pitch of the monitor. Some of the older VGA monitors respond to only one VGA resolution setting, but today most VGA monitors support resolutions from 640×480 to 1024×768. As the resolution on the monitor increases, the size of each pixel (smallest control point on the screen) decreases because the size of the video screen is fixed. Popular monitor sizes are 14, 15, 17, 19, and 21 inches, and dot pitch ranges from 0.42 mm to 0.26 mm (the smaller the dot pitch, the better and more costly the video monitor), while resolution ranges from 640×480 to 1024×768 (1024×1024 on 17-inch and larger monitors). Most VGA monitors today contain four or five ICs, assorted power transistors, and passive support circuitry. The larger and more expensive monitors contain their own microcontroller (special single-chip microprocessor) to automatically adjust the video screen size when video resolution changes.

The VGA color monitor (Fig. 8-8) most closely resembles the RGB color monitor in terms of circuitry. The differences are an absence of an intensity input and the fact that the red, green, and blue inputs accept and respond to analog signals instead of just TTL signals. The intensity input is not needed because the color signals are analog signals, which allows them

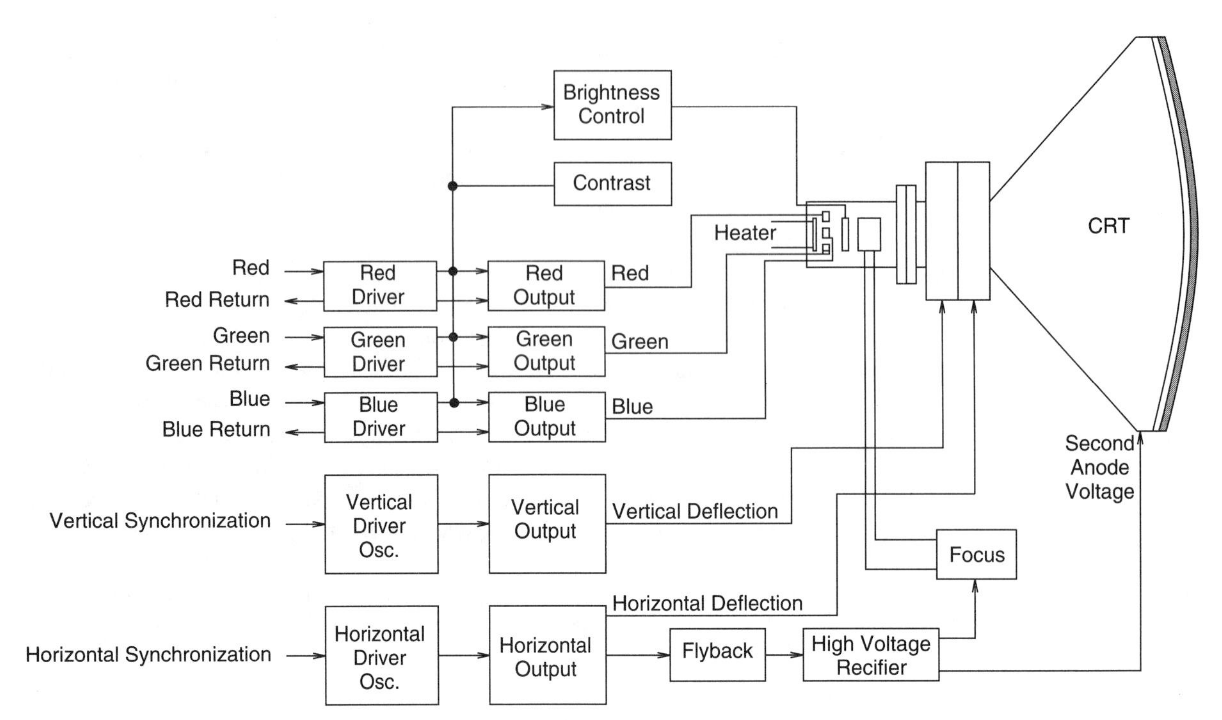

Figure 8-8 VGA color monitor block diagram.

to produce different amounts of each color. The vertical and horizontal synchronization signals are still at TTL levels. The VGA monitor still provides brightness and contrast controls. In the most expensive computers, additional controls may be provided to adjust the size and position of the video image.

The color drivers (red, green and blue) and outputs all perform the same function as in the RGB color monitor except that these circuit blocks respond to analog input voltages that range up to 1.4 volts peak to peak and can change as fast as 25 to 75 million times per second.

The vertical driver/oscillator and output, horizontal driver/oscillator and output, vertical-horizontal deflection coils, flyback transformer, high voltage rectifier, and focus sections in the RGB monitor perform their functions much as they do in the RGB color monitor. The differences are as follows: The horizontal frequency ranges from 37.5 KHz to 48.4 KHz and retrace takes 20% of the time period; and the vertical frequency ranges from 60 Hz to 72 Hz and retrace takes 6% of the time period, depending on the resolution setting, which can range from 640 × 480 to 1024 × 768.

8.3.4 Troubleshooting Video Monitors

Before a technician can determine if a problem is in the monitor or the video adapter, the technician should make sure that the outputs of the adapter are correct.

Horizontal line in center of screen. No vertical deflection.

1. Check vertical synchronization input.
2. Check output of the vertical oscillator.
3. Check vertical output circuitry.
4. Check for possible bad vertical deflection coil.

Vertical line in center of screen. No horizontal deflection.

1. Since there is a raster on the screen, some horizontal frequency must be getting to the high-voltage section of the monitor. Check for a bad horizontal deflection coil.

No raster, but signal from adapter is good. No high voltage from high-voltage section.	1. Check input of horizontal oscillator. 2. Check output of horizontal driver. 3. Check second anode voltage (high-voltage probe only).
Screen is shrunken, out of focus, or dim. Low value on second anode.	1. Check horizontal output. 2. Check flyback transformer. 3. Check high-voltage rectifier. 4. Check position of yoke.
Screen tilted.	1. Check for lose yoke; readjust and tighten.
Contents of screen rolls from from bottom to top or top to bottom. Vertical frequency drift.	1. Check output of integrator. 2. Check components of vertical oscillator circuit. This is usually a capacitor going bad (leaky).
Loss of screen display, slanted horizontal lines or high-pitched sound from monitor. Horizontal frequency drift.	1. Check output of AFC (if available). 2. Check output of differentiator. 3. Check horizontal oscillator circuit.

8.4 VIDEO ADAPTERS

This chapter discusses the details of the MDA, CGA, and VGA video adapters. There are other types of video adapters, but they function basically the same way and use the same basic principles. Video adapters are divided into two types: Digital video adapters produce digital signals for the video monitor, and analog video adapters produce analog signals for the video monitor. See Table 8-3 for video adapter specifications for each type of video adapter listed.

8.4.1 Digital Video Adapters

MDA. The MDA (monochrome display adapter) video adapter was the first type of display adapter used with the IBM PC. It was developed to produce very readable text on a monochrome video display. The MDA adapter does not support graphics, although it does display graphic symbols from the extended ASCII code character set on the video screen. The MDA adapter is still used in data entry applications, is the least costly video adapter, and provides an on-board printer adapter. The MDA adapter is a digital video adapter that contains 4K of video memory. In the mid 1980s a company called Hercules Graphics developed a monochrome video adapter that produced text as readable as the MDA adapter but allowed limited graphic capability, provided that the software supported the Hercules Graphics standard. In order to allow graphics to be displayed, the Hercules Graphics adapter contained 64K of video memory.

CGA. The CGA (color-graphics adapter) video adapter was IBM's first color graphics video adapter. It allowed color graphics to be displayed on the video screen, but at the cost

TABLE 8-3
Video Adapter Specifications

Adapter	*Type*	*Resolution*	*Characters Column/Row*	*Character Box Horizontal/ Vertical*	*Colors*	*Memory*
MDA	Digital	720 × 350	80 × 25	9 × 14	1	4K
CGA	Digital	320 × 200	40 × 25	8 × 8	2 –16	16K
		640 × 200	80 × 25	8 × 8	2 –16	
EGA	Digital	All MDA	All MDA	All MDA	MDA	64K–256K
		All CGA	All CGA	All CGA	CGA	
		320 × 350	40 × 25	8 × 14	16	
		640 × 350	80 × 25	8 × 14	16	
VGA	Analog	All MDA	All MDA	All MDA	MDA	256–1MB
		All CGA	All CGA	All CGA	CGA	
		All EGA	All EGA	All EGA	16–256	
		360 × 400	40 × 25	8 × 16	16	
		640 × 400	80 × 25	8 × 16	16	
		640 × 480	80 × 25	8 × 16	16–256	
SVGA	Analog	All MDA	All MDA	All MDA	MDA	512K–2MB
with 8514A		All CGA	All CGA	All CGA	CGA	
(mode)		All EGA	All EGA	All EGA	16–256	
		All VGA	All VGA	All VGA	16–16M	
		800 × 600	80 × 25	8 × 16	16–64K	
		1024 × 768	80 × 25	8 × 16	16–256	

of text that was not as readable as with the MDA adapter. The CGA has two text modes and three graphics modes (one of which requires special software set-up) and allows up to 16 different colors depending on the resolution mode selected. The CGA adapter is a digital video adapter and contains 16K of video memory.

EGA. The EGA (enhanced graphics adapter) video adapter, the next step in video adapters, could produce color graphics with higher resolution, display more colors at higher resolutions, and produce text that was as readable as that with the MDA adapter but with color. The EGA adapter supports all MDA and CGA modes, is a digital type video adapter, and contains 64K to 256K of video memory. The resolution of this adapter was increased to 640 × 350 at 16 colors. This was the first video adapter that allowed PCs reasonable graphic capabilities, which helped GUIs (graphical user interfaces) to become popular.

8.4.2 Analog Video Adapters

PGC. The PGC (professional graphics controller) video adapter was IBM's first attempt to provide higher resolution and more than 16 colors. Although this adapter provided 640 × 480 resolution at 256 colors simultaneously, it never became very popular and was quickly replaced by the VGA.

VGA. The VGA (video graphics array) video adapter was first used in IBM's PS/2 (Personal System 2) computer systems. The VGA circuitry was built into the system board of the PS/2. VGA became so popular because ICs were developed to provide 640 × 480 resolution with 256 colors. Since the introduction of the VGA, it has become the standard for video systems and is available on video adapters that are not built into the system board. VGA supports all MDA, CGA, and EGA standards to allow compatibility with older software. The VGA video adapter is an analog video adapter and contains from 256K to 1MB of video memory.

SVGA. The SVGA (super video graphics array) video adapter is a modified version of the VGA video adapter. SVGA video adapters support higher resolutions of 800 × 600 to

1024 × 768 (8514-A video standard) and, depending on the amount of memory on the video adapter, colors ranging from 16 to 16 million. The SVGA adapter is an analog video adapter and can have memory that ranges from 512K to 2MB. SVGA adapters may use DRAM or VRAM for video memory. VRAM is a special form of video memory that allows much faster access than DRAM, but VRAM costs about four times more than DRAM. Some SVGA adapters provide graphic accelerators or graphic coprocessors, which are special ICs used with the video controller to speed up graphic operations.

8.4.3 Video Display Modes

Video adapters operate in one of two different display modes. In the text (alphanumeric) mode the video adapter displays information on the screen in the form of a predefined number of dots in a row and column matrix that makes up a set of characters and graphic symbols. The text mode resolution ranges from a minimum of 40 × 25 (40 characters per row and 25 rows per screen) to over 80 × 25 (80 characters per row and 25 rows per screen) depending on the video system used and the software support of the video adapter. The text mode allows the fastest method of displaying data on the video screen, because all the codes needed to create the text are programmable into a character generator. The characters and graphic symbols are made up of a set of dots laid out in a row and column matrix. The number of rows and columns of dots that make up a character or graphic symbol varies with the type of video system. The more rows and columns (Fig. 8-9) available to create the character and graphic symbol, the greater the number of dots. As the number of dots increases, the characters and graphic symbols produced become shaper and easier to view. Each character on the screen requires 2 bytes of data. One byte of data is the ASCII code used to create the raster code by the character generator. The other byte of data, called the attribute code, provides information on how the character is displayed on the screen (bold face, underline, blinking, character color, and background color are all attributes).

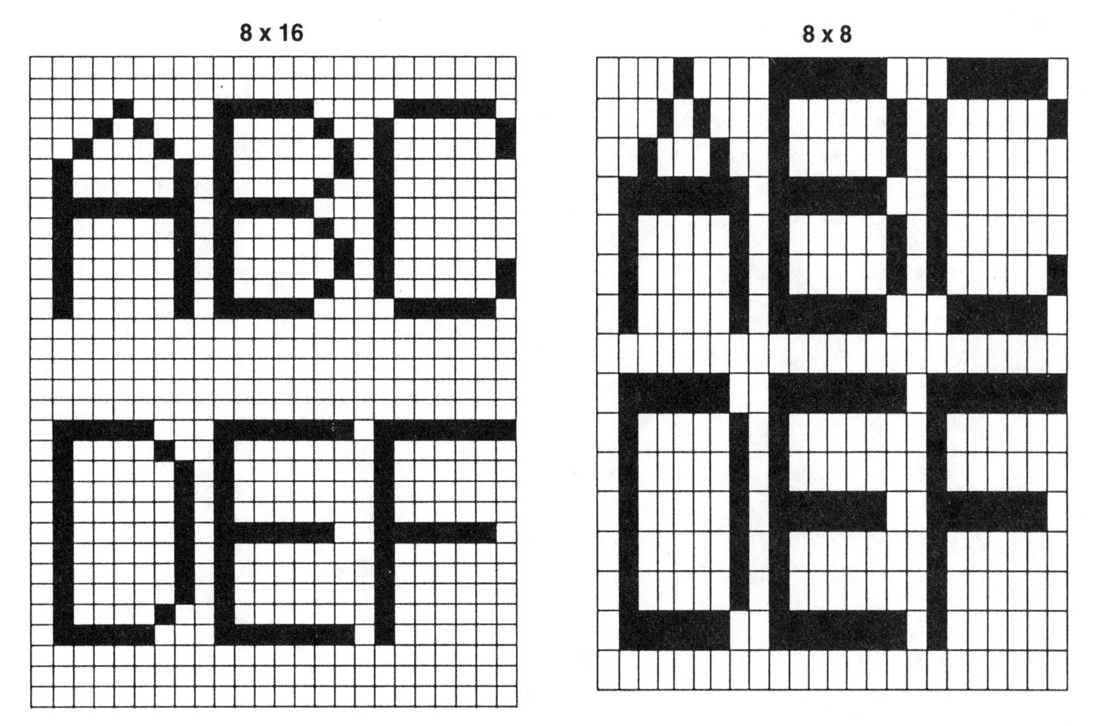

Figure 8-9 Character matrix examples.

The second video display mode is called the graphics (all-points-addressable) mode. While in the graphics mode, the video adapter turns on and off one dot or a group of dots at a time. The smallest unit of control in the video adapter is called a pixel, which changes with different graphic modes. Because in the graphics mode the video adapter controls more than one character location (a number of columns and rows of dots used as one unit), the process of producing the image on the screen is slower than the text mode. The number of colors in which each pixel can be displayed also affects the speed of the video system and the amount of video memory needed for the display. If a video adapter has a resolution of 640 × 480 and is in the graphics mode, it controls 307,200 dots (pixels in this example). In order to display the dots in either of two colors (on or off), 307,200 bits (38.4K) of video memory is needed. If each dot can be displayed in one of four colors, 76.8K of video memory is needed, for 16 colors 153.6K, and for 256 colors 307.2K. Therefore, as the number of colors for each dot or the number of dots that can be controlled on the video screen increases, more video memory is needed.

8.5 6845 CATHODE RAY TUBE CONTROLLER

The heart of the IBM PC monochrome or color graphics video controller board is the 6845 (cathode ray tube controller). The 6845 (Fig. 8-10) is a 40-pin NMOS smart IC that contains 19 programmable registers, which determine how the display appears on the screen and allows this IC to be used with just about any computer system. The IBM PC, XT, AT, and all IBM-compatible systems use this IC for their displays.

The purpose of the CRTC is to control the flow of data from screen memory to the character or graphic generator logic and then to the CRT monitor. This IC also controls horizontal-vertical synchronization pulses, cursor size and display type, display blanking, and light pen logic.

8.5.1 Main Features of the 6845

Screen memory addressing logic
Light pen logic
Limited cursor logic

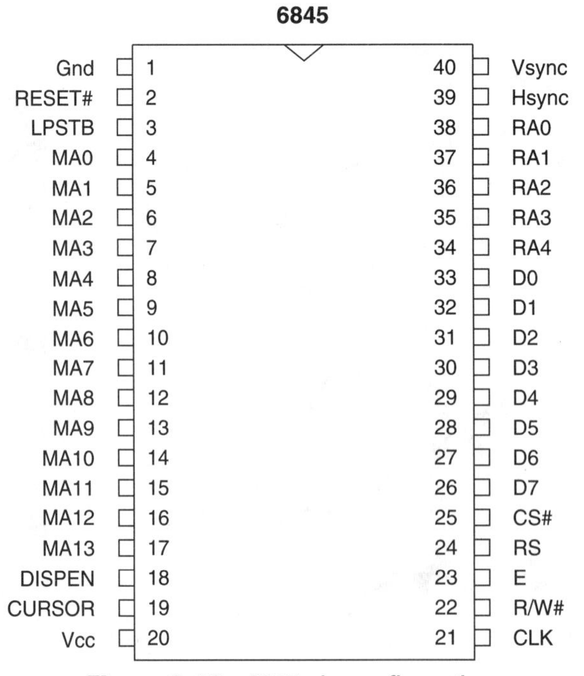

Figure 8-10 6845 pin configuration.

Scan line counters
Scrolling logic
Programmable video scanning
Limited blanking logic
Horizontal synchronization generation logic
Vertical synchronization generation logic
19 internal programmable formatting registers

8.5.2 6845 Pin Functions

MPU Interface Signals

CLK — The CLocK input is the clock input for internal and external operations in the CRTC. This signal synchronizes all 6845 signals to produce the proper timing sequence in the video adapter. The CLK signal must operate at least the character clock rate; this signal is responsible for addressing and for video memory raster addressing timing.

CS# — The Chip Select input allows the computer to select the CRTC when low, for programming or reading the status of certain registers.

D0–D7 — These eight bidirectional data lines are used to transfer data between the MPU and the 19 internal registers of the 6845. The only times the MPU accesses the CRTC are during formatting and the reading of the status of the internal registers. Screen data does not enter the 6845; instead the 6845 is used like a DMA for the control of data flow.

E — The enable input pin allows the CRTC to be controlled by the 8088 when high. When low, the CRTC refreshes the data on the screen from the screen memory. The level on this pin provides synchronization between the MPU bus operations and the CRTC. This signal does not affect the refreshing of the screen memory.

RESET# — The RESET input pin is used to reset all registers inside the CRTC so that it can be initialized. The reset stops the video display but does not affect the program accessible counters; the program can resume when the pin goes high.

RS — The Register Select input, when low, allows the data on the data bus to be loaded into the address register. When this line is high, the value in the address register selects the register specified.

R/W# — The Read/Write input pin determines whether data is written to or read from the selected internal register. When low, a write operation may be preformed, provided that the enable and chip select lines are at their proper levels. When high, a read operation may be performed, provided that the enable and chip select lines are at their proper levels.

Supply Pins

Vcc — This pin is connected to a 5-volt DC power supply and supplies the necessary potential for proper operation of the CRTC.

Vss — This is the ground pin that supplies all necessary current for the proper operation of the CRTC.

Screen Memory Pins

MA0–MA13 — The 14 screen Memory Address lines form the screen memory address bus, which allows either the MPU 8088 or the 6845 CRTC to control which screen memory location is accessed. Screen memory is accessed by the MPU only when a character is added or changed on the screen. The majority of the time, the CRTC 6845 controls screen memory, so that it can refresh the data on the video screen. The device that controls whether the 8088 or 6845 controls the screen memory address bus is a discrete device. Although it is on the video adapter, it is not part of the 6845.

RA0–RA4 — The Raster Address outputs are used to drive the raster scan inputs to the character generator. The five lines allow up to 32 raster scan lines per character. In the monochrome display, 14 raster scan lines are used to form the character box. In the color graphics display, 8 raster scan lines are used to form the character box. How the internal registers in the 6845 are programmed determines how many counts are supplied to the raster address outputs.

Monitor Signals

CURSOR — The CURSOR enable pin is used to create a steady stream of dots on the CRT screen, when and only when the screen memory address matches the cursor address. A high on the cursor pin produces the cursor at the specified time. They type of cursor produced depends on how the registers associated with the cursor are programmed.

DISPEN — The DISPlay ENable pin is used to enable the electron beam during trace time (output is high). When this pin is low, during retrace time the electron beam is turned off (blanked). Horizontal trace time occurs when the electron beam moves from the left to the right side of the display screen. Vertical trace time occurs when the electron beam moves from the top to the bottom of the screen. Horizontal retrace time occurs when the electron beam moves from the right to the left side of the display screen. Vertical retrace time occurs when the electron beam moves from the bottom to the top of the screen.

Hsync — The Horizontal synchronization signal causes the internal horizontal oscillator in the CRT monitor to become synchronized with the horizontal synchronization output of the video controller. This signal increases or decreases the horizontal scan rate of the monitor to obtain synchronization. In some monitors, the signal is the horizontal frequency needed to drive the horizontal circuitry in the monitor itself.

LPSTB — The Light Pen STroBe input is used with external circuitry to cause the value of current screen memory addressed to be stored in the light pen register. A high level on this pen performs this function. The address can then be read by the MPU and can subsequently determine the position at which the light pen was detected, after which the program must determine what to do with the data at the screen memory address.

Vsync — The Vertical synchronization signal causes the video monitor's vertical oscillator to synchronize onto the vertical synchronization (vertical scan rate) of the video controller board. This signal either increases or decreases the vertical scan rate to synchronize the monitor with the CRTC. In some monitors, the vertical synchronization signal is the actual vertical frequency of the monitor itself.

8.5.3 6845 Registers

There are 19 registers in the 6845, one of which is used for addressing the other 18 registers, which are called parameter registers. These are used for formatting the operation of the CRTC or reading certain specified status.

Register Addressing. To address one of the 18 registers of the 6845, the following levels are required. A 5-bit address must be loaded into the address register. This occurs when CS# = 0, RS = 0, E = 1, and R/W# = 0. At this time the five LSBs of data on the data bus are placed in the address register. Once the data is loaded into the address register, the contents of the selected register may be read or written to, depending on the type of register selected.

Register Grouping. R0 to R3 is the horizontal format and timing parameter group. It determines the number of characters per line and the horizontal operating frequency.

R4 to R7 is the vertical format and timing parameter group. It determines the number of character rows on the display screen and the vertical operating frequency.

R0 to R11 must be loaded when the system is first started; they are not accessed afterwards, unless the display mode or type of cursor is changed.

R8 to R17 controls the cursor characteristics, screen memory addressing, and light pen interface group. They determine the type and size of the cursor, the number of raster scans per character, screen memory addressing values, and light pen address value.

R12 and R13 establish a 14-bit starting (top of the page) address for screen memory. The data in these registers are manipulated when scrolling the screen.

R14 and R15 establish a 14-bit cursor address which is used by the CRTC to indicate the screen location of the next character to be input on the screen (into screen memory).

R16 and R17 are used to indicate at which screen memory location the light pen was triggered.

Horizontal Registers. The horizontal register group is used by the 6845 to determine how raster lines are displayed on the screen (Fig. 8-11, Table 8-4).

R0 — The horizontal total register is an 8-bit write-only register that determines the horizontal synchronization frequency by defining the Hsync period in character times. This value is equal to the total displayed characters and nondisplayed character (retrace) times plus one.

R1 — The horizontal display register is an 8-bit write-only register that determines the number of displayed characters per line. The value of this register must be less than the value in register R0 because this value only represents the total number of displayed characters per line. This value essentially determines trace time.

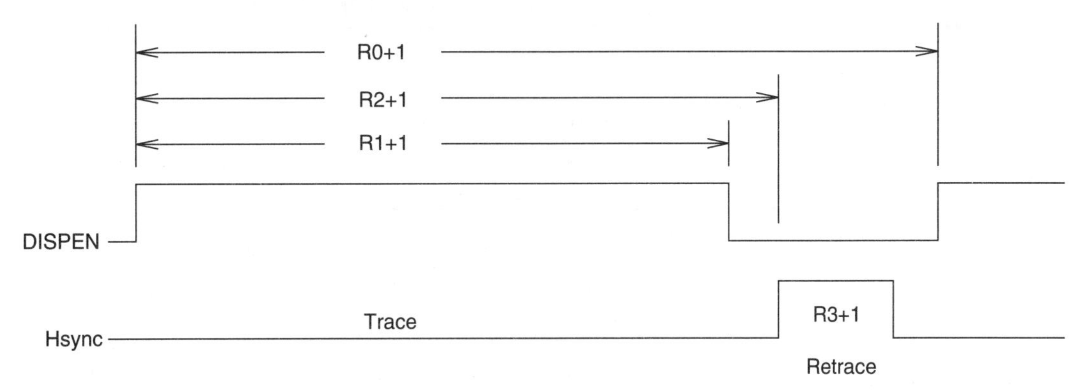

Figure 8-11 Horizontal register waveforms.

R2 The horizontal synchronization position register is an 8-bit write-only register that controls whether the character is displayed to the left or right side of the screen. If the value of the data is increased, the display on the screen is shifted to the left, and if the value is decreased, the display is shifted to the right. The value in this register determines when retrace begins. The value must be less than R0 and R1.

R3 The synchronization width register is an 8-bit write-only register that determines the width of the horizontal synchronization pulse. The value can be anywhere from 1 to 15 character clock periods, which allows different types of monitors to be used. This value determines how long it takes to perform a retrace.

The difference between R0 and R1 is the horizontal blanking interval, which allows the electron beam to retrace and return to the left side of the screen. The retrace time is determined by the monitor's horizontal scanning period. If the electron beam overscans, the retrace time should be increased.

Vertical Registers (Table 8-5)

R4–R5 The vertical total register (R4) is a 7-bit write-only register, and the vertical total adjust register (R5) is a 5-bit write-only register. They are used together to determine the vertical scan rate. The total number of character line times is an integer plus a fraction that produces a 60-Hz vertical refresh rate. The integer number minus one is stored in R4 and the fractional number is stored in R5. The values in these two registers determine the vertical trace and retrace times of the display screen.

R6 The vertical display register (Fig. 8-12) is a 7-bit write-only register that contains the number of displayed row times on the screen. This number must be less than R4. The value of this register determines how many rows of characters are displayed on the video screen. This

TABLE 8-4
Horizontal Registers

Name	*Register*	*R/W*	*Bits*
Horizontal total	00	W	8
Horizontal display	01	W	8
Horizontal sync. position	02	W	8
Horizontal sync. position	03	W	4

TABLE 8-5
Vertical Registers

Name	*Register*	*R/W*	*Bits*
Vertical total	04	W	7
Vsync adjust	05	W	5
Vertical display	06	W	7
Vertical sync position	07	W	7

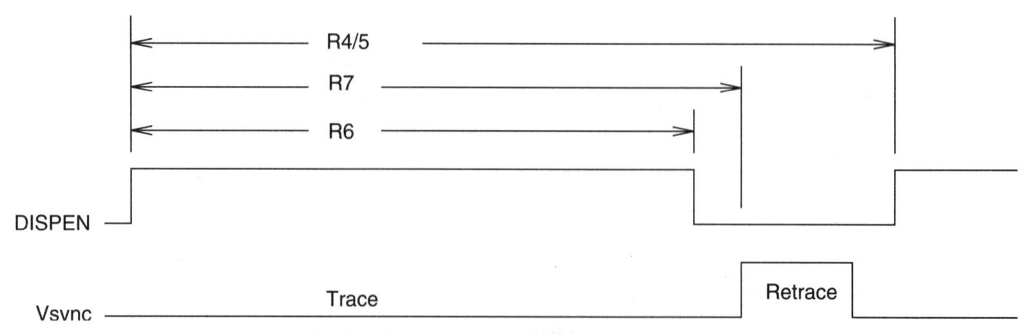

Figure 8-12 Vertical register waveforms.

TABLE 8-6
Display Mode

Bit 1	*Bit 0*	*Mode*
0	0	Normal synchronization (non-interlaced)
0	1	Interlace synchronization mode
1	0	Normal synchronization (non-interlaced)
1	1	Interlace synchronization and video

TABLE 8-7
Non-interlaced Display

Scan	*Data*	*Frame Display*
0	000 1000	— — — X — — —
1	001 0100	— — X — X — —
2	010 0010	— X — — — X —
3	100 0001	X — — — — — X
4	111 1111	X X X X X X X
5	100 0001	X — — — — — X
6	100 0001	X — — — — — X
7	000 0000	— — — — — — —

register determines the trace time from the top of the screen to the bottom of the screen.

R7 The vertical synchronization position register is a 7-bit write-only register that sets the position reference for the vertical row position. This value is one less than the number of computed character line (row) times. If the value increases, the characters on the display move up, and if the value decreases, the characters on the display move down. This value must be less than or equal to R4. The value of this register determines when retrace occurs. The retrace time is fixed.

Display Format Registers

R8 The interlace mode and skew register is a 6-bit write-only register. Bits 0 and 1 control the type of interlace (Tables 8-6 and 8-7).

Normal synchronization mode (non-interlaced, also known as raster scan) displays one entire character or picture during each scan field. Each scan field is refreshed at the vertical scan frequency of 60 Hz. This display mode allows less flickering because of the refresh rate; therefore, it is used in most computer monitors.

Interlace synchronization mode displays every other line of the character or picture during each scan field (field for short); during the next field, the missing lines of the character or picture are filled, giving one frame or the entire display. Thus, character or pictures (frames) are fully refreshed 30 times a second (one-half the vertical frequency), which sometimes causes flickering in displays. These are not used except for TVs and are not discussed more fully.

Interlace synchronization and video mode operates the same as interlace synchronization mode, except that alternating lines of the characters are displayed in the even and odd fields, which doubles the given bandwidth of the CRT monitor. Because of flickering, this method of scanning the monitor is not used and not discussed more fully.

Bits 4 and 5 of R8 determine the level on the DISPEN pin; the skew rate is the amount of delay added to the time the display is enabled during the printing of a character. Because

TABLE 8-8
Character Skew

Bit 5	*Bit 4*	*Mode*
0	0	No character skew
0	1	One character skew
1	0	Two character skew
1	1	No available skew

TABLE 8-9
Cursor Skew

Bit 7	*Bit 6*	*Mode*
0	0	No character skew
0	1	One character skew
1	0	Two character skew
1	1	No available skew

TABLE 8-10
Cursor Type

Bit 6	*Bit 5*	*Cursor mode*
0	0	Nonblinking cursor
0	1	Nondisplay cursor
1	0	Blinking at 1/16 field rate
1	1	Blinking at 1/32 field rate

TABLE 8-11
Cursor Display

Scan	*R10 = 0 R11 = 7*	*R10 = 2 R11 = 6*
0	XXXXXXXX	— — — — — — — —
1	XXXXXXXX	— — — — — — — —
2	XXXXXXXX	X X X X X X X X
3	XXXXXXXX	X X X X X X X X
4	XXXXXXXX	X X X X X X X X
5	XXXXXXXX	X X X X X X X X
6	XXXXXXXX	X X X X X X X X
7	XXXXXXXX	— — — — — — — —

non-interlace is used, there is no need to enable the DISPEN pin longer than the time it is being accessed; therefore no character skew is used (Table 8-8).

Bits 6 and 7 of R8 control the cursor skew or the amount of extra delay added to the time the cursor stays lit during each frame. Because non-interlace scanning is used, there is no need to enable the cursor pin longer than the time it is being accessed; therefore no character skew is used (Table 8-9).

R9 The maximum scan line address register is a 5-bit write-only register that determines the number of scan lines per character row including the spacing. The value in this register is the total number of rows per character box. The contents of the register controls the levels on the RA (Raster Address) outputs.

R10–R11 The cursor start register (R10, 7-bits write-only) and the cursor stop register (R11, 5-bits write-only) are used to determine the size of the cursor and how it is displayed on the screen. Bits 5 and 6 of R10 are used to determine how the cursor is displayed, while bits 0 to 4 are used to indicate the beginning raster address of the character row. The 5 LSBs of R11 are used to determine the raster address where the cursor ends. The values of these registers change any time the cursor changes its size or how it is seen on the display screen (Tables 8-10 and 8-11).

R12–R13 The start address register is a 14-bit write-only register that controls the first address output by the CRTC after vertical blanking. R12 acts as the lower 8 bits and R13 acts as the upper 6 bits. The value of this register changes during scrolling of the screen. This register points to the first character in the first row to be displayed on the screen.

TABLE 8-12
Display Format Registers

Name	*Register*	*R/W*	*Bits*
Interlace mode	08	W	6
Scan line address	09	W	5
Cursor start	0A	W	7
Cursor stop	0B	W	5
Start address (LSB)	0C	W	8
Start address (MSB)	0D	W	6
Cursor address (LSB)	0E	R/W	8
Cursor address (MSB)	0F	R/W	6
Light pen (LSB)	10	R	8
Light pen (MSB)	11	R	6

TABLE 8-13
6845 Format Work Sheet

6845 Format Work Sheet		*Mono*	*Color*	
Displayed characters-row		80	80	Char.
Displayed character rows-screen		25	25	Rows
Character matrix	a columns	7	7	Col.
	b rows	9	7	Rows
Character box	a columns	9	8	Col.
	b rows	14	8	Rows
Frame refresh rate		50	60	Hz
Horizontal oscillator frequency		18	15.75	KHz
Active scan lines (line 2 × line 4b)		350	200	Lines
Total scan lines (line 6 × line 5)		360	262	Lines
Total rows per screen (line 8 × line 4b)		25	32	Rows
		10	6	Lines
Vertical sync. delay (char. rows)		0	0	Rows
Vertical sync. width (scan lines 16)		16	16	Lines
Horizontal sync. delay (char. times)		8	16	Char.
Horizontal sync. delay (char. times)		9	17	Char.
Total character times (lines 1 + 12 + 13)		97	113	Char.
Character rate (lines 6 × 15)		1.746	1.779	MHz
Dot clock rate (lines 4a × 16)		15.71	14.23	MHz

R14–R15 The cursor register is a 14-bit read-write register that holds the address of the cursor on the screen. R14 acts as the lower 8 bits and R15 acts as the upper 6 bits. The 8088, when adding new characters to screen memory, uses this address for the address of the new character to be entered and increments this register to point to the new address. The value of this register is also used when any type of modification is made to the contents of the screen when the cursor is involved.

R16–R17 The light pen register is a 14-bit read-only register that places the current value of the screen memory address in the light pen register, when the LPSTB input pin is pulled high. Where R16 is the lower 8 bits of the address and R17 contains the upper 6 bits of the address, the programmer or program must decide what to do with the address once it has been transferred.

Register Values at Startup

TABLE 8-14
Monochrome Display Startup Values

Register	*Register Name*	*Unit Value*	*Hex. Value*
R0	Hor. total	Characters	61
R1	Hor. display	Characters	50
R2	Hor. sync. position	Characters	52
R3	Hor. sync. width	Characters	0F
R4	Vert. total	Character rows	19
R5	Vert. total adjust	Scan lines	06
R6	Vert. display	Character rows	19
R7	Vert. sync. position	Character rows	19
R8	Interlace mode	Code value	02
R9	Max. scan line	Scan lines/char.	0D

TABLE 8-14
Continued

Register	*Register Name*	*Unit Value*	*Hex. Value*
R10	Cursor start	Scan line	0B
R11	Cursor stop	Scan line	0C
R12	Start address (high)	Address value	00
R13	Start address (low)	Address value	00
R14	Cursor address (high)	Address value	00
R15	Cursor address (low)	Address value	00
R16	Light pen add. (high)	NA	NA
R17	Light pen add. (low)	NA	NA

NA = Not applicable

Table 8-15
Color Graphics Display Startup Values

Register	*Register Name*	*Unit Value*	*Alphanumeric*		*Graphic*
			40 × 25	*80 × 25*	
R0	Hor. total	Characters	38	71	38
R1	Hor. display	Characters	28	50	28
R2	Hor. sync. pos.	Characters	2D	5A	2D
R3	Hor. sync. width	Characters	0A	0A	0A
R4	Vert. total	Char. rows	1F	1F	7F
R5	Vert. total adj.	Scan line	06	06	06
R6	Vert. display	Char. rows	19	19	64
R7	Vert. sync. pos.	Char. rows	1C	1C	70
R8	Interlace mode	Code value	02	02	02
R9	Max. scan line	Scans/char.	07	07	07
R10	Cursor start	Scan line	06	06	06
R11	Cursor stop	Scan line	07	07	07
R12	Start add. (high)	Add. value	00	00	00
R13	Start add. (low)	Add. value	00	00	00
R14	Cursor add.(high)	Add. value	00	00	00
R15	Cursor add. (low)	Add. value	00	00	00
R16	Light pen add. (high)	Add. value	—	—	—
R17	Light pen add. (low)	Add. value	—	—	—

— = No Set Value

8.5.4 6845 Monitor Timing Diagram

The timing diagram in Figure 8-13 shows how the CRTC controls the screen memory and raster scan addressing during a typical scan of the screen.

The MA0 to MA13 lines are used to access the video screen memory. The beginning address for the first character displayed on the screen is determined by the value of the start-of-page register, just after vertical retrace has occurred. The electron beam is in the upper left of the screen. At this time, the raster scan counters inside the CRTC are reset to zero, so the first raster scan can be produced. The horizontal synchronization lines are low and the display enable line goes high, which enables the electron beam.

As each CLK signal is received, the memory address counters inside the CRTC are incremented by one. Once the first 80 (if in the 80 character display mode) memory address locations are accessed by the memory address registers, the raster scan address counter incre-

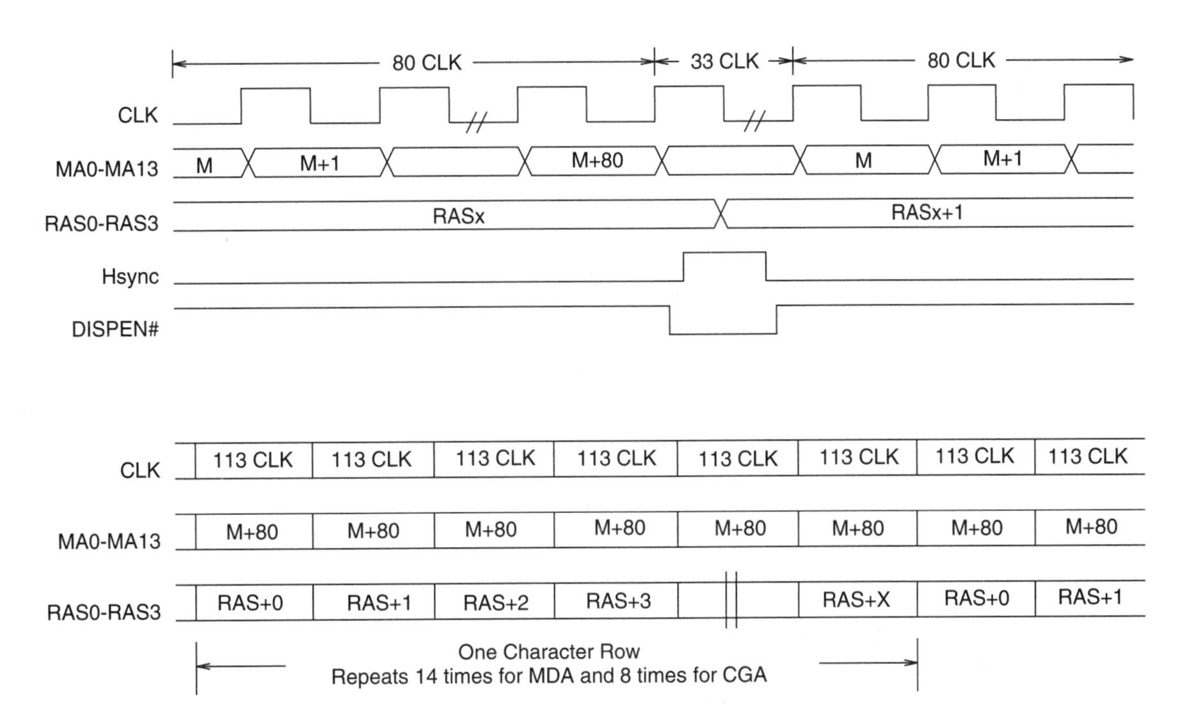

Figure 8-13 Horizontal timing.

ments by one. During this time, the horizontal synchronization line goes high to produce a retrace in the monitor (moves the electron beam from the right to left side of the screen). Before and a little after the time the horizontal synchronization line goes high, the CRTC brings the display enable lines low, which disables the electron beam (turns it off), so that the horizontal retrace is not seen on the screen.

After the retrace time period has ended, the horizontal synchronization line goes low again, indicating the end of the retrace sequence. This causes the CRTC to enable the display enable pin by bringing it high. The memory address counters begin accessing the video screen memory and the same 80 memory locations are accessed. After the first 80 memory locations have been accessed, the raster scan counter is incremented and the retrace begins.

The above process occurs until the first 80 screen memory address locations have been accessed a total of 8 times for CGA and 14 times for MDA. At that point, the next 80 screen memory locations are accessed 8 times for CGA or 14 times for MDA. This process continues until all character rows on the display have been printed on the screen. Then a vertical retrace occurs.

8.5.5 CRTC Read/Write Timing Diagram

The timing diagram in Figure 8-14 shows how the MPU can address the address register and then program or read the status of the selected register.

In order to select a register inside the CRTC, the MPU must bring the CS# low, RS line low, R/W# line low, and E line high; at this point the data on the data bus is written to the internal address register of the CRTC. When the enable line is high, the CRTC is refreshing the screen. The RS line, when low, allows the data on the data bus to enter the address register.

Once a register has been selected by the value in the address register, the MPU can write data into the selected register by bring the CS# and R/W# lines low. The data on the data bus is transferred into the selected register. The RS line should stay high during this cycle.

Once a register has been selected by the value in the address register, the MPU can read the data in the selected register by bringing the CS# line low and the R/W# line high. A copy of the data in the selected register is seen on the data bus and transferred to the device specified. The RS line should stay high during this cycle.

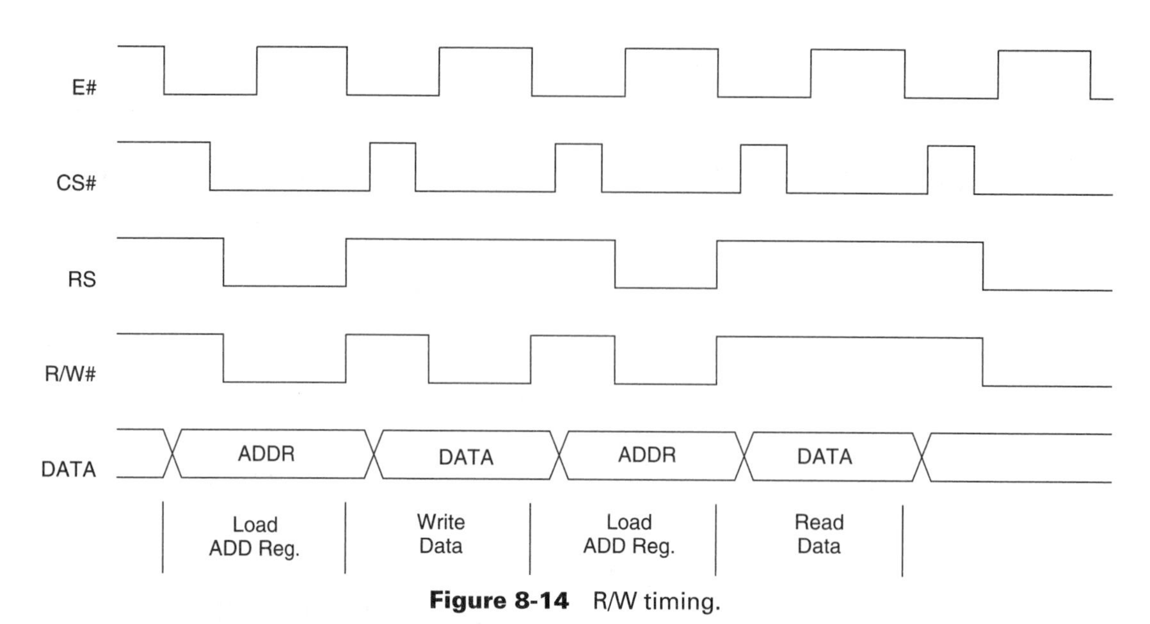

Figure 8-14 R/W timing.

8.6 IBM MONOCHROME VIDEO ADAPTER

The block diagram in Figure 8-15 shows the video section of the IBM monochrome video adapter. Many of the blocks of the diagram operate the same as in the color graphic block diagram, which is discussed later.

8.6.1 Block Functions

The purpose of the 6845 programmable CRTC is to produce the horizontal-vertical synchronization timing, screen refreshing, raster scanning, and display during the trace time. It also produces and locates the cursor and provides light pen locations or trigger location (on color displays only).

The MPU accesses this device during the following times and conditions.

1. When the computer is reset or first turned on, the 8088 MPU formats this device so that normal operations can take place.
2. Every time a character is to be displayed, the computer must read the location of the cursor in screen memory. The 8088 MPU then stores the ASCII code at the address of the cursor in screen memory. The cursor address then increments by one.
3. When changing the attribute values of the screen, it is accessed.
4. When changing the type of cursor that is displayed, it is accessed. At all other times the CRTC controls the display on the screen.

The memory address multiplexer is used to determine whether the MPU or the CRTC accesses screen memory.

The MPU supplies 12 address lines to the memory address multiplexer, of which 10 are used to actually address the screen memory. Of the two extra lines, one is used to determine if the data on the data bus goes into character or attribute memory. The other line is used to determine whether the control (003B8) or status (003BA) port is accessed. The control port must be loaded at startup with 01 (high-resolution output) to enable the monochrome display adapter. If the control port is not loaded, BIOS does not recognize the adapter. The status port indicates the status of the horizontal synchronization (if horizontal pulses are present) and video enable (indicates when the dots are being created). The status of these two bits may be

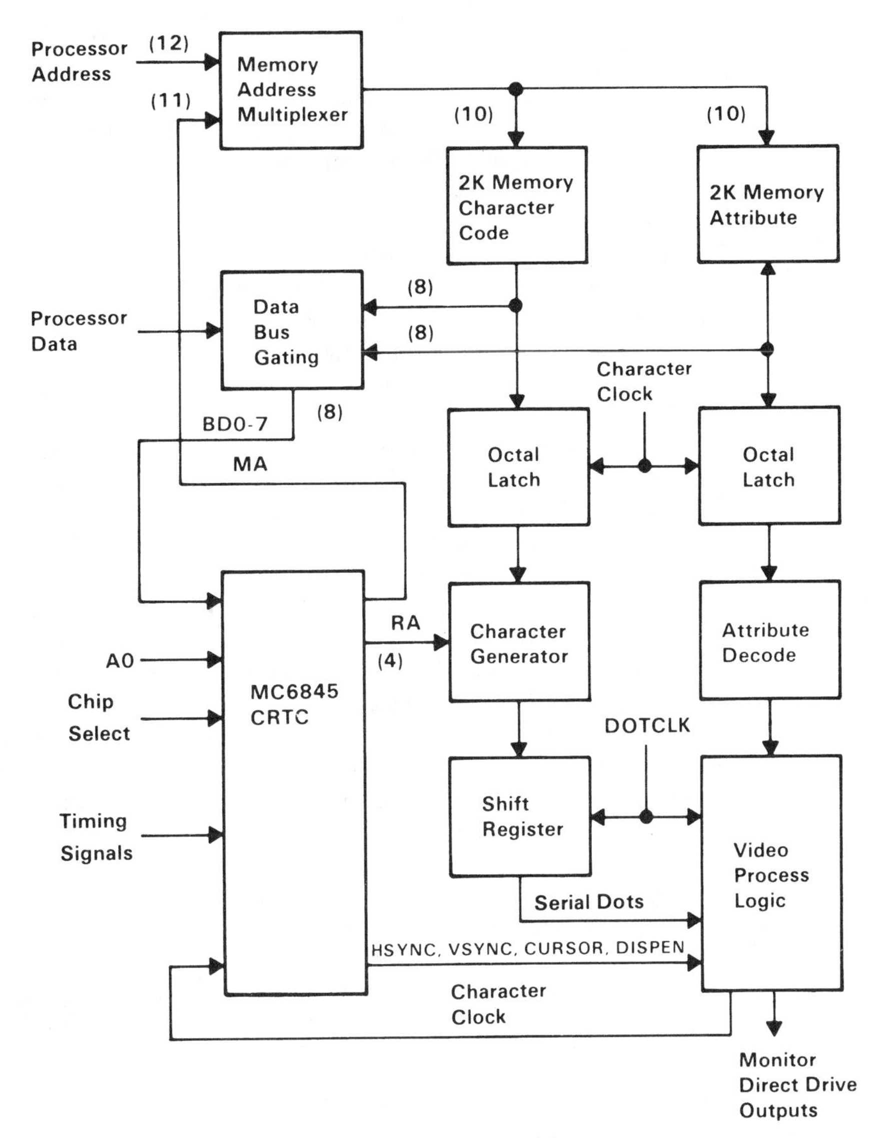

Figure 8-15 IBM monochrome block diagram. (Courtesy of IBM)

used to determine if the adapter is working. In most cases, the status port is never read. The 8088 MPU has control over the screen memory address bus only when storing character or attribute data in a screen location.

Whenever the 8088 MPU does not control the screen memory address bus, the CRTC controls it. The CRTC outputs 11 address lines to the memory address multiplexer. Of the 11 address lines, 10 are used to address screen memory, and the last line is used to make sure both the character and attribute memory blocks are enabled at the same time for each screen address accessed.

The data bus gating circuit is a bidirectional switching network. If the CRTC is to be programmed or its registers read, the data flow between the 8088 MPU data bus and the CRTC data bus. If the MPU writes or reads character or attribute values into screen memory, data flow to or from the 8088 MPU data bus to screen memory; otherwise bus gating is disabled. When the data bus gating is disabled, the CRTC is allowed to refresh the screen display from screen memory.

The screen memory on the monochrome display adapter consists of 4K static RAM. The static memory was chosen because of the small size; screen memory could be placed on four memory ICs and no refresh is needed. Screen memory is divided into two 2K memory blocks. During screen refresh, both blocks of memory are enabled at the same time, producing the character and attribute codes at the same time. Whenever the MPU updates information in screen memory, only one block of memory becomes enabled at any one time, allowing the MPU to write or read data to or from character or attribute screen memory.

The character code screen memory block contains the ASCII codes for the characters displayed on the screen. The size of the character code memory block is 2048 bytes, even though only 2000 (80 × 25) memory locations are needed to store one screen of characters. This memory is accessed at the same time attribute memory is accessed. The data bus of the character screen memory block is applied to the data bus gating and the octal latch for character codes. When the 8088 MPU has control of the data bus gating, the data is applied to the data lines of the character screen memory block. During refresh, the data is applied to the octal latch for character codes.

The attribute screen memory block contains 2K of attribute data. Each piece of attribute data corresponds to a character code in the character code part of screen memory and shares the same address location. When the character code is addressed, the attribute for that character code is also accessed. The data from the attribute part of screen memory is applied to the attribute octal latch during refresh. When the MPU has control over the data bus gating, the data is applied to the attribute memory block. The attribute byte determines how the character code is displayed on the screen Table 8-16 gives the purpose of each bit in the attribute byte.

Screen memory is located in the IBM PC memory map beginning at address B0000.

There are two octal latches (actually flip-flops), one for the character codes and one for the attribute byte. The octal latch for the character code holds the ASCII code from character screen memory until the ASCII code can be processed by the character generator. The output of the character octal latch is used as the eight LSBs of the address in the character generator.

TABLE 8-16

Attribute Bit Definition. (Courtesy of IBM)

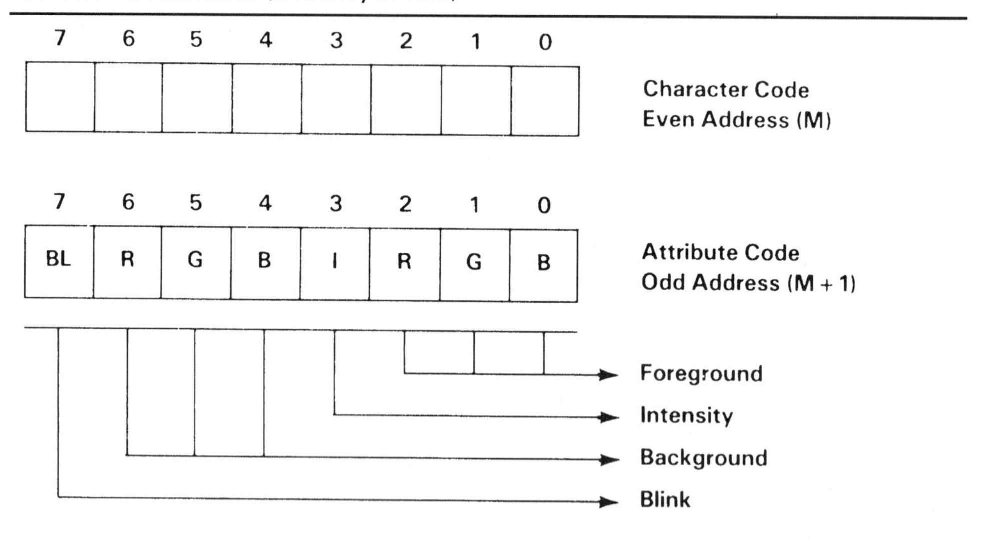

Background R G B	Foreground R G B	Function
0 0 0	0 0 0	Non-display
0 0 0	0 0 1	Underline
0 0 0	1 1 1	White character/black background
1 1 1	0 0 0	Reverse video

There are a total of 256 different extended ASCII codes (127 standard ASCII codes). The octal latch for the attribute byte is used to hold the attribute data from the attribute screen memory block until the attribute decode circuitry can produce the proper code for the video process logic. The latching of both occurs at the character clock rate (1.746 MHz) but only during screen refresh. The character and attribute codes are latched on the leading edge of the character clock.

The character generator is nothing more than an 8K ROM/EPROM that contains 7168 raster scan codes for two sets of extended ASCII characters. One set of 3584 raster scan codes (14 per character) is for extended ASCII characters. The other 4K is used for the color graphics adapter character set.

The character generator has two sets of inputs. The inputs from the character octal latch (one byte wide) act as the lower 8 bits of the address to the character generator. The other input is from the CRTC. The monochrome adapter uses four of the CRTC's raster scan address lines to act as the upper four bits of the address to the character generator. There are a total of 3584 memory locations in the character generator. Each character in the character generator produces an 8-bit raster code.

Each raster code is eight bits wide. The first bit is always low, which produces the one dot horizontal spacing to the left of each character. A special circuit produces an additional (ninth) bit for the shift register. This X bit is low for all ASCII codes from 00 to CF and E0) to FF, which produces one dot horizontal spacing to the right of the character. If the ASCII code falls between B0 and DF (graphic characters), the circuit causes the X bit to copy the eighth bit of the raster scan code from the character generator. The reason for the X bit copying the eighth bit of the character generator is that graphic characters do not require spacing.

The first two raster scan codes for characters are normally blanked for superscript displays. Raster scan lines 12 and 13 are used for subscript displays, and the very last line is used for underlining (if enabled) (Table 8-17).

Tables 8-18 and 8-19 show the IBM quick reference character set used in the IBM PC, with either a monochrome or color graphic display adapter.

The attribute decoder circuitry sends the attribute levels to the video process logic to cause one or more of the following functions for each character displayed on the screen. Reversed video occurs when the background is light and foreground is darker. Blinking causes a character to flash at 1/16 the field rate. High intensity makes the foreground (character) lighter than normal. Underlining causes the last raster line to light up in all dot positions for the character.

TABLE 8-17
Monochrome Character Display

Raster	*X*	*Code*	*Display*
0	0	0000 0000	— — — — — — — — — —
1	0	0000 0000	— — — — — — — — — —
2	0	0001 0000	— — — — X — — — —
3	0	0011 1000	— — — X X X — — —
4	0	0110 1100	— — X X — X X — —
5	0	1100 0110	— X X — — — X X —
6	0	1100 0110	— X X — — — X X —
7	0	1111 1110	— X X X X X X X —
8	0	1100 0110	— X X — — — X X —
9	0	1100 0110	— X X — — — X X —
A	0	1100 0110	— X X — — — X X —
B	0	0000 0000	— — — — — — — — — —
C	0	0000 0000	— — — — — — — — — —
D	*	**** ****	— — — — — — — — — —

TABLE 8-18
Quick Reference Character Set—Part 1. (Courtesy of IBM)

DECIMAL VALUE	➡	0	16	32	48	64	80	96	112
⬇	HEXA DECIMAL VALUE	0	1	2	3	4	5	6	7
0	0	BLANK (NULL)	►	BLANK (SPACE)	0	@	P	‘	p
1	1	☺	◄	!	1	A	Q	a	q
2	2	☻	↕	"	2	B	R	b	r
3	3	♥	‼	#	3	C	S	c	s
4	4	♦	¶	$	4	D	T	d	t
5	5	♣	§	%	5	E	U	e	u
6	6	♠	▬	&	6	F	V	f	v
7	7	•	↨	'	7	G	W	g	w
8	8	◘	↑	(	8	H	X	h	x
9	9	○	↓	)	9	I	Y	i	y
10	A	◙	→	*	:	J	Z	j	z
11	B	♂	←	+	;	K	[	k	{
12	C	♀	∟	,	<	L	\	l	¦
13	D	♪	↔	-	=	M	]	m	}
14	E	♫	▲	.	>	N	^	n	~
15	F	☼	▼	/	?	O	_	o	△

TABLE 8-19
Quick Reference Character Set—Part 2. (Courtesy of IBM)

DECIMAL VALUE	➡	128	144	160	176	192	208	224	240
⬇	HEXA DECIMAL VALUE	8	9	A	B	C	D	E	F
0	0	Ç	É	á	░	└	╨	α	≡
1	1	ü	æ	í	▒	┴	╤	β	±
2	2	é	Æ	ó	▓	┬	╥	Γ	≥
3	3	â	ô	ú	│	├	╙	π	≤
4	4	ä	ö	ñ	┤	─	╘	Σ	⌠
5	5	à	ò	Ñ	╡	┼	╒	σ	⌡
6	6	å	û	ª	╢	╞	╓	µ	÷
7	7	ç	ù	º	╖	╟	╫	τ	≈
8	8	ê	ÿ	¿	╕	╚	╪	Φ	°
9	9	ë	Ö	⌐	╣	╔	┘	Θ	•
10	A	è	Ü	¬	║	╩	┌	Ω	·
11	B	ï	¢	½	╗	╦	█	δ	√
12	C	î	£	¼	╝	╠	▄	∞	ⁿ
13	D	ì	¥	¡	╜	═	▌	φ	²
14	E	Ä	₧	«	╛	╬	▐	∈	■
15	F	Å	ƒ	»	┐	╧	▀	∩	BLANK 'FF'

The dot shift register is a 9-bit parallel-in serial-out shift-right shift register. The first bit shifted out is the LSB from the character generator. This is followed by the rest of the bits from the character generator. The last bit shifted out is the X bit. The bits of the dot shift register are shifted out serially at the rate of the dot clock (15.714 MHz). These bits turn the electron beam on and off, causing the raster scan line to be produced for the character that is displayed on the screen.

The video process logic receives inputs from four different sources, attribute decode circuitry, serial data from the shift register, signals from the CRTC (HSync, VSync, cursor enable, and display enable), and the dot clock signal from the dot clock. The combination of these signals produces output signals from the video process logic. There are two groups of output signals, the character clock signal and the monitor drive output signals.

The attribute decode circuitry is used to determine how the display is seen: blinking, reversed video, underlined, or high intensity. The circuitry causes the intensity pin go to high for the characters that are being displayed using high intensity. The circuitry causes the video data signal to become disabled once every 1/16 of a field when a blinking character is displayed. The circuitry causes the data bits to become inverted when a character is displayed with reversed video. The circuitry also causes the last raster line to become enabled when an underlined character is displayed.

The serial data is the dot information from the shift register. This signal turns the electron beam on or off during the horizontal and vertical trace. If high, the electron beam is on; if low the electron beam turns off (no dot appears on the screen for that time period).

The CRTC signal group provides four signals. The first signal produces the horizontal synchronization pulse for the monitor; when high, it causes the monitor to start its retrace. The second signal produces the vertical synchronization pulse for the monitor; when high, it causes the monitor to start its vertical retrace. The third signal produces the cursor when the cursor address matches the address the CRTC is currently addressing in screen memory. The last signal is the display enable output from the CRTC; the line goes high during the horizontal and vertical trace and turns off the display during retrace (which contains no useful screen information).

The dot clock signal input makes sure that all signals change at their proper times.

The character clock output signal is established by dividing the counter produced by the dot clock input by N. This signal is used to clock the CRTC to address screen memory. It also keeps small frequency drifts form causing a loss of display synchronization.

8.6.2 Screen Refresh Operation

The following assumes that the computer is turned on, keys are pressed (keyboard buffer is empty), and the scanning of the screen begins at the top of the screen.

1. The CRTC uses the top of page register to find the first screen memory location to be addressed.
2. The CRTC sets its raster address scan address at 0 in hexadecimal.
3. Memory address lines from the CRTC address the first screen memory location. Screen memory releases the first character and its associated attribute code to the screen memory data bus.
4. The data on the screen memory data bus is latched to the character octal buffer and attribute octal buffer on the rising edge of the character clock.
5. The data from the character latch and the outputs from the CRTC raster address outputs address the raster scan code in the character generator. The output of the character generator supplies its raster scan code to the parallel input of the dot shift register along with the data from the X bit code. At the same time, the attribute data from the attribute latch is decoded to produce the necessary signals for the video process logic.
6. The dot clock loads the dot shift register and begins shifting out the nine bits of data through its serial output into the video process logic.
7. As the electron beam sweeps horizontally across the screen, the video process logic causes the level on the video output to turn the electron beam on and off. If high, a dot lights up on the screen; if low, no dot appears.
8. As the first nine dots of the first raster scan line are applied to the monitor for the first character scan row, the CRTC goes to the next memory location. With the raster scan address still at hexadecimal zero, steps 3 through 7 are repeated, until 80 characters have been addressed.
9. After the first 80 characters have been addressed, the CRTC changes its raster scan address to 1 in hexadecimal and steps 3 through 8 are repeated. This process continues until the raster scan address has processed 14 raster scan lines. At that point, one complete character row is displayed on the screen. Therefore, to produce one complete character row, the CRTC and character generator must address the first 80 screen memory locations 14 times (a total of 1120 address cycles).
10. The sequence in steps 2 through 9 continues until all 25 character rows have been displayed on the screen, which occurs 60 times per second. Then the process repeats.

8.6.3 Writing Into Screen Memory

1. The 8088 MPU, when ready to write an extended ASCII code to screen memory, accesses the CRTC and reads the cursor address position from the cursor address register.
2. The MPU disables the CRTC from controlling screen memory and addresses the memory location of the cursor. Next the MPU writes data to the character location of

the character screen memory block and gives control back to the CRTC. The CRTC performs a memory refresh. The CRTC increments the cursor location by one to point to the next screen location.

3. If the character produced needs an attribute other than normal, the MPU uses the same screen memory address to address the screen memory again. This time it writes data to the attribute data location. If the normal attribute is needed, the memory location does not have to be changed.

8.7 IBM COLOR-GRAPHICS VIDEO ADAPTER

The IBM color-graphics video adapter is much like the monochrome video adapter, except that it has more screen memory and some additional circuitry for the decoding of color graphics.

8.7.1 Alphanumeric Display Mode

When the color-graphics adapter is in the alphanumeric display mode, it can display the IBM extended ASCII code set, which contains 256 characters. There are two display sizes. The 40 × 25 format produces characters in a double dot format, 7 × 7 in an 8 × 8 character box. The frequency from the CRTC changes to allow full screen display. This format is not the default format. This format requires 2K of memory, 1000 bytes for the character codes and 1000 bytes for the attribute data. The second display size is the 80 × 25 format, which produces characters in a double dot 7 × 7 format in an 8 × 8 character box. This format requires 4K of memory, 2000 bytes for the character codes and 2000 bytes for the attribute data.

IBM's extended ASCII character set contains 96 standard ASCII characters, 48 block graphics symbols, 48 foreign language characters, 16 Greek symbols, 15 scientific notation symbols, 16 game symbols, and 15 word processing control characters (nonprintable).

Both formats in the alphanumeric mode allow the following attributes. In black and white, the attributes are reverse video, blinking, and highlighted characters. The color attributes are 16 foreground (character) colors, 8 background colors, 16 border colors, and character blinking.

8.7.2 All-Points-Addressing Display Mode

There are three levels of resolution that can be used with the color-graphics video adapter (Fig. 8-16). The low-resolution graphics mode is not used except when a TV is used as a video monitor. This mode requires additional software routines to operate. It is not normally supported by this adapter; therefore, this book does not go into any great detail on this display format.

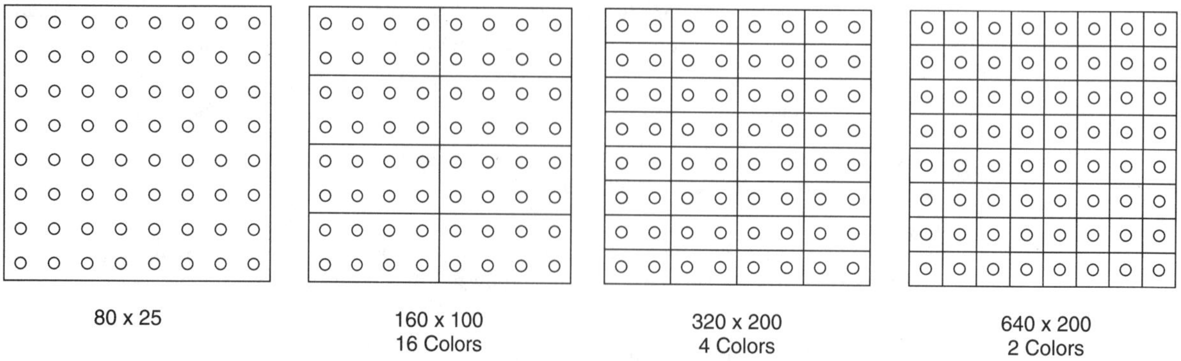

80 x 25

160 x 100
16 Colors

320 x 200
4 Colors

640 x 200
2 Colors

Figure 8-16 Character box and graphics display.

The medium-resolution format in the all-points-addressing display mode allows the adapter to control 320 × 200 points on the screen. Each point that can be controlled on the screen is a pixel. Each pel (or pixel) can be in one of four colors. One of the colors is the background color, the other three colors can be from one of the following color sets (green, red, brown) or (cyan, magenta, white). There are 16 different background colors. Each pixel in this format contains two dots (2 dots wide and 1 dot high) which means that there are 32 different pixels in each character location on the screen that can be controlled independently of each other. Because each pixel can be one of four colors, each pixel requires 2 bits of memory to be displayed, and because there are 320 × 200 pixels on the screen (a total of 64,000 control points), 16,000 bytes of memory are required to store this information.

In the high-resolution mode there are 640 × 200 pixels on the screen. Each pixel in this display format can be in one of two colors (black or white). Each pixel controls an area 1 dot wide and 1 dot high. Therefore, there are 64 pixels for each character location on the screen. There are a total of 128,000 pixels on every screen, and this requires 16,000 bytes of memory to store the information, because each pixel is defined by one bit of memory.

The all-points-addressing graphics mode also allows the video adapter to add alphanumeric characters to graphic display. The graphics mode supports all IBM extended ASCII character codes. These are produced by the ROM character generator.

The character display shown in Table 8-20 is the type that will be seen on the color-graphics adapter in any mode if an alphanumeric character is to be displayed. This display is called a double dot character; all vertical lines are two dots wide and all horizontal lines are a single dot wide.

8.7.3 Types of Color-Graphics Video Displays

The IBM color-graphics video display supports three types of video display signals. From an external connector on the video adapter board an output can be attached to an external RF modulator. The output of the RF modulator is applied to a standard black and white or color TV that can produce only low-resolution graphics and display characters in a 40 × 25 mode.

There is a monochrome-color composite video output, which is accessible from the back of the adapter through an RCA jack. This signal is a combination of the red-blue-green color signals modulated on a 3.59-MHz color subcarrier, horizontal synchronization, vertical synchronization, and intensity signals. The output is one signal that looks much like a composite video signal from a TV set. The monitor must demodulate the one signal to separate it into the proper signals and apply it to the proper circuitry. The only difference between this signal and a standard TV signal is that this signal is not modulated at very high frequencies. Consequently, the signal does not have to be demodulated as much. This allows sharper video displays.

TABLE 8-20
Alphanumeric Display Character

Raster Number	*Raster Code*	*Character Display*
0	0001 0000	— — — X — — — —
1	0011 1000	— — X X X — — —
2	0110 1100	— X X — X X — —
3	1100 0110	X X — — — X X —
4	1111 1110	X X X X X X X —
5	1100 0110	X X — — — X X —
6	1100 0110	X X — — — X X —
7	0000 0000	— — — — — — — —

The last output from the back of the video display adapter is the RGB output. The signals leave a 9-pin jack. This output directly drives the video, horizontal, vertical, green color gun, blue color gun, and red color gun circuits, with all outputs TTL compatible. The outputs are as follows.

1. The intensity output controls the overall intensity of the video signal.
2. The horizontal synchronization output provides a synchronization pulse to the horizontal circuitry of the monitor. This causes the horizontal sweep to increase or decrease in order to maintain lock.
3. The vertical synchronization output provides a synchronization pulse to the vertical circuitry of the monitor. This causes the vertical sweep to increase or decrease in order to maintain lock.
4. The red, blue, and green outputs are each applied to their respective color circuitry which controls the intensity of each of the three color guns in the color CRT.

8.7.4 Functions of the Block Diagram

Address Latches. There are two address latches. One latch is enabled when the 8088 MPU is required to access the display buffer (screen memory). The 8088 MPU has control of the screen memory address bus only when it writes data to or reads data from either the character or attribute part of the display buffer. The other address latch is enabled whenever the 6845 CRTC is refreshing the display screen, which occurs more often than the MPU accesses screen memory.

6845 CRTC. The 6845 CRTC is used to control screen refreshing. The CRTC produces the following signals: vertical and horizontal synchronization signals, display enable signal (enables the display during trace times), cursor producing (cursor size and type), cursor addressing (cursor location on the screen), and light pen position addressing (location on the screen where the light pen was enabled) signals.

Palette and Overscan. The palette and overscan circuitry is used to modify the graphics code in the color encoder to determine how the pixels appear on the screen. This circuitry also keeps the screen in vertical synchronization on vertical retrace because of the odd number of retrace lines produced on the screen.

This circuitry contains a color select register, which is a 6-bit write-only register that contains information on the colors used in the selected mode (Fig. 8-17).

Model Control. The model control register is used to provide extra control data needed by the timing generator and control circuitry to control the modulation in the composite color generator.

Timing Generator and Control. This timing circuitry provides the necessary additional frequencies and controls to modulate the data from the color encoder and horizontal and vertical synchronization signals in the composite color generator.

Composite Color Generator. This circuitry is a modulator that mixes the signals from the color encoder (red, blue, green, and intensity signals) and horizontal and vertical synchronization. The output of this signal is similar to a TV signal except it is at a lower frequency.

Input Buffer-Output Latch. The input buffer and output latch is used by 8088 MPU to control the display buffer (screen memory) data bus and how it moves data to and from the MPU's data bus.

Display Buffer. The display buffer, also called screen memory, is a 16K block of memory. This memory is dynamic RAM and is refreshed by support refresh circuitry on the

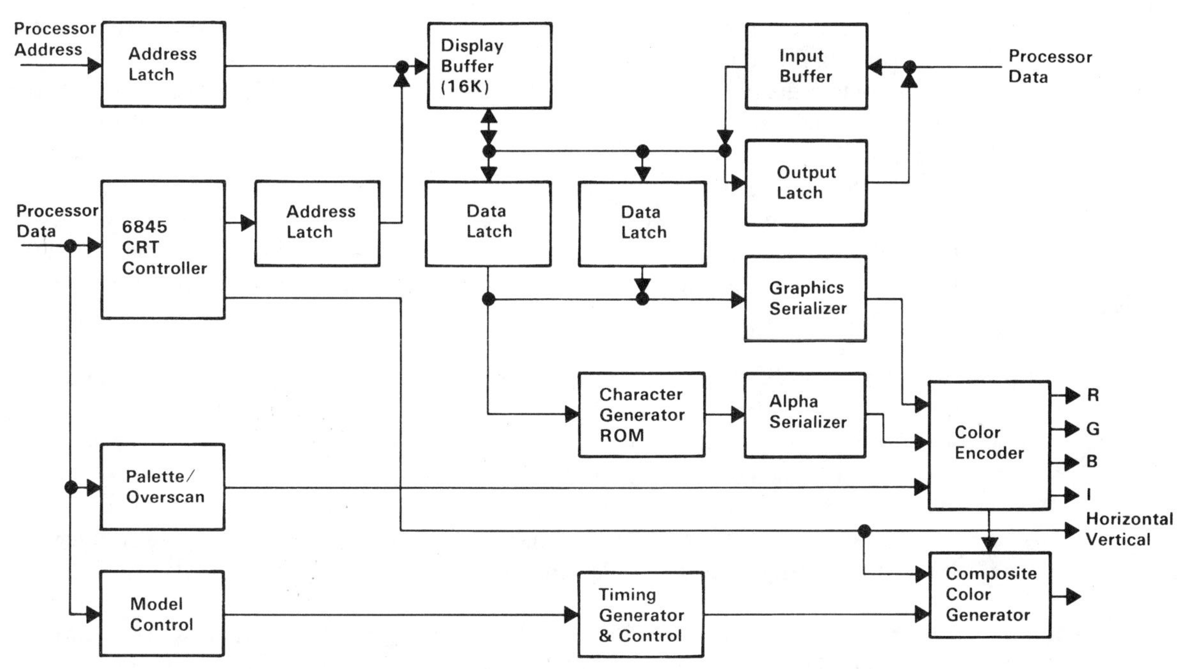

Figure 8-17 IBM color-graphics adapter block diagram. (Courtesy of IBM)

TABLE 8-21
Attribute Color Data Table

Bit	*7 BL*	*6 R*	*5 G*	*4 B*	*3 IN*	*2 R*	*1 G*	*0 B*
Color	*NA*		*Background*			*Foreground*		
B-W								
Normal	BL	0	0	0	IN	1	1	1
Rev Video	BL	1	1	1	IN	0	0	0
Black (ND)	BL	0	0	0	IN	0	0	0
White (ND)	BL	1	1	1	IN	1	1	1

Key:
BL = Blinking display only in monochrome (B/W).
R = Red color.
G = Green color.
B = Blue color.
IN = Intensity of display. (IN = 1 color set 1, IN = 0 set 2)
Normal display (black background, white foreground).
Reverse Video (black foreground, white background).
Black (ND)—nondisplay in black.
White (ND)—nondisplay in white.

color-graphics adapter itself. If the motherboard is not set up to use the color adapter, the computer does not use the proper screen memory location and the screen displays no readable characters or may display nothing at all.

In the alphanumeric mode the lower 8K of the display buffer are used for character data, memory locations B8000 through B9F3F. The upper 8K of the display buffer are used for attribute data, memory locations BA000 through BBF3F. This display buffer can store eight screens (pages) of characters in the 40 × 25 format and four screens of characters in the 80 × 25 format, but only under software control (Table 8-21).

In the medium-resolution graphics mode the lower 8K of the display buffer contain the pixel data for all even character locations on the screen. The upper 8K of the display buffer contain the pixel data for all odd character locations on the screen. Each character location contains four pixels for every raster scan line and is in one of four colors (Table 8-22).

In the high-resolution graphics mode, the lower 8K of the display buffer contain the palette data for all even character locations on the screen. The upper 8K of the display buffer contain the palette data for all odd character locations. Each character location contains eight palettes for every raster scan line and is in either black or white (Table 8-23).

Data Latch. There are two data latches. One holds the character data and the other holds the attribute data when the adapter is in the alphanumeric mode. When the adapter is in the graphics mode, one data latch holds the data for the even character location palettes and the other for the odd character location palettes.

Graphics Serializer. The graphics serializer (also called the graphics shift register) is used to convert parallel palette data into serial data for the color encoder.

Character Generator ROM. The character generator ROM is 8K. The ROM used in the monochrome adapter is used in the color-graphics adapter. The lower 4K produce the raster codes for the monochrome adapter. The next 2K are used to produce the 2048 raster codes needed for the double dot (standard format) character format, and the last 2K are used to produce the 2048 raster codes for single dot character format. There are eight 8-bit raster codes for each character using the color graphics adapter.

TABLE 8-22
Color Bit Table

Pixel bits	*7*	*6*	*5*	*4*	*3*	*2*	*1*	*0*
Color bits	*C1*	*C0*	*C1*	*C0*	*C1*	*C0*	*C1*	*C0*
Pixel number	—1—		—2—		—3—		—4—	

C1	*C0*	*Color Type*
0	0	Background color (1 of 16)
0	1	Color 1 of the selected color set
1	0	Color 2 of the selected color set
1	1	Color 3 of the selected color set

Color	*Color Set 1*	*Color Set 2*
1	Green	Cyan
2	Red	Magenta
3	Brown	White

TABLE 8-23
Color Pixel Table

Pixel bits	7	6	5	4	3	2	1	0
Color bit	C	C	C	C	C	C	C	C
Pixel number	1	2	3	4	5	6	7	8

If C = 0, color = black; if C = 1, color = white.

Alpha Serializer. The alphanumeric serializer converts parallel raster scan codes into serial dot codes and applies them to the color encoder.

Color Encoder. The color encoder is an IC that uses data from the palette-overscan circuitry, graphics serializer, and alphanumeric serializer to produce the necessary output signals for the monitor connected to the circuit. It produces four output signals, and the output of these signals is determined by the display mode the adapter is in at the time. The four output signals are red, green, blue, and intensity. Table 8-24 shows how the three colors signals and the intensity signal produce the colors on the screen of the monitor.

8.7.5 Alphanumeric Screen Refresh

The operation of alphanumeric screen refresh in the color-graphics adapter is the same as the screen refresh operation in the monochrome adapter. The only difference is that the attribute data is color information applied to the color encoder.

8.7.6 Screen Refresh and Input

Alphanumeric Screen Input. The alphanumeric screen input operation in the color-graphics adapter is the same as in the monochrome adapter.

Graphic Screen Refresh. The graphic screen refresh operation in the color-graphics adapter is the same as in the alphanumeric screen refresh operation, except that instead of having the data in the display buffer go to the character generator, it is applied to the graphics serializer. The data from the serializer is applied to the color encoder, which causes the screen to display a pixel in one of the available colors at the proper screen location.

Graphic Screen Input. The graphic screen input operation in the color-graphics adapter is the same as in the alphanumeric screen input operation, except that there are more pixel memory locations in the display buffer. The 320 × 200 or 640 × 200 graphics mode requires four bytes to control the same area on the screen as one byte of character information.

TABLE 8-24
Color Display Table

Intensity	*Red*	*Green*	*Blue*	*Color*	*Display Location*
0	0	0	0	Black	fore/back/border
0	0	0	1	Blue	fore/back/border
0	0	1	0	Green	fore/back/border
0	0	1	1	Cyan	fore/back/border
0	1	0	0	Red	fore/back/border
0	1	0	1	Magenta	fore/back/border
0	1	1	0	Brown	fore/back/border
0	1	1	1	White	fore/back/border
1	0	0	0	Gray	back/border
1	0	0	1	Light Blue	back/border
1	0	1	0	Light Green	back/border
1	0	1	1	Light Cyan	back/border
1	1	0	0	Light Red	back/border
1	1	0	1	Light Magenta	back/border
1	1	1	0	Yellow	back/border
1	1	1	1	White (High Int.)	back/border

8.8 VIDEO ADAPTER OUTPUTS

The following pages contain the output signals that would be seen on the outputs of the IBM monochrome adapter and color-graphics adapter.

Figure 8-18 shows the outputs from the IBM TTL monochrome adapter. In the horizontal section note what happens to the video, intensity, and vertical synchronization signals when horizontal retrace occurs. In the vertical section, note what happens to the video, intensity, and horizontal synchronization signals during vertical retrace. The video signal is used to create the characters on the screen. The intensity output controls the brightness of the character displayed on the screen.

Figure 8-19 shows the outputs from the color-graphics adapter using the RGB output. Again, note all signals during horizontal retrace and during vertical retrace. The RGB outputs determine the color of background and foreground, which changes with the type of character being displayed. The intensity output determines which color set is used (standard colors or bright colors).

Figure 8-20 shows the outputs from the color-graphics adapter using the composite output. In the composite waveform the video, horizontal-vertical synchronization pulses, intensity, and color information are modulated together to produce the waveform in Figure 8-20. During the horizontal blanking pedestal, the display is blanked and retrace lines do not appear on the screen. At the center of the horizontal blanking pedestal is the horizontal synchronization pulse, which causes the electron beam to move from the right to left side of the screen. On the trailing edge of the horizontal blanking pedestal are eight cycles of the chrominance signal (color information), which is used with the video information between the blanking pedestals to produce the proper color display. The signals between the blanking pedestals are the video information, which produces the characters. The color composite signal is just like the signals found in a TV.

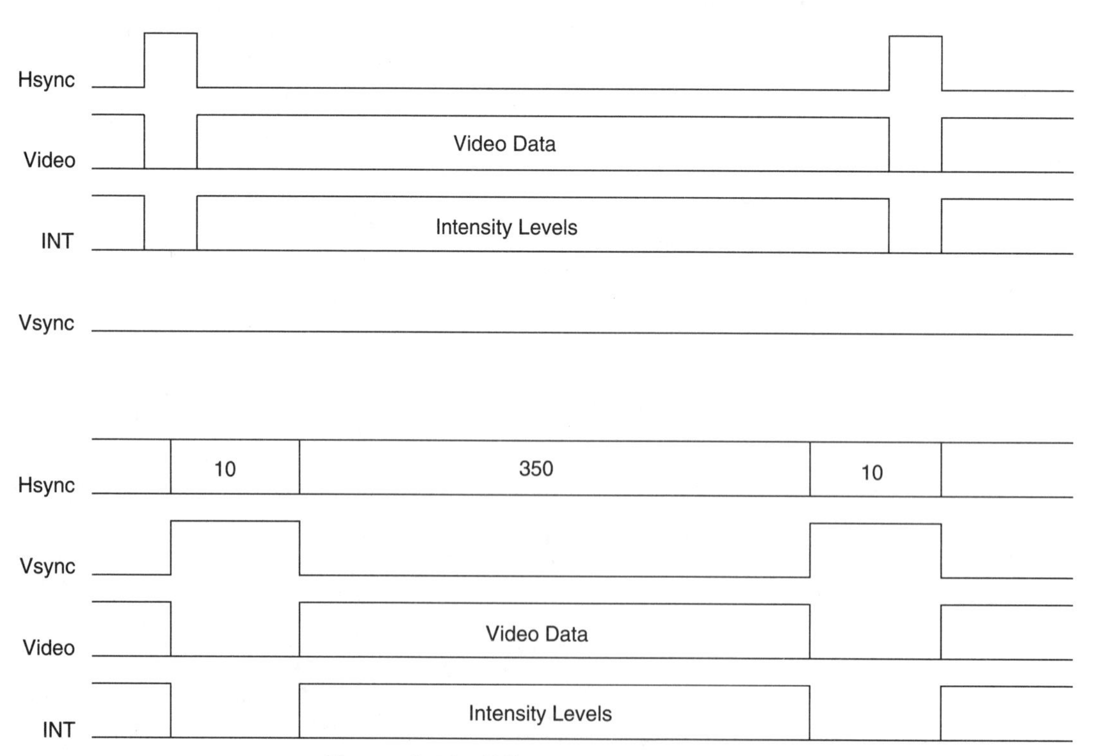

Figure 8-18 TTL monochrome adapter outputs.

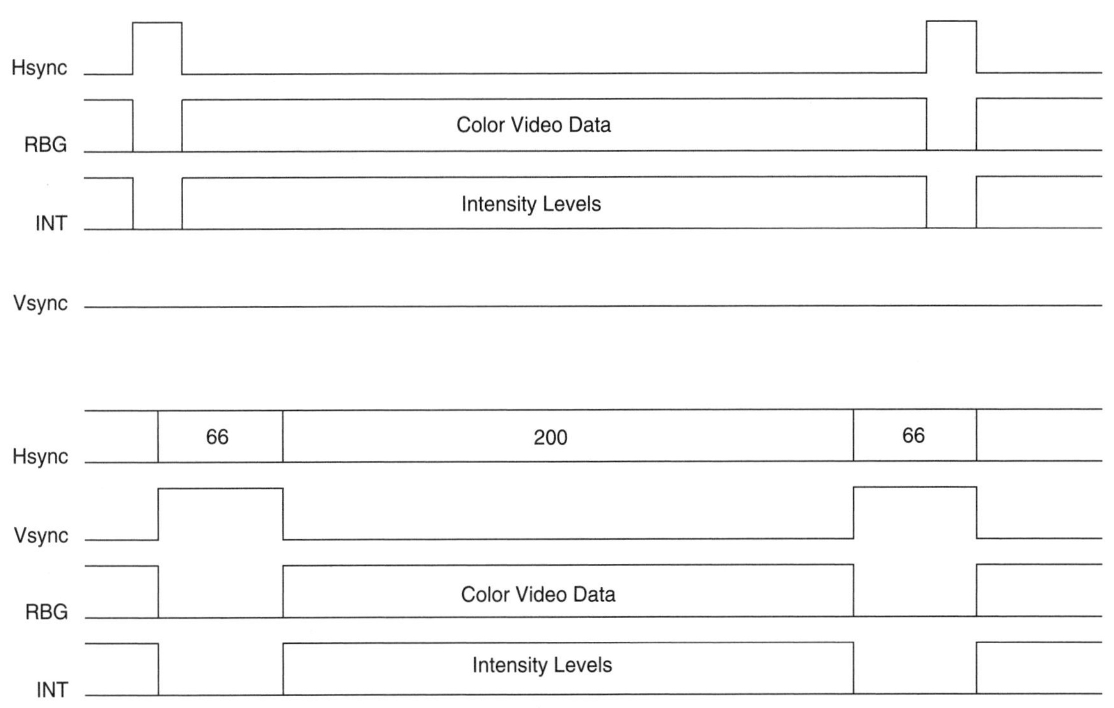

Figure 8-19 RGB color-graphics adapter outputs.

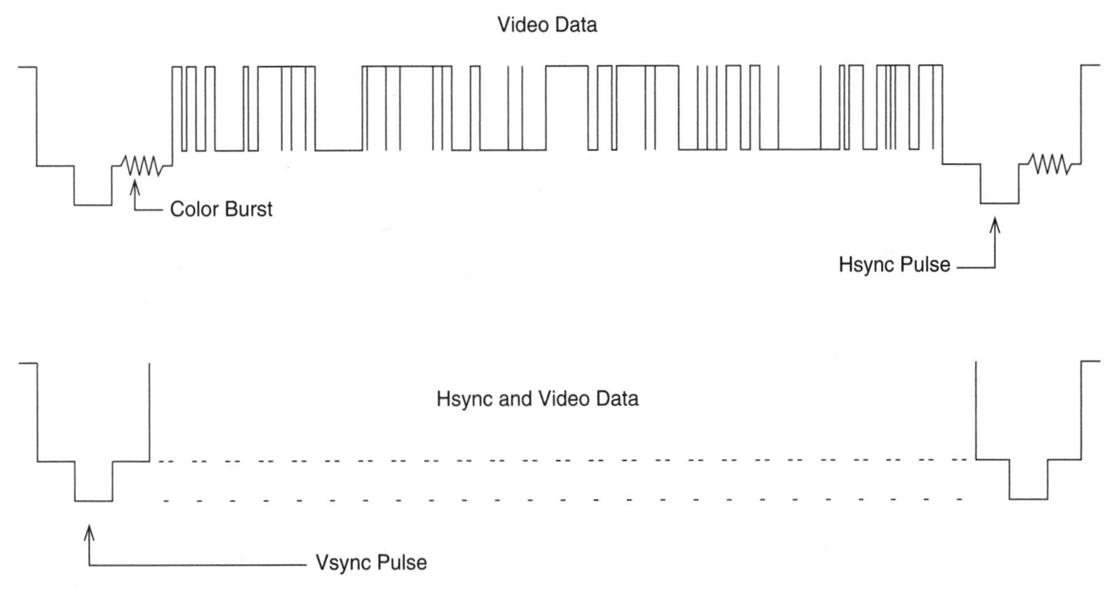

Figure 8-20 Composite color-graphics adapter output.

8.8.1 Troubleshooting the Video Adapter

No display and no raster

1. Verify that monitor has power.
2. Check connections to monitor.
3. Check equipment DIP switches on the system board; make sure they are correct for the type of adapter in the system.
4. Check outputs from the video adapter against one of the waveforms in Figures 8-18 to 8-20. If a signal is present, troubleshoot the monitor. If no signal is present, make the following checks.
   a. Check character clock frequency (1.746 MHz for MDA or 1.779 MHz for CGA).
   b. Check dot clock frequency (15.71 MHz for MDA or 14.23 MHz for CGA).
   c. Check supply to adapter and 6845.
   d. Check the Hsync pin on the 6845, for a pulse at 18 kHz on MDA or 15.75 kHz on CGA.
   e. Check the Vsync on the 6845, for a pulse at 50 Hz on MDA or 60 Hz on CGA.
   f. Check the DISPEN pin on the 6845, for a pulse at 18 kHz and 50 Hz on MDA or 15.75 kHz and 60 Hz on CGA.
   g. Check the enable pin on the 6845 for a pulse during POST.
   h. Check MA0–MA10 and RA0–RA3 (on CGA) or RA0–RA4 (on MDA) for changes. If no changes occur, the 6845 may not have been formatted.

   If any of the above checks is not right, the 6845 may be bad or not formatted.

   If the above checks are fine, check the video processing logic in the MDA or the color encoder logic in the CGA.

Every other character is missing or wrong, or a character is wrong or missing in locations on the screen that are powers of two.

1. Check video RAM address bus for a stuck address line.

8.9 VGA DISPLAY ADAPTER

The VGA video controller IC (Fig. 8-21) in the block diagram of a typical VGA video adapter operates somewhat differently than the 6845 CRTC used in the MDA and CGA video adapters. The VGA video adapter basically contains the CGA video controller, video memory, VGA BIOS, and a palette/DAC (digital to analog converter). The more advanced VGA and SVGA video adapters that contain graphic accelerators or coprocessors have additional support circuitry.

The VGA BIOS is a ROM or EPROM chip that contains the programming information to configure and modify the configuration of the VGA adapter upon boot-up. This IC contains not only the programming information for BIOS but also character set raster coding information needed by the VGA controller IC to display text information while the VGA adapter is in

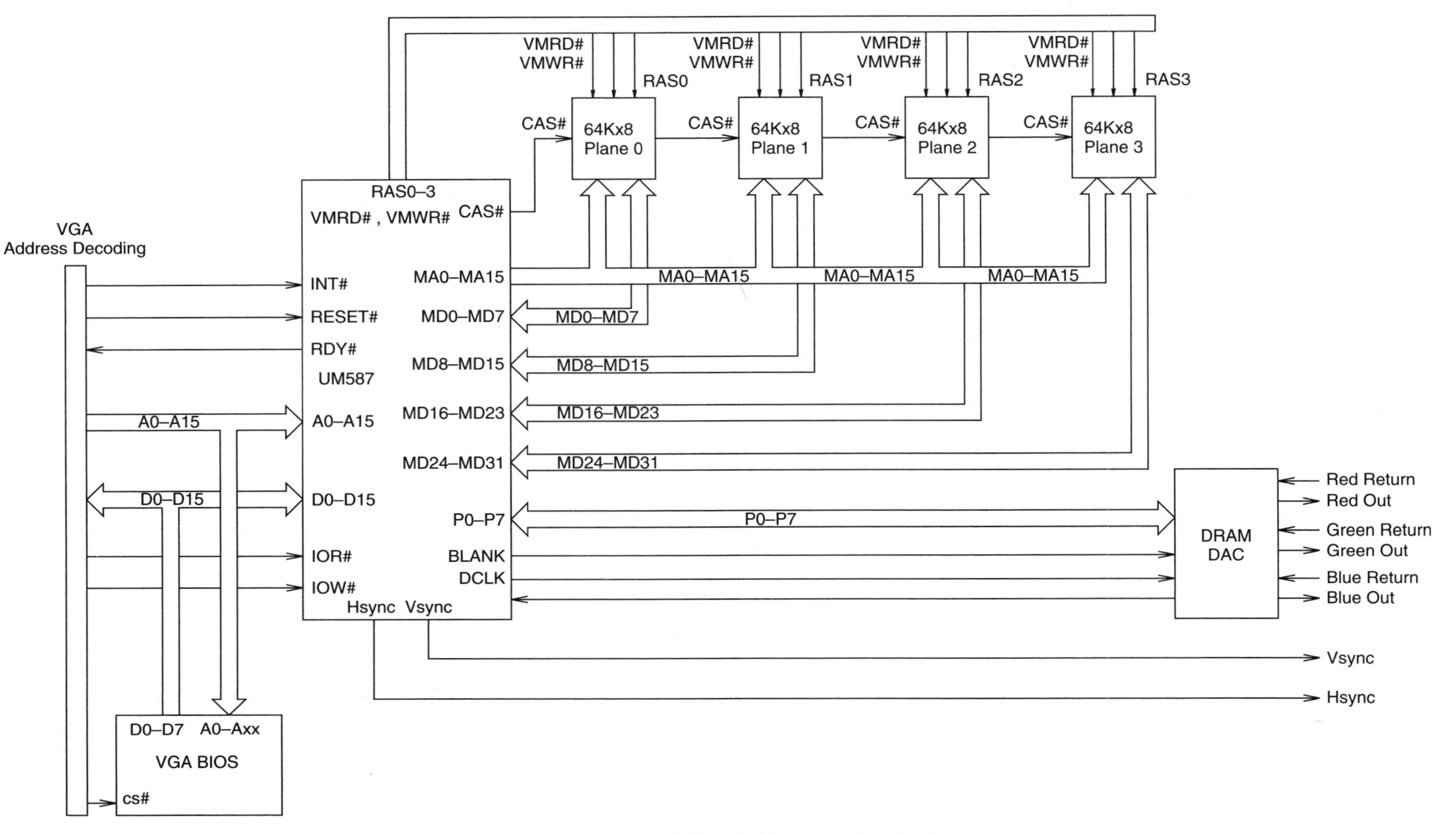

Figure 8-21 VGA adapter block diagram.

the text mode. Upon boot-up VGA BIOS is one of the first devices the computer accesses, in order to enable and test the video adapter.

The video memory is divided into four banks (also referred to as planes) of 64K × 8 DRAM. Each bank controls 8 bits of the 32-bit video memory data bus of the VGA controller. The address bus for each bank of video memory is connected in parallel to the VGA controller's video memory address bus. Connected in parallel to each bank of video memory is the CAS# (column address strobe) line, which performs the same function as the column address strobe lines in any DRAM IC. The VMRD# and VMWR# lines are connected in parallel to video memory and select with the read or write functions of the DRAM. Each bank of video memory has a separate RASx# (row address strobe) line connected to it from the VGA controller. The controller uses these lines to determine which of the banks of video memory is addressed during any video memory operation. The VGA controller has total control over the bank or banks of video memory accessed during any VGA operation.

The VGA controller (such as the Intel 82706 or UM587) has many of the same responsibilities as the 6845 plus some additional functions. As opposed to the MDA and CGA adapters, the host MPU has no direct access to the video memory of the VGA adapter; all data that goes into video memory is processed by the VGA controller. The standard interface lines (RESET, IOW#, IOR#, INTR#, D0–D15, A0–A15, RDY#) operate as they do on the system board. The SENSE input gets its input signal from the palette/DAC, which indicates where a color or monochrome video display monitor is connected to the VGA adapter.

On the video memory side of the VGA controller, the MA0 to MA15 (video memory address bus), MD0 to MD31 (video memory address bus), VMRD#, VMWR#, CAS#, and RAS0# through RAS3# lines are used to control the video memory on the VGA adapter (refer to video memory section for pin operations).

Once the data is processed into a format needed by the palette/DAC, it is sent to the palette/DAC on the P0 to P7 (pixel) lines. When the VGA adapter is in the graphics mode, the pixel information represents the color of each pixel; in the text mode the data represents dot code, color, and attribute information. The VGA controller automatically converts the data from video memory into dot codes (raster scan codes) and sends the data out serially to the palette/DAC while in the text mode. Also during this time the VGA controller sends out attribute information. The BLANK output is used to disable the output of the palette/DAC IC during retrace. The DCLK (dot clock) output is also applied to the palette/DAC IC to ensure that the DAC converter is synchronized with the VGA controller.

The VGA controller also produces the horizontal and vertical synchronization pulses for the video monitor. The frequency of these pulses varies with the VGA resolution mode.

The outputs of the VGA adapter are applied to a 15-pin, 3-row female connector. The name of each pin and its function are listed in Table 8-1, and the position of each pin is displayed in Figure 8-3. The vertical and horizontal synchronization pulses are TTL levels, while the red, green, and blue outputs are analog levels with a maximum of 1.4 volts peak to peak.

8.9.1 Troubleshooting VGA Adapters

The process of troubleshooting a VGA video adapter is the same as the MDA and CGA video adapters, with the exception that the red, green, and blue signals are analog in nature.

CHAPTER SUMMARY

1. The video adapter is the interface between the computer and video monitor. It converts the data from the computer and supplies control signals to the video monitor to display.
2. The video monitor is the display for the video system; information provided by the video adapter controls the image on the video screen.

3. The CRT (cathode ray tube) is the video screen for the video monitor. The CRT uses an electron beam and magnetic fields set up by the deflection coils to position and display the image on the video screen.
4. Horizontal scanning causes the electron beam to move from left to right and right to left on the video screen.
5. Vertical scanning causes the electron beam to move from top to bottom and bottom to top on the video screen.
6. During trace time the electron beam turns on and off to produce images on the video screen.
7. During retrace time the electron beam is returned to either the left side or the top of the screen; the electron beam is keep off during retrace.
8. MDA = monochrome display adapter
   CGA = color graphics adapter
   RGB = red, green, blue
   EGA = enhanced graphics adapter
   VGA = video graphics array
   SVGA = super video graphics array
9. Digital monitors use only TTL level signals for control of the monitor, while analog monitors use both TTL and analog level signals for control of the monitor.
10. Resolution is the amount of independent control over the areas of the video screen.
11. Dot pitch refers to the amount of space between each phosphor dot or triad on the video screen.
12. MDA monitors provide highest resolution characters on the video screen in one of two colors. MDA monitors cannot perform graphic commands.
13. CGA monitors were the first type of monitor to allow color graphics to be displayed on the video screen. Today they offer no advantages because of their very low resolution, few colors, poor character resolution, and slow speed.
14. VGA monitors today provide high-resolution graphics, good character resolution, and up to 256 colors at a low cost.
15. SVGA monitors supply higher-resolution graphics, more colors, and greater speed than a VGA display. The disadvantage of SVGA monitors is that they cost more than a VGA display.
16. The video driver increases the voltage of the video signal and supplies the signal to the video output.The video output provides the voltage and current necessary to control the CRT and supporting circuitry.
17. The vertical driver increases the voltage of the vertical synchronization signal and supplies the signal to the vertical output.The vertical output provides the voltage and current necessary to control the vertical deflection coils on the CRT.
18. The horizontal driver increases the voltage of the horizontal synchronization signal and supplies the signal to the horizontal output. The horizontal output provides the voltage and current necessary to control the horizontal deflection coils on the CRT and supplies the input signal for the flyback transformer, which is used to produce the focus and second anode voltage.
19. The vertical driver/oscillator is used to key in the vertical frequency of the oscillator to the vertical synchronization signal from the video adapter. When no vertical synchronization signal is being generated, the oscillator provides the vertical frequency for the monitor. The driver increases the voltage of the vertical synchronization signal and supplies the signal to the vertical output. The vertical output provides the voltage and current necessary to control the vertical deflection coils on the CRT.

20. The horizontal driver/oscillator is used to key in the horizontal frequency of the oscillator to the horizontal synchronization signal from the video adapter. When no horizontal synchronization signal is being generated, the oscillator provides the horizontal frequency for the monitor. The driver increases the voltage of the horizontal synchronization signal and supplies the signal to the horizontal output. The horizontal output provides the voltage and current necessary to control the horizontal deflection coils on the CRT and supplies the input signal for the flyback transformer, which is used to produce the focus and second anode voltage.
21. The flyback section contains the flyback transformer, which converts the horizontal output voltage (55 to 130 volts) into the high-voltage value (9 to 15 Kv). This section contains the high-voltage section, which uses a rectifier to convert the horizontal frequency into a pulsating DC voltage, which is then filtered to produce the second anode voltage. The high-voltage section also provides the voltage for the focus grids.
22. The color drivers increase the voltage of the color signal and supply the signal to the color output. There are three color drivers, one for each of the primary colors (red, green, blue). The color output provides the voltage and current necessary to drive the cathodes of the CRT. There are three cathodes in a color CRT; the red cathode controls the electron beam that strikes the red phosphor, and the green and blue cathodes control the electron beams that strike the green and blue phosphors on the video screen.
23. The 6845 provides screen memory addressing logic, raster scanning logic, cursor logic, screen blanking, screen scrolling, and horizontal and vertical synchronization generation logic.
24. Each character on the video screen occupies a memory location. When the 6845 accesses the screen memory location, the ASCII code for the character is applied to the character generator, along with the raster address produced by the 6845. Each character contains eight or more raster codes, which represent each scan line that makes up a character. The character generator produces the raster scan code, which is converted into a serial format to produce the dot codes necessary to turn the electron beam on and off. The 6845 processes each screen memory location at least eight times for each character location on the screen. The first raster scan codes for each character in the character row being processed before the 6845 accesses the second raster scan code for each character in the character row.
25. Character codes are 8-bit binary values that represent all available characters in the extended ASCII set used by the IBM PC. Character codes are located in character screen memory.
26. Attribute codes are 8-bit binary values used to represent how the character associated with it is displayed. There is one attribute code for each character location on the video screen. Attribute codes are found in the attribute part of screen memory.
27. The character generator is a ROM that contains all the raster scan codes for each character that can be displayed on the video screen in the text mode. The ASCII code from screen memory and the raster address provide the address of the character generator. The output of the character generator supplies the raster scan codes, which are used to produce the dot codes for the video adapter.
28. The attribute decoder is used to control how the video processor logic processes the dot code.
29. The shift register of the MDA video adapter converts the 8-bit raster scan code into nine serial dot codes, which are used to turn the electron beam on and off.
30. To determine the character that will be displayed on the screen from the extended ASCII code, use the most significant hex value to determine which column will be used. Follow the column down to the least significant hex value to determine which row is used. The character located at the selected column and row is the character that will be displayed. Note: There are two parts to the character set; part 1 is used for all hex values between 00 and 7F and part 2 for 80 to FF.

31. The alphanumeric mode displays only characters from the extended ASCII character set. Each character takes up one space on the video screen.
32. In the all-points-addressing mode, each screen character location is divided into smaller groups that the computer can control independently of each other. Each group of dots in the all-points-addressing mode is referred to as a pixel. The number of dots in each pixel depends on the resolution of the video screen. Each pixel can be in one of two or more colors. The number of colors and pixels depends on the type of video adapter used and the amount of video memory on the video adapter.
33. The UM587 is the VGA processor of the video adapter; this device is responsible for controlling all signals and timing for the video adapter. The VGA processor controls video memory addressing and video memory data translation and produces blanking, clocking, and horizontal and vertical synchronization signals.
34. VGA BIOS is a ROM or EPROM used to provide the boot-up configuration of the VGA processor. This device also provides programming information that is used to change the configuration of the VGA processor when switching between different VGA modes.
35. There are four banks (planes) of video memory on the VGA adapter, and each bank provides 64K of memory. The VGA memory bus is connected to each plane of video memory. Plane 0 of memory supplies 8 bits of data while planes 1 to 3 supply 8 additional bits of memory for the VGA processor. The 32-bit data bus for the VGA processor allows the VGA adapter to use slower DRAM memory by stacking the data on the data bus until the VGA processor can process the data.
36. The DRAM DAC (digital to analog converter) is used to convert the TTL levels from the 8-bit palette bus into an analog signal on the red, green, and blue outputs of the DAC. With an 8-bit palette bus, the DAC can produce 256 different voltage levels among the three color outputs.

REVIEW QUESTIONS

1. Which part of the video system contains the CRT?
   a. Video adapter
   b. Video monitor
2. Which part of the video system contains the flyback transformer?
   a. Video adapter
   b. Video monitor
3. Which part of the video system contains the video memory?
   a. Video adapter
   b. Video monitor
4. Which part of the CRT narrows the electron beam as it moves toward the video screen?
   a. Cathode
   b. 2nd anode
   c. Control grid
   d. Focus grids
5. The horizontal synchronization signal causes the electron beam to move from the top to the bottom of the video screen and back to the top.
   a. True
   b. False
6. The electron beam is turned ________ during vertical retrace.
   a. On
   b. Off

7. The electron beam is enabled during the horizontal synchronization pulse.
  a. True
  b. False

8. Which video adapter produces the highest resolution characters on the video screen?
  a. MDA
  b. CGA
  c. EGA
  d. VGA

9. Which video system has the lowest graphics resolution in its high-resolution mode?
  a. MDA
  b. CGA
  c. EGA
  d. VGA

10. Which video system uses analog signals?
  a. MDA
  b. CGA
  c. EGA
  d. VGA

11. The horizontal and vertical synchronization signals are TTL regardless of the type of video system.
  a. True
  b. False

12. The higher the resolution, the ________ control you have over the video screen.
  a. More
  b. Less

13. A dot pitch of 0.39 mm produces a better-looking image than a dot pitch of 0.26 mm.
  a. True
  b. False

14. Which part of a video monitor drives the vertical deflection coil?
  a. Video output
  b. Vertical output
  c. Horizontal output

15. If no signal is applied to the flyback transformer, what happens to the video screen?
  a. A single vertical line appears in the center of the screen.
  b. A single horizontal line appears in the center of the screen.
  c. No raster appears.
  d. None of the given answers

16. Which one of the following functions is the 6845 not responsible for?
  a. Horizontal synchronization pulse
  b. Cursor size and type
  c. Raster scan code
  d. Display blanking
  e. None of the given answers

17. Which part of the MDA video adapter produces the raster scan code?
  a. 6845 CRTC
  b. Attribute memory
  c. Shift register
  d. Character generator
  e. None of the given answers

18. Using the extended ASCII code chart, determine which character is produced with an ASCII code of 41(hex).
    a. "a"
    b. "A"
19. Which video mode must the video adapter be in to perform graphic commands?
    a. Alphanumeric mode
    b. All-points-address mode
20. Which part of the VGA adapter is responsible for converting TTL levels into analog levels?
    a. VGA controller
    b. VGA video memory banks
    c. DRAM DAC
    d. Address decoding
    e. None of the given answers

chapter 9

Mass Storage Systems

OBJECTIVES

Upon completion of this chapter, you should be able to perform the following tasks:

1. List and define the purpose of the four parts of any mass storage system.
2. Specify which of the following mass storage systems use magnetic or optical storage methods: floppy disk drive, hard drive, floptical, CD drive, and tape back-up.
3. List the differences between FM and MFM encoding.
4. List and define the functions of tracks and logical sectors.
5. Define the purpose of the sector ID field.
6. List the functions of the 765A floppy disk controller.
7. List the functions of each pin of the 765A floppy disk controller.
8. List and define the command sequence of the 765A.
9. Define the basic purpose of each command for the 765A.
10. List and define the purpose of each block in the block diagram of the floppy disk controller.
11. List the duties of the disk drive unit.
12. List and define the purpose of each block in the block diagram of the floppy disk drive unit.
13. Define how DOS uses the diskette to control the data on the diskette.
14. List the differences between double-density and high-density floppy disk systems.
15. List the differences between floppy and hard disk systems.
16. List the steps necessary to prepare a disk system for data storage.
17. List the advantages, disadvantages, and differences between the different types of hard disk systems.

9.1 OVERVIEW

Mass storage systems fall into five basic types, each with its own advantages and disadvantages. Mass storage systems are normally made up of four different items—controller adapter, drive unit, storage medium, and some type of controlling software.

The controller adapter converts data formats between the computer and drive unit. The adapter converts commands from the computer into the signals necessary to control the drive unit and uses feedback information provided by sensors on the drive unit to verify the operation of the commands issued to the drive unit.

The drive unit contains the motors, sensors, and electronic interface circuitry that converts TTL levels into levels needed by the R/W heads. Most drive units contain two motors. One motor rotates or moves the storage medium in order to place the data at different locations on it. Some drive units contain only one motor, which is used to physically move the storage medium while the R/W heads remain fixed. In drive units that contain two motors, the second motor is used to change the position of the R/W heads while the first motor moves the storage medium. Drive units that contain fixed R/W heads access data sequentially, which allows very high-density storage (large amount of data in a very small area) but slow access time. Drive units that contain movable R/W heads access data in random fashion, which allows high-density storage but rapid access. In a mass storage system that uses magnetic fields, the R/W heads are made up of a set of small coils. The R/W heads either create or detect a pattern of magnetic fields on the surface of the storage medium with a particular alignment pattern, which represents the data stored in that area of the medium. The head creates the magnetic fields during a write operation and detects them during a read operation. In optical mass storage systems, the R/W head is made up of a laser beam and light sensor. As the light from the laser shines on the reflective surface of the storage medium, the light sensor either detects or does not detect the reflected laser light. If the reflective surface is not pitted the light is detected, but if the surface is pitted the light is deflected and the light sensor detects no light from the laser.

A mass storage system may use magnetic fields, optical scanning, or a combination of both as the method of storing data on the surface of the medium. The medium may be a disk-shaped object (flexible or nonflexible) or a tape placed in a cartridge. The storage medium is the part of the mass storage system where the data is stored. In a mass storage system that uses magnetic fields, the surface of the storage medium has a magnetic coating in order to store the data in the form of small, specially aligned magnetic fields. Optical mass storage systems use a medium made of a reflective material sandwiched between two clear protective plastic coatings. The data is stored by distorting (pitting) or not distorting the surface of the reflective material. Pitted areas of the storage medium do not reflect the light from the laser, while areas not pitted do reflect the light. The floptical storage medium uses an optical track to position the R/W heads, but uses the magnetically coated area to actually store the data.

The software part of the mass storage system programs the computer to control the hardware. The only standardized mass storage system today is the disk (floppy or hard) system; therefore, the software used to control these mass storage systems is called DOS (disk operating system). The primary purpose of DOS is to provide easy control of the disk (floppy or hard) drive system; DOS also provides control over most other standard hardware in the computer system. Other forms of mass storage such as optical drives, flopticals, and tape back-up systems require a program called a driver, which allows DOS to access the hardware of the mass storage system. If the correct driver is not used, the results of the operation may be incorrect.

9.1.1 Floppy Drives

The floppy disk system was the first reliable form of mass storage for the IBM PC. The first floppy disk system replaced cassette tape storage, which was very unreliable and slow. Basically all computer systems today contain at least one floppy drive unit, with the 3.5-inch microfloppy being the most popular. Over the years the cost of the floppy drive system has decreased, while the reliability, storage capabilities, and speed of the system have increased. Today floppy drive systems are the most popular and common way of loading and transferring software between computer systems. Figure 9-1 shows a typical 5.25-inch half height and 3.5-inch floppy disk drive unit.

9.1.2 Hard Drives

Today most computer systems use one or more hard drives to store large amounts of data and software. Over the years the cost of hard drive systems has decreased from $10 per megabyte to $ .50 per megabyte, while increasing reliability by 1000%, storage capacity by over 100 times, and speed by more than 20 times. As opposed to the floppy system, the interface adapter and the format of the data stored on the hard drive unit differ from system to system. Even through actual methods of storing the data on the hard drive system differ, DOS accesses the interface adapters in the same fashion. Figure 9-2 shows an older 5.25-inch MFM (modified frequency modulation) (20 to 80MB) hard drive and a newer 3.5-inch IDE (integrated drive electronics) (100MB to 1.2GB) hard drive.

9.1.3 Magnetic/Optical Drives

The floptical drive is one of the newer types of mass storage systems. A floptical system uses both optical and magnetic methods to store data. It can store up to 21MB of data on one 3.5-inch floptical diskette, which is about 15 times more data than a standard high-density diskette holds. The storage capability of the floptical system comes from using an optical track used to position the R/W heads, while data is stored in the form of magnetic fields. A floptical drive looks just like a standard 3.5-inch floppy disk drive unit. The only visible difference is where

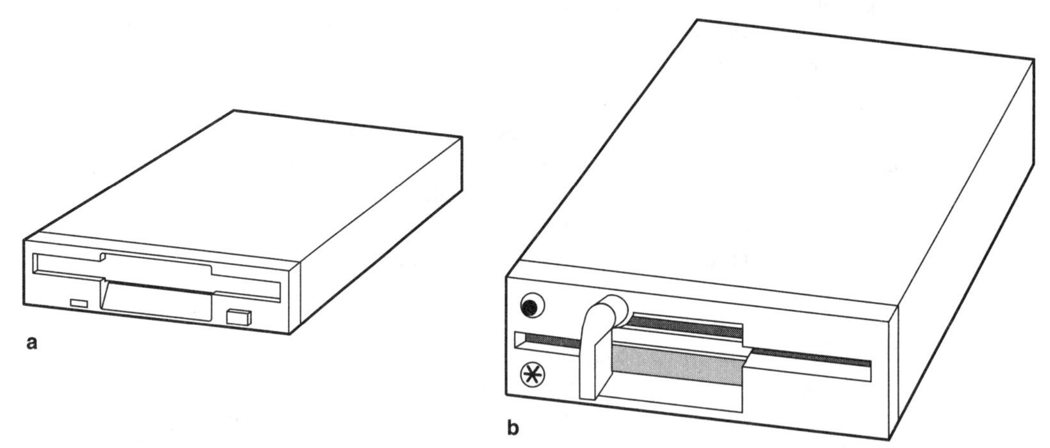

Figure 9-1 **(a)** 3 1/2-inch and **(b)** 5 1/4-inch floppy disk drive units.

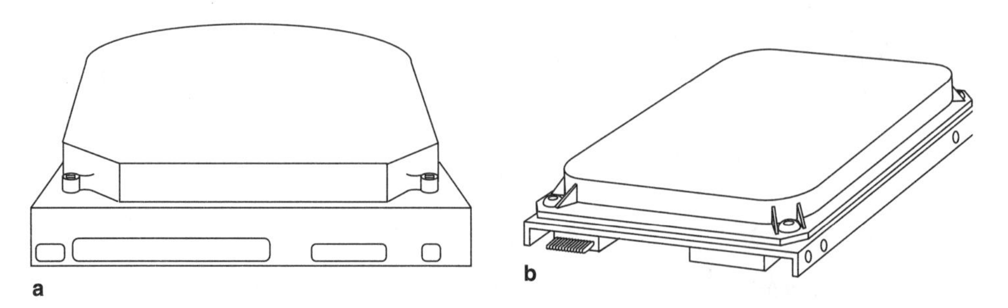

Figure 9-2 **(a)** 5 1/4-inch MFM and **(b)** 3 1/2-inch IDE hard drive units.

the cable connects to the drive unit: Floptical drives use an SCSI (small computer system interface), which is commonly used with CD drives and is a 50-pin ribbon cable.

9.1.4 Optical Drives

Optical drives allow the user to store very large amounts of data (100MB to 500MB) on a special optical diskette that is normally between 3.5 and 5.25 inches in size. Optical drives use a laser beam and light sensor to read and write data from and to the optical diskette. Because the data is not stored magnetically, it cannot be erased by magnetic fields and is therefore less proned to the problems of magnetic storage over long periods of time. The problems with optical drives are their slow speed (slower than a hard drive) and very high cost per megabyte of storage. An optical drive looks much like a standard 3.5-inch floppy disk drive unit but uses the SCSI.

Another type of optical drive is called the CD (compact disc) ROM. CD drives normally do not allow the user to write data to the disc itself; instead the user purchases a CD which contains 500MB or more of data that can either be accessed from the CD during the execution of the program or be installed on the hard drive system. Although there are master CD units that are used to write the data to the CD, these systems are not commonly available to the individual user. If a mistake is made while writing to the CD, corrections cannot be made and the CD may not be usable. CDs allow software companies to protect their software because copying a CD is not very cost effective. A standard 5.25-inch CD costs only about $3.00, which is equal to the cost of about 10 high-density diskettes (which weigh more and take up more shipping space), and a CD can contain about 500 times more data than a high-density diskette. For these reasons it is less expensive to sell large software programs on CD than on high-density diskettes. Most CD drives (Fig. 9-3) use a SCSI, either from a SCSI adapter or from a sound card, but some of the latest CD drives use the IDE interface common to hard drive systems.

9.1.5 Tape Drives

Tape drive (Fig. 9-4) systems (also called tape back-ups) are used primarily to archive large amounts of data in case a problem occurs on the hard drive system. Tape drive systems, although very slow, allow very large amounts of data to be stored on a small cassette cartridge in one simple package. Tape drive systems are normally used to back up data stored on a hard drive. Tape drives are very slow because the data is stored on the drive asynchronously. Most

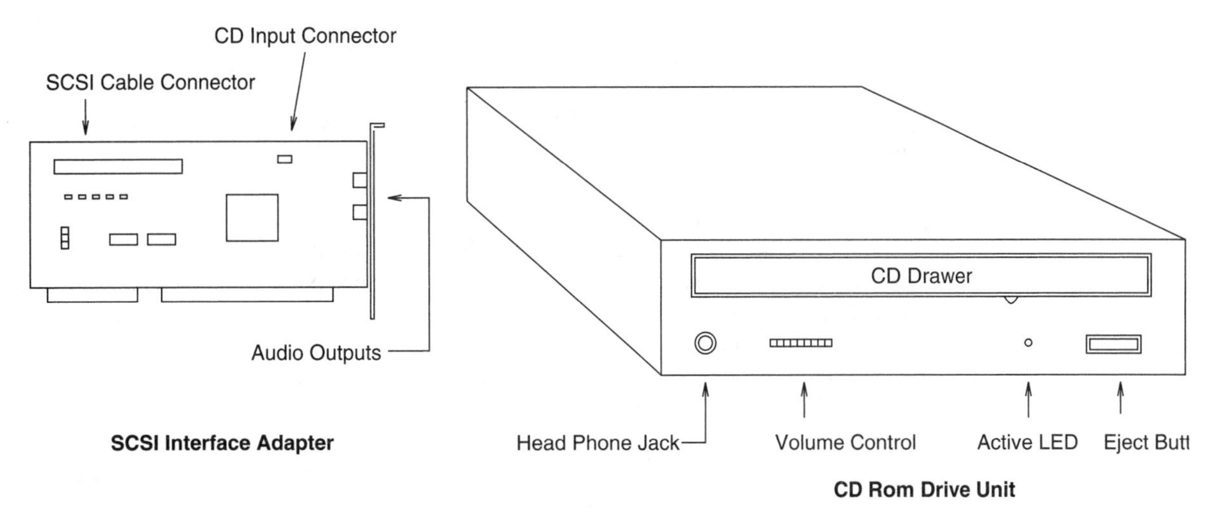

Figure 9-3 CD system.

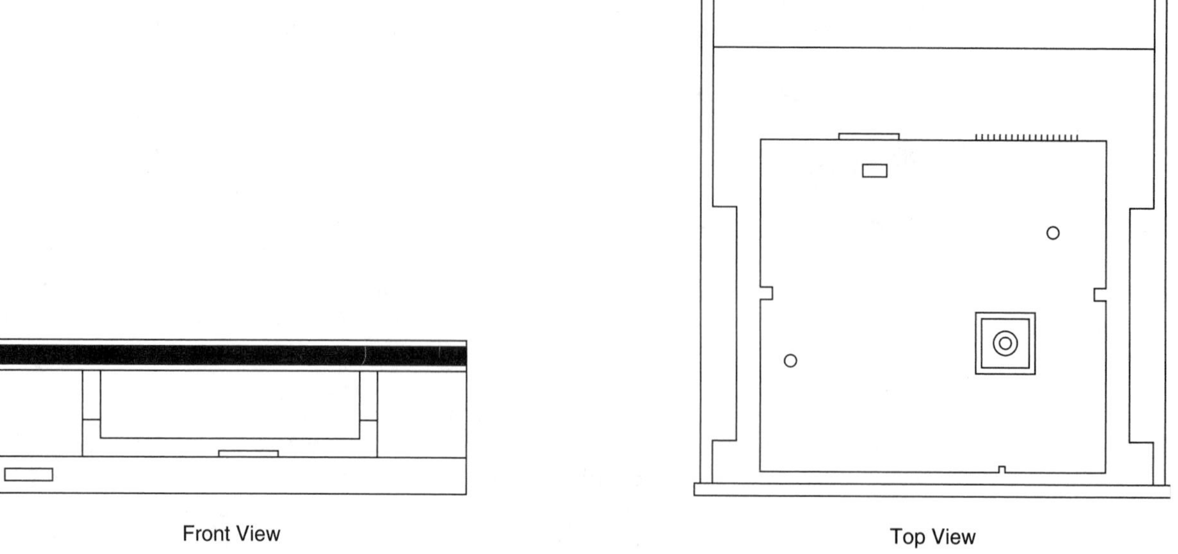

Figure 9-4 Tape back-up system.

tape drives connect to the floppy disk adapter for standard operations, but if a high speed is required a special tape interface adapter is used instead of the floppy disk adapter.

9.2 FLOPPY DISK SYSTEMS

Because all computers today have at least one floppy disk drive, and the floppy disk system is the least complex system, it is the first system discussed in this chapter and is covered in greatest detail.

9.2.1 Disk Recording Techniques

The IBM PC uses an IBM System 34 format called MFM (modified frequency modulation) to store data on a double-density diskette. The older PCs used an IBM 3740 FM (frequency modulation) single-density format.

When storing data using FM, the adapter produces a 2-μs clock pulse with a 50% duty cycle, followed by a 2-μs time period for a data pulse signal (Fig. 9-5); therefore, it takes 4 μs to store one bit of data. When writing a logic one, the FM adapter produces a 2-μs clock pulse with a 50% duty cycle followed by a 2-μs data pulse with a 50% duty cycle. When writing a logic- 0, the FM adapter produces a 2-μs clock pulse with a 50% duty cycle followed by 2 μs of no pulses. Therefore, if one pulse is detected in a 4-μs time period, a logic-0 bit is indicated; if two pulses are detected in a 4-μs time period, a logic-1 bit is indicated.

During a read operation, if the disk controller circuitry sees two pulses in a 4-μs time period, the data is seen as a logic-1. If only one pulse is seen (just the clock pulse) during a 4-μs time period, the data is seen as a logic-0. On the adapter the data is applied to the data separator, which is a PLL (phase lock loop) that separates the clock pulses from the data and sends the data to the floppy disk control IC. The PLL maintains proper tracking of the clock and data pulses even with variations in rotation speed and flux density.

In MFM (double-density recording), the clock signal is incorporated with the data pulse. Therefore, if a logic-1 is stored on the disk, one clock-data pulse is seen every 2 μs. If a logic-0 is stored on the disk, no pulse is seen during the 2-μs time period (Fig. 9-5).

When writing a logic-1 to the disk, the write circuit sends a 0.25-μs pulse to the disk, 1 μs after the new bit time has started. If a logic-0 is written to the disk, no pulse is seen during the 2-μs time period.

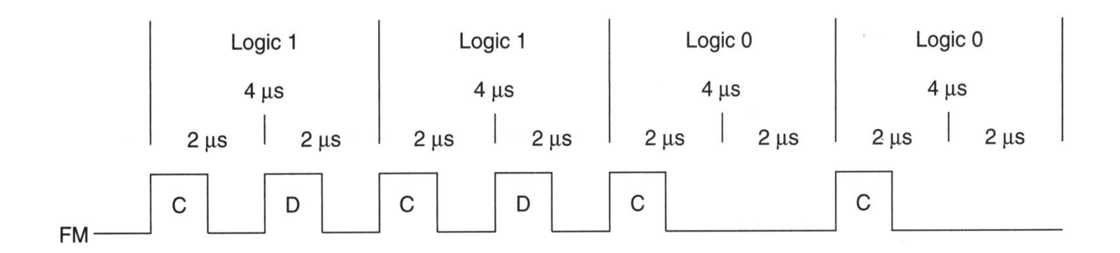

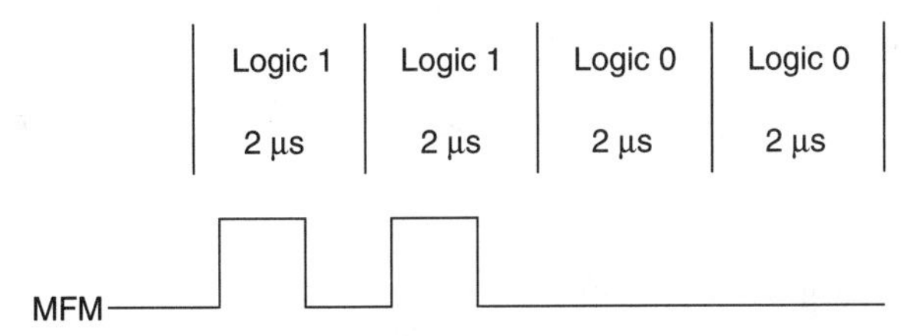

Figure 9-5 FM and MFM encoding.

When reading the MFM data, the PLL is used to maintain a lock on the data being read. If no pulse is seen during a full bit time period, a logic-0 is indicated. If a pulse occurs during a 2-µs time period, a logic-1 is indicated.

Therefore, because each bit time does not have a clock to maintain synchronization, all factors in the operation of the diskette must be closely regulated in order to maintain data accuracy. The PLL is used to maintain the lock frequency of the input combined clock and data signals. The data separator separates the clock signals from the data.

9.2.2 Tracks and Logical Sectors

Each side of a diskette is divided into a number of storage areas called records, which are positioned on a track and sector. The number of sectors and tracks depends on the type of diskette and its formatting (Table 9-1).

A track is an area on the diskette where the R/W heads can be positioned. In the PC the number of tracks (head positions) is either 40 or 80, depending on the formatted size of the diskette. The track at the outermost edge of the diskette is called track 0 and the track at the innermost edge is either track 39 or 79, depending on the formatted size.

As the diskette rotates, it is divided into timing rotation divisions; each division is called a sector. Because the diskette rotates at 300 rpm, the diskette makes one complete rotation every 200 ms. The floppy disk adapter uses timing to determine which sector the R/W heads are currently located on. The timing varies with the formatted size of the diskette.

Track or Logical Sector Allocation. Each track contains three packets of information—sector information, sector data, and track information. This information is placed on the diskette during the formatting process. The number of packages of sector information and data varies with the format size of the diskette. See Table 9-1 for specifications and Figure 9-6 for the track and logical sector layouts.

Each track has an *index mark gap* (gap 5), which contains 16 bytes of information (FF in hexadecimal). Also found on each track is one index address mark. This contains one byte of information (FC). Track information requires 17 bytes of storage space on the diskette, leaving 6233 bytes of storage space on the track. The track information allows the disk controller circuitry time to synchronize itself with the write precompensation circuitry only when formatting the diskette.

There is one packet of sector information for each sector formatted on the diskette. The data found on each track and in each sector is often called a record.

TABLE 9-1
Floppy Disk Type Specifications

Description	*360K*	*1.2MB*	*720K*	*1.44MB*	*2.88MB*
Form factor	5.25-inch	5.25-inch	3.5-inch	3.5-inch	3.5-inch
Density	Double	High	Double	High	EHD
Recording method	MFM	MFM	MFM	MFM	MFM
Number of sides	2	2	2	2	2
Bytes/sector	512	512	512	512	512
Number of tracks	40	80	80	80	80
Number of sectors/track	9	15	9	18	36
Number of records/disk	720	2,400	1,440	2,880	5,760
Formatted size	368,640	1,228,800	737,280	1,474,560	2,949,120
Storage size	362,496	1,222,656	731,136	1,468,416	2,936,832
Type of PC	PC/XT/AT	AT	PC/XT/AT	AT	AT

The *post index gap* (gap 1) contains 17 bytes of information (11 bytes of FF and 6 bytes of 00). It is used to synchronize the diskette with the data separator for the reception of the serial data bits. This data is placed on the diskette during formatting.

The *sector ID field* contains 7 bytes of information. The information in the sector ID field is written into the field when the diskette is formatted and does not change unless the diskette is reformatted.

1. Byte 1 is the *ID address mark* (data FE).
2. Byte 2 is the *track address* (data 00–28 hexadecimal).
3. Byte 3 is the *head address.* It states which side of the disk the R/W head is using (0/1).
4. Byte 4 is the *sector address* (00–08).
5. Byte 5 is *sector length* information that states the size of the data field in the sector (10 = 512 bytes of data).
6. Bytes 6 and 7 are the *CRC* (Cyclic Redundancy Check) error detection code.

The *post ID field gap* (gap 2) contains 14 bytes (10 bytes = FF and 4 bytes = 00). In a read operation, this data is used to synchronize the data separator for the data field. During a write operation, the first 10 bytes are used to turn on the write circuitry, and the last 4 bytes are rewritten in order to synchronize the write circuitry. This information is written to the diskette during formatting, and the last 4 bytes are rewritten during each write operation.

The *sector data field* in the IBM PC contains three packets of information. This field contains a total of 515 bytes of data.

The *data address mark* contains one byte (FB hexadecimal) that indicates to the floppy disk controller that data will follow. This address mark is placed on the diskette during formatting.

The *data stream* contains 512 bytes. This is the actual data stored in the sector.

The *CRC code* represents the sum of all logical 1s in the data stream. The two LSBs of the sum are inverted and stored following the data stream. When the data is read, the floppy disk controller takes the sum of all the data in the data stream plus one and creates a 2-byte CRC code. The new CRC code created by the FDC is added to the CRC code read from the disk. If the sum of two LSBs equals zero, no error has occurred and the data is good. If the sum is not equal to zero, an error has occurred and the data is bad.

The *post data field gap* (gap 3) contains 16 bytes (FF hexadecimal) and is used to separate the old data field from the next sector ID field.

The total number of data bytes in the sector area is 559 bytes.

The very last part of the track contains the *final gap* (gap 4). The size of this gap varies according to the number of sectors formatted on the track. This gap contains the bytes left on

Logical Sector Layout

Post Index Gap (gap 1) 11 bytes FF 6 bytes 00	Sector ID Address Mark (FE)	Track Address	Head Address	Sector Address	Sector Length	ID CRC 2 bytes	Post ID Field Gap (gap 2) 10 bytes FF 4 bytes 00	Data Address Mark (FB)	Data Stream 512 bytes	Data CRC 2 bytes	Post Data Field Gap (gap 3) 16 bytes FF

Track Layout

Track Mark Gap (gap 5) 16 bytes FF	Index Mark 1 byte (FC)	Logical Sector 1 559 bytes	Logical Sector 2 559 bytes	Logical Sector 3 559 bytes	Logical Sector 4 559 bytes	Logical Sector 5 559 bytes	Logical Sector 6 559 bytes	Logical Sector 7 559 bytes	Logical Sector 8 559 bytes	Logical Sector 9 559 bytes	Final Gap (gap 4) 1202 bytes FF

Figure 9-6 Standard floppy diskette track format.

the track until the index mark gap (gap 5) is reached. Gap 4 on the IBM PC contains 1202 bytes of FF, which takes up the remaining space on the track.

Track Layout. The number of tracks and sectors per track depends on the formatted size of the diskette. Figure 9-6 shows a typical 360K 5.25-inch track layout.

9.3 FLOPPY DISK ADAPTER

We begin with the floppy disk adapter found in the IBM PC/XT computer system. Although this particular adapter does not support high-density diskettes, it forms the basis for any type of floppy disk adapter and even the hard disk adapter.

9.3.1 Disk Controller Adapter

The IBM PC floppy disk controller adapter (Fig. 9-7) uses the NEC FD765A (floppy disk controller) as the main controlling element of the disk system. The NEC FD765A conforms to the IBM 3740 single-density format and the IBM System 34 double-density format. In many of the compatible PCs other floppy disk controllers are used, such as the Standard Microsystems Corporation FDC 9266 and the Intel 8272 FDC, which is hardware and software compatible with the NEC 765AC.

Features of the FDC 765A

Controls up to 4 disk drive units
Accepts data transfers from MPU or DMA
8-MHz clock, single phase
40 pin LSI DIP
+5 V DC supply
Single- or double-density format
Multisector-track transfer
Programmable record lengths: 128, 256, 512, 1024
Compatible with most microprocesssors

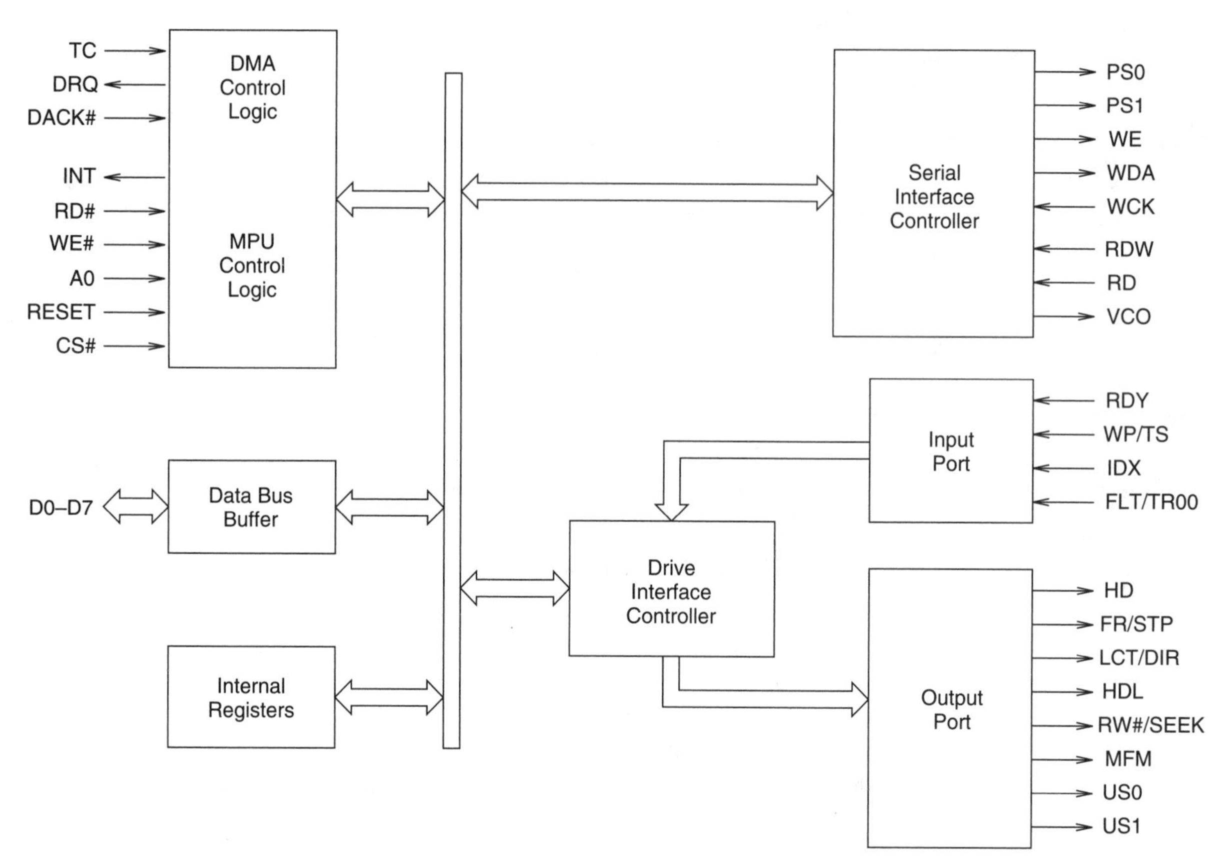

Figure 9-7 NEC FD765A block diagram.

FDC 765AC Pin Configuration

MPU Interface Bus

RESET	The RESET active high input is used to reset all FDD (Floppy Disk Drive unit) output lines of the FDC to a logic-low, while not affecting the last command. This condition is known as the idle state in the FDC.
RD#	The ReaD data active low input is used to cause the FDC to perform a read operation where data is transferred from the FDC to the data bus (see Fig. 9-7)
WR#	The WRite data active low input is used to make the FDC perform a write operation. Data is transferred from the data bus to the FDC.
CS#	The Chip Select active low input is used by the MPU to select or activate the FDC during the programming or status reading of the FDC. The FDC is selected during a read operation (to check the status of the FDC) or during a write operation (to program one of the FDC registers).
A0	The Address line 0 input is used by the MPU to determine whether the data or main status registers will be accessed by the read or write operation.
INT	The INTerrupt active high output, when high, is used by the FDC to indicate to the MPU (via the PIC) that an interrupt has occurred in the FDC.

DMA Interface Bus

DRQ — The DMA ReQuest active high output indicates to the DMA that the FDC needs to be serviced if the FDC is using a DMA for data transfers. The DMA is only used when data is being transferred to or from the FDC during the execution phase of the command. This means that this input is used only when data is being sent to or received from the disk. During a read operation, this line informs the DMA that new data is ready to be transferred to memory. During a write operation, this line informs the DMA that new data may be transferred to the FDC for storage on the disk. In the IBM PC, this line is connected to the DRQ2 line of the DMA.

DACK# — The DMA ACKnowledge active low input is used by the DMA to inform the FDC that the DMA is performing a DMA data transfer cycle. Before this line goes low, the DMA increments its current address register and decrements its current word count register.

TC — The Terminal Count active high input is used by the DMA to inform the FDC that the DMA cycle has ended and no more data is to be transferred on the data bus.

MPU/DMA Interface Bus

DB0–DB7 — The 8 Data Bus bidirectional lines are used by either the MPU or DMA to transfer data to and from the FDC. During a read operation, data is transferred from the FDC to the data bus. During a write operation, data is transferred from the data bus to the FDC.

Supply Bus

Vcc — The Vcc line provides the necessary voltage potential for the FDC.

GND — The GrouND line provides the necessary current and reference for the FDC.

CLK — The CLocK input line supplies the necessary clock signals for the FDC. In the IBM PC, the input clock frequency is 4 MHz.

FDD (Floppy Disk Drive Unit) Interface

IDX — The InDeX input line, when high, informs the FDC that the index circuitry on the floppy disk drive unit has detected the index hole. The index hole indicates that the R/W heads are on the first sector of the diskette.

WCK — The Write ClocK input supplies the proper clock frequency during a write operation. The time period is 4 μs with a 25% duty cycle.

RDW — The Read Data Window input is used by the PLL to supply a data sample to the FDC. This signal causes the FDC to change when the data bit from the data separator is sampled. This input allows the data bit to be sampled in the middle of bit cell, even if there are variations in the speed at which the bit is read.

RD — The ReaD data input receives the clock and data bits from the data separator. Internally, the FDC sends the correct bits to their proper locations.

VCO	The Voltage Controlled Oscillator output enables or disables the VCO in the PLL during different types of FDC operations. The VCO is disabled during a write operation, which turns off the data separator.
WE	The Write Enable output enables the write circuitry on the floppy disk drive unit during a write operation.
MFM	The density select output supplies density information to the PLL. Because the IBM PC uses only the MFM double density format, this output is not connected in the circuit.
HD	The HeaD select output informs the floppy disk drive unit which R/W head should be used to perform the specified operation.
US0–US1	These two Unit Select output lines are used to select which one of the four disk drive units is to be enabled during a specified operation. In the standard IBM PC disk controller adapter, these lines are not used and not connected in the circuit. Instead, the IBM PC uses a separate control port to determine which one of the drives is used in the specified operation.
WDA	The Write DAta output sends the clock and data bits from the FDC to the write precompensation circuitry on the disk controller adapter.
PS0–PS1	The Precompensation Select output lines are used to select the type of delay the R/W head uses in storing the data and clock bits on the diskette during a write operation. 00 = 0 ns, 0 1 = 125 ns, 10 = 250 ns, and 11 = 375 ns. These outputs compensate for differences in the R/W heads in different disk drive units.
FLT/TR0	The FauLT/TRack 0 input line informs the FDC that the R/W head is on track 0 during a recalibrate command. This line also indicates a fault if the R/W head is on track 0 when it should be on a different track.
WP/TS	The Write Protect/Two Side input line informs the FDC that the diskette is write protected. The second function is not used in the IBM PC.
RDY	The ReaDY input informs the FDC that the disk drive unit is ready to send or receive data.
HDL	The HeaD Load output is used to inform the disk drive unit to engage the R/W heads near the surface of the diskette, so that a read or write operation can be performed. This output is not used in the IBM PC because, when the door latch to the drive unit is closed, the R/W heads are loaded automatically. This output is normally used in 8-inch drive units.
FR/STP	The Fault Reset/STeP output line sends a step pulse to the disk drive unit. The step pulse causes the R/W heads to move toward or away from the center of the diskette by one track per pulse.
LCT/DIR	The Low CurrenT/DIRection output line has two functions. When the FDD is in the seek mode, the output indicates which direction the heads should move when a step pulse is received. When the FDD is in the R/W mode, this line lowers the current to the R/W head during operations on the inner tracks of the diskette.
RW#/SEEK	The Read Write/SEEK output line determines whether the disk drive unit is in the R/W mode when low or in the SEEK mode (head movement) when high.

FDC 765AC Block Functions

DATA BUS BUFFER The data bus buffer is an 8-bit bidirectional parallel-in parallel-out buffer that temporarily holds data to be transferred to and from the internal data bus of the FDC.

REGISTER There are two directly accessible registers in the block diagram, the main status register and the data register. These allow the FDC to be programmed and provide status information on the commands performed.

The main status register is read only. It supplies FDC status information about the overall status of the FDC and all its disk drives. This register can be read at any time during any phase of a command.

Main Status Register (CS#, A0, RD# = 0, WR# = 1)

- Bit 0 Disk drive 0 is busy (R/W or seek mode).
- Bit 1 Disk drive 1 is busy (R/W or seek mode).
- Bit 2 Disk drive 2 is busy (R/W or seek mode).
- Bit 3 Disk drive 3 is busy (R/W or seek mode).
- Bit 4 FDC R/W operation is in progress.
- Bit 5 The non-DMA mode is used to give the status of a DMA operation. If high, the FDC does not use the DMA lines. This bit sets during the execution phase of a command, when not in the DMA mode, and resets at the end of the execution phase. This is not used in the IBM PC because the IBM uses the DMA.
- Bit 6 The data I/O bit indicates the direction of data flow. When high, data flows out of the data register; when low, data flows into the data register.
- Bit 7 The request for master bit indicates that the data register is ready to receive or transmit data from or to the outside world, through the data bus.

The second accessible register is an 8-bit data register, which is made up of three sections of registers, the command register set, parameter register set, and drive status set. To perform an operation, all register sets in the data register must be accessed in a certain order, specified by the command.

The command register section is a 16-bit write-only register. It requires two bytes of data to load the command. It is the first of the registers to be loaded for an operation. This register can be accessed only during the command phase.

The parameter register section contains nine 8-bit registers that may or may not be needed for the selected command. These registers are R/W registers and can be accessed only during the command phase or result phase. Not all registers are used with every command.

Parameter Register

- C This register contains the track address of the command specified.
- H This register contains the head number (0/1) that is used in the command specified.
- R This register contains the sector address of the command specified.
- N This register contains the sector size of the command specified.
- EOT This register contains the sector address of the last sector on the track in the command specified.
- GPL This register contains the length of gap 3 (the spacing between sectors) of the command specified.
- DTL This register contains information on the sector size. This information temporarily alters the sector size of the disk during special operation.
- D This register contains information that determines the pattern of data written to the data field during formatting.
- STP The value in this register is used to modify the current sector number in order to determine the next sector to be scanned.

The drive status register section contains four registers that specify detailed information on the drives. These registers can be read only during the result phase of the command. Which status registers are read is determined by a specified command.

Status Register 0

Bit 0
Bit 1 These two bits specify the drive used during the last interrupt.
Bit 2 This bit indicates the head being used at the time of the interrupt.
Bit 3 This bit indicates that the drive unit was not ready for a read or write operation.
Bit 4 This bit indicates a hardware track disk drive unit error.
Bit 5 This bit indicates when a seek command is complete or the R/W head is correctly positioned.
Bit 6
Bit 7 These two bits indicate four different types of disk interrupts 00 = termination of command without error. 01 = termination of command prior to completion. 10 = invalid command. 11 = termination of command because disk drive unit was not ready during the execution of the command.

Status Register 1

Bit 0 This bit indicates a missing data address mark.
Bit 1 This bit indicates a write protect error.
Bit 2 This bit indicates that the sector could not be found on the specified track.
Bit 3 Not used (low).
Bit 4 This bit indicates that the data in the data bus buffer has not been read and that new data has been written into the buffer; old data is lost (overrun error).
Bit 5 This bit indicates that a bad CRC code was found in the ID field in the sector.
Bit 6 Not used (low).
Bit 7 This bit indicates that the FDC has attempted to access a sector that is beyond the range of the specified track.

Status Register 2

Bit 0 This bit indicates that the FDC cannot find a data address mark or deleted data address mark on the specified track.
Bit 1 This bit indicates that the track number in the ID field does not match the track number in the FDC (bad track error).
Bit 2 This bit indicates that the FDC could not find the sector information specified on the track during a scan command (scan not satisfied).
Bit 3 This bit indicates that the FDC found the sector information on the specified track during a scan command (scan hit).
Bit 4 This bit indicates that the track (cylinder) address in the ID field does not match the track address register in the FDC (cylinder address error).
Bit 5 This bit indicates that a CRC error has been detected in the ID field.
Bit 6 This bit indicates that a deleted address mark was found during a read or scan command. It also indicates if an address mark was found during a deleted read command.
Bit 7 Not used (low).

Status Register 3

Bit 0
Bit 1 These bits indicate the currently selected disk drive unit.
Bit 2 This bit indicates the address of the currently selected R/W head.
Bit 3 This bit indicates the status of a double sided disk drive unit.

Bit 4 This bit indicates that the selected R/W is on track 0.
Bit 5 This bit indicates that the selected drive is ready for an operation.
Bit 6 This bit indicates that the selected drive is write protected.
Bit 7 This bit indicates that the selected drive has a hardware error.

SERIAL INTERFACE CONTROLLER The purpose of this section is to convert parallel data from the internal data bus to serial data to be applied to the write precompensation circuitry on the disk controller adapter during a write operation. This section also determines the value of the delay shift in a write operation and sends the data and clock information out at the proper times in a write operation.

In a read operation, this section converts serial data from the data separator and sends the data part of the signal to a serial-in parallel-out shift register. When the shift register is full, this section informs the DMA to read the data. If the data is not read before new bits are shifted into the shift register, an overrun error occurs. This section varies the VCO frequency of the data separator, which corrects phase differences between the data and clocks bits.

DRIVE INTERFACE CONTROLLER This control section is used in controlling the disk drive units, to determine the command levels to be supplied to the output port or to define how status information on the input port is to be used.

INPUT PORT The input port contains sensor status information for the disk drive unit: Ready, Write Protect, Index mark, and Track 0.

OUTPUT PORT The output port contains control information for the disk drive unit. The output of this port determines which operation the disk drive unit will perform.

READ/WRITE DMA CONTROL LOGIC This section controls how the MPU or DMA accesses the registers inside the FDC. It also provides handshaking between the MPU or DMA and FDC.

Command Operation Sequence: There are three phases in any command operation when the FDC is used—the command phase, execution phase, and result phase.

COMMAND PHASE In the command phase, which is the first sequence of an operation, the actual command instruction is stored in the FDC by the MPU, followed by the necessary parameter register values (different parameters are needed for different commands). The FDC is in the command phase after the completion of the last command or after a system reset.

Before each byte of data is stored in the necessary parameter registers, the MPU must read the main status register. Bit 6 must be low and bit 7 must be high before the next parameter can be read. After loading a register, the MPU must wait at least 12 μs before reading the main status register. To program the parameter or command registers, the MPU must use the WR# and CS# lines.

EXECUTION PHASE The execution phase automatically begins after the last parameter has been loaded in the parameter registers. This phase of the command ends if an error occurs or at the end of the last data transfer to or from the FDC. Below is the communications sequence between the DMA and 765.

1. After the parameter registers are loaded and data is available in the 765, it generates a high on the DRQ line (which is connected to the DRQ2 input on the 8237 DMA).
2. Once the DMA gains control of the bus, the DMA brings either its IOR# or IOW# line low (depending on the operation) as well as its DACK2 line. (The DACK2 line is connected to the DACK input line of the 765.) This informs the 765 to release its data during a read operation or accept the data on the bus during a write operation. The DRQ line is reset, once the data is transferred.
3. If the TC (terminal count) line from the DMA is still low or an EOT code has not been read or written onto the disk, the 765 brings the DRQ line high again. This signal

causes the 765 to ask for another DMA transfer, which causes step 2 to be repeated until the following condition occurs. Once the DMA has determined that the DMA cycle is completed (execution phase), the TC line goes high, causing the 765 to end the execution phase. This resets the DRQ line, and no data is transferred to or from the 765, ending the execution phase of the command.

RESULT PHASE During the result phase, the main status register must be read by the MPU before any data can be transferred. After each transfer, the MPU must wait 12 μs and then check to make sure bits 6 and 7 are high in the main status register before proceeding.

Once all specified status registers have been read, the result phase is completed and the next command phase can begin. Status registers Ø through 3 may be read only at this time. If all specified registers are not read, the 765 does leave the result phase. The MPU must use the RD# and CS# lines to read the data. If an error occurs, the operation is terminated.

NEC FD 765A Instructions: There are a total of 15 instructions that the FDC can perform. It is normally the purpose of DOS to supply the necessary instructions to perform the selected tasks. One DOS instruction may require more than one FDC instruction.

READ DATA The READ DATA command performs a read operation on the specified track, sector, and head address. Prior to this, the head must be positioned over the proper track with the SEEK command. As soon as the index mark is read, the 765 starts reading the ID fields. When the ID field track, sector, and head numbers match the values in the parameter registers, the data from that data stream is transferred to the computer's data bus. The data transfer is controlled by the DMA.

Command Phase

Command	B7	B6	B5	B4	B3	B2	B1	B0
Byte 0	MT	MFM	SK	0	0	1	1	0
Byte 1	0	0	0	0	0	HDS	DS1	DS0

Parameter Bytes

Byte 2 Cylinder address
Byte 3 Head address
Byte 4 Sector address
Byte 5 Sector size
Byte 6 End of track
Byte 7 Gap 3 length
Byte 9 Special sector size

Execution Phase

Data in the selected sector is transferred from the FDC to the data bus.

Result Phase

Read status register 0.
Read status register 1.
Read status register 2.
Read cylinder address register.
Read head address register.

Read sector address register.

Read sector size register.

Where

MT = multitrack operation (double sided disk only).

MFM = high for MFM, low for FM.

SK = skip flag, for skipping deleted sector.

HDS = head address

DS1–DS0 = drive select address.

READ DATA Sequence

We assume that the R/W heads are already on the proper track.

Command Sequence: The command register is loaded. Then the following registers are loaded in this order: cylinder address, head address, sector address, sector size, end of track, Gap 3, and special sector size register.

Execution Sequence: After a maximum time delay of 50 ms (head load and head settling times), the 765 reads the ID field data and compares the data in its parameter register with the data read from the ID fields. When the values match, the DRQ line goes high, requesting a DMA cycle. Once the DMA acknowledges the requests, the first byte of data is transferred from the FDC to the data bus and is stored in memory. This resets the DRQ line. If, during the last DACK, the TC line goes high, it indicates to the FDC that the cycle is completed and the execution phase is over. If the TC line is low during the DACK, the DRQ line goes high after the next transfer, causing the DMA to increment its current address register to select the next memory location for the transfer. The DMA also decrements the current word count register, which controls the TC line. When the count reaches zero, the TC line goes low, ending the execution phase. After the actual data is read, the FDC stops transferring data to the data bus, but it reads the CRC codes to verify the read operation. This ends the execution phase.

Result Sequence: The MPU reads status register 0 through 2, the cylinder address, read head address, sector address, and sector size registers to determine if the read operation was successful.

READ DELETED DATA The READ DELETED DATA command performs the same task as the READ DATA command except that it begins its task after reading the data address mark when it sees a skip flag.

Command Phase

Command	B7	B6	B5	B4	B3	B2	B1	B0
Byte 0	MT	MFM	SK	0	1	1	0	0
Byte 1	0	0	0	0	0	HDS	DS1	DS0

Parameter Bytes

Byte 2 Cylinder address

Byte 3 Head address

Byte 4 Sector address

Byte 5 Sector size

Byte 6 End of track

Byte 7 Gap 3 length

Byte 9 Special sector size

Execution Phase

Data from the selected sector is transferred from the FDC to the data bus, while the skip flag is checked.

Result Phase

Read status register 0.
Read status register 1.
Read status register 2.
Read cylinder address register.
Read head address register.
Read sector address register.
Read sector size register.

WRITE DATA This command causes the data in the data bus buffer to be converted into serial data and stored on the diskette in the selected area of the diskette. The FDC first finds the proper location, and then the data is transmitted until the sector is written with 512 bytes, followed by the 2-byte CRC code.

Command Phase

Command	B7	B6	B5	B4	B3	B2	B1	B0
Byte 0	MT	MFM	0	0	0	1	0	1
Byte 1	0	0	0	0	0	HDS	DS1	DS0

Parameter Bytes

Byte 2	Cylinder address
Byte 3	Head address
Byte 4	Sector address
Byte 5	Sector size
Byte 6	End of track
Byte 7	Gap 3 length
Byte 9	Special sector size

Execution Phase

Data is transferred from the data bus to the FDC and then to the selector sector.

Result Phase

Read status register 0.
Read status register 1.
Read status register 2.
Read cylinder address register.
Read head address register.
Read sector address register.
Read sector size register.

WRITE DELETED DATA This command performs the same task as the WRITE DATA command except that it checks for the skip flag at the time it sees the data address mark.

Command Phase

Command	B7	B6	B5	B4	B3	B2	B1	B0
Byte 0	MT	MFM	0	0	1	0	0	1
Byte 1	0	0	0	0	0	HDS	DS1	DS0

Parameter Bytes

Byte 2 Cylinder address
Byte 3 Head address
Byte 4 Sector address
Byte 5 Sector size
Byte 6 End of track
Byte 7 Gap 3 length
Byte 9 Special sector size

Execution Phase

Data is transferred from the data bus to the FDC and then to the selected sector while the status of the skip flag is checked.

Result Phase

Read status register 0.
Read status register 1.
Read status register 2.
Read cylinder address register.
Read head address register.
Read sector address register.
Read sector size register.

READ A TRACK The READ A TRACK command reads the entire data filed from sector to sector as one complete block of data. All sectors on the track are read.

Command Phase

Command	B7	B6	B5	B4	B3	B2	B1	B0
Byte 0	MT	MFM	SK	0	0	0	1	0
Byte 1	0	0	0	0	0	HDS	DS1	DS0

Parameter Bytes

Byte 2 Cylinder address
Byte 3 Head address
Byte 4 Sector address
Byte 5 Sector size
Byte 6 End of track
Byte 7 Gap 3 length
Byte 9 Special sector size

Execution Phase

Data from all sectors on the selected track are transferred from the FDC to the data bus.

Result Phase

Read status register 0.
Read status register 1.
Read status register 2.
Read cylinder address register.
Read head address register.
Read sector address register.
Read sector size register.

READ ID The READ ID command reads the contents of the first valid sector on the selected track and transfers this information to the data bus.

Command Phase

Command	B7	B6	B5	B4	B3	B2	B1	B0
Byte 0	0	MFM	0	0	1	0	1	0
Byte 1	0	0	0	0	0	HDS	DS1	DS0

Execution Phase

Data from the first valid sector on the selected track is transferred from the FDC to the data bus.

Result Phase

Read status register 0.
Read status register 1.
Read status register 2.
Read cylinder address register.
Read head address register.
Read sector address register.
Read sector size register.

FORMAT TRACK The FORMAT TRACK command initializes a diskette by inserting the following information on each track: track mark, index mark, ID address mark, ID field information, data address mark, data field, and final mark.

Command Phase

Command	B7	B6	B5	B4	B3	B2	B1	B0
Byte 0	0	MFM	0	0	1	1	0	1
Byte 1	0	0	0	0	0	HDS	DS1	DS0

Parameter Bytes

Byte 2 Number of bytes per sector
Byte 3 Number of sectors per track
Byte 4 Gap 3 length
Byte 5 Data pattern

Execution Phase

Data is transferred from the data bus to the FDC and then written to the diskette until the entire track is written.

Result Phase

Read status register 0.

Read status register 1.

Read status register 2.

Read cylinder address register.

Read head address register.

Read sector address register.

Read sector size register.

SCAN EQUAL The SCAN EQUAL command reads sector data and compares this data with the data supplied by the DMA. When these match, this command ends and the scan hit flag is set. The scan equal command also ends when the last sector on the track is read or if the DMA terminal count is reached.

Command Phase

Command	B7	B6	B5	B4	B3	B2	B1	B0
Byte 0	MT	MFM	SK	1	0	0	0	1
Byte 1	0	0	0	0	0	HDS	DS1	DS0

Parameter Bytes

Byte 2	Cylinder address
Byte 3	Head address
Byte 4	Sector address
Byte 5	Sector size
Byte 6	End of track
Byte 7	Gap 3 length
Byte 9	Scan Sector Increment register

Execution Phase

Data being read is compared with data from the DMA. If they match, the command ends and the scan hit flag is set.

Result Phase

Read status register 0.

Read status register 1.

Read status register 2.

Read cylinder address register.

Read head address register.

Read sector address register.

Read sector size register.

SCAN LOW OR EQUAL The SCAN LOW OR EQUAL command compares the data read from the sector with the data supplied from the DMA; if less than or equal to the data, the command ends. The command also ends when the terminal count from the DMA or the last sector of the specified track is reached.

Command Phase

Command	B7	B6	B5	B4	B3	B2	B1	B0
Byte 0	MT	MFM	SK	1	1	0	0	1
Byte 1	0	0	0	0	0	HDS	DS1	DS0

Parameter Bytes

Byte 2 Cylinder address
Byte 3 Head address
Byte 4 Sector address
Byte 5 Sector size
Byte 6 End of track
Byte 7 Gap 3 length
Byte 9 Scan Sector Increment register

Execution Phase

Data is read and compared with data from the DMA. If it is less than or equal to the DMA data, the commands ends, and the scan hit flag sets.

Result Phase

Read status register 0.
Read status register 1.
Read status register 2.
Read cylinder address register.
Read head address register.
Read sector address register.
Read sector size register.

SCAN HIGH OR EQUAL The SCAN HIGH OR EQUAL command compares the data read from the sector with the data supplied from the DMA; if greater than or equal, the command ends. The command also ends when the terminal count from the DMA or the last sector of the specified track is reached.

Command Phase

Command	B7	B6	B5	B4	B3	B2	B1	B0
Byte 0	MT	MFM	SK	1	1	1	0	1
Byte 1	0	0	0	0	0	HDS	DS1	DS0

Parameter Bytes

Byte 2 Cylinder address
Byte 3 Head address
Byte 4 Sector address
Byte 5 Sector size
Byte 6 End of track
Byte 7 Gap 3 length
Byte 9 Scan Sector Increment register

Execution Phase

Data is read and compared with data from the DMA. If it is greater than or equal to the DMA data, the command ends, and the scan hit flag sets.

Result Phase

Read status register 0
Read status register 1.

Read status register 2.

Read cylinder address register.

Read head address register.

Read sector address register.

Read sector size register.

RECALIBRATE The RECALIBRATE command causes the R/W heads to position themselves at track 0.

Command Phase

Command	D7	D6	D5	D4	D3	D2	D1	D0
Byte 0	0	0	0	0	0	1	1	1
Byte 1	0	0	0	0	0	0	DS1	DS0

Execution Phase

The R/W head moves to home position, track 0.

Result Phase

None.

SENSE INTERRUPT STATUS The SENSE INTERRUPT STATUS command supplies interrupt status of the selected drive unit. This command is normally used to check the status of interrupts after a read, write, scan or format command is processed.

Command Phase

Command	D7	D6	D5	D4	D3	D2	D1	D0
Byte 0	0	0	0	0	1	0	0	0

Execution Phase

None.

Result Phase

Reads status register 0 and cylinder address register.

SPECIFY The SPECIFY command is used to set the initial value of the three internal timers of the FDC. The head-unload time defines the amount of time it takes to unload the heads from the diskette surface. The step rate time determines the time intervals between adjacent step pulses. The head-load time specifies the amount of time it takes to load the heads onto the diskette surface. This is not used in the IBM PC.

Command Phase

Command	D7	D6	D5	D4	D3	D2	D1	D0
Byte 0	0	0	0	0	1	0	1	1

Parameter Bytes

Byte 1 ______SPT______ ______HUT______

Byte 2 ____________HLT____________ NA

SPT = Step rate time code, 4-bits
HUT = Head-unload time code, 4-bits
HLT = Head-load time code, 7-bits

SENSE DRIVE STATUS The SENSE DRIVE STATUS command is used to obtain status of any of the floppy disk drive units.

Command Phase

Command	D7	D6	D5	D4	D3	D2	D1	D0
Byte 0	0	0	0	0	0	1	0	0
Byte 1	0	0	0	0	0	HDS	DS1	DS0

Result Phase

Read status register 3.

SEEK The SEEK command moves the R/W heads to another track. The direction of the movement is determined by the value of the direction pin.

Command Phase

Command	D7	D6	D5	D4	D3	D2	D1	D0
Byte 0	0	0	0	0	1	1	1	1
Byte 1	0	0	0	0	0	HDS	DS1	DS0

Parameter Byte

Byte 2 Cylinder address register

Execution Phase

The R/W heads are moved from one track to another track. The direction is controlled by the level on the direction pin of the FDC.

9.3.2 IBM PC Floppy Disk Controller Adapter

Bus Buffer. The bus buffer section of the floppy disk controller adapter in Figure 9-8 is an 8-bit parallel-in parallel-out buffer. The buffer temporarily holds data for the 765 during a read or write operation or holds data for the digital control port during its programming. This buffer speeds up the data transfers of the floppy disk controller (765) during a read or write operation. This buffer holds one data byte while the floppy disk controller is processing another data byte.

FD 765A. The floppy disk controller is responsible for the following functions.

1. Determines the position of the R/W heads on the floppy diskette.
2. Determines which one of the two R/W heads is used for the instruction.
3. Detects hardware problems in the drive unit.
4. Converts parallel data into serial data and applies it at the proper times to the R/W heads during a write operation.
5. Converts serial data into parallel data and applies it to the data bus at the proper times during a read operation.

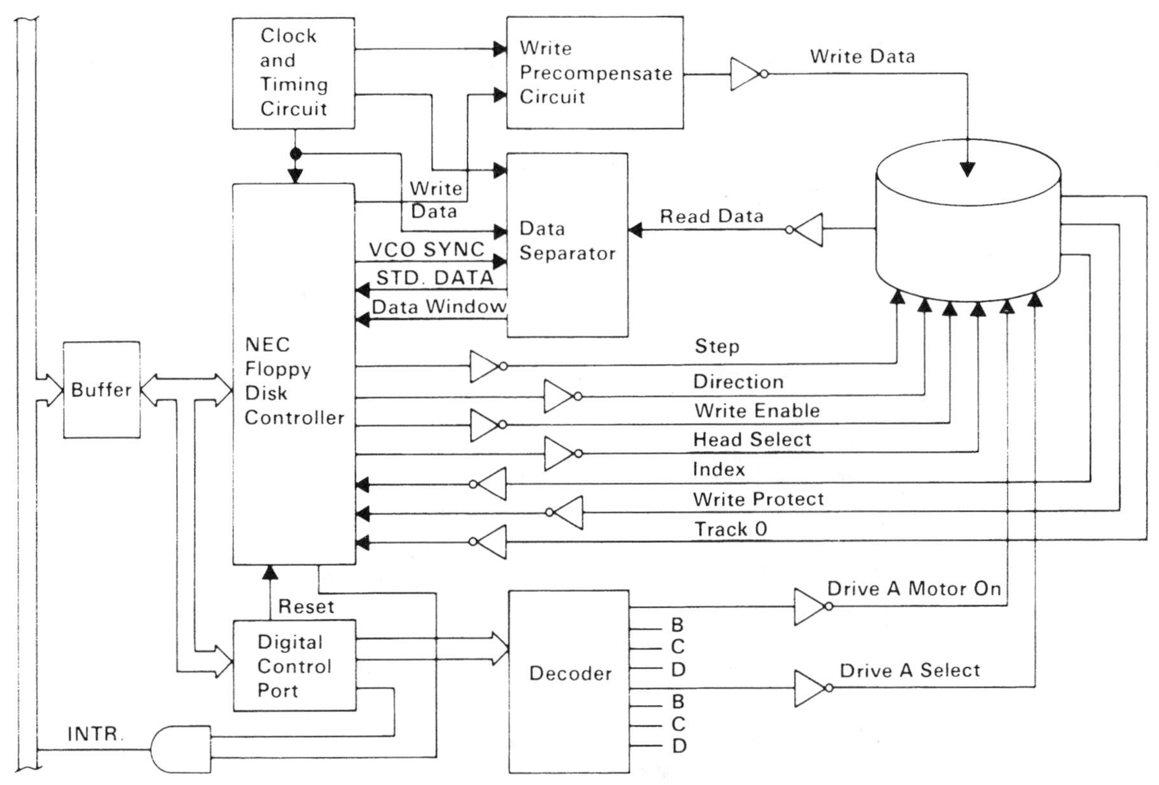

Figure 9-8 Floppy disk controller adapter. (Courtesy of IBM)

6. Detects the location of the R/W heads and senses if the diskette is write protected.
7. Modifies data being read by the data separator circuitry and turns off the data separator circuitry during a write operation (VCO).

It also performs the following 15 low level disk commands.

1. READ DATA: reads from the diskette at the selected sector, track, and side.
2. READ DELETED DATA: reads deleted data from the diskette at the selected sector, track, and side where space is deleted.
3. WRITE DATA: writes data to the diskette at the selected sector, track, and side.
4. WRITE DELETED DATA: writes deleted data to the diskette at the selected sector, track, and side when space is deleted.
5. READ A TRACK: reads all data on the selected track and side.
6. READ ID: reads the ID sector field on the selected track and side.
7. FORMAT TRACK: writes sector format information on the selected track and side.
8. SCAN EQUAL: scans the diskette beginning at the selected sector, track, and side. When the data supplied by the DMA matches the data stored on the diskette, the process stops and the scan hit flag is set.
9. SCAN LOW OR EQUAL: scans the diskette beginning at the selected sector, track, and side. When the data that is supplied by the DMA matches or has a value less than the data read on the diskette, the command ends and the scan hit flag is set.
10. SCAN HIGH OR EQUAL: scans the diskette beginning at the selected sector, track, and side. When the data supplied by the DMA matches or has a greater value than the data read on the diskette, the command ends and the scan hit flag is set.

11. RECALIBRATE: causes the R/W heads to position themselves at track 0.
12. SENSE INTERRUPT STATUS: supplies the interrupt status of the selected drive unit.
13. SPECIFY: sets the initial values of the three internal timers of the FDC. Only the step rate time is changed on the IBM.
14. SENSE DRIVE STATUS: obtains the status of any of the floppy disk drive units.
15. SEEK: moves the R/W heads to another track.

Clock and Timing. The clock and timing circuitry is used to supply three signals for the floppy disk controller adapter. One signal is the 10-MHz clock for the write precompensation circuit and two 500-kHz clocks are for the data separator and for the floppy disk controller (765).

Digital Control Port. The digital control port is a six-bit buffer parallel-in parallel-out. Two output lines are used to select one of disk drive units (motor), which is controlled by a dual one-to-four drive unit decoder. Another set of two output lines is used to select one of the disk drive units (control bus, enables sensors, and R/W heads for the unit selected). One of the last two outputs is used to reset the floppy disk drive unit control bus from the floppy disk controller (765), and the other line is used to gate the interrupt output from the floppy disk controller, which is applied to the PIC.

FDD Decoder. The floppy disk drive unit decoder is used to select one of the floppy disk drive units that can be connected to the floppy disk controller adapter. The decoder is a separately controlled one-to-four decoder. One decoder selects the motor, and the other decoder selects the other control circuitry.

Write Pre-comp Circuit. The purpose of the write precompensation circuitry (Fig. 9-9) is to delay the data going to the R/W heads, to compensate for the differences in the operation of different types of diskettes and in the heads themselves. The circuitry supplies three different delay times, 0 ns, 125 ns, and 250 ns, which are controlled by the floppy disk controller (765). The floppy disk controller determines which delay to use when reading gap 2.

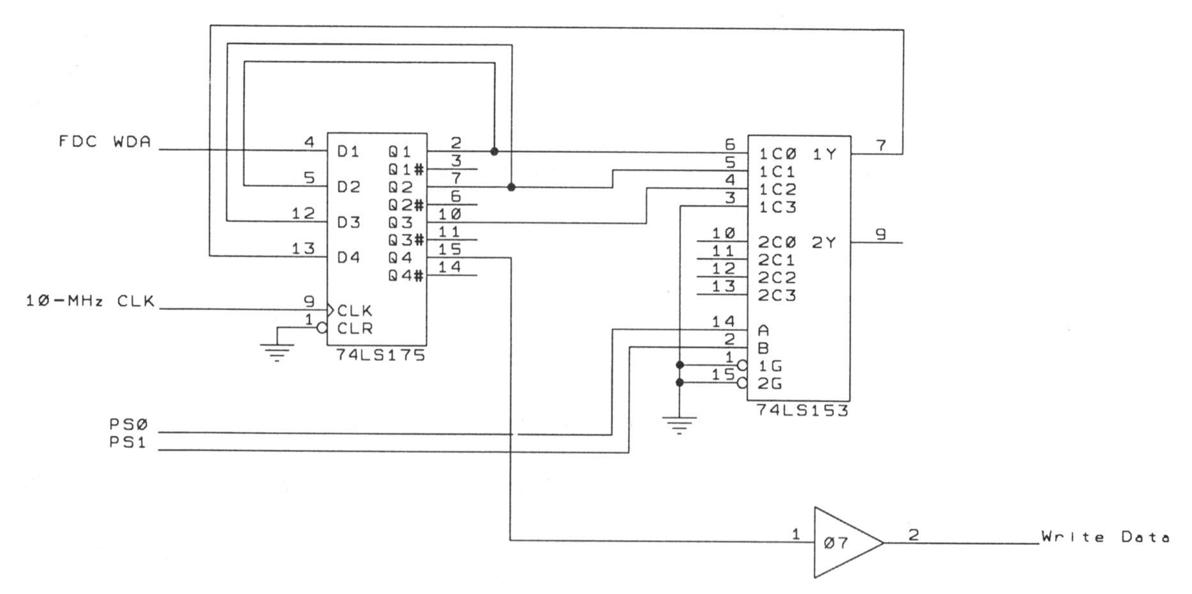

Figure 9-9 Write precompensation circuit.

Data Separator. The data separator separates the data from the clock. In the double density format these are combined together. The circuit contains a PLL, Schmitt trigger, and some control logic.

A Schmitt trigger section is used for wave shaping of the raw read data from the selected R/W head.

The control logic serves two purposes. When a write operation is being performed, it disables the VCO in the PLL. During a read operation when no data is being read, the control logic supplies a 500-kHz clock signal to the VCO in the PLL for faster synchronization.

The PLL in Figure 9-10 produces a frequency that matches the input frequency from the Schmitt trigger. The difference is in the phase shift between the VCO and the data read. This difference is used by the floppy disk controller to determine when the data is to be shifted to the serial-in parallel-out shift register.

Drivers. There is one output driver for each of the output lines from the floppy disk controller adapter. There is also one input driver for each input line from the floppy disk drive unit to the floppy disk controller adapter. The drivers are used to supply extra drive current for the disk drive unit and the disk controller adapter.

IBM I/O Channel Bus Interface

A0–A9 — The ten address lines are used to address the floppy disk controller during formatting, reading the status of the FDC, and transferring data.

D0–D7 — These eight bidirectional data lines are used to transfer data to and from the floppy disk adapter.

IOW# — This I/O Write enable line allows the MPU or DMA to control the write operation of the floppy disk adapter when low.

IOR# — The I/O Read enable line allows the MPU or DMA to control the read operation in the floppy disk adapter when low.

AEN — The Address ENable line is used to enable the address latches on the floppy disk adapter when high.

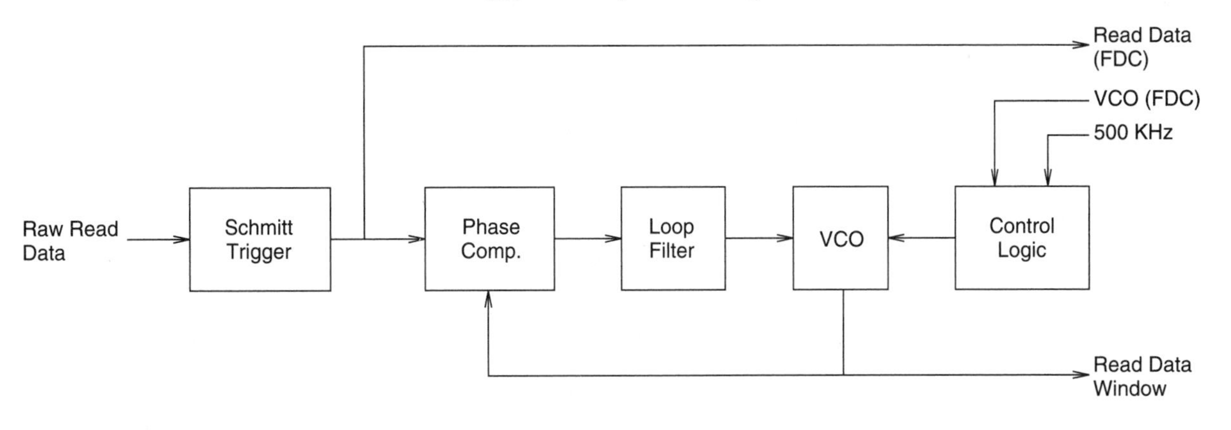

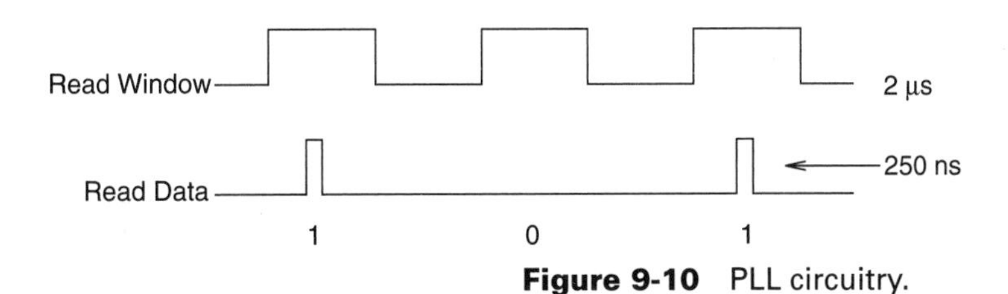

Figure 9-10 PLL circuitry.

DRQ2 The DMA ReQuest 2 line lets the FDC request a DMA transfer during execution phase of a FDC command when high.

DACK2# The DMA ACKnowledge 2 line is used by the DMA to inform the FDC to transfer data when low.

TC The Terminal Count line from the DMA is used to inform the FDC when the last DMA transfer has occurred when high.

IRQ6 The Interrupt ReQuest 6 line is controlled by the digital control port and informs the PIC that the FDC has ended its cycle. This causes the MPU to read the status register in the FDC.

Disk Drive Interface Cable. A 34-conductor ribbon cable connects the floppy disk adapter to the floppy drive unit. On one end of the cable is a 34-pin dual row header that connects to the floppy drive adapter. On the other end of the cable are two 34-pin edge connectors that are attached to the floppy drive units. Table 9-2 lists the function of each connection. In the IBM PC, the edge connector at the end of the cable that connects to the floppy disk drive unit is for drive A and the edge connector near the center of the cable is for drive B.

Drive Select 1 and 2 These two active low lines allow the floppy disk adapter to select (enable) the electronics on the selected disk drive unit.

Motor Enable 1 and 2 These two Motor Enable lines enable the spindle motor and allow the stepper motor to work when a pulse is produced. The motor is enabled when the line is low.

DIR The DIRection line is controlled by the direction pin of the FDC and is used to determine the direction of the stepper motor once it is pulsed. The level determines the direction and must be stabilized before the step pulse is produced.

STEP The STEP line is controlled by the STEP pin of the FDC and causes the stepper motor to move. The direction in which the motor moves is determined by the level on the direction pin. The signal on the line is a pulse, and for each pulse, the R/W heads move once.

TABLE 9-2
34-Pin Floppy Disk Cable Connections

Pin Number	*Description*	*Pin Number*	*Description*
1	Ground	2	Not used
3	Ground	4	Not used
5	Ground	6	Not used
7	Ground	8	Index
9	Ground	10	Motor enable 1
11	Ground	12	Drive select 2
13	Ground	14	Drive select 1
15	Ground	16	Motor enable 2
17	Ground	18	Direction
19	Ground	20	Step pulse
21	Ground	22	Write data
23	Ground	24	Write gate
25	Ground	26	Track 00
27	Ground	28	Write protect
29	Ground	30	Read data
31	Ground	32	Side 1 select
33	Ground	34	Not used

Side 1 Select	The Side 1 Select line, controlled by the FDC, determines which R/W head the operation is to use. The lower R/W head is active when high, and the upper R/W head is active when low.
WRITE GATE	This line enables the current in the selected R/W head.
WRITE DATA	This line is used by the FDC to transfer data from the floppy disk adapter to the R/W head circuitry on the disk drive unit.
READ DATA	This line is used by the FDC to transfer data from the disk drive unit to the floppy disk adapter.
INDEX	The INDEX line is used by the disk drive unit to transfer the index pulse to the floppy disk adapter. When the R/W heads are on the first sector of the diskette, this line uses an active low, 100–500 μs pulse.
WRITE PROTECT	The WRITE PROTECT line is used by the disk drive unit to inform the floppy disk adapter when the diskette is write protected. When low, a write operation is not performed on the diskette.
TRACK 00	The TRACK zero line is used by the disk drive unit to inform the FDC when the R/W heads are on track zero of the disk drive unit. This line goes low when the R/W heads are on track 0.

9.3.3 Disk Drive Unit

One of the most used floppy disk drive units for the IBM PC is the TANDON TM-100-2 double-sided, double-density, half-height drive (Fig. 9-11). The disk drive unit in Figure 9-11 contains a DC spindle motor, which turns at 300 rpm and rotates the diskette in the drive unit. The second motor is a 48 position stepper motor, which determines the position of the R/W head assembly. Also contained on the drive unit are two optical sensors; one senses if the diskette is write protected and the other senses the index hole. The disk drive unit also contains a drive board that contains digital and analog circuits that control the levels applied to the R/W heads, motors, and sensors.

The Electronics of the Floppy Disk Drive Unit

Pulse Shaping Network: The purpose of the pulse shaping network is to supply a pulse from the index hold optical sensor that is wide enough and strong enough for the floppy disk controller adapter to use.

Write Protect Switch: The write protect switch is an optical sensor that sends a logic high to the floppy disk controller adapter if the diskette in the drive unit is write protected. If the diskette is not write protected, the level on the line is low.

Track 00: The track 00 switch is a mechanical switch that indicates the R/W heads are positioned on track zero.

Spindle Drive Control: The spindle drive control circuitry contains a tachometer, which maintains the speed of the spindle motor. The difference between the reference frequency to the tachometer and the signal from the DC motor increases or decreases the speed of the spindle motor to maintain a 300 rpm speed. This speed is obtained in 250 ms.

Positioner Control: The positioner control circuitry controls the R/W head stepper motor. There are two input signals from the floppy disk controller adapter; one input determines the direction the stepper motor takes once the step pulse is given. The step input uses a pulse (the width of the pulse is determined by the 765 FDC) that causes the R/W head to move one step in the direction indicated by the level on the direction input. There will be a delay of 20 ms before a R/W operation can be performed.

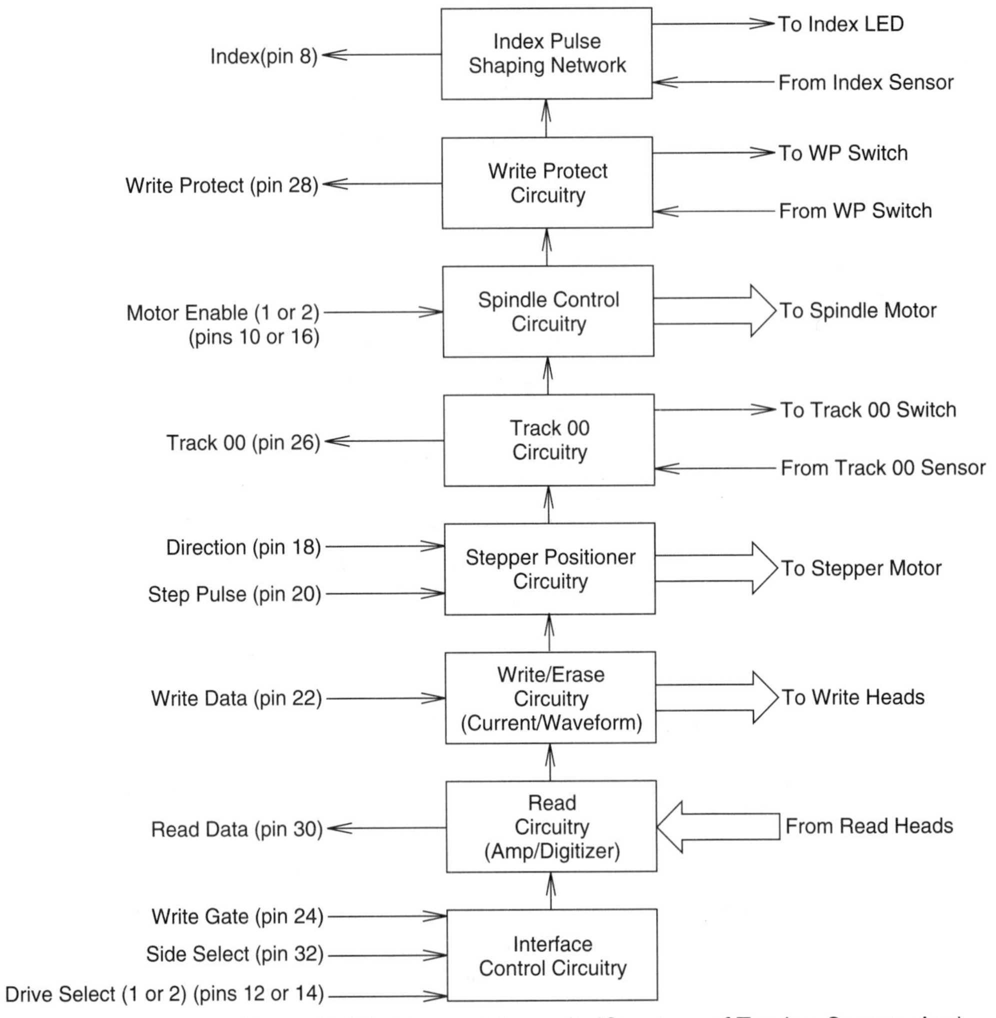

Figure 9-11 Floppy drive unit. (Courtesy of Tandon Corporation)

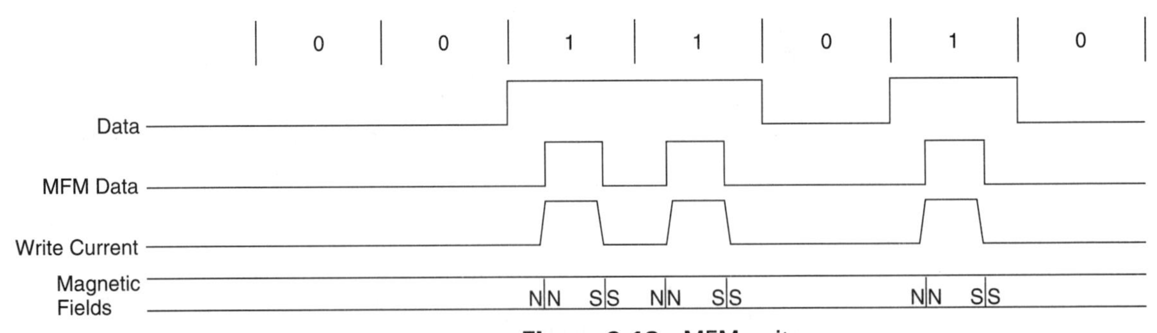

Figure 9-12 MFM write.

Write/Erase Current Source and Waveform Generator: The purpose of the write/erase current source and waveform generator (Fig. 9-12) is to perform the following functions. During a write operation the R/W head selected is supplied with an alternate magnetic field when the bit is a logic one. If the data bit to be stored is a logic zero, the magnetic field does not alternate during that time period. The erase current supplied to the R/W head (on the falling edge of the data bit) is used to provide a small demagnetized area on the diskette to provide spacing between the magnetic fields recorded on the diskette. This area allows the flux in the R/W

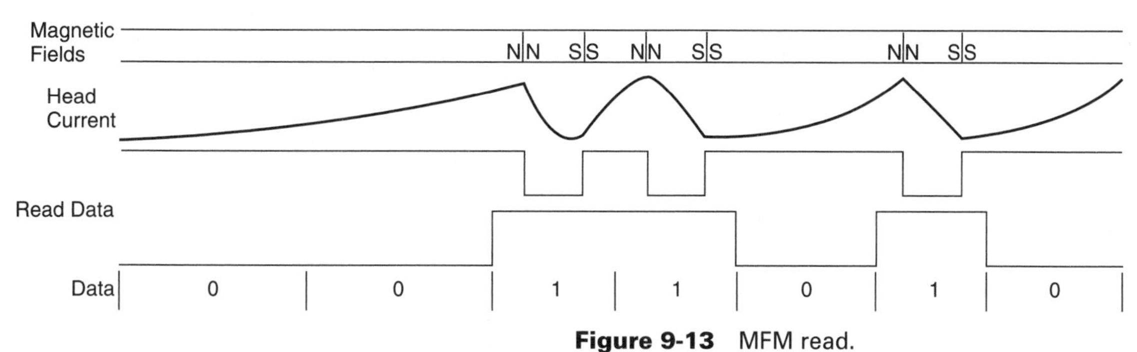

Figure 9-13 MFM read.

head time to collapse before the next magnetic field is read during a read operation. This increases the accuracy of the read operation.

Read Amplifier and Digitizer: The read amplifier and digitizer circuitry (Fig. 9-13) converts the magnetic field picked up in the R/W heads from analog signals into digital signals that can be applied to the floppy disk controller adapter.

The signals that are picked up by the selected R/W head during a read operation are applied to a high gain amplifier and a differentiator circuit. The differentiator circuit is an operation amplifier circuit that is used to eliminate any stray magnetic noise that may be picked up by the R/W heads. The signal from the differentiator circuit is applied to a digitizer, which converts the waveform into a digital signal, which is applied to the floppy disk controller adapter. The digitizer circuitry is an EX OR gate and a 74222 multivibrator. The signal from the differentiator is applied directly to one of the inputs. The other input receives the same signal except that it is first applied to a low pass filter, which removes the low frequency element from the signal. The output from the EX OR is then applied to a multivibrator, which operates at the bit time frequency and does the final clean up of the signal (which is digital at this time).

Troubleshooting the Floppy Disk System. In most cases, a problem with a floppy disk system lies within the mechanical parts of the disk drive unit. The following are some key points to check when troubleshooting a floppy disk system.

Read/Write Heads: One of the biggest problems in disk drive systems are the R/W heads. The heads are close to the surface of the diskette. If a user allows the surface of the diskette to become contaminated or uses cheap floppy diskettes, damaging particles may build up on the R/W heads.

When enough particles have built up on the R/W heads, the heads may become scratched (the surface of the head that comes closest to the surface of the diskette is made of glass). A scratch on the head surface may scratch the magnetic material off the diskette, which ruins the diskette. If enough particles build up, the heads may crash, causing damage to the heads themselves, or may cause the heads to go out of alignment. During a head crash, heads dig into the diskette, which damages the diskette and may bend the heads themselves. Normally head crashes occur on hard disk systems.

The technician should always clean the R/W heads with a swab and alcohol before trying to perform any type of alignment. Simply open the door latch, remove the diskette, and gently clean the heads by rubbing the swab with the alcohol on the surfaces of the heads, until no more dirt or other contaminants come off. This should be done as part of maintenance on a monthly basis, or sometimes weekly basis, depending on how much a drive unit is used. A cleaning diskette can be used occasionally but should not be relied on exclusively.

Inability to Read: If data cannot be read off a diskette, a few things may be wrong with the system.

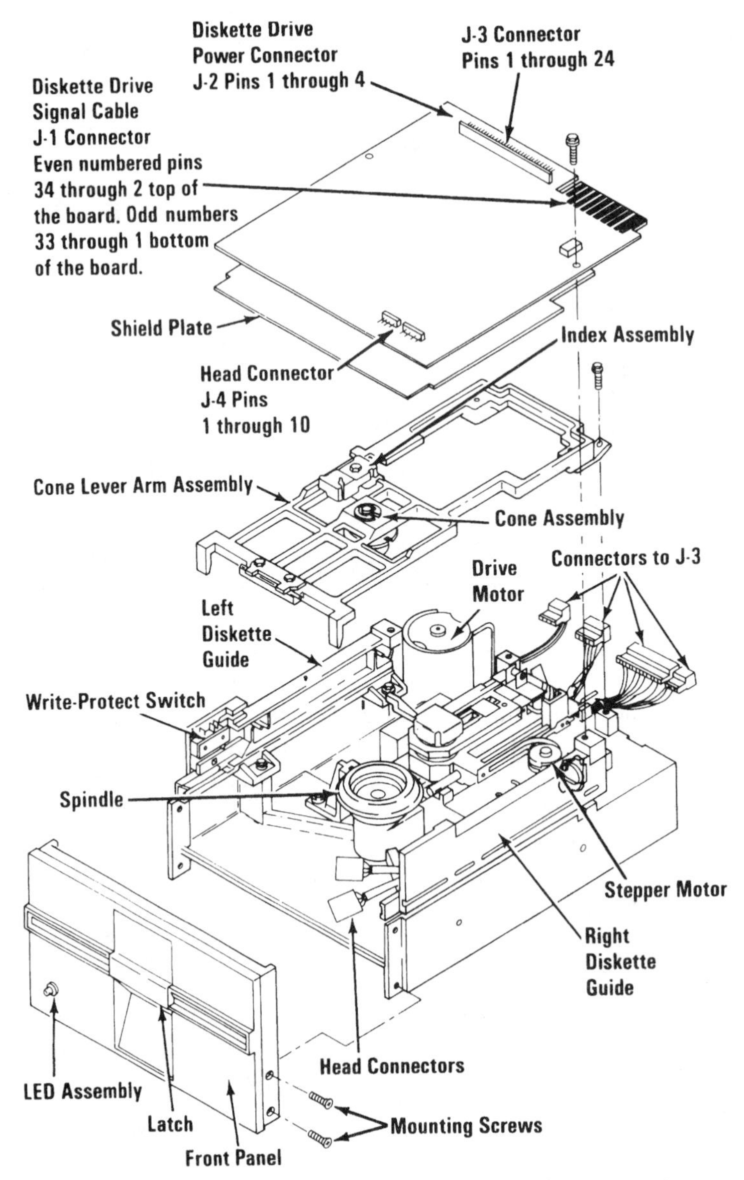

Figure 9-14 Type 2 diskette drive assembly. (Courtesy of IBM)

1. Make sure the diskette used is good.
2. Check the speed of the diskette using the IBM diagnostic.

   On older disk drive units, there is an adjustment pot on the disk speed board. This pot controls the amount of current supplied to the spindle motor, and controls the speed of the motor. Adjust this until the speed is correct. If the correct speed cannot be obtained, check the drive belt and bushings. If nothing is wrong with the belt or bushings, load the spindle motor (apply pressure to it) and check the voltage from the PLL speed correction circuit. This voltage should go up. If it does not, check the PLL circuitry. If all else fails, change the spindle motor.

   On direct drive units, there are no adjustments for spindle speed. Check to make sure the voltages to the motor are correct. If they are, change the spindle motor; otherwise troubleshoot the circuits driving the spindle motor.
3. The radial alignment may be off. The radial alignment is a measurement of how well the R/W head is centered over the diskette. If the radial alignment is off, the R/W

TABLE 9-3
Mechanical and Electrical Specifications for a Floppy Disk System

Description	*5.25-inch*	*5.25-inch*	*3.5-inch*	*3.5-inch*
Density	DSDD	DSHD	DSDS	DSHD
Tracks per inch	48	96	135	135
Number of tracks	40	80	80	80
Height size	Full	Half	NA	NA
Dimension				
Height	85 mm	41 mm	26 mm	26 mm
Width	149 mm	146 mm	101 mm	101 mm
Depth	203 mm	196 mm	129 mm	129 mm
Weight	2 kg	1.2 kg	390 g	395 g
Temperature				
Operating	10°–44° C	5°–45° C	5°–45° C	5°–45° C
Nonoperating	-40°–71° C	-40°–65° C	-40°–65° C	-40°–65° C
Relative humidity				
Operating	22%–80%	20%–80%	20%–80%	20%–80%
Nonoperating	5%–95%	5%–90%	5%–90%	5%–90%
Power	+12 v/900 ma	+12 v/480 ma		
	+5 v/60 ma	+5v/210 ma	+5 v/125 ma	+5 v/125 ma
Seek time	5 ms	3ms	3ms	3ms
Head setting time	15 ms	15ms	15ms	15ms
Error rate	1 per 10^9	1 per 10^9	1 per 10^9	1 per 10^9
Head life	20,000 hr.	25,000 hr.	30,000 hr.	30,000 hr.
Media life (passes)	3.6×10^6	3.6×10^6	3.6×10^6	3.6×10^6
Disk speed	300 rpm	300 rpm	300 rpm	300 rpm
Inst. speed varition	+/- 3.0%	+/-1.5%	+/-1.0%	+/-1.0%
Start/stop time	250 ms	100 ms	100 ms	100 ms
Transfer rate				
MFM DSDD	250K bps	250K bps	250K bps	250K bps
MFM DSHD	NA	500K bps	500K bps	500K bps
Bytes/disk	500K	1.6K	1MB	2MB
Recording modes	MFM	MFM	MFM	MFM

heads may not read the correct track or may not pick up the proper field in full. Radial adjustment normally affects the entire assembly in relation to the drive unit's frame.

4. The track zero reference may be off. Using a diagnostic diskette, position the R/W heads at the mechanical track zero. Then adjust the track zero indicator. The sensor may be mechanical or optical. If track zero is not mechanically correct, all other tracks on the drive will be off, and disk access may be impossible.
5. The azimuth on the R/W heads refers to the center angle of the heads compared with the center of each track. This angle should be 90 degrees. If the magnetic fields are at right angles to the gap in the R/W heads, maximum error free reading and writing should be achieved. If the azimuth of the heads is off, a disk alignment diskette or tester is needed to readjust the azimuth. Normally, the azimuth of the heads changes if diskettes are forcibly inserted or removed from the disk drive unit; sometimes they change if the computer is moved from one location to another.
6. The index signal is used in the rotational timing of the diskette to ensure that the R/W heads are on the first sector of the diskette at the time the FDC acknowledges the signal. By adjusting the prototransducer, the time it takes to recognize the index mark may be adjusted to make up for any mechanical deficiencies that occur as the drive becomes older. A scope and alignment tool and/or diskette is needed for this adjustment.
7. As the R/W heads move toward and away from the center of the diskette, the stepper motor step-widths become important. If the steps of the motor are not the same, data cannot be written to or read from the innermost tracks of the diskette. The reason for this is that the outermost track, track zero, is used as the starting reference. Using the

diagnostic diskette or alignment tool, verify valid steps by the stepper motor. Normally, one of two things causes improper steps: a worn-out stepper motor (as the motor is used, the permanent magnets inside break down causing the steps to be unequal) or an incorrect stepping pulse from the floppy disk adapter.

8. Make sure the operation of the R/W circuitry has not changed over time. This may be done using an alignment tool and diskette. Many times alignment problems look like circuitry problems.

Floppy Disk Adapter: If the floppy disk adapter is suspected of causing a read or write problem in the disk system, which is very rare, check the signals.

1. Vcc to card and 765 FDC.
2. Clock input to the 765 FDC.
3. Reset input to the 765 FDC.

Using a scope, check the write data output line from the floppy disk adapter when performing a write operation to the diskette. If a data pattern is present, troubleshoot the disk drive unit. If no data pattern is present, troubleshoot the precompensation circuitry on the disk drive adapter.

Using a scope, check the read data input line from the disk drive interface edge connector when performing a read operation. If no signal is present, troubleshoot the disk drive unit. If the signal is not present, troubleshoot the PLL and data separator on the floppy disk adapter.

If the floppy disk adapter is suspected of causing head movement or lack of head movement problems, make the following checks:

1. Check the level on the direction pin of the 765. This level should change when the movements of the heads change. This level must be stable before the step pulse is produced.
2. Check the step pulse width from the 765 against manufacturer's parameters.
3. Check the write protect line on the 765 and verify the proper level.
4. Check the ready line on the 765 and verify proper level and operation.

9.3.4 DOS and the Disk System

There are two different DOS formats, PC-DOS (written for IBM) and MS-DOS (written for IBM compatibles). Both work in any IBM PC or compatible. The main differences between the two formats are in the way the displays look on the screen and a few commands. At the time of printing, DOS 6.0 is still popular while DOS 6.2 is fast becoming the standard.

Formatting a Diskette with DOS. When DOS is used to format a 360K diskette, the following default parameters are used.

40 tracks (cylinders) per side
9 logical sectors per track
2 sides
360 logical sectors per side
720 logical sectors per diskette
512 bytes per logical sector (called a record)
Total unformatted disk space of 368,640 bytes
Total formatted disk space of 362,496 bytes

DOS stores its data in clusters that consist of two logical sectors (two-records, 1024 bytes). By storing data in clusters of 1024 bytes instead of a single logical sector (512 bytes), MicroSoft increases the speed of the operating system. The only problem is that even if the data to be stored is only one byte, DOS allocates 1024 bytes of space. This reduces the amount of data that can be stored on the diskette.

Table 9-4 shows the relationship between the logical sectors, track numbers, and the two sides of the diskette as far as DOS uses it. Each track of the disk contains 18 logical sectors which DOS uses as 9 sectors (clusters). Each value of C per track (9 total per track) is used to store one block of data. Cluster 5 is divided in two parts on every track; one half is located on side one and the other half is on side two. Therefore, on every track, cluster 5 equals logical sector 9 on side 1, and logical sector 0 on side 2 is treated as one unit. DOS stores its information in this way to reduce head movement and increase the speed of any disk operation.

TABLE 9-4

Logical Sector and Cluster Table

Logical Sector Track	*Side 1*					*Side 2*			
	1/2	*3/4*	*5/6*	*7/8*	*9/1*	*2/3*	*4/5*	*6/7*	*8/9*
00	C-1	C-2	C-3	C-4	C-5	C-6	C-7	C-8	C-9
01	C-1	C-2	C-3	C-4	C-5	C-6	C-7	C-8	C-9
02	C-1	C-2	C-3	C-4	C-5	C-6	C-7	C-8	C-9
03	C-1	C-2	C-3	C-4	C-5	C-6	C-7	C-8	C-9
04	C-1	C-2	C-3	C-4	C-5	C-6	C-7	C-8	C-9
05	C-1	C-2	C-3	C-4	C-5	C-6	C-7	C-8	C-9
06	C-1	C-2	C-3	C-4	C-5	C-6	C-7	C-8	C-9
07	C-1	C-2	C-3	C-4	C-5	C-6	C-7	C-8	C-9
08	C-1	C-2	C-3	C-4	C-5	C-6	C-7	C-8	C-9
09	C-1	C-2	C-3	C-4	C-5	C-6	C-7	C-8	C-9
0A	C-1	C-2	C-3	C-4	C-5	C-6	C-7	C-8	C-9
0B	C-1	C-2	C-3	C-4	C-5	C-6	C-7	C-8	C-9
0C	C-1	C-2	C-3	C-4	C-5	C-6	C-7	C-8	C-9
0D	C-1	C-2	C-3	C-4	C-5	C-6	C-7	C-8	C-9
0E	C-1	C-2	C-3	C-4	C-5	C-6	C-7	C-8	C-9
0F	C-1	C-2	C-3	C-4	C-5	C-6	C-7	C-8	C-9
10	C-1	C-2	C-3	C-4	C-5	C-6	C-7	C-8	C-9
11	C-1	C-2	C-3	C-4	C-5	C-6	C-7	C-8	C-9
12	C-1	C-2	C-3	C-4	C-5	C-6	C-7	C-8	C-9
13	C-1	C-2	C-3	C-4	C-5	C-6	C-7	C-8	C-9
14	C-1	C-2	C-3	C-4	C-5	C-6	C-7	C-8	C-9
15	C-1	C-2	C-3	C-4	C-5	C-6	C-7	C-8	C-9
16	C-1	C-2	C-3	C-4	C-5	C-6	C-7	C-8	C-9
17	C-1	C-2	C-3	C-4	C-5	C-6	C-7	C-8	C-9
18	C-1	C-2	C-3	C-4	C-5	C-6	C-7	C-8	C-9
19	C-1	C-2	C-3	C-4	C-5	C-6	C-7	C-8	C-9
1A	C-1	C-2	C-3	C-4	C-5	C-6	C-7	C-8	C-9
1B	C-1	C-2	C-3	C-4	C-5	C-6	C-7	C-8	C-9
1C	C-1	C-2	C-3	C-4	C-5	C-6	C-7	C-8	C-9
1D	C-1	C-2	C-3	C-4	C-5	C-6	C-7	C-8	C-9
1E	C-1	C-2	C-3	C-4	C-5	C-6	C-7	C-8	C-9
1F	C-1	C-2	C-3	C-4	C-5	C-6	C-7	C-8	C-9
20	C-1	C-2	C-3	C-4	C-5	C-6	C-7	C-8	C-9
21	C-1	C-2	C-3	C-4	C-5	C-6	C-7	C-8	C-9
22	C-1	C-2	C-3	C-4	C-5	C-6	C-7	C-8	C-9
23	C-1	C-2	C-3	C-4	C-5	C-6	C-7	C-8	C-9
24	C-1	C-2	C-3	C-4	C-5	C-6	C-7	C-8	C-9
25	C-1	C-2	C-3	C-4	C-5	C-6	C-7	C-8	C-9
26	C-1	C-2	C-3	C-4	C-5	C-6	C-7	C-8	C-9
27	C-1	C-2	C-3	C-4	C-5	C-6	C-7	C-8	C-9
28	C-1	C-2	C-3	C-4	C-5	C-6	C-7	C-8	C-9

One may also ask why there is a difference of 6144 bytes between the unformatted disk space and the formatted disk space. The answer is that DOS allocates this space for three special files that are not normally seen by the user when performing a directory command in DOS. This space is used to hold special information that DOS needs to use the diskette. The names of these files are boot record, file allocation table (FAT), and the directory. If any of these files are not on the diskette, DOS does not allow the diskette to be used and displays one of the following messages: NON-SYSTEM DISK or NON-DOS DISK.

Boot Record: The boot record contains a short program and causes the computer to use the disk and I/O control routines in BIOS. The boot record is the very first file that is stored on the diskette. It is located in cluster 1 (logical sectors 1 and 2, side 0, track 0) of every diskette that is formatted using DOS. If the boot record is not found on a diskette, the computer produces a non-system disk error message. The last word of the boot record is called the signature. For the boot record to be valid, the signature must equal 55AA in hexadecimal.

File Allocation Table (FAT): The FAT is used by DOS to convert cluster locations into logical sectors and maps the allocated and available disk space. Its entries are 12 bits for floppy and 16 bits for hard disk systems and they are used to map all clusters on the diskette, except the cluster for the boot record. The FAT also contains the information that determines which clusters are tied (chained) together and the order in which they are accessed to perform the specified operation.

The first two entries in the FAT are (1) OFD hexadecimal, which represents the size of the diskette (40 tracks, 9 sectors on 2 sides) and (2) 002 hexadecimal, which represents the beginning of the mapped area of the diskette (cluster 2).

The FAT is always located in clusters 2 and 3 (2048 bytes [logical sectors 3, 4, 5, 6, track 0, side 0]). Each 12-bit entry represents one cluster except for cluster 1 (which is not mapped). Cluster 2 contains one copy of the FAT, and cluster 3 contains the second copy of the FAT. DOS makes two copies, in case one FAT is damaged. If both copies of the FAT become damaged, it is very hard to correct the problem, if not impossible.

The following are allocation codes entries contained in the FAT and what they mean.

Data (hexadecimal)	Description
000	Unused, available cluster.
XXX	This is the next cluster number in file. The first cluster of a file is located in the directory. Valid values are 002 to 168 hexadecimal. (168 hexadecimal = 360 decimal, total number of clusters on diskette).
FF7	This value has two meanings. If the cluster is not part of an allocation chain, this value represents a bad cluster (bad area on the disk). If the cluster is part of an allocation chain, this indicates a reserved cluster, which is used for the file specified in the allocation chain.
FF8–FFE	This is used to mark the end of a file. It is normally used with random or sequential files.
FFF	This indicates the last cluster number in a allocation chain (file).

The FAT is updated whenever a write to the diskette is performed.

Directory. The directory contains a listing of all files, their names, extensions, attributes, allocation values, date and time of creation, and size (including hidden files that are not normally displayed with a directory command in DOS). Each entry in the directory contains 32 bytes of data. The directory is always located in clusters 4, 5, 6 (logical sectors 7, 8, 9, side 0, track 0 and logical sectors 1, 2, 3, side 1, track 0 [requires 3072 bytes of space]). The directory is updated after each write operation.

The following is a list of the bytes that make up this information.

Byte	Description
0	First byte of file name or one of special following codes. 00 indicates a file name that is never used. 05 indicates that the first character in the file name is an escape character. 2E indicates that the file name is a subdirectory. E5 is a file name that has been erased. XX is any valid file name character that represents the first character of the standard file name.

Bytes (ASCII character codes) must be valid file name characters.

1	Second byte of file name, if used.
2	Third byte of file name, if used.
3	Fourth byte of file name, if used.
4	Fifth byte of file name, if used.
5	Sixth byte of file name, if used.
6	Seventh byte of file name, if used.
7	Eighth byte of file name, if used.
8	First byte of extension, if used.
9	Second byte of extension, if used.
A	Third byte of extension, if used.
B	Attribute codes (characteristics of file). 01 indicates a read-only file. 02 indicates a hidden file (can be executed but not listed). 04 indicates a system file (can be listed but is used only by the system). 08 indicates a volume label (root directory only). 10 indicates a subdirectory. 20 indicates an archive bit (bit used in back-up and restore DOS commands). 00 is used if none of the above attribute codes is used.
C-16	DOS reserved area (no access).
17	Time (mmmxxxxx) where m = lower three bits for the number of minutes and x = binary number for two-second increments.
18	Time (hhhhhmmm) where h = binary number for hours and m = high three bits for the number of minutes.
19	Date (mmmddddd) where m = three lower bits for the month number and d = binary numbers days of the month.
1A	Date (yyyyyyym) where y = binary number of years and m = high bit for the number of the month.
1B	Lower byte of starting cluster number.
1C	Upper byte of starting cluster number.
1D	Least significant byte of file size.
1E	Next least significant byte of file size.
1F	Most significant byte of file size.

When the disk is formatted, the 6144 bytes of disk space that make up the boot record, FAT, and directory are not listed in the directory and cannot be viewed using normal DOS commands.

When the diskette is formatted with the system on it, DOS creates three additional files, two of which are hidden and cannot be listed using the directory command. The third is a normal file that is listed with the directory command. The following list shows the three files that are created.

IBMBIO.SYS: The IBM BIOS file is used to load the other system files on the diskette, once the IBM PC is reset (hardware or software). This is the first file. It is a hidden, system, read only, archive file. The file is not listed when a directory command is given. The system attribute means that it is accessed only by the system and not by the user. Read only means that the computer does not normally write into this area. The archive attribute means that it was written into during the formatting of the diskette and closed properly. The file size is 4736 bytes and is stored in the first 5 clusters available after the directory area.

IBMDOS.SYS: The IBM DOS file is used to modify the devices and parameters of the computer system. This file contains the same attributes as the IBMBIO file. The size of this file is 17,024 bytes, and it is stored in the first 17 clusters following the IBMBIO file.

COMMAND.COM: The COMMAND.COM file contains the PC-DOS internal commands (the commands loaded into the computer when it is booted). The attributes of this file are that it is a normal file (listed during a directory command) and is archived (see the IBMBIO archive attribute). The size of this file is 17,192 bytes, and it is stored in the first 17 clusters following the IBMDOS.SYS file.

DOS always stores files beginning at the first empty cluster or erased cluster it finds nearest the outside track of the disk. When DOS deletes a file from the disk, it does not erase the data stored in the file. Instead it changes the first character in the file in the directory to the delete character (E5 in hex). The delete character informs DOS that the file space is available for new data. With some disk utilities, the user can go into the directory and replace the delete character with another valid character to indicate and activate data.When the character is inserted, the data can be used normally again.

9.3.5 AT Floppy Disk System

The floppy disk system for an AT computer functions basically the same as in an XT computer. With the advent of AT computer systems came high-density floppy disk systems. The difference between a double-density and a high-density floppy disk system lies in the following: a high-density diskette, a disk drive unit capable of using high-density diskettes, and a floppy disk adapter capable of supporting high-density drives.

High-Density Floppy Diskette. There are two sizes of high-density floppy diskettes, both share the same characteristics. The magnetic coating on a high-density diskette allows the magnetic fields that represent each bit of data to be formed on it in a much shorter time because it can be magnetized easily. Because the rotation speed of the diskette remains constant, if it only takes half the time to form a magnetic field that represents a data bit, then twice as much data can be stored on each track, provided that the circuitry of the drive unit and adapter can handle the speed. In the early days of high-density floppy diskettes, they were referred to as quad-density diskettes.

The only way to differentiate a 5.25-inch high-density minidiskette from a double-density diskette is that the high-density diskette does not have a reinforcement ring around the spindle hole. High-density 3.5-inch microdiskettes are identified by a notch cut in the case on the side opposite the write-protect notch. The high-density notch cannot be covered.

Because the density of a high-density diskette is much greater than that of a double-density diskette, if the user tries to format it as a double-density diskette or tries to use it on a double-density drive, data errors occur. If a double-density diskette is formatted as a high-density diskette, bad sectors will most likely be indicated during the formatting process, and the integrity of the data is not constant.

High-Density Floppy Drive Units. The differences between a double-density and high-density drive are as follows: The width (head gap) of the R/W heads in a high-density drive is about half that of the double-density drive unit. Because the magnetic coating on the diskette allows a magnetic field to be created more rapidly, the fields are smaller, resulting in more data bits stored in the same amount of surface area of the diskette. When formatting or writing a double-density diskette, the circuitry of the floppy drive unit increases the current to the R/W heads and reduces the bit timing, as opposed to formatting or writing a high-density diskette. When reading a double-density diskette, the circuitry on the drive unit causes the PLL to slow down to capture the data bits (250K bps), as opposed to the 500K bps on a high-density diskette.

In a 5.25-inch high-density floppy drive unit, the second difference is that because more data bits are stored on each track of the diskette, the rotation speed of the diskette must be more closely regulated. In the older style double-density diskette, the rotation of the spindle was generated by a separate DC motor, which turned a spindle flywheel using a rubber belt. This set-up leads to speed variations that cause data errors in a high-density drive unit. In all high-density drive units, the rotation of the diskette is controlled by a direct drive motor. No belts are used, and attached to the flywheel are permanent magnets. The flywheel is then attached to the spindle, and the spindle is attached to a circuit board (Fig. 9-15). Underneath the flywheel attached to the circuit board are 20 or more small coils of wire. At the end of each coil is a small piece of metal, which forms a small electromagnet when current is flowing through the coil. In order to cause the motor to turn, the circuit on the circuit board causes current to flow through the wire in one or more coils, causing the permanent magnet on the flywheel to start turning. As the flywheel turns, the circuitry causes the current to flow in different coils to keep the flywheel turning. A light sensor or hall effect transistor (a transistor designed to change its characteristics with changes in magnetic-field changes) is used to monitor the speed of the flywheel (spindle) so that the circuitry on the circuit board can vary the current and speed at which the current switches between coils to maintain the speed of the

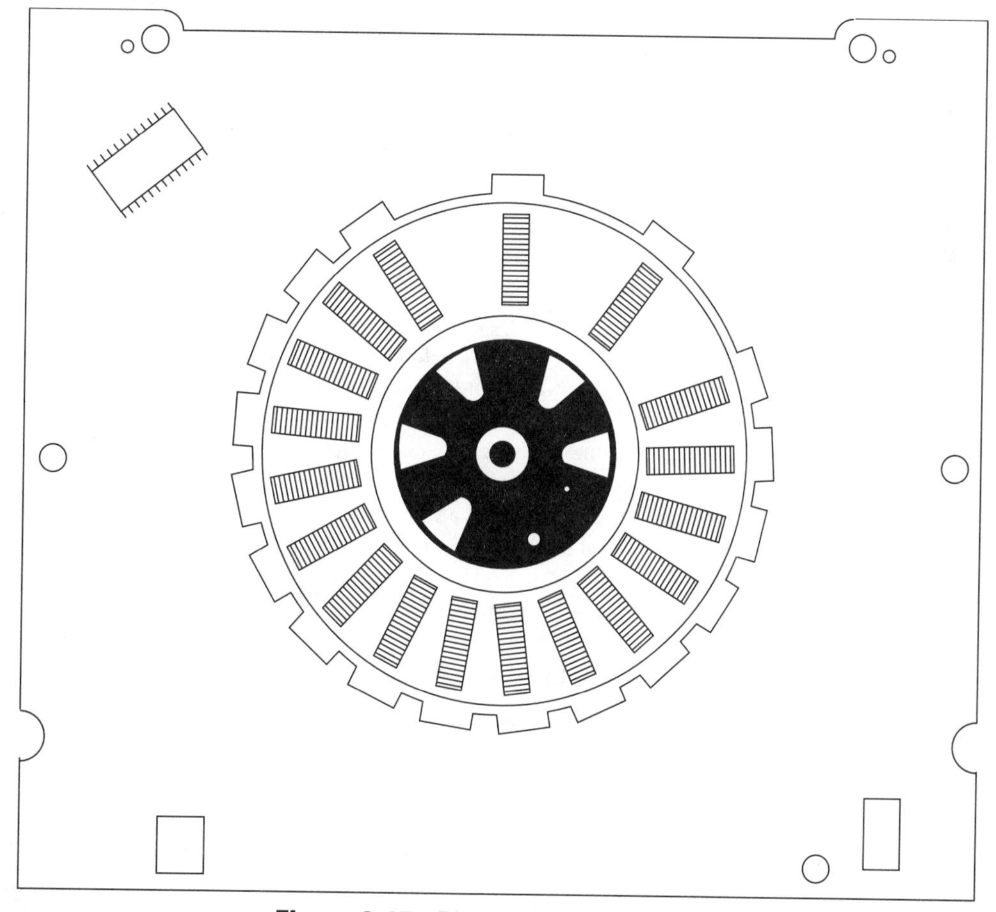

Figure 9-15 Direct drive spindle motor.

spindle under varying loads. Because direct drive motors are so accurate and do not vary with age under normal conditions, there are no speed adjustments on direct drive motors. If the motor goes bad, the motor board is simply replaced.

On a 5.25-inch high-density floppy drive unit, the stepper motor for the R/W heads has 96 steps rather than the 48 steps of a double-density drive unit. Only the first 80 steps are used in an IBM system. When the controller formats a double-density drive, two step pulses are used to move the R/W heads to the next track location, as opposed to one step pulse when formatting a high-density diskette.

Because a 3.5-inch floppy drive unit always contains direct drive motors, there is no difference in the motors. The number of tracks for double-density or high-density drive units is the same, so there is no difference in the stepper motor for the R/W heads. In a high-density drive, a sensor detects whether the diskette in the drive is high density or double density. The high-density sensor may use light or may be mechanical.

The ribbon cable for both the double-density and high-density unit is the same. The 5.25-inch drive uses a 34-pin card edge connector, and the 3.5-inch drive uses a 34-pin header connector (see Table 9-2) for the pin configuration of each connection.

AT Floppy Diskette Adapter. One of the ICs commonly found in AT floppy disk adapters that support high-density diskettes is the INTEL 82077 CHMOS single-chip floppy disk controller. This IC not only performs all the same functions as the 765 floppy disk controller but also integrates all the external circuitry normally found on the floppy disk adapter into one 68-pin PLCC IC package. Besides being able to accept bit transfer rates for double-density diskettes, it also provides for high-density bit transfer rates (500K bps), and, to reduce data bus delays, it incorporates a 16-byte first-in first-out data buffer. Figure 9-16 shows a typical application of the 82077 as a floppy disk adapter. Only external address decoding is needed to create the floppy disk adapter.

Using High-Density Floppy Disk Systems. To use a high-density floppy disk drive unit in an AT computer system, the CMOS set-up must be programmed to indicate which type of drive is connected. If the incorrect setting is selected, the drive does not function. DOS uses the CMOS settings in an AT system to determine how to communicate with the drive being used. DOS makes the necessary changes in the FAT and root directory to accommodate the difference in drive size when formatting the diskette. The boot record remains the same.

Troubleshooting High-Density Floppy Disk Systems. Because high-density floppy systems use higher bit transfer speeds and smaller head gap spacing, it is important to make sure the R/W heads are kept clean. Cleaning of the R/W heads is the same as in a double-density system. Dirty heads are the most common cause of problems with floppy systems and even more so with high-density systems.

In most cases disk drive alignments cannot be performed reliably on a high-density drive unit unless special test and alignment equipment is available. Because the cost of floppy drive units ranges from $32 to $60, very few repairs are made. Before replacing a drive unit, the following checks should be made:

1. Clean the R/W heads.
2. Verify the operation of the drive with a known good floppy diskette.
3. Check CMOS settings and verify that they are correct.
4. Check all connections to the drive unit, 34-pin ribbon cable, and power connections.

Most AT systems use a combination floppy and hard disk adapter. These adapters range in cost from $15 to $69 for a standard ISA bus; unless a special floppy/hard disk adapter is needed, replacement rather than repair is normally done.

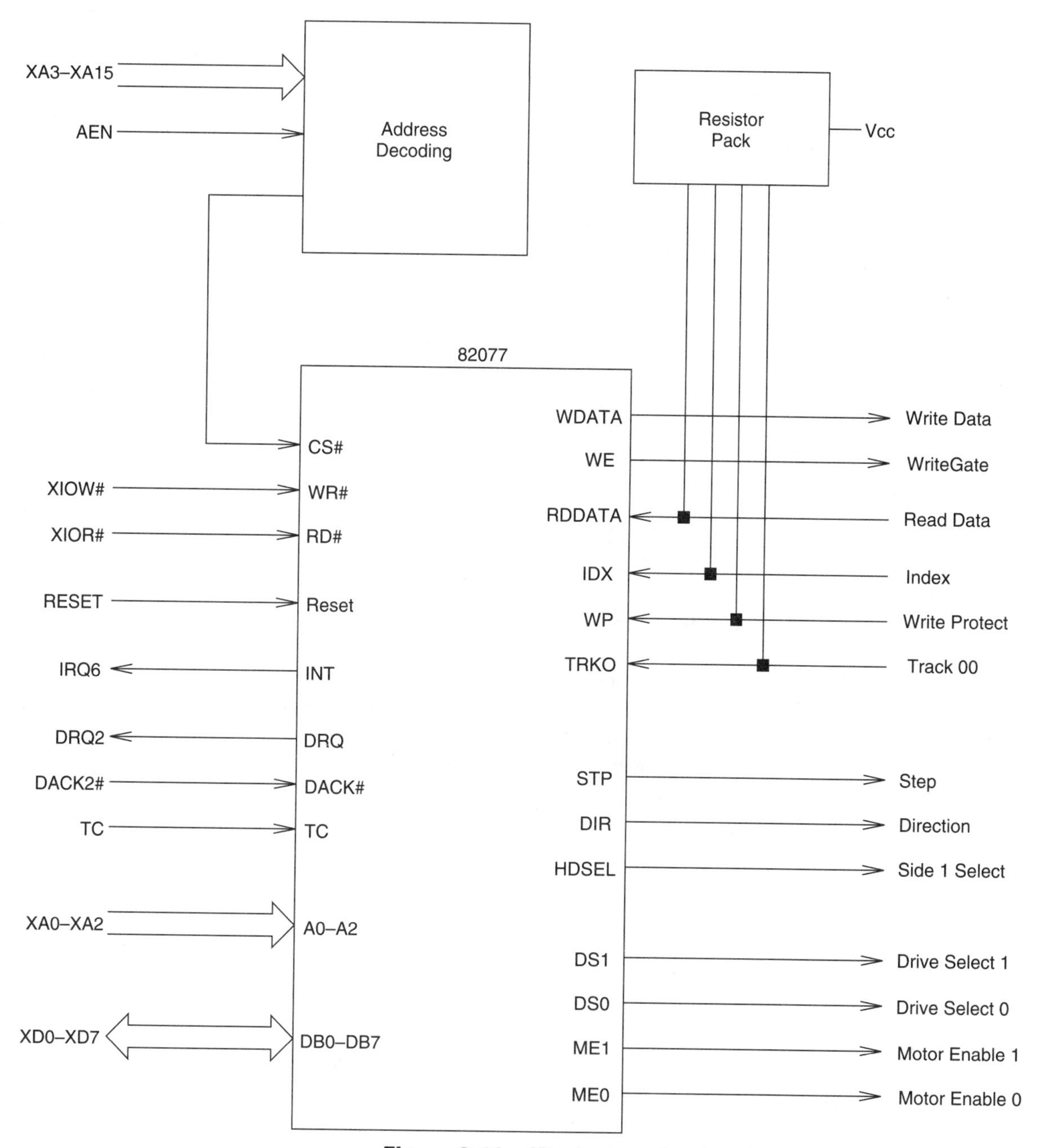

Figure 9-16 AT adapter using the 82077 IC.

9.4 HARD DRIVE SYSTEMS

Hard drive systems are made up of three parts:

1. The hard drive adapter, which performs the same function as the floppy disk adapter except that it processes data 10 to 100 times faster than a floppy disk adapter.
2. The hard drive unit, which contains two motors; one motor rotates the disks and another positions the R/W heads over the correct track (cylinders). The disks are rigid platters that are used like a floppy disk to store the data.
3. DOS, the software that is used to interface the hard drive system with the computer system.

9.4.1 Hard Drive Adapter

The hard drive adapter contains a hard drive controller IC, which functions much the same as a floppy disk controller. A hard drive controller IC processes the data much faster than a floppy controller. Whereas a floppy disk transfers data bits to and from the floppy drive unit at 250 to 500K bps, the hard drive controllers transfer data at 5 to 15M bps. The hard drive controllers also have larger internal registers because of the amount of data that can be stored on the hard drive unit. In an HDC (hard drive controller) the data is buffered to allow 0-wait state data transfers. Most HDCs also provide on-chip error detection and correction circuitry. While the 765 floppy disk controller supports 15 high-level commands, hard drive controllers support only 8 high-level commands (read sector, write sector, write format, compute correction, restore, seek, scan ID, and set parameter). Most HDCs require a fair amount of additional support circuitry such as address decoding, clocking circuitry, write precomp circuitry, PLL, data separator, and drive select circuitry. As with the floppy disk controller, NRZ (non-return to zero) encoding (converting parallel data to serial data bits) and decoding (converting serial data bits into parallel data) is used. Some HDCs use NRZI (non-return to zero inverse) for encoding and decoding data. Once the data is serialized, the HDC uses a format method to store or retrieve the data to or from the disk drive unit. Popular forms are MFM (modified frequency modulation), RLL (run length limited), or a specialized recording method that varies from drive to drive unit.

9.4.2 Hard Drive Unit

As opposed to a floppy disk drive unit, which uses a removable storage medium, most hard drive systems use one or more nonremovable disks (platters). Each platter has two recording sides and its own R/W head. The platters (Fig. 9-17) are stacked on a single spindle with only enough space to allow the R/W heads to move back and forth without interfering with each other. As the spindle rotates, all platters in the hard drive rotate at the same speed. The rotation

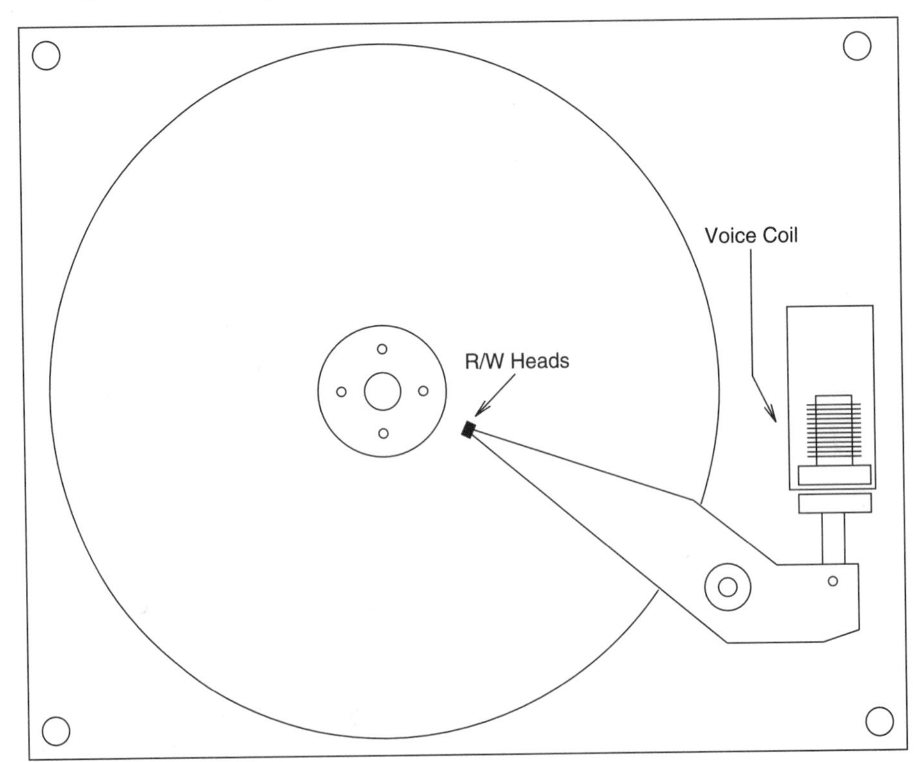

Figure 9-17 Hard drive, internal view.

speed of most hard drives is 3600 rpm. All hard drives use direct drive motors to drive the spindle. A typical hard drive unit contains one to eight platters.

The platters vary in size, thickness, and types of magnetic coatings and therefore in the amount of data that can be stored on each one. The platters range from 5.25 inches in diameter for older drives to 3.5 inches, 2.5 inches, and 1 inch for newer drives. The thickness is normally between 1/16 and 3/16 of an inch. Older platters use iron oxide coatings, while newer drives use plated, thin film, and sputtered platters.

In the hard drive unit, the R/W heads never come in direct contact with the surface of the platters; the heads float on a cushion of air about 4 to 10 microinches from the surface. Because of the construction of the platters (not flexible) and the speed at which they rotate, it is necessary that the hard drive unit be sealed against any type of pollutants. If the seal is broken in any type of environment other than a clean room, the drive most likely will not function properly because pollutants will cause the R/W heads to scratch the surface of the platter. If the R/W heads dig into the surface too much, a head crash could occur and destroy the platters and/or R/W heads. Never take the cover off a hard drive unit. If a head crash occurs or the drive breaks down and data on the disk is needed, the following steps should be taken. Back up any data on the disk if possible; next use one of the third-party software packages to recover data. If the data cannot be recovered or has not been backed up, send the hard drive to a data recovery company that specializes in recovering data from defective drives. These companies have specialized equipment that can usually recover the data; although the cost of data recovery is high.

Because there are two sides to each platter, each platter requires two R/W heads, one for each side. All R/W heads are stacked onto a shaft, that causes all heads to move at the same time whenever a different track (cylinder) is called for. In older hard drive units, stepper motors were used to move the R/W heads, while the newer hard drive units use voice coils to position the R/W heads. Hard drive units that use stepper motors are slow, the motor takes up a lot of space, and as the motor wears the step position changes, which may cause errors because the tracks are no longer aligned. Voice coils are used today in most hard drive units because they are very fast, their alignment does not change, they are very small, and they do not wear out as fast as stepper motors.

Because hard drive units are sensitive to vibration, it is necessary to park the R/W heads before the drive unit is moved or transported. The parking process causes the R/W heads to position themselves to an area of the platter that is not used to store data. In older drivers this process was performed with a software command; in today's hard drive units, this process is performed automatically when power is removed from the drive unit.

Inside each hard drive unit is a micron filter used to trap any loose. particles. As the platters spin, air is circulated in the drive unit and the filter traps the particles. Any loose particles may cause the R/W heads to crash. The loose particles may be bits of magnetic coating that come off the platters as they rotate or with age.

Each hard drive in a hard drive system must be identified; most hard drive adapters control two or more hard drive units. The first hard drive that is accessed upon boot-up is called the master, and second hard drive unit is called the slave. A jumper is provided on the drive unit to designate the drive unit as either a master or a slave.

9.4.3 Preparing a Hard Drive

A hard drive must be prepared to accept data. Because of the different types, sizes, and platters used in hard drive units, DOS cannot format a hard drive like a floppy diskette (which has predefined parameters). In all hard drive systems except IDE systems, three steps are required to prepare the drive unit for data storage. When DOS formats a floppy diskette, it automatically performs all three steps.

Once the hard drive unit is installed and the computer is booted with a floppy diskette, the first step is to perform a lower level format. During that process, the hard drive controller writes the data and verifies all tracks and sector locations on the hard drive unit. A lower-level format can be performed in one of three different ways. Regardless of the method, the user must know the number of cylinders, heads, and sectors per track. The user can access ROM BIOS on the hard drive adapter that contains a lower-level format routine. The second method uses BIOS, if available, to perform a lower-level format; this method may not be available with all computer BIOS. When this method is used, the data from the CMOS set-up is used to perform the lower-level format. The third method uses a third-party software program to perform the lower-level format. During the format the software asks for a bad track map, which is supplied by the hard drive manufacturer (located on a label on the drive unit); the software also verifies each surface and indicates any bad locations. Any bad surface locations are marked and are not used when the disk drive is formatted.

The second step in preparing the hard drive is to partition the drive, which allows the user to subdivide the physical hard drive into one or more logical drives. Prior to DOS 5.0, the size of any single hard drive partition was limited to 32MB. If the hard drive was larger than 32MB, the user had to partition it. Each partition used by DOS is given a drive letter; the primary DOS partition is always drive C. Any space in the extended DOS partition can be logically divided into as many drives as the user wants as long as there is space remaining. Logical extended DOS drives are specified beginning with drive letter D. Drive C is the only DOS partition that can be used to boot DOS. FDISK (Fixed DISK) is a DOS command used to partition the hard drive unit.

The final step in preparing the hard drive is to perform high-level formatting on the hard drive unit. During the high-level format, DOS creates the boot record, FAT, and root directory. The user can have the format command make the hard drive bootable by transferring the system files to the hard drive after the formatting process is complete. The time it takes perform a high-level format depends on the size and speed of the hard drive. Each partition must be formatted separately.

9.4.4 Types of Hard Drive Systems

There are basically five types of hard drive systems, each with its advantages and disadvantages. The systems use different methods of data transfer to and from the hard drive units and adapter.

Modified Frequency Modulation. Modified frequency modulation (MFM) drives were the first type of hard drive available. MFM drives follow ST506/412 specifications. MFM specifies how the data bits are stored to and retrieved from the hard drive unit. An MFM hard drive is controlled by two ribbon cables—a 34-pin control cable and a 20-pin data cable (Table 9-5). The drive unit uses an edge card connector for both cables, while the hard drive adapter uses a header connector for both cables. Normal sizes for MFM drives were 10MB, 20MB, 40MB, and 80MB, with access time ranging from 100 ms to 28 ms. The form factor size was 5.25 inches.

This type of drive used stepper motors and was not very reliable after 2 years of use. To maintain data integrity the data had to be pulled off the drive and the drive lower-level formatted, after which the data could be restored to the hard drive.

Run Length Limited. The run length limited (RLL) method of modulation also follows ST506/412 specifications. RLL differs from MFM by reducing the intervals between flux changes to increase the amount of data on the surface of the platter. RLL provides 50% more storage space on the same size hard drive unit. RLL drives use the same cable connections as MFM drives. Normal sizes for RLL drives were 30MB, 60MB, and 120MB, with access time ranging from 65 ms to 28 ms; the form factor size was 5.25 inches. These types of drives used stepper motors and were not very reliable after 2 years of use. The advantage of RLL drives is greater storage space on the same size drive unit. The disadvantage of RLL drives is that they are more prone to data errors. They also share the disadvantages of MFM drives.

TABLE 9-5
MFM/RLL Drive Cable Connections

34-Pin Hard Drive Control Cable			
Pin Number	*Description*	*Pin Number*	*Description*
1	Ground	2	Reduce Write Current#
3	Ground	4	Not used
5	Ground	6	Write Gate#
7	Ground	8	Seek Complete#
9	Ground	10	Track 00#
11	Ground	12	Write Fault#
13	Ground	14	Head Select 1# (Drive 0)
15	Ground	16	Not used
17	Ground	18	Head Select 2# (Drive 1)
19	Ground	20	Index#
21	Ground	22	Ready#
23	Ground	24	Step#
25	Ground	26	Drive Select 1#
27	Ground	28	Drive Select 2#
29	Ground	30	Not used
31	Ground	32	Not used
33	Ground	34	Direction#
20-Pin Hard Drive Cable			
Pin Number	*Description*	*Pin Number*	*Description*
1	Drive Select#	2	Ground
3	Not used	4	Ground
5	No pin	6	Ground
7	Not used	8	Ground
9	Spare	10	Spare
11	Ground	12	Ground
13	MFM Write Data	14	MFM Write Data#
15	Ground	16	Ground
17	MFM Read Data	18	MFM Read Data#
19	Ground	20	Ground

Enhanced Small Device Interface. Enhanced small device interface (ESDI) systems use hard drive adapters that allow more intelligent communications between the adapter and the drive unit itself. ESDI drives are used in high-speed applications and large hard drive systems. Typical data transfer rates range from 10 to 15M bps, a great improvement over MFM and RLL drives. Table 9-6 shows the ESDI drive cable connections. ESDI cables use 34-pin and 20-pin connectors, with edge connectors on the drive unit and headers on the adapter. ESDI systems range in size from 100MB through 1GB, with access times ranging from 40 ms to 20 ms. The form factor size is 5.25 inches. The advantages of ESDI systems are high data transfer rates, fast access times, and large sizes. The disadvantages of ESDI systems are very high cost and some incompatibility.

Small Computer System Interface. Small computer system interface (SCSI) systems have the hard drive controller located on the drive unit itself. The SCSI adapter is a simple gate that controls the flow of data between the data bus and the drive unit. The controller on the drive unit performs all the encoding and decoding of data. Data is passed between the adapter and the controller on the drive unit in a parallel format. SCSI provides only 8-bit data transfers, while the new SCSI-2 provides 8-, 16-, and 32-bit data transfers. Up to 16 different devices can be daisy chained together using one SCSI adapter. SCSI adapters are also used to control other devices such as floppy drives (non-IBM drives), CD ROMs, optical scanners, printers, and optical R/W drives. SCSI drives use a 50-pin header connector (Table 9-7) on both the drive unit and adapter. Each

TABLE 9-6
ESDI Drive Cable Connections

34-Pin Control Cable			
Pin Number	*Description*	*Pin Number*	*Description*
1	Ground	2	Head select 2#
3	Ground	4	Head select 2#
5	Ground	6	Write gate#
7	Ground	8	Config#/Status data#
9	Ground	10	Transfer ACK#
11	Ground	12	Attention#
13	Ground	14	Head select 1#
15	Ground	16	Sector#
17	Ground	18	Head select 2#
19	Ground	20	Index#
21	Ground	22	Ready#
23	Ground	24	Transfer request#
25	Ground	26	Drive select 1#
27	Ground	28	Drive select 2#
29	Ground	30	Drive select 3#
31	Ground	32	Read gate#
33	Ground	34	Direction#

20-Pin Data Cable			
Pin Number	*Description*	*Pin Number*	*Description*
1	Drive select#	2	Sector#
3	Command complete#	4	Address mark enable#
5	Reserved for step mode#	6	Ground
7	Write clock	8	Write clock#
9	Reserved for step mode#	10	Read/Reference clock
11	Read#/Reference clock	12	Ground
13	NRZ write data	14	NRZ write data#
15	Ground	16	Ground
17	NRZ read data	18	NRZ read data
19	Ground	20	Index#

drive connected must be addressed by using a jump to identify the drive number. SCSI drives range from 40MB to 2.2GB, access times range from 65 ms to 10 ms, and data transfer rates range from 1 to 15M bps. The advantages of SCSI drives are their fast speeds, large disk sizes, and high transfer rates. The disadvantages of SCSI drives are high cost, incompatibility between SCSI adapters and drive units, and incompatibility with some software.

Integrated Drive Electronics. Integrated drive electronics (IDE) is the newest type of drive interface. The IDE standard was developed for AT systems. Like the SCSI, the controller adapter is built into the hard drive unit itself, and the IDE adapter is simply a gate that buffers the data between the bus and the drive unit. Because the IDE standard was developed for AT systems, it has access to 16 bits of data directly to the computer's data bus. Many system boards incorporate the IDE adapter on the board itself. IDE drive units also have large cache buffers to help maintain the throughput of data transfers. In the beginning IDE drives were small, but today IDE drives are as large as SCSI drives, data transfer rates are as high as 16M bps, and access times range from 28 ms to 2 ms. IDE drives are already lower-level formatted. IDE drives also contain error correction circuitry to maintain data integrity. IDE drives provide the lowest cost per megabyte of storage of any drive. IDE drives are also the easiest drives to install and configure. The IDE drive uses a 40-pin header connector (Table 9-8) on both the

TABLE 9-7
SCSI Drive Cable Connections

Pin Number	*Description*	*Pin Number*	*Description*
1	Ground	2	DB0#
3	Ground	4	DB1#
5	Ground	6	DB2#
7	Ground	8	DB3#
9	Ground	10	DB4#
11	Ground	12	DB5#
13	Ground	14	DB6#
15	Ground	16	DB7#
17	Ground	18	Data parity (odd)#
19	Ground	20	Ground
21	Ground	22	Ground
23	Ground	24	Ground
25	Not Used	26	Not Used
27	Ground	28	Ground
29	Ground	30	Ground
31	Ground	32	Attention#
33	Ground	34	Ground
35	Ground	36	BSY#
37	Ground	38	ACK#
39	Ground	40	Reset#
41	Ground	42	Message#
43	Ground	44	SEL#
45	Ground	46	C/D#
47	Ground	48	Request#
49	Ground	50	I/O#

TABLE 9-8
IDE Drive Cable Connections

Pin Number	*Description*	*Pin Number*	*Description*
1	Host reset#	2	Ground
3	+ Host data 7	4	+ Host data 8
5	+ Host data 6	6	+ Host data 9
7	+ Host data 5	8	+ Host data 10
9	+ Host data 4	10	+ Host data 11
11	+ Host data 3	12	+ Host data 12
13	+ Host data 2	14	+ Host data 13
15	+ Host data 1	16	+ Host data 14
17	+ Host data 0	18	+ Host data 15
19	Ground	20	Key
21	Not used	22	Ground
23	Host IOW#	24	Ground
25	Host IOR#	26	Ground
27	+ IO Ch Rdy	28	+ Host ALE
29	Not used	30	Ground
31	+ Host IRQ14	32	Host IOCS 16#
33	+ Host Addr 1	34	Host PDIAG#
35	+ Host Addr 0	36	+ Host ADDR 2
37	Host CS 0#	38	Host CS 1
39	Host SLV/ACT	40	Ground

adapter and the drive unit. IDE drives provide large storage in small space with form factor sizes ranging from 3.5 inches to 1 inch. With the popularity of the IDE interface, more devices are becoming IDE comparable. Currently IDE CD ROMs are available, and optical scanners will be available soon. IDE drives are the current standard for AT interfaces. The advantages of IDE drives are large size, high data transfer rates, fast access times, large drive cache buffers, error correction circuitry, lower-level formatting, ease of installation, and low cost. The only disadvantage is the lack of additional devices other than hard drives and CD ROMs.

CHAPTER SUMMARY

1. The controller adapter converts data formats between the computer and drive unit. The adapter converts commands from the computer into the signals necessary to control the drive unit and uses feedback information provided by sensors on the drive unit to verify the operation of the commands issued to the drive unit.
2. The drive unit contains the motors, sensors, and electronic interface circuitry that will convert TTL levels into levels needed by the R/W heads. The drive unit also converts levels provided by the R/W heads into TTL levels for the computer.
3. Mass storage systems use magnetic fields, optical scanning, or a combination of both as the method of storing data on the surface of the media.
4. DOS is the software part of the mass storage system that is used to program the computer. It allows the computer to control the hardware of the mass storage system.
5. Tape, floppy, and hard disk systems use magnetic storage methods. CD drives use optical storage methods. Floptical systems use both magnetic and optical storage methods.
6. FM encoding requires two pulses (clock and data pulse) to represent a logic-1 and one pulse for a logic-0. MFM encoding requires one pulse (clock and data combined) to represent a logic-1 and no pulse for a logic-0.
7. A logical sector is a package of data that contains 512 bytes of user data. A track is a location on the surface of the diskette where the R/W heads are positioned. Each track contains a number of logical sectors. The number of logical sectors varies depending on the formatted size of the diskette.
8. The 765 floppy disk controller is responsible for converting parallel data in serial data during a write operation and converting serial data into parallel data during a read operation. The 765 is also responsible for interpreting DOS instructions to perform the selected task on the disk drive unit.
9. The following defines the function of each pin of the 765:

RESET	The RESET line is used to reset the 765 to starting conditions.
RD#	The Read Data line is used to cause the FDC to perform a read operation.
WR#	The Write Data line is used to cause the FDC to perform a write operation.
CS#	The Chip Select line is used to access the FDC by the MPU for programming or reading the status of one or more of the registers.
A0	Address line zero is used by the MPU to select whether the data or main status registers will be accessed by the read or write operation.

INT	The INTerrupt output line is used by the FDC to indicate to the MPU that an interrupt has occured in the FDC.
DRQ	The DMA ReQuest output indicates to the DMA that the FDC needs servicing.
DACK#	The DMA ACKnowledge input is used by the the DMA to inform the FDC that the DMA is performing a DMA data transfer cycle.
TC	The Terminal Count input is used by the DMA to inform the FDC that the DMA cycle has ended.
D0 to D7	The bidirectional data lines are used by either the MPU or DMA to transfer data to and from the FDC.
Vcc	Vcc provides the necessary votlage potential for the FDC.
GND	The GrouND line provides the necessary current and reference for the FDC.
CLK	The CLocK input line supplies the necessary clock signals for the FDC.
IDX	The InDeX input informs the FDC that the index circuitry on the floppy disk drive unit has detected the index hole.
WCK	The Write ClocK input supplies the proper clock frequency to the FDC during a floppy disk write operation.
RDW	The Read Data Window input is used by the PLL to supply a data sample to the FDC.
RD	The Read Data input receives the clock and data bits from the data separator.
VCO	The Voltage Controller Oscillator output enables or disables the VCO in the PLL during different types of FDC operations.
WE	The Write Enable output enables the write circuitry on the floppy disk drive unit during a write opertion.
MFM	The density select output supplies density information to the PLL.
HD	The HeaD select output informs the floppy disk drive unit which R/W head should be used to perform the specified operation.
US0 to US1	The Unit Select output lines are used to select which one of the four disk drive units will be enabled during a FDC operation.
WDA	The Write DAta output sends the clock and data bits from the FDC to the write precompensation circuitry on the adapter.
PS0 to PS1	The precompensation select output lines are used to select the type of delay the R/W heads will use in storing the data and clock bit on the diskette during a FDC write operation.
FLT/TR00	The FauLT/Track 00 input lines inform the FDC that the R/W heads are on track) during a recalibrate command.
WP/TS	The Write Protect and Two Side input lines inform the FDC that the diskette is write protected.

RDY	The ReaDY input informs the FDC that the disk drive unit is ready to send or receive data.
HDL	The HeaD Load output is used to inform the disk drive unit to engage the R/W heads near the surface of the diskette.
FR/STP	The Fault Reset and STeP output line sends a step pulse to the disk drive unit.
LCT/DIR	The Low CurrenT and DIRection output is used to indicate to the disk drive unit to lower its current to the R/W heads or to determine the direction the R/W heads will move.
RW#/SEEK	The read/write and seek output determines whether the disk drive unit will be in the R/W mode or the seek mode.

10. All floppy disk operations require three phases. During the command phase the command is issued to the FDC and the necessary parameters are loaded into the FDC registers. During the execution phase, the operation is performed by the drive unit and adapter. The last phase is the result phase, which verifies the operation of the execution phase by reading different registers of the FDC.
11. 765 FDC commands:

    READ DATA: reads data from the diskette at the selected sector, track, and side.

    READ DELETED DATA: reads deleted data from the diskette.

    WRITE DATA: writes data to the diskette at the selected sector, track, and side.

    WRITE DELETED DATA: writes data to a location where data was deleted.

    READ A TRACK: reads all data on the selected track and side.

    READ ID: reads the ID sector field on the selected track and side.

    FORMAT TRACK: writes sector format information on the selected track and side.

    SCAN TRACK: scans the diskette beginning at the selected sector, track, and side.

    SCAN LOW OR EQUAL: scans the diskette beginning at the selected sector, track, and side.

    SCAN HIGH OR EQUAL: scans the diskette beginning at the selected sector, track, and side.

    RECALIBRATE: causes the R/W heads to position themselves at track 0.

    SENSE INTERRUPT STATUS: supplies the interrupt status of the selected drive unit.

    SPECIFY: sets the initial values of the three internal timers of the FDC.

    SENSE DRIVE STATUS: obtains the status of any of the disk drive units.

    SEEK: moves the R/W heads to another track.
12. The purpose of each block of the block diagram of the floppy disk drive adapter follows:

    Bus Buffer: The data bus buffer is a buffer that is used to hold data until either the FDC or DMA can use the data.

    765 FDC: The FDC converts data passed between the MPU and disk drive unit. The FDC also interprets the command from DOS and controls the hardware to complete the command task.

    Clock and Timing: The clock and timing circuitry is used to supply clock signals to the FDC, write precompensation circuitry, and the data separator.

    Digital Control Port: Supplies information to the FDD decoder to indicate which disk drive unit will be used during the selected operation.

    FDD Decoder: The floppy disk drive decoder selects which drive unit will be enabled and disabled.

Write Pre-comp Circuit: The write precompensation circuitry is used to determine which time delay will be used during a write operation.

Data Separator: The data separator separates the data from the clock signals during a read operation.

Drivers: The open collector drivers are used to provide the necessary voltage and current for the disk drive unit and FDC.

13. The disk drive unit is used to move the storage media so that data can be either read or written onto the diskette.
14. The following summarizes the electronics of the disk drive:

    The pulse shaping network is used to create the index pulse, when the index mark passes the index sensor.

    The write protect switch is used to indicate when the diskette is write protected.

    The Track 00 switch indicates the R/W heads are on track zero.

    Spindle drive control circuitry is used to rotate the floppy diskette at the proper speed.

    Positioner control circuitry is used to position the R/W heads on the proper track.

    Write/Erase current source and waveform generator circuitry is used to supply current and frequency necessary for the R/W heads to write data on the diskette.

    The read amplifier and digitizer is used to amplify and digitize the signals being supplied by the R/W heads during a read operation.
15. When a disk operation is called for, DOS will convert the DOS syntax into the necessary hardware command sequence to perform the operation on the floppy diskette. One DOS command may represent 100 or more hardware commands.
16. A high-density floppy disk system basically uses the same items that a double-density floppy disk system uses, except when a high-density disk is used, the disk drive unit has higher quality and faster R/W heads and circuitry, which allows more data to be stored on each track of the disk. A high-density floppy disk system can read and write data onto double-density disks. In some cases, the data written onto a double-density diskette written by a high-density disk system may not be used reliably on a double-density floppy system because the magnetic fields placed on the disk from a high density system are smaller in size and not as strong.
17. A hard drive system is made up of three parts as opposed to the four parts of a floppy disk system. The hard drive adapter performs the same functions as a floppy disk adapter, except that it operates much faster. The hard drive unit performs the same functions as the floppy disk drive unit and floppy disk, except the hard drive disks (platters) are nonremovable, turn at much higher speeds, contain from one to eight hard drive platters per drive unit with one pair of R/W heads per platter, and the density of the coating on each platter is much more dense. DOS for both the floppy disk and hard drive systems performs the same functions, except the size of the directory and file allocation table are larger for the hard drive systems.
18. All hard drive systems, except IDE systems, require three steps before they can be used for data storage. The first step in preparing a hard drive is to perform a low-level format (sometimes performed by the manufacture). (This step is not needed with IDE systems.) The second step is to partition the hard drive using DOS's FDISK command. The third step is to perform a high-level format using DOS's FORMAT command.
19. MFM hard drive systems were the first type of hard drive system. Their sizes range from 10MB to 80 MB with access speeds ranging from 100 ms to 28 ms, and a 5.25-inch form factor. They were the first type of hard drive system. To maintain data integrity a low-level format must be performed periodically because the internal parts of the drive unit fall out of calibration.
20. RLL hard drive systems provide 50% more storage space on the same size disk as a MFM drive. Their sizes range from 30MB to 120MB and can access speeds from 65 ms to

28 ms with a 5.25-inch form factor. To maintain data integrity, a low-level format must be performed periodically. These types of hard drive systems are more prone to data errors because of how the data was stored on the disk drive.

21. ESDI hard drive systems have sizes that range from 100MB to 1GB, access speeds from 40 ms to 20 ms, data transfer rates of 10 to 15 M bps, and a 5.25-inch form factor. Advantages to using these kinds of systems are high data transfer rates, fast access times, and large sizes. Some disadvantages are high cost and some incompatibility.
22. SCSI hard drive systems have sizes that range from 40MB to 2.2GB, access speeds from 65 ms to 10 ms, data transfer rates from 1 to 15 M bps, and a 5.25- to 3.5-inch form factor. Advantages include a high speed, large size, and high transfer rates. Disadvantages include a high cost, incompatibility between some SCSI adapters and drive units, and incompatibility with some software.
23. IDE hard drive systems have sizes that range from 20MB to 2.2GB, access speeds from 28 ms to 2 ms, data transfer rates as high as 16M bps, and 3.5-, 2.5-, and 1-inch form factor sizes. Advantages include a large size, very low cost, high data transfer rates, fast access times, large drive cache buffers, error correction circuitry, and ease of installation. Disadvantages include a lack of additional devices that use the interface other than hard drives and CD ROMs.

REVIEW QUESTIONS

1. Which part of any mass storage system acts as the interface between the mass storage system and the MPU?

a. Disk adapter

b. Disk drive unit

c. DOS

2. Which part of any mass storage system allows the user to issue commands to the mass storage system?

a. Disk adapter

b. Disk drive unit

c. DOS

3. Which of the following mass storage systems is normally referred to as a read only storage system?

a. Floppy drive system

b. Hard drive system

c. CD drive system

d. Tape back-up system

e. Floptical system

4. A tape back-up system uses optical methods to store data on the tape.

a. True

b. False

5. A hard drive system is faster than a floppy disk system.

a. True

b. False

6. Which type of encoding method allows more data to be stored on a disk?

a. FM

b. MFM

7. Each position the R/W heads stop at is called a ________.

a. Sector

b. Track

c. Cluster

8. Each track contains one sector.
   a. True
   b. False
9. The IBM PC/XT has a floppy disk sector size of ________.
   a. 128 Bytes
   b. 256 Bytes
   c. 512 Bytes
   d. 1024 Bytes
10. The chip select of the 765 must go high in order to access the 765.
    a. True
    b. False
11. Which 765 input is used to inform the 765 that the R/W heads are on the first sector of the track?
    a. WCK
    b. IDX
    c. VCO
    d. MFM
    e. None of the given answers
12. Which 765 input is used to input read data from the data separator?
    a. RDW
    b. VCO
    c. RD
    d. HD
    e. None of the given answers
13. Which 765 output is used to output data to the write precompensation circuitry?
    a. WDA
    b. WCK
    c. HDL
    d. RDY
    e. None of the given answers
14. Which of the following phases is the first phase in a floppy disk operation?
    a. Execution phase
    b. Command phase
    c. Result phase
15. The write-deleted data command is used when data is written onto an area of the disk where old data was once stored.
    a. True
    b. False
16. The digital control port is used to separate the clock and data being read from the disk drive unit.
    a. True
    b. False
17. The step line on the floppy diskette ribbon connector is used to tell the floppy disk drive unit how large a step it should take each time the R/W heads move.
    a. True
    b. False
18. Which part of the floppy disk drive unit is used to inform the floppy disk drive adapter that the R/W heads are retracted to their starting position?
    a. Index pulse shaping network
    b. Write protect switch
    c. Track zero switch

d. Head positioner control

e. None of the given answers

19. The most common problem of a floppy disk drive unit is dirty R/W heads.

a. True

b. False

20. Which part of the DOS format contains the format signature?

a. Boot record

b. File allocation table

c. Directory

d. COMMAND.COM

e. None of the given answers

21. The spindle speed of a high-density disk unit is faster than a double-density disk unit.

a. True

b. False

22. Hard drive controllers support more high-level commands than floppy disk controllers.

a. True

b. False

23. The spindle speed of a hard drive system is faster than the spindle speed of a floppy disk system.

a. True

b. False

24. The part of preparing a hard drive that allows the user to electronically divide a physical hard drive unit into smaller drives is called ________.

a. Low-level formatting

b. Partitioning

c. High-level formatting

25. Which hard drive standard is the least expensive and easy to install?

a. MFM

b. ESDI

c. SCSI

d. IDE

e. RLL

chapter 10

I/O Systems and Devices

OBJECTIVES

Upon completion of this chapter, you should be able to perform the following tasks:

1. List the differences between parallel and serial communications.
2. List the advantages and disadvantages of parallel and serial communications.
3. List the function of each Centronics printer interface signal as used in the IBM PC.
4. List and define the function of each block in the block diagram for the printer interface used in the IBM PC.
5. List the operational sequence for the printer interface.
6. List and define the three different methods of data communications.
7. List the operational sequence of a typical NRZ asynchronous transmission.
8. List and define the purpose of each block in the block diagram for the asynchronous adapter.
9. List and define the purpose for each signal pin of RS-232-C and 20-ma current loop interface standards.
10. Define basic printer terminology.
11. List and define the purpose for each block in a block diagram of a typical printer.
12. List the operational sequence for a typical printer.
13. List and define how different printers produce print.
14. List the advantages and disadvantages of different types of printers.
15. List the basic operation of a laser printer.
16. List the basic operation of an ink jet printer.
17. List the advantages and disadvantages of each type of mouse.
18. List the basic operation of each type of mouse.

10.1 PARALLEL DATA COMMUNICATIONS

Parallel data transmission is normally restricted to the internal bus of the computer or to peripherals that require fast transmission of data. In parallel data communications, data is transferred from one device to another device in a parallel format: More than one bit of data is transmitted at the same time, which requires one transmission line for each bit transmitted. In most peripherals that use parallel communications, one byte is transmitted during each cycle and a total of eight transmission lines are used for the data. Additional lines are used to control the data transmission. The additional control lines establish smart communications (called handshaking) between the computer and the peripheral device. During handshaking information is shared between the two devices regarding how the data is transferred. The Centronics parallel printer interface is the only parallel standard. If the computer is to communicate to a parallel device other than a printer, there is no set standard.

Advantages

It is fast because more than one bit is transmitted at a time.

Normally no additional hardware is needed.

Disadvantages

Cost is high because of the number of transmission lines (wires) needed.

It has a tendency to pick up stray noise, because of the number of transmission lines.

The distance between devices is normally kept below 20 feet, because of the two disadvantages listed above.

10.1.1 Centronics Printer Interface

The Centronics parallel printer interface was established in the early 1970s by one of the largest printer companies at the time, Centronics, Inc. The interface uses TTL levels to transfer data and perform the handshaking between the computer and printer. The original Centronics interface uses a 36-pin connector, which is standard for all printers with a parallel printer interface. Of the 36 pins, IBM and most other computer companies only use 25 pins. The following list gives the pin numbers, type of I/O, and signal name for the standard Centronics interface. These signals are at the printer.

Pin 1	The STROBE# line is an active low input for the printer. This line, when low, informs the printer that the data on the data inputs is valid.
Pins 2–9	The eight data inputs (D1–D8) are used by the printer interface to supply the data byte of the printer; the data is usually in ASCII.
Pin 10	The ACK# (acknowledge) output from the printer, when low, informs the printer interface that the data is being processed by the printer.
Pin 11	The BUSY output from the printer, when high, informs the printer interface that the printer is busy and is not ready to receive any new data.
Pin 12	The PE (paper empty) output line from the printer, when high, informs the printer interface that the printer is out of paper. This line may not be monitored by all interfaces and therefore may not be used by all interfaces.
Pins 13, 35	These pins are used to supply a 5-volt reference level between the printer interface and printer itself.
Pin 14	The AUTO FEED XT# input line causes a printer line feed when this line goes low.

Pins 15, 18, 34	These lines are not used.
Pin 16	This line is the logic signal ground.
Pin 17	This line is the chassis ground.
Pins 19–30	These lines act as return lines for the data inputs.
Pin 31	The INT (initialize) input line is used by some printer interfaces to enable the printer, when high. This line is not used by all printers.
Pin 32	The ERROR# output line from the printer, when low, informs the printer interface that a problem has occurred in the printer and data should not be sent.
Pin 33	This is the signal return ground line for the paper empty signal (PE pin 12).
Pin 36	The SLCT IN# (select input) input line, when low, selects if the printer is to be on line (enabled).

10.1.2 IBM Printer Adapter

This section covers the IBM PC parallel printer interface board (Fig. 10-1) and how the IBM uses the Centronics parallel printer interface.

Address Decoding. The majority of the address decoding (Fig. 10-2) is performed by an eight-input NAND gate. In order for this chip to be addressed, the following lines must be within the following address ranges.

$$A3 = 1, A4 = 1, A5 = 1, A6 = 1, A7 = 0, A8 = 0, A9 = 1$$

$$AEN\# = 0, (IOR\# = 0 \text{ or } IOW\# = 0)$$

Hexadecimal address 00278–0027F

Command Decoder. The command decoder (Fig. 10-2) is a 74155 dual two-to-four multiplexer. The purpose of the command decoder is to generate the read and write command sequence for the five different registers of the printer adapter (Table 10-1).

When an output is enabled, it goes from a high to a low state. The transmission causes the data to be latched into the selected latch or stored in one of the read buffers. All disabled outputs stay high.

Control Latch. The control latch is used to latch to the necessary command level output lines for the printer adapter. The data is latched to this latch when address 0027A (write control logic) is selected during a write operation. The conditions of the output lines are controlled by the software that is being used and not by the hardware in this adapter.

Bit 0	STROBE# line bit.
Bit 1	AUTO FD XT# line bit.
Bit 2	INIT line bit.
Bit 3	SLCTIN# line bit.
Bit 4	IRQEN line bit (applied to control buffer).
IRQEN	(use to generate IRQ7).

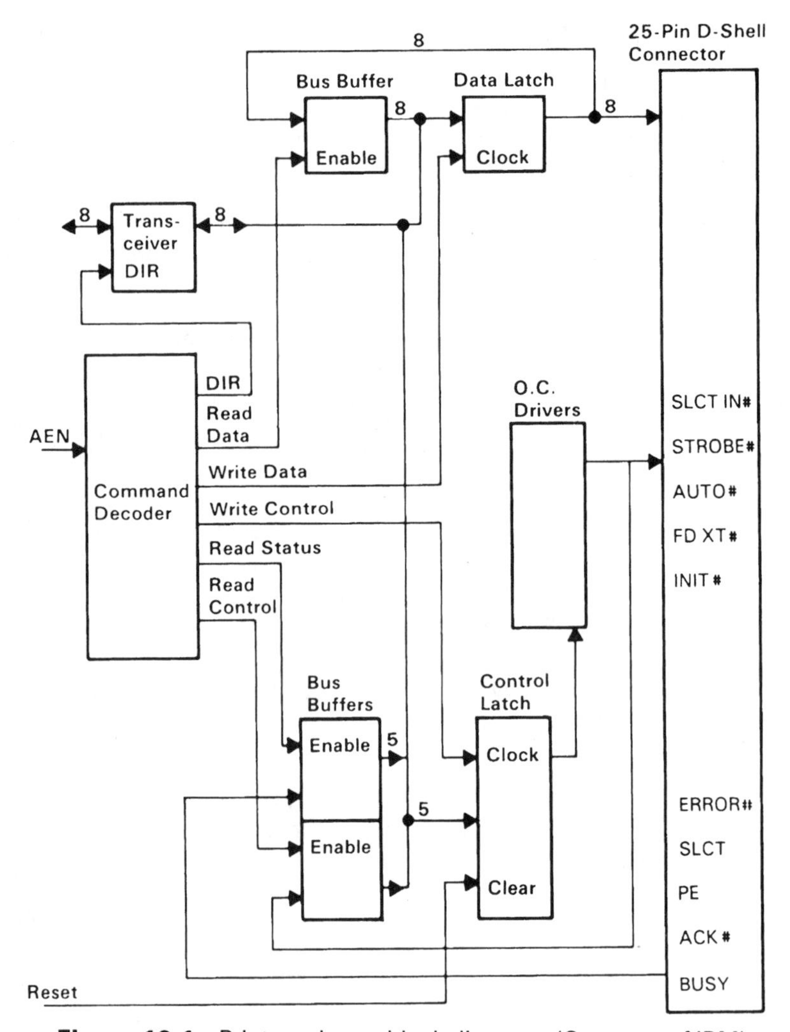

Figure 10-1 Printer adapter block diagram. (Courtesy of IBM)

These outputs, as well as being applied to the printer, are applied to the control buffer.

Control Buffer. The purpose of the control buffer (bottom section of the bus buffer) is to supply a positive response of the condition of the control latch outputs. The software performing the printer operation may or may not check these levels during the operation. The following five bits are used to verify the levels on the control latch:

Bit 0	STROBE# bit.
Bit 1	AUTO FD TX# bit.
Bit 2	INIT bit.
Bit 3	SLCT IN# bit.
Bit 4	IRQ EN bit.

Status Buffer. The status buffer (upper section of bus buffer) holds the following five bits of information about the status of the printer. The bits in this buffer are read when I/O

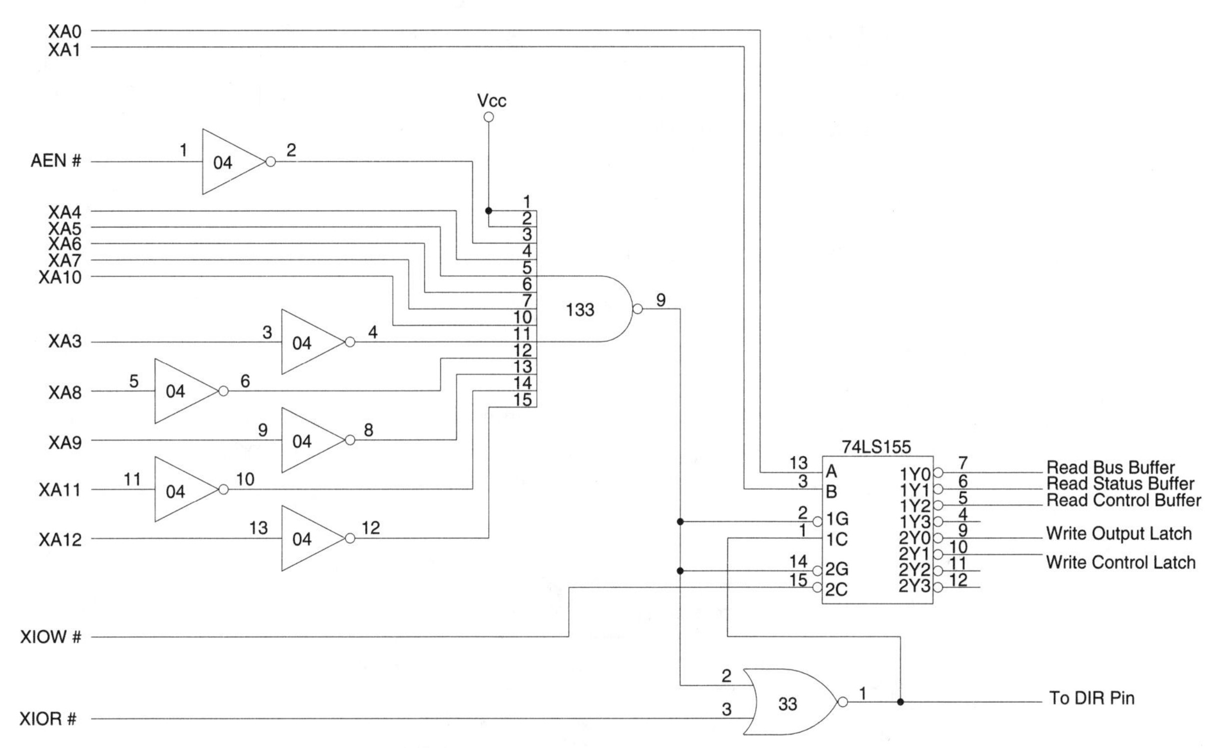

Figure 10-2 Address decoding for printer interface.

TABLE 10-1
Printer Port Addresses

Address	*IOR#*	*IOW#*	*Enabled Output*
00278	0	1	Read bus buffer
00279	0	1	Read status buffer
0028A	0	1	Read control buffer
00278	1	0	Write output latch
0027A	1	0	Write control latch

address 00279 is accessed. The levels of these bits are determined by the output lines from the printer and show the operational condition of the printer at the time.

Bit 0 ERROR# bit.

Bit 1 SLCT bit.

Bit 2 PE bit.

Bit 3 ACK# bit.

Bit 4 BUSY bit.

Data Latch. The output data latch holds the eight bits of data until the printer can use the information. The data is latched when the address 00278 is written into.

Data Buffer. The input latch is used by some software to verify that the data on the eight data outputs from the output data latch is good. The output data is latched when address 00278 is read.

Bus Transceiver. The data bus transceiver is used to control the flow of data to and from the printer adapter. When the card is properly addressed and an IOR# is performed, data flows from the printer card to the data bus. When the card is properly addressed and an IOW# is performed, data flows from the data bus to the printer card.

Open Collector Drivers. The open collector drivers are used to supply the necessary current for the drive control signals to the printer. Each open collector driver uses a 4.7 K ohm pull-up resistor to supply.

IBM Centronics Pin Out. Table 10-2 is a comparison of the IBM and Centronics interface pin out.

IBM PC Printer Connector. The following diagram and pin functions represent the connections of the IBM PC printer connector (Fig. 10-3). The parallel printer port uses a DB-25 connector to connect the printer to the printer adapter.

TABLE 10-2
IBM to Centronics Table

IBM Pin	*Centronics Pin*	*Signal Name*
1	1	STROBE#
2–9	2–9	DB0-DB7
10	10	ACKNOWLEDGE#
11	11	BUSY
12	12	PAPER EMPTY
13	36	SELECT OUTPUT
14	14	AUTO FEED#
15	32	ERROR#
16	31	INITIALIZED PRINTER
17	36	SELECT INPUT#
18–25	16	GROUNDS

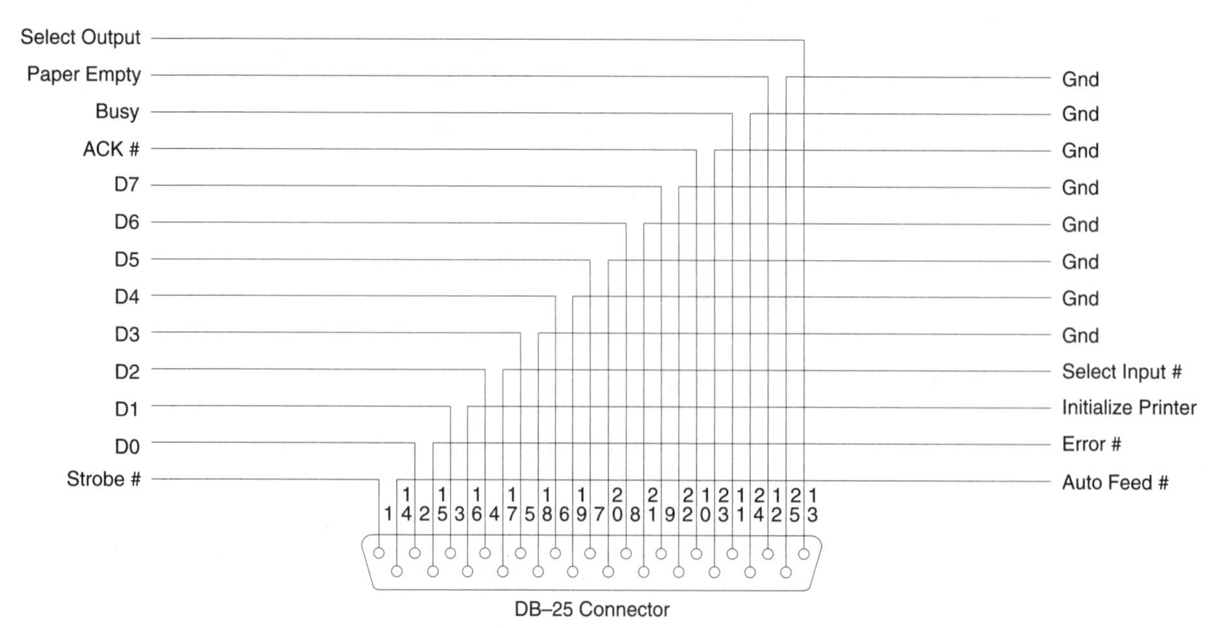

Figure 10-3 DB-25 pin configuration.

Pin 1	The STROBE# is used to inform the printer that the printer adapter has placed data on the data bus for the printer. This output is an active low and must stay low for a minimum of 5 μs.
Pins 2–9	These eight output lines (DB0–DB7) supply the ASCII codes for the printer. The data must stay active at least 10 μs after the BUSY line goes high.
Pin 10	The ACK# (ACKnowledge) line informs the adapter when the printer is ready for more data. This input is an active low.
Pin 11	The BUSY line informs the adapter when the printer is busy and cannot accept data. The printer is busy when the printer is off line, reading data on the data bus outputs, printing, and if an error state is active. This is an active high input.
Pin 12	The paper empty (PE) input goes high when the printer is out of paper.
Pin 13	The select input is connected to the select input at the printer. When this line goes high the printer is on line (selected). Not all printers use this input.
Pin 14	The AUTO FEED# output line produces a line feed at the end of paper. This line is controlled by software. The printer can be set to disable this function, in which case the printer produces its own line feed at the end of the paper. Most software supports this output; therefore, most printers have their auto line feed function disabled internally. This output is an active low.
Pin 15	The ERROR# input is used by some software to determine some type of printer error state. This goes low with most printers when PE goes high, the printer is off line, or a hardware error is detected in the printer.
Pin 16	Initialize printer output is used to enable some printers. This function is not available on all printers. When low, this output line resets the printer. It is normally used to change character and graphics sets available in the printer.
Pin 17	SELECT INPUT# is an output used to select which printer is to be used if more than one printer is connected to the printer adapter. If used, this line is an active low.
Pins 18–25	Ground lines are connected together to supply return current and offer a reference point between adapter and printer.

Typical Printer Adapter Operation

1. The MPU reads the status buffer (top part of bus buffer) and checks to make sure the following levels are present: ACK# = 1, BUSY = 0, PE = 0, ERROR# = 1. If the levels are correct, the MPU continues to the next step. If the levels are not correct, the MPU may keep reading the status buffer until the levels are correct or may do some other task, depending on the software.
2. The MPU places data in data latch. The data becomes valid on DB0–DB7.
3. After at least 5 μs, the MPU loads the control latch with the code to bring the STROBE# line low.
4. After the STROBE# line is low for at least 5 μs, the MPU reads the status buffer. If the busy bit is high, the MPU sets the STROBE# line. If the BUSY line stays low, the MPU continues to read the status buffer until the BUSY line goes high. In some software, if the BUSY line does not go high after a certain period of time, software error occurs.

5. Once the BUSY line goes high, the MPU brings the STROBE# line high and then polls (continuously reads) the status buffer until the ACK# line goes low. Once the data has been accepted by the printer, the ACK# line goes low for a minimum of 10 μs.
6. After the ACK# line goes low, the MPU polls the status buffer until the BUSY line goes low. A low on the BUSY line, when the ACK# line is low, indicates that the printer is ready to accept another byte of data. At this point the sequence starts over again (Fig. 10-4).

10.2 SERIAL DATA COMMUNICATIONS

Serial data communications occur when one bit of data is transmitted at a time. Because only one bit is transmitted at a time, only one line is needed to transfer the data. Because only one transmitted line is used, it take eight clock cycles to transfer one byte of data. There are two forms of serial data communications—serial asynchronous and serial synchronous. In serial asynchronous communications, data is transmitted in character packages (8 to 12 bits). During serial synchronous communications, data is sent in blocks (each block contains a large number of bytes, either defined or undefined), with each bit transmitted separately. There are many communication standards for serial data transmission.

Advantages

It requires fewer wires to complete the transmission.
It costs less because fewer wires are needed.
It carries less induced noise, because of fewer wires.

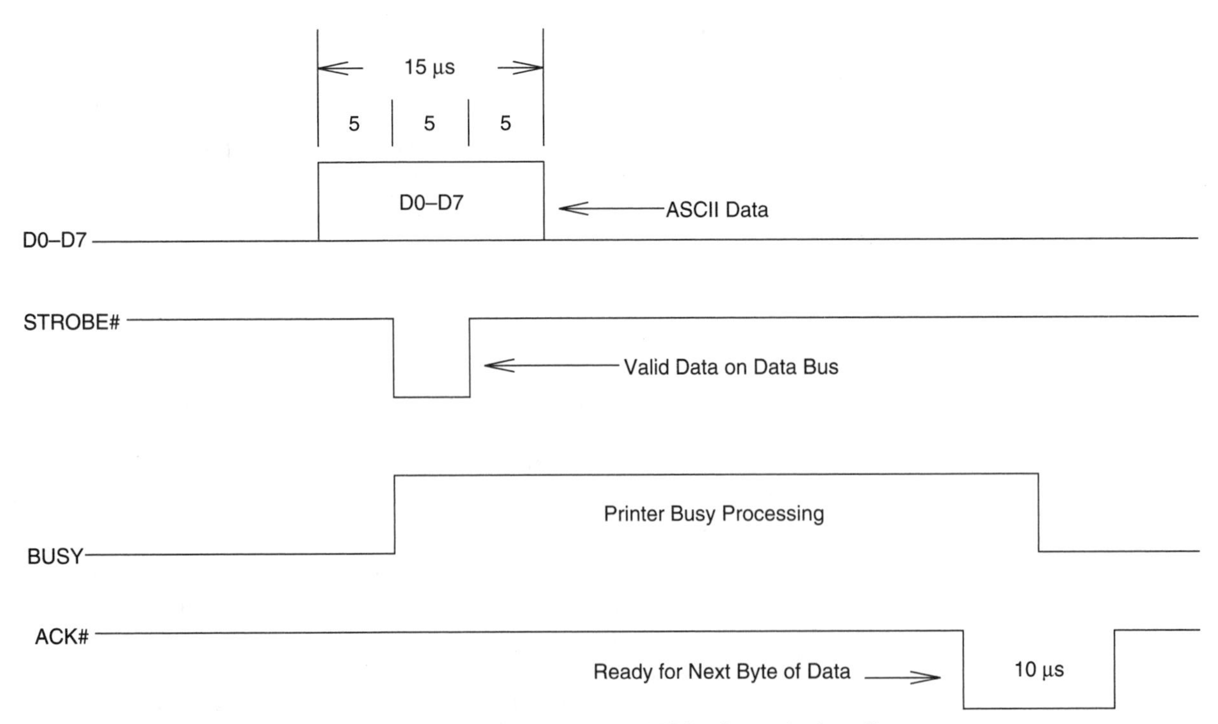

Figure 10-4 IBM printer timing diagram.

Disadvantages

It normally requires some type of additional hardware.

Transmission is slow because data is sent one bit at a time.

10.2.1 Data Communication Methods

There are three methods by which data communications can take place—simplex, half-duplex, and full-duplex.

In simplex communications (Fig. 10-5), data is transferred in only one direction. Simplex can take place with either parallel or serial data transmission. Simplex communication is the oldest type of data communications: The data is transferred without regard to whether the device receiving the information is ready or not. This type of communication is called dumb communication, because the device transmitting the data assumes the data is being received. In simplex communications, no handshaking occurs. The transmitting device never knows what the receiving device is doing. One example of this type of communication is the communication between the computer and the keyboard.

Half-duplex (Fig. 10-5) allows communications between two devices over the same line or lines but in only one direction at a time. Most computers internally use half-duplex for the data bus. During a read operation, the data flows in one direction, and during a write operation, the data flows in the opposite direction. Half-duplex can be used in both parallel and serial data communications. Another example of half-duplex is the communications of a tape back-up system connected to a computer.

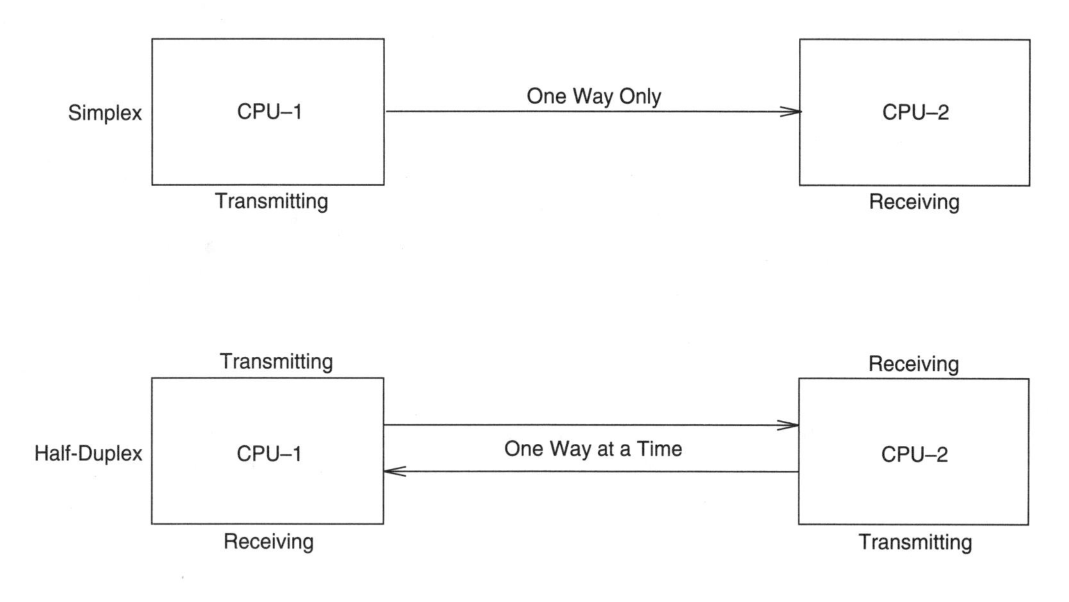

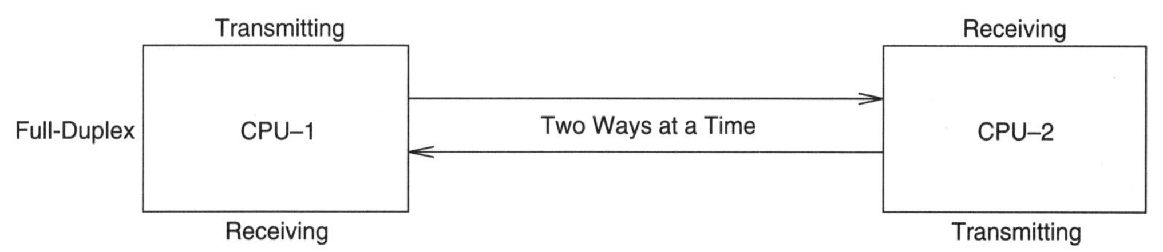

Figure 10-5 Simplex, half-duplex, and full-duplex communications.

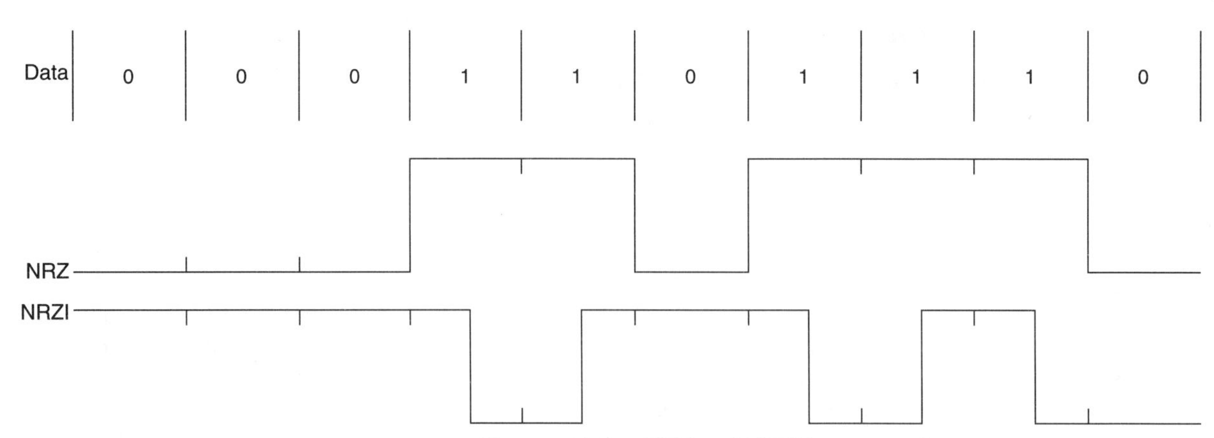

Figure 10-6 NRZ and NRZI interpretation.

When full-duplex communications occur, as in Figure 10-5, data is transferred in both directions at the same time over a pair of lines. Full-duplex is normally used only with serial data communications. The most common place full-duplex used is in communication between two modems. A modem is a device that converts TTL data into frequencies that can be sent over some type of transmission medium. The modem at the other end converts the frequencies into TTL data for the other device. The sending modem transmits at one frequency while the receiving modem transmits at a different and unrelated frequency. Because the two frequencies are different, data can flow in both directions at the same time.

10.2.2 Methods of Interpretation

The two most popular ways of interpreting asynchronous data are by using NRZ (non-return to zero) and NRZI (non-return to zero inverse) methods.

The NRZ method of interpreting data (Fig. 10-6) uses voltage or current levels to determine the type of data on the data line. Data is usually sampled in the center of the bit time (the amount of time the data stay valid on the data line). This type of interpretation requires the smallest amount of hardware and therefore costs less.

The NRZI method of interpretation in Figure 10-6 determines the type of data on the data line as voltage or current transitions during any bit time. If a transition occurs during any bit time, the data for that bit time is seen as a logic-1. If no transition occurs during any bit time, the data for that bit time is seen as a logic-0. This type of interpretation requires complex circuitry, which can be costly. NRZI circuitry requires some type of memory circuitry as well as timing circuitry.

10.2.3 NRZ Asynchronous Circuitry Example

Most of this chapter is concerned with how NRZ circuitry works. Figure 10-7 shows NRZ asynchronous circuitry.

In the example circuit, the 74322 is the main controlling element. The 74322 is an eight-bit SPISPO shift-right shift register, the same IC used in the serial to parallel conversion of keyboard data on the system board of the IBM PC. The following descriptions indicate the function of the inputs and outputs used in this example.

A/QA–H/QH These are the eight parallel inputs and outputs of the 74322. The combination of levels on OE# and S/P# lines determines whether they act as inputs or outputs.

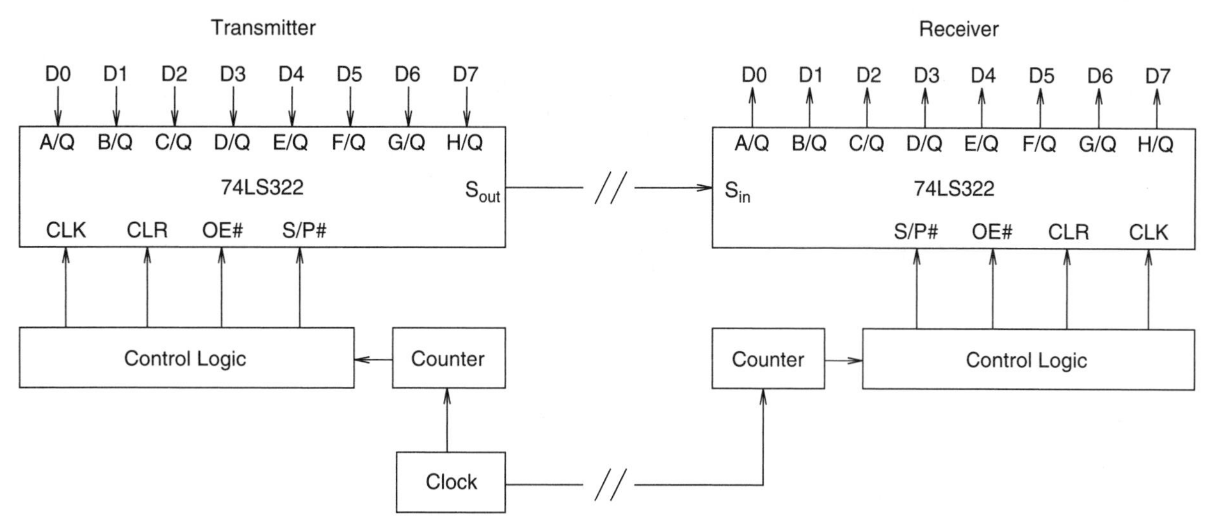

Figure 10-7 Example of NRZ circuitry.

CLK	The clock input is used to load either serial or parallel data into and out of the 74322. One clock is needed to load data in parallel format. Eight clocks are needed to read or write serial data into the shift register. Once the shift register is loaded, no clocks are required to read that data on the parallel outputs.
OH	The most significant bit output (output H) is used as the serial output from the 74322. This pin is used only as an output.
D0	This pin is used as the serial input (which happens to be the LBS of the data byte). It is used only as a serial input and cannot be used as a parallel input.
CLR	The clear line resets all data to zero in the shift register.
OE#	The output enable line allows A/QA through H/QH to act as outputs when low. When high, these lines act as inputs.
S/P#	The serial-parallel line control determines if serial data is processed in the shift register or if parallel data is processed. If low, the shift register processes parallel data. If high, serial data is processed.

The counter circuit controls how the clock signals are interpreted by both the transmitting and receiving circuitry. The circuit in this example is a mod-9 counter, which gets its input from a clock circuit. There are four outputs from this circuit, which are applied to the control logic.

The control logic is used to control when clock signals are applied to both the transmitter and receiver and also controls the control lines of the circuit and when they change.

Transmit Operation

CLOCK 1	The transmitter is reset by control logic. OE# goes high and S/P# goes low for a parallel load of the data on A/QA through H/QH. The loading occurs on the rising edge of the clock input to the counter.
	Control logic delays the clock pulse to the receiver control logic by one-half bit time from the time the data is loaded.
CLOCK 2	S/P# goes high for serial transmission, enabling the HO output. The MSB is seen on the data line.

CLOCK 3	74322 is clocked. The data from the LSB in the shift register is shifted toward the MSB. Bit A moves to B, B to C, C to D, and so forth, until bit G moves to the location of bit H. The new value for HO is seen on the data line.
CLOCK 4	74322 is clocked. The data in the shift register is shifted toward the MSB and a new bit is seen on the data line.
CLOCK 5	74322 is clocked. The data in the shift register is shifted toward the MSB and a new bit is seen on the data line.
CLOCK 6	74322 is clocked. The data in the shift register is shifted toward the MSB and a new bit is seen on the data line.
CLOCK 7	74322 is clocked. The data in the shift register is shifted toward the MSB and a new bit is seen on the data line.
CLOCK 8	74322 is clocked. The data in the shift register is shifted toward the MSB and a new bit is seen on the data line.
CLOCK 9	74322 is clocked. The data in the shift register is shifted toward the MSB and a new bit is seen on the data line. At this time, the data that was in bit location A (LSB) is in the location of H (MSB).

Receive Operation

CLOCK 1	The clock from the transmitter, which is delayed by one half bit time, is applied to the receiver counter. The receiver counter causes the receiver's control logic to clear all data bits in the receiver shift register. The control logic brings OE# high and S/P# high.
CLOCK 2	The clock from the transmitter causes the control logic in the receiver to read the data on D0. The data read is loaded into the LSB location of the shift register A/QA.
CLOCK 3	The clock from the transmitter causes the control logic in the receiver to read the next bit of data on D0. The data read is loaded into the LSB location of the shift register and all successive bits are shifted toward the MSB location.
CLOCK 4	The receiver is clocked and the data line is read. The data bit read is shifted into the LSB location and all successive bits are shifted toward the MSB location.
CLOCK 5	The receiver is clocked and the data line is read. The data bit read is shifted into the LSB location and all successive bits are shifted toward the MSB location.
CLOCK 6	The receiver is clocked and the data line is read. The data bit read is shifted into the LSB location and all successive bits are shifted toward the MSB location.
CLOCK 7	The receiver is clocked and the data line is read. The data bit read is shifted into the LSB location and all successive bits are shifted toward the MSB location.
CLOCK 8	The receiver is clocked and the data line is read. The data bit read is shifted into the LSB location and all successive bits are shifted toward the MSB location.

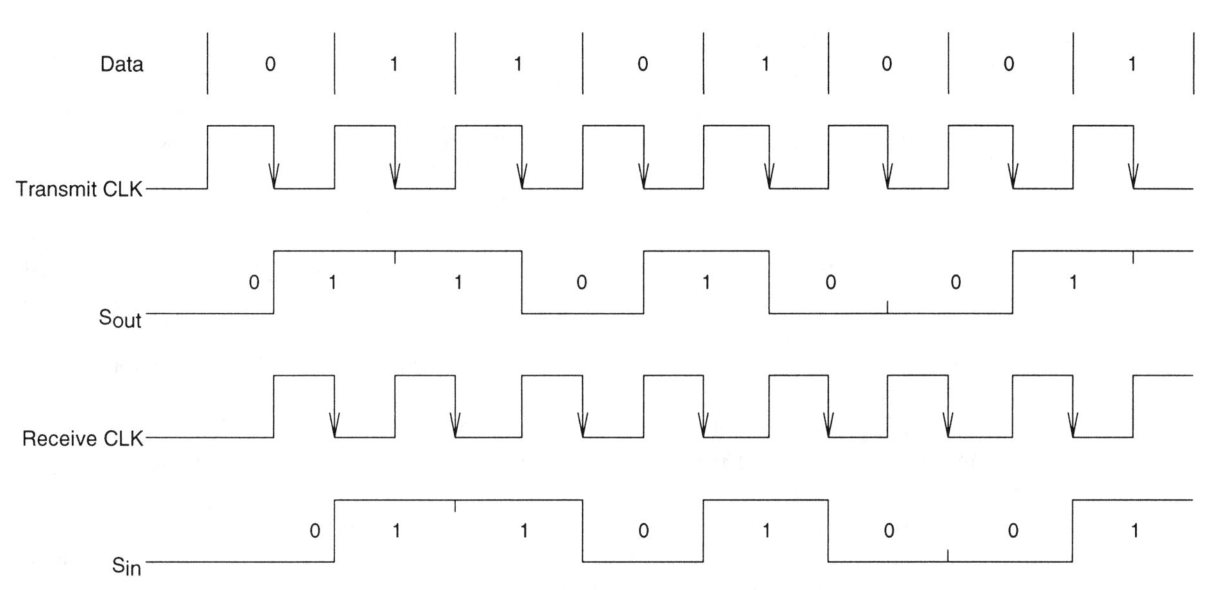

Figure 10-8 Example of NRZ timing.

TABLE 10-3
Asynchronous Data Packaging

Start Bit	*Data Field*	*Parity Bit*	*Stop Bit/s*
1-Bit	5–8 Bits	1-Bit	1/2-Bits

CLOCK 9 The receiver is clocked and the data line is read. The data bit read is shifted into the LSB location and all successive bits are shifted toward the MSB location. At this time, OE# is brought low and S/P# is brought low, enabling the parallel outputs and allowing the data to be read in a parallel format.

The timing diagram in Figure 10-8 shows the operation of the NRZ example circuit in Figure 10-7. Notice the phase relationship between the transmission and receiving clocks.

10.2.4 Asynchronous Data Packaging

In order for the transmission to have meaning, asynchronous data is sent in bit packages. The packages consists of the actual data along with control bits. The control bits are used by the receiver to determine when the transmission is beginning and ending and if the data has been transmitted properly (Table 10-3).

The idle state on most transmission lines is normally high because the logic-high requires less current. Keeping the idle state always at one level allows the receiver to easily identify the start of the transmission.

The start bit is identified when a high to low transition is seen on the data line after the previous transition is completed. After the circuit sees the transition, it waits one-half bit time before reading the data line again. If the data line is still low, the receiver assumes that the start bit is valid and begins reading the data line once every bit time. The reading of the data line occurs half-way through the bit time. This allows the data line to stabilize. If the data line is high, the start bit is not valid and the circuitry waits for the next high to low transition.

The data field is the actual data that is begin transmitted. This block of data must be predefined by both the transmitter and receiver. Otherwise errors occur. The number of bits can vary from five to eight, with seven being the most common, because ASCII is a seven-bit code. The next most popular code is the IBM EBCDIC code which uses eight-bits. Because 90% of all asynchronous data communications use ASCII, this is the code we will discuss.

ASCII (American Standard Code for Information Interchange) represents all characters in the English alphabet, as well as punctuation marks and numeric characters. There are also special characters for teletypes, like record separator, group separator, form feed, start of heading, and end of text. The IBM PC uses an extended ASCII code, which uses eight bits. When the MSB is low, the code produces standard ASCII, and when the MSB is high, special extended codes are transmitted. These extended codes represent graphic, scientific, foreign, and mathematical characters.

The parity bit is optional. It is used to determine if the proper number of logic-1s were received. The parity bit is not part of the data field. It is used only by the parity circuitry to check for proper transmission.

To end the data transmission the data package contains what is known as the stop bit. There are one or two stop bits, usually two. Stop bits are always high. Therefore, after the receiver receives the proper number of data field bits and the parity bit if used, the last bits should be high to indicate the end of the transmission. If the last bits received are not logic-high, the receiver generates a framing error. This usually indicates a timing problem or an interruption of data on the data line.

10.2.5 Asynchronous Data Transmission Examples

The following examples (Figs. 10-9 and 10-10) use ASCII, even parity, and two stop bits.

The data in Figure 10-9 are the letters M and a.

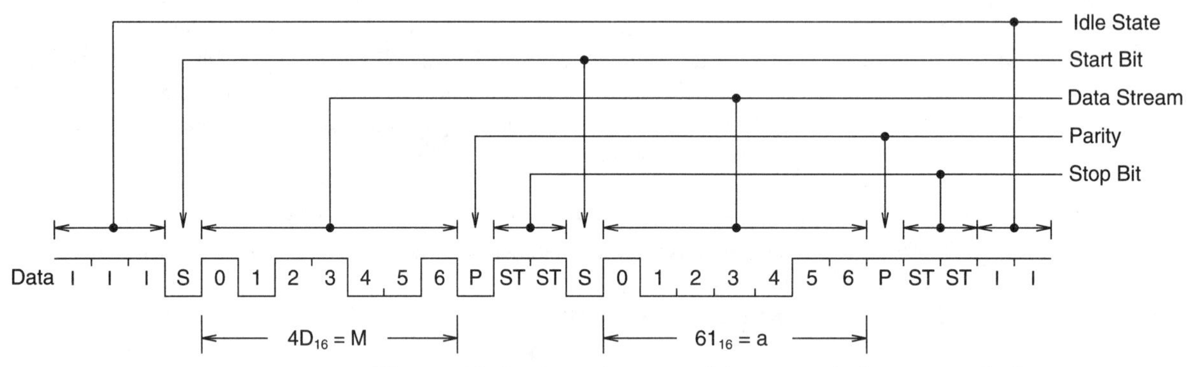

Figure 10-9 Asynchronous data transmission, example 1.

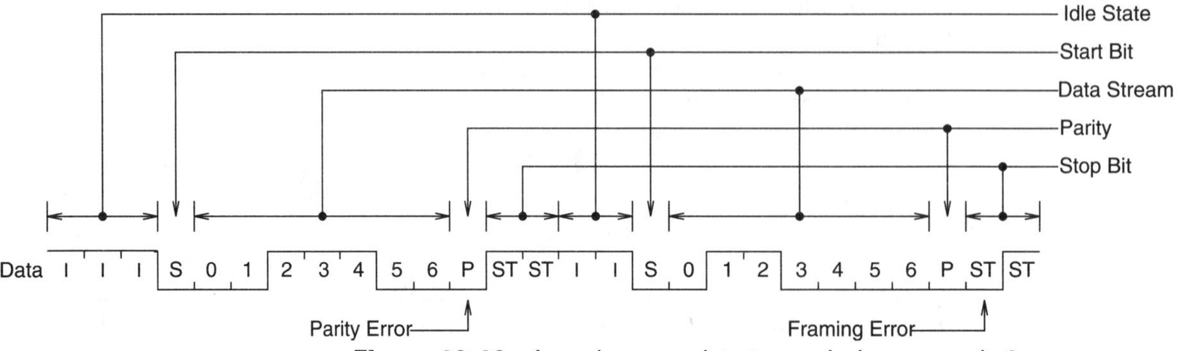

Figure 10-10 Asynchronous data transmission, example 2.

The data in Figure 10-10 are both invalid. The first transmission has a parity error. The reason for the parity error is that the total number of logic-1s is not even.

The second transmission has a framing error. The reason for the framing error is that the first stop bit was not a logic-1.

10.2.6 Baud Rate

In asynchronous data communication each bit of data transferred must stay on the data line for a predetermined amount of time, called bit time. In order to determine how many bits are transferred over a given data line in one second, the bit time is inverted. The result of this math operation gives the baud rate. Therefore, baud rate is the number of bits that can be transferred over a given data line in one second without compression.

Bit time = 1/Baud rate (in seconds)

Baud rate = 1/Bit time (in bits per seconds)

10.2.7 8250 Asynchronous Communication Element

The IBM PC uses Intel's 8250 ACE as the main controlling element in asynchronous communication between the IBM PC and any smart device. The 8250 ACE is programmable through BIOS and application software. BIOS calls this port COM1 or COM2 depending on whether the primary communication port (COM1) or the secondary communication port (COM2) is being accessed (Fig. 10-11).

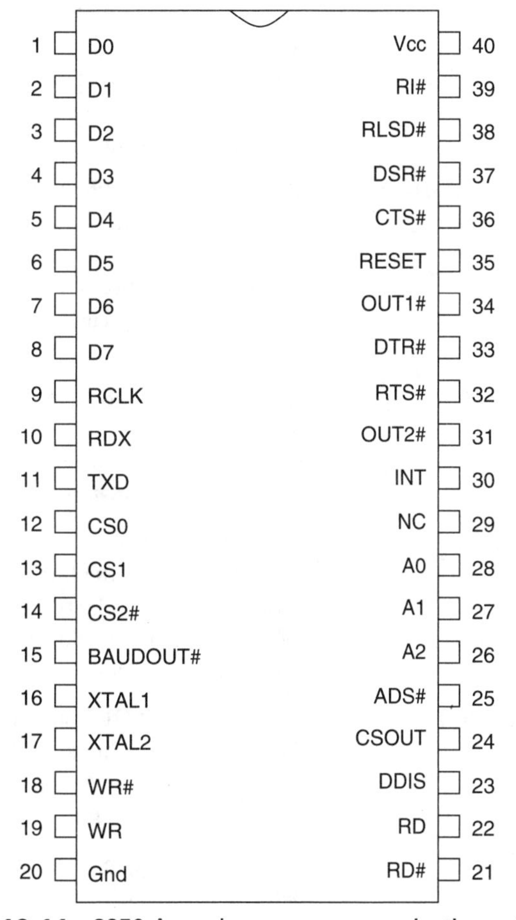

Figure 10-11 8250 Asynchronous communications element.

8250 ACE Features

10 programmable registers
40-pin LSI IC
Special modem control and status
Programmable baud rates
Maximum operating frequency 3.1 MHz
Operates off single + 5-volt supply

8250 Pin Configuration

MPU Bus

D0–D7 The 8-bit bidirectional Data bus is used to program or read the status of the ACE. During a write operation these lines act as inputs, allowing the programmer to configure the ACE, or they transfer data to the ACE for transmission. During a read operation, these lines act as outputs, allowing the status of the selected registers to be read by the MPU on the transfer of data out of the ACE.

CS0–1–CS2# There are three Chip-Select inputs used to access the ACE. To access the ACE CS0 and CS1 both have to be high, while CS2# must be low. The chip selects enable only this IC. The address lines are used to access one of the ten programmable registers inside the IC.

DOSTR–DOSTR# There are two input Data Output STRobes; one is an active high while the other is an active low. Only one of these lines is used in a computer system. The input that is used replaces the write enable line in most computers and allows the computer to write into the selected register. The input that is not used must be disabled by tying the line high or low.

DISTR–DISTR# There are two Data Input STRobes; one is an active high while the other is an active low. Only one of these lines is used in a computer system. The input that is used replaces the read enable line in most computers and allows the computer to read the contents of one of the selected registers. The input that is not used must be disabled by tying the line high or low.

A0–A2 Address lines A0 to A2 allow the MPU to access the registers inside the ACE. A0 to A2 by themselves allow access of only eight registers, but by using the DOSTR and DISTR lines, the MPU can access the additional two registers.

ADS# The ADdress Strobe line is an input used to latch in the address and chip selects when the line goes low. This line is used to verify that the address and chip selects are valid when the ACE is being programmed by a device other than an MPU. Because the IBM PC uses the 8088 MPU, this line is always enabled because it is grounded.

CSOUT The Chip Select OUTput indicates to the device that is programming the ACE that the chip selects are valid, when this line is high. This line is used when the device that is programming the ACE is not an MPU. In the IBM PC this line is left open (no connection).

DDIS The Data DISplay output line is used to indicate to an external device that one of the registers inside the ACE is being read. During a read operation, this line goes low. This line is used when the device that is programming the ACE is not an MPU. In the IBM PC this line is left open.

INTRPT	The INTeRruPT output line is used to generate an interrupt signal to the MPU when high. In the IBM PC, this line is connected to the PIC.
MR	The Master Reset input, when high, resets all registers inside the ACE to a starting value.
Vcc	The Vcc supplies the necessary voltage potential for the ACE.
GND	This supplies the necessary current for the ACE.
XTAL1-2	The two crystal inputs supply the necessary frequency to the ACE. If a frequency source other than a crystal is used, the signal is applied to the XTAL1 input.
RCLK	The Receive CLocK input determines when the ACE samples the data being seen on the serial input line. The input frequency is divided by 16 to determine the sample rate of the data line. Dividing by 16 minimizes any frequency difference between the transmission clock and receiving clock.
BAUDOUT#	The BAUD OUTput is a frequency that is 16 times greater than the transmission baud rate. This line is tied to the RCLK input in most computer applications, including the IBM PC.

Serial Interface Bus

SOUT	The Serial OUTput (also called the TX or transmit) line is used to output serial data one bit at a time at the selected baud rate.
SIN	The Serial INput (also called the RX or receive) line is used to input serial data one bit at a time to the ACE.

The following lines are used to communicate with the device connected to the ACE, in a handshaking mode.

RI#	The Ring Indicator input is used with modems to indicate when the telephone is ringing. RI# is an active-low signal. This signal is used when the modem is in the autoanswer mode.
RLSD#	The Receive-Line Signal-Detect input is a modem signal. RLSD is an active-low signal. This signal indicates when a carrier is detected by a modem, indicating that another modem is on the line.
DSR#	The Data Set Ready input line is used to indicate to the ACE that the device that it is communicating is on and working. DSR# is an active-low signal.
CTS#	The Clear-To-Send input line is used to indicate to the ACE that the device has processed the data and is ready to receive new data. CTS# is an active-low signal.
DTR#	The Data-Terminal Ready output line allows the ACE to control an external device. DTR# is an active-low output. DTR# can also be used to turn on and off any device connected to the ACE.
RTS#	The Ready-To-Send output line indicates to the device that the ACE has processed the data and is ready to receive more data. RTS# is an active-low output.
OUT1-2#	OUTputs 1 and 2# are extra output lines that can be programmed to go low when a command is given. These two lines are active lows and

have no predefined functions. In the IBM PC, OUT2# is used to enable interrupts created by the INTRPT. OUT1# is not used and is not connected.

8250

Line Control Register: The line control register (Fig. 10-12) is an 8-bit, write-only register that is used to format how the ACE transmits and receives data.

Bits 0–1	These bits determine the word lengths transmitted and received. Word lengths vary from five to eight bits.
Bit 2	This bit determines the number of stop bits transmitted and received. The number of stops can be set at 1, 1 1/2, or 2. The 1 1/2 stop is only used when the word length is set at five.
Bit 3	This bit is used to enable parity checking and generation.
Bit 4	This bit is used to select even parity.
Bit 5	This bit determines how the parity error flag is interpreted, whether a parity error produces a high or low level in the parity error flag.
Bit 6	The set break control bit is used to disable the SOUT output by sending out a break signal (continuous logic-low). The break signal interrupts the device that the ACE is communicating with.
Bit 7	The divisor latch access bit (DLAB) accesses the baud rate generator when high. When low, the receive buffer, transmitter holding register, or interrupt enable register can be accessed.

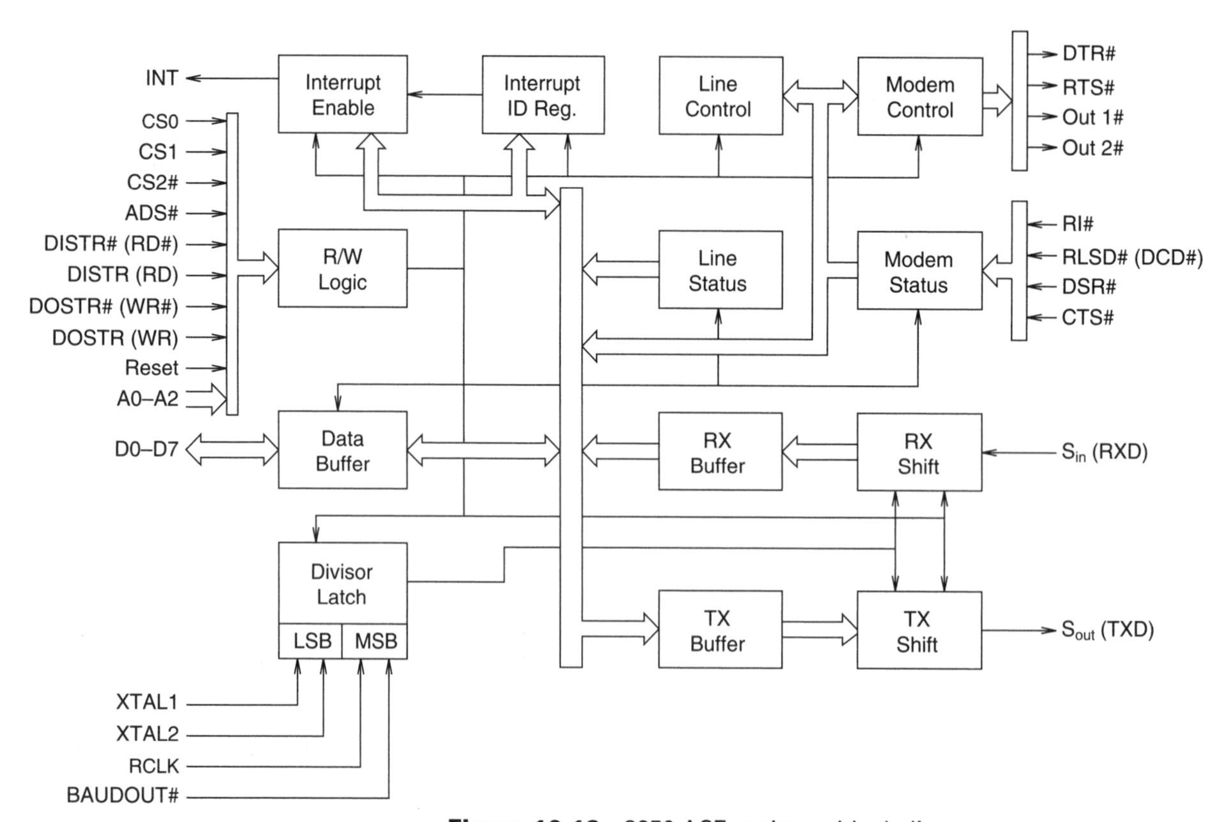

Figure 10-12 8250 ACE registers block diagram.

Divisor Latch LSB and MSB: These two 8-bit write-only registers contain the numerical value of the divisor that is used to divide the clock frequency to produce the baud rate. These values control the programmable baud rate generator. The maximum input frequency is 3 MHz. The IBM PC uses a 1.8432-MHz input frequency to produce the baud rate. The following formula can be used to determine the divisor value for any given frequency input and baud rate value.

$$\text{divisor} = (\text{frequency input})/(\text{baud rate} \times 16)$$

Table 10-4 shows the values of divisors for the most popular baud rate values for the IBM PC. The baud rates are in decimal and the divisor values are in binary.

Line Status Register: The line status register is an 8-bit R/W register that is used to provide the overall status of the ACE.

Bit 0	The data-ready bit sets when the data byte has been transferred into the receiver buffer. This bit resets when the receiver buffer is read or when a logic-0 is moved into the bit. Therefore, this bit indicates that new data has been received and should be read by the MPU.
Bit 1	The overrun error bit sets whenever new data is received before the data in the receiver buffer has been read. The data in the receiver buffer is lost.
Bit 2	The parity error bit sets whenever the total number of logic-1s received plus the parity bit is not equal to the parity selected.
Bit 3	The framing error bit sets if the proper number of stop bits has not been received.
Bit 4	The break interrupt bit sets whenever the data received is held low (spacing level) for longer than one complete transmission.
Bit 5	The transmitter holding register empty bit indicates when the data in the transmitter holding register has been sent to the transmit shift register. When this bit sets, it indicates that a new byte can be stored in the transmitter holding register without affecting the transmission. Once the data in the transmitter holding register is transferred to the transmit shift register, the data is automatically transferred out through the SOUT.
Bit 6	The transmitter shift register empty bit sets whenever the transmitter shift register is empty or idle. When the data is transferred from the transmitter holding register to the transmitter shift register, this bit resets.
Bit 7	Bit 7 is not used and is always reset.

TABLE 10-4
Baud Rate Divisor Table

Baud Rate	*Divisor*
110	0000 0100 0001 0111
300	0000 0001 1000 0000
1200	0000 0000 0110 0000
2400	0000 0000 0011 1010
3600	0000 0000 0011 0000
4800	0000 0000 0001 1000
9600	0000 0000 0000 1100

Modem Control Register: The modem control register is a R/W, 8-bit register that is used to control the serial interface outputs of the ACE. Five of the eight bits are used; the other bits are inactive.

Bit 0	This bit controls the DTR# output line.
Bit 1	This bit controls the RTS# output line.
Bit 2	This bit controls the OUT1# output line.
Bit 3	This bit controls the OUT2# output line.
Bit 4	This bit enables the loop-back diagnostic testing function of the 8250.
Bits 5–6	These bits are not used and should be reset.

Modem Status Register: The modem status register is an 8-bit, R/W register that is used to indicate the status of the serial interface input lines.

Bit 0	Indicates the status of the CTS# (clear-to-send) input line.
Bit 1	Indicates a change in the status of the DSR# (data-set-ready) input line. This bit is called DDSR# (delta data set ready).
Bit 2	This bit indicates a change in the status of the RI# (ring indicator) input. This bit is called DRI# (delta ring indicator).
Bit 3	This bit indicates a change in the status of the RLSD# (receive-line signal-detect) input. This bit is called DRLSD# (delta receive line signal detect).
Bit 4	This bit indicates the current status of the CTS# (clear-to-send) input.
Bit 5	This bit indicates the current status of the DSR# (data-set-ready) input.
Bit 6	This bit indicates the current status of the RI# (ring indicator) input.
Bit 7	This bit indicates the current status of the RLSD# (receive-line signal-detect) input.

Interrupt Identification Register: This 8-bit register is used to determine the priority levels of up to four types of interrupts. Bits 0 to 2 set up the priorities of the interrupts. Bits 3 to 7 are always reset. In the IBM PC the following priorities are used.

The high priority is the receiver line status interrupt. This interrupt occurs whenever any one of the following conditions exists: overrun, parity, framing error, or a break interrupt signal. The interrupt is reset by reading the line status register.

The second highest priority is the received data available interrupt. This interrupt occurs whenever the data ready bit in the line register is set. The interrupt is reset by reading the receiver buffer.

The third highest priority is the transmitter-holding register-empty interrupt. This interrupt occurs whenever the transmitter-holding register-empty bit in the line status register is set. The interrupt is reset by writing data into the transmitter holding register or reading this bit in the interrupt identification register.

The lowest priority is the modem status interrupt. This interrupt occurs whenever one of the following inputs goes low: CTS#, DSR#, RI#, or RLSD#. The interrupt is reset by reading the status of the modem status register.

Interrupt Enable Register: This register is used to enable one or more of the interrupts selected in the interrupt identification register. Only bits 0 to 3 are used. Bits 4 to 7 are always reset when used in the IBM PC.

Bit 0 This bit enables the highest priority interrupt.

Bit 1 This bit enables the second highest priority interrupt.

Bit 2 This bit enables the third highest priority interrupt.

Bit 3 This bit enables the lowest priority interrupt.

Transmitter Holding Register: This register is a write-only register that contains the data byte to be transmitted. The LSB of the register is the first bit transferred after the start bit.

Receiver Buffer Register: This register is a read-only register that contains the data byte that has just been received, after the start, parity, and stop bits are removed. The LSB is the first bit received after the start bit.

Typical Transmit Operation

Step 1 Reset DTR#.

Step 2 Read modem status register. If DSR# and CTS# are low, go to step 3. If DSR# and CTS# are not low, repeat step 2 until they go low.

Step 3 Read the transmitter-holding register-empty bit in the line status register. If it is set, go to step 4; if reset, repeat step 3.

Step 4 Reset RTS#.

Step 5 Write data into transmitter-holding register.

Step 6 Read transmitter-shift register-empty bit in the line status register. If it is set, go to step 2 to transmit next byte of data or end program; if reset, repeat step 6.

Typical Receive Operation

Step 1 Reset the DTR#.

Step 2 Read the line status register and check the condition of the following bits.

If BI bit is set, repeat step 2.

If DR bit is set, continue with step 2; otherwise repeat step 2.

If the OR, PE, or FE bit is set, go to the appropriate error routines; otherwise go to step 3.

Step 3 Read data in receiver buffer register. If the reception of data is finished, end the program; if it is not completed, go to step 2.

10.2.8 IBM Asynchronous Communications Adapter

The IBM PC asynchronous communication adapter in Figure 10-13 allows the IBM PC to communicate with asynchronous serial devices. The main controlling element is the 8250 asynchronous communication element (also known as the ACE). There are RS-232-C drivers and receivers on the card to supply and receive RS-232-C levels from outside devices. Also located on the adapter card is a 20-ma current loop circuit for external devices that require current loop interfacing.

The main purpose of the 8250 is to take parallel data from the MPU bus and transmit it serially at the proper baud rate to an external device during a transmit operation. During a receive operation, serial data is converted into parallel data for the MPU bus. During both operations, the status of the operations is determined and can be used by the program to indicate if the operation was successful.

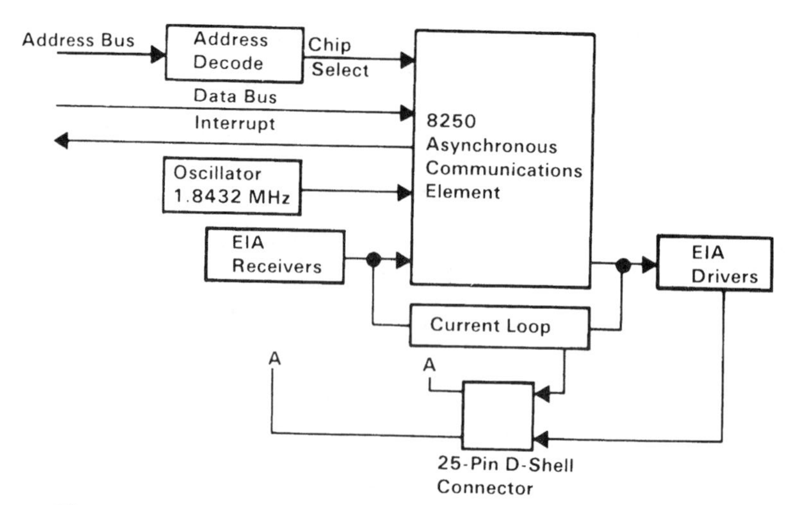

Figure 10-13 IBM asynchronous communications adapter. (Courtesy of IBM)

TABLE 10-5
Serial Port Register Address Table

COM1	*COM2*	*Register*	*DLAB Bit*
3F8	2F8	TX holding buffer	0 WRITE
3F8	2F8	RX buffer	0 READ
3F8	2F8	Divisor latch LSB	1
3F9	2F9	Divisor latch MSB	1
3F9	2F9	INT enable register	0
3FA	2FA	INT identification register	0
3FB	2FB	Line control register	0
3FC	2FC	Modem control register	0
3FD	2FD	Line status register	0
3FE	2FE	Modem status register	0

The address decoder circuitry is used to select one of the registers inside the 8250 to perform the selected operation. The address map of the IBM PC allows two asynchronous communication ports to be placed in the system, but only one port can be used at any one time. Table 10-5 shows the address of the different registers inside the 8250.

The input frequency to the 8250 is derived from an oscillator running at 18.432 MHz. going through a circuit that divides by 10. Therefore, the input frequency is actually 1.8432 MHz.

EIA (RS-232-C) Interface Standard. In the late 1960s the electronic industry association (EIA) set up an asynchronous communication standard. This standard is still used today in asynchronous communications between computers and other devices. The RS in RS-232 stands for receive and send. The following describes the electrical and mechanical specifications. Any line that indicates DTE (Data Terminal Equipment) refers to the computer itself or the device controlling the communications. Any line that indicates DCE (Data Communication Equipment) refers to the device with which the computer or DTE is communicating.

RS-232 Electrical Specifications

MARK (LOGIC 1) any voltage between −3 V and −24 V.

SPACE (LOGIC 0) any voltage between 3 V and 24 V.

All lines must be able to drive a 3K to 7K resistive load in parallel with 0.0025 uf load without interfering with the signal level.

All lines must be able to take a direct short between any other line or ground, without destroying the circuit.

Maximum length of transmission line in worst case conditions is 50 feet, without a repeater.

Maximum baud rate under worst case conditions is 20K baud.

Mechanical Specifications

ITT	DB-25P or DB-25S connector
TWR	DB-25P or DB-25S connector

Line Specifications: Data Signals

Pin 2	Primary high-speed transmit data
Pin 3	Primary high-speed receive data
Pin 14	Secondary low-speed transmit data
Pin 16	Secondary low-speed receive data

Line Specifications: Control Signals

Pin 4	RTS (Ready-To-Send)
Pin 5	CTS (Clear-To-Send)
Pin 6	DSR (Data-Set-Ready)
Pin 8	RLSD (Receive-Line Signal-Detect)
Pin 12	SRLSD (Secondary-Receive Line Signal-Detect)
Pin 13	SCTS (Secondary Clear-To-Send)
Pin 19	SRTS (Secondary Ready-To-Send)
Pin 20	DTR (Data-Terminal-Ready)
Pin 21	SQD (Signal-Quality-Detect)
Pin 22	RI (Ring Indicator)
Pin 23	RXDR (Data-Signal Rate-Select)

Line Specifications: Grounds

Pin 1	Chassis ground
Pin 2	Signal ground
Pin 15	TxCLK (Transmit clock from source DCE)
Pin 17	RxCLK (Receive clock)
Pin 24	TxCLK (Transmit clock from DTE)

EIA Receivers and Drivers. The IBM PC, as well as most other computer systems, uses the 1488 RS-232-C line driver and the 1489 RS-232-C line receiver (Fig. 10-14) to communicate with other devices that require RS-232-C signal levels.

The 1488 RS-232-C line driver converts TTL levels from the 8250 into RS-232-C signal levels. The IC contains four DTL NAND gates which produce a −6-volt logic-1 level and +6 volts for a logic-0 level, when the supply to the IC is (+ and −) 12 volts DC. One of the NAND gates has only one input while the other three NAND gates have two inputs. If only one input is needed, tie the other input to the supply.

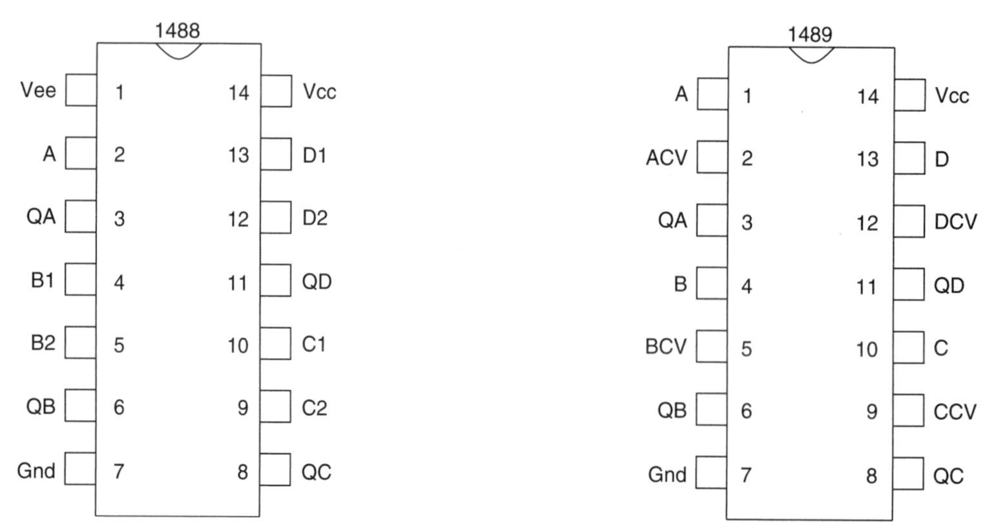

Figure 10-14 1488 RS-232 transmitter and 1489 RS-232 receiver.

The 1489 RS-232-C line receiver converts RS-232-C levels received from the external device into TTL levels. This IC contains four gated DTL single input NAND gates. The gate (control voltage input) is used to disable the output of any of the NAND gates. These inputs are normally left open to enable the gate. When the IC is operated using a +5-volt supply, the following TTL levels are produced: TTL logic-1 is any voltage on the input between −.75 and −1.25 volts, and TTL logic-0 is any voltage on the input between 1 and 1.5 volts.

The IBM also contains a 20-ma current loop circuit for any device that requires this type of interface standard. In the early 1960s the EIA set up an asynchronous standard for devices, such as teletypes and other devices that ran their inputs using relays. Because relays are current devices, the EIA specified the communication between the devices in terms of current levels rather than voltage levels. This standard is still used in some devices, but RS-232-C is used in most devices. The following are the electrical, mechanical, and signal specifications for 20-ma current loop interfacing.

Electrical Specifications

MARK (LOGIC-1) any current between 16 ma and 24 ma.
SPACE (LOGIC-0) current less than 4 ma.
Maximum transmission length 25 feet.
Maximum baud rate 4800 bps.
Low impedance transmission lines must be used.

Mechanical Specifications

Amphonal 4-pin connector or terminal block.

Signal Specifications

RxD(+)	Receive current (DTE or DCE).
RxD(−)	Receive current ground (DCE).
TxD(−)	Transmit current (DTE or DCE).
TxD(+)	Transmit current source (DCE).

The circuit in Figure 10-15 is used in the IBM PC for producing the 20 ma current loop interface standard.

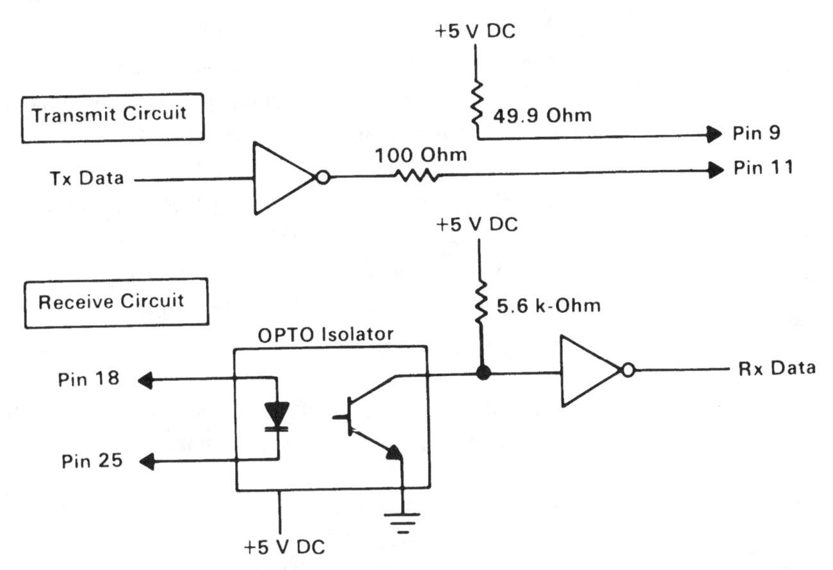

Figure 10-15 IBM 20-ma current loop interface. (Courtesy of IBM)

10.2.9 Other Forms of Serial Data Communication

Besides asynchronous serial data communication, which is easy to use and costs less, there are two other forms of serial communication. Both involve synchronous serial data transfers. Synchronous serial data communication transmits a block of characters or bits rather than one character or 8 bits per package. Each block may contain from 126 to 10,240 bytes of data or 256 to 40,960 bits of data. Each block contains a start sequence, parity sequence, and ending sequence.

Therefore, to transmit 10,240 bytes of data using synchronous serial communications requires 89,952 bit times, plus 32 bit times for the start, parity, and ending sequence, a total of 89,984 bit times. If the same number of bytes were transmitted using asynchronous serial communications, it would require 112,640 bit times because each byte transmitted requires 11 bits of data. Consequently, synchronous serial communication needs less time to transmit and receive the same amount of information.

The problem with synchronous serial data communications is that the support circuitry and programming are more complex and therefore cost more. Synchronous serial data communication is normally used between high-speed computers and networks (LANs) rather than when computers are communicating with peripheral devices.

SDLC (Synchronous Data Link Control) is a protocol that IBM uses for transmission of blocks of data bits. BSC (Binary Synchronous Communications) is a protocol that IBM uses for the transmission of blocks of characters. More information on both of these protocols can be found in IBM publications.

Troubleshooting the Printer Adapter. If the problem is in printing, check the following.

1. Make sure the printer is working correctly by running the built-in printer test. If the printer passes the test, the problem is in one of the following steps. If the printer fails the test, the problem is in the printer.
2. With a diagnostic program, make sure the computer can communicate with the adapter. If the computer cannot establish communications with the adapter, make the following checks.
   a. Check power to adapter and all major ICs.

b. Verify the address decoding section of the adapter.
c. Check the operation of the command decoder.

If the computer can establish communications with the adapter but there is no printing on the printer, make the following checks.

a. Using a break-out box, verify the levels on the printer cable. If the cable is bad, correct the problem. Usually the problem is a broken wire in the connector. If the cable checks out, check the timing for the printer adapter and correct any problem.
b. Check for stuck levels on the open collector drivers.
c. Check for proper status signals for stuck levels.

3. If no hardware problems can be found, check the software drivers.

Troubleshooting the Asynchronous Communications Adapter. If asynchronous communications cannot be established, make the following checks.

1. Using a diagnostic diskette, try to establish communication with the asynchronous communication adapter. To verify that the communication adapter is operational, the serial out is fed back into the serial input. If communication can be made with the adapter and the data being sent out is being received, the problem is not in the adapter. It is in either the cabling or the asynchronous device connected to the adapter.
2. If communication cannot be made with the adapter, check the address decoding of the adapter and all control lines that are used to select the adapter.
3. If communications with the adapter can be made but the data being received is not the data being transmitted, transmit the same character repeatedly, checking the serial output with a scope. The waveform on the oscilloscope should match the proper data package programmed to be transmitted. If the serial output from the adapter is bad, check the output of the 8250. If the signal is good, the trouble is with the RS-232 interface. If the signal is bad, the 8250 is bad.

10.3 PRINTERS

The main difference among printers is the way each printer produces the character or symbol. The following is common terminology related to printers.

Fully formed printers use a character that is permanently formed on a print head. The print head may be made of metal, plastic, or hard rubber. Fully formed printers give letter quality print, but the user cannot change the type of characters on the print head, although in some cases one can change the print head to achieve a different character set. These printers use mechanical force to make the print head strike the ribbon and produce the character on the paper. Because mechanical force is used, these printers are normally slow (unless a line printer is used), noisy, and large and require a fair amount of maintenance.

Dot matrix printers use an MPU inside the printer to create characters on the paper using a series of dots in rows and columns. Because most dot matrix printers have their own MPU, they are usually programmable. The user may use one of the preprogrammed character and symbol sets or create their own character-symbol set. Dot matrix printers are the most popular type of printer because of their programmability, speed, and low cost. Dot matrix printers may use mechanical force, light, heat, or electrostatic force to print the characters on the paper.

Most printers today use bidirectional *print head movement* to print the characters on the paper. A bidirectional printer prints characters from left to right for one line and from right to left for the next line. This means that the print head does not have to move to the left hand side of the paper before starting a new line. This increases the speed of the printer by reducing the amount of head movement. Any head movement that does not produce a character is a waste of print time. Although a bidirectional printer is faster, it requires much more control logic. In older printers, unidirectional print head movement was used because of its simple control circuitry and because it does not require an MPU. Unidirectional printing repositions the head after each line of print, which decreases the number of characters the printer can print in a given time period.

There are two basic types of *paper feed* in printers. In a friction feed printer, the paper is moved through the printer by use of friction between the paper and special rollers (as in a typewriter). The problem with friction feed printers is that if the rollers are not set properly the paper can slip.

The second type of paper feed is called tractor (or pin) feed. In this type, special paper with holes on both sides is used, along with sprockets on the printer itself, to move the paper through the printer. The sprockets use holes in the paper to move the paper through the printer and keep it from slipping from side to side. After the document is printed, the edges of the paper that contain the holes for the pin feed mechanism can be removed, leaving a clean, standard size piece of paper. Pin feed paper is normally fan folded, which means the paper is continuous, allowing the printer to print long documents without the need to reload paper after each page is printed. The pages are separated after printing is finished.

Impact printers are printers that use electromechanical force to print the characters or graphics on paper. Fully formed printers are always impact printers, and some dot matrix printers fall into this category.

Nonimpact printers are printers that use nonmechanical force to cause characters to be printed on the paper. This type of printer usually requires special paper. Nonimpact printers normally use heat, light, or special liquid inks to produce the characters.

Some printers contain a keyboard, so that the printer can also be used as a video terminal (using the paper like a video screen). This type of printer, called a *KSR* (Keyboard Send and Receive) printer, allows the user to input data to the computer as well as receive data from the computer. This type of printer is very useful where space is limited or only one remote communication device is acceptable. Printers that only receive data from a host computer are called *receive only* (RO) printers. These cost much less than KSR printers.

Printers may receive data in either a serial or parallel form. Printers that receive data in a serial form are normally called serial printers; those that receive in parallel form are called parallel printers. Today, most printers have both serial and parallel interfaces built into them.

A *font* refers to the type style (how the characters look) of a printer. Typical fonts are courier (standard typewriter style print), geneva, helvetica, New York, Athens, Chicago, London, and Los Angeles.

The *pitch* of the printer refers to the size of the character. The standard pitches for courier are pica (10 characters/inch), elite (12 characters/inch), semicompressed (14 characters/inch), and compressed (17 to 18 characters/inch).

10.3.1 Printer Electronics

Figure 10-16 shows a typical block diagram for a dot matrix printer; the diagram contains five basic sections.

8049. The 8049 is an 8-bit, single-chip MPU from Intel. This is the same type of chip used in the IBM PC keyboard. The 8049 has an 8-bit data bus, 27 programmable I/O port lines, 64×8 RAM, and a 1K or $2K \times 8$ ROM-EPROM. In the block diagram, this MPU (also called a microcontroller) is used as the master controller. The main function of this controller in this printer is as follows.

The 8049 is used to control all handshaking between the printer and host computer (IBM PC or compatible). Handshaking is the process of sending signals back and forth between the host computer and master MPU in the printer, to allow data transmission to occur in the proper sequence.

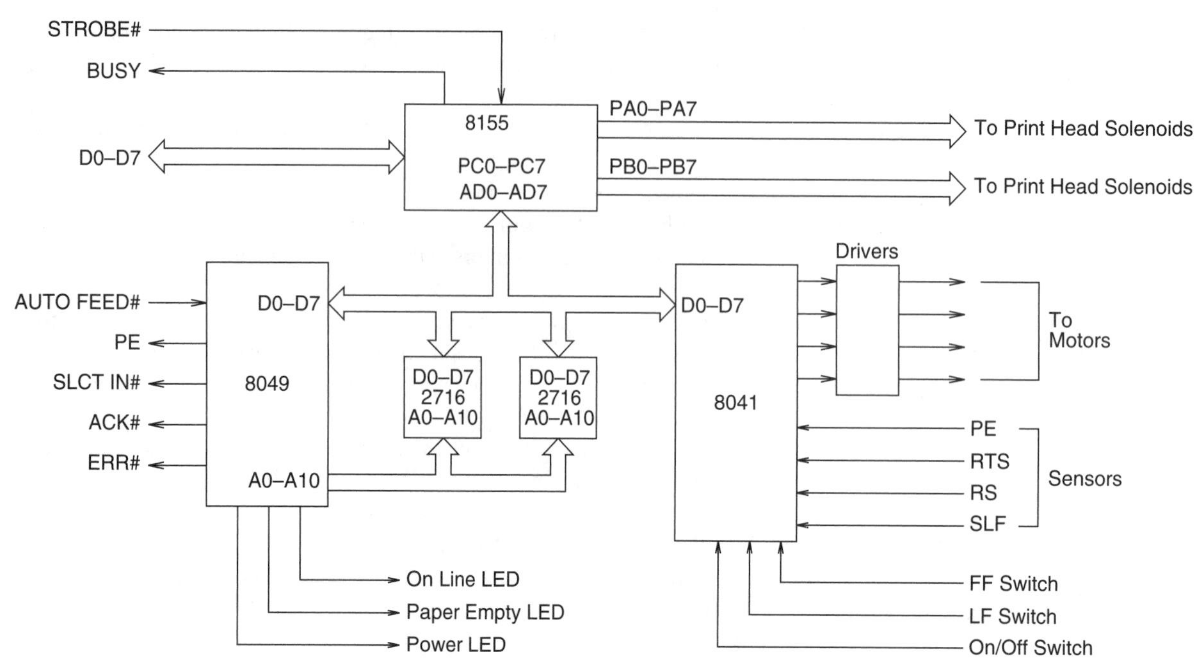

Figure 10-16 Dot matrix printer block diagram.

The 8049 is also used to transfer character column codes to the print heads. The 8049 reads the ASCII code in the print buffer (RAM) and uses the code to address the character column codes in the EPROMs. It transfers the character column codes to energize the print head solenoids to print the character on the paper.

It is also used to format and or program the 8041 8-bit microcontroller which is used to determine the position of the print head and the way the paper moves through the printer. The 8049 reads the DIP switches to format the 8041 with the selected hard-wired character and font set and other printer characteristics. Different character and font sets may be selected when the 8049 reads a special control code received by the print buffer. Once the special code is read, the 8049 reprograms the 8041 to change how the print head and paper are positioned during printer operation.

The 8049 is also responsible for the control of the 8155 RAM (print buffer and I/O port controller). As each byte of data is received in a parallel format, the data is placed in the printer buffer (RAM). Once one entire row of character codes is received, the 8049 reads the print buffer. When the printer prints from left to right, the 8049 accesses the print buffer as a first-in first-out buffer. When the printer prints from right to left, it accesses the print buffer as a first-in last-out buffer.

The 8049 also controls the status LEDs on the printer. The LEDs indicate the following conditions to the user: paper empty and the power and on-line status.

8155. The Intel 8155 is a 256-byte RAM with three 8-bit I/O ports and an 8-bit address-data bus. In the printer this device is used to perform the following.

As the 8049 communicates with the host computer, parallel data from the host computer is placed in the 8155's memory (print buffer). The 8155 controls the transfer of the parallel data with the permission of the 8049, which is communicating with the host computer. The 8155 is also used to transfer the character column codes from the EPROMs to the print head. The codes are used to energize the solenoids in the print head.

Because the 8155 has I/O ports and timing circuits, it generates the busy signals for the printer to tell the host computer that the data on the parallel data lines has been processed. The 8155 is also responsible for receiving the strobe signal, which is used by the host computer to tell the 8155 when a new data byte is on the data bus.

8041. The 8041 Universal Peripheral Interface, an 8-bit slave microcontroller (similar to the 8049), is used to control print head and paper movement motors in the printer. This controller contains 1K × 8 ROM, 64 × 8 RAM, and two 8-bit I/O ports which are all TTL compatible. The 8041 is used instead of the 8049 because the 8041's I/O ports are all TTL compatible and it contains special I/O instructions, designed for controlling peripheral devices. The 8041 performs the following tasks.

Some I/O ports are used to control the carriage return motor, which is used to position the print head at different locations on the paper.

Some I/O ports are used to control the line feed motor. The line feed motor is used to position the paper in the printer.

Some I/O ports are used as status inputs for the printer switches. These switches perform operations on the printer.

Some I/O ports are used as inputs for different status sensors in the printer. These sensors sense the status of mechanical sections of the printer.

2716. The 2716 EPROMs are 2K × 8 and hold the character column codes for the different character and font sets available for the printer. The 8049 uses its address bus to control which character column code is placed on the data bus to be transferred to the print head through the 8155. Different addresses are selected when the 8049 reads a change in the DIP switches or if the 8049 reads an escape control code from the print buffer.

Driver. The driver section is normally made of SCRs, or triacs or optical drivers (with a Darlington pair or SCR or triac outputs). These drivers are used to buffer the TTL outputs of the 8041 from any surges in the motors and to supply extra current to control the motors. The circuitry is also used in some computers to control the solenoids in the print head.

10.3.2 Printer Operations

The following is a sample operation of the printer in the block diagram in Figure 10-16.

1. When power is first applied to the printer, the 8049 reads the DIP switches to determine the font, pitch of the characters, and number of characters per line.
2. The 8049 then formats the 8155 and 8041 with the proper parameters from the information supplied by the DIP switches.
3. The 8049 reads the 8041 to determine if the printer is on or off line. If the printer is on line, it can accept codes from the host computer, and the next step is performed. If the printer is off line, the 8049 continues to read the 8041 to determine what to do next. When the printer is off line the user, through printer switches, can control paper movement or select a different print function (if available on the printer).
4. The 8049 checks the SLCT IN# input line, which indicates if the printer should receive codes from the host computer. If the line is high, the printer is deselected and does not operate from the host computer. The AUTO FEED# line may or may not be checked, depending on the configuration of the printer.
5. The host computer places its data byte on the eight data lines.
6. 5 μs after the computer places the data on the data lines, the host computer brings the STROBE# lines low.
7. The STROBE# line causes the 8155 to store the data byte on the data lines into RAM (print buffer). The 8155 selects at what address the data is stored. The 8155 brings the BUSY line high. This high indicates to the host computer that the 8155 is busy and a new data byte should not be placed on the data lines.
8. Once the data byte is placed in the memory of the 8155, the 8049 checks the print buffer.

If the buffer is full, the 8049 starts reading the print buffer. As each data code is read, the 8049 uses the ASCII code to determine which character column code is produced by the EPROM. Once the character column code is available on the data bus, the 8049 transfers this code through the 8155 to the print head solenoids. This code is also used by the 8041 to determine head positioning while the character is being printed. The 8049 keeps the host computer from sending any more characters until the buffer is empty. Once a line of characters is printed, the printer either generates an internal line feed or waits for a line feed from the host computer.

If the buffer is not full, the 8049 brings the ACK# line low, which does two things. First, it tells the computer that the printer has accepted the code and to get ready to send another character. Once the 8155 is ready, the 8049 brings the BUSY line high, which informs the host computer it may send the next character. This process continues until the print buffer is full.

9. If at any time during the operation of the printer the PE line goes high or the ERR# line goes low, the host computer will not send any more data until the problem is corrected.

10.3.3 Printer Mechanical Hardware

The print head is the part of the printer that actually forms the character. In a dot matrix printer the print head contains from 9 to 24 solenoids, in the format of a column. As the print head is moved across the paper, the solenoids are energized and the tip of the solenoid strikes a ribbon to cause a dot to be placed on the paper. When enough dots are placed side by side, a character is formed.

The carriage return motor controls the position of the print head as it moves across the paper. This motor may be a stepper motor or a standard synchronous motor and is controlled by the 8041 through the 8049.

The line feed motor controls the position of the paper as it moves through the printer. This motor is usually a stepper motor and is controlled by the 8041 through the 8049.

The print head reference switch determines when the print head is at the far left side of the paper. This switch is usually optical, but sometimes it is mechanical.

The paper empty switch is used to determine if the printer is out of paper. If the switch shows the printer is out of paper, the printer stops operating until paper is loaded.

10.3.4 Types of Printers

The most used type of printer is an impact dot matrix printer. The reason for their popularity is their low cost, high speed and the ability to program different character sets. Dot matrix printers normally have the ability to print characters and/or graphics.

Dot Matrix Printers. Dot matrix printers (see Figs. 10-17 and 10-18) print character and graphics using a row and column matrix. The print head has one solenoid for each row in each column of the character. The standard print head contains nine print solenoids, one for each row that is to be printed. As the printer moves the print head across the page of paper the solenoids are energized. Each character requires six column codes. Each column position contains a row code. As the column is printed the row code causes the proper solenoids to energize, producing the dots needed for that column. The more rows and columns the character has, the better looking the character is. Print heads for dot matrix printers contain 9, 18, or 24 print solenoids; the more print solenoids there are, the more expensive the printer is, but the resolution of the characters is better. The character matrix (Fig. 10-19) can vary from printer to printer. Typical character matrixes are 5×7, 6×9, 10×10, or 18×24.

Most of today's dot matrix printers also provide an NLQ (near letter quality) printing mode. In the NLQ mode each character is created by a double pass of the print head. During the second pass the second set of dots that make up the character are offset to fill in any gaps

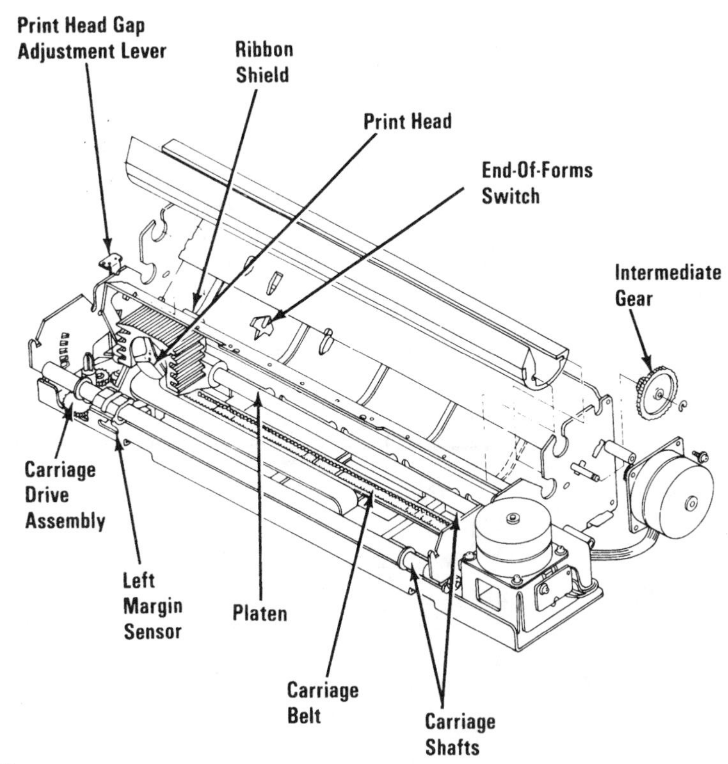

Figure 10-17 IBM graphics printer, print mechanism assembly (front view). (Courtesy of IBM)

in the first set of dots. Figure 10-19 shows typical draft mode and NLQ mode characters. Because NLQ printing requires two passes of the print head, it takes more than twice the time to print the same character.

Many dot matrix printers also provide a color option. The color option contains a special ROM to reprogram the printer, new software printer drivers, and a three- or four-color ribbon.

Advantages

Programmable character and graphic symbols
Speeds of 90–600 characters per second (cps) in draft mode
Speeds of 36–100 cps in NLQ mode
Color printing optional
Cost: $100–$2000

Disadvantages

NLQ only at low speeds
Graphics printer speed very slow
High noise level

Nonimpact Dot Matrix Printers: The thermal dot matrix printer uses a dot matrix in which each dot location on the print head is a heating element. If a dot is to be printed on the row, the element heats up, causing the special paper to discolor and produce a dark gray or black dot. The major problem with this type of printer is that it is slow because the print head must have time to cool down before it can be moved to the next character location. Today thermal dot matrix printers have been replaced by ink jet printers.

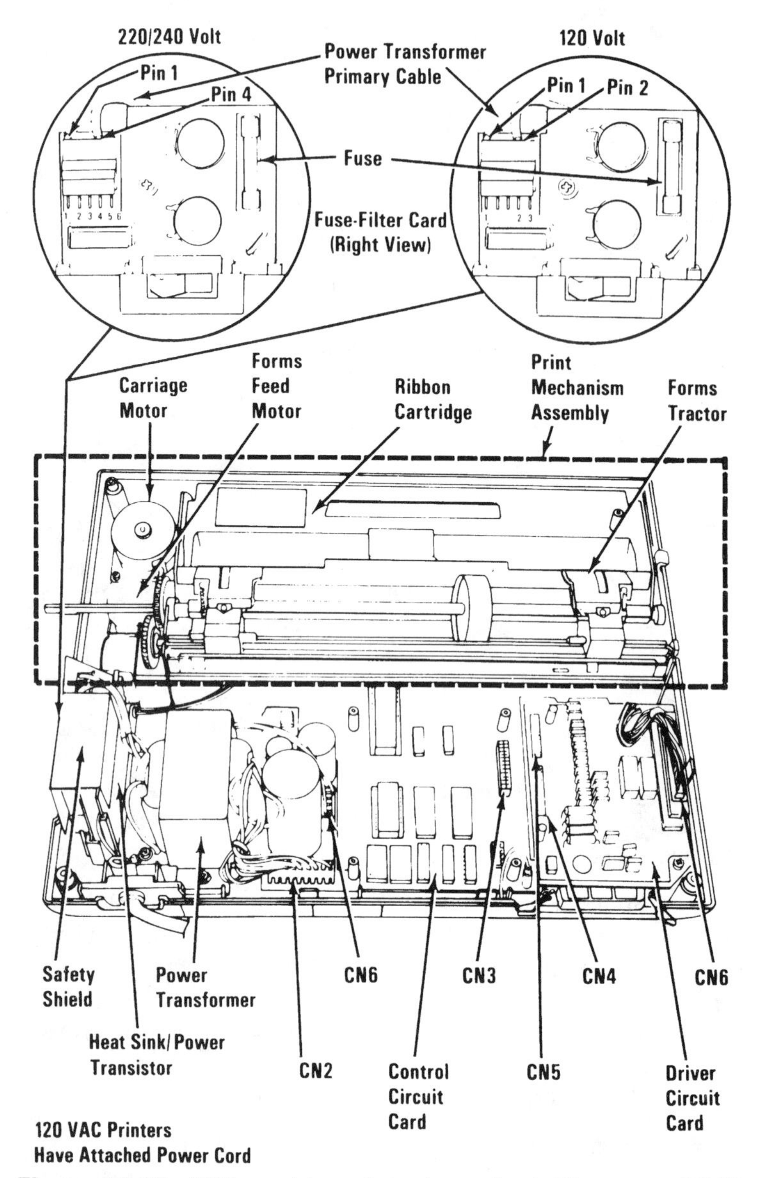

Figure 10-18 IBM graphics printer (rear view). (Courtesy of IBM)

Advantages

Very quiet

No mechanical parts in the print head

Light weight

Disadvantages

Speed of 16–65 cps

High cost of special paper

Limited paper sizes

Cost $200–$1500

Laser Printers. One of the most popular types of printers today is the laser printer because of its very high resolution, programmability, high speed, and new low cost (as little as

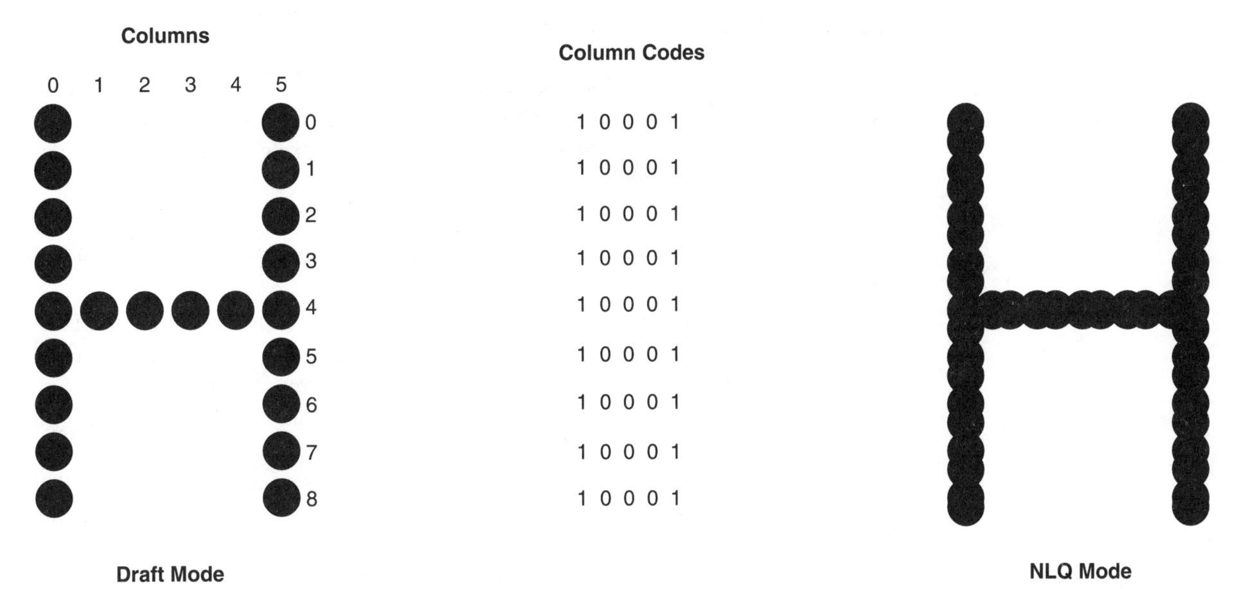

Figure 10-19 6 × 9 character matrix example.

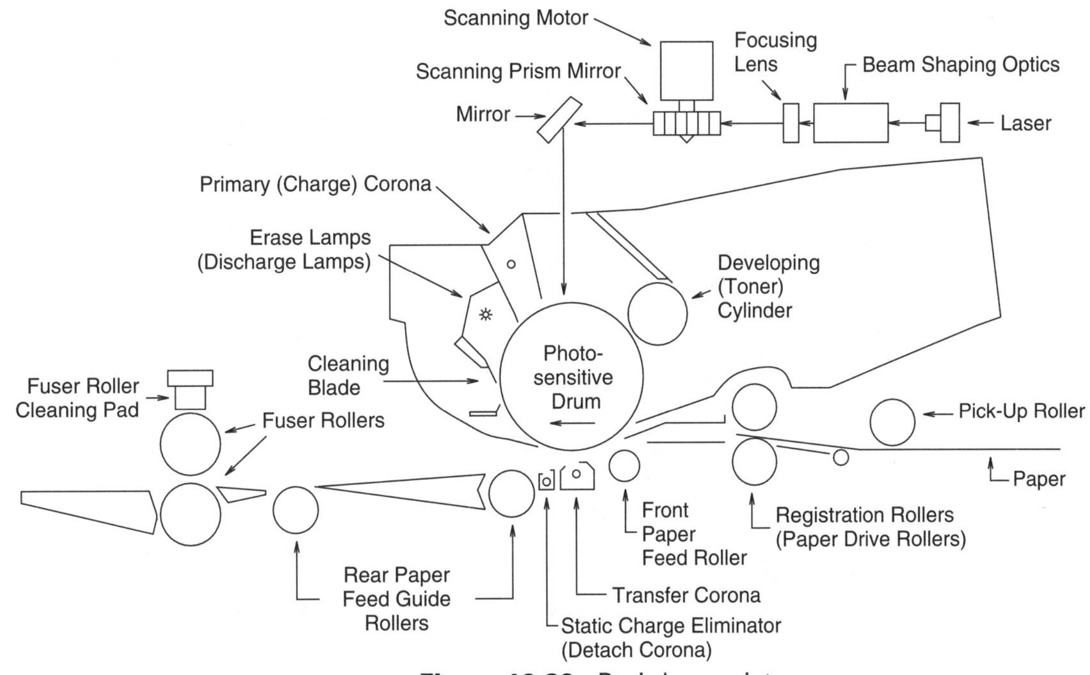

Figure 10-20 Basic laser printer.

$400). Laser printers contain high-speed large microcomputers in order to provide the control needed to print images on paper. Because the printing process cannot start until an entire page of data is downloaded into the printer, laser printers require a minimum of 512K RAM to print text and a minimum of 1.5MB RAM to print simple graphics. The more complex the image, the more RAM is needed.

Figure 10-20 shows a basic side view of a laser printer. The actual printing process does not begin until one page of data has been downloaded and processed by the MPU inside the laser printer. The MPU and its support circuitry convert the data from the printer interface and program any special functions into the laser printer. Once one page of data is downloaded into RAM, the printing process begins.

Laser Printing Process

1. The printing process begins with cleaning of the photosensitive drum. The drum is rotated while the erase lamps (also known as discharge lamps) remove any charges on the drum. During this process a blade in the toner cartridge mechanically cleans any toner off the drum.
2. The drum is conditioned for the printing process. During this step the primary (charge) corona places a uniform negative charge on the surface of the drum of about −600 volts.
3. As the drum passes the primary corona, the MPU of the laser printer starts to turn the laser on and off as the the (six-sided) scanning prism mirror scans the drum. The scanning produces the horizontal component of the photo image on the drum. Rotating the drum produces the vertical component of the photo image. When the laser light strikes the photosensitive drum, the negative charge on that area of the drum is discharged to about −100 volts. The system can turn the laser beam on and off so that it can produce between 300 and 600 dots per inch of the drum. The rotation of the drum also allows 300 to 600 rows of dots per inch. **Caution:** *Because the laser beam is invisible and can cause eye damage, laser units are enclosed in a sealed chamber. Check the repair manual before working on a laser printer.* The pick-up roller pulls a piece of paper from the paper tray toward the paper drive rollers (registration rollers) to prepare the paper for the fifth step of the printing process.
4. As it rotates, the drum contains an electrostatic image created by step 3. As the drum passes the toner-covered developing (toner) cylinder (toner and developer are mixed in some systems, while other systems use separate toner and developer), the toner is attracted to the areas of the drum that the laser has discharged. The image has now changed from an electrostatic image to a visible image on the drum. Toner is made up of microscopic iron particles encased in black plastic which are attracted to the less negative charges on the drum. The voltage applied to the toner cylinder controls the amount of toner drawn to the drum. This voltage allows for contrast and density control of the printed image.
5. As the drum continues to turn and the visible image begins moving toward its lowest point in the rotation, the paper drive rollers (registration rollers) move the paper close to the drum. The paper moves 0.5 inch past the bottom part of the drum before the image reaches the bottom of the drum rotation. The transfer corona places a positive charge on the back side of the paper as the paper passes between the drum and the transfer corona. The positive charge causes the toner on the surface of the drum to be attracted to the paper. The image is transferred to the paper as the drum turns and the paper moves. After passing the transfer corona, the paper passes across a detach corona (static charge eliminator) to remove any static charge on the paper and keep the paper from sticking to the drum. As the process continues and the paper and drum turn, the cleaning blade cleans any extra toner off the drum, while the erase lamps discharge the drum so that it can accept a new image after it has been conditioned by the primary corona. The drum rotates three times to print one letter-size page; larger paper requires more rotations.
6. The last step of the printing process occurs as the paper passes the detach corona and over the rear paper feed guide rollers and enters the fuser. The image at this time is held to the paper by gravity. The fuser roller applies heat and pressure to the paper. The heat melts the toner while the pressure from the special nonstick fuser rollers presses the toner into the grain of the paper. The fuser assembly is heated to about 180° C during the printing process and is held at a lower temperature in the standby mode. The fuser roller cleaning pad removes any excess toner from the upper fuser roller. As the paper exits the fuser, many laser printers employ a static discharge grid to remove any charges on the paper. The drum continues to turn until it is completely clean.

Advantages

Very quiet

Speeds from 4 to 22 pages per minute once the data is downloaded

No mechanical parts in the print head

Excellent print contrast, density, and quality

Very high character and graphic resolution—300 to 600 dots per inch (dpi)

Single- and double-sided (requires two printing passes) printing available

New lower prices on personal laser printers of about $400

Disadvantages

Business size and speed laser printers cost over $1000.

Cost of toner cartridges

Must be kept clean to reduce mechanical problems

Ink Jet Printer. Over the last 8 years ink jet printers have made changes in the way they print images on paper. In the early days of ink jet printers, pressure from a pump or created by the flexing of crystals was used to force ink from the print head to produce the dots that made up characters or graphics on the paper. Maintenance on the first ink jet printers was high because the ink would dry up in the print heads if they were not used for a period of time. Also control of the amount of ink in the dots was poor, causing overspray or leakage. Today's ink jet printers (Fig. 10-21) have the print head built into the ink cartridge, so the integrity of the print quality is maintained.

An ink jet printer functions like both a dot matrix printer and a laser printer. It acts like a dot matrix printer in that it starts printing before an entire page of text has been downloaded

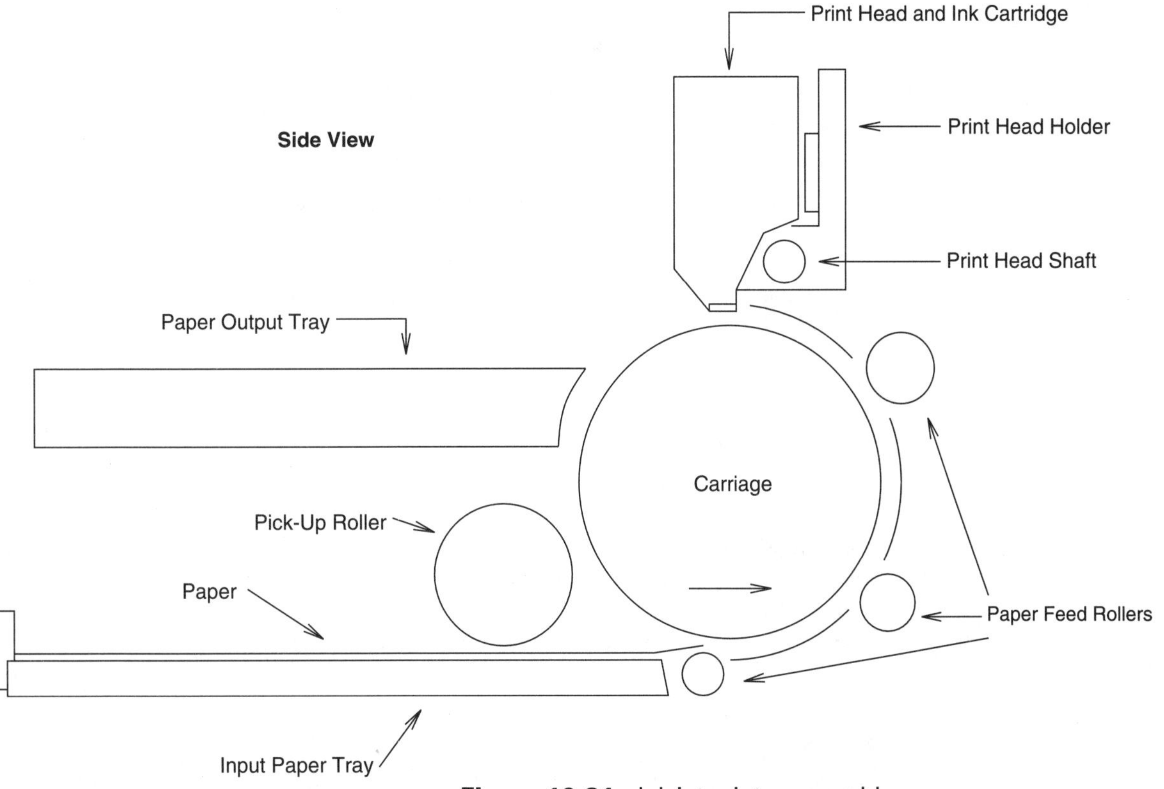

Figure 10-21 Ink jet printer assembly.

into the printer and it prints one row of characters at a time. It acts like a laser printer in that it has the ability to produce high-resolution images, from 300 × 300 dpi to 300 × 600 dpi, and it does not require mechanical force to produce the image on the paper.

Most ink jet printers today also have the ability to print in both color and black and white. The computer inside the ink jet printer allows the mixing of colors to produce photo quality outputs on special high-gloss paper. Ink jet printers can also print on transparencies and different forms of special paper.

Ink Jet Printing Process: The printing process is much like the process used in a dot matrix printer. The MPU of the ink jet printer controls the flow of data between the printer interface and the internal RAM of the printer. The RAM ranges from 64K to 256K. As soon as enough data has been received by the printer, the MPU starts the printing process. The pick-up roller grabs the paper from the input paper tray and passes it between the print carriage and paper feed rollers. The pressure on the paper between the rollers and the carriage turning cause the paper to move vertically.

Once the paper reaches the top of the print carriage, the print head holder causes the print head to scan the paper horizontally. The print head in the ink cartridge or ink cartridges (in a color printer) produces dots on the paper. The MPU causes the resistor in one or more of the 50 cavities to heat the ink in the cavity. The ink expands until it is pushed out of the cavity. The hot ink dries almost immediately after hitting the paper. In color printing colors can be mixed to create new colors. Because the hot ink dries immediately, there is no overspray and no leakage.

As each horizontal scan is completed, the paper is moved vertically to the next position. The paper exits the printing area onto a paper output tray, which contains a holding mechanism that holds recently printed paper off the bottom of the output tray until the next page is printed.

Advantages

Very quiet

Speeds from 120 cps to 400 cps

No mechanical parts in the print head

Very simple and reliable paper movement mechanics

Low cost: $200 to $2000.

Color printing options

Low power, small size, and light weight (can be used on portables)

High resolution: 300 × 300 to 300 × 600 dpi

Disadvantages

The higher the paper quality, the better the print quality.

Output is not quite as good as that of a laser printer.

10.3.5 Printer Troubleshooting

Normally, a problem with a printer is due to one of three things—lack of proper maintenance, mechanical breakdown, or lack or proper interfacing. Make the following checks to determine the problem. Serious printer problems usually are handled by authorized dealers.

If the printer does not print, run the self-test and check printer manual for sequence. If it fails the test, the problem is within the printer itself. Make the following checks.

1. Check AC line to printer.
2. Check DC power supply; correct if it is bad.
3. Check the voltage to major components on the printer control board.

4. Check the status sensors on mechanical parts.
5. Check the driver circuit to motors and the print head; a short circuit may have occurred.

If the printer produces print but it is not correct, do the following.

1. Check DC voltage to all major components on the printer control board.
2. Check for stuck data or address or control lines in the printer control board.
3. Check for jammed mechanical parts.

If the printer passes the test, the problem is not in the printer. Do the following.

1. Check the printer adapter (use diagnostic diskette) and correct the problem. If printer adapter is good, check the printer cable.
2. Check the printer cable by placing a break-out box on line (normally at the ends of the printer cable). A break-out box is a device that monitors the levels on the printer cable. A broken line at one of the ends of the printer cable is likely.

If none of the above is applicable, check the printer with different software. Lack of maintenance usually causes the biggest problems. The following procedures should be performed once a month under average use, to maintain maximum performance.

1. Clean out all paper dust. Dust can cause gears or the print head to jam, paper to move incorrectly, and characters to print irregularly.
2. Keep the carriage shafts clean and lightly oiled; these shafts allow the print head to move back and forth across the paper.
3. Clean the print head to reduce ink build-up and to reduce possible solenoid jams.
4. Clean the platen. Although a dirty platen does not normally cause any mechanical problems, it can cause poor print quality.

10.4 THE MOUSE

The most popular pointing device used in computers today is the mouse. The mouse is a device that moves the position of the cursor in a high-speed fashion that cannot be replicated by the functions of the keyboard. There are basically three different types of mice that use different methods to determine the X- and Y-axis movements of the mouse. Each uses an MPU to translate movements into serial data information that is applied to a serial port. The serial data received from the serial port is converted into mouse movements by the driver software downloaded into the comupter. Most mice today have either two or three buttons, which are used like the enter key to make the selection to which the mouse is pointing.

10.4.1 Electromechanical Mouse

Figure 10-22 shows the internal structure of a typical electromechanical mouse. As the mouse ball moves, the two rollers are used to detect movements. The X-axis roller supplies the horizontal movements of the mouse, and the Y-axis roller supplies the vertical movements of the mouse. At the end of the shaft connected to each of the rollers is a wheel on which are found small metal contacts. A mechanical sensor determines the amount of movememt as the wheel turns. The small electrical spikes sent from the mechanical sensor are applied to the mouse processor, which converts the signals from both the X and Y mechanical sensors into the correct data to be sent to the serial port in an asynchronous fashion. The mouse processor also sends out any switch closures from the mouse buttons when they are depressed.

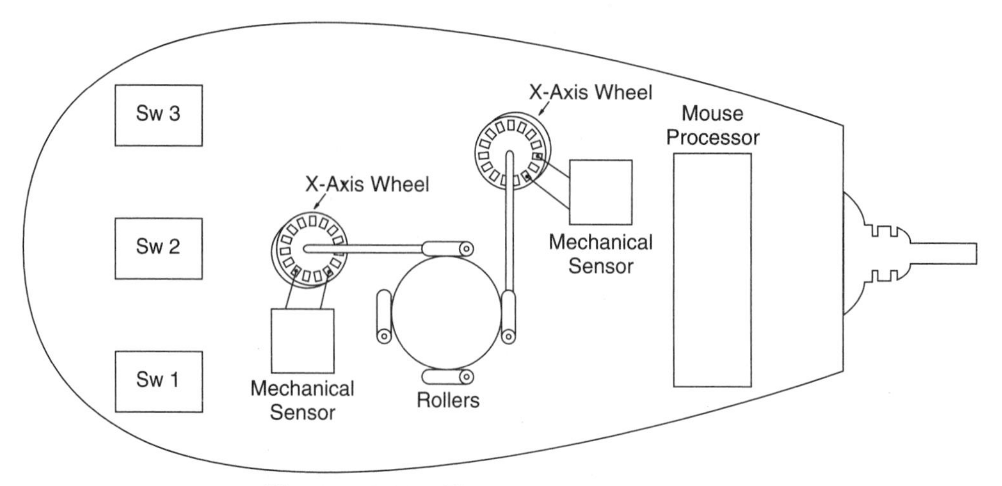

Figure 10-22 Electromechanical mouse.

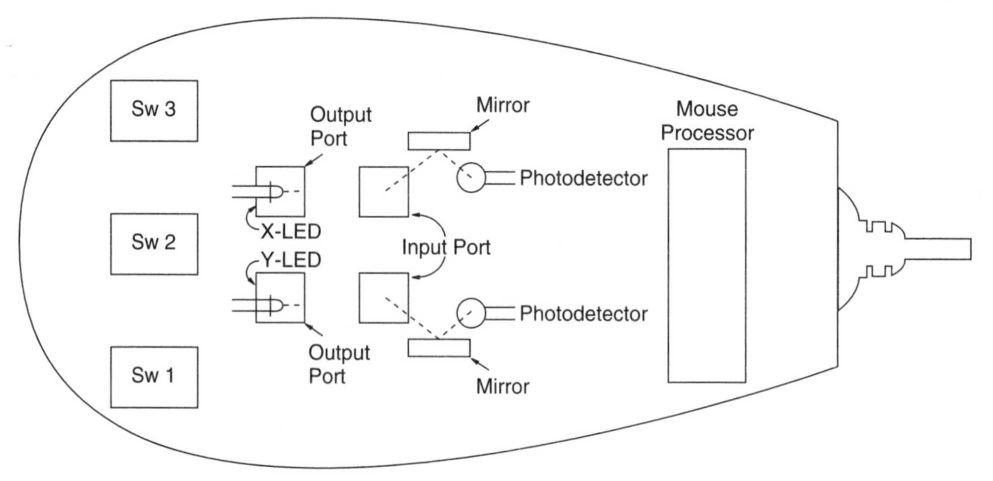

Figure 10-23 Optical mouse.

The problem with a electromechanical mouse is that the contacts on the wheels to the mechanical sensors can get dirty and impede tracking. Also as the mouse ball rolls it has a tendency to pick up dirt, which also impedes the correct movement of the mouse.

10.4.2 Optical Mouse

Figure 10-23 shows the internal structure of a typical optical mouse. An optical mouse does not use a mouse ball to determine the movement but instead uses a special mouse pad that contains markings to determine the movement. Two LEDs, one for each axis, are directed through two holes in the bottom of the mouse. The light emitted from the two LEDs is reflected from the mouse pad into two input openings on the bottom of the mouse. The light reflected back into the mouse housing is then reflected by two mirrors onto a photo detector. As the mouse moves, the light is broken up into pulses, which are used to determine the X- and Y-axis movements of the mouse. The photodetectors send their signals to the mouse processor, where they are converted into the proper format for the serial port. The mouse processor performs the same functions as that of the electromechanical mouse.

The advantage of an optical mouse is that, other than the mouse buttons, it contains no mechanical parts. Therefore, dirt affects the mouse only if the dirt is interperted as an axis line on the mouse pad. Optical mice tend to be expensive.

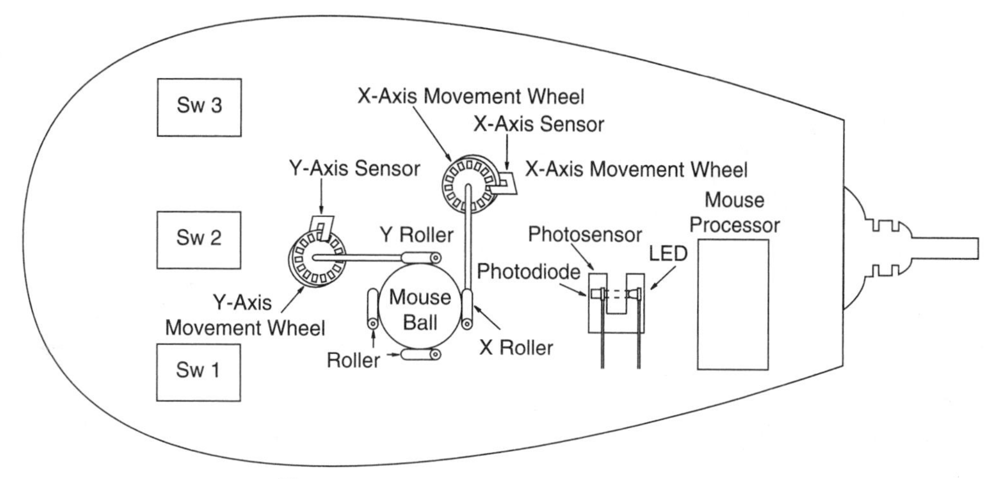

Figure 10-24 Optomechanical mouse.

10.4.3 Optomechanical Mouse

Figure 10-24 shows the internal structure of a typical optomechanical mouse. As the mouse ball moves, the two rollers detect movements. The X-axis roller supplies the horizontal movements of the mouse, and the Y-axis roller supplies the vertical movements of the mouse. At the end of the shaft connected to each of the rollers is a wheel, which contains many small slits. An optical photosensor containing an LED and photodiode detects the movement of the wheel as the light passes through the slits. The changes in light are converted into small electrical pulses. These are applied to the mouse processor, which converts the signals from the X and Y photosensors into the correct data to be sent to the serial port in an asynchronous fashion. The mouse processor also sends out any switch closures from the mouse buttons when they are depressed.

As the mouse ball rolls, it tends to pick up dirt, which may impede the correct positioning of the mouse. Most problems with mice result from dirt forming on the mouse ball and/or rollers. The mouse ball may be cleaned with warm soapy water and any dirt or fuzz removed from the rollers. Other common problems are using the wrong mouse driver and driver software that is bad or incompatible with the software of the system. If the serial port is bad, the mouse does not function correctly.

CHAPTER SUMMARY

1. Parallel communication allows the transfer of one byte of data at a time.
2. Serial communication allows the transfer of one bit of data at a time.
3. The advantages of parallel communication is that one byte of data is transferred at a time and normally no additional hardware is needed. The disadvantages of parallel communication are the high cost of the cabling, a tendency to pick up stray noise, and limitation of the distance between devices.
4. The advantages of serial communication are that fewer wires are needed for transmission, the cable costs less, and it tends to induce less noise. The disadvantages are that it requires additional hardware for the transmission, and transmission is slower than with parallel communication.
5. Printer signals:

STROBE#	Used to inform the printer that the printer adapter has placed data on the data bus for the printer.

D0–D7	Used to supply a byte of data to the printer.
ACK#	Informs the adapter when the printer is ready for more data.
BUSY	Informs the adapter when the printer is busy and cannot accept new data.
PE	Goes high when the printer is out of paper.
AUTO FD#	Produces a line feed.
ERR#	Used by some software to determine some type of printer error state.
INIT#	Used to enable some printers.
SEL#	Used to select which printer is to be used.

6. Print adapter block diagram functions:

   Address decoding: Used to select the printer adapter and its different registers.

   Command decoder: Used to send commands to the different registers of the printer interface adapter.

   Control latch: Used to issue the command for the printer adapter. The command is latched into the latch until software changes the level.

   Control buffer: Used to hold the condition of the control latch information and to verify command lines to the printer.

   Status buffer: Holds the status levels from the printer to the printer adapter.

   Data latch: Holds the data byte for the data bus.

   Data buffer: Holds the data byte from the data latch.

   Bus transceiver: Used to control the flow of data between the printer adapter and the computer I/O bus.

   Open collector drivers: Used to supply the necessary current for the drive control signals to the printer. Each control line uses an open collector driver.

7. The printer checks the status of the printer; if it is ready, the printer adapter places a new byte of data into the data latch. The adapter then causes the STROBE# line to go low for 5 μs, which informs the printer that a new byte of data is on the data bus. The printer causes the busy line to go high as it starts processing the data from the data bus. Once the printer can accept new data, it causes the ACK# line to go low, indicating that the printer is ready to accept another byte of data. When the printer finishes processing the data, the BUSY line goes low.
8. Simplex communication is one-way communication.

   Half-duplex allows two-way communication, but not at the same time.

   Full-duplex allows two-way communication to occur at the same time.
9. NRZ interpretation uses the level of the data line over a given time period.

   NRZI interpretation uses a transitional change in levels in a given time period to represent a logic-1 and no transitional change in levels to represent a logic-0.
10. Asynchronous communication adapter function:

    Address decoder: Provides the chip select to enable the ACE for programming or writing data and reading status or data.

    EIA receivers: Convert RS-232-C levels into TTL levels.

    EIA drivers: Convert TTL levels into RS-232-C levels.

    8250 Asynchronous communications element (ACE): Converts parallel data into serial data during a transmission and converts serial data into parallel data while receiving data. This device also verifies the control and status of the serial communications.

Current loop: Converts TTL levels into 20-ma current loop levels and 20-ma current loop levels into TTL levels.

11. RS-232-C specifications:

    Transmit data line: Used to output data.

    Receive data line: Receives input data from transmitting device.

    RTS#: The ready-to-send line indicates that the transmitting device is ready to send data to the receiving device.

    CTS#: The clear-to-send line indicates that the receiving device is ready to receive new data.

    DSR#: The data-set-ready line indicates that the receiving device is on line.

    RLSD#: The receive-line signal-detect line indicates that the receive line has a proper level on it.

    RI#: The ring indicator informs the serial port that the modem has detected a carrier on the phone line.

12. Printer terminology:

    Fully formed printer: Uses a character that is permanently formed on a printer head.

    Dot matrix printer: Uses an MPU inside the printer to create characters made up of a series of dots in rows and columns.

    Bidirectional printer: Prints characters in both directions.

    Impact printer: Uses electromechanical force to print the characters.

    Nonimpact printer: Uses nonmechanical force to print the characters.

    Font: The type style of a character.

    Pitch: The size of the character.

13. Dot matrix printer block diagram:

    The 8049 single-chip MPU is used to perform all handshaking between the printer and host computer. The 8049 reads the ASCII code in the printer buffer (RAM) and uses the code to address the character column codes in the EPROMs. This IC also transfers the character column codes to energize the print head solenoids to print the character.

    The 8041 microcontroller determines the position of the print head and controls the way paper moves through the printer.

    The 8155 controller accepts the data from the printer adapter and generates the BUSY and STROBE# signals. This IC also holds the data in RAM until it can be used to control the printer solenoids of the print head.

    The 2716 EPROMs hold the character column codes for the different character and font sets available for the printer.

    The motor drivers supply the necessary voltage and current to drive the motors of the 8041 microcontroller.

14. In a dot matrix printer, as the print head moves horizontally across the paper the MPU in the printer energizes the solenoids in the print head to produce the selected column of dots to form the character. Each character is made up of six or more column codes.
15. The least expensive printer is the dot matrix impact printer, which costs as little as $150, but it is slow and noisy. Today the ink jet printer produces near laser quality output from a printer that starts at about $200, is much faster than an impact dot matrix printer, and is very quiet. Ink jet printers normally cost more per page to print than an impact dot matrix or laser printer but provide good color printing. Currently ink jet printers are the most common printers sold. If speed and high-resolution printing are needed, a laser printer is

the best choice. With prices starting at about $400, this type of printer is becoming the standard for small and large offices and many home offices.

16. Once one page of data is downloaded into the laser printer, the printing process starts. First the photosensitive drum is cleaned and erased. Next the primary corona conditions the drum. As the drum turns, the laser and its scanning prism mirror produce an electrostatic image on the photosensitive drum. As the electrostatic image passes the developing cylinder, the areas that were struck by the laser light attract the toner to produce the visual image on the drum. The paper starts moving. As the rotation of the drum places the visual image of the drum at its lowest position, the transfer corona causes the toner from the drum to be attracted to the paper, which is passing between the corona and the drum. The visual image is thereby transferred to the paper. The paper then goes into the fuser assembly, where the toner on the paper is heated and pressed into the grain of the paper, completing the printing process.
17. Once enough data has been downloaded into the ink jet printer, the printer causes the paper to start moving in the printer. As the paper reaches the proper position of the print heads, the MPU in the printer causes the print head to start scanning the paper horizontally. To print a character the MPU causes a resistor in one or more of 50 print head cavities to heat the ink in the cavity; as the ink heats up, it expands and exits the cavity. Characters are formed in much the same way as in a dot matrix printer. Once one row of characters is printed, the carriage motor moves the paper to the next vertical row position.
18. Electromechanical and optomechanical mice provide the best performance for the dollar. No special mouse pad is needed to operate these mice. The main disadvantage is that they are susceptible to dirt and fuzz on the mouse ball and rollers, so frequent cleaning is necessary. Optical mice, although not as prone to the effects of dirt and fuzz, cost more and require a special mouse pad.
19. The electromechanical and optomechanical mice use a mouse ball and rollers to determine mouse movement in X and Y axes. The difference between the two mice is how the movement is converted from mechanical to some type of electrical signal. Optical mice use a light source, reflections, and light detection to determine the movement of the mouse. In all types of mice the mouse processor converts the mouse movements into a serial format for the serial port of the computer. The mouse driver software converts the serial data into cursor movements.

REVIEW QUESTIONS

1. Which type of data communication requires only two wires for the transfer of data?
   a. Parallel data communications
   b. Serial data communication

2. In general, which type of data communication is faster?
   a. Parallel
   b. Serial

3. What is the function of the STROBE# line in the Centronics printer interface?
   a. Informs the printer that new data is on the data bus
   b. Informs the printer adapter that the printer has accepted the data
   c. Informs the printer adapter that the printer is ready for more data
   d. Informs the printer adapter that the printer is busy
   e. None of the given answers

4. How many data bits are transferred at a time using the Centronics printer interface?
a. 1
b. 8
c. 16
d. None of the given answers

5. The purpose of the command decoder in a printer adapter is to provide address decoding for the adapter.
a. True
b. False

6. The printer adapter places the data on D0–D7 25 μs after the STROBE# line goes low.
a. True
b. False

7. Which method of data communication is most intelligent?
a. Parallel
b. Serial

8. Which method of data communication is fastest?
a. Parallel
b. Serial

9. NRZ asynchronous transmissions are more reliable than NRZI transmissions.
a. True
b. False

10. Which part of the asynchronous adapter converts TTL signals into RS-232-C levels?
a. 8250 ACE
b. 1488
c. 1489
d. None of the given answers

11. Which type of printers uses mechanical force to print characters on paper?
a. Ink jet printers
b. Laser printers
c. Impact dot matrix printers
d. Both a and b
e. None of the given answers

12. Printers print characters vertically as the print head scans the paper.
a. True
b. False

13. A bidirectional printer is slower than a unidirectional printer.
a. True
b. False

14. What is the major cause of problems in most printers?
a. Mechanical
b. Electronic

15. What is the major cause of problems in electromechanical and optomechanical mice?
a. Mechanical
b. Electronic

chapter 11

Data Communications/ Networks

OBJECTIVES

Upon completion of this chapter, you should be able to perform the following tasks:

1. Define data communications.
2. List the basic methods for data communications with modems.
3. Define the different methods of modulation used in data communications.
4. List the operational sequence of an FSK modem.
5. List the differences between mainframe networks and LANs.
6. List and define the two different types of network access methods.
7. List and define the differences in topologies.
8. Define the basic functions of network interface adapters.
9. Define network throughput.
10. List and define the four types of data encoding.
11. List and define the parameters of the four types of transmission lines discussed in this chapter.
12. List the parameters of Ethernet, Token Ring, FDDI, and ARCnet networks.
13. Define the function of protocols.
14. Define the difference between dedicated server and peer-to-peer networks.

11.1 DATA COMMUNICATIONS

Data communications is the transfer of data between computers separated by long distances. Today the term *information superhighway* expresses the exchange of information between computers, computer networks, and other devices. The most common method of connecting the computer to the information superhighway is a phone modem. Phone modems are used because phone lines are available just about everywhere in the world. An alternative is to connect a modem to a cellular phone system or some other type of system that allows communication over radio airwaves. Data communication can also take place between devices other than a computer, such as between fax machines. The important feature of data communications is the exchange of information between two or more different locations.

Modems are the most common device today for data communications. Speed, which has always been a factor in using modems, is becoming less of a problem. Five years ago 2400 baud (2.4K bits per second) was commonplace; this speed was improved to 9600 baud, followed by 14.4K baud. Today modems reach the speed of 28.8K baud over standard phone lines. Not only do today's modems have higher baud rates, but they contain special circuitry that allows the data to be compressed, and with the help of software, provide effective transfer rates of greater than 60K baud. In the following discussion the basics of modems and phone systems are covered, followed by an example of a typical FSK 2400-baud modem.

11.1.1 Modems

Modem is an acronym for modulator and demodulator. A modem is a device that changes the characteristics of data from one communication medium into those of another communication medium.

The modulator part of the modem converts data from the computer, usually TTL (transistor/transistor logic) or CMOS (complementary metal oxide semiconductor) levels, into some level that is compatible with the transmission medium.

The demodulator part of the modem converts the modulated signal from the transmission line back into the original form, which can be used by the computer or other device.

Modem Transmission Media. Modems use many different media for communication: telephone lines, RF coaxial cables, station-to-station or station-to-satellite microwaves, and fiber optics.

Telephone communication is the most common medium for modems because of its availability and relatively low cost. Because of the characteristics of the phone lines, however, data transmission is slow. There are two types of phone lines—standard and dedicated phone lines.

Standard lines are the regular lines used for voice communications. These lines may be switched from one line path to another path for trafficking calls. This switching occurs during normal voice communication but is never heard because of the high frequency of the switching. Switching may cause errors during high-speed data transmission.

Dedicated phone lines are not switched by the phone company; rather the signal is reconditioned every time it goes through a switching exchange. The problem with dedicated lines is their relatively high cost, but they do allow faster and more reliable communications.

RF coaxial cable communication modulates digital data onto a RF carrier and transmits the signal along a coaxial cable. At the receiver, the RF is demodulated and the digital data is recovered. This type of communication is used where distances of up to 10 miles are involved and high-speed transmission and noise immunity are needed.

Station-to-station microwave transmission involves modulating and demodulating digital data to microwave carrier frequencies. The advantage of this medium is its speed. The disadvantages are the initial high cost and the fact that the distances between stations are limited to about 150 miles because of the curvature of the earth.

Station-to-satellite microwave transmission operates the same as station-to-station microwave *communication* except that the transmission is beamed to a satellite in outer space, back to a receiving station on earth, and then to the other device. The advantages of this medium are that it is very fast and communication around the world can be achieved without going through many repeating stations. The disadvantage is the high cost of equipment and of the rental of a communication link code for the satellite.

Fiber optic transmission is the newest medium. It will probably become the most common type of communication in the future. Digital data is modulated to a light beam and then transmitted over fiber optic cable. Fiber optic cable is either clear glass or plastic strands that have a reflective coating on the outside to prevent light from escaping from the cable. Any light that leaves the cable decreases the intensity of the light and cuts down the distance the data can be transmitted. The light from the cable at the receiving end is demodulated and the data is recovered for use with the computer or other device. This method is fast and offers the most resistance to noise. The major disadvantage is the initial cost of the equipment.

Telephone System. Because phone modems are the most common form of telecommunication, we will discuss phone systems and phone modems in detail. Figure 11-1 shows the basic phone hook-up and how it works.

The phone is connected to the phone system by four wires. The wire labeled G is the ground for the system. The wire labeled R is the line that causes the phone to ring. The Tx (transmit) line allows information from the microphone to be placed on the phone line. The Rx (receive) line receives the data from the phone line.

When the handset is on the phone, the hook switch is closed. When the handset is picked up, CT1 turns off, indicating that the line is being used. CT2 turns on when data is received, blocking feedback. At this time the special hybrid transformer allows information to transfer to and from the handset.

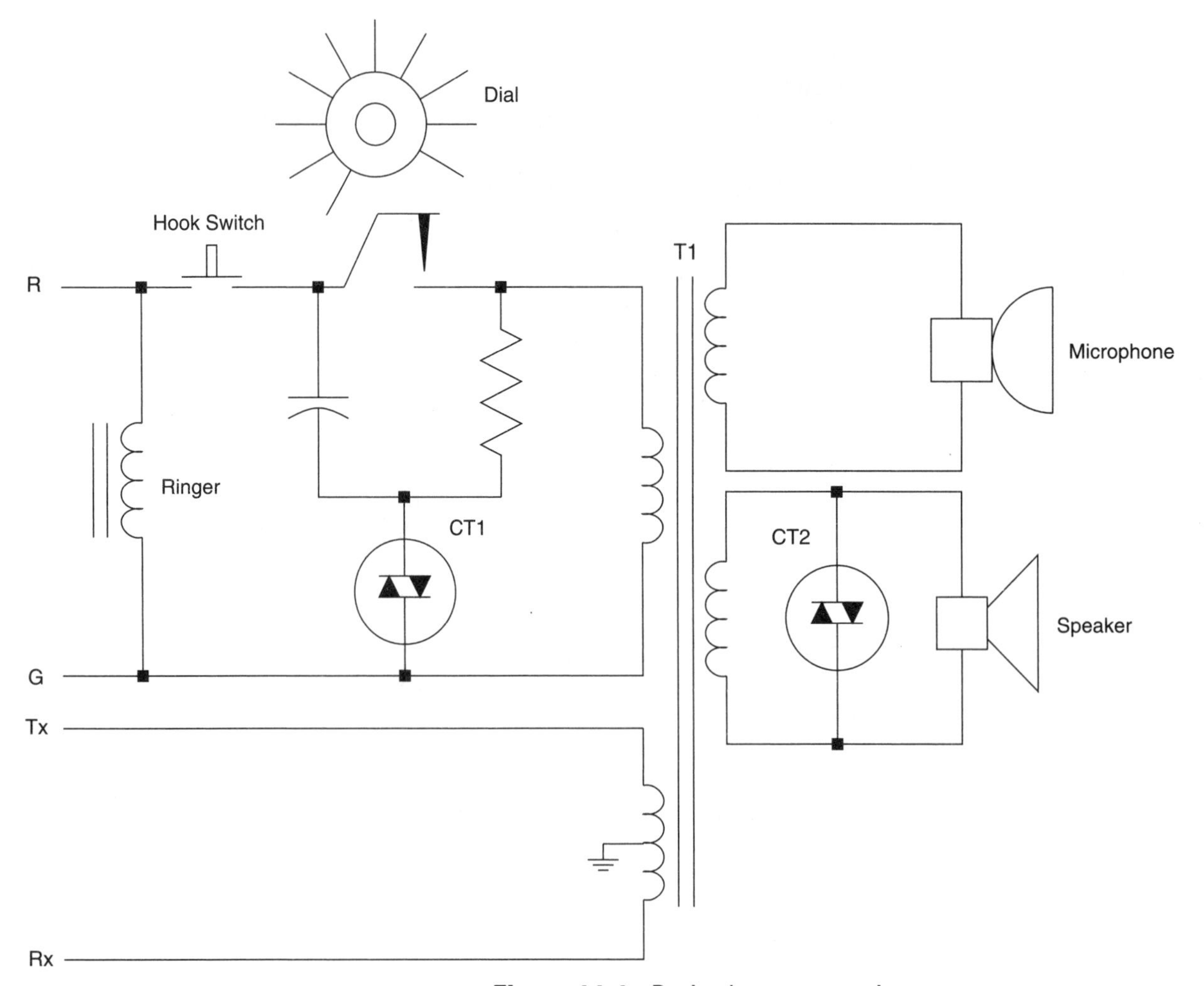

Figure 11-1 Basic phone connection.

When the phone is not in use, the reactance of the phone line is approximately 5 K ohms. CT1 and CT2 are used to determine if the phone is in use and to prevent any excessive signal levels from going through the line.

Dialing occurs at approximately 10 Hz or 100 msec. The off time of the pulse is 60 msec and the on time lasts 40 msec for proper decoding of the dial code.

Telephone lines and equipment are made to transmit voice data in the audible frequency range of 30 Hz to 3.4 KHz.

The bandwidth (Fig. 11-2) is so small because of the capacitance, inductance, and the equipment used in a phone system. Also, because phone lines cover great distances, signals are attenuated by the resistance in the cable. Phone companies must recondition the signal periodically. This reconditioning is usually handled by small repeaters in remote locations and at switching exchanges.

Distortion in Phone Systems: Distortion occurs in many forms in a phone system; the three most common forms are amplitude distortion, white noise, and impulse noise.

Amplitude distortion occurs because of the capacitive and inductive reactance in the phone system. This type of distortion limits the bandwidth from location to location.

White noise is the hissing sound heard on phone lines. This type of distortion is caused by thermal noise in switching and amplifying equipment.

Impulse noise is the static sound heard on phone lines. This type of distortion is caused by storms, switching equipment, and other types of electrical generating equipment.

Therefore, any type of data communication that uses phone lines must be able to distinguish between noise (distortion) and data.

Types of Modulation: Because the bandwidth is narrow in a telephone system, standard TTL changes cannot be transmitted on the phone line. For example, one cannot transmit a large number of logic-1s. Because the phone system cannot handle straight DC levels, the signal would never reach the other modem. Also, because a square wave at the time of transition has, in theory, no time delay, the frequency would be too high for the phone system to transmit. Therefore, the rising and falling edges of all digital signals would be attenuated. If a frequency of the transmission is too high, the attenuation of the waveforms might be such that the information would be interpreted incorrectly. Because of this bandwidth problem, the digital signals (data) must be modulated to a frequency that the phone system can handle (Fig. 11-3).

Figure 11-2 Phone system bandwidth.

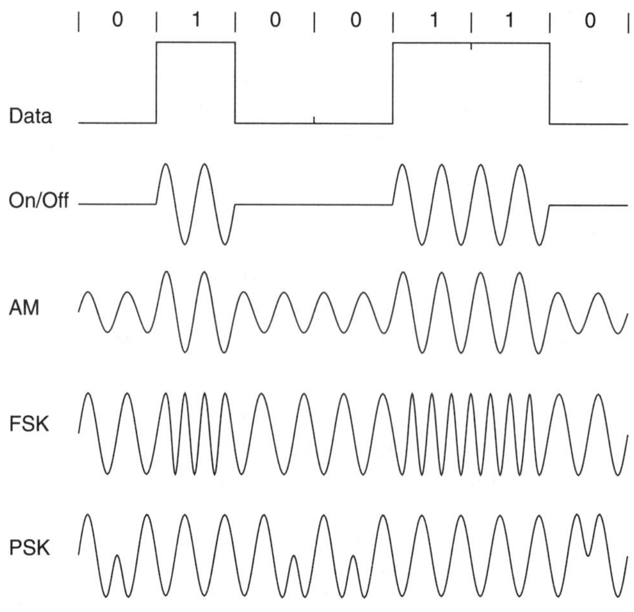

Figure 11-3 Types of modulation.

On–off keying modulation turns on and off a certain carrier frequency for digital data. The carrier is on to represent a logic-1 and off for logic-0. This carrier frequency must be within the bandwidth of the phone system—30 Hz to 3 KHz.

Amplitude modulation causes the amplitude of the carrier frequency to change in level. High-amplitude levels represent a logic-1, whereas small amplitude levels represent a logic-0.

In *frequency-shift keying modulation,* two different frequencies are used to represent different logic levels. The higher frequency normally represents a logic-1 (mark). The lower of the two frequencies represents a logic-0 (space). The frequencies must be within the bandwidth of the phone system and must not be harmonically related.

Phase-shift keying modulation uses the phase shift between two carriers to represent digital data. A 0-degree phase shift represents a mark and a 180-degree phase shift represents a space. The receiver in this system determines the phase relationship.

Differential phase-shift keying modulation operates the same way as phase shift keying, except that at the receiver, if the new data has a phase shift of 0 degrees from the previous phase shift, the data is a mark. If there is a 180-degree phase shift from the last piece of data received, the data represents a space.

FSK Modem: The most popular types of modulation in modems are FSK (frequency shift keying) and PSK (phase shift keying). In Figure 11-4 FSK modulation is used to transmit and receive data.

The modulator part of the FSK modem in the above example is simply a VCO (voltage controlled oscillator). Its purpose is to convert digital levels (logic-0 or -1) into two frequencies. The output of the VCO is decoupled by a capacitor to block any DC offset from the VCO. The signal is then applied to an audio amplifier to increase both the voltage and the current. The output of the audio amplifier is applied to a speaker that is coupled to the microphone on the handset.

The demodulator part of the modem gets its frequency input from the speaker on the handset. The speaker is coupled to a microphone on the modem. The output of the microphone is run through a decoupling network to filter out any DC level that might be picked up by the microphone. The output of the decoupling network is applied to a preamplifier, which increases the output voltage. The output of the preamplifier is applied to a bandpass filter with a narrow bandwidth. The bandpass filter only allows a narrow range of frequencies to pass

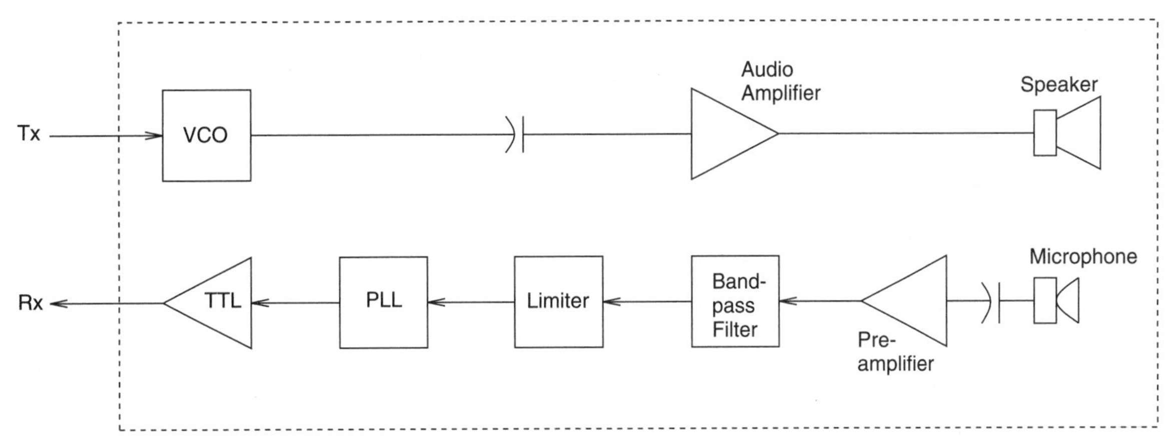

Figure 11-4 FSK modem.

TABLE 11-1
Originating and Answering Modem Frequencies

Mode	*Transmit Frequency*	*Receive Frequency*	*Bandwidth*
Originate	0–1070 Hz	0–2025 Hz	1950–2300 Hz
Originate	0–1270 Hz	1–2225 Hz	1950–2300 Hz
Answer	0–2025 Hz	0–1070 Hz	950–1400 Hz
Answer	1–2225 Hz	1–1270 Hz	950–1400 Hz

through the network, and all unwanted frequencies are attenuated. The output of the bandpass filter is applied to a limiter circuit. The limiter removes the upper and lower portion of the signal from the filter. Because noise is normally seen in the amplitude of a waveform, the limiter removes the noise from the circuit. The limiter also makes sure the signal to the PLL (phase lock loop) is the same amplitude if there are any frequency changes. The output of the limiter is applied to the frequency input of the PLL. The PLL tracks the incoming frequency and produces a DC output voltage that is proportional to the tracking frequency. The lower the frequency, the lower the DC output voltage; the higher the frequency, the higher the DC output voltage. The output of the PLL is then applied to a Schmitt trigger buffer. Its function is to make sure the levels applied to the computer are at proper TTL or CMOS levels.

Connecting a Modem to a Phone System: There are two ways modems can be connected to a phone line, acoustic and direct coupled.

Acoustic-coupled modems are the type in which the handset sits on the modem unit and a microphone and speaker are used to transfer the data over the phone. This type of modem is used in locations where the type of phone is not known. Acoustic modems can be used with any type of phone, even payphones.

Direct-coupled modems connect directly to the phone line through a phone jack and do not require the handset be connected to the phone. The advantage of this type is that it cannot pick up any environmental noise that may cause data errors. The disadvantage is that connections must be made to the phone line.

U.S. Modem Frequencies: Modems can be in one of two different operational modes. The modem is in the originating mode when it is making contact with the other modem. A modem is in the answering mode when it is answering the modem originating the communication link. Because modems operate on full-duplex, two different and unrelated frequencies are used for the communication link. Full-duplex allows the transmission and reception of data over the phone line at the same time without losing any information. Table 11-1 shows the originating and answering modem frequencies used in the United States.

In the previous example of an FSK modem, the baud rate of the modem would normally be limited to 600 baud, because of the frequencies used. If a high baud rate is required for data communication, a modem using PSK or differential PSK (DPSK) must be used along with more complex circuitry (usually including a microcontroller). This type of modem circuitry is beyond the scope of this book.

Modem Hardware Support: The modem hardware is only as good as the software support used to make the communications possible. The software is responsible for the transmission of data at the proper baud rate (number of bits per second), type of parity, size of data package, and format of the data.

The data format (also called a protocol) is a set of rules (laws) that determine how the transmission takes place. The most popular protocols today are KERMIT, XModem, and Z-Modem. These protocols govern how many bytes are transmitted in a package of information and how the handshaking takes place.

Troubleshooting Modems: To troubleshoot a modem or modem system, make the following checks.

1. First verify that the asynchronous communication adapter is operating correctly; use the diagnostic diskette.
2. Verify that the asynchronous cable is good by using a break-out box.
3. Run the internal self-test for the modem, making sure that the adapter can access the modem. If the modem fails the test, the problem is in the modem. If the modem passes the test but communications cannot be made, make the following checks.
   a. Verify that the phone line is good.
   b. Use an oscilloscope and diagnostic diskette to verify that the proper signals are getting to the modem.
   c. Verify that the software is good and is configured for the correct equipment.
   d. Check the phone line with the oscilloscope while running a modem test program; the data being sent by the modem should be transmitted to the phone lines. Remember, the data will be in the form of different frequencies not logic levels.
4. If the proper frequencies are being transmitted and received, and the modem is still not communicating properly, check the following.
   a. Verify that the software is correct for both systems.
   b. Verify that the communication parameters in the software match the parameters of both computer systems.
   c. Verify that the proper protocols are being used by both systems.

11.2 NETWORKS

The idea of networking has been around since the first mainframe computer systems. Networking allows two or more users or computers to share the resources of the network. The resources of a mainframe system include the processing power of the mainframe itself, RAM, and software. Resources also include peripherals such as hard drive units, printers, tape drives, and plotters. In a mainframe computer system (Fig. 11-5) each user uses a terminal to communicate with the network. In most mainframe systems the terminals are dumb. A dumb terminal contains no independent processing power; to process data it must be connected to the mainframe network. The terminal allows the user to input data and commands through a keyboard and to view the communications with the mainframe using a video display. Smart terminals contain their own processing power and use a network interface emulator to access the mainframe as a centralized source of data and contact with peripheral resources.

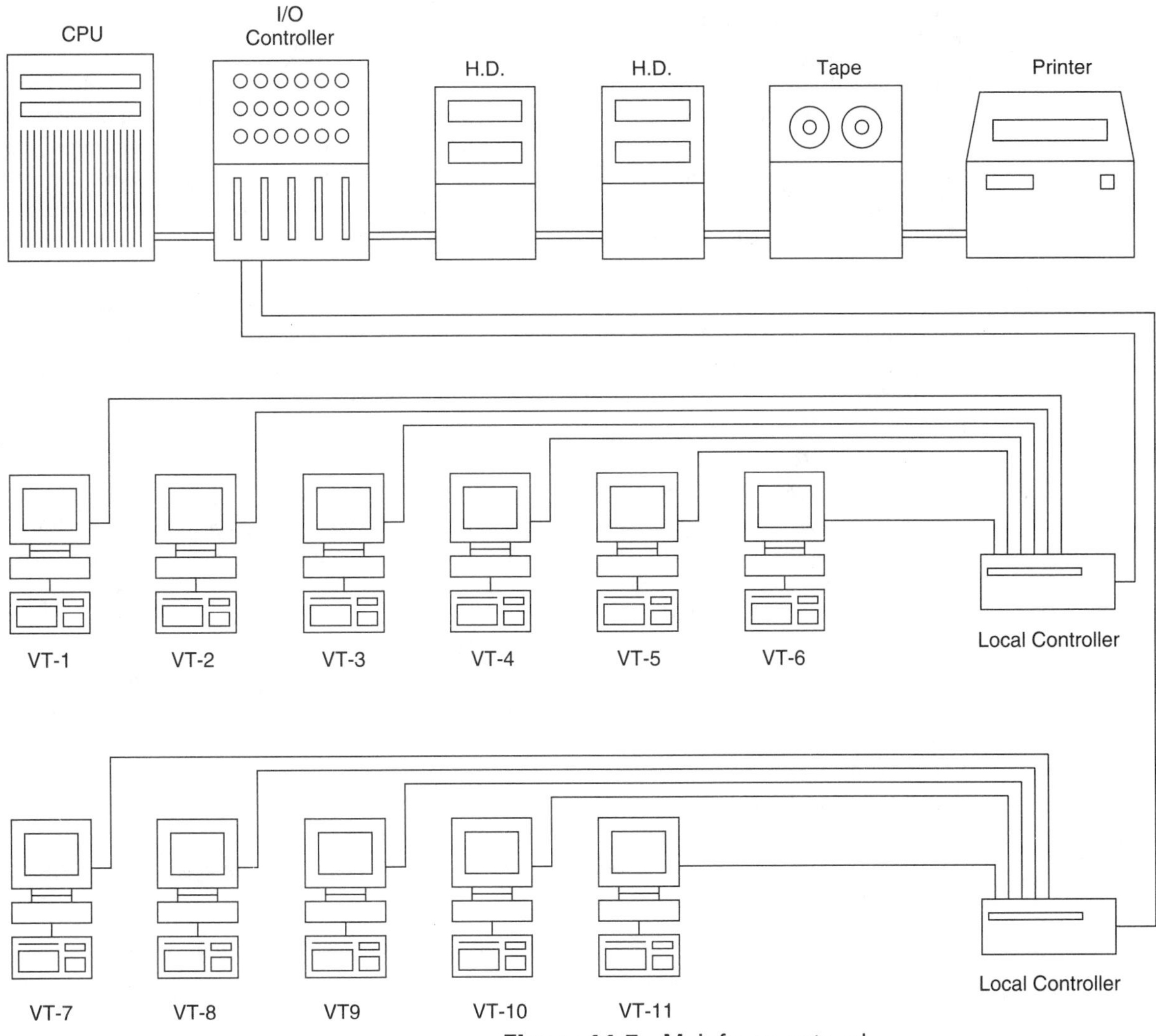

Figure 11-5 Mainframe network.

As each user is connected to the mainframe system, he is given a time slice. The mainframe switches between users, giving each user a certain amount of time to control their operations on the network. Only one user receives service from the mainframe at any one time. Because the mainframe is so large and fast, it looks like each user has control of the system, but in reality the mainframe is switching among users. As the number of users connected to the mainframe increases, each user's time slice gets smaller. Users connected to a mainframe may also experience a slowdown when one or more users request a process that ties up the mainframe's processing power. If the mainframe goes down, the users connected to the mainframe are not able to process any data.

Today many companies use mainframes as a centralized source of data and control and as a means by which users can gain access to peripherals otherwise too expensive to use on a single-user system. In this type of configuration each user uses a personal computer with a network interface card to make the connection to the mainframe. The personal computer processes the data and uses the mainframe as a source of data. If the mainframe goes down, each user is able to process data locally as long as the data is in the personal computer. Also, because the mainframe is not responsible for processing data, users connected to the mainframe do not experience the same amount of slowdown as with a mainframe using dumb terminals. As the number of users connected to the mainframe increases, the response time also decreases.

11.2.1 Local Area Networks (LANs)

With the advent of high-power personal microcomputer systems, networking on a local level has become popular. LANs allow two or more computers to be connected to each other, with each computer executing application software locally. Depending on the configuration of the system, each computer can share the resources of the other computers on the network. By networking personal computers, expensive peripherals can be shared by the computers connected to the network without the expense of having the peripheral connected to each computer directly. LANs also provide a central source for data and control.

Network computers require three items: the network adapter, network software, and the transmission line. The network adapter connects to one of the I/O interface slots of the computer, and the software sets up the rules to allow the data or resources to be shared. The transmission line is the physical wire or other transmission medium that connects the computers in the network.

LAN Architecture. In this chapter three of the most popular types of LAN architecture are described. The Ethernet system was first developed in 1972 by Digital Equipment Corporation, Intel, and Xerox and is the most popular system for LANs using 100 or fewer computers (300 maximum). The first ARCnet (attached resource computer network) system was developed in 1977 by Datapoint Corporation. The first Token Ring system was developed in 1985 by IBM. The FDDI (fiber distributed data interface) standard was specified by ANSI in 1985 and operates much like Token Ring networks. The LAN architecture defines the design of the LAN, hardware requirements, software requirements, and protocols (rules for communications) of the network. Before discussing the details of the three types of LAN architecture, the basic terminology is explained.

Access Method. Access method describes how the communications take place on the network. Ethernet architecture uses what is called carrier sense multiple access with collision detection (CSMA/ CD) for accessing the network. ARCnet and Token Ring architectures use token-passing media access.

CSMA /CD: CSMA /CD works on the principle of first come, first served. Before a network interface card (node) tries to communicate on the network, it first monitors the transmission line for activity. If no activity is sensed, the transmission begins. If activity is sensed on the transmission line, the node waits a predetermined time before it tries again.

Once the transmission starts, the node sends out a packet of data while monitoring the transmission line. If no collision occurs during the transmission, the data packet is processed by the select node. If during the transmission there is a loss of the carrier or a sudden change in average voltage on the transmission line, a collision (two or more nodes transmitting data at the same time) has occurred. Once a collision has been detected, all nodes transmitting data stop transmitting. The nodes detecting the collision then transmit a jabber code and then stop transmitting data for a time period determined by TBEBA (truncated binary exponential back-off algorithm).

The jabber signal ensures that all nodes on the network go into a time delay before trying to access the transmission line. TBEBA is a routine built into the Ethernet controller IC and is used to provide different types of time delays for each node before any node can try to transmit data.

Token Passing: In the token-passing access method, a token signal is sent from one station to another in the network. Each station connected to the network has a unique address. The token is passed in a predetermined order. In order to transmit data the station must have the token that gives it control of the network. Therefore in a token-passing network only one station can transmit data, and no collisions can occur.

Topology. Topology describes both the flow of data and the physical connections of the computers in the network. Logical topology describes the flow of data, and physical topology describes the physical connection in the network.

Logical Topology: Although ARCnet and Token Ring both use the token-passing access method to determine which station may transmit the data, the actual data packet transmission differs. In ARCnet all stations in the network hear the transmission but only the destination station processes the data packet. In the IBM Token Ring network, only the destination station hears the data packet transmission.

Physical Topology (Fig. 11-6)

BUS In bus topology, each node is connected to a common main bus transmission line. In this type of network the bus must be terminated on each end of the bus line with a terminating resistor. If a break occurs on the bus, no communication can take place because the transmission line is open. To troubleshoot a bus topology network, start at one end of the bus and connect two computers together with terminators on each end. Start the network software and try to establish communication. If communication can be established, the cable between the two computers is good. Remove the terminator from one end of the cable and connect the next segment of cable to the next computer and terminate. Start the network software and try to establish communication. If communication can be established, the new segment of cable added to the network is good; otherwise the cable is bad. Continue this process until the break in the cable is found. If the layout of the network is available, you can divided the network in half and terminate each end of the cable segment to be tested. Test the segment with the network software; if communication can be made that cable segment is good. Then test the other half of the network. If communication cannot be made, the cable segment is bad. Then divide the bad cable segment in half, terminate each end, and test each cable segment. Continue this process until the break in the network cable is found.

STAR-WIRED RING In a star-wired ring topology network, each station is directly linked only to stations on each side of it, and each station is connected to a multistation access unit (MAU) through a lobe cable. The lobe cable contains two pairs of wires, one for transmission and one for receiving. The lobe cable with its pair of twisted wires form the

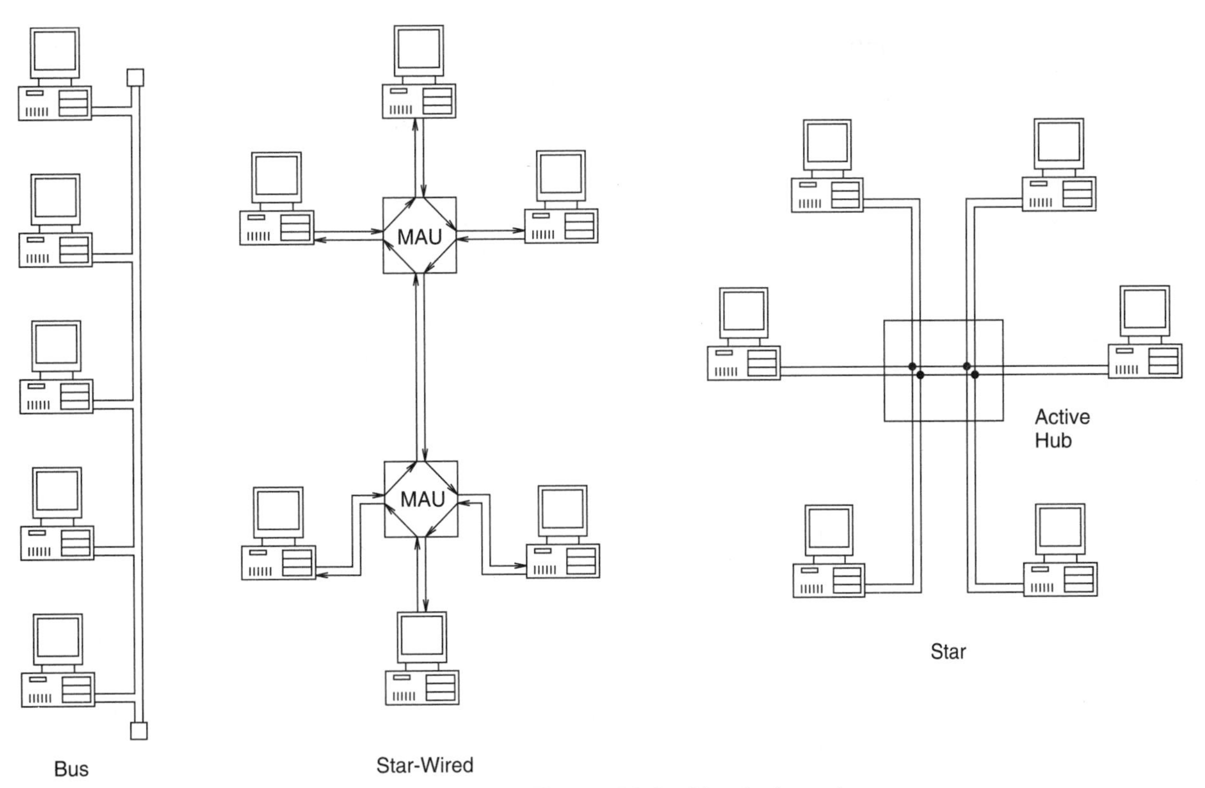

Figure 11-6 Physical topology.

ring in the star-wired ring network. If a lobe cable or main cable connecting the MAUs together fails, the entire network goes down because there is a break in the ring.

STAR/DISTRIBUTED STAR In a star or distributed star topology network, each station in the network is connected in parallel through a hub. All stations in a star or distributed star network hear data transmissions at the same time. There are two types of hubs in a star or distributed star network. An active hub amplifies and splits the signal between the stations, whereas a passive hub simply splits the signal. If one or more of the connections to the hub fail or break, the network can still function.

Network Interface Adapter. Basically all network interface adapters (Figure 11-7) do the same thing—they act as synchronous serial interfaces to provide high-speed serial communications on some type of transmission line. Basically, the adapter during transmission must convert parallel data from the I/O bus of the PC into serial data that is formatted in a synchronous data package ranging from 40 bytes to about 18K per package, depending on the network architecture used. Once the data has been converted and packaged, the network adapter must convert the TTL levels into levels that can be used on the transmission line and send the data out at the proper speed. During the reception of data, the adapter must convert the signals provided by the transmission line into TTL levels. The data is then unpacked and checked for proper reception and then converted from a serial format into a parallel format so that it can be transferred onto the I/O bus of the PC. The network adapter must also verify and control the protocols for the network in question. Each data packet contains not only the data that is being transferred but also other additional information that is used to verify the transfer, addressing, and other software and protocol information.

In an ARCnet system each character package is 11 bits, which sounds like asynchronous serial transfers. That is true except that in ARCnet between 1 and 508 character packages are sent as a data packet group, which makes it operate more like a standard synchronous serial transmission.

Network Throughput. Throughput defines the speed at which data is transferred through the network. Throughput is a combination of the transmission speed and the amount

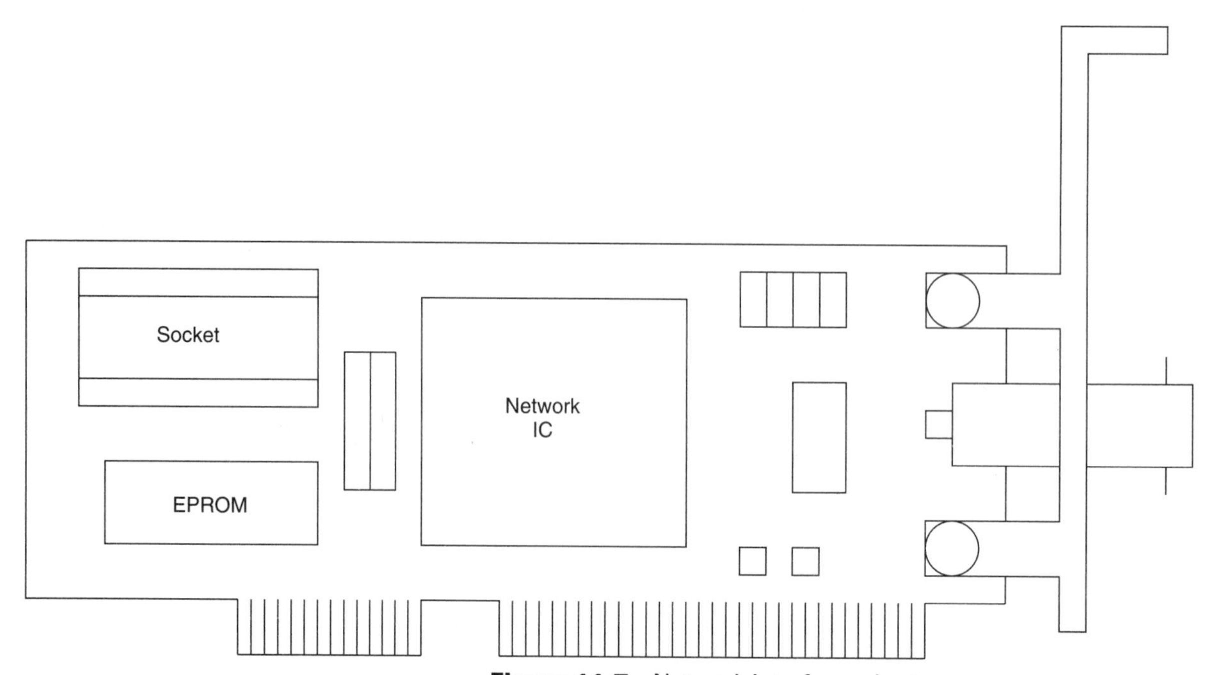

Figure 11-7 Network interface adapter.

of time it takes to process the transmission. The transmission speed is specified in the number of bits per second (Mb/sec = millions of bits per second) that are carried on the transmission line. The type of transmission line is a major factor in determining the maximum transmission speed. Table 11-2 specifies the speeds of the three networks discussed in this chapter.

Data Encoding. In order to transmit the data on the transmission line, the TTL data that has been serialized and packaged into the proper format must be converted into a signal that is compatible with the transmission line. The process of converting the signals is called encoding (Fig. 11-8).

Ethernet encodes data using the Manchester standard. In Manchester encoding the clock and data bits are combined and data is sent out on a square wave with voltage levels from 0 volts for a high to −2.0 volts for a low. A transition from 0 volts (high) to −2.0 volts (low) occurs during the middle of the bit time, and the data represents a logic-0. A transition from −2.0 volts to 0 volts occurs during the middle of the bit time, and the data represents a logic-1. Any transitions at the beginning of the bit time are used to set up the next data bit.

Token Ring encodes data using the differential Manchester standard. The Token Ring system uses two pairs of shielded twisted pair (STP) wires. One STP wire is used to transmit data and the other pair to receive data. If the data bit transmitted is a logic-0, there are two voltage swings, one at the beginning and one in the middle of the bit time. If the data bit transmitted is a logic-1, there is only one voltage swing in the middle of the bit time. The voltage swing between either one of the wires in the twisted pair and the shield is between 0 and 1.5 volts. The other wire in the twisted pair also has a swing between 0 and −1.5 volts. Therefore the difference in voltage during a transmission ranges from 0 volts to 3 volts peak to peak (1.5 volts to −1.5 volts).

In an FDDI system, the encoded data is converted into light during transmission and the light received by the adapter is converted into voltage during reception. FDDI systems use NRZI (non–return-to-zero inverse) to encode the data bits. If a logic-1 is transmitted, the voltage level that controls the light source or the voltage received by the detector toggles (it does not matter in which direction). If a logic-0 is transmitted or received, the voltage level does not toggle (it does not matter what the voltage level was).

ARCnet systems do not use square waves during encoding. A logic-0 is seen when the voltage level on the line is at 0 volts during the entire bit time. A logic-1 produces a sine wave pulse with a 50% duty cycle that swings from 0 volts to 10 volts down to −10 volts and back to 0 volts in a given bit time.

Transmission Lines. The transmission lines (Figure 11-9) are used to connect the different computers in the network together. The network interface card must convert TTL data from the computer into signals that can be sent down the transmission line during the transmission of data. The network interface card must also convert the signals from the transmission line into TTL signals for the computer while receiving data. Each type of transmission line has certain characteristics that govern the speed of the network. The following are the different types of transmission lines used with networks.

TABLE 11-2
Network Transmission Speeds

Network Architecture	*Transmission Speed*
FDDI Token Ring	100 Mb/sec
Ethernet (modified Ethernet)	10 Mb/sec (100 Mb/sec)
ARCnet (ARCnet Plus)	2.5 Mb/sec (20 Mb/sec)
IBM Token Ring (new IBM Token Ring)	4 Mb/sec (16 Mb/sec)

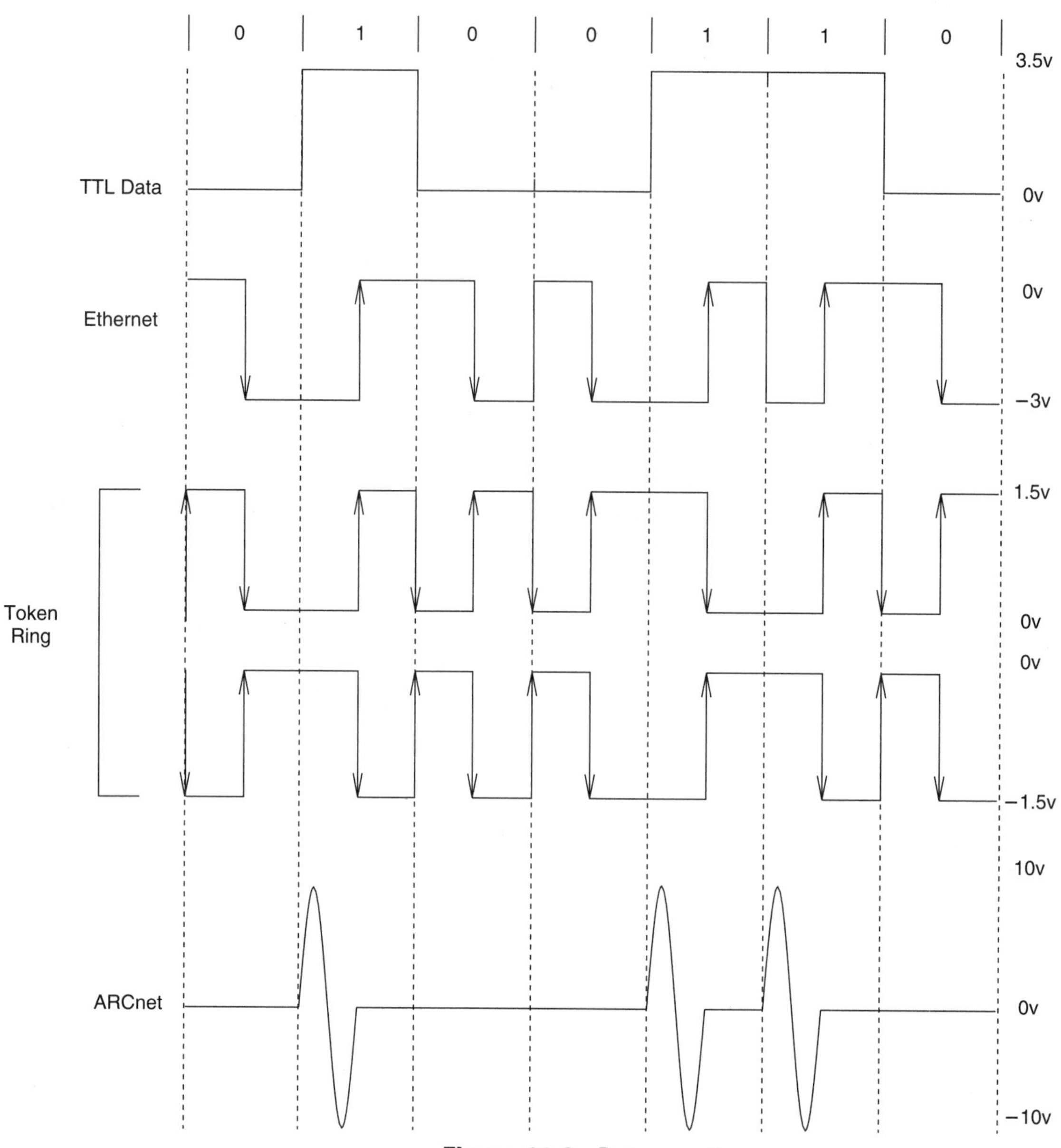

Figure 11-8 Data encoding.

Coaxial Cable: There are many types of coaxial cables, each with its own characteristics that govern the speed of the transmission line. Coaxial cables also differ in impedance values, insulation, shielding, and types of conductors. A coaxial cable contains a solid or stranded center conductor surrounded by an air (polyethylene tube larger than the conductor separated by a spiral plastic spacer and air), cellular, or solid polyethylene insulator. The insulator is then surrounded by a wire mesh and/or foil shield that is covered in fire-retardant polyethylene. Table 11-3 lists the common coaxial cable specifications for networking.

Twisted Pair: There are two types of twisted pair cables. STP (shielded twisted pair) cables contain one or two sets of twisted pairs of wires, each pair surrounded by a foil shield. The shielded pairs are placed within a wire mesh shield covered with a polyethylene outer coating. The advantages of the STP are higher transmission speeds than with unshielded twisted pair (UTP) and less likelihood of electrical noise. The disadvantages of the STP are greater cost but narrower bandwidth than with coaxial and fiber optic cables. The UTP cable contains one or more pairs of twisted wires covered with a polyethylene coating. The UTP is similar to phone

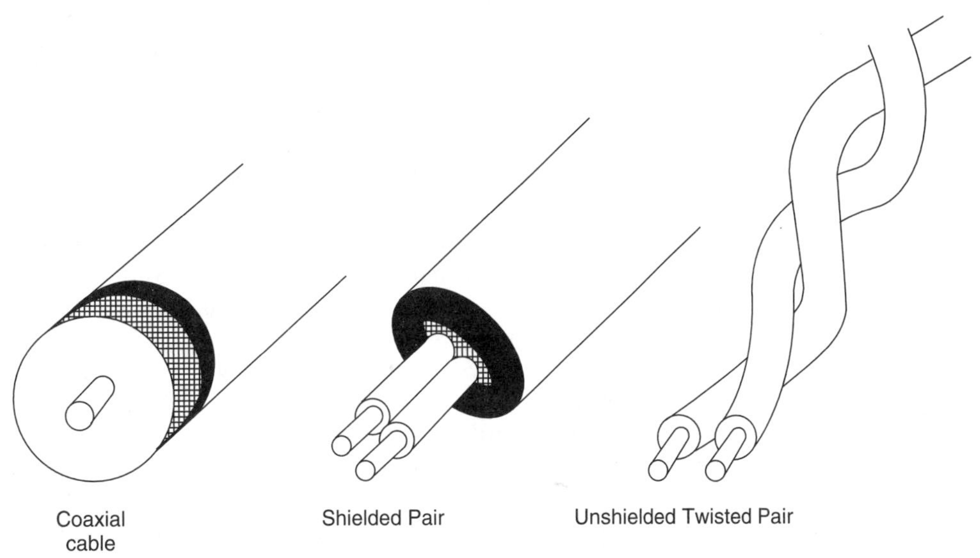

Figure 11-9 Transmission lines.

TABLE 11-3
Coaxial Cable Specifications for Networking

Coaxial Cable Number	*Impedance*	*Polyethylene Insulator*	*Diameter*	*Notes*
RG-8A/U	52 ohms	Solid		
RG-8/U	50 ohms	Cellular	10.3 mm	Ethernet ThickNet
RG-58A/U	50 ohms	Solid	4.95 mm	Ethernet ThinNet
RG-59B/U	75 ohms	Solid	6.15 mm	Used in older systems
RG-62A/U	93 ohms	Air	6.15 mm	ARCnet

wires. The advantages of UTP are the low cost of cabling and, in many circumstances, the presence already of cabling in the building. The disadvantages of UTP are its narrow bandwidth and its susceptibility to noise.

Fiber Optics: Fiber optic cables contain one or more very thin strands of specially made glass fiber (50 to 100 microns); each fiber contains a clad coating to eliminate loss of the light signal, which keeps the light traveling along the length of the fiber. Over the coated glass fiber, a protective plastic surrounds the cable to protect the clad coating and provide an easy way to handle the cable. The diameter and type of glass and clad coating used determine the bandwidth and distance the fiber optic cable can transmit the signal. The biggest advantage of fiber optic cable networks is very high speeds; also no electromagnetic field is created (provides maximum security), and the cross talk and noise associated with metal cables is eliminated. The disadvantages of fiber optic cables are that they are more expensive (cable and equipment to operate the network) and harder to repair when damaged.

Ethernet Networks. Ethernet networks are the most popular type of networking architecture today. In this section we discuss media access and cable encoding. Today three different cabling formats are available for Ethernet networks. Of the three different cabling formats, ThinNet is the most popular, with 10BaseT being the second most popular and ThickNet the least popular.

In an Ethernet network using ThinNet cabling, RG-58A/U (50-ohm coaxial cable) is used to connect the Ethernet interface network adapter. Each adapter has a BNC connector.

Connected to this BNC connector is a BNC T-adapter, which allows two cables to be connected to the back of the network adapter. Typically one segment of coaxial cable is connected to one end of the T-connector and another coaxial segment is connected to the other end of the T-connector. On each end of the coaxial bus, the coaxial cable is terminated by a 50-ohm BNC (bayonet Neill-Concelman) terminating resistor. The T-connector allows each Ethernet card to tap into the coaxial bus. The speed of a ThinNet Ethernet network is 10 Mb/sec, with a new enhanced version operating at 100 Mb/sec. The maximum length of the cable segments is limited to 300 meters without a repeater. Up to five cable segments can be connected together to increase the maximum length to over 1000 meters with the use of repeaters.

Ethernet networks using 10BaseT have a modular type (like an RJ-11 phone jack) connector to join the network adapter to the passive or active hub. The four-wire cable (two UTPs) is used to make the connection. One pair of twisted wires is used for transmitting data, and the other pair is used to receive data from the hub. Hubs normally control up to 16 different Ethernet adapters. To increase the size of the network, hubs can be networked together. STP cables can be used to increase the distance between the computer and the hub. Active hubs not only connect the network adapters to the main bus, but also monitor the status of the network and clean and re-establish the signal levels on the network. The hubs are needed so that the 10BaseT network can operate at 10 Mb/sec. The 10BaseT format allows a star configuration, which makes troubleshooting easier. The maximum length of any one of the stars is about 100 meters.

Ethernet networks using ThickNet use RG-8/U coaxial cable, which must be terminated at each end just as in ThinNet. The connection between the network adapter and coaxial cable is made through AUI (attachment unit interface) cable. The cable can be up to 50 meters in length and has a DB-15 connected to each end. The AUI cable then is connected to a transceiver which clamps onto the coaxial cable, and a pin makes the connection precisely to the center conductor of the coaxial cable. The transceiver performs collision checking, transmitting, and receiving operations. The only advantage to this cable format is that it allows up to 500 meters of coaxial cable to be used, but it limits the number of stations per cable segment to 100. If a repeater is used, the number of stations can be increased to 1024. ThickNet is not as popular because of the cost of the coaxial cable and transceivers used in the system.

Token Ring Networks. Token Ring networks use two pairs of STP wires to connect the network. One pair of STP wires is used to transmit data and the other pair is used to receive data. Each network adapter is connected to the MAU, which controls up to eight adapters to form the ring. The MAU controls whether the transmit or receive twisted pair is used for the selected adapter. Each lobe cable contains eight wires (four wires of the two twisted pairs and four shields) and has a nine-pin connector on each end. The lobe cable connects the adapter to the MAU. Multiple MAUs can be connected to form a large ring of up to 240 adapters.

FDDI Networks: FDDI uses a Token Ring format for communications, but instead of using wires and voltage to transfer data it converts voltage levels into light or light into voltage levels. With single attached stations a pair of fibers is used, one to transmit data and the other to receive data. With dual attached stations two pairs of fibers are used, one to transmit and one to receive data.

ARCnet Network: ARCnet systems can use either UTP wires or coaxial cable. The type of physical topology used determines which type of cable to use. In a star topology configuration, each network adapter is connected to a hub (active or passive) using UTP cable. ARCnet can also be configured using a bus topology which uses coaxial (RG-62) cable to connect the adapters to each other. In this type of configuration the coaxial cable must be terminated (93 ohms) on both ends or a low-impedance card used as the terminator. All other adapters connected to the bus must be high-impedance adapters in order to balance the transmission line.

Protocols. Protocols specify the communication rules for each type of network. Protocols are specified in the OSI (open standard interconnection) seven-layer specifications.

Layer 1 (physical) specifies the physical handshaking rules, connections, cables, voltage levels, and time for the type of network. Characteristics in this layer vary with the different types and speeds of the different network architecture.

Layer 2 (data link) formats the data from layer 3 during transmission and adds any additional information to form proper data packets for transmission. When receiving the data packets, this layer is used to verify data and unpack the information from the data packet.

Layer 3 (network) is used by the network to determine paths between the transmitter and receiver. During the transmission process this layer packages the data from the transport layer for layers 1 and 2. During the receiving process this layer converts the data received into data for the transport layer.

Layer 4 (transport) is used to control the flow of data between the layers of the protocol and the network. It also determines the status of the network for the network interface card.

Layer 5 (session) deals with network management functions. This layer allows the conversion of station addresses into station names and checks log-on and log-off, password verification, and network reporting.

Layer 6 (presentation) deals with file format transfers along with network security.

Layer 7 (application) deals with network application interfacing, including server and printer software.

LAN Configurations. There are basically two types of LAN configurations. In a dedicated server system, the server acts as the mainframe. The server in a dedicated LAN is not used by a user but is used only for controlling the network. The server controls all access to the network. The data and application software is stored on the server, but the application software is executed on each workstation. Because the server is the main controlling element of the system, it contains large amounts of memory in order to provide fast network operations. Just as in a mainframe system, as the number of users connected to the network increases the speed of the network decreases. Complex data processing operations also slow down the network speed. When a workstation wishes to use an application program, the server downloads the program files to the workstation where it will be executed. This cuts down on the amount of processing power the server must have. The advantages of a dedicated server LAN are high level of software security (all application software is loaded on the server), centralized data (data located on the server), and network access security. The disadvantages of a dedicated server LAN are that the server must be powerful and contain large amounts of memory, and if the server goes down the workstations cannot process any data. Dedicated server LANs are used primarily in networks that contain a large number of workstations (25 or more) because of the cost of the server and networking software.

The second type of LAN configuration is called a peer-to-peer LAN. Each computer in a peer-to-peer LAN system contains its own application software. Peer-to-peer LANs contain one or more servers (each computer in the network can be a server) and workstations. If a computer is configured as a server, the other computers connected to the network can use its resources (peripherals). Each server has the ability to enable or deny other servers or workstations use of its resources. If a computer is configured as a workstation, it has the ability to use the resources of the server, but no computer in the network can use the workstation resources. A peer-to-peer LAN contains one or more servers, and the servers act as a centralizing point for data and peripherals. Because the server is not dedicated for the entire network, the server does not have to be as powerful or as fast as in a dedicated LAN. The advantages of peer-to-peer LANs are that network software is less complex, no special high power computers are needed, it is easy to use, and network operations are normally transparent to the users on the networks. The disadvantages of peer-to-peer LANs are that they are limited to the number of computers connected to the network (normally 300 or fewer computers), total network control is lacking (each server controls its own resources), and application software cost is higher because each computer requires its own software. Peer-to-peer LANs are the newest entry to the networking field. They are very popular in small networks (100 or fewer computers) because of their very low cost and ease of use and the fact that in most cases existing computers can be used as servers.

11.2.2 Network Software

The last choice or perhaps the first choice is the network software, which is one of the most important parts of the network. There are basically two types of network software—server and peer-to-peer. The most popular type of server-dedicated software is Novel. With this type of software one of the computers connected to the network is set up as the server. The computer that is set up as the server is the computer that controls the network. All other computers in the network (remotes or clients) share the resources and software of the server. In a server-type system, the server must be a very large and fast computer because the remotes use the server for their data, application software, and peripherals.

In a peer-to-peer system, any or all computers on the network can be servers. The difference between a server and a remote (client) is that a remote can share the resources of the server, but a server cannot share the resources of the remote. A peer-to-peer system allows the most flexibility and in small systems allows the network to operate more effectively than a server-type network. Popular peer-to-peer network software packages are Lantastic, Invisible LAN, and Windows for Workgroups.

Also included in network software are different types of operating systems that directly support networking; these include Windows NT, OS/2, and Unix.

CHAPTER SUMMARY

1. Data communication is the transfer of data between two or more devices over some type of transmission medium.
2. A telephone system (standard or dedicated line) is the most common transmission medium when using a modem because of its low cost and availability. RF coaxial cable is popular where short distance, high speed, and reliability are important. Station-to-station, station-to-satellite, and fiber optic communications provide the greatest speeds, highest volume, and least interference with communications but at a somewhat higher cost.
3. On-off keying transmits data by turning on and off a carrier frequency. If the carrier frequency is turned on during a data bit time period, the data represents a logic-1; if no carrier is detected during that time period, the data represent a logic-0.

   In amplitude modulation the amplitude of the carrier frequency is used to transfer the data. Low-amplitude levels represent logic-0s, whereas high-amplitude levels represent logic-1s.

   In frequency shift keying the change in carrier determines the data being sent. Two different frequency sources are used. If the higher frequency source is controlling the carrier frequency during a time period, the data represents a logic-1. If during the time period the lower frequency source is controlling the carrier, the data is a logic-0.

   Phase shift keying uses the phase shift between two carriers to represent digital data. No phase shift represents a logic-1, whereas a 180-degree shift represents a logic-0.

   Differential phase shift keying compares the current phase shift with the previous phase shift. If there is a 180-degree phase shift from the last piece of data received, the data represents a logic-0, whereas no phase shift from the previous piece of data received represents a logic-1.
4. During the transmission of data the modulator part of the modem is used. TTL data from the computer or RS-232-C data from the serial port is applied to a voltage controlled oscillator (VCO). The logic level voltages applied to the VCO produce two different output

frequencies, depending on the logic level voltages. The output of the VCO is decoupled through a capacitor and is then applied to an audio amplifier. The audio amplifier amplifies the frequencies of the VCO to drive the speaker. The speaker is acoustically coupled to the microphone of the handset.

During the reception of data the speaker of the handset uses sound waves to drive the microphone of the acoustically coupled modem. The movement of the air on the microphone produces a small voltage that varies with the change in air pressure. The voltage produced by the microphone is decoupled and applied to an audio pre-amplifier. The pre-amplifier amplifies the signal to a level that can be used by the bandpass filter. The bandpass filter is used to allow only the correct frequencies to pass to the limiter. The limiter circuit is used to eliminate any amplitude noise. The output of the limiter is applied to a phase lock loop (PLL). The change in error voltage created by the PLL during the reception of the data frequencies is applied to a Schmitt trigger to create a TTL-compatible logic level that is applied to the computer. The output of the Schmitt trigger may also be converted into an RS-232-C level, which is then applied to the computer's serial port.

Table 11-4 provides the parameters of the networks discussed in this chapter.

Table 11-4

Description	*Ethernet*	*Token Ring*	*FDDI*	*ARCnet*
Development date	1972	1985	1985	1977
Developed by	DEC, Intel, Xerox	IBM	ANSI	Datapoint
Access method	CSMA/CD	Ring passing	Ring passing	Ring passing
Physical topology	Bus	Star-wired ring	Ring	Star, distributed star
Data packet size (bytes)	46-1.5K	1-18K	1-4.5K	1-508
Transmission speed (Mb/sec)	10 or 100	4 or 16	100	2.5 or 20
Data encoding	Manchester	Differential Manchester	NRZI	Analog
Voltage range	0 to −2.0	1.5 to −1.5	NA	10 to −10
Waveform type	Square	Square	Square	Sinewave
Transmission line	ThickNet ThinNet UTP	STP	Fiber optic	UTP ThinNet

REVIEW QUESTIONS

1. Which type of transmission medium is the least expensive?

a. Telephone lines
b. RF coaxial cable
c. Station-to-station microwave
d. Station-to-satellite microwave
e. Fiber optic cable

2. Which type of transmission medium provides the greatest noise protection?

a. Telephone lines
b. RF coaxial cable
c. Station-to-station microwave
d. Station-to-satellite microwave
e. Fiber optic cable

3. What is the purpose of the Tx line in a phone system?
   a. Allows data to exit the phone
   b. Allows data to enter the phone
4. The bandwidth of a basic phone system is 100 Hz to 2.4 KHz.
   a. True
   b. False
5. Which type of noise is created by switching equipment?
   a. Amplitude distortion
   b. White noise
   c. Impulse noise
   d. None of the given answers
6. Data transmitted through a phone system must be modulated because of the noise generated by the phone system.
   a. True
   b. False
7. Which type of modulation is most affected by noise?
   a. On–off keying modulation
   b. Amplitude modulation
   c. Frequency-shift keying modulation
   d. Phase-shift keying modulation
   e. Differential phase-shift keying modulation
8. Which type of modulation uses two different frequencies?
   a. On–off keying modulation
   b. Amplitude modulation
   c. Frequency-shift keying modulation
   d. Phase-shift keying modulation
   e. Differential phase-shift keying modulation
9. What is the purpose of the VCO in an FSK modem?
   a. Converts frequencies into voltage levels
   b. Converts voltage levels into frequencies
   c. Reduces the noise from the phone line
   d. Amplifies the transmit data for the phone line
   e. None of the given answers
10. What is the purpose of the bandpass filter in an FSK modem?
    a. Converts frequencies into voltage levels
    b. Converts voltage levels into frequencies
    c. Reduces the noise from the phone line
    d. Amplifies the transmit data for the phone line
    e. None of the given answers
11. A dumb terminal processes data internally, independent of the server.
    a. True
    b. False
12. Which network normally uses dumb terminals?
    a. Mainframe network
    b. LAN
13. What does LAN stand for?
    a. Logically Arranged Network
    b. Local Area Network

c. Large Area Network
d. Local Asynchronous Network
e. None of the given answers

14. Which type of network architecture uses CSMA /CD?
    a. Ethernet
    b. Token Ring
    c. ARCnet
15. When a collision occurs on an Ethernet network, the network stops operating and a message is sent to all users that the network must be reset.
    a. True
    b. False
16. A Token Ring adapter must take control of the token before it can transmit data.
    a. True
    b. False
17. In a bus topology, the network continues to function when the coaxial cable is not terminated.
    a. True
    b. False
18. In a network using star-wired ring topology, one pair of wires is used for transmitting data and the other pair is used for receiving data.
    a. True
    b. False
19. The function of a hub is to route the transmission of data to only the proper adapter.
    a. True
    b. False
20. During transmission, the network interface adapter will convert the data from the transmission media into logic levels for the computer.
    a. True
    b. False
21. The type of network adapter and transmission line determines network throughput.
    a. True
    b. False
22. Which type of network interface is the fastest?
    a. Ethernet
    b. ARCnet
    c. Token Ring
    d. FDDI Token Ring
23. Which type of network interface uses analog voltages on the transmission line?
    a. Ethernet
    b. ARCnet
    c. Token Ring
    d. FDDI Token Ring
24. Which type of transmission line is less likely to pick up noise?
    a. Coaxial cable
    b. STP
    c. UTP
    d. Fiber optic cable
25. Which network interface is the least expensive?
    a. Ethernet
    b. ARCnet

c. Token Ring
d. FDDI Token Ring

26. The rules for communications in a network are defined as ________.
    a. Protocols
    b. Logical topology
    c. Access method
    d. Architecture
    e. None of the given answers

27. Which layer of the OSI specification is responsible for network management functions?
    a. Layer 1
    b. Layer 2
    c. Layer 3
    d. Layer 4
    e. Layer 5

28. High-level software security is one advantage of a dedicated server network.
    a. True
    b. False

29. Less expensive computer systems is one advantage of a peer-to-peer network.
    a. True
    b. False

30. Novel is the most popular type of dedicated server network.
    a. True
    b. False

chapter 12

Test Equipment and Troubleshooting

OBJECTIVES

Upon completion of this chapter, you should be able to perform the following tasks:

1. List the basic functions, advantages, and disadvantages of each type of test equipment discussed in this chapter.
2. List and define microcomputer signals and specifications.
3. List and define the two different types of computer problems.
4. List the basic boot-up sequence of a typical computer.
5. Using the troubleshooting information on hardware and software problems in this chapter, select the possible cause of a problem.
6. List and define the purpose and functions of diagnostic software and hardware used in troubleshooting a computer system.
7. List and define the purpose of a logical approach to system troubleshooting.
8. List the three different methods of repairing hardware problems.
9. List the importance of maintaining logs.

12.1 TEST EQUIPMENT

Because most signals in a computer are digital in nature, some of the test equipment used in troubleshooting computer circuitry is different. In this section we look at basically two types of test equipment. The first type of test equipment can monitor only one signal at a time, whereas the other type can monitor multiple signals at any one time.

DMM. A *digital multi-meter* is a device that can monitor voltage (AC/DC), resistance, and current (AC/DC). The DMM may monitor only one signal at a time and then can monitor only the present level. Because of the above factors, DMMs are used mainly for checking supply voltage levels and resistances in a computer. The DMM is not used to measure digital levels in the computer because these levels normally change faster than the DMM can operate, unless the computer is in a standby state and not switching.

When a DMM is used in a computer circuit, it is very important that the DMM have as high an input impedance as possible, to minimize loading effects on the IC that is being tested. Also, because threshold levels are so small, the resolution should be at least 3 1/2 digits on the low scale with an accuracy of 0.1% plus or minus the LSD.

Digital Logic Probe. A digital probe is a device that is more suited for looking at logic levels in their true state. Digital logic probes come in two forms—memory and nonlatching. Most logic probes look like large felt-tipped pens, with at least two LEDs to indicate logic levels. The latching probes normally contain a switch to reset the memory in the probe and an additional LED to indicate when the memory has changed.

A nonlatching digital-logic probe is nothing more than a digital buffer and inverter, which are used to power two LEDs. One of the LEDs represents a logic low when lit; the other represents a logic high when lit. If both LEDs are on or both are off, the digital-logic probe is probably showing a bad logic level or the device being tested is in tristate. Both LEDs are on or off when testing a tristate device only when no other output is connected to the output device being tested.

The problems with a nonlatching digital-logic probe are that it can only monitor one digital signal at a time, and it checks the logic state only at that moment in time.

A memory (latching) digital-logic probe stores the last change in logic states and does not allow the output to change until the probe is reset. This type of probe allows you to check a fast logic change and indicate the change.

Frequency Counter. A frequency counter is a device that displays the frequency at the input of the device, in a numerical format. Although a frequency counter provides an easy way to determine the frequency of a digital clock, it lacks the ability to display other important information about the clock signal (like duty cycle, phase relationship between two or more digital clocks, clock levels, and noise associated with the clock signal). Sometimes frequency counters are integrated in an oscilloscope. Normally, most frequency counters look very much like a DMM, except that the display has more characters.

Oscilloscope. A more useful device in some digital circuits is an oscilloscope. An oscilloscope allows the technician to check digital clock signals for proper frequency and levels and for noise levels within the clock signals. Oscilloscopes are used to monitor logic levels in a standby state, noise in the power supply, video display signals, and disk drive signals. The oscilloscope is an important tool that is used with a disk drive exerciser to align disk drive units. If a multitrace oscilloscope is used, the technician can monitor delays between two or more repeated signals.

One of the major problems with oscilloscopes is that unless the input changes occur at a regular rate, the user may not notice the change. Another problem with oscilloscopes is that they normally do not monitor more than four inputs at any one time.

Digital Storage Oscilloscope. A digital storage oscilloscope solves one of the major problems with a standard oscilloscope. Unless the signal is repetitive, the person using the standard oscilloscope may miss seeing a nonrepetitive change occur. The digital storage oscilloscope solves this problem by adding memory to a standard oscilloscope. The memory retains and displays the waveform on the display. The technician can then analyze the nonrepetitive waveform in greater detail, to find the problem in the waveform.

The digital storage oscilloscope can monitor the same conditions as a standard oscilloscope; it can also monitor simple nonrepetitive signals.

The only problem with the digital storage scope is that it is limited in the number of inputs it can monitor and it lacks the ability to perform multilevel triggering on the inputs.

Logic Analyzer. A logic analyzer is the most useful piece of test equipment for troubleshooting serious computer problems. A logic analyzer is used to identify timing and sequencing problems that other types of test equipment cannot detect. Although a logic analyzer is the last word in troubleshooting equipment, it is not often used in the field to troubleshoot common computer problems. Instead it is used during training to teach trainees the timing and signal sequencing in a computer system, to assist in research and development, and to troubleshoot complex computer systems in which standardized replacement boards are not readily available or are not a cost-effective option.

Basically, a logic analyzer is a device that samples (looks at and remembers) 8 to 64 different digital inputs periodically and stores the logic levels in memory. The user determines the time period between samples. Once memory is filled the user can display the information on a video display in a variety of different formats. The user also has the ability to cause the logic analyzer to trigger on a predicator event (a multilevel set of logic signals on one or more inputs) and to retain the number of sample levels before and following the triggered event. By setting the trigger event, the user can identify the logic levels on the inputs before (leading up to) and following the selected event.

Logic analyzers come in three forms. A *self-contained logic analyzer* (Fig. 12-1) contains all parts of the logic analyzer and the video display. This type is the most expensive because the user must purchase the entire unit. Normally it takes longer to learn to operate this type of logic analyzer. The cost ranges from $7,000 to $45,000; therefore, these logic analyzers are used only in design and where in-house repair of the computer system is not possible. The self-contained logic analyzer looks much like an oscilloscope, except that a keypad usually replaces most of the switches and buttons on the oscilloscope. The latest version of self-contained logic analyzers looks like a portable computer, with an LCD display and floppy and hard drives.

An *external logic analyzer* contains all parts needed for the logic analyzer except the display device. Normally, this type of logic analyzer uses a standard oscilloscope for the display device. Because the user does not have to purchase the video display, the cost of this type of logic analyzer is much less ($2,000 to $7,000). This type of logic analyzer is used for less crit-

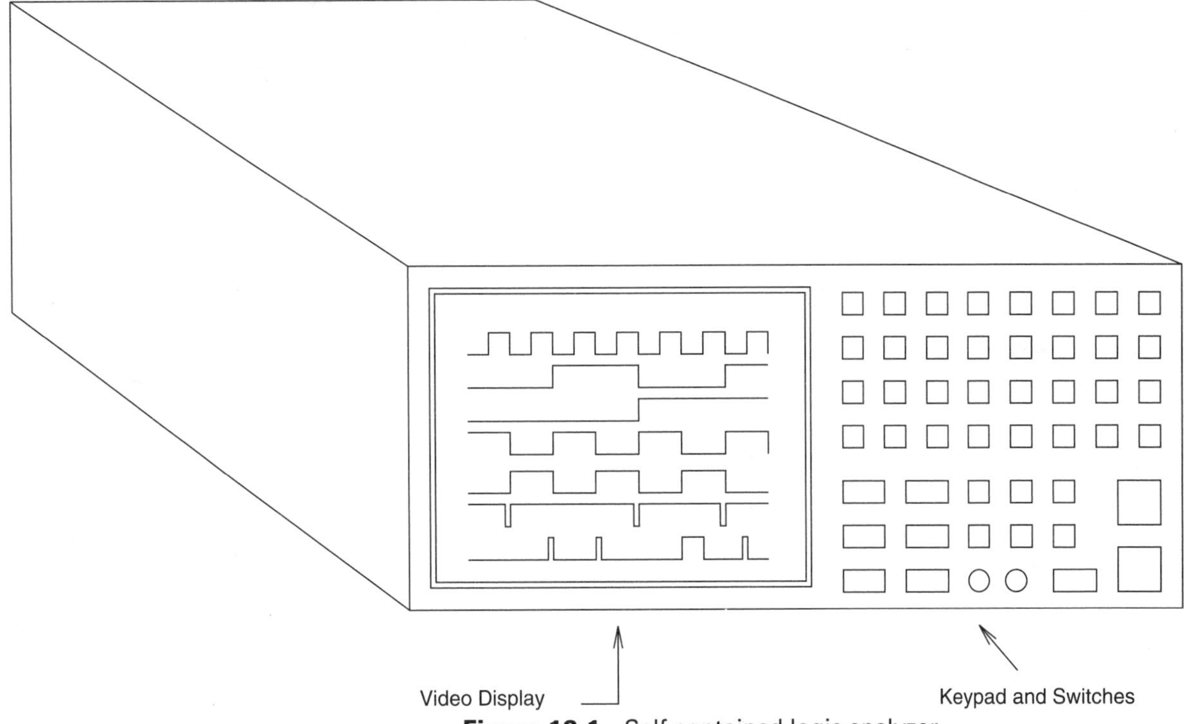

Figure 12-1 Self-contained logic analyzer.

ical repair projects. It normally takes the form of a bench type DVM or frequency counter, except that a keypad is used to set and determine the modes of operation.

A *fully integrated logic analyzer* (Fig. 12-2) usually contains a high-speed memory, logic control, and multiplexing circuitry on a computer I/O expansion card. The expansion card uses some of the circuitry of the computer, along with the display, to convert the computer into a fully functional logic analyzer. Because much of the circuitry of the logic analyzer is supplied by the computer, the cost of this type of logic analyzer is low ($600 to $4,000), and its versatility is very high. The versatility comes from the fact that the user can store the traces on diskette and print the traces on paper to be used for analysis. Because the logic analyzer is incorporated with the computer, it does not take long to learn how to use the software needed to make the computer function as a logic analyzer.

Logic Analyzer Terminology and Functions: The following is a summary of the basic functions of most types of logic analyzer.

The *input pod* is a buffering circuit placed between the circuit being tested and the logic analyzer circuitry itself to protect it from damaging input voltage levels. Each input pod normally monitors eight input (channel) levels, a ground line, and an input clock.

Sample rate refers to how often the logic analyzer looks at all the inputs being monitored. Each time a sample is taken, the logic levels on all channels are normally set to sample on the leading or falling edge of the sampling clock. The sample rate may be controlled by an internal clock or an external clock.

Internal clock sampling (asynchronous sampling) is used for analyzing timing problems. Because this type of sampling is not synchronized with the clock of the circuit being tested, the sampling rate must be four to six times faster than the fastest changing input signal. Sampling rates of less than four times the clock speed cause inaccurate timing representations of the channels being monitored, duty cycles may not be accurate, and signal changes that have a shorter time period than the sampling rate may not be seen. A sampling rate greater than six times the fastest changing input signal allows more accurate timing, but because the logic analyzer has a limited number of memory locations only a small portion of a timing diagram is seen.

External clock sampling (synchronous sampling) is used to show the state (logic level) of each channel just after the clock signal is received. External clock sampling requires that a digital clock signal be applied to the clock input of the input pod. Synchronous sampling is normally used in debugging new hardware and software.

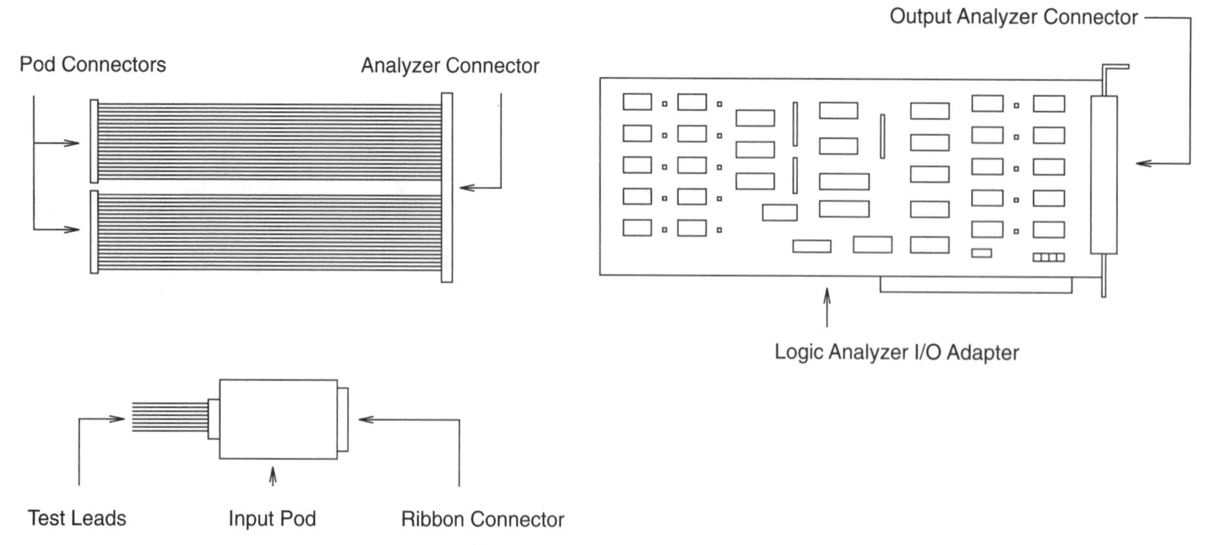

Figure 12-2 Fully integrated logic analyzer.

The *trigger word* defines the logic levels needed on the inputs of the logic analyzer which specifies the *trigger event*. The logic analyzer continuously monitors all selected channels at the sampling rate and places the logic levels in a first-in, first-out memory circuit. The memory of the logic analyzer, therefore, is continuously updated until a trigger event occurs. After the trigger event occurs and memory is full, the analyzer displays the levels on the video display. The triggered event occurs when the logic levels on the selected channels match the trigger word. Once the memory is filled, the logic analyzer stops monitoring the channels and displays the information on the video display. Depending on the trigger position value, the user can select the number of samples prior to or following the trigger event. The lower the trigger position value, the more samples following the trigger event will be displayed. The higher the trigger position, the more samples preceding the trigger event will be displayed. Therefore, to determine what causes the trigger event to occur, the trigger position should be set to a high value. To determine what happens to the computer after the trigger event, the trigger position should be set to a low value. The trigger word is selected by the user of the analyzer; each bit in the word can be set to one of the following three levels: logic-1, logic-0, or X (don't care—either 0 or 1). Each sampling of the channels requires one memory location; the number of memory locations normally ranges from 256 to 1024.

There are two display modes in most logic analyzers. The *memory display mode* allows the user to display the contents of memory in binary, octal, decimal, hexadecimal, and ASCII (optional) and with optional software mnemonics. The memory display mode normally displays information on a video display. This mode is very useful in looking for certain types of data information on the channels being monitored.

The second display mode is called the *timing (trace) display mode*. In this mode, the video monitor is used to create a timing diagram for the triggered event. The timing diagram is used to determine timing delays and problems of the circuit being tested.

Figure 12-3 illustrates what might be seen on the video display after the trigger event has occurred. The timing diagram represents a four-bit ripple counter. It shows the relationship between the clock input (CLK) being applied to the counter and the four outputs (Q0, Q1, Q2, and Q3) from the counter. The Q0 line is the LSB of the counter, and the Q3 is the MSB of the counter. The trigger word was set for a low on the CLK, Q0, Q1, Q2, and Q3 lines. From the display the user can determine the proper operation of the circuit.

Because fully integrated logic analyzers use the computer as part of the analyzer, the data stored in memory can be stored on diskette or hard disk, and the trace can also be printed on paper using a printer. Waveforms of good working circuitry stored on disk can be compared with waveforms generated by a circuit not functioning correctly to help the technician determine the device causing the problem.

Signature Analyzer. A signature analyzer is a device like a logic analyzer in its ability to store the logic level on its one input in memory for every sample. This group of samples is called the signature of the circuit. Therefore, during a specified operation in a digital circuit, a device produces a unique signature. The signature for that circuit is stored in the signature analyzer and compared with a new signature from the circuit being tested. If the two signatures do not match, the circuit is bad. Because the user does not need to monitor as many signals, this device is easy to use.

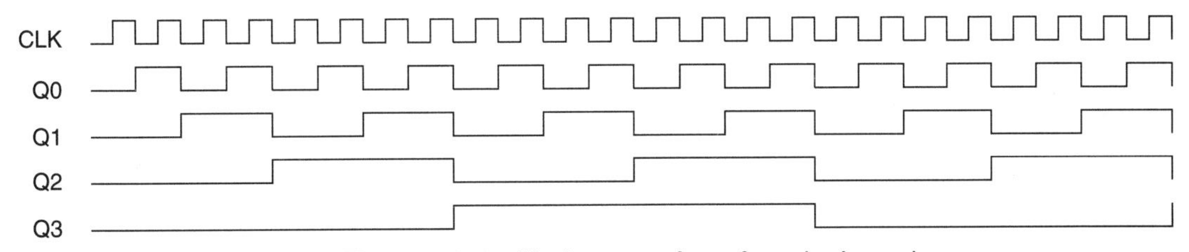

Figure 12-3 Timing waveform from logic analyzer.

One of the disadvantages of a signature analyzer is that it has only one input; therefore multiple testing must be performed to test a circuit containing more than one test point. Another disadvantage is that a signature for a good circuit must be produced before the comparison can take place.

Logic Monitor. A logic monitor is a simple device that clips onto a dip package IC. This device contains one LED for each pin on the IC. The LED is on if the logic level on that pin is a logic-1. If the LED is off, the logic level is 0. The advantage of this type of device is that its cost is very low ($30), and it allows the user to monitor multiple logic signals at one time. The disadvantage of this type of device is that it monitors only logic states and does not monitor fast-changing logic states effectively.

A logic monitor looks like a dual inline test clip, with LEDs representing the levels on each pin of the IC being tested.

12.2 MICROCOMPUTER SIGNALS AND SPECIFICATIONS

Most of the signals developed in today's computers are digital; only about 10% of the signals are analog. Of the analog signals, only about 1% are analog by nature; the other 9% are digital signals converted into analog.

Logic Signals. Digital logic signals are seen by the device to have only two valid levels (Fig. 12-4), even though the voltage level of the signal may vary within a range. All small computer systems today use positive logic. This means that the highest voltage level represents a logic high and the lowest voltage level represents a logic low.

Personal computers are made up of mostly MOS (LSI and VLSI ICs) and TTL (SSI and MSI ICs). In order to reduce the load on MOS ICs, they normally have TTL-compatible inputs and outputs. Therefore, levels over about 2.2 volts may be considered as a logic high. Voltages below 0.8 volt are a logic low. Signals between these levels are considered to be improper levels, as long as the circuit uses TTL levels.

Because most ICs in the computer share the same output lines (bus), standard TTL outputs cannot be used. Standard TTL outputs fight each other to control the bus, and one of the outputs burns up. In order to eliminate this problem and to reduce loading of ICs inside the computer, most ICs that share the same bus have tristate outputs.

In a tristate device, when the device is in its third state, the output normally shows a low level or a bad level, if no other device is connected to the output. If the device is in the third state and another device is connected to the output, the level of the other device is the output level for the ICs in question. The only difference is that the output does not have any current flowing into or out of the output. When an IC with a tristate output is enabled, the output produces a logic low or high.

When replacing ICs in a computer, make sure that the same logic family and/or subfamily is used; otherwise loading may take place, which results in improper operations, especially at high frequencies.

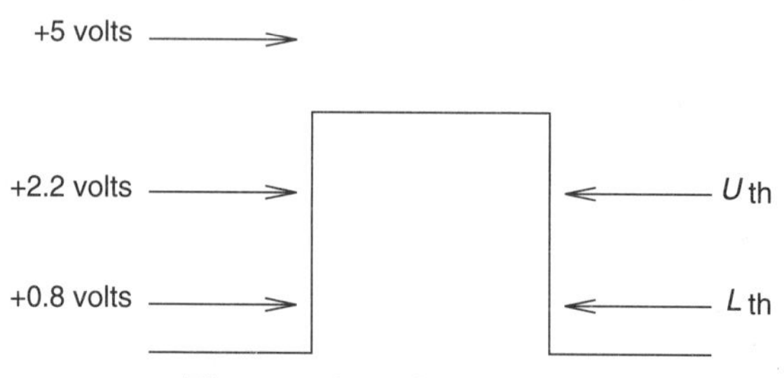

Figure 12-4 Logic levels.

Because at high frequencies the internal capacitance becomes high, a good digital signal may become distorted, and noise may be induced into the logic level. Therefore, some areas of the computer contain ICs with Schmitt trigger inputs. A Schmitt trigger input (Fig. 12-5) does not switch states until the opposite threshold level is reached by the IC. This eliminates most problems caused by levels that fall into the don't-care (bad) region of the logic level.

If the input voltage falls within the don't care (bad) range of TTL, the output of the TTL ICs with a standard input causes the output to switch when it should have stayed low. If the same condition is applied to a TTL circuit with a Schmitt trigger input, the output does not change, because the noise did not exceed the upper threshold voltage level of 2.2 volts DC.

Analog Signals. An analog signal is a signal with many different levels. Devices that interpret analog signals that are analog in nature are called analog devices. Most of the analog signals found inside the computer are converted into digital signals at some point. Also, most of the analog signals developed inside the computer were digital at some time. Figure 12-6 is an example of an analog type signal.

Normally, an analog signal is found in the computer only when the computer needs to control an analog type device like the motor speed of the spindle. Another example of an analog signal found in a computer occurs when analog signals from a tape or disk read-write head are converted into digital data.

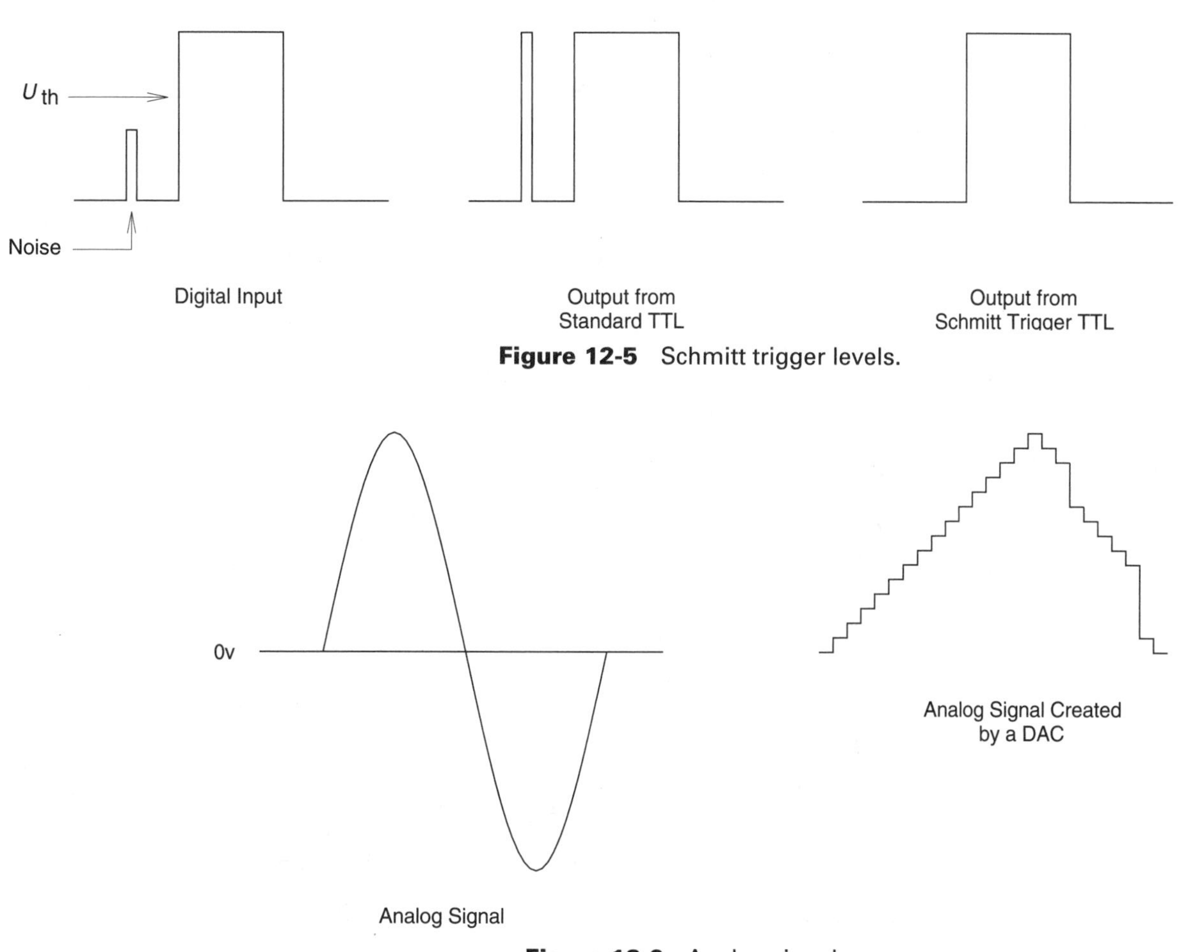

Figure 12-5 Schmitt trigger levels.

Figure 12-6 Analog signals.

12.3 MANUFACTURER'S SPECIFICATIONS

Supply Voltages. Because most ICs in the computer are either TTL or MOS (which are TTL compatible), the majority of the ICs require a +5-volt DC supply. Normal operation continues as long as the voltage stays within + or −10% of the +5-volt supply, unless the manufacturer's specifications state otherwise.

Certain ICs in the computer system require more than one voltage; these ICs are specified in the diagram. The computer supplies more than just +5 volts DC for a few special ICs within the computer and for any special adapters that may be added to the computer at a later date.

Noise created by gates inside the computer switching at high frequencies also causes problems. To reduce any noise that may feed back into the computer's power supply, small capacitors are normally placed from Vcc to ground for every five TTL ICs.

Noise created by the power supply is normally not a problem in the computer unless the noise causes the supply voltage to leave the guaranteed operating level of the IC. Power supply noise normally comes in two forms. Ripple voltage is the noise that gets through the power supply from the AC source or is generated from a device that operates from an AC source. This type of noise normally occurs at 60 Hz or 120 Hz. The second type of noise is generated by the output impedance of the switching power supply, which changes in order to regulate the output voltage. This noise has a frequency that normally varies with the changes in load resistance.

Another form of noise in a computer system is developed by the ICs switching from one level to another level. During this transition, the ICs require a great deal of power. The fluctuation causes the voltage of the supply to vary. The voltage regulator in the power supply tries to compensate for these changes, which may cause the power supply to create noise. If the noise becomes too great, the ICs may start to oscillate or cause the supply voltage temporarily to leave the guaranteed operating range of the IC, causing data errors.

Threshold Logic Levels (Table 12-1). Voltage levels in digital circuits are interrupted only twice. These levels are called threshold levels. The upper threshold level is the minimum amount of input voltage necessary to cause the output to recognize a logic-1. The lower threshold level is the maximum amount of input voltage necessary to cause the output to recognize a logic-0. Voltages that fall between the upper and lower thresholds may be interrupted as a logic-1 or logic-0, which may cause errors in the output of the circuit.

Noise. Noise margin is defined as the minimum amount of input noise that will cause the output to switch states. This value is normally the value of the smallest threshold value. If the noise margin is exceeded, the output may switch and cause a bad signal. Normally, only the worst case noise margin is given in manufacturer's specifications.

TTL . . . . 1-volt noise margin
MOS with TTL in/out . . . . 1-volt noise margin
CMOS . . . . 30% of Vcc

Operating Speed. The maximum operating speed of a digital IC is a function of propagation delay in the device. Propagation delay is the amount of time it takes the output to switch when the input level switches.

TABLE 12-1
Threshold Logic Levels

Logic Family	Logic-0	Logic-1	Supply
TTL	>.8 volt	<2.2 volts	+5 volts
MOS with TTL in/out	>.8 volt	<2.2 volts	+5 volts
CMOS	>35% Vcc	<60% Vcc	+5 volts

There are three factors that govern the maximum operating speed of a digital IC, propagation delay, power dissipation, and interelectrode capacitance. It is very important that you replace defective ICs with exactly the same number from the same subfamily; otherwise the circuit may not work properly. Each logic family and subfamily have different operating speeds, and if a slower logic subfamily is placed in a circuit, timing errors may occur. Table 12-2 is a general list of TTL and MOS propagation delays and maximum speeds. The delays and speeds listed in this table are for single gates and vary from IC to IC, depending on the number of digital gates in the IC.

Power Dissipation. Power dissipation in a computer is critical, because of the number of ICs within the computer. Power dissipation is the maximum amount of power dissipated in an IC. Because the IC requires the greatest amount of power when it switches states, power dissipation is normally at a higher value during operation than when the IC is in a standby state and not switching.

Again it is very important that the proper logic family and subfamily be used when replacing ICs, because each IC differs in power dissipation. The following is a general list of power dissipation in ICs. It varies with the number of digital gates within the IC. Table 12-3 represents power dissipation per gate.

12.4 MICROCOMPUTER PROBLEMS

One of the first things a computer technician must determine is whether the problem with the computer is caused by hardware or software. The most important part of determining the cause of the problem is getting as much information as possible about the problem from the person who was operating the computer when the problem occurred. This information should include

TABLE 12-2
Propagation Delays and Maximum Speeds

Logic Family or Subfamily		*Propagation Delay*	*Maximum Speed*
Standard TTL	7400s	10 nsec	35 MHz
High-speed TTL	74H00s	5 nsec	65 MHz
Low-power TTL	74L00s	30 nsec	9 MHz
Schottky TTL	74S00s	2 nsec	130 MHz
Low-power Schottky TTL	74LS00s	10 nsec	40 MHz
NMOS	xxxxx	50 nsec	20 MHz
PMOS	xxxxx	65 nsec	15 MHz
CMOS	74C00-4000s	40 nsec	25 MHz

TABLE 12-3
Logic Power Dissipation

Logic Family or Subfamily		*Power Dissipation*
Standard TTL	7400s	10 mw
High-speed TTL	74H00s	22 mw
Low-power TTL	74L00s	2 mw
Schottky TTL	74S00s	15 mw
Low-power Schottky TTL	74LS00s	2 mw
NMOS	xxxxx	1 mw
PMOS	xxxxx	1 mw
CMOS	74C00–4000s	.2 mw

the time it happened, what was done before and after the problem occurred, and any other problems that have occurred in the system. This information can help the technician determine if the problem is hardware or software related. Although such information is very valuable and may save time, it is not always available.

Most computer problems come from misuse of equipment or software. Hardware problems are normally caused by environmental factors (bad AC power, dirt, liquids, heat, lack of ventilation, and moving parts that wear out or become misaligned). Most software problems are caused by mishandling diskettes, users who do not know how to operate the software properly, software masters that are not write protected, trying to operate software on a computer that does not have the minimum requirement of hardware for the software package, and software bugs and viruses.

12.4.1 Hardware Problems

Normally hardware problems found in a computer system are not in the computer circuitry itself. The problems are normally found in the circuits that interface the computer with the outside world. In fact, most hardware problems are caused by the outside world.

If the problem is related to hardware, the next step is to use your senses to determine where the problems may be. Perform the following checks first.

1. Check all cables outside of the computer.
2. Look the computer over to see if there is anything out of place and check for the presence of foreign material.
3. Use your sense of smell to determine if any part of the computer has been burned.
4. Verify that the computer is receiving proper AC voltage.

If no obvious problem is found, it is time to run the internal self-test of the computer. In the IBM PC and compatibles, the test is known as the Power-On Self-Test (POST). When the computer is first powered up, it runs POST as part of the monitor program. POST checks the power supply, major LSI ICs on the system board (8088, 8259, 8284, 8237, 8253, 8288 and 8255), system RAM, ROM, keyboard, video adapter, floppy disk, and the expansion unit (found only in older IBM PCs).

All peripherals should be connected and powered up prior to determining if POST is performed correctly. Generally the disk operating system is placed in the default drive (drive A).

Running POST

1. Turn on the power to the computer and peripherals.
2. Normally, the power supply fan motor is heard.
3. The cursor should appear on the screen after about 4 to 15 seconds.
4. In a compatible computer, a number appears in the upper left corner of the screen. This number, which represents the RAM in the system that is being checked, increments. IBM PCs do not show the number of kilobytes of RAM that are being checked, but after POST, the total number of bytes tested is displayed.
5. After about 40 to 75 seconds, the user should hear one beep, which indicates that everything has checked out. How long it takes for the beep is determined by the amount of RAM in the computer.
6. The computer tries to read floppy disk drive A and boot the computer from the software in the drive. If no diskette is in the drive in an IBM PC or XT and no hard disk is being used, the computer boots into cassette BASIC or boots the software in the drive. If a compatible is being used and no diskette or hard drive is in the system, the computer constantly tries to read the floppy drive. This probably locks up the computer: The keyboard has no effect and the computer usually needs to be started from a cold boot.

POST Problems	Location of Error
1. No beep and no display	power supply
2. Continuous beeping	power supply
3. Repeated short beeps	power supply
4. One long and two short beeps, with incorrect or no display	display adapter
5. One long and one short beep	system board
6. Two short beeps	keyboard
7. Normal beep, no display	display monitor
8. Normal beep, drive is not accessed	disk adapter or drive

9. Normally, if the problem with the computer is not significant enough to keep POST from completing, one of the following error codes is displayed on the screen. The letter x represents a number; the last two digits cannot both be zero.

Error Code	Location of Error
02x	Power supply
1xx	System board
20x or xxxxx xx20x	RAM
30x or xx30x	Keyboard
4xx	Monochrome adapter
5xx	Color adapter
6xx	Floppy disk adapter or unit
7xx	Math co-processor
9xx	Primary printer adapter
11xx	Primary Async adapter
12xx	Alternate Async adapter
13xx	Game adapter
14xx	Alternate printer adapter
15xx	SDLC adapter
17xx	Fixed disk adapter or unit
18xx	Expansion unit (old PCs)
xxxxx ROM	ROM

If the hardware problem is not found by POST, it may be a minor problem and related to the type of software package that is being used. To verify that the problem is in the hardware, additional testing must be done. Normally the next step is to use the IBM Advanced Diagnostic diskette or compatible type of diagnostic software, which performs a detailed analysis on each part of the computer and the peripherals connected to it. This software verifies that all modes of each part of the computer are working correctly. The following is a list of tests and procedures that most diagnostic software performs.

1. Complete system board check
2. System RAM test
3. Complete keyboard test. Checks for stuck keys and key codes being received.
4. Display test. Checks the display adapter, display attributes, maximum number of characters on a display, printer adapter (monochrome adapter only), video, and sync. If a color and graphic adapter is attached, the test also checks the 40×25 display mode, the three different graphics modes, the color bar, and the eight pages of video RAM.

5. Floppy disk drive adapter and unit test. Checks the sequential access, random seek, read/write function, and speed.
6. Asynchronous and Alt Asynchronous communication adapter test. Verifies the proper operation of the asynchronous communication port for which a special loop-back adapter is needed.
7. Printer adapter test. Checks the adapter and printer connected to the printer.
8. Game control adapter. Checks the joystick buttons and the X and Y axes.
9. Fixed disk test. Checks the fixed disk adapter, read/write, disk surface, seek, and head selection.

Using this type of advanced diagnostic software, most problems can be solved. Purchase of the hardware maintenance and service package from IBM is often a wise investment.

If the diagnostic software does not locate the problem, the next solution is to use a computer emulator. A computer emulator is essentially another computer that is designed to send specific signals to the defective computer to trace the problem. Emulators are very expensive and seldom used except by repair specialists.

If the problem is within the floppy disk drive units, a disk drive exerciser is normally needed to recalibrate the drives.

If the computer fails during the execution of POST, a POST card can be used. A POST card is an I/O adapter that monitors the BIOS ERROR port. If the computer fails POST, an ERROR code appears on the BIOS error port, which indicates where in POST the computer stopped with an error. The technician can then lookup the error code for the specific BIOS to determine which devise in the computer caused the failure.

Intermittent Hardware Problems. There are two common intermittent hardware problems. Because these rarely seem to occur when the technician is present, it is very important to ascertain from the user exactly when and under what conditions the problems occur. The technician must then try to recreate the same conditions to be able to correct the problem.

Parity Error and Parity Check One: Depending on the computer that is producing the error, either parity error or parity check one (error) may be displayed on the screen during the normal operation of the computer. This indicates that a parity error has been detected when system RAM is read. Normally this type of error is not fatal; it simply results in a data error. It is not uncommon to occasionally see these errors. If the problem occurs often, there is probably a problem with power supply, refresh circuitry, or the DRAM itself. Whenever a parity error or parity check one occurs, the software and data should be reloaded and checked.

I/O Parity Error and Parity Check Two: Depending on the computer that is producing the error, either I/O parity error or parity check two (error) may be displayed on the screen during the normal operation of the computer. This refers to a parity error in external RAM or in an I/O device. This type of error usually locks the computer if it comes from an I/O device. If the error comes from external RAM, it is normally not fatal to the computer. After the I/O parity error or parity check two message is displayed, an error code is given. This code indicates where the problem is located.

If these error codes do not occur often, there is no major problem. If they are frequent, the first thing to check is power. Then seek the device causing the error. This diagnostic procedure usually finds the problem as long as it is not a power supply or AC power problem.

12.4.2 Software Problems

In most cases, if the computer successfully boots in DOS, the problem is not in the hardware, but in the software. Sometimes software problems are very hard to remedy. Below is a list of common software problems and some possible causes.

1. Software fails to boot.
   a. Diskette may be defective; check it on a different computer.
   b. Diskette recorded on a disk drive unit that is misaligned; check it on a different computer.
   c. Diskette properly recorded; disk drive unit being used is misaligned.
   d. Software may require hardware different from the computer in use. Often the incorrect type of video display adapter is used, or there is not enough memory.
   e. Files on diskette may be missing or bad or have been changed.
2. Software boots but fails to operate.
   a. Files on diskette may be missing or bad or have been changed.
   b. Software may require hardware different from the computer in use.
   c. Software bug; check software updates.
   d. Improper use of software.

12.5 LOGICAL APPROACH TO SYSTEM TROUBLESHOOTING

After determining that the problem is caused by hardware, a technician must mentally divide the computer into manageable sections and define the purpose of each section to troubleshoot the problem effectively.

System Block Diagram. The highest level of block diagram (Fig. 12-7) is sometimes referred to as the system block diagram. In this diagram, all devices connected to the computer are specified, and the adapters located inside the system unit are specified separately. The following is such a block diagram.

The technician determines the function of each block in the system block diagram and decides which block is causing the problem by checking the input and output of each block outside the system unit. If the input signals are not there, the technician must look at the block in the system unit that is supplying the signal for the device outside the system unit. If the input is correct and the output is not correct, the device outside of the system is bad.

The following are the functions of each block in the system block diagram.

System Unit: The *system board* is responsible for performing all timing, data transfer, memory-related, and controlling operations for devices located inside the system unit. The system board is also responsible for receiving and decoding key codes.

The *power supply* is responsible for supplying the power for all equipment inside the system unit. The voltages are ± 5 volts and ± 12 volts DC.

The *video display adapter* receives ASCII codes from the system board and converts them into the signals required to display the characters they represent on the video monitor. The output from this adapter is Hsync, Vsync, video data (red, blue, green data in a color adapter), and an intensity signal.

The *floppy disk adapter* is responsible for transferring data to and from the diskette and converting binary codes from the MPU into the signals necessary to control the operation of the disk drive unit. The adapter also uses signals from the hardware of the disk drive unit to determine if the operation called for is complete.

The *floppy disk drive unit* is responsible for transferring serial data to and from the floppy diskette, converting analog signals into TTL/CMOS levels, controlling mechanical parts of the drive unit, and sending status information to the floppy disk drive adapter.

The *printer adapter* is responsible for transferring ASCII data from the MPU to the printer and supervising software-controlled printer functions.

The *asynchronous adapter* is responsible for converting parallel data from the data bus into serial data for the outside. It also converts serial data from the outside world into parallel

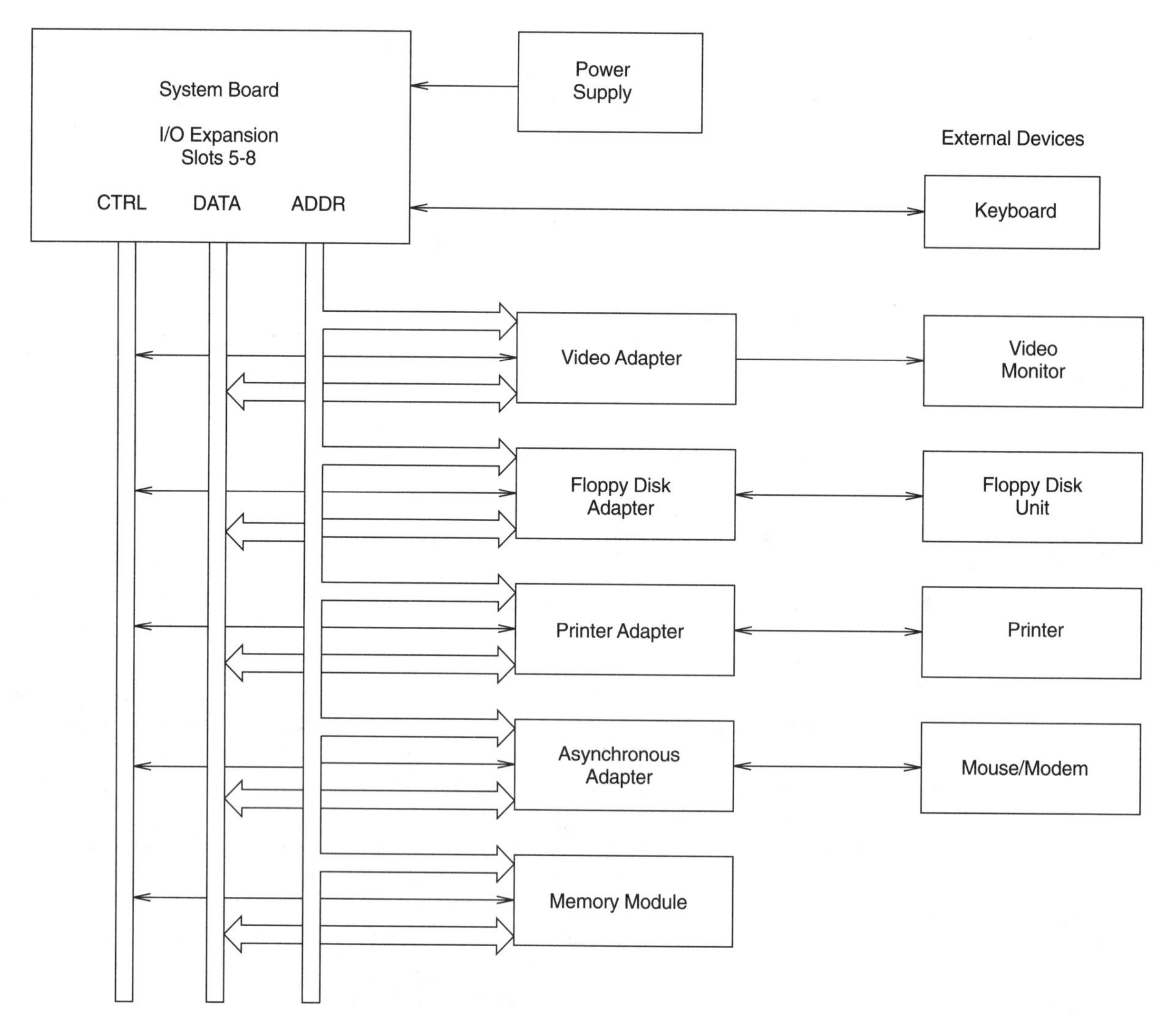

Figure 12-7 System block diagram.

data for the data bus. The adapter checks for transfer errors and controls and reads the status of the serial device connected to its output.

The *memory expansion* expands system-board memory beyond the 256K limit of older PCs and compatibles and the 640K limit in newer PCs and compatibles. Memory beyond the 640K limit must be supported by software.

Devices Outside the System Unit: The *keyboard* sends key code information about the key pressed to the computer.

The *video monitor* receives serial data information from the video display adapter and turns on and off the electron beam at the proper times to produce the proper ASCII characters on the screen.

The *printer* produces characters on paper from the ASCII codes sent to it from the printer adapter.

The *modem* converts TTL/CMOS levels received from the asynchronous adapter into signals that can be sent over selected transmission lines. The modem also converts data from the transmission lines into TTL/CMOS levels that can be used by the asynchronous adapter and checks the data for accuracy.

The *mouse* is a device that takes mechanical or optical movement and converts them into digital code, indicating the type and amount of movement. The code is sent to the asynchronous adapter where, with the help of software, it is translated into cursor movement.

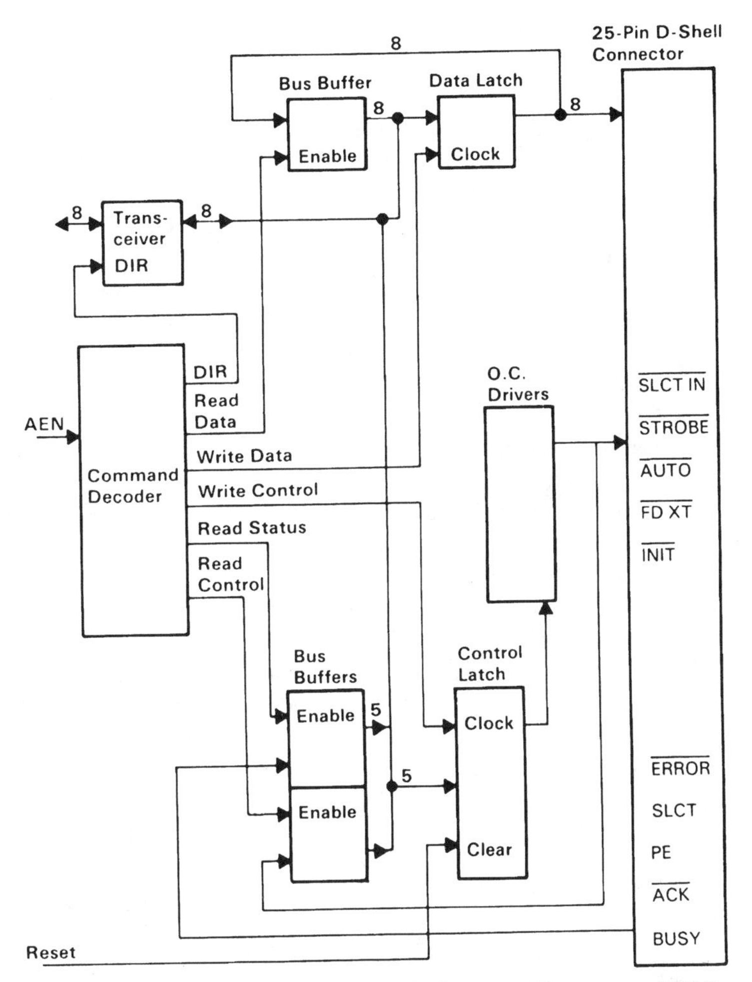

Figure 12-8 Printer adapter block diagram. (Courtesy of IBM)

Once the problem is traced to one of the blocks in or out of the system block diagram, the technician must divide the block into subsystem blocks to trace the problem further.

Subsystem Block Diagram. A subsystem block diagram divides one of the blocks from the system block diagram. The subsystem block diagram normally specifies the major components of the selected block, without going into pin-outs or specifying all the support hardware needed to make the card work. Each block in the subsystem block diagram may contain one IC or many ICs.

The purpose of the subsystem block diagram is to give the technician an easier method of determining which small block is causing the problem. In most cases, the subsystem block diagram gives enough information to locate the problem without going to the schematic diagram.

Figure 12-8 is the subsystem block diagram for a printer adapter in the IBM PC. Again, it is important to understand the function of each block inside the diagram.

The *command decoder* is used to address the five registers of the printer adapter during the read and write command sequence. This operation is performed by a 74155 dual 2 to 4 decoder-multiplexer.

The *data bus transceiver* controls the direction of data to and from the printer adapter. The data flow is determined by the command decoder.

The control-status *bottom bus buffers* perform two functions. Some of the bits in the control-status bus buffers are used as control bits. These bits help produce the output command for the printer. The information in this section of the buffer is applied to the control latch, which generates the output control signals for the printer. Information from these bits may also be sent to the data bus of the system board to aid the computer in making a decision on the operation of the adapter. The status part of the buffers holds the status information from the printer until it can be used by the adapter or computer. The information found inside the status part of the buffer allows the computer to decide if the printer is ready to receive new information. This information is also applied to the control latch to help determine when the command in the control latch is to be performed.

The *control latch* generates the actual control signals supplied to the printer. The sequence is determined by the command decoder. The information from the control-status bus buffer determines when the sequence occurs; in some cases, the computer also has a part in determining when the sequence takes place.

The *open-collector drivers* supply enough drive current for the control signals from the adapter to control the printer.

The *upper bus buffer* or input data latch is used by some software to verify that the data from the eight data outputs from the output data latch is good.

The *output data latch* holds eight bits of data from the data bus for the printer until they can be used by the printer. New data is loaded after the old has been accepted by the printer. The progress of the data transfer is indicated by the data found in the status part of the control-status buffers.

Because the information on the operation of this block is specified in more detail, the technician can verify the input and output signals for each block. Again, if the block has the proper input but does not have the proper output, the problem is within that block. If the proper input signals are not found at the input of the block, the technician goes to the block that is controlling the input.

Normally this type of troubleshooting is not done at the customer site unless necessary. It is more efficient to change the printer adapter if it is found to be defective and repair the defective card elsewhere.

Schematic Diagram. Of all the diagrams of a computer system, the schematic diagram gives the greatest detail of everything in the computer. Even though the schematic diagram gives all pin-out and signal levels, it is still necessary for the technician to know how each IC in the computer works. The hard part for most technicians is to relate how one component affects another. This sequence is normally made easier by the use of a timing diagram. A timing diagram shows the technician how the circuit should operate.

Because each IC in the IBM is covered in greater detail in other chapters of this book, we do not go into any detail on the example in Figure 12-9 of the printer adapter schematic diagram. Be sure you understand that you, as a technician, need to know the operation of each and every IC before you can use it effectively to troubleshoot the problem. The more you work on circuitry, the easier it becomes to troubleshoot the circuitry.

12.6 REPAIRING HARDWARE PROBLEMS

When working in field service, it is important to get the computer system operating as soon as possible. To do this, the service technician must first determine which possible block in the system block diagram could be causing the problem. The technician must then decide how to correct the problem, using one of the following methods.

Card Swapping. Most field service technicians exchange an I/O card known to be good with the possibly bad card to determine if the card is bad. If the new card corrects the problem, the technician may leave the good card in the computer and repair the bad card back at the office. This procedure occurs normally, unless the technician does not have an extra set

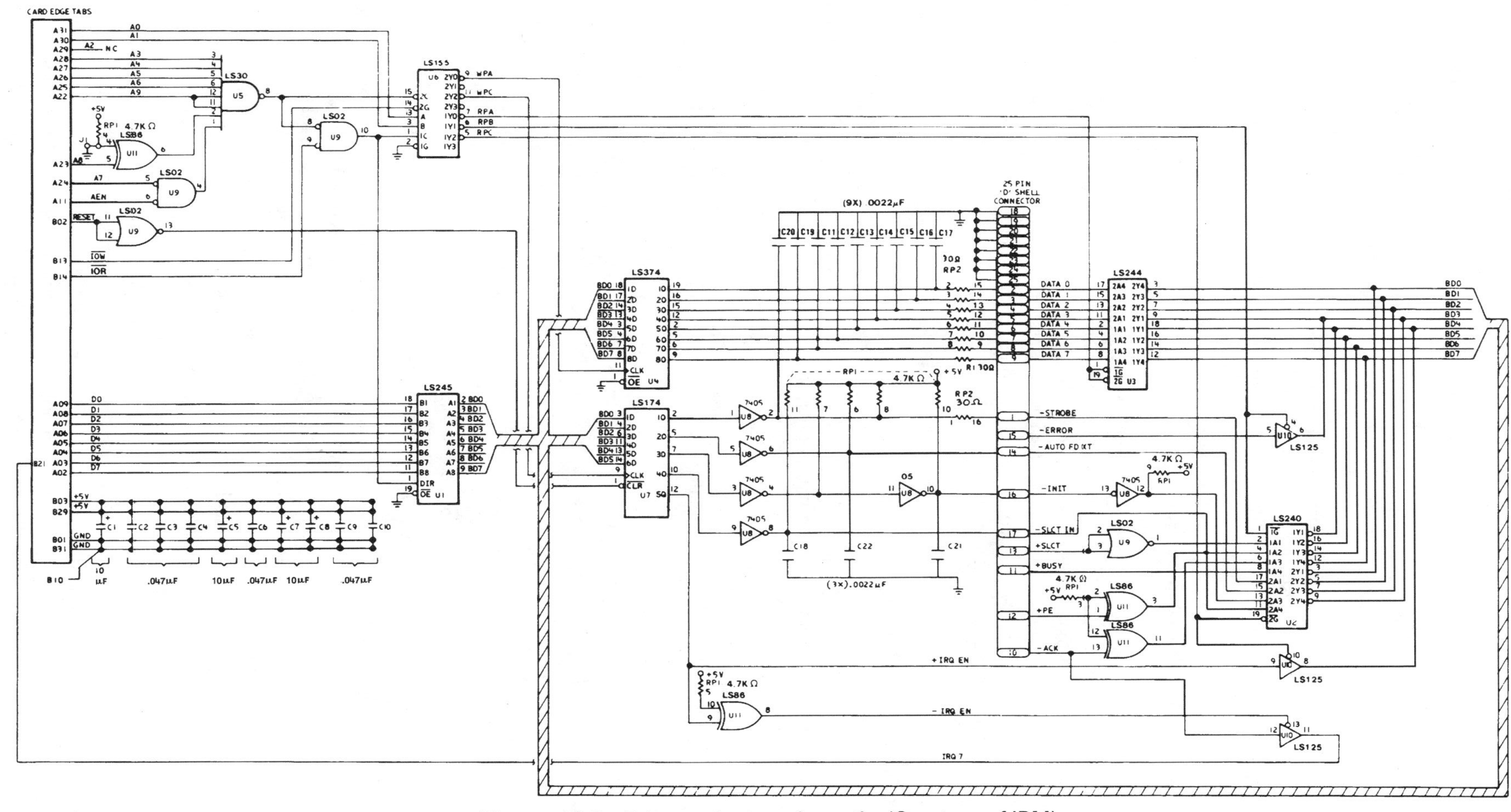

Figure 12-9 Printer adapter schematic. (Courtesy of IBM)

of I/O cards. In some cases, the technician may just tag the problem and replace it at a later time.

Card swapping requires very little work and usually solves most problems. Normally card swapping requires no measurement of the equipment.

Substituting Parts. If card swapping is not possible, the technician goes to the next level of troubleshooting. This involves using a subsystem block diagram to try to locate the problem. At this level, simple test equipment (DMM, logic probe, and oscilloscope) measurements help narrow the problem. Once the suspected component is found, the technician can substitute a good component for the old one. Most of the time a field service technician does not do this out in the field.

Field Engineering Changes. Another task that a field-service technician may perform is called a field engineering change, in which the technician replaces a card or circuit board. Although this task is not normally a troubleshooting task, it often corrects the problems that might have been built into the circuit. Field engineering changes may occur in the field or the changes may be performed at the office. Simple field engineering changes are normally performed at the customer site.

12.7 TROUBLESHOOTING TIPS

Maintenance and Repair Logs. Because most computer problems are not caused by electronic circuit failures, much of the technician's time is spent performing preventive maintenance or repairs relating to the lack of maintenance. If maintenance is not the problem, the problem often involves improper use of the computer equipment and/or software. The most common hardware problems are those involving moving parts that wear out and fall out of calibration.

A technician, especially a new technician, should keep a record of the symptoms and cures for any repair made. The repair log should list the type of equipment (version, model number), problems reported, any type of testing used, and what was done to correct the problem. The technician will find that many times the same problem occurs in the same type of equipment, and by keeping a log of these problems, repair time decreases because the probable cure has been done before.

Some customer sites already have a maintenance-repair log that must be updated each time the technician works on the equipment. The technician should also keep a personal log.

Investigate. The greatest tool the technician may have is common sense. Use it before jumping into any project or repair. Remember three things, **think**, **look**, and **listen**.

1. Get as much information about the equipment and problem as possible. Before going out on the call, check maintenance and repair records for the customer's site if they are available. From these records, you can often get a good idea of what the problem is and if additional test equipment and/or parts might be needed.
2. Once on site, try to talk to the person who was using the equipment when it went down. Get as much information from that person as possible about the problem.
3. Look at the equipment and make sure it looks normal from the outside. Check the area for any possible outside influence that could have caused the problem (excessive heat, dirt, and bad power supply). At this time you may also want to use your sense of smell and touch to determine if anything has been burned or is loose.
4. At this point, if you have a repair record for the equipment, there may be enough information to correct the problem without performing any tests. Otherwise, from the information obtained, you should be able to determine the next step to be taken.

5. If the problem is not related to the line power or burning in the computer, the next step is to turn on the computer and use POST to help locate the problem.
6. From POST, in most cases, you can determine the section of the system that is causing the problem. If card changing is to be done to solve the problem, change the card that is suspect. If the firm you work for prefers part substitution, you will have to use some test equipment to trace the problem.
7. After the repair has been made, it is important, if possible, to have the user of the equipment verify that the equipment is working satisfactorily. This ensures that the problem is not caused by improper use of the equipment.

CHAPTER SUMMARY

1. Test equipment:

   DMM—used to measure DC and AC current and voltage, resistance, and in some cases transistor gain, continuity, and capacitance. The DMM is not good for measuring fast-changing levels or for determining noise.

   Digital logic probes—good for looking at the logic state of a digital circuit. They are not good for looking at fasting-changing logic levels.

   Frequency counters—good for measuring the exact frequency of a signal; noise may be seen as a change in frequency.

   Oscilloscopes—used to measure voltage in reference to time; they are good for measuring changing voltages. If the signal is not repeated the user may not see the image on the oscilloscope. Oscilloscopes are also good for determining frequency and duty cycle.

   Digital storage oscilloscopes—similar to oscilloscopes with the added feature of storing the image on the screen in digital memory for display at a later time. This device is good for measuring waveforms that are not repeated.

   Logic analyzers—the best tool for analyzing problems with microcomputer waveforms which require monitoring of many different digital signals. The logic analyzer can display a timing diagram, which can be used to determine frequency, time delays, noise, and sequencing problems.

   Signature analyzer—like a signal trace logic analyzer except that it can also measure and store analog signals. Signature analyzers are used in data communications where a test signal can be analyzed and stored for later comparison.

   Logic monitor—allows the user to monitor the state logic level of a digital circuit; it does not function well with fast-changing digital signals.
2. Microcomputer signals and specifications:

   Logic levels—a range of voltages that a digital circuit responds to in a logical way. Any voltage between 0 and the lower threshold level is seen as a logic-0. Any voltage between the upper threshold level and supply is seen as a logic-1. Any voltage higher than the lower threshold level but less than the upper threshold level may be seen as either a logic-1 or logic-0 or may toggle. This voltage is not a valid digital voltage level.

   Schmitt trigger input—used to reduce the effect of noise on a digital gate. In order for a gate with a Schmitt trigger input to see a logic level change, the input voltage must go either below the lower threshold level or above the upper threshold level.

   Analog signal—a signal that has more than two usable voltage levels; analog signals may be positive, negative, or positive and negative voltage levels.

   Supply voltage—the voltage needed by the device to operate properly.

   Noise margin—the greatest amount of noise a device can withstand before the noise is used as data or a proper signal.

   Threshold levels—the voltage levels that digital circuits respond to. The upper threshold level is the least amount of voltage that causes a digital circuit to see a valid logic-1. The lower threshold is the maximum amount of voltage that causes a digital circuit to see a valid logic-0.

Operating speed—the maximum speed at which a digital IC can switch.

Power dissipation—the amount of power dissipated by the gate during a typical operation.

3. Hardware problems occur when a hardware component fails. Software problems occur while operating particular types of software programs. Sometimes hardware problems look like software problems and software problems look like hardware problems. In order to determine the type of problem, you must isolate the hardware or software and test the hardware or software independently.
4. Once power is applied to the computer and peripherals, the fan motor on the power supply of the computer should be heard. The cursor should appear on the screen after about 4 to 15 seconds. In many computers the copyright screen should be seen on the video monitor and a number should start counting in the upper left hand corner of the screen as memory (RAM) is checked. If everything is operating correctly and POST cannot find a problem in about 40 to 75 seconds, a beep should be heard. The amount of time depends on the speed of the computer and the amount of memory in the computer. The computer then tries to read floppy drive unit A to perform a boot-up. If no boot disk is in drive A, the computer tries to boot up in drive C (hard drive). If no hard drive is present or the hard drive does not contain the system boot files, the computer responds with some type of error message.
5. POST produces error codes in a visual or audio format to indicate problems that POST has found during the booting of the computer system. POST does not perform a detailed test of the equipment in the computer system; additional test diagnostic equipment or software may be required.
6. If the computer fails to boot, diagnostic hardware may be used to track down the problem with the computer system. A POST card is an I/O card that supplies information about the computer system without the need for the computer to be in operating condition. Diagnostic software works as long as the computer can boot up. Diagnostic software is used to test and find noncritical problems with the computer hardware or software.
7. The logical approach to system troubleshooting is to divide and subdivide the computer system into manageable blocks and parts to help track down problems. This method starts with the block diagrams of the system, after determining which block is bad by determining proper input and output signals. Once the block is identified as bad, the bad block is subdivided into smaller blocks. Once the smaller bad block is identified, the next step is to go to the component level to determine the cause of the problem.
8. Hardware repairs can fall into three methods. Card swapping sometimes corrects the problem by swapping known good cards with the cards in the computer until the problem goes away. Once the card or component is suspected, the part can be substituted and the circuit tested to determine if the replaced component was bad. Field engineering changes occur when problems in the design occur after the equipment has been out in the field under actual operating conditions.
9. Keeping maintenance and repair logs is very important to provide information to the repair technician about what type of service and repairs were made on the equipment in question. Many times a pattern develops that allows the technician to determine the problem in the computer system without performing complex testing.

REVIEW QUESTIONS

1. Which piece of test equipment is best for viewing the logic levels on many differing digital inputs?

a. Logic probe

b. Oscilloscope

c. Logic analyzer

2. Which piece of test equipment is best for measuring duty cycle?

a. Oscilloscope

b. DMM

c. Logic probe

3. Which piece of test equipment is the easiest to use to determine the logic state of a digital circuit?

a. Logic probe

b. Oscilloscope

c. Logic analyzer

4. The maximum amount of voltage that can be seen on the input of a digital circuit and be seen as a valid logic low is called ________.

a. Upper threshold level

b. Lower threshold level

5. The maximum amount of noise that a digital circuit can withstand before seeing the noise as valid data is called noise margin.

a. True

b. False

6. A hardware problem is a problem that occurs regardless of the type of software being used.

a. True

b. False

7. If the number 301 appears on the video screen, what may be causing the problem?

a. Power supply

b. System board

c. Keyboard

d. Floppy disk controller

e. Hard disk controller

8. During the booting of the computer system, the computer fails to boot but one long and two short beeps are heard. What is the problem?

a. Power supply

b. System board

c. Keyboard

d. Video adapter

e. None of the given answers

9. If the computer fails to boot up, diagnostic software can be used to find the problem.

a. True

b. False

10. What is the most common intermittent computer hardware problem?

a. Keyboard

b. Memory parity errors

c. 8088 errors

11. Which step in a logical approach to troubleshooting contains the most amount of detail?

a. System block diagram

b. Subsystem block diagram

c. Schematic diagram

12. Manufacture specifications are important because they specify the operational characteristics of the computer hardware.

a. True

b. False

13. If POST does not locate the problem, the problem is not the hardware.

a. True

b. False

14. Microcomputer problems are normally caused by ________.

a. Software

b. Hardware breakdown

c. Environmental conditions

15. POST does not verify the operation of the ________.

a. MPU

b. RAM

c. ROM

d. Printer

e. None of the given answers

Glossary

8088: A 16-bit Intel MPU that contains an 8-bit data bus and a 20-bit address bus, used in the PC and PC/XT. The 8088 is software compatible with the 8086.

8086: A 16-bit Intel MPU that contains a 16-bit data bus and a 20-bit address bus.

80286: A 16-bit Intel MPU that contains a 16-bit data bus, a 24-bit address bus, and memory management. The 80286 (also known as the 286) was the first MPU used in AT-type PCs and is software compatible with the 8088/8086.

80386SX: A 32-bit Intel MPU that contains a 16-bit data bus, a 32-bit address bus, and advanced memory management. The 80386SX (also known as the 386SX) is used in AT-type PCs and is software compatible with the 8088/8086/80286.

80386DX: A 32-bit Intel MPU that contains a 32-bit data bus, a 32-bit address bus, and advanced memory management. The 80386DX (also known as the 386DX) is used in AT-type PCs and is software compatible with the 8088/8086/80286/80386SX.

80486DX: A 32-bit Intel MPU that contains a 32-bit data bus, a 32-bit address bus, advanced memory management, internal cache memory, and numeric coprocessor. The 80486DX (also known as the 486DX or DX) is used in AT-type PCs and is software compatible with the 8088/8086/80386SX/80386DX.

80486DX2: A 32-bit Intel MPU that contains a 32-bit data bus, a 32-bit address bus, advanced memory management, internal cache memory, and numeric coprocessor and operates at twice the speed of the DX internally. The 80486DX2 (also known as the 486DX2 or DX2) is used in AT-type PCs and is software compatible with the 8088/8086/80386SX/80386DX/80486SX/80486DX.

80486DX4: A 32-bit Intel MPU that contains a 32-bit data bus, a 32-bit address bus, advanced memory management, internal cache memory, and numeric co-processor and operates at three times the speed of the DX internally. The 80486DX4 (also known as the 486DX4 or DX4) is used in AT-type PCs and is software compatible with the 8088/8086/80386SX/80386DX/80486SX/80486DX.

80486SX: A 32-bit Intel MPU that contains a 32-bit data bus, a 32-bit address bus, advanced memory management, and internal cache memory. The 80486SX (also known as the 486SX or SX) is used in AT-type PCs and is software compatible with the 8088/8086/80386SX/80386DX.

80486SX2: A 32-bit Intel MPU that contains a 32-bit data bus, a 32-bit address bus, advanced memory management, and internal cache memory and operates at twice the speed of the SX internally. The 80486SX (also known as the 486SX2 or SX2) is used in AT-type PCs and is software compatible with the 8088/8086/80386SX/80386DX/80486SX/80486DX.

Accumulator: A register inside the computer that is used by some instructions to hold the results of an ALU operation.

Alphanumeric mode: A display mode in a computer system that allows only ASCII characters and graphic symbols to be displayed on the video screen.

ALU: Arithmetic Logic Unit, a digital device found inside the MPU that is responsible for performing all math and logic operations on the data inside the MPU.

Architecture: Specifies the design of an MPU and how its internal elements function with each other. Architecture can also be applied to how the support circuitry of the MPU works with the MPU.

AT: Advance Technology, refers to a PC that uses one of the following Intel MPUs (80286, 80386SX, 80386DX, 80486SX, 80486DX, Pentium, or more advanced MPU).

ASCII: American Standard Code for Information Interchange, a 7-bit binary code that is used to exchange data between computer systems. The code represents all letters, numbers, and punctuation symbols common to computers. The IBM PC uses extended ASCII, which is an 8-bit binary code. When bit eight is reset, the extended ASCII code represents normal ASCII characters. When bit eight is set,a subset group of graphic and special characters is represented.

Binary: A base 2 number system that allows each character to be represented by either a zero or a one.

BIOS: Basic Input Output System, a monitor program used in the IBM PC. The programming of BIOS configures the computer system and tells the computer how to act like a computer.

Bit: A single piece of binary data.

Boot record: A short program located on each usable diskette that enables disk routines inside BIOS, which, in turn, allows the computer to use the diskette.

Break-out box: A piece of computer test equipment that allows the technician to easily access the different signal lines of the computer interface cable.

Bus cycle: The amount of time it takes to transfer a piece of data from one location to another location in the computer system.

Byte: A group of eight binary bits that are used as one data package.

Cache: A form of high-speed memory that is used to buffer slower memory from the MPU.

CGA: Color Graphics Adapter, the first type of color video adapter available for the IBM PC.

CMOS: Complementary Metal Oxide Semiconductor, a type of low-power digital logic family.

Compatible PC: A microcomputer system that executes most, if not all, software written for the IBM PC. Compatibles also accept I/O devices for the IBM PC.

Default drive: The disk drive unit that the computer uses whenever a drive parameter is not specified.

Diagnostic software: A special type of software used to identify trouble areas of the computer system.

DMM: Digital Multi Meter, a piece of test equipment used to measure resistance, DC/AC voltage, and current. Some DMMs can measure transistor gain, frequency, and capacitance and verity the operation of a good or bad diode.

DOS: Disk Operating System, a group of programs used to control the hardware in the computer system.

DOS commands:

BUFFERS—Specifies the number of buffers used in a disk operation.

CALL—Allows a new batch program to take control without stopping the first batch program.
CD—Changes the working directory.
CHKDSK—Analyzes the files on the disk and produces a report.
CHOICE—Prompts the user during the execution of a batch file.
CLS—Clears the video screen of data and places the cursor in the upper-left corner of the screen.
COMP—Compares the contents of two files.
COPY—Transfers a copy of a file(s) to another location.
DATE—Allows the user to change or view the current date.
DBLSPACE—Compresses the files on a disk to gain space.
DEBUG—A single-pass assembler.
DEFRAG—Allows DOS to defragment a disk.
DEL—Deletes a file(s) from a disk.
DELTREE—Deletes, files, and removes the specified directory(ies).
DEVICE—Loads a device driver into memory.
DEVICEHIGH—Loads a device drive into upper memory.
DIR—Displays a listing of the files and other related file information.
DISKCOMP—Compares the information on two disks.
DISKCOPY—Makes an exact copy of another disk.
DOS—Loads part of MS-DOS into high memory.
DOSSHELL—Allows the user to process most common DOS commands using a menu interface.
ECHO—Controls whether text in a batch file is displayed or not.
EDIT—A menu-driven text processor.
EMM386—Expanded memory manager for DOS.
ERASE—Deletes a file(s) from a disk.
FC—Compares the contents of two files.
FCBS—Specifies the number of file control blocks.
FDISK—Configures a hard drive for use with MS-DOS.
FILES—Specifies the number of files DOS can access at one time.
FORMAT—Initializes a new disk or reinitializes a used disk.
GRAPHICS—Allows MS-DOS to print graphical information on the printer.
HELP—Provides help information on all DOS commands.
HIMEM—Extended memory manager for DOS.
LABEL—Allows the user to change, create, or delete the volume label name on a disk.
LASTDRIV—Specifies the maximum number of drives in DOS.
LH—Loads the program file into upper memory.
MD—Makes a subdirectory within a directory.
MEM—Displays a listing of programs in memory.
MEMMAKER—Optimizes memory configurations.
MORE—A filter used to display information one video page at a time.
MOVE—Moves a file(s) from one location to another location.
MSAV—Enables the MicroSoft Anti-Virus program.
MSBACKUP—A menu program that controls the backing up of the disk system.
MSCDEX—A MicroSoft CD-ROM command interface.
MSD—Enables the MicroSoft Diagnostics program.
PATH—Specifies a path for DOS to search when calling up an executable file.
PAUSE—Causes the computer to suspend processing of a batch file until the user responds.
PRINT—Prints the contents of a file on a printer.
RD—Removes a directory.
REM—Allows the user to place remarks in a batch file.
RENAME—Changes the name of a file.
SET—Sets the environment variables in a batch file.

SETVER—Sets up the DOS version table.
SHELL—Specifies the location of the command interpreter.
SMARTDRV—MicroSoft's disk caching system.
STACKS—Specifies the number of data stacks.
SYS—Transfers the system files to a disk, making it bootable.
TIME—Allows the user to change or view the current time.
TREE—Displays the directory structure.
TYPE—Displays the ASCII characters in a file on the video screen.
UNDELETE—Restores files that were previously deleted by using the DEL (ERASE) command.
UNFORMAT—Restores the disk information that was erased by the format command.
VERIFY—Controls the condition of the VERIFY switch.
VOL—Displays the volume label.
VSAFE—A memory resident anti-virus monitor.
XCOPY—Transfers a copy of a file(s) to another location, including files within subdirectories.

Double word: A group of 32-bits that are used as one unit.

Download: A computer process that takes the data stored in a secondary storage unit and places that data into the main memory of the computer.

DRAM: Dynamic Random Access Memory, a type of read/write memory that stores its data in the capacitance of the IC. Because data is stored in the form of charges in the IC, the charge levels (data) must be re-established within a certain time period, which is known as refresh.

DSDD: Double-Sided Double-Density, a type of floppy diskette.

DSHD: Double-Sided High-Density, a type of floppy diskette.

DVM: Digital Volt Meter, a piece of test equipment used to perform the same tasks as a DMM (see DMM).

EGA: Enhanced Graphics Adapter, a type of color graphics video display system.

EIA: Electronic Industries Association, a group of electronics manufacturers formed to standardize electronic hardware and signal specifications.

EISA: Extended Industry Standard Architecture, a 32-bit high-speed I/O interface standard.

EPROM: Erasable Programmable Read Only Memory, a ROM-type digital storage device that allows data to be programmed into its memory cells. Once those data have been programmed into the EPROM, the data cannot be changed under normal conditions; they can only be read when the IC is accessed. The data are erased by applying ultraviolet light to the erase window for a certain amount of time,which erases all data in the EPROM.

ESDI: Enhanced Small Device Interface, a type of hard drive system.

Floppy disk adapter: An I/O interface adapter that is used to control the interrupts in the instructions from the MPU and perform the operations on the floppy disk system. This adapter also converts parallel data from the MPU into serial data for the floppy disk drive unit during a write operation and converts serial data from the floppy disk drive unit into parallel data for the MPU duringa read operation.

Floppy disk drive unit: A device that contains TTL and analog circuitry, two motors, and assorted sensors, needed to store data onto the floppy diskette.

Floppy disk system: A floppy disk system consists of four parts: the floppy disk adapter, floppy disk drive unit, diskette, and operating system.

Floppy diskette: The magnetic storage medium for the floppy disk system (5.25-inch, called a minifloppy, or 3.5-inch, called a microfloppy). Data is stored on the diskette in the form of small magnetic fields.

GUI: Graphical User Interface, a type of software interface that allows the user to use graphical symbols to perform different types of software operations. In the PC, Windows, Windows for Workgroups, Windows NT, and OS/2 are GUIs.

Hard drive system: A system that operates much like a floppy disk system, except that the storage medium is not removable, it operates at a much higher speed, and it can store 20 to 1000 times more data. The hard disk system consists of three parts: the hard drive adapter, the hard drive unit (which contains the nonremovable storage medium), and an operating system.

Hexadecimal: Also known as hex, a base-16 numbering system in which each character can have one of 16 different values. Characters in hex from 0 to 9 represent the first 10 values of the numbering system and A to F represent the last 6 values, where A equals the number 10 in decimal and F equals the number 15 in decimal.

High byte: In a word (16-bit) operation, the eight most significant bits of the value.

IBM: International Business Machine Inc., developer of the IBM PC and the largest manufacturer of personal computer systems.

IDE: Intelligent Drive Electronics, a type of hard drive system.

Integrated motherboard: A type of system board that contains most of the I/O interface adapters built into the system board itself.

I/O: An acronym for the words Input and Output. It is sometimes used to refer to signals and hard-ware devices that are used to transfer data between the computer and the outside world.

ISA: Industry Standard Architecture, a 16-bit AT I/O interface standard.

K: Also known as Kbyte or kilobyte or KB, a group of 1000 bytes of data.

Keyboard: An input device with between 83 and 101 keys (switches) which allows the user to input data into the computer. Each key represents a different type of character or control function.

Logic one: A digital voltage level that represents the most positive logic state, also known as a logic high.

Logic zero: A digital voltage level that represents the most negative logic state, also known as a logic low.

Low byte: In a word (16-bit) operation, the lower or eight least significant bits of the data.

LSB: Least Significant Bit or Byte, depending on the context.

Mark: A logic one, with reference to data communications.

MCA: Micro Channel Architecture, a high-speed IBM I/O interface standard.

MDA: Monochrome Display Adapter, the first type of video display adapter used on the IBM PC. The adapter cannot perform graphic commands.

MFM: Modified Frequency Modulation, a type of hard drive system.

MHz: A frequency unit of measurement of 1 million hertz or cycles per second. Also known as megahertz.

Mnemonic: An English acronym for a computer instruction.

Motherboard: The main circuit board of the computer that contains the MPU, its support circuitry, ROM, RAM, and I/O interface bus. Also known as a systemboard.

MPU: MicroProcessor Unit, the main controlling element of a microcomputer system.

MSB: Most Significant Bit or Byte, depending on the context.

Nibble: Four binary bits used together as one data package.

Operand: Specifies a value or result for a computer instruction.

PC: Personal Computer, a microcomputer system that is normally found in the home or in a business.

PCI: Peripheral Component Interconnect, Intel's high-speed 32/64-bit Pentium-based I/O interface standard.

Pentium: A 64-bit Intel MPU that contains a 64-bit data bus, 32-bit address bus, large internal cache, integrated smart superscaler pipeline, and floating point unit (FPU, also known as a math coprocessor). The Pentium is software compatible with all Intel 80xxx MPUs.

Peripheral: An electronic device that is connected to a computer system. A peripheral may be a input, output, or input/output device.

PLA: Programmable Logic Array, a device that is programmable to emulate one or more gate circuits within one IC.

Port: A point in a circuit or computer circuit where signals may enter (input) or exit (output).

POST: Power On Self-Test, a test program that is executed as soon as the IBM PC is turned on. This program tests all major components within the computer system.

Power supply: An electronic device that converts AC from a wall outlet into DC voltage to power the computer. Personal computers use a switching power supply to perform this task.

Printer: A device that takes information (character or graphic) from the computer and makes a permanent record on paper (hard copy).

RAM: A type of read and write digital memory that is used to temporarily store data in the computer. This is a volatile type of memory device, which means that if the power is lost the data is lost.

Refresh: The process of re-establishing the data in a DRAM.

Refresh cycle: The amount of time it takes to perform a refresh operation on DRAM.

Register: A digital device that is used to store one or more bits of binary data inside the computer.

RGB: A term used when talking about a video monitor; it represents the three primary colors used in a color monitor display (red, green, and blue).

RLL: Run Length Limited, a type of hard drive system.

ROM: Read Only Memory, a memory device that is programmed during the time the device is made and can never be reprogrammed. This is a nonvolatile type of memory.

Sample: The process of obtaining data, information, or levels from a circuit in a predetermined way.

Schmitt trigger: A special type of digital input that increases the noise margin of a circuit.

SCSI: Small Computer System Interface, a type of mass storage system interface that includes hard drive systems, CD ROMs, and other devices.

SSDD: Single-Sided Double-Density, a type of floppy diskette.

Static RAM: A type of read and write memory that does not require refresh to retain its data. In static RAM the part of memory cell that is conducting determines the data stored in the memory cell.

Stepper motor: A type of motor that has a shaft which rotates a certain number of degrees each time the motor is pulsed.

SVGA: Super Video Graphics Array, a high resolution color graphics video display system.

System board: The main circuit board of the computer that contains the MPU, its support circuity, ROM, RAM, and I/O interface bus. Also known as a motherboard.

System unit: A computer case that contains the system board, I/O interface adapters, floppy drive units, hard drive units, and power supply for the computer.

Timing diagram: A visual diagram that shows the time relationship between two or more signals at any time.

Timing error: A condition that occurs when a signal level changes at the incorrect time.

TTL: Transistor Transistor Logic, a type of digital logic family that uses multiemitter transistors as the main input element.

UART: Universal Asynchronous Receiver Transmitter, an electronic device that converts data back and forth between serial and parallel formats.

VGA: Video Graphics Array, a color graphics video display system.

Video display: An output device that provides the user with a visual indication of the operation the computer is performing. A video display operates much like a TV.

VLB: Video Local Bus, a high-speed 32-bit AT I/O bus standard.

Word: A group of 16 bits that are used together as one unit.

XT: An acronym for eXtra Technology, which refers to the second generation of PCs developed by IBM. XTs use the 8088 MPU from Intel.

References

The following are excellent technical references:

IBM Corporation, Boca Raton, Florida
IBM Guide to Operations, Personal Computer
IBM PC DOS Manual
IBM Technical Reference Personal Computer
IBM Technical Reference Options and Adapters, Vols. 1 & 2
IBM Hardware Maintenance and Service Manual, Vols. 1 & 2
IBM DOS Technical Reference Manual

Intel Corporation, Santa Clara, California
Intel Microprocessors, Vols. I, II, and III, 1994
Intel 82340 High-Performance ISA-compatible Chip Set Manual, 1990
82350 EISA Chip Set Manual

Microsoft Corporation, Redmond, Washington
Microsoft MS-DOS User Manual

Micro Technology Inc., Boise, Idaho
MOS Data Book

Texas Instruments Inc., Dallas, Texas
TI TTL Data Manual
TI Programmable Logic Data Book

Tandon Magnetics Corporation, Chatsworth, California
Tandon Disk Drive Operating and Service Manual

Answers to Review Questions

Chapter 1

1. b. 65,536
2. c. 80386SX
3. a. System unit
4. e. None of the given answers (equipment DIP switches)
5. a. XT
6. b. Integrated motherboard
7. c. +9 volts and d. −9 volts
8. a. DSHD
9. a. MDA
10. a. XT
11. b. Parallel
12. a. One plus the background color
13. e. Delete
14. c. Shift
15. d. Application
16. b. False
17. b. False
18. a. True
19. b. False
20. a. ROM/EPROM

Chapter 2

1. b. Disk Operating System
2. c. Directory
3. b. File allocation table
4. b. False
5. a. True
6. d. F3
7. a. True
8. b. External
9. a. True
10. d. COMMAND.COM
11. a. True
12. b. False
13. b. False
14. b. False
15. b. /S
16. d. VOL
17. c. /P
18. b. False
19. b. XCOPY
20. b. FC
21. b. False
22. b. False
23. a. MEM
24. a. True
25. b. False

Chapter 3

3.1 Microprocessor Subsystem

1. b. OSC
2. d. Nothing is wrong with the computer; this is a normal condition.
3. a. The computer operates normally but a little faster.
4. c. NMI
5. b. False
6. b. False
7. b. The status lines from the 8088 goes passive.
8. a. True
9. b. False
10. b. False

3.3 Address Decoding

1. e. None of the above answers is correct. (G1 = 1, $\overline{G2A}$ = 0, $\overline{G2B}$ = 0)
2. b. Low
3. a. A = 1, B = 1, C = 0
4. e. None of the above answers is correct. ($\overline{WRT\ DMA\ PG\ REG}$)
5. b. False
6. a. True
7. b. $\overline{RAS1}$ and $\overline{CAS1}$
8. e. $\overline{CS7}$
9. c. Hard disk adapter
10. b. False (003D0–003DF)

3.4 DMA Section

1. c. DREQx
2. b. $\overline{MEMR}$
3. c. ADSTB
4. b. HLDA
5. e. $\overline{EOP}$
6. d. $\overline{DACKx}$
7. a. True
8. a. True
9. b. False (it is always low during a DMA cycle)
10. c. 74670

3.5 ROM/EPROM

1. a. $\overline{CS7}$
2. b. Disabled
3. b. From side B to side A
4. b. DMA (8237)
5. a. $\overline{CE}$ = 0, $\overline{OE}$ = 0

3.6 DRAM

1. c. $\overline{RAS2}$
2. b. A logic high
3. a. A logic low
4. a. Low
5. a. Low
6. b. High
7. c. XU69
8. b. False (A parity error can occur only after a memory read cycle.)
9. a. Read and b. Write
10. e. None of the banks of DRAMs

3.7 8253

1. a. Low
2. b. High
3. b. OUT1
4. a. OUT0
5. c. OUT2

3.8 8255

1. a. Bad equipment DIP switch setting
2. a. True
3. a. True
4. a. PC0 to PC3
5. c. PB1

3.9 I/O Interface Bus Channel

1. d. $\overline{\text{DACK0}}$
2. c. $\overline{\text{IOR}}$
3. a. Side A
4. b. False
5. b. CLK

Chapter 4

1. c. 8088
2. d. None of the given answers (6 MHz)
3. a. 82284
4. b. 82288
5. a. 74245
6. b. IRQ1
7. a. 8237
8. d. 8254
9. d. RTC
10. a. 8042
11. c. 2
12. c. 16
13. d. AU
14. c. MMU
15. a. Real
16. a. Real
17. c. 16
18. b. Flags
19. d. MSWR
20. b. False
21. d. GDTR
22. e. None of the given answers (IDTR)
23. c. DX
24. b. False
25. d. $\overline{\text{BHE}}$
26. c. $\overline{\text{READY}}$
27. e. HLDA
28. b. False
29. a. 82230
30. b. 82231
31. a. True

Chapter 5

1. a. Bus interface
2. b. Protected mode
3. c. 16
4. d. 32
5. b. False
6. b. False
7. a. $\overline{\text{BHE}} = 0$, $\overline{\text{BLE}} = 0$
8. a. True
9. a. True
10. a. True
11. a. True
12. b. HOLD
13. b. False
14. a. True
15. a. True
16. b. False
17. b. 82231
18. c. Page mode interlace
19. b. False
20. d. SRAM
21. a. Direct
22. b. False
23. b. False
24. b. False
25. d. ISA controller

Chapter 6

1. d. a, b, and c
2. c. 33 MHz
3. a. 486SX
4. c. Pentium
5. c. Pentium
6. a. Bus interface unit
7. e. Memory management unit
8. b. BE0# to BE3#
9. e. None of the given answers
10. b. False
11. b. False
12. a. True
13. a. True
14. c. 8
15. a. Low
16. b. False
17. a. True
18. a. True
19. b. False
20. a. True
21. d. PCI
22. c. MCA
23. b. 82304
24. c. 82358
25. b. 82434LX

Chapter 7

1. b. AT keyboard
2. a. XT keyboard
3. b. AT keyboard
4. b. IRQ1
5. a. Capacitive keys
6. b. AT
7. a. XT
8. d. The key closure and/or contacts
9. a. Short
10. b. Undervoltage
11. d. Series pass transistor
12. b. Switching power supply
13. b. Increase
14. b. Increase
15. a. Decrease
16. a. Decrease
17. a. Decrease
18. c. Stays the same
19. c. 50%
20. c. +12 V

Chapter 8

1. b. Video monitor
2. b. Video monitor
3. a. Video adapter
4. d. Focus grids
5. b. False
6. b. Off
7. b. False
8. a. MDA
9. b. CGA
10. d. VGA
11. a. True
12. a. More
13. b. False
14. b. Vertical output
15. c. No raster appears
16. c. Raster scan code
17. d. Character generator
18. b. "A"
19. b. All-points-address mode
20. c. DRAM DAC

Chapter 9

1. a. Disk adapter
2. c. DOS
3. c. CD drive system
4. b. False
5. a. True
6. b. MFM
7. b. Track
8. b. False
9. c. 512 bytes
10. b. False
11. b. IDX
12. c. RD
13. a. WDA
14. b. Command phase
15. a. True
16. b. False
17. b. False
18. c. Track zero switch
19. a. True
20. a. Boot record
21. b. False
22. b. False
23. a. True
24. b. Partitioning
25. d. IDE

Chapter 10

1. b. Serial data communications
2. a. Parallel
3. a. Informs the printer that new data is on the data bus
4. b. 8
5. a. True
6. b. False
7. b. Serial
8. a. Parallel
9. b. False
10. b. 1488
11. c. Impact dot matrix printers
12. a. True
13. b. False
14. a. Mechanical
15. a. Mechanical

Chapter 11

1. a. Telephone lines
2. e. Fiber optic cable
3. a. Allows data to exit the phone
4. b. False
5. c. Impulse noise
6. b. False
7. b. Amplitude modulation
8. c. Frequency-shift keying modulation
9. b. Converts voltage levels into frequencies
10. c. Reduces the noise from the phone line
11. b. False
12. a. Mainframe network
13. b. Local Area Network
14. a. Ethernet
15. b. False
16. a. True
17. b. False
18. a. True
19. b. False
20. b. False
21. a. True
22. d. FDDI Token Ring
23. b. ARCnet
24. d. Fiber optic cable
25. a. Ethernet
26. a. Protocols
27. e. Layer 5
28. a. True
29. a. True
30. a. True

Chapter 12

1. c. Logic analyzer
2. a. Oscilloscope
3. a. Logic probe
4. b. Lower threshold level
5. a. True
6. a. True
7. b. System board
8. d. Video adapter
9. b. False
10. b. Memory parity errors
11. c. Schematic diagram
12. a. True
13. b. False
14. c. Environmental conditions
15. d. Printer

Index

A

AC power, 407. *See also* Power supplies
  distortion, 408–9
  protection, 409
Advanced AT buses, 375–83
  82311 chip set, 376–78
  EISA, 378–80
    82350 chip set, 379–80
  microchannel architecture, 375–76
  PCI, 381–83
  VESA VLB, 380–81
Advanced Technology. *See* Advanced AT buses; AT entries
Alternating current. *See* AC power; Power supplies
Analog signals, 595
ARCnet network, 582
Asynchronous adapter, 545–48
  function of, 601–2
  troubleshooting, 550
Asynchronous data. *See also* Serial data communications
  communications element. *See* 8250 Asynchronous communications element
  methods of interpretation, 534. *See also* NRZ asynchronous circuitry
  packaging, 537–38
  transmission
    baud rate, 539
    examples, 538–39
AT chip set. *See* 80286 System board using 82230/82231 AT chip set; 82230 AT chip; 82231 AT chip; 82340 AT chip set
AT floppy disk, 508–10
AT keyboards, 397–98
  system board key encoding, 402–6
AT system boards, 6–7, 242–43
  key encoding circuitry, 402–3
  key encoding operation, 403–6
AUTOEXEC. BAT file, 93–95
  example, 102–3

B

Batch files, 91–93
Baud rate, 539
Block diagrams
  dot matrix printer, 552
  8253 programmable interval timer, 227–28
  8255 programmable peripheral interface, 233–34
  80286 system board, 243
  80386 system board, 301–2
  80486 system board, 341–43
  82385 32-bit cache controller, 325–27
  IBM PC XT system board
    DRAM section, 210–12
    8253 programmable interval timer, 227–28
    8255 programmable peripheral interface 233–34
  Pentium, 363–64
  video adapters
    IBM color graphics, 458–61
    IBM monochrome, 451
    VGA, 465
Branch prediction, 365
BUFFERS= command (CONFIG.SYS), 95
Bus arbitration
  80486 MPU, 352–53
  Pentium processor, 369–70

C

CAD. *See* Computer-aided drafting
CALL command, 91
Carrier sense multiple access with collision detection (CSMA/CD), 576
Cathode ray tube (CRT), 426
Cathode ray tube controller (CRTC). *See* 6845 CRTC
CD command, 78
CD ROM, 475
Central processing unit (CPU), 302, 344
Centronics printer interface, 526–27
CGA. *See* Color-graphics adapter
Character matrix, 557
CHKDSK command, 72–73
CHOICE command, 91
Clear screen. *See* CLS command
Clock
  8284 generator/driver, 121, 268, 276
    pin configuration, 122–26

Clock, 8284 generator/driver *(cont.)*
timing diagrams, 126–27
80486 MPU, pin configurations, 354
82284 clock generator/driver, 259–61, 272–73
82335 IC, 318
external sampling, 592
internal sampling, 592
Pentium processor, pin configuration, 372
6818 RTC, 245, 264, 275
CLS command, 46
Coaxial cable, in LANs, 580
Color-graphics adapter (CGA), 14, 437–38
COMMAND. COM file, 508
Command decoder, function of, 603
COMP command, 63
Computer-aided drafting (CAD), 26
CONFIG. SYS file, 95–103
buffers, 95
device driver, 95
device high, 95–96
EMM386, 97
examples, 98–102
HIMEM, 97
last drive, 96
menu commands, 97–100
set version, 97
shell, 96
stacks, 97
COPY command, 60–61
CPU. *See* Central processing unit
CRT. *See* Cathode ray tube
CRTC (Cathode ray tube controller). *See* 6845 CRTC
CSMA/CD. *See* Carrier sense multiple access with collision detection

D

Data base, 26
Data bus transceiver, function of, 603
Data communications, 568–74. *See also* Asynchronous data; Parallel data communications; Serial data communications
ports, 17
Data integrity, in Pentium processor, 365
DATE command, 44–45
DBLSPACE command, 105
DEBUG commands, 88–90
DEFRAG command, 104–5
DELETE command, 64–65
DELTREE command, 79–86
Desktop publishing, 26
DEVICE= command (CONFIG. SYS), 95, 97
DEVICEHIGH= command (CONFIG. SYS), 95–96
Diagnostics, 75–78
Digital logic probe, 590
Digital multi-meter (DMM), 589–90
Digital storage oscilloscope, 590
DIR command, 53–56
Directory, 506–7
root, 38–40
Directory commands, 53–56, 77–78
DISKCOMP command, 59–60
DISKCOPY command, 57–59
Diskettes. *See* Floppy disk
Disk operating system (DOS), 12, 34
command prompt, 44
commands, 42, 44–90
list of, 114–16
directory commands, 77–78
disk system and, 35–37
file types, 37–38
formatting diskette with, 504–8
function of, 35
loading, 43–44
root directory, 38–40
special keys, 40–41
syntax, 40–41
versions of, 41–42, 97–99
wild cards, 40
DMA controller. *See* 8237 Direct memory access controller
DMM. *See* Digital multi-meter
DOS. *See* Disk operating system
DOS= command (CONFIG. SYS), 96
DOSSHELL command, 86–88
Dot matrix printers, 23, 550, 554–56
advantages and disadvantages of, 555
block diagram, 552
nonimpact, 555–56
advantages and disadvantages of, 556
Doublespace, 105
Drivers, printer, 553
Dynamic random-access memory (DRAM)
80286 system board, 246
using 82230/82231 AT chip set, 287–90
80386DX system board, 335
80386SX system board, 324
82335 IC, 320–23
IBM PC XT system board, 209–25
internal operation, 212–13
MT4264 block diagram, 210–12
pin assignments, 213–15
timing diagrams, 216–23
troubleshooting, 223

E

ECHO command, 91
EDIT command, 90
EGA. *See* Enhanced graphics adapter
EIA. *See* Electronic industry association
8041 Universal peripheral interface, 553
8042 Universal peripheral interface microcontroller, 245, 287, 291, 324, 335
8048/8049 Microcontrollers, 390–93
pin configurations, 391
in printer, 551–52
8087 Math coprocessor, 143–44
pin configurations, 144–45
troubleshooting, 145
8088 Microprocessor, 4, 128–43
addressing modes, 132
architecture, 129–30
bus, 128–29
data formats, 131
effective address, 131
instructions, 132
pin configuration, 132–41
timing diagrams, 141–43
8155 RAM, 552
8237 Direct memory access (DMA) controller, 185–201, 245, 268–69, 277
internal registers and control logic, 187–88
operational modes, 191
pin configuration, 189–90
reading page register, 194
timing diagrams, 194–98
troubleshooting, 198–99
types of transfers, 191–92
writing to page register, 192
8250 Asynchronous communications element, 539–45
divisor latch LSB and MSB, 543
features, 540
interrupt enable register, 544–45
interrupt identification register, 544
line control register, 542
line status register, 543
modem control register, 544
modem status register, 544
MPU bus, 540–41
pin configuration, 540–42
receive operation, 545
receiver buffer register, 545
serial interface bus, 541–42
transmit operation, 545
transmitter holding register, 545
8253 Programmable interval timer (PIT), 225–31
block diagram, 227–28
functions of, 229–30
operational modes, 228–29
troubleshooting, 230–31
8254 Programmable interval timer (PIT), 245, 268, 276
8255 Programmable peripheral interface (PPI), 231–38
block diagram, 233–34
functions of, 235–37
operational modes, 234–35
pin configuration, 234
troubleshooting, 237
8259 Programmable interrupt controller (PIC), 156–58, 245, 263–64, 274–75
cascaded control lines, 161
pin configurations, 158–61
timing diagrams, 161–63

8284 Clock generator/driver, 121, 268, 276
pin configuration, 122–26
timing diagrams, 126–27
8288 Bus controller, 145–46
pin configurations, 146–51
timing diagrams, 151–56
80286 Microprocessor, 246–55, 270–71
base architecture, 246–48
instructions, 249–51
memory management, 248
modes of, 248–49
pin configurations, 251–55
registers, 249–51
80286 System board, 242–99
block diagram, 243
DRAM section, 246
8042 universal peripheral interface microcontroller, 245
8237 DMA controller, 245
8254 programmable interval timer, 245
8259 programmable interrupt controller, 245, 263–64, 274–75
math coprocessor. *See* 80287 Math coprocessor
microprocessor subsystem, 243–45
MPU. *See* 80286 Microprocessor
ROM section, 246
6818 real time clock, 245, 264, 275
80286 System board using 82230/82231 AT chip set, 258–77
address/data bus control, 264–65
bus control logic, 262–63, 273
command delay logic, 261, 273
coprocessor interface logic, 263, 274
CPU shutdown logic, 262–63, 273–74
DRAM, 287–90
8042 keyboard microcontroller, 287, 291
8237 DMA controller, 268–69, 277
8254 counter timer, 268, 276
8259 programmable interrupt controller, 263–64, 274–75
8284 clock generator/driver, 268, 276
82284 clock generator/driver, 259–61, 272–73
82288 bus controller, 261–62, 273
extended data bus, 265, 270, 277
I/O interface bus slots, 291–92
miscellaneous logic, 267–68, 275–76
parity check logic, 270, 277
peripheral select decode, 266–67
read/write logic, 266–67, 275
refresh and DMA arbitration and timing, 269, 277
ROM, 286–87
74612 memory mapper, 268, 276
6818 real time clock in, 264, 275
system board timing, 277–86
I/O write bus cycle (word), 284–86
memory read bus cycle (low byte), 277–81
memory read bus cycle (word), 282
memory write bus cycle (high byte), 283–84
80287 Math coprocessor, 255–58, 272
operational sequence, 257–58
pin configuration, 255–57
80386DX system board, 325–29
DRAM section, 335
8042 keyboard controller, 335
82340 AT chip set, 333–35
82344 ISA controller, 334
82345 data buffer, 333
82346 system controller, 333–34
82385 32-bit cache controller, 325, 327
block diagram, 325–27
direct mapped cache memory, 328
pin configurations, 329–33
two-way set associative cache memory, 329
I/O interface bus slots, 335
ROM section, 334
80386 Microprocessor, 307–14
address bus, 308–10
bus control, 313
bus cycle, 312–13
coprocessor, 314
data bus, 312
DMA bus, 314
interrupt bus, 313–14
pin configurations, 307–11
power bus, 314
80386SX system board
DRAM section, 324
8042 keyboard microcontroller, 324
82330/82231 AT chip set, 324
82335 IC
address mapper and decoder, 318–20
bus cycle translator, 323
clock generator and reset synchronizer, 318
DRAM, 320–23
interface device, 318
math coprocessor interface, 324
parity generator and checker, 318
ready generator, 324
I/O interface bus slots, 324–25
ROM section, 324
80386 System board, 300–340
addressing, 304–6
block diagrams, 301–2
bus interface, 301
central processing unit, 302
80387 numeric coprocessor, 315–17
instructions, 306
memory management unit, 302
MPU. *See* 80386 Microprocessor
operating modes, 302–6
registers, 303–4
debug, 306
flags, 303–4
general, 303
global descriptor table, 305
interrupt descriptor table, 305
local descriptor table, 306
segment, 306
task, 305–6
test, 306
word control, 305
80387 Numeric coprocessor, 315–17
80486 Microprocessor, 349–63
bus cycles, 356–63
basic 2-2, 356–58
basic 3-3, 358
burst cacheable, 361–63
non-burst cacheable, 359–61
non-cacheable burst, 358
versus Pentium, 365–66
pin configurations, 349–56
address bit 20 Mask, 356
address bus, 350
burst control, 353
bus arbitration, 352–53
bus control, 354
bus cycle definition, 352
bus size control, 353
cache control, 355
cache invalidation, 355
clock, 354
data bus, 354
interrupt signals, 354–55
numeric error reporting, 356
page caching control, 355
parity, 353–54
80486 System board, 341–63
block diagram, 341–43
bus interface unit, 341–43
cache section, 344–45
central processing unit, 344
floating point unit, 344
instructions, 349
instruction unit, 343–44
internal cache, 348–49
memory management unit, 344
microarchitecture, 341–45
MPU. *See* 80486 Microprocessor
registers, 345–48
addressing, 347
base architecture, 345–47
debug and test, 348
flags, 345
floating point, 347–48
global descriptor table, 347
instruction pointer, 345
interrupt descriptor table, 347
segment, 345
system address, 347
system level, 347
task state segment, 347
word control, 347
82077 Floppy disk controller, 378
82230 AT chip, 258–65, 270, 272–75, 324
address/data bus control, 264–65
bus control logic, 262–63, 273

command delay logic, 261, 273
coprocessor interface logic, 263, 274
CPU shutdown logic, 262–63, 273–74
8259 programmable interrupt controller, 263–64, 274–75
82284 clock generator/driver, 259–61, 272–73
82288 bus controller, 261–62, 273
extended data bus, 265
6818 real time clock in, 264, 275
82231 AT chip, 258–59, 266–70, 275–77, 324
8237 DMA controller, 268–69, 277
8254 counter timer, 268, 276
8284 clock generation and logic, 268, 276
extended data bus, 270, 277
miscellaneous logic, 267–68, 275–76
parity check logic, 270, 277
peripheral select decode, 266–67
read/write logic, 266–67, 275
refresh and DMA arbitration and timing, 269, 277
74612 memory mapper, 268, 276
82284 Clock generator/driver, 259–61, 272–73
82288 Bus controller, 261–62, 273
82303 Local I/O support chip, 376
82304 Local I/O support chip, 377
82307 DMA/CACP controller, 377
82308 Microchannel bus controller, 377
82309 Address bus controller, 377
82311 Microchannel architecture (MCA) chip set, 376–78
bus slots and edge connectors, 378
82077 floppy disk controller, 378
82303 local I/O support chip, 376
82304 local I/O support chip, 377
82307 DMA/CACP controller, 377
82308 microchannel bus controller, 377
82309 address bus controller, 377
82706 VGA graphics controller, 377
82335 Integrated circuit
address mapper and decoder, 318–20
bus cycle translator, 323
clock generator and reset synchronizer, 318
DRAM, 320–23
interface device, 318
math coprocessor interface, 324
parity generator and checker, 318
ready generator, 324
82340 AT chip set, 333–35
82344 ISA controller, 334
82345 Data buffer, 333
82346 System controller, 333–34
82350 EISA chip set, 378–80
82375EB PCEB, 383
82378 System I/O, 383
82385 32-Bit cache controller, 325, 327
82430 PCIset, 382–83
82375EB PCEB, 383
82378 system I/O, 383
82433LX local bus accelerator, 383
82434LX PCI/cache/memory controller, 382–83
82706 VGA graphics controller, 377
EISA. *See* Extended industry standard architecture
Electronic industry association (EIA)
receivers and drivers, 547–48
electrical specifications, 548
mechanical specifications, 548
signal specifications, 548
standard interface. *See* RS-232-C interface
EMM386 Device driver, 97
Enhanced graphics adapter (EGA), 15, 438
Enhanced small device interface (ESDI), 13, 515
ERASE command, 64–65
Error detection, in Pentium processor, 365
ESDI. *See* Enhanced small device interface
Ethernet networks, 581–82
Extended industry standard architecture (EISA), 378–80
82350 chip set, 378–80
External clock sampling, 592
External memory cards, 15–17

F

FAT. *See* File allocation table
FCBS= command (CONFIG. SYS), 96
FC command, 63–64
FDDI networks, 582
FDISK program, 106–8
Fiber optics in LANs, 581
Fiber optic transmission, 570
File allocation table (FAT), 506
File control blocks, 96
Files, comparing, 63–64
FILES= command (CONFIG. SYS), 96
Fixed disk program, 106
Floating point unit (FPU)
80486, 344, 347–48
Pentium, 365
Floppy disk, 9–11, 473–74, 476–79
adapter, 9, 479–510. *See also* NEC FDC 765A controller
function of, 601
AT system, 508–10
controller, 378
drive unit, 9–10, 499–504
electronics of, 499–501
function of, 601
troubleshooting, 501–4
formatting, 46–51
with DOS, 504–8
handling, inserting, and removing, 26–27
high-density, 508–10
troubleshooting, 510
recording techniques, 476–77
tracks and logical sectors, 477–79
Floptical drives, 474–75
Font, 551
FORMAT command, 46–51
FPU. *See* Floating point unit
Frequency counter, 590
FSK modem, 572–73
Fully formed printers, 550
Function keys, 20

G

GRAPHICS command, 71–72
Graphics software, 26

H

Hard disk, 12–14, 474, 511–18
adapter, 12–13, 512
drive unit, 13–14, 512–13
enhanced small device interface, 515
formatting, 46–51
integrated drive electronics, 516–18
modified frequency modulation, 514
preparing, 513–14
run length limited, 514
setting up, 106–13
small computer system interface, 515–16
types of, 514–18
Hardware, 3
intermittent problems, 600
repairing, 604–6
card swapping, 604–6
field engineering changes, 606
substituting parts, 606
troubleshooting, 598–600
HELP command, 44
High-density floppy disk, 508–10
HIMEM device driver, 97

I

IBM Advanced Diagnostic diskette, 599
IBM Asynchronous communications adapter, 545–48
troubleshooting, 550
IBMBIO. SYS file, 507–8
IBMDOS. SYS file, 508
IBM PC XT keyboards, 396–97
system board key encoding, 399–402
IBM PC XT system board, 4–6, 119–241
address decoding, 176–84
determining, 178
differences in, 182
RAM timing, 179–82
74138 Pin configuration, 178
troubleshooting, 182–84
DMA. *See* IBM PC XT system board, 8237 DMA controller
DRAM section, 209–25. *See also* IBM PC XT system board, RAM section
internal operation, 212–13
MT4264 block diagram, 210–12
pin assignments, 213–15
timing diagrams, 216–23
troubleshooting, 223

8087 math coprocessor, 143–44
pin configurations, 144–45
troubleshooting, 145
8088 microprocessor, 4, 128–43
addressing modes, 132
architecture, 129–30
bus, 128–29
data formats, 131
effective address, 131
instructions, 132
pin configuration, 132–41
timing diagrams, 141–43
8237 DMA controller, 185–201
internal registers and control logic, 187–88
operational modes, 191
pin configuration, 189–90
reading page register, 194
timing diagrams, 194–98
troubleshooting, 198–99
types of transfers, 191–92
writing to page register, 192
8253 programmable interval timer, 225–31
block diagram, 227–28
functions of, 229–30
operational modes, 228–29
troubleshooting, 230–31
8255 programmable peripheral interface, 231–38
block diagram, 233–34
functions of, 235–37
operational modes, 234–35
pin configuration, 234
troubleshooting, 237
8259 programmable interrupt controller, 156–58
cascaded control lines, 161
pin configurations, 158–61
timing diagrams, 161–63
8284 clock generator/driver, 121
pin configuration, 122–26
timing diagrams, 126–27
8288 bus controller, 145–46
pin configurations, 146–51
timing diagrams, 151–56
extended control logic, 173–76
troubleshooting, 176
I/O interface bus channel, 238–40
functions of, 238–40
troubleshooting, 240
key-encoding circuitry, 399–401
key-encoding operation, 401–2
microprocessor subsystem, 119–73
timing diagrams, 166–70
troubleshooting, 170–71
PIT section. *See* IBM PC XT system board, 8253 programmable interval timer (PIT)
PPI section. *See* IBM PC XT system board, 8255 programmable peripheral interface (PPI)
RAM section, 209–25. *See also* IBM PC XT system board, DRAM section
integrated circuits in, 215–16
ROM section, 201–8
8364 ROM, 201
integrated circuits in, 206–7
pin configuration, 201–3, 205–6
timing diagram, 207
2764 EPROM, 203–6
sections of, 119
74245 data bus transceiver, 164–65
74373 address latch, 163–64
IBM PS/2, 375–76
Impact printers, 551
Information superhighway, 568
Ink jet printers, 25, 559–60
advantages and disadvantages of, 560
printing process, 560
Input/output systems, 525–67. *See also* Parallel data communications; Serial data communications
Input pod, 592
Integrated system board, 7
Internal clock sampling, 592
I/O parity error, 600

K

Keyboard matrix, 394
Keyboards, 20–23, 390–407, 602
AT, 397–98
system board key encoding, 402–6
8048/8049 microcontrollers, 390–93
pin configurations, 391
function keys, 20
IBM PC XT, 396–97
system board key encoding, 399–402
key types, 393
numeric keypad, 22–23
troubleshooting, 406–7
KSR printer, 551

L

LABEL command, 52
LANs. *See* Local area networks
Laser printers, 24, 556–59
advantages and disadvantages of, 559
printing process, 558
LASTDRIVE= command (CONFIG. SYS), 96
LH command, 92
Local area networks (LANs), 576–84
access method, 576
architecture, 576
configurations, 583–84
data encoding, 579–80
interface adapter, 578
protocols, 582–83
software, 584
throughput, 578–79
topology, 576–78
logical, 577
physical, 577–78
transmission lines, 579–81
transmission speeds, 579
Logic analyzer, 591–93
external, 591–92
fully integrated, 592
functions of, 592–93
self-contained, 591
Logic monitor, 594
Logic signals, 594–95

M

Magnetic/optical drives, 474–75
Mainframe network, 575
Maintenance-repair logs, 606
Make directory, 78
Manufacturer's specifications, 596–97
Mass storage, 8–14, 472–524. *See also* Floppy disk; Hard disk
types of, 473–76
Math coprocessor
8087, 143–44
pin configurations, 144–45
troubleshooting, 145
80287, 255–58, 272
operational sequence, 257–58
pin configurations, 255–57
80387, 315–17
82335 IC interface, 324
Maximum speeds, 597
MCA. *See* Microchannel architecture
MCA chip set. *See* 82311 Microchannel architecture chip set
MDA. *See* Monochrome display adapter
MD command, 78
MEM command, 74–75
MEMMAKER, 105–6
Memory
analyzing, 74–75
display mode, 593
DMA. *See* 8237 Direct memory access controller; 82307 DMA/CACP controller
DRAM. *See* Dynamic random-access memory
82385 32-bit cache controller
direct mapped cache, 328
two-way set associative cache, 329
82434LX PCI/cache/memory controller, 382–83
EPROM, 203–6, 553
expansion, 602
external cards, 15–17
internal cache
80486 system board, 348–49
management
80286 system board, 248
80386 system board, 302
80486 system board, 344
page sizing, Pentium, 365
RAM. *See* Random access memory
ROM. *See* Read-only memory
screen, writing into, 455–56
74612 memory mapper, 268, 276
MENUCOLOR= command (CONFIG. SYS), 99–100
Menu commands, 97–100

MENUDEFAULT= command (CONFIG. SYS), 100
MENUITEM= command (CONFIG. SYS), 100
MFM. *See* Modified frequency modulation
Microchannel architecture (MCA), 375–76
82311 chip set, 376–78
Microcomputer signals and specifications, 594–95
Microprocessors (MPUs). *See* 8088 Microprocessor; 80286 Microprocessor; 80386 Microprocessor; 80486 Microprocessor; Pentium processor
Microprocessor subsystems
80286, 243–45
IBM PC XT, 119–73
timing diagrams, 166–70
troubleshooting, 170–71
MicroSoft Anti-Virus, 104
MicroSoft Diagnostics (MSD), 75–78
Modems, 569–74, 602
connecting to phone system, 573
FSK, 572–73
hardware support, 574
transmission media, 569–70
troubleshooting, 574
U.S. frequencies for, 573–74
Modified frequency modulation (MFM), 12–13
Monitors. *See* Video monitors
Monochrome display adapter (MDA), 430, 437
MORE command, 56–57
Motherboard. *See* System board
Mouse, 561–63, 602
electromechanical, 561–62
optical, 562
optomechanical, 563
MOVE command, 70
MSBACKUP, 88
MSAV command, 104
MSBACKUP command, 88
MSCDEX command, 92
MSD. *See* MicroSoft Diagnostics
MSD command, 75–78
Multiprocessing, Pentium, 365

N

NEC FDC 765A controller, 479–94, 479–99
block functions, 483–85
command operation sequence, 485–86
features, 479
instructions, 486–94
pin configuration, 480
DMA interface bus, 481
FDD interface, 481–82
MPU/DMA interface bus, 481
MPU interface bus, 480
supply bus, 481
Networks, 574–84
ARCnet, 582
Ethernet, 581–82
FDDI, 582
local area. *See* Local area networks
mainframe, 575
token ring, 582
Noise
manufacturer's specifications for, 596
Nonimpact printers, 551
NRZ asynchronous circuitry
example, 534–37
receive operation, 536–37
transmit operation, 535–36
NRZI asynchronous circuitry, 534
Numeric coprocessor. *See* Math coprocessor
Numeric keypad, 22–23

O

Open-collector drivers, function of, 604
Operating speed, manufacturer's specifications for, 596–97
Operating system, 12, 25. *See also* DOS; OS2; Unix; Windows
Optical drives, 475
OS2, 12
Oscilloscope, 590
Output data latch, function of, 604

P

Paper feed, 551
Parallel data communications, 526–32
advantages and disadvantages of, 526
Centronics printer interface, 526–27
IBM printer adapter, 527–32
address decoding, 527
bus transceiver, 530
command decoder, 527
control buffer, 528
control latch, 527–28
data buffer, 529
data latch, 529
open collector drivers, 530
status buffer, 528–29
typical operation, 531–32
Parity check one, 600
Parity check two, 600
Parity error, 600
PATH command, 78, 92
PAUSE command, 92–93
PCI bus. *See* Peripheral component interconnect bus
PC XT. *See* IBM PC XT
Pentium processor, 363–75
block diagram, 363–64
branch prediction, 365
code and data caches, 364
data integrity and error detection, 365
versus 80486, 365–66
floating point unit, 365
memory page sizing, 365
multiprocessing support, 365
performance monitoring, 365
pin configuration, 366–75
address bit 20 Mask, 374
burst control, 371
bus arbitration, 369–70
bus control, 372
bus cycle definition, 369
cache control, 373
cache invalidation, 372–73
clock, 372
data bus, 369
error detection, 371–72
external bus write status, 374
interrupt signals, 372
numeric error reporting, 373–74
operational control, 374
page caching control, 373
system management, 374
testing bus, 374–75
32-bit address bus, 366
64-bit data bus, 365
superscalar architecture, 363–64
Performance monitoring, Pentium, 365
Peripheral component interconnect (PCI) bus, 381–83
82430 PCIset, 382–83
Personal computer(s)
turning on and off, 27–29
types of, 1–3
PGC. *See* Professional graphics controller
PIC. *See* 8259 Programmable interrupt controller
Pin configurations
8048/8049 microcontrollers, 391
8087 math coprocessor, 144–45
8088 microprocessor, 132–41
8237 DMA controller, 189–90
8250 asynchronous communications element, 540–42
8255 programmable peripheral interface, 234
8259 programmable interrupt controller, 158–61
8284 clock generator/driver, 122–26
8288 bus controller, 146–51
80286 microprocessor, 251–55
80287 math coprocessor, 255–57
80386 microprocessor, 307–11
80486 microprocessor, 349–56
82385 32-bit cache controller, 329–33
NEC FDC 765A controller, 480–82
Pentium processor, 366–75
PIT. *See* 8253 Programmable interval timer; 8254 Programmable interval timer
Pitch, 551
Ports, data communication, 17
POST. *See* Power-on self-test
Power back-up systems, 419–20
Power dissipation, manufacturer's specifications for, 597
Power-on self-test (POST), 29
running, 598–600
Power supplies, 7–8, 407–20. *See also* AC power
linear regulated, 410–11

7805 voltage regulator, 411
SG3525A PWM switching regulator, 412–14
operation, 414
switching, 411–12
PWM, 414–18
troubleshooting, 418–19
Power supply, function of, 601
PPI. *See* 8255 Programmable peripheral interface
PRINT command, 72
Printer, 23–25, 602
KSR, 551
Printer adapter
function of, 601
troubleshooting, 549–50
Printer drivers, 553
Printers, 550–61
dot matrix. *See* Dot matrix printers
electronics, 551–53
fonts, 551
fully formed, 550
impact, 551
ink jet. *See* Ink jet printers
laser. *See* Laser printers
mechanical hardware, 554
nonimpact, 551
operations, 553–54
paper feed, 551
pitch, 551
print head movement, 550
receive only, 551
troubleshooting, 560–61
types of, 554–60
Professional graphics controller (PGC), 438
Programming languages, 25
Propagation delays, 597
Pulse width modulation (PWM), 411–18

R

Random access memory (RAM). *See also* Dynamic random-access memory
8155 RAM, 552
IBM PC XT system board, 209–25
integrated circuits in, 215–16
timing, 179–82
Read-only memory (ROM)
CD, 475
80286 system board, 246
using 82230/82231 AT chip set, 286–87
80386DX system board, 334
80386SX system board, 324
EPROM, 203–6, 553
IBM PC XT system board, 201–8
8364 ROM, 201
integrated circuits in, 206–7
pin configuration, 201–3, 205–6
timing diagram, 207
2764 EPROM, 203–6
Receive only printer, 551
Registers
8237 DMA controller, 187–88, 192, 194
8250 asynchronous communications element, 542, 544–45
80286 MPU, 249–51
80386 system board, 303–6
80486 system board, 345–48
6845 CRTC, 443–47
REM command, 93
RENAME command, 70
RF coaxial cable communication, 569
RGB color monitors, 433–35
Root directory, 38–40
RS-232-C interface, 546–47
electrical specifications, 546–47
line specifications
control signals, 547
data signals, 547
grounds, 547
mechanical specifications, 547
RTC. *See* 6818 Real time clock

S

Sample rate, 592
SCANDISK command, 73–74
Schematic diagram, 604–5
SCSI. *See* Small computer system interface
Serial data communications, 532–50, 549–50. *See also* Asynchronous data
advantages and disadvantages of, 532–33
asynchronous data packaging, 537–38
baud rate, 539
interpretation, 534
methods, 533–34
NRZ asynchronous circuitry, 534–37
receive operation, 536–37
transmit operation, 535–36
NRZI asynchronous circuitry, 534
SET command, 93
SETUP program, 107–13
SETVER= command (CONFIG. SYS), 97
7805 Voltage regulator, 411
74245 Data bus transceiver, 164–65
74373 Address latch, 163–64
74612 Memory mapper, 268, 276
SG3525A PWM switching regulator, 412–14
operation, 414
SHELL= command (CONFIG. SYS), 97
Signature analyzer, 593–94
6818 Real time clock (RTC), 245, 264, 275
6845 CRTC, 440–49
features of, 440–41
pin functions, 441–42
registers, 443–47
addressing, 443
display format, 445–47
grouping, 443
horizontal, 443–44
values at startup, 447
vertical, 444–45
timing diagrams
monitor, 448–49
read/write, 449
Small computer system interface (SCSI), 12, 515–16
SMARTDRV command, 93
Software, 25–26
troubleshooting, 600–601
Spread sheet, 26
SPS. *See* Standby power system
STACKS= command (CONFIG. SYS), 97
Standby power system (SPS), 420
Star/distributed star, 578
Star-wired ring, 577–78
Station-to-satellite microwave transmission, 569
Station-to-station microwave transmission, 569
Storage. *See* Floppy disk; Hard disk; Mass storage
Subsystem block diagram, troubleshooting, 603–4
Superscalar architecture, 363–64
Super video graphics array (SVGA), 15, 438–39
Supply voltages, manufacturer's specifications for, 596
SVGA. *See* Super video graphics array
Switching power supplies, 411–12
SYS command, 53
System block diagram, troubleshooting, 601–3
System board (motherboard), 4–7
AT, 6–7, 242–43. *See also* Advanced AT buses; 80286 System board
function of, 601
integrated, 7
PC XT. *See* IBM PC XT system board
System unit, 3–4
cases for, 17–18
functions of, 601–2

T

Tape drives, 475–76
Telephone communication, 569
Telephone system, 570–74
bandwidth, 571
distortion in, 571
modulation in, 571–72
amplitude, 572
differential phase-shift keying, 572
frequency-shift keying, 572
on-off keying, 572
phase-shift keying, 572
Test equipment, 589–94
Threshold logic levels, manufacturer's specifications for, 596
TIME command, 45–46
Timing, 80286 system board using 82230/82231 AT chip set, 277–86

Timing diagrams
  8088 MPU, 141–43
  8237 DMA controller, 194–98
  8259 programmable interrupt controller, 161–63
  8284 clock generator/driver, 126–27
  8288 bus controller, 151–56
  IBM PC XT system board
    DRAM section, 216–23
    microprocessor subsystem, 166–70
    ROM section, 207
  6845 CRTC
    monitor, 448–49
    read/write, 449
Timing (trace) display mode, 593
Token passing, 576
Token ring networks, 582
TREE command, 78–79
Trigger event, 593
Trigger word, 593
Troubleshooting, 597–607
  asynchronous adapters, 550
  8087 math coprocessor, 145
  8237 DMA controller, 198–99
  8253 programmable interval timer, 230–31
  8255 programmable peripheral interface, 237
  floppy disk
    drive unit, 501–4
    high-density, 510
  hardware, 598–600
  IBM PC XT system board
    address decoding, 182–84
    DRAM section, 223
    extended control logic, 176
    I/O interface bus channel, 240
    microprocessor subsystem, 170–71
  keyboards, 406–7
  logical approach to, 601–4
  modems, 574
  power supplies, 418–19
  printer adapters, 549–50
  printers, 560–61
  software, 600–601
  subsystem block diagram, 603–4
  system block diagram, 601–3
  tips for, 606–7
  video monitors, 436–37
TTL monochrome monitors, 430–33
Twisted pair cables, in LANs, 580–81
2716 EPROM, 553
TYPE command, 71

U

UNDELETE command, 65–70
UNFORMAT command, 51–52
Uninterruptible power system (UPS), 420
Unix, 12
Upper bus buffers, function of, 604
UPS. *See* Uninterruptible power system
Utilities, 25

V

VERIFY command, 61–62
VGA. *See* Video graphics array
VGA color monitors, 435–36
Video adapters, 14–15
  analog, 438–39
    PGC, 438
    SVGA, 15, 438–39
    VGA, 438, 464–66
  digital, 437–38
    CGA, 14, 437–38
    EGA, 15, 438
    MDA, 430, 437
  display modes, 439–40
  IBM color graphics, 456–61
    all-points-addressing display mode, 456–57
    alphanumeric display mode, 456
    alphanumeric screen refresh, 461
    block diagram, 458–61
    types of displays, 457–58
  IBM monochrome, 450–56
    block diagram, 451
    screen refresh operation, 455
    writing into screen memory, 455–56
  outputs, 462–64
  troubleshooting, 464, 466
Video display adapter, function of, 601
Video display screen, 426–28
Video Electronic Standard Association (VESA)
  video local bus, 380–81
Video graphics array (VGA), 438, 464–66
Video image, producing, 428–29
Video local bus (VLB), 380–81
Video monitors, 18–20, 429–37, 602
  RGB color, 433–35
  troubleshooting, 436–37
  TTL monochrome, 430–33
  VGA color, 435–36
Video systems, 425–71
Virus protection, 104
VLB. *See* Video local bus
VOL command, 53
Voltage regulators, 411
VSAFE command, 93

W

Windows, 12, 25
Word processor, 26

X

XCOPY command, 62–63
XT. *See* IBM PC XT